Springer Handbook of Crystal Growth

Govindhan Dhanaraj, Kullaiah Byrappa,
Vishwanath Prasad, Michael Dudley (Eds.)

Springer Handbook of Crystal Growth
Organization of the Handbook

使 用 说 明

1.《晶体生长手册》原版为一册，分为A～H部分。考虑到使用方便以及内容一致，影印版分为6册：第1册—Part A，第2册—Part B，第3册—Part C，第4册—Part D、E，第5册—Part F、G，第6册—Part H。

2.各册在页脚重新编排页码，该页码对应中文目录。保留了原书页眉及页码，其页码对应原书目录及主题索引。

3.各册均给出完整6册书的章目录。

4.作者及其联系方式、缩略语表各册均完整呈现。

5.主题索引安排在第6册。

6.文前介绍基本采用中英文对照形式，方便读者快速浏览。

材料科学与工程图书工作室

联系电话 0451-86412421

0451-86414559

邮　　箱 yh_bj@yahoo.com.cn

xuyaying81823@gmail.com

zhxh6414559@yahoo.com.cn

Springer 手册精选系列

晶体生长手册

熔体法晶体生长技术

【第2册】

Springer Handbook of *Crystal Growth*

〔美〕Govindhan Dhanaraj 等主编

（影印版）

黑版贸审字08-2012-047号

Reprint from English language edition:
Springer Handbook of Crystal Growth
by Govindhan Dhanaraj, Kullaiah Byrappa, Vishwanath Prasad
and Michael Dudley

图书在版编目（CIP）数据

晶体生长手册. 2, 熔体法晶体生长技术 = Handbook of Crystal Growth. 2, Crystal Growth from Melt Techniques : 英文 / (美) 德哈纳拉 (Dhanaraj,G.) 等主编. — 影印本. — 哈尔滨 : 哈尔滨工业大学出版社,2013.1
（Springer手册精选系列）
ISBN 978-7-5603-3867-5

Ⅰ.①晶… Ⅱ.①德… Ⅲ.①熔体晶体生长 – 手册 – 英文 Ⅳ.①O78-62

中国版本图书馆CIP数据核字(2012)第292345号

材料科学与工程
图书工作室

责任编辑 杨 桦 许雅莹 张秀华
出版发行 哈尔滨工业大学出版社
社 址 哈尔滨市南岗区复华四道街10号 邮编 150006
传 真 0451-86414749
网 址 http://hitpress.hit.edu.cn
印 刷 哈尔滨市石桥印务有限公司
开 本 787mm × 960mm 1/16 印张 24.5
版 次 2013年1月第1版 2013年1月第1次印刷
书 号 ISBN 978-7-5603-3867-5
定 价 76.00元

序 言

多年以来，有很多探索研究已经成功地描述了晶体生长的生长工艺和科学，有许多文章、专著、会议文集和手册对这一领域的前沿成果做了综合评述。这些出版物反映了人们对体材料晶体和薄膜晶体的兴趣日益增长，这是由于它们的电子、光学、机械、微结构以及不同的科学和技术应用引起的。实际上，大部分半导体和光器件的现代成果，如果没有基本的、二元的、三元的及其他不同特性和大尺寸的化合物晶体的发展则是不可能的。这些文章致力于生长机制的基本理解、缺陷形成、生长工艺和生长系统的设计，因此数量是庞大的。

本手册针对目前备受关注的体材料晶体和薄膜晶体的生长技术水平进行阐述。我们的目的是使读者了解经常使用的生长工艺、材料生产和缺陷产生的基本知识。为完成这一任务，我们精选了50多位顶尖科学家、学者和工程师，他们的合作者来自于22个不同国家。这些作者根据他们的专业所长，编写了关于晶体生长和缺陷形成共计52章内容：从熔体、溶液到气相体材料生长；外延生长；生长工艺和缺陷的模型；缺陷特性的技术以及一些现代的特别课题。

本手册分为七部分。Part A介绍基础理论：生长和表征技术综述，表面成核工艺，溶液生长晶体的形态，生长过程中成核的层错，缺陷形成的形态。

Part B介绍体材料晶体的熔体生长，一种生长大尺寸晶体的关键方法。这一部分阐述了直拉单晶工艺、泡生法、布里兹曼法、浮区熔融等工艺，以及这些方法的最新进展，例如应用磁场的晶体生长、生长轴的取向、增加底基和形状控制。本部分涉及材料从硅和Ⅲ–Ⅴ族化合物到氧化物和氟化物的广泛内容。

第三部分，本书的Part C关注了溶液生长法。在前两章里讨论了水热生长法的不同方面，随后的三章介绍了非线性和激光晶体、KTP和KDP。通过在地球上和微重力环境下生长的比较给出了重力对溶液生长法的影响的知识。

Part D的主题是气相生长。这一部分提供了碳化硅、氮化镓、氮化铝和有机半导体的气相生长的内容。随后的Part E是关于外延生长和薄膜的，主要包括从液相的化学气相淀积到脉冲激光和脉冲电子淀积。

Part F介绍了生长工艺和缺陷形成的模型。这些章节验证了工艺参数和产生晶体质量问题包括缺陷形成的直接相互作用关系。随后的Part G展示了结晶材料特性和分析的发展。Part F和G说明了预测工具和分析技术在帮助高质量的大尺寸晶体生长工艺的设计和控制方面是非常好用的。

最后的Part H致力于精选这一领域的部分现代课题，例如蛋白质晶体生长、凝胶结晶、原位结构、单晶闪烁材料的生长、光电材料和线切割大晶体薄膜。

我们希望这本施普林格手册对那些学习晶体生长的研究生，那些从事或即将从事这一领域研究的来自学术界和工业领域的研究人员、科学家和工程师以及那些制备晶体的人是有帮助的。

我们对施普林格的Dr. Claus Acheron，Dr. Werner Skolaut和le-tex的Ms Anne Strobach的特别努力表示真诚的感谢，没有他们本书将无法呈现。

我们感谢我们的作者编写了详尽的章节内容和在本书出版期间对我们的耐心。一位编者（GD）感谢他的家庭成员和Dr. Kedar Gupta(ARC Energy 的CEO)，感谢他们在本书编写期间的大力支持和鼓励。还对Peter Rudolf, David Bliss,Ishwara Bhat和Partha Dutta在A、B、E部分的编写中所给予的帮助表示感谢。

Nashua, New Hampshire, April 2010 G. Dhanaraj
Mysore, India K. Byrappa
Denton, Texas V. Prasad
Stony Brook, New York M. Dudley

Preface

Over the years, many successful attempts have been made to describe the art and science of crystal growth, and many review articles, monographs, symposium volumes, and handbooks have been published to present comprehensive reviews of the advances made in this field. These publications are testament to the growing interest in both bulk and thin-film crystals because of their electronic, optical, mechanical, microstructural, and other properties, and their diverse scientific and technological applications. Indeed, most modern advances in semiconductor and optical devices would not have been possible without the development of many elemental, binary, ternary, and other compound crystals of varying properties and large sizes. The literature devoted to basic understanding of growth mechanisms, defect formation, and growth processes as well as the design of growth systems is therefore vast.

The objective of this Springer Handbook is to present the state of the art of selected topical areas of both bulk and thin-film crystal growth. Our goal is to make readers understand the basics of the commonly employed growth processes, materials produced, and defects generated. To accomplish this, we have selected more than 50 leading scientists, researchers, and engineers, and their many collaborators from 22 different countries, to write chapters on the topics of their expertise. These authors have written 52 chapters on the fundamentals of crystal growth and defect formation; bulk growth from the melt, solution, and vapor; epitaxial growth; modeling of growth processes and defects; and techniques of defect characterization, as well as some contemporary special topics.

This Springer Handbook is divided into seven parts. Part A presents the fundamentals: an overview of the growth and characterization techniques, followed by the state of the art of nucleation at surfaces, morphology of crystals grown from solutions, nucleation of dislocation during growth, and defect formation and morphology.

Part B is devoted to bulk growth from the melt, a method critical to producing large-size crystals. The chapters in this part describe the well-known processes such as Czochralski, Kyropoulos, Bridgman, and floating zone, and focus specifically on recent advances in improving these methodologies such as application of magnetic fields, orientation of the growth axis, introduction of a pedestal, and shaped growth. They also cover a wide range of materials from silicon and III–V compounds to oxides and fluorides.

The third part, Part C of the book, focuses on solution growth. The various aspects of hydrothermal growth are discussed in two chapters, while three other chapters present an overview of the nonlinear and laser crystals, KTP and KDP. The knowledge on the effect of gravity on solution growth is presented through a comparison of growth on Earth versus in a microgravity environment.

The topic of Part D is vapor growth. In addition to presenting an overview of vapor growth, this part also provides details on vapor growth of silicon carbide, gallium nitride, aluminum nitride, and organic semiconductors. This is followed by chapters on epitaxial growth and thin films in Part E. The topics range from chemical vapor deposition to liquid-phase epitaxy to pulsed laser and pulsed electron deposition.

Modeling of both growth processes and defect formation is presented in Part F. These chapters demonstrate the direct correlation between the process parameters and quality of the crystal produced, including the formation of defects. The subsequent Part G presents the techniques that have been developed for crystalline material characterization and analysis. The chapters in Parts F and G demonstrate how well predictive tools and analytical techniques have helped the design and control of growth processes for better-quality crystals of large sizes.

The final Part H is devoted to some selected contemporary topics in this field, such as protein crystal growth, crystallization from gels, in situ structural studies, growth of single-crystal scintillation materials, photovoltaic materials, and wire-saw slicing of large crystals to produce wafers.

We hope this Springer Handbook will be useful to graduate students studying crystal growth and to researchers, scientists, and engineers from academia and industry who are conducting or intend to conduct research in this field as well as those who grow crystals.

We would like to express our sincere thanks to Dr. Claus Acheron and Dr. Werner Skolaut of Springer and Ms Anne Strohbach of le-tex for their extraordinary efforts without which this handbook would not have taken its final shape.

We thank our authors for writing comprehensive chapters and having patience with us during the publication of this Handbook. One of the editors (GD) would like to thank his family members and Dr. Kedar Gupta (CEO of ARC Energy) for their generous support and encouragement during the entire course of editing this handbook. Acknowledgements are also due to Peter Rudolf, David Bliss, Ishwara Bhat, and Partha Dutta for their help in editing Parts A, B, E, and H, respectively.

Nashua, New Hampshire, April 2010 G. Dhanaraj
Mysore, India K. Byrappa
Denton, Texas V. Prasad
Stony Brook, New York M. Dudley

About the Editors

Govindhan Dhanaraj is the Manager of Crystal Growth Technologies at Advanced Renewable Energy Company (ARC Energy) at Nashua, New Hampshire (USA) focusing on the growth of large size sapphire crystals for LED lighting applications, characterization and related crystal growth furnace development. He received his PhD from the Indian Institute of Science, Bangalore and his Master of Science from Anna University (India). Immediately after his doctoral degree, Dr. Dhanaraj joined a National Laboratory, presently known as Rajaramanna Center for Advanced Technology in India, where he established an advanced Crystal Growth Laboratory for the growth of optical and laser crystals. Prior to joining ARC Energy, Dr. Dhanaraj served as a Research Professor at the Department of Materials Science and Engineering, Stony Brook University, NY, and also held a position of Research Assistant Professor at Hampton University, VA. During his 25 years of focused expertise in crystal growth research, he has developed optical, laser and semiconductor bulk crystals and SiC epitaxial films using solution, flux, Czochralski, Bridgeman, gel and vapor methods, and characterized them using x-ray topography, synchrotron topography, chemical etching and optical and atomic force microscopic techniques. He co-organized a symposium on Industrial Crystal Growth under the 17th American Conference on Crystal Growth and Epitaxy in conjunction with the 14th US Biennial Workshop on Organometallic Vapor Phase Epitaxy held at Lake Geneva, WI in 2009. Dr. Dhanaraj has delivered invited lectures and also served as session chairman in many crystal growth and materials science meetings. He has published over 100 papers and his research articles have attracted over 250 rich citations.

Kullaiah Byrappa received his Doctor's degree in Crystal Growth from the Moscow State University, Moscow in 1981. He is Professor of Materials Science, Head of the Crystal Growth Laboratory, and Director of the Internal Quality Assurance Cell of the University of Mysore, India. His current research is in crystal engineering of polyscale materials through novel solution processing routes, particularly covering hydrothermal, solvothermal and supercritical methods. Professor Byrappa has co-authored the Handbook of Hydrothermal Technology, and edited 4 books as well as two special editions of Journal of Materials Science, and published 180 research papers including 26 invited reviews and book chapters on various aspects of novel routes of solution processing. Professor Byrappa has delivered over 60 keynote and invited lectures at International Conferences, and several hundreds of colloquia and seminars at various institutions around the world. He has also served as chair and co-chair for numerous international conferences. He is a Fellow of the World Academy of Ceramics. Professor Byrappa is serving in several international committees and commissions related to crystallography, crystal growth, and materials science. He is the Founder Secretary of the International Solvothermal and Hydrothermal Association. Professor Byrappa is a recipient of several awards such as the Sir C.V. Raman Award, Materials Research Society of India Medal, and the Golden Jubilee Award of the University of Mysore.

Vishwanath "Vish" Prasad is the Vice President for Research and Economic Development and Professor of Mechanical and Energy Engineering at the University of North Texas (UNT), one of the largest university in the state of Texas. He received his PhD from the University of Delaware (USA), his Masters of Technology from the Indian Institute of Technology, Kanpur, and his bachelor's from Patna University in India all in Mechanical Engineering. Prior to joining UNT in 2007, Dr. Prasad served as the Dean at Florida International University (FIU) in Miami, where he also held the position of Distinguished Professor of Engineering. Previously, he has served as a Leading Professor of Mechanical Engineering at Stony Brook University, New York, as an Associate Professor and Assistant Professor at Columbia University. He has received many special recognitions for his contributions to engineering education. Dr. Prasad's research interests include thermo-fluid sciences, energy systems, electronic materials, and computational materials processing. He has published over 200 articles, edited/co-edited several books and organized numerous conferences, symposia, and workshops. He serves as the lead editor of the Annual Review of Heat Transfer. In the past, he has served as an Associate Editor of the ASME Journal of Heat. Dr. Prasad is an elected Fellow of the American Society of Mechanical Engineers (ASME), and has served as a member of the USRA Microgravity Research Council. Dr. Prasad's research has focused on bulk growth of silicon, III-V compounds, and silicon carbide; growth of large diameter Si tube; design of crystal growth systems; and sputtering and chemical vapor deposition of thin films. He is also credited to initiate research on wire saw cutting of large crystals to produce wafers with much reduced material loss. Dr. Prasad's research has been well funded by US National Science Foundation (NSF), US Department of Defense, US Department of Energy, and industry.

Michael Dudley received his Doctoral Degree in Engineering from Warwick University, UK, in 1982. He is Professor and Chair of the Materials Science and Engineering Department at Stony Brook University, New York, USA. He is director of the Stony Brook Synchrotron Topography Facility at the National Synchrotron Light Source at Brookhaven National Laboratory, Upton New York. His current research focuses on crystal growth and characterization of defect structures in single crystals with a view to determining their origins. The primary technique used is synchrotron topography which enables analysis of defects and generalized strain fields in single crystals in general, with particular emphasis on semiconductor, optoelectronic, and optical crystals. Establishing the relationship between crystal growth conditions and resulting defect distributions is a particular thrust area of interest to Dudley, as is the correlation between electronic/optoelectronic device performance and defect distribution. Other techniques routinely used in such analysis include transmission electron microscopy, high resolution triple-axis x-ray diffraction, atomic force microscopy, scanning electron microscopy, Nomarski optical microscopy, conventional optical microscopy, IR microscopy and fluorescent laser scanning confocal microscopy. Dudley's group has played a prominent role in the development of SiC and AlN growth, characterizing crystals grown by many of the academic and commercial entities involved enabling optimization of crystal quality. He has co-authored some 315 refereed articles and 12 book chapters, and has edited 5 books. He is currently a member of the Editorial Board of Journal of Applied Physics and Applied Physics Letters and has served as Chair or Co-Chair for numerous international conferences.

List of Authors

Francesco Abbona
Università degli Studi di Torino
Dipartimento di Scienze Mineralogiche
e Petrologiche
via Valperga Caluso 35
10125 Torino, Italy
e-mail: *francesco.abbona@unito.it*

Mohan D. Aggarwal
Alabama A&M University
Department of Physics
Normal, AL 35762, USA
e-mail: *mohan.aggarwal@aamu.edu*

Marcello R.B. Andreeta
University of São Paulo
Crystal Growth and Ceramic Materials Laboratory,
Institute of Physics of São Carlos
Av. Trabalhador Sãocarlense, 400
São Carlos, SP 13560-970, Brazil
e-mail: *marcello@if.sc.usp.br*

Dino Aquilano
Università degli Studi di Torino
Facoltà di Scienze Matematiche, Fisiche e Naturali
via P. Giuria, 15
Torino, 10126, Italy
e-mail: *dino.aquilano@unito.it*

Roberto Arreguín-Espinosa
Universidad Nacional Autónoma de México
Instituto de Química
Circuito Exterior, C.U. s/n
Mexico City, 04510, Mexico
e-mail: *arrespin@unam.mx*

Jie Bai
Intel Corporation
RA3-402, 5200 NE Elam Young Parkway
Hillsboro, OR 97124-6497, USA
e-mail: *jie.bai@intel.com*

Stefan Balint
West University of Timisoara
Department of Computer Science
Blvd. V. Parvan 4
Timisoara, 300223, Romania
e-mail: *balint@math.uvt.ro*

Ashok K. Batra
Alabama A&M University
Department of Physics
4900 Meridian Street
Normal, AL 35762, USA
e-mail: *ashok.batra@aamu.edu*

Handady L. Bhat
Indian Institute of Science
Department of Physics
CV Raman Avenue
Bangalore, 560012, India
e-mail: *hlbhat@physics.iisc.ernet.in*

Ishwara B. Bhat
Rensselaer Polytechnic Institute
Electrical Computer
and Systems Engineering Department
110 8th Street, JEC 6031
Troy, NY 12180, USA
e-mail: *bhati@rpi.edu*

David F. Bliss
US Air Force Research Laboratory
Sensors Directorate Optoelectronic Technology
Branch
80 Scott Drive
Hanscom AFB, MA 01731, USA
e-mail: *david.bliss@hanscom.af.mil*

Mikhail A. Borik
Russian Academy of Sciences
Laser Materials and Technology Research Center,
A.M. Prokhorov General Physics Institute
Vavilov 38
Moscow, 119991, Russia
e-mail: *borik@lst.gpi.ru*

Liliana Braescu
West University of Timisoara
Department of Computer Science
Blvd. V. Parvan 4
Timisoara, 300223, Romania
e-mail: *lilianabraescu@balint1.math.uvt.ro*

Kullaiah Byrappa
University of Mysore
Department of Geology
Manasagangotri
Mysore, 570 006, India
e-mail: *kbyrappa@gmail.com*

Dang Cai
CVD Equipment Corporation
1860 Smithtown Ave.
Ronkonkoma, NY 11779, USA
e-mail: *dcai@cvdequipment.com*

Michael J. Callahan
GreenTech Solutions
92 Old Pine Drive
Hanson, MA 02341, USA
e-mail: *mjcal37@yahoo.com*

Joan J. Carvajal
Universitat Rovira i Virgili (URV)
Department of Physics and Crystallography
of Materials and Nanomaterials (FiCMA-FiCNA)
Campus Sescelades, C/ Marcel·lí Domingo, s/n
Tarragona 43007, Spain
e-mail: *joanjosep.carvajal@urv.cat*

Aaron J. Celestian
Western Kentucky University
Department of Geography and Geology
1906 College Heights Blvd.
Bowling Green, KY 42101, USA
e-mail: *aaron.celestian@wku.edu*

Qi-Sheng Chen
Chinese Academy of Sciences
Institute of Mechanics
15 Bei Si Huan Xi Road
Beijing, 100190, China
e-mail: *qschen@imech.ac.cn*

Chunhui Chung
Stony Brook University
Department of Mechanical Engineering
Stony Brook, NY 11794-2300, USA
e-mail: *chuchung@ic.sunysb.edu*

Ted Ciszek
Geolite/Siliconsultant
31843 Miwok Trl.
Evergreen, CO 80437, USA
e-mail: *ted_ciszek@siliconsultant.com*

Abraham Clearfield
Texas A&M University
Distinguished Professor of Chemistry
College Station, TX 77843-3255, USA
e-mail: *clearfield@chem.tamu.edu*

Hanna A. Dabkowska
Brockhouse Institute for Materials Research
Department of Physics and Astronomy
1280 Main Str W.
Hamilton, Ontario L8S 4M1, Canada
e-mail: *dabkoh@mcmaster.ca*

Antoni B. Dabkowski
McMaster University, BIMR
Brockhouse Institute for Materials Research,
Department of Physics and Astronomy
1280 Main Str W.
Hamilton, Ontario L8S 4M1, Canada
e-mail: *dabko@mcmaster.ca*

Rafael Dalmau
HexaTech Inc.
991 Aviation Pkwy Ste 800
Morrisville, NC 27560, USA
e-mail: *rdalmau@hexatechinc.com*

Govindhan Dhanaraj
ARC Energy
18 Celina Avenue, Unit 77
Nashua, NH 03063, USA
e-mail: *dhanaraj@arc-energy.com*

Ramasamy Dhanasekaran
Anna University Chennai
Crystal Growth Centre
Chennai, 600 025, India
e-mail: *rdhanasekaran@annauniv.edu; rdcgc@yahoo.com*

Ernesto Diéguez
Universidad Autónoma de Madrid
Department Física de Materiales
Madrid 28049, Spain
e-mail: *ernesto.dieguez@uam.es*

Vijay K. Dixit
Raja Ramanna Center for Advance Technology
Semiconductor Laser Section,
Solid State Laser Division
Rajendra Nagar, RRCAT.
Indore, 452013, India
e-mail: *dixit@rrcat.gov.in*

Sadik Dost
University of Victoria
Crystal Growth Laboratory
Victoria, BC V8W 3P6, Canada
e-mail: *sdost@me.uvic.ca*

Michael Dudley
Stony Brook University
Department of Materials Science and Engineering
Stony Brook, NY 11794-2275, USA
e-mail: *mdudley@notes.cc.sunysb.edu*

Partha S. Dutta
Rensselaer Polytechnic Institute
Department of Electrical, Computer
and Systems Engineering
110 Eighth Street
Troy, NY 12180, USA
e-mail: *duttap@rpi.edu*

Francesc Díaz
Universitat Rovira i Virgili (URV)
Department of Physics and Crystallography
of Materials and Nanomaterials (FiCMA-FiCNA)
Campus Sescelades, C/ Marcel·lí Domingo, s/n
Tarragona 43007, Spain
e-mail: *f.diaz@urv.cat*

Paul F. Fewster
PANalytical Research Centre,
The Sussex Innovation Centre
Research Department
Falmer
Brighton, BN1 9SB, UK
e-mail: *paul.fewster@panalytical.com*

Donald O. Frazier
NASA Marshall Space Flight Center
Engineering Technology Management Office
Huntsville, AL 35812, USA
e-mail: *donald.o.frazier@nasa.gov*

James W. Garland
EPIR Technologies, Inc.
509 Territorial Drive, Ste. B
Bolingbrook, IL 60440, USA
e-mail: *jgarland@epir.com*

Thomas F. George
University of Missouri-St. Louis
Center for Nanoscience,
Department of Chemistry and Biochemistry,
Department of Physics and Astronomy
One University Boulevard
St. Louis, MO 63121, USA
e-mail: *tfgeorge@umsl.edu*

Andrea E. Gutiérrez-Quezada
Universidad Nacional Autónoma de México
Instituto de Química
Circuito Exterior, C.U. s/n
Mexico City, 04510, Mexico
e-mail: *30111390@escolar.unam.mx*

Carl Hemmingsson
Linköping University
Department of Physics, Chemistry
and Biology (IFM)
581 83 Linköping, Sweden
e-mail: *cah@ifm.liu.se*

Antonio Carlos Hernandes
University of São Paulo
Crystal Growth and Ceramic Materials Laboratory,
Institute of Physics of São Carlos
Av. Trabalhador Sãocarlense
São Carlos, SP 13560-970, Brazil
e-mail: *hernandes@if.sc.usp.br*

Koichi Kakimoto
Kyushu University
Research Institute for Applied Mechanics
6-1 Kasuga-kouen, Kasuga
816-8580 Fukuoka, Japan
e-mail: *kakimoto@riam.kyushu-u.ac.jp*

Imin Kao
State University of New York at Stony Brook
Department of Mechanical Engineering
Stony Brook, NY 11794-2300, USA
e-mail: *imin.kao@stonybrook.edu*

John J. Kelly
Utrecht University,
Debye Institute for Nanomaterials Science
Department of Chemistry
Princetonplein 5
3584 CC, Utrecht, The Netherlands
e-mail: *j.j.kelly@uu.nl*

Jeonggoo Kim
Neocera, LLC
10000 Virginia Manor Road #300
Beltsville, MD, USA
e-mail: *kim@neocera.com*

Helmut Klapper
Institut für Kristallographie
RWTH Aachen University
Aachen, Germany
e-mail: *klapper@xtal.rwth-aachen.de;*
helmut-klapper@web.de

Christine F. Klemenz Rivenbark
Krystal Engineering LLC
General Manager and Technical Director
1429 Chaffee Drive
Titusville, FL 32780, USA
e-mail: *ckr@krystalengineering.com*

Christian Kloc
Nanyang Technological University
School of Materials Science and Engineering
50 Nanyang Avenue
639798 Singapore
e-mail: *ckloc@ntu.edu.sg*

Solomon H. Kolagani
Neocera LLC
10000 Virginia Manor Road
Beltsville, MD 20705, USA
e-mail: *harsh@neocera.com*

Akinori Koukitu
Tokyo University of Agriculture and Technology
(TUAT)
Department of Applied Chemistry
2-24-16 Naka-cho, Koganei
184-8588 Tokyo, Japan
e-mail: *koukitu@cc.tuat.ac.jp*

Milind S. Kulkarni
MEMC Electronic Materials
Polysilicon and Quantitative Silicon Research
501 Pearl Drive
St. Peters, MO 63376, USA
e-mail: *mkulkarni@memc.com*

Yoshinao Kumagai
Tokyo University of Agriculture and Technology
Department of Applied Chemistry
2-24-16 Naka-cho, Koganei
184-8588 Tokyo, Japan
e-mail: *4470kuma@cc.tuat.ac.jp*

Valentin V. Laguta
Institute of Physics of the ASCR
Department of Optical Materials
Cukrovarnicka 10
Prague, 162 53, Czech Republic
e-mail: *laguta@fzu.cz*

Ravindra B. Lal
Alabama Agricultural and Mechanical University
Physics Department
4900 Meridian Street
Normal, AL 35763, USA
e-mail: *rblal@comcast.net*

Chung-Wen Lan
National Taiwan University
Department of Chemical Engineering
No. 1, Sec. 4, Roosevelt Rd.
Taipei, 106, Taiwan
e-mail: *cwlan@ntu.edu.tw*

Hongjun Li
Chinese Academy of Sciences
R & D Center of Synthetic Crystals,
Shanghai Institute of Ceramics
215 Chengbei Rd., Jiading District
Shanghai, 201800, China
e-mail: *lh_li@mail.sic.ac.cn*

Elena E. Lomonova
Russian Academy of Sciences
Laser Materials and Technology Research Center,
A.M. Prokhorov General Physics Institute
Vavilov 38
Moscow, 119991, Russia
e-mail: *lomonova@lst.gpi.ru*

Ivan V. Markov
Bulgarian Academy of Sciences
Institute of Physical Chemistry
Sofia, 1113, Bulgaria
e-mail: *imarkov@ipc.bas.bg*

Bo Monemar
Linköping University
Department of Physics, Chemistry and Biology
58183 Linköping, Sweden
e-mail: *bom@ifm.liu.se*

Abel Moreno
Universidad Nacional Autónoma de México
Instituto de Química
Circuito Exterior, C.U. s/n
Mexico City, 04510, Mexico
e-mail: *carcamo@unam.mx*

Roosevelt Moreno Rodriguez
State University of New York at Stony Brook
Department of Mechanical Engineering
Stony Brook, NY 11794-2300, USA
e-mail: *roosevelt@dove.eng.sunysb.edu*

S. Narayana Kalkura
Anna University Chennai
Crystal Growth Centre
Sardar Patel Road
Chennai, 600025, India
e-mail: *kalkura@annauniv.edu*

Mohan Narayanan
Reliance Industries Limited
1, Rich Branch court
Gaithersburg, MD 20878, USA
e-mail: *mohan.narayanan@ril.com*

Subramanian Natarajan
Madurai Kamaraj University
School of Physics
Palkalai Nagar
Madurai, India
e-mail: *s_natarajan50@yahoo.com*

Martin Nikl
Academy of Sciences of the Czech Republic (ASCR)
Department of Optical Crystals, Institute of Physics
Cukrovarnicka 10
Prague, 162 53, Czech Republic
e-mail: *nikl@fzu.cz*

Vyacheslav V. Osiko
Russian Academy of Sciences
Laser Materials and Technology Research Center,
A.M. Prokhorov General Physics Institute
Vavilov 38
Moscow, 119991, Russia
e-mail: *osiko@lst.gpi.ru*

John B. Parise
Stony Brook University
Chemistry Department
and Department of Geosciences
ESS Building
Stony Brook, NY 11794-2100, USA
e-mail: *john.parise@stonybrook.edu*

Srinivas Pendurti
ASE Technologies Inc.
11499, Chester Road
Cincinnati, OH 45246, USA
e-mail: *spendurti@asetech.com*

Benjamin G. Penn
NASA/George C. Marshall Space Flight Center
ISHM and Sensors Branch
Huntsville, AL 35812, USA
e-mail: *benjamin.g.penndr@nasa.gov*

Jens Pflaum
Julius-Maximilians Universität Würzburg
Institute of Experimental Physics VI
Am Hubland
97078 Würzburg, Germany
e-mail: *jpflaum@physik.uni-wuerzburg.de*

Jose Luis Plaza
Universidad Autónoma de Madrid
Facultad de Ciencias,
Departamento de Física de Materiales
Madrid 28049, Spain
e-mail: *joseluis.plaza@uam.es*

Udo W. Pohl
Technische Universität Berlin
Institut für Festkörperphysik EW5-1
Hardenbergstr. 36
10623 Berlin, Germany
e-mail: *pohl@physik.tu-berlin.de*

Vishwanath (Vish) Prasad
University of North Texas
1155 Union Circle
Denton, TX 76203-5017, USA
e-mail: *vish.prasad@unt.edu*

Maria Cinta Pujol
Universitat Rovira i Virgili
Department of Physics and Crystallography
of Materials and Nanomaterials (FiCMA-FiCNA)
Campus Sescelades, C/ Marcel·lí Domingo
Tarragona 43007, Spain
e-mail: *mariacinta.pujol@urv.cat*

Balaji Raghothamachar
Stony Brook University
Department of Materials Science and Engineering
310 Engineering Building
Stony Brook, NY 11794-2275, USA
e-mail: *braghoth@notes.cc.sunysb.edu*

Michael Roth
The Hebrew University of Jerusalem
Department of Applied Physics
Bergman Bld., Rm 206, Givat Ram Campus
Jerusalem 91904, Israel
e-mail: *mroth@vms.huji.ac.il*

Peter Rudolph
Leibniz Institute for Crystal Growth
Technology Development
Max-Born-Str. 2
Berlin, 12489, Germany
e-mail: *rudolph@ikz-berlin.de*

Akira Sakai
Osaka University
Department of Systems Innovation
1-3 Machikaneyama-cho, Toyonaka-shi
560-8531 Osaka, Japan
e-mail: *sakai@ee.es.osaka-u.ac.jp*

Yasuhiro Shiraki
Tokyo City University
Advanced Research Laboratories,
Musashi Institute of Technology
8-15-1 Todoroki, Setagaya-ku
158-0082 Tokyo, Japan
e-mail: *yshiraki@tcu.ac.jp*

Theo Siegrist
Florida State University
Department of Chemical
and Biomedical Engineering
2525 Pottsdamer Street
Tallahassee, FL 32310, USA
e-mail: *siegrist@eng.fsu.edu*

Zlatko Sitar
North Carolina State University
Materials Science and Engineering
1001 Capability Dr.
Raleigh, NC 27695, USA
e-mail: *sitar@ncsu.edu*

Sivalingam Sivananthan
University of Illinois at Chicago
Department of Physics
845 W. Taylor St. M/C 273
Chicago, IL 60607-7059, USA
e-mail: *siva@uic.edu; siva@epir.com*

Mikhail D. Strikovski
Neocera LLC
10000 Virginia Manor Road, suite 300
Beltsville, MD 20705, USA
e-mail: *strikovski@neocera.com*

Xun Sun
Shandong University
Institute of Crystal Materials
Shanda Road
Jinan, 250100, China
e-mail: *sunxun@icm.sdu.edu.cn*

Ichiro Sunagawa
University Tohoku University (Emeritus)
Kashiwa-cho 3-54-2, Tachikawa
Tokyo, 190-0004, Japan
e-mail: *i.sunagawa@nifty.com*

Xu-Tang Tao
Shandong University
State Key Laboratory of Crystal Materials
Shanda Nanlu 27, 250100
Jinan, China
e-mail: *txt@sdu.edu.cn*

Vitali A. Tatartchenko
Saint – Gobain, 23 Rue Louis Pouey
92800 Puteaux, France
e-mail: *vitali.tatartchenko@orange.fr*

Filip Tuomisto
Helsinki University of Technology
Department of Applied Physics
Otakaari 1 M
Espoo TKK 02015, Finland
e-mail: *filip.tuomisto@tkk.fi*

Anna Vedda
University of Milano-Bicocca
Department of Materials Science
Via Cozzi 53
20125 Milano, Italy
e-mail: *anna.vedda@unimib.it*

Lu-Min Wang
University of Michigan
Department of Nuclear Engineering
and Radiological Sciences
2355 Bonisteel Blvd.
Ann Arbor, MI 48109-2104, USA
e-mail: *lmwang@umich.edu*

Sheng-Lai Wang
Shandong University
Institute of Crystal Materials,
State Key Laboratory of Crystal Materials
Shanda Road No. 27
Jinan, Shandong, 250100, China
e-mail: *slwang@icm.sdu.edu.cn*

Shixin Wang
Micron Technology Inc.
TEM Laboratory
8000 S. Federal Way
Boise, ID 83707, USA
e-mail: *shixinwang@micron.com*

Jan L. Weyher
Polish Academy of Sciences Warsaw
Institute of High Pressure Physics
ul. Sokolowska 29/37
01/142 Warsaw, Poland
e-mail: *weyher@unipress.waw.pl*

Jun Xu
Chinese Academy of Sciences
Shanghai Institute of Ceramics
Shanghai, 201800, China
e-mail: *xujun@mail.shcnc.ac.cn*

Hui Zhang
Tsinghua University
Department of Engineering Physics
Beijing, 100084, China
e-mail: *zhhui@tsinghua.edu.cn*

Lili Zheng
Tsinghua University
School of Aerospace
Beijing, 100084, China
e-mail: *zhenglili@tsinghua.edu.cn*

Mary E. Zvanut
University of Alabama at Birmingham
Department of Physics
1530 3rd Ave S
Birmingham, AL 35294-1170, USA
e-mail: *mezvanut@uab.edu*

Zbigniew R. Zytkiewicz
Polish Academy of Sciences
Institute of Physics
Al. Lotnikow 32/46
02668 Warszawa, Poland
e-mail: *zytkie@ifpan.edu.pl*

Acknowledgements

B.7 Indium Phosphide: Crystal Growth and Defect Control by Applying Steady Magnetic Fields
by David F. Bliss

The author would like to thank Prof. Michael Dudley for helpful guidance in x-ray topography and crystallographic analysis, G.G. Bryant and R. Lancto for their skilled assistance in crystal growth and data acquisition, and Dr. G. Iseler for many discussions. Topography was carried out at the Stony Brook Synchrotron Topography Facility, beamline X19C, at the National Synchrotron Light Source, at Brookhaven National Laboratory, which is supported by the US Department of Energy. This research effort was supported by the US Air Force Office of Scientific Research.

B.9 Czochralski Growth of Oxide Photorefractive Crystals
by Ernesto Diéguez, Jose Luis Plaza, Mohan D. Aggarwal, Ashok K. Batra

The authors gratefully acknowledge the support of the present work through NSF RISE grant # HRD-0531183. Special thanks are due to Mr. G. Sharp for fabrication of crystal growth system components. One of the authors (M.D.A.) would like to acknowledge support from NASA Administrator's Fellowship Program (NAFP) through United Negro College Fund Special Program (UNCFSP) Corporation under their contract #NNG06GC58A.

B.11 Growth and Characterization of Antimony-Based Narrow-Bandgap III–V Semiconductor Crystals for Infrared Detector Applications
by Vijay K. Dixit, Handady L. Bhat

The authors are grateful to the collaborators: B. Bansal and V. Venkataraman, Departments of Physics Indian Institute of Science, Bangalore; and B.M. Arora and K.S. Chandrasekaran, Solid State Group, TIFR Mumbai. The authors are also grateful to the many authors whose work is included in this review.

B.13 Laser-Heated Pedestal Growth of Oxide Fibers
by Marcello R.B. Andreeta, Antonio Carlos Hernandes

The authors would like to thank the Brazilian agencies CNPq, FAPESP, and CAPES for financial support and also all the editors and journals that kindly allowed the reproduction of the figures that illustrate this chapter.

B.14 Synthesis of Refractory Materials by Skull Melting Technique
by Vyacheslav V. Osiko, Mikhail A. Borik, Elena E. Lomonova

We wish to thank the following for help in the preparation of this chapter: Dr. O. M. Borik, Dr. V. A. Panov, and S. Semyova.

目 录

缩略语

Part B 熔体生长晶体技术

Contents

List of Abbreviations

μ-PD micro-pulling-down
1S-ELO one-step ELO structure
2-D two-dimensional
2-DNG two-dimensional nucleation growth
2S-ELO double layer ELO
3-D three-dimensional
4T quaterthiophene
6T sexithienyl
8MR eight-membered ring
8T hexathiophene

A

a-Si amorphous silicon
A/D analogue-to-digital
AA additional absorption
AANP 2-adamantylamino-5-nitropyridine
AAS atomic absorption spectroscopy
AB Abrahams and Burocchi
ABES absorption-edge spectroscopy
AC alternate current
ACC annular capillary channel
ACRT accelerated crucible rotation technique
ADC analog-to-digital converter
ADC automatic diameter control
ADF annular dark field
ADP ammonium dihydrogen phosphate
AES Auger electron spectroscopy
AFM atomic force microscopy
ALE arbitrary Lagrangian Eulerian
ALE atomic layer epitaxy
ALUM aluminum potassium sulfate
ANN artificial neural network
AO acoustooptic
AP atmospheric pressure
APB antiphase boundaries
APCF advanced protein crystallization facility
APD avalanche photodiode
APPLN aperiodic poled LN
APS Advanced Photon Source
AR antireflection
AR aspect ratio
ART aspect ratio trapping
ATGSP alanine doped triglycine sulfo-phosphate
AVT angular vibration technique

B

BA Born approximation
BAC band anticrossing
BBO BaB_2O_4
BCF Burton–Cabrera–Frank
BCT $Ba_{0.77}Ca_{0.23}TiO_3$
BCTi $Ba_{1-x}Ca_xTiO_3$
BE bound exciton
BF bright field
BFDH Bravais–Friedel–Donnay–Harker
BGO $Bi_{12}GeO_{20}$
BIBO BiB_3O_6
BLIP background-limited performance
BMO $Bi_{12}MO_{20}$
BN boron nitride
BOE buffered oxide etch
BPD basal-plane dislocation
BPS Burton–Prim–Slichter
BPT bipolar transistor
BS Bridgman–Stockbarger
BSCCO Bi–Sr–Ca–Cu–O
BSF bounding stacking fault
BSO $Bi_{20}SiO_{20}$
BTO $Bi_{12}TiO_{20}$
BU building unit
BaREF barium rare-earth fluoride
BiSCCO $Bi_2Sr_2CaCu_2O_n$

C

C–V capacitance–voltage
CALPHAD calculation of phase diagram
CBED convergent-beam electron diffraction
CC cold crucible
CCC central capillary channel
CCD charge-coupled device
CCVT contactless chemical vapor transport
CD convection diffusion
CE counterelectrode
CFD computational fluid dynamics
CFD cumulative failure distribution
CFMO Ca_2FeMoO_6
CFS continuous filtration system
CGG calcium gallium germanate
CIS copper indium diselenide
CL cathode-ray luminescence
CL cathodoluminescence
CMM coordinate measuring machine
CMO $CaMoO_4$
CMOS complementary metal–oxide–semiconductor
CMP chemical–mechanical polishing
CMP chemomechanical polishing

COD	calcium oxalate dihydrate
COM	calcium oxalate-monohydrate
COP	crystal-originated particle
CP	critical point
CPU	central processing unit
CRSS	critical-resolved shear stress
CSMO	$Ca_{1-x}Sr_xMoO_3$
CST	capillary shaping technique
CST	crystalline silico titanate
CT	computer tomography
CTA	$CsTiOAsO_4$
CTE	coefficient of thermal expansion
CTF	contrast transfer function
CTR	crystal truncation rod
CV	Cabrera–Vermilyea
CVD	chemical vapor deposition
CVT	chemical vapor transport
CW	continuous wave
CZ	Czochralski
CZT	Czochralski technique

D

D/A	digital to analog
DBR	distributed Bragg reflector
DC	direct current
DCAM	diffusion-controlled crystallization apparatus for microgravity
DCCZ	double crucible CZ
DCPD	dicalcium-phosphate dihydrate
DCT	dichlorotetracene
DD	dislocation dynamics
DESY	Deutsches Elektronen Synchrotron
DF	dark field
DFT	density function theory
DFW	defect free width
DGS	diglycine sulfate
DI	deionized
DIA	diamond growth
DIC	differential interference contrast
DICM	differential interference contrast microscopy
DKDP	deuterated potassium dihydrogen phosphate
DLATGS	deuterated L-alanine-doped triglycine sulfate
DLTS	deep-level transient spectroscopy
DMS	discharge mass spectroscopy
DNA	deoxyribonucleic acid
DOE	Department of Energy
DOS	density of states
DPH-BDS	2,6-diphenylbenzo[1,2-*b*:4,5-*b*']diselenophene
DPPH	2,2-diphenyl-1-picrylhydrazyl
DRS	dynamic reflectance spectroscopy
DS	directional solidification
DSC	differential scanning calorimetry
DSE	defect-selective etching
DSL	diluted Sirtl with light
DTA	differential thermal analysis
DTGS	deuterated triglycine sulfate
DVD	digital versatile disk
DWBA	distorted-wave Born approximation
DWELL	dot-in-a-well

E

EADM	extended atomic distance mismatch
EALFZ	electrical-assisted laser floating zone
EB	electron beam
EBIC	electron-beam-induced current
ECE	end chain energy
ECR	electron cyclotron resonance
EDAX	energy-dispersive x-ray analysis
EDMR	electrically detected magnetic resonance
EDS	energy-dispersive x-ray spectroscopy
EDT	ethylene dithiotetrathiafulvalene
EDTA	ethylene diamine tetraacetic acid
EELS	electron energy-loss spectroscopy
EFG	edge-defined film-fed growth
EFTEM	energy-filtered transmission electron microscopy
ELNES	energy-loss near-edge structure
ELO	epitaxial lateral overgrowth
EM	electromagnetic
EMA	effective medium theory
EMC	electromagnetic casting
EMCZ	electromagnetic Czochralski
EMF	electromotive force
ENDOR	electron nuclear double resonance
EO	electrooptic
EP	EaglePicher
EPD	etch pit density
EPMA	electron microprobe analysis
EPR	electron paramagnetic resonance
erfc	error function
ES	equilibrium shape
ESP	edge-supported pulling
ESR	electron spin resonance
EVA	ethyl vinyl acetate

F

F	flat
FAM	free abrasive machining
FAP	$Ca_5(PO_4)_3F$
FCA	free carrier absorption
fcc	face-centered cubic
FEC	full encapsulation Czochralski

FEM finite element method
FES fluid experiment system
FET field-effect transistor
FFT fast Fourier transform
FIB focused ion beam
FOM figure of merit
FPA focal-plane array
FPE Fokker–Planck equation
FSLI femtosecond laser irradiation
FT flux technique
FTIR Fourier-transform infrared
FWHM full width at half-maximum
FZ floating zone
FZT floating zone technique

G

GAME gel acupuncture method
GDMS glow-discharge mass spectrometry
GE General Electric
GGG gadolinium gallium garnet
GNB geometrically necessary boundary
GPIB general purpose interface bus
GPMD geometric partial misfit dislocation
GRI growth interruption
GRIIRA green-radiation-induced infrared absorption
GS growth sector
GSAS general structure analysis software
GSGG $Gd_3Sc_2Ga_3O_{12}$
GSMBE gas-source molecular-beam epitaxy
GSO Gd_2SiO_5
GU growth unit

H

HA hydroxyapatite
HAADF high-angle annular dark field
HAADF-STEM high-angle annular dark field in scanning transmission electron microscope
HAP hydroxyapatite
HB horizontal Bridgman
HBM Hottinger Baldwin Messtechnik GmbH
HBT heterostructure bipolar transistor
HBT horizontal Bridgman technique
HDPCG high-density protein crystal growth
HE high energy
HEM heat-exchanger method
HEMT high-electron-mobility transistor
HF hydrofluoric acid
HGF horizontal gradient freezing
HH heavy-hole
HH-PCAM handheld protein crystallization apparatus for microgravity
HIV human immunodeficiency virus
HIV-AIDS human immunodeficiency virus–acquired immunodeficiency syndrome
HK high potassium content
HLA half-loop array
HLW high-level waste
HMDS hexamethyldisilane
HMT hexamethylene tetramine
HNP high nitrogen pressure
HOE holographic optical element
HOLZ higher-order Laue zone
HOMO highest occupied molecular orbital
HOPG highly oriented pyrolytic graphite
HOT high operating temperature
HP Hartman–Perdok
HPAT high-pressure ammonothermal technique
HPHT high-pressure high-temperature
HRTEM high-resolution transmission electron microscopy
HRXRD high-resolution x-ray diffraction
HSXPD hemispherically scanned x-ray photoelectron diffraction
HT hydrothermal
HTS high-temperature solution
HTSC high-temperature superconductor
HVPE halide vapor-phase epitaxy
HVPE hydride vapor-phase epitaxy
HWC hot-wall Czochralski
HZM horizontal ZM

I

IBAD ion-beam-assisted deposition
IBE ion beam etching
IC integrated circuit
IC ion chamber
ICF inertial confinement fusion
ID inner diameter
ID inversion domain
IDB incidental dislocation boundary
IDB inversion domain boundary
IF identification flat
IG inert gas
IK intermediate potassium content
ILHPG indirect laser-heated pedestal growth
IML-1 International Microgravity Laboratory
IMPATT impact ionization avalanche transit-time
IP image plate
IPA isopropyl alcohol
IR infrared
IRFPA infrared focal plane array
IS interfacial structure
ISS ion-scattering spectroscopy
ITO indium-tin oxide
ITTFA iterative target transform factor analysis
IVPE iodine vapor-phase epitaxy

J

JDS	joint density of states
JFET	junction FET

K

K	kinked
KAP	potassium hydrogen phthalate
KDP	potassium dihydrogen phosphate
KGW	$KY(WO_4)_2$
KGdP	$KGd(PO_3)_4$
KLYF	$KLiYF_5$
KM	Kubota–Mullin
KMC	kinetic Monte Carlo
KN	$KNbO_3$
KNP	$KNd(PO_3)_4$
KPZ	Kardar–Parisi–Zhang
KREW	$KRE(WO_4)_2$
KTA	potassium titanyl arsenate
KTN	potassium niobium tantalate
KTP	potassium titanyl phosphate
KTa	$KTaO_3$
KTaN	$KTa_{1-x}Nb_xO_3$
KYF	KYF_4
KYW	$KY(WO_4)_2$

L

LACBED	large-angle convergent-beam diffraction
LAFB	L-arginine tetrafluoroborate
LAGB	low-angle grain boundary
LAO	$LiAlO_2$
LAP	L-arginine phosphate
LBIC	light-beam induced current
LBIV	light-beam induced voltage
LBO	LiB_3O_5
LBO	$LiBO_3$
LBS	laser-beam scanning
LBSM	laser-beam scanning microscope
LBT	laser-beam tomography
LCD	liquid-crystal display
LD	laser diode
LDT	laser-induced damage threshold
LEC	liquid encapsulation Czochralski
LED	light-emitting diode
LEEBI	low-energy electron-beam irradiation
LEM	laser emission microanalysis
LEO	lateral epitaxial overgrowth
LES	large-eddy simulation
LG	$LiGaO_2$
LGN	$La_3Ga_{5.5}Nb_{0.5}O_{14}$
LGO	$LaGaO_3$
LGS	$La_3Ga_5SiO_{14}$
LGT	$La_3Ga_{5.5}Ta_{0.5}O_{14}$
LH	light hole
LHFB	L-histidine tetrafluoroborate
LHPG	laser-heated pedestal growth
LID	laser-induced damage
LK	low potassium content
LLNL	Lawrence Livermore National Laboratory
LLO	laser lift-off
LLW	low-level waste
LN	$LiNbO_3$
LP	low pressure
LPD	liquid-phase diffusion
LPE	liquid-phase epitaxy
LPEE	liquid-phase electroepitaxy
LPS	$Lu_2Si_2O_7$
LSO	Lu_2SiO_5
LST	laser scattering tomography
LST	local shaping technique
LT	low-temperature
LTa	$LiTaO_3$
LUMO	lowest unoccupied molecular orbital
LVM	local vibrational mode
LWIR	long-wavelength IR
LY	light yield
LiCAF	$LiCaAlF_6$
LiSAF	lithium strontium aluminum fluoride

M

M–S	melt–solid
MAP	magnesium ammonium phosphate
MASTRAPP	multizone adaptive scheme for transport and phase change processes
MBE	molecular-beam epitaxy
MBI	multiple-beam interferometry
MC	multicrystalline
MCD	magnetic circular dichroism
MCT	HgCdTe
MCZ	magnetic Czochralski
MD	misfit dislocation
MD	molecular dynamics
ME	melt epitaxy
ME	microelectronics
MEMS	microelectromechanical system
MESFET	metal-semiconductor field effect transistor
MHP	magnesium hydrogen phosphate-trihydrate
MI	morphological importance
MIT	Massachusetts Institute of Technology
ML	monolayer
MLEC	magnetic liquid-encapsulated Czochralski

MLEK	magnetically stabilized liquid-encapsulated Kyropoulos
MMIC	monolithic microwave integrated circuit
MNA	2-methyl-4-nitroaniline
MNSM	modified nonstationary model
MOCVD	metalorganic chemical vapor deposition
MOCVD	molecular chemical vapor deposition
MODFET	modulation-doped field-effect transistor
MOMBE	metalorganic MBE
MOS	metal–oxide–semiconductor
MOSFET	metal–oxide–semiconductor field-effect transistor
MOVPE	metalorganic vapor-phase epitaxy
mp	melting point
MPMS	mold-pushing melt-supplying
MQSSM	modified quasi-steady-state model
MQW	multiple quantum well
MR	melt replenishment
MRAM	magnetoresistive random-access memory
MRM	melt replenishment model
MSUM	monosodium urate monohydrate
MTDATA	metallurgical thermochemistry database
MTS	methyltrichlorosilane
MUX	multiplexor
MWIR	mid-wavelength infrared
MWRM	melt without replenishment model
MXRF	micro-area x-ray fluorescence

N

N	nucleus
N	nutrient
NASA	National Aeronautics and Space Administration
NBE	near-band-edge
NBE	near-bandgap emission
NCPM	noncritically phase matched
NCS	neighboring confinement structure
NGO	$NdGaO_3$
NIF	National Ignition Facility
NIR	near-infrared
NIST	National Institute of Standards and Technology
NLO	nonlinear optic
NMR	nuclear magnetic resonance
NP	no-phonon
NPL	National Physical Laboratory
NREL	National Renewable Energy Laboratory
NS	Navier–Stokes
NSF	National Science Foundation
nSLN	nearly stoichiometric lithium niobate
NSLS	National Synchrotron Light Source
NSM	nonstationary model
NTRS	National Technology Roadmap for Semiconductors
NdBCO	$NdBa_2Cu_3O_{7-x}$

O

OCP	octacalcium phosphate
ODE	ordinary differential equation
ODLN	opposite domain LN
ODMR	optically detected magnetic resonance
OEIC	optoelectronic integrated circuit
OF	orientation flat
OFZ	optical floating zone
OLED	organic light-emitting diode
OMVPE	organometallic vapor-phase epitaxy
OPO	optical parametric oscillation
OSF	oxidation-induced stacking fault

P

PAMBE	photo-assisted MBE
PB	proportional band
PBC	periodic bond chain
pBN	pyrolytic boron nitride
PC	photoconductivity
PCAM	protein crystallization apparatus for microgravity
PCF	primary crystallization field
PCF	protein crystal growth facility
PCM	phase-contrast microscopy
PD	Peltier interface demarcation
PD	photodiode
PDE	partial differential equation
PDP	programmed data processor
PDS	periodic domain structure
PE	pendeo-epitaxy
PEBS	pulsed electron beam source
PEC	polyimide environmental cell
PECVD	plasma-enhanced chemical vapor deposition
PED	pulsed electron deposition
PEO	polyethylene oxide
PET	positron emission tomography
PID	proportional–integral–differential
PIN	positive intrinsic negative diode
PL	photoluminescence
PLD	pulsed laser deposition
PMNT	$Pb(Mg, Nb)_{1-x}Ti_xO_3$
PPKTP	periodically poled KTP
PPLN	periodic poled LN
PPLN	periodic poling lithium niobate
ppy	polypyrrole
PR	photorefractive
PSD	position-sensitive detector
PSF	prismatic stacking fault

PSI phase-shifting interferometry
PSM phase-shifting microscopy
PSP pancreatic stone protein
PSSM pseudo-steady-state model
PSZ partly stabilized zirconium dioxide
PT pressure–temperature
PV photovoltaic
PVA polyvinyl alcohol
PVD physical vapor deposition
PVE photovoltaic efficiency
PVT physical vapor transport
PWO $PbWO_4$
PZNT $Pb(Zn, Nb)_{1-x}Ti_xO_3$
PZT lead zirconium titanate

Q

QD quantum dot
QDT quantum dielectric theory
QE quantum efficiency
QPM quasi-phase-matched
QPMSHG quasi-phase-matched second-harmonic generation
QSSM quasi-steady-state model
QW quantum well
QWIP quantum-well infrared photodetector

R

RAE rotating analyzer ellipsometer
RBM rotatory Bridgman method
RC reverse current
RCE rotating compensator ellipsometer
RE rare earth
RE reference electrode
REDG recombination enhanced dislocation glide
RELF rare-earth lithium fluoride
RF radiofrequency
RGS ribbon growth on substrate
RHEED reflection high-energy electron diffraction
RI refractive index
RIE reactive ion etching
RMS root-mean-square
RNA ribonucleic acid
ROIC readout integrated circuit
RP reduced pressure
RPI Rensselaer Polytechnic Institute
RSM reciprocal space map
RSS resolved shear stress
RT room temperature
RTA $RbTiOAsO_4$
RTA rapid thermal annealing
RTCVD rapid-thermal chemical vapor deposition
RTP $RbTiOPO_4$
RTPL room-temperature photoluminescence
RTR ribbon-to-ribbon
RTV room temperature vulcanizing
R&D research and development

S

S stepped
SAD selected area diffraction
SAM scanning Auger microprobe
SAW surface acoustical wave
SBN strontium barium niobate
SC slow cooling
SCBG slow-cooling bottom growth
SCC source-current-controlled
SCF single-crystal fiber
SCF supercritical fluid technology
SCN succinonitrile
SCW supercritical water
SD screw dislocation
SE spectroscopic ellipsometry
SECeRTS small environmental cell for real-time studies
SEG selective epitaxial growth
SEM scanning electron microscope
SEM scanning electron microscopy
SEMATECH Semiconductor Manufacturing Technology
SF stacking fault
SFM scanning force microscopy
SGOI SiGe-on-insulator
SH second harmonic
SHG second-harmonic generation
SHM submerged heater method
SI semi-insulating
SIA Semiconductor Industry Association
SIMS secondary-ion mass spectrometry
SIOM Shanghai Institute of Optics and Fine Mechanics
SL superlattice
SL-3 Spacelab-3
SLI solid–liquid interface
SLN stoichiometric LN
SM skull melting
SMB stacking mismatch boundary
SMG surfactant-mediated growth
SMT surface-mount technology
SNR signal-to-noise ratio
SNT sodium nonatitanate
SOI silicon-on-insulator
SP sputtering
sPC scanning photocurrent
SPC Scientific Production Company
SPC statistical process control
SR spreading resistance
SRH Shockley–Read–Hall
SRL strain-reducing layer
SRS stimulated Raman scattering

SRXRD spatially resolved XRD
SS solution-stirring
SSL solid-state laser
SSM sublimation sandwich method
ST synchrotron topography
STC standard testing condition
STE self-trapped exciton
STEM scanning transmission electron microscopy
STM scanning tunneling microscopy
STOS sodium titanium oxide silicate
STP stationary temperature profile
STS space transportation system
SWBXT synchrotron white beam x-ray topography
SWIR short-wavelength IR
SXRT synchrotron x-ray topography

T

TCE trichloroethylene
TCNQ tetracyanoquinodimethane
TCO thin-film conducting oxide
TCP tricalcium phosphate
TD Tokyo Denpa
TD threading dislocation
TDD threading dislocation density
TDH temperature-dependent Hall
TDMA tridiagonal matrix algorithm
TED threading edge dislocation
TEM transmission electron microscopy
TFT-LCD thin-film transistor liquid-crystal display
TGS triglycine sulfate
TGT temperature gradient technique
TGW Thomson–Gibbs–Wulff
TGZM temperature gradient zone melting
THM traveling heater method
TMCZ transverse magnetic-field-applied Czochralski
TMOS tetramethoxysilane
TO transverse optic
TPB three-phase boundary
TPRE twin-plane reentrant-edge effect
TPS technique of pulling from shaper
TQM total quality management
TRAPATT trapped plasma avalanche-triggered transit
TRM temperature-reduction method
TS titanium silicate
TSC thermally stimulated conductivity
TSD threading screw dislocation
TSET two shaping elements technique
TSFZ traveling solvent floating zone
TSL thermally stimulated luminescence
TSSG top-seeded solution growth
TSSM Tatarchenko steady-state model
TSZ traveling solvent zone
TTV total thickness variation
TV television
TVM three-vessel solution circulating method
TVTP time-varying temperature profile
TWF transmitted wavefront
TZM titanium zirconium molybdenum
TZP tetragonal phase

U

UC universal compliant
UDLM uniform-diffusion-layer model
UHPHT ultrahigh-pressure high-temperature
UHV ultrahigh-vacuum
ULSI ultralarge-scale integrated circuit
UV ultraviolet
UV-vis ultraviolet–visible
UVB ultraviolet B

V

VAS void-assisted separation
VB valence band
VB vertical Bridgman
VBT vertical Bridgman technique
VCA virtual-crystal approximation
VCSEL vertical-cavity surface-emitting laser
VCZ vapor pressure controlled Czochralski
VDA vapor diffusion apparatus
VGF vertical gradient freeze
VLS vapor–liquid–solid
VLSI very large-scale integrated circuit
VLWIR very long-wavelength infrared
VMCZ vertical magnetic-field-applied Czochralski
VP vapor phase
VPE vapor-phase epitaxy
VST variable shaping technique
VT Verneuil technique
VTGT vertical temperature gradient technique
VUV vacuum ultraviolet

W

WBDF weak-beam dark-field
WE working electrode

X

XP x-ray photoemission
XPS x-ray photoelectron spectroscopy
XPS x-ray photoemission spectroscopy
XRD x-ray diffraction
XRPD x-ray powder diffraction
XRT x-ray topography

Y

YAB $YAl_3(BO_3)_4$
YAG yttrium aluminum garnet
YAP yttrium aluminum perovskite
YBCO $YBa_2Cu_3O_{7-x}$
YIG yttrium iron garnet
YL yellow luminescence
YLF $LiYF_4$
YOF yttrium oxyfluoride
YPS $(Y_2)Si_2O_7$
YSO Y_2SiO_5

Z

ZA Al_2O_3-$ZrO_2(Y_2O_3)$
ZLP zero-loss peak
ZM zone-melting
ZNT ZN-Technologies
ZOLZ zero-order Laue zone

Crystal Gr

Part B

Part B Crystal Growth from Melt Techniques

7. Indium Phosphide: Crystal Growth and Defect Control by Applying Steady Magnetic Fields

David F. Bliss

The application of steady magnetic fields during crystal growth of indium phosphide is described, and the effect of the magnetic fields on crystal properties is analyzed. The use of magnetic fields is one of many engineering controls that can improve homogeneity and crystal quality. This method is especially relevant to InP because of the high pressure requirement for crystal growth. Under high pressure, fluid flows in the melt and in the gas environment can become uncontrolled and turbulent, with negative effects on crystal quality and reproducibility. If properly configured, a steady magnetic field can reduce random oscillatory motion in the melt and reduce the likelihood of defect formation during growth. This chapter presents the history and development of magnetic-field-assisted growth of InP and an analysis of the effects of applied fields on crystal quality.

7.1 Historical Overview

The development of bulk InP crystal growth has been historically driven by the commercial telecommunications market for solid-state lasers. InP was chosen because it has a lattice match with the semiconductor compounds that serve as laser materials for fiber-optic communications. Small 2 in diameter wafers were sufficient from the outset to fabricate large volumes of discrete lasers with commercially viable yields. Because of the small size of the lasers, the demand for larger-diameter wafers was slow to gain momentum. Later, during the 1990s, new applications for optoelectronic devices requiring larger-area InP substrates led to the development of 4 in diameter InP crystals. As recently as 2006, a new application for high-speed laser modulation was developed using 6 in InP wafers bonded to silicon [7.1].

Crystal growth of large-diameter InP boules places extreme control demands on the crystal growth environment, which becomes increasingly chaotic as the size of the melt increases. At the melting temperature of 1060 °C, the equilibrium vapor pressure of phosphorus over InP is 27 atm. Controlled crystal growth must be performed in a pressure vessel exceeding 30 atm. Two principal issues arising from this pressure dependence are the synthesis of a stoichiometric InP compound and the containment of a highly volatile melt. Experiments on compound semiconductor crystal growth from volatile compounds in the 1960s focused on two possibilities: either the crystal must be grown in a completely sealed hot-wall ampoule (Bridgman or gradient freeze method), or the crystal can be pulled using an encapsulant under pressure. Using a pressure-balancing technique, a breakthrough came in 1962 when *Metz* and a group at Westinghouse [7.2] developed the liquid-encapsulated Czochralski (LEC) method for growth of PbSe and PbTe. The LEC method was later adapted for growth of InP using B_2O_3 as an encapsulant [7.3–5].

The state of the art today for bulk InP crystal growth is still divided between two competing technologies: top-seeded crystal pulling with liquid encapsulation and vertical growth in a container with bottom seeding. Various names for these techniques have arisen from each unique laboratory research effort. In general, the pulling method has been the most cost-effective, but with high thermal gradients and a high level of strain during growth, the crystals may have high dislocation densities. On the other hand, vertical container growth offers a very low dislocation density because of its low-stress environment, but it is plagued by yield problems due to twinning and interface breakdown in heavily doped crystals. Two recent reviews [7.6, 7] have compared the growth, characteristics, and applications of InP crystals grown by both techniques. This chapter focuses on the development of techniques to control the turbulent melt environment using applied magnetic fields, and characterization to determine the electrical and optical properties of InP crystals after growth and thermal treatment.

7.2 Magnetic Liquid-Encapsulated Growth

Growth of InP in an axial magnetic field has two main advantages: melt stability contributes to reduced incidence of twinning and improved dopant uniformity. On the other hand, the radial temperature gradient is increased by an axial magnetic field because convective heat transfer in the melt is reduced. The resulting convex interface contributes to large hoop stresses during crystal growth, contributing to dislocation multiplication and propagation. On the other hand, if a cusped magnetic field is employed (with two opposing vertical magnetic fields), the radial thermal gradients are predicted to be flatter, since the nonuniform field will not affect radial fluid flow near the surface. Flattening the melt isotherm should reduce the dislocation density of InP crystals. In this chapter we will compare the cusped with the axial magnetic field for InP bulk crystal growth, and evaluate the practicality of imposing a strong axial magnetic field on the melt during growth. To determine if there are advantages in terms of process control, we compare crystallographic defects such as twins, dislocations, and striations under either controlled or uncontrolled environments during crystal growth. A major goal is to determine the effect on melt convection of magnetic field configuration, either axial or cusped.

7.2.1 Evolution of Crystal Growth Under Applied Magnetic Fields

In the 1960s several authors [7.8, 9] pointed out that the application of an external magnetic field would be useful to damp the time-dependent turbulent convective flows in melts during the growth of semiconductors. Experimental support for this concept did not appear in research laboratories until the 1980s [7.10, 11], and more recently it has attracted commercial interest for growth of InP crystals. There are two main advantages which result from magnetic field growth:

1. Thermal fluctuations are reduced, thereby stabilizing the microscopic growth rate, and consequently
2. The diffusion boundary layer is increased, enhancing the uniformity of dopant distribution.

Natural convection and forced convection are two coexisting forces during crystal pulling from the melt, and the magnitude and direction of fluid flow plays a crucial role in heat and mass transfer during growth. In Czochralski growth, both the thermal field and the solute distribution are dominated by convection. However, since most semiconductor melts are metallic in nature, changes to the flow dynamics can be made by applying a magnetic field. With the application of a static magnetic field, convective forces in the liquid are restricted when the flow crosses the magnetic lines of flux by the Lorentz force $\boldsymbol{L}$

$$\boldsymbol{L} = \boldsymbol{j} \times \boldsymbol{B} , \tag{7.1}$$

where $\boldsymbol{j}$ is the ionic current density vector and $\boldsymbol{B}$ is the applied magnetic field. The characteristic ratio between the electromagnetic (EM) body force and the viscous force is Ha^2, where the Hartmann number is

$$\mathrm{Ha} = BR\left(\frac{\sigma}{\mu}\right)^{1/2} , \tag{7.2}$$

where B is the magnetic flux density of the magnetic field, R is the characteristic dimension of the melt, σ is the electrical conductivity of the melt, and μ is the dynamic viscosity of the melt. As the Hartmann number increases, the characteristic velocity of the melt is reduced. Although it is not practical to produce a magnetic flux density sufficiently large to eliminate all melt motion due to natural and forced convection, a strategy of controlling melt motion for dopant uniformity can be achieved with moderate magnetic fields. The model of *Burton* et al. [7.12] reveals the effect of melt convection on crystal growth. A diffusion boundary layer δ is assumed, beyond which the melt composition is maintained uniform by convection and inside of which transport is by diffusion only. The boundary layer is thus confined by the characteristic melt velocity. After steady state is reached, for an infinite liquid, one can define an effective distribution coefficient k_{eff}

$$\frac{C_{\mathrm{S}}}{C_{\mathrm{L}}} \equiv k_{\mathrm{eff}} = \frac{k_0}{k_0 + (1 - k_0)\mathrm{e}^{R\delta/D}} , \tag{7.3}$$

where R is the growth rate and D is the diffusion coefficient for the solute in the melt. This is a useful expression because it relates the composition of the growing crystal to convection conditions and the growth rate. It describes the dopant distribution provided that the thickness of the boundary layer is small compared with the extent of the crucible. For example, an increase in the growth rate or the boundary-layer thickness tends to enhance the effective distribution coefficient towards unity. In the case of an applied magnetic field, the diffusion boundary-layer thickness increases, and hence k_{eff} is closer to 1, resulting in a more uniform incorporation of the dopant. The dopant distribution in magnetic field growth has been modeled by several authors [7.13, 14]. Experimental confirmation of this model will be discussed in Sect. 7.3.2.

7.2.2 Crystal Shaping Measures

The prevailing crystal pulling method is liquid-encapsulated Czochralski growth, named after *Czochralski* [7.15], a Polish scientist who developed a pulling method for growth of metallic rods. Although his work was focused on thin metal rods, Czochralski's name has become associated with the characteristic shape of large single crystals pulled from the melt. Magnetic liquid-encapsulated Czochralski (MLEC) growth is a variant of Czochralski growth with improved control over melt turbulence by application of a static magnetic field. Crystal pulling is initiated immediately after seeding, and as the melt temperature is reduced, a sloped shoulder emerges from the melt with a grow-out angle between 60 and 80°. The phenomenon of growth twinning is a problem in the shoulder region because the shoulder angle may traverse a critical angle where twinning is likely to occur. As will be shown in Sect. 7.3.2, to avoid the problem of growth twinning, the shape of the crystal must be controlled reproducibly. In Fig. 7.1 we see an example of an InP MLEC crystal grown with a controlled shape to avoid the critical angle

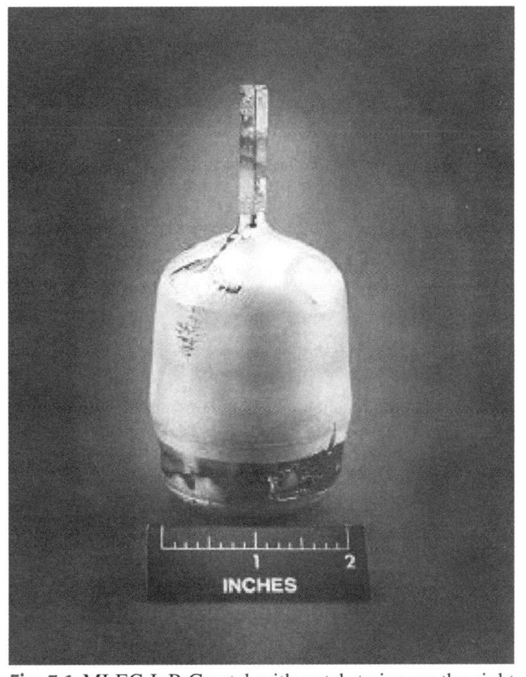

Fig. 7.1 MLEC InP Crystal with patch twins on the right shoulder (grown at the US Air Force Research Laboratory, Hanscom AFB)

Fig. 7.2 Twin-free MLEK InP crystal (grown at the US Air Force Research Laboratory, Hanscom AFB)

for twinning. Despite the controlled angle, *patch* twins are visible on the shoulders. These twins are relatively benign, as most of the crystal remains usable.

Research at several laboratories has taken advantage of crystal shaping as a means to minimize the incidence of twinning [7.16]. Magnetically stabilized liquid-encapsulated Kyropoulos (MLEK) [7.17–19] growth is one such technique, where the crystal is grown with a flat top before initiation of pulling, in order to reduce the incidence of twins (Fig. 7.2). When pulling commences, the crystal has already reached full diameter, and the growth angle changes from perpendicular to parallel to the growth axis. At that moment the growth angle must pass through the critical range, but facet formation is reduced because thermal gradients near the periphery are steeper than at the center. This method relies on magnetic stabilization of the melt to suppress turbulent flows at the solid–liquid interface. Without the magnetic field, the flat top would become unstable because of perturbations due to the temperature fluctuations of natural convection.

It is worthwhile to note here a peculiar historical twist. The Kyropoulos method is named after Spyro Kyropoulos, a German scientist (1911–1967) who later emigrated to the USA. His technique for growth of large alkali halide crystals was conceived to avoid the cracking problems arising from container growth [7.20]. The Kyropoulos method was further developed during the 1930s and 1940s by several groups to grow large optical crystals of alkali halides for spectroscopy. Until World War II (WWII), it was considered to be one of the leading techniques for growing large single crystals. In a survey of crystal growth techniques from 1930 to 1946, *Wells* [7.21] referred to only three general methods for obtaining single crystals: Bridgman growth, Kyropoulos growth, and solid-state recrystallization. The Czochralski method was not considered as a method for large crystal growth. Nevertheless, after WWII when germanium and silicon crystal growth were demonstrated at Bell Labs and Texas Instruments, the name of the new process was assigned to Czochralski. Coincidentally, the language of science was shifting at that time from German to English, a fact that may explain the misunderstanding. Because of this historical twist, some clarification is in order to distinguish

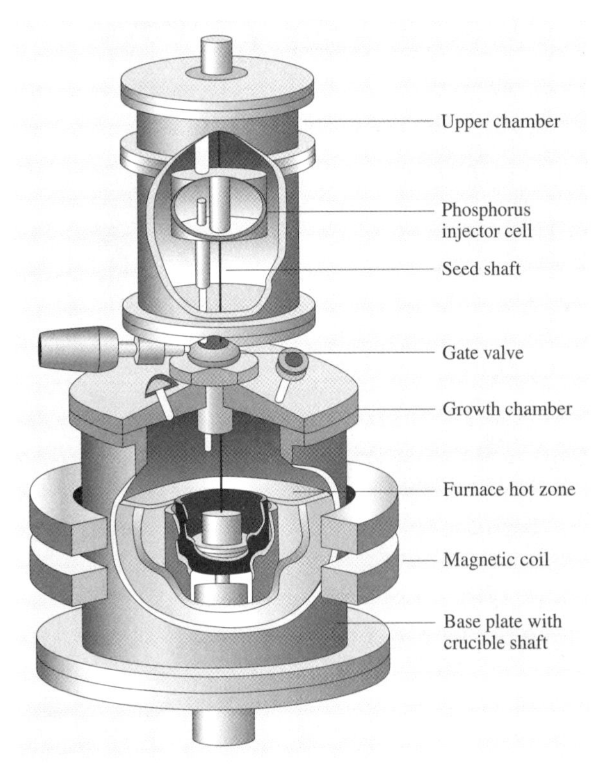

Fig. 7.3 Schematic diagram of a magnetic crystal growth furnace

the two techniques. The main distinction is the pronounced difference between the crystal shapes, and the consequent differences between the curvatures of the solid–liquid interface. With the Kyropoulos method, the crystal forms an ellipsoid of rotation with its center deep in the melt. On the other hand, with the sloping shoulders of Czochralski growth, the solid–liquid interface is flatter.

7.2.3 Apparatus for Magnetically Stabilized Crystal Growth

The schematic diagram in Fig. 7.3 shows the basic components required for MLEC or MLEK growth. This custom furnace design was realized and is now in operation at the author's laboratory. Positioned around the outside of the growth chamber is a large Helmholtz coil consisting of two toroidal magnets to provide an axisymmetric field up to 0.4 T. The two coils may be operated in tandem to provide an axial field, or in opposition to provide a cusped field. For a cusped field, the north pole of one coil points in the opposite direction to the other. The growth chamber is fabricated from nonmagnetic stainless steel to minimize screening losses and to allow the magnetic field to interact with the molten semiconductor. The water-cooled chamber is designed to operate at 20–40 atm ambient pressure. The upper chamber can be raised to allow access to the seed rod and a phosphorus ampoule. The upper chamber is separated by an isolation valve from the lower chamber, so that the phosphorus ampoule can be removed after in situ compounding of InP is completed. A seed is then attached to the seed rod, which can then be lowered into the growth chamber for crystal growth.

7.3 Magnetic Field Interactions with the Melt

The choice of magnetic field configuration and crystal shape are two types of engineering controls that can be exploited to improve the quality of InP crystals, with respect to twins, dislocations, and striations. In this chapter, the properties of InP crystals grown using these engineering controls are evaluated by synchrotron white-beam x-ray topography (SWBXT), chemical etching, and photoluminescence. Other controls, such as thermal gradient control and vapor pressure control, involving hardware modifications to the hot zone, are also considered for their practical application.

7.3.1 Hydrodynamic Principle

Although realizable magnetic flux densities are not large enough to eliminate melt motion altogether, a moderate magnetic field of 0.1–0.3 T can be used to tailor the melt motion for practical control of the crystal growth process. There are infinitely many ways to tailor the strength and configuration of the externally applied magnetic field, as well as the rotation rates, distribution of heat flux into and out of the melt, etc., so that models to accurately predict the system behavior are needed for process optimization. *Hurle* and *Series* [7.23] and *Walker* [7.24] reviewed the literature on the use of magnetic fields during bulk growth of semiconductors. Various components of fluid flow in InP melts have been modeled: melt-depletion flow [7.22], forced convection resulting from the rotational effects of the crystal and crucible [7.25], and natural buoyant convection [7.26]. A global model combining the effects of pressure, thermal flux, magnetic field, and stress in the crystal was developed by *Zhang* and *Prasad* [7.27,28]. The characteristic velocity U_c of melt motion in the magnetically

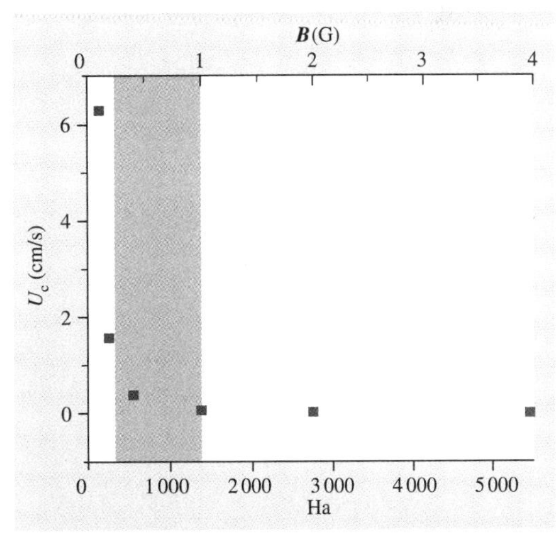

Fig. 7.4 Average melt velocity as a function of applied axial magnetic field and Hartmann number, calculated from the model of [7.22]. The *shaded area* shows the range where EM damping is sufficient for effective crystal growth control

damped melt is a critical indicator of EM suppression. This is shown graphically in Fig. 7.4, where calculated values of the average velocity are plotted against the Hartmann number [7.26]. The shaded area is the region where practical values of the magnetic flux can be used to control crystal shape and doping concentration for a given crucible design.

7.3.2 Effect of Magnetic Field on Crystal Twinning

The appearance of growth twins during the bulk crystal growth of InP is a recurrent yield-limiting problem. Because of their low stacking fault energy, indium-containing III–V compounds such as InP and InSb may have a higher tendency to form twins than other zincblende crystals. However, the general problem of twins in III–V crystal growth has been studied by experimentalists for many years [7.31, 32]. The first full theoretical treatment of twinning in III–V compound semiconductors was offered by *Hurle* in 1995 [7.29]. He observed that there were certain conditions under which internal (111) edge facets could be anchored to the three-phase boundary (TPB) at the solid–liquid–encapsulant interface. In addition, the appearance of an external shoulder facet coincides in many instances with the point of internal (111) facet anchoring. A diagram of the Hurle model in Fig. 7.5 shows the shouldering angle α and the angle ν between the edge facets and the extension of the crystal surface.

From a calculation of thermodynamic conditions for each orientation, Hurle derived a range of shoulder angles for which twinning was likely to occur. The model compared the undercooling at a high-index high-surface-energy external facet with that of a low-index low-surface-energy facet to determine if formation of a twin nucleus was energetically favorable. Hurle used this approach to predict the most dangerous shouldering angles for several zincblende crystals, including InP grown in the $\langle 001\rangle$ and $\langle 111\rangle$ directions. An experimental verification of this model was later published [7.33], where modifications to the Hurle theory were reported. These modifications were based on direct observations of the nucleation of twins on edge facets anchored to the TPB in S-doped MLEC-grown $\langle 001\rangle$ InP single crystals using SWBXT and optical microscopy.

Undercooling at a facet during crystal pulling is determined by the nucleation process for a particular crystal face. *Brice* [7.30] pointed out that, to a simple approximation, the temperature gradient G in the solid at the interface depends on the radius of the crystal. If the curve of the interface passes through any low-index face, a facet will develop with undercooling ΔT. The crystal diameter is reduced by a chord formed by the facet, reducing the radius by a small amount (dimension c). As shown in Fig. 7.6, the undercooling at the center of the facet is given by

$$\Delta T = Gc\,. \tag{7.4}$$

The growth rate at the facet is controlled by this undercooling. The geometric relation between the facet half-width w and the temperature gradient can then be derived as

$$w^2 = c(2R - c)\,, \tag{7.5}$$

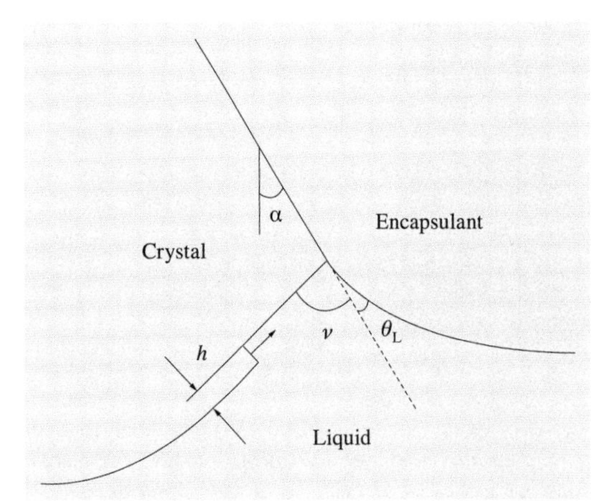

Fig. 7.5 Diagram showing the attachment of an internal edge facet to the three-phase boundary (TPB) (after [7.29])

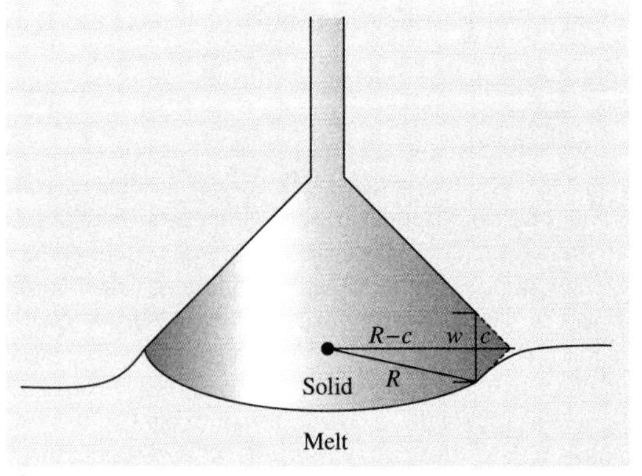

Fig. 7.6 Cross section (viewed from an angle slightly above the melt) schematic of a crystal with a shoulder facet (after [7.30])

and since the value of $R \gg c$ we can combine (7.3) and (7.4) to get

$$w^2 = \frac{2\Delta TR}{G} . \tag{7.6}$$

For a given crystal radius and a fixed growth rate, the undercooling is proportional to the square of the facet size. As the crystal diameter increases, so will the facet size, unless the undercooling is too small to sustain the facet. If undercooling is too large, the system becomes inherently unstable, and seeks a lower free energy.

In verifying the predicted relationship between twinning and growth angle, several [001] InP crystals were grown by the author with various shoulder angles from a 35° sloped shoulder to a 90° flat top. The incidence of twinning is generally higher in S-doped crystals than in crystals grown with other dopants, further motivating an investigation of the twinning defect. Study of the twins in these crystals by visual observation reveals a familiar pattern. Two kinds of twins exist: *patch twins*, which grow like epaulets on the shoulder, and *standard twins*, which grow on {111} planes passing through the boule center. Both types of twins nucleate on [110] ridge lines, but the patch twins are somewhat more benign in terms of wafer yield because they soon grow out of the crystal. A trend was observed after studying many crystals: twinning seems to occur more often on broad-shouldered crystals, but not so frequently on either low-angle shoulders or flat-top crystals. However, these experiments show a marked difference in twinning probability depending on the field configuration.

Axial Field Growth

Magnetic stabilization using an axial magnetic field is a useful engineering control for controlling the InP growth process. As illustrated in Fig. 7.4, the melt velocity is reduced by an order of magnitude with an applied field of 0.2 T. Because of melt stabilization, dopant incorporation is more uniform, and shape control becomes reproducible. From the point of view of doping uniformity, it was possible to determine the effect on the solute boundary layer by comparing magnetically stabilized growth and standard LEC growth of InP [7.17]. To test the effect of magnetic fields on interface shape and dopant uniformity, two tin-doped crystals were grown, one without an applied magnetic field, and the other with an axial field of 0.2 T. Both were grown from melts weighing about 1 kg with tin concentrations of $5 \times 10^{19}\,\mathrm{cm}^{-3}$. After growth, these crystals were sliced into samples representing by weight each fraction of the grown crystal. The slices were then analyzed by glow-discharge mass spectrometry (GDMS). A plot of the tin concentration versus fraction grown was used to determine the effective distribution coefficient k_{eff} for each process. A comparison of the axial doping profiles revealed that k_{eff} is 2.5 times closer to unity for the magnetically stabilized growth. The

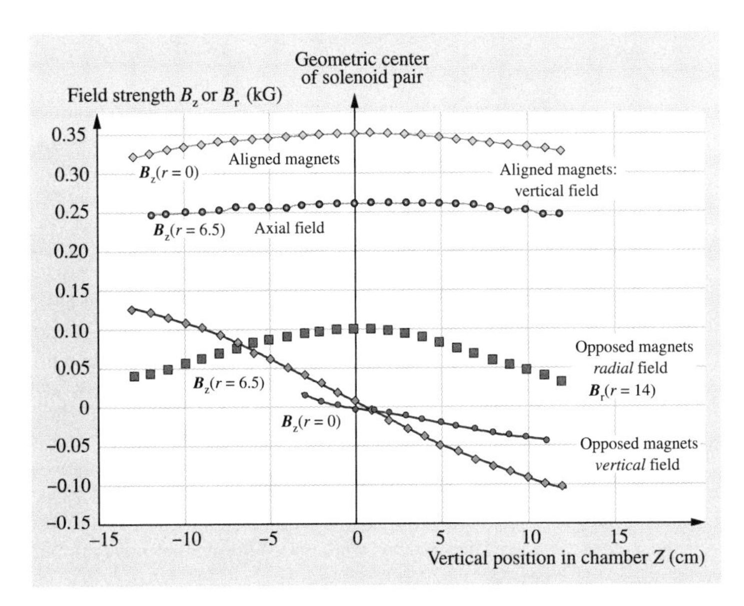

Fig. 7.7 A map of magnetic field strength for axial and cusped fields inside the crystal growth chamber at constant power

difference in k_{eff} indicates that static magnetic fields increase the boundary-layer thickness δ at the melt–solid interface. Considering the Burton, Prim and Schlichter (BPS) analysis in (7.3), one can see that k_{eff} is an exponential function of δ, since the growth rate R was the same in both cases, and the value of diffusivity D_L is assumed to be independent of the applied magnetic field. Using this analysis, the boundary-layer thickness is found to be 3.5 times larger for 0.2 T axial magnetic field growth compared with standard LEC growth.

Figure 7.7 shows the measured field strength within the chamber for cusped and axial fields. The axial field with aligned magnet pairs produced the highest magnetic field in the z-direction. Vertical scans were measured at the chamber center, and at 6.5 cm away from the center. The opposed magnet fields were measured in both the radial and axial directions. At the center, the vertical component B_z was zero, but small magnetic fields were measured at positions offset either vertically or radially.

Time-dependent temperature fluctuations in the melt were recorded with a sapphire optical thermometer and two thermocouples to reveal the effect of an axial magnetic field configuration on turbulent flow. For these measurements, the sensors were placed near the center of the melt at 2 and 12 mm from the crucible bottom (total InP depth approximately 31 mm and melt diameter 95 mm) with no crucible rotation. After the temperature reached steady state the magnet power was turned on and the field remained constant at 0.2 T. Figure 7.8 shows the off–on transient thermal behavior for the axially aligned field, where the temperature near center the drops rapidly to a 20 °C lower level and remains steady with minor fluctuations.

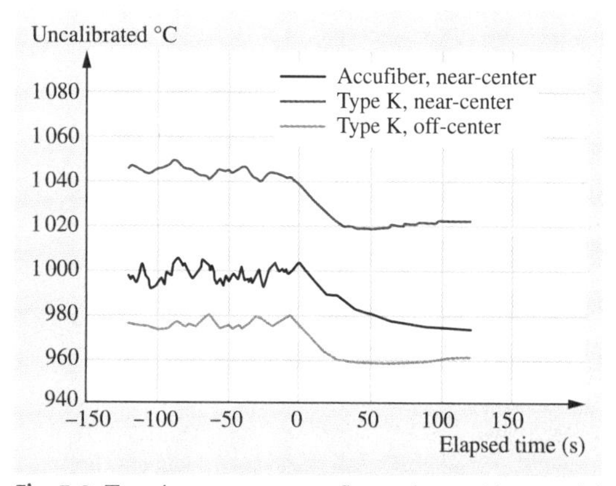

Fig. 7.8 Transient temperature fluctuations with no axial magnetic field ($t < 0$), and after axial magnetic field is turned on ($t > 0$)

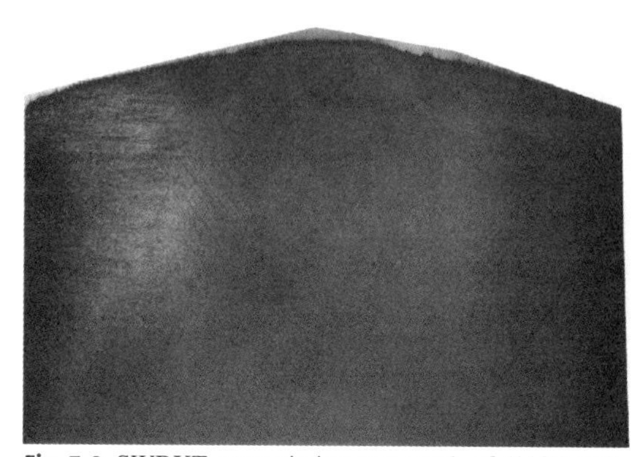

Fig. 7.9 SWBXT transmission topograph of (110)InP:Sn cross-sectional slice showing dislocations and darker slip bands but no edge facets or striations. $g = 004$, $\lambda = 0.5$ Å, scale: crystal diameter = 55 mm

A topograph of an InP:Sn crystal grown under an axial magnetic field is shown in Fig. 7.9. Here the conical growth angle is $\alpha = 82°$, and the presumed angle ν would be 115° if there were edge facets attached to the TPB. Technically, the angle ν is the angle between the (111) edge facets and the extension of the crystal surface, as defined in Fig. 7.5. In this case there are no edge facets to be seen either by x-ray or by infrared (IR) topography. Both iron- and tin-doped crystals were grown without twins in the axial magnetic field using the flat-topped configuration.

Cusped Field Growth

For InP growth in a cusped field [7.34], the magnetic stabilization effect is much weaker. Since the two magnets are opposed, much of the field strength is canceled, and in fact the axial field goes to zero at the center. Although the radial component of the magnetic field is about 0.1 T, this is not strong enough for convective damping. Without a strong applied magnetic field, random oscillatory flow still exists in the melt, and flat-top growth is not possible because of twin formation or dendritic growth. In order to obtain a single crystal under these conditions, the crystal must be pulled as in LEC growth, i. e., avoiding the critical 74° cone angle. A few crystals were successfully grown in this fashion, but with occasional twin formation. The appearance of (111) edge facets attaching to the TPB are observed in IR transmission micrographs, as shown in Fig. 7.10. A twin line extends from one of the smaller facets at the shoulder.

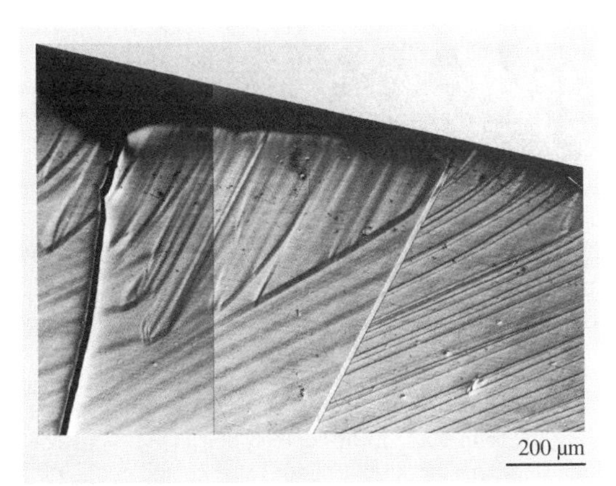

Fig. 7.10 Nomarski micrograph of the shoulder of an InP crystal grown in a cusped field, showing edge facets attached to the TPB and a twin line parallel to the (111) plane

A measurement of the transient temperature behavior in a cusped field shows only a small effect on turbulence at the moment when the magnet is turned on. Using the same temperature sensors as described above, time-dependent temperature fluctuations were recorded, again with no crucible rotation. After the temperature reached steady state the magnet power was turned on and the field remained constant at 0.2 T. Figure 7.11 shows the off–on transient thermal behavior for the cusped field, with opposed magnets. There is no discernible change in the temperature near the center, and a slight increase near the crucible edge. Thermal fluctuations are reduced by $\approx 25\%$ in amplitude, with more

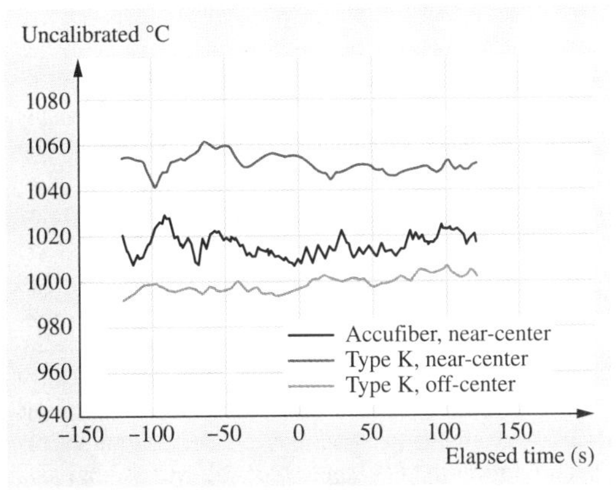

Fig. 7.11 Transient temperature fluctuations before ($t < 0$) and after ($t > 0$) cusped magnetic field is turned on

regular frequency, and there is a slight shift in average temperature.

Examination of the growth striations by IR transmission microscopy revealed some interesting differences between the axial and cusped magnetic fields [7.34]. For flat-topped crystals grown under axial field conditions, regular striations were observed which appeared to follow the rotational frequency. Calculating the rotational growth period by dividing the growth rate by the rotation rate gives $V/R = 50\,\mu\text{m}$ as the length per rotational period. The observed striations showed a regular spacing corresponding to approximately 50 μm. No facets were observed to be anchored to the TPB.

On the other hand, crystals grown in a cusped field exhibited a random striation pattern (Fig. 7.10), with striae periods as small as 30 μm and as large as 150 μm. A large variation from center to edge was also observed, indicating that the interface shape is locally variant. Edge facets on (111) planes were observed near the top surface of the crystal, coinciding with the plane of twin formation.

The interface shape was examined with infrared transmission images taken with the IR camera for the two different magnetic configurations. For axial field growth the interface was convex at the seed end (radius of curvature $r = 5$ cm). On the other hand, for crystals grown with a cusped field, the interface shape is sigmoidal, i. e., convex with a radius of 9 cm and deflected edges.

Figure 7.12 shows a detail from a scanned SWBXT image, recorded in transmission Laue geometry, from a (110) wafer cut from a S-doped, [001]-grown InP single crystal in which a twin was formed in a region where $(1\bar{1}1)$ edge facets came into contact with the external shoulder of the crystal (i. e., were anchored at the TPB during crystal growth). On the right-hand side of this image a twin nucleates in a region containing anchored edge facets at the point where the shoulder angle reaches 74.21°. When viewed from above, a $(1\bar{1}\bar{5})$ external shoulder facet appears on the shoulder at the same point. After twinning, this $(1\bar{1}\bar{5})$ facet is replaced by a $(\bar{1}11)$ facet. In this experiment, the external shoulder facets present before and after twinning were precisely identified using synchrotron white-beam back-reflection spot patterns [7.33].

Considering the left-hand side of this image, we observe anchored edge facets in several regions of differing shoulder angle, but none close enough to 74.21° to enable the creation of the $(\bar{1}1\bar{5})$ external shoulder facet required for the nucleation of a twin. Consequently no twinning is observed in this region, a confirmation

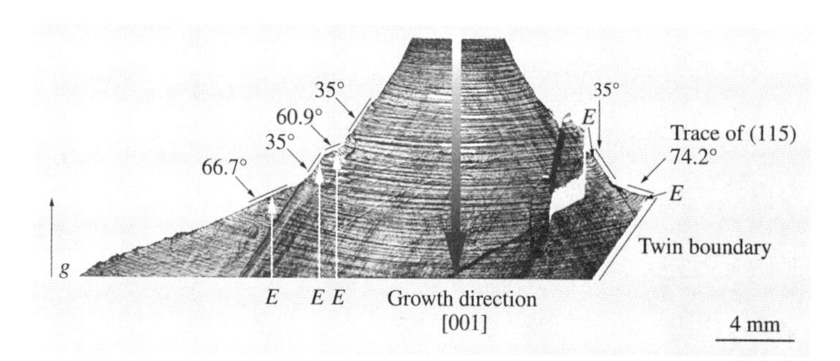

Fig. 7.12 SWBXT image recorded in transmission Laue geometry ($g = 004$, $\lambda = 0.45$ Å) showing the shoulder region of crystal. *E* indicates anchored edge facets. Local shoulder angles are indicated

of the importance of shoulder geometry. In the same boule, twinning was also observed to occur on the $(\bar{1}\bar{1}\bar{1})$ planes, leading to the conversion of a $(\bar{1}\bar{1}\bar{4})$ external shoulder facet to a $(\bar{1}\bar{1}0)$ one. In this case the twin, although nucleated in the shoulder region, grows out of the crystal. The local shoulder angle, in this case, must become equal to 70.53°, i. e., the shoulder must be parallel to $(\bar{1}\bar{1}\bar{4})$. Again, the presence of the various shoulder facets in the actual crystal was confirmed using synchrotron white-beam back-reflection spot patterns.

The concept of facet conversion resulting in a lower free energy is consistent with the Hurle model, but the model was modified by the authors [7.33] to agree with the observed geometry. For twinning in $\langle 001 \rangle$ growth of InP, the most dangerous shouldering angle was changed from 35.5 to 74.21°; and the range of shoulder angles over which edge facets are thermodynamically favored to be anchored to the TPB was changed from $31° < \nu < 86.5°$ to $31° < \nu < 112°$, as illustrated in Fig. 7.13. A significant reduction in the estimated undercooling required to promote twinning was changed from ≈ 15 to ≈ 2 °C.

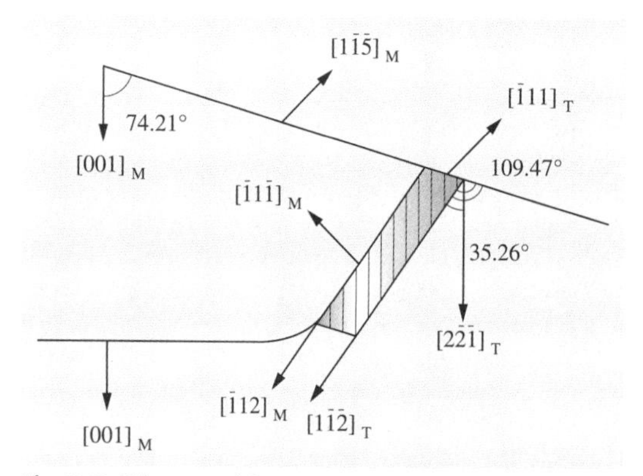

Fig. 7.13 Diagram of facets at TPB and a twin at the most likely growth angle for twin formation

Polarity of Twinning

Since the twin plane in InP is the polar {111} family of planes, another important factor to be considered is the polarity of the observed twin planes. The question is: of the two types of edge facets, either $\{111\}_{In}$ or $\{\bar{1}\bar{1}\bar{1}\}_P$, anchored to the three-phase boundary, which has the higher propensity for generation of twins? Choosing the polarity of the seed face which is in contact with the melt in order to avoid twinning in $\langle 111 \rangle$-grown InP has long been recognized [7.32]. However, an understanding of the influence of polarity on facet formation and twinning is not available for all zincblende structure crystals. In particular such understanding is lacking for InP.

In order to study the influence of polarity on faceting and twinning, it is important to be able to determine polarity unambiguously. Traditional practice has been to use chemical etching to distinguish between the two polar opposite faces of {111} InP. The etching behavior of the In and P faces of InP are clearly different, but questions have been raised as to which surface is P- or In-terminated. *Mullin* et al. [7.35] report on the use of 6 : 3 : 1 and 6 : 6 : 1 H_2O : HNO_3 : HCl in order to distinguish P faces from In faces in InP. *Bachmann* and *Buehler* [7.36], using the opposite indexing convention, confirm the usefulness of the 6 : 6 : 1 etch in producing etch pits on the In face to discriminate between polar faces.

An experiment to confirm the etching method and to determine the crystallographic polarity of the {111} edge facets by x-ray anomalous scattering was reported by *Dudley* [7.37]. To generate significant anomalous scattering for the case of InP, a characteristic radiation with a wavelength close to the In absorption edge ($\lambda_K = 0.444$ Å) was used. In this SWBXT study, plots of diffracted intensity as a function of Bragg angle corresponding to the $(1\bar{1}1)$ and $(\bar{1}1\bar{1})$ reflections, and also a plot of the ratio of the intensities of these reflections, were used to verify the etching results (Fig. 7.14a,b).

It can be seen that, for both reflections, the diffracted intensity changes abruptly at the absorption edge due to the drastic variation of the absorption coefficient. More important, the intensity values of the two reflections are significantly different in the vicinity of the absorption edge. The calculated ratios $\left|F^{P}_{\bar{1}\bar{1}1}/F^{In}_{\bar{1}1\bar{1}}\right|^2$ and $\left|F^{P}_{\bar{1}\bar{1}1}/F^{In}_{\bar{1}1\bar{1}}\right|$ are also shown in Fig. 7.14. These two curves represent the theoretical diffracted intensity ratios according to the kinematical and dynamical x-ray diffraction models, respectively. One would expect that, for a single crystal containing defects, the diffraction generally would contain both dynamical and kinematical contributions. It can clearly be seen that the profile of the ratios of the measured intensities is similar in shape to the theoretical predictions. Therefore, synchrotron x-ray anomalous scattering shows unambiguously that $(1\bar{1}1)$ is a P face while $(\bar{1}1\bar{1})$ is an In face. These results are consistent with etching figures from either Br/methanol or 6 : 6 : 1 etchants that produce a smooth surface on the P face, while producing triangular pits on the In face.

With the polarity of $(1\bar{1}1)$ and $(\bar{1}1\bar{1})$ being unambiguously defined, we can now define the polarity of the twins in InP, to say that twinning is observed to nucleate on only $\{\bar{1}\bar{1}\bar{1}\}_P$ facets. Quite clearly, the preference for twinning on $\{\bar{1}\bar{1}\bar{1}\}_P$ facets is related to the high incidence of twinning observed by *Bonner* [7.38] when growing InP in the $[111]_{In}$ direction. The preferred seed orientation is with the P face in contact with the melt. Similar results were reported by *Steinemann* and *Zimmerli* [7.39] for GaAs, which exhibited a preference for twinning on $\{\bar{1}\bar{1}\bar{1}\}_{As}$ planes. The effect of seed polarity on twinning incidence could be accounted for if the surface energy of the In-terminating facet was significantly higher than the P-terminating one. *Hurle* [7.29] cited the results of the model calculations of *Oshcherin* [7.40], who reports a value for the surface energy of the Ga-terminating facet that is $\approx 12\%$ higher than that of the As-terminating one. Surface energy values were reported by Oscherin also for InP. Calculated values of the surface energy of $\{111\}_{In}$ faces were reported to be more than 30% larger than those for $\{\bar{1}\bar{1}\bar{1}\}_P$ faces.

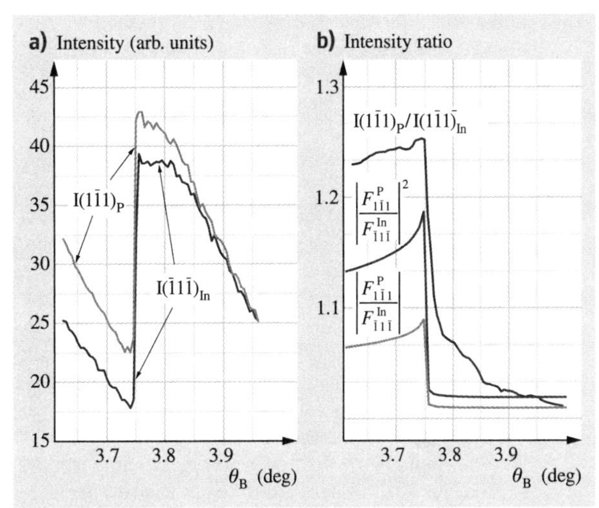

Fig. 7.14a,b Plots of **(a)** observed intensity versus Bragg angle plots for the two reflections, and **(b)** observed ratios of diffracted intensity as a function of Bragg angle from the $(1\bar{1}1)_P$ and $(\bar{1}1\bar{1})_{In}$ planes, and calculated ratios of diffracted intensity according to kinematical theory, i. e., $\left|F^{P}_{\bar{1}\bar{1}1}/F^{In}_{\bar{1}1\bar{1}}\right|^2$ and dynamical theory, i. e., $\left|F^{P}_{\bar{1}\bar{1}1}/F^{In}_{\bar{1}1\bar{1}}\right|$

Twinning Summary

As a result of these studies, we have a better understanding of the fundamental causes of, and methods to control, twinning in InP crystal growth. Specifically, the magnetic field configuration, the dopant species, and the conical growth angle are three important parameters for control of twinning. Trade-offs can be made among these variables to minimize the occurrence of twins.

Although there are many possible causes of twinning in InP, two major causes which can be influenced by engineering controls are:

1. Uncontrolled crystal shape. In shaping the crystal from a narrow seed to a full-diameter body, the conical growth angle is a function of the imposed thermal gradient and the crystal pulling rate.
2. Sudden perturbations in the melt. Because InP crystals are grown with the encapsulant surface exposed to high pressure, temperature gradients in the liquid are steep (50–110°/cm). Uncontrolled oscillatory flows can cause severe thermal perturbations in the melt, resulting in rapid changes in the solid–melt interface shape.

The modified Hurle model, which predicts the incidence of twinning in the growth of III–V semiconductor compounds, focuses on uncontrolled shape as the primary cause of twinning. However, this is not the only factor in controlling twin nucleation in InP. The choice of dopant and magnetic field configuration are also important. A comparison between the theoretical predictions of the incidence of twinning and experimental observation of the incidence of twinning in MLEC-grown [001] InP reveals the following:

1. X-ray topographic observations, etching studies, and IR transmission microscopy revealed the occurrence of twins nucleated at {111} edge facets anchored to the TPB. For ⟨100⟩ seeded InP crystals, this condition is most likely to occur when the angle ν between the edge facet and the extension of the crystal surface, falls in the range $31° < \nu < 112°$. For crystals grown with the angle $\nu > 112°$ in which edge facets do not appear to be attached to the TPB, twin-free crystals can be grown by using the axial magnetic field configuration.
2. The conical growth angle α where twinning is most likely is 74.2°, where a {115} external shoulder facet is converted to a {111} shoulder facet. The most favorable angle for conical growth is 35.3°, where diagonal twins are minimized, for ⟨100⟩-oriented InP crystals. However, this angle allows the possibility to produce patch twins. At a growth angle of 35.3°, the dominant {111} matrix facets typically appear on the crystal shoulders and these facets are favored because of their low surface energy. On the other hand, at a growth angle of 74°, the $\{115\}_M$–$\{111\}_T$ transformation follows exactly the crystallographic orientation relationships of 180° rotation twins. In other words, a {111} external facet would be introduced by twinning if the external surface of the growing crystal becomes parallel to a {115} lattice plane just prior to twinning.
3. Of all the impurity elements used as dopants in this series of experiments, sulfur appears to be the most likely to cause twins. S-doped crystals also exhibit large shoulder facets during growth, an indication that sulfur doping increases undercooling. Other dopants, such as tin, even at the same high concentration, are not as likely to cause twinning.
4. The axial magnetic field configuration is preferred over the cusped field for controlling twins. With the axial field, there are more options to take advantage of crystal shaping as a means of reducing twins. However, because of the increased radial thermal gradient, the axial magnetic field increases thermal stress in the crystal.
5. Close examination of {110} cross-sectional views of S-doped InP crystals reveals that the edge facets may actually increase and then decrease in size before twinning nucleates, suggesting that a critical undercooling may not control the twinning process. Rather, the critical point at which the twin nucleates seems to be associated with production of the {115} shoulder facet, which upon twinning, converts to a {111} shoulder facet.

7.4 Dislocation Density

The primary cause of the high dislocation density in pulled crystals of InP is thermoelastic stress during crystal growth. A high density of dislocations can degrade the performance of photodetectors [7.41] and it affects the properties and performance of many other InP-based devices [7.42]. Dislocations may also change the mechanical properties of InP and contribute to point defect and impurity migration [7.43]. To reduce the dislocation density, sulfur doping is a preferred method, because of its lattice-hardening effect. The need for crystals with low dislocation density is increasing, because these structural defects can limit the performance of advanced monolithic microwave integrated circuits (MMICs) and optoelectronic integrated circuits (OEICs).

High dislocation density is less of a problem with vertical gradient freeze (VGF) or vertical Bridgman (VB) crystals, where the temperature gradient is significantly reduced compared with LEC growth. Densities of less than $500\,\mathrm{cm}^{-2}$ are generally reported for VGF crystals. Thermoelastic stresses are lower in the vertical configuration of container growth because the temperature gradients are so low. Thermal gradients at the solid–melt interface are on the order of 10–15 K/cm in the VGF process, very low compared with typical LEC growth at 60–80 K/cm or higher. For pulled crystals, the steep temperature gradient is the price one must pay to provide stability to the crystal growth environment. Without such steep gradients, diameter control is more difficult, twinning probability is increased, and surface decomposition occurs as the crystal grows out of the encapsulant.

Despite conditions of very steep thermal gradients, for Czochralski silicon growth it is possible to grow large crystals that are virtually dislocation free because the use of *Dash* [7.44] seeding eliminates dislocations

Table 7.1 Results of dislocation reduction experiments

Crystal/ dopant	Seeding condition	Growth rate (mm/h)	Clearance length (mm)	Interface shape	Mechanism	Origin of dislocation
InP:S crystal A	Hot seed	0–27	12	Convex	Glide	Stress induced
InP:S crystal B	Cold seed	0–54	18	Convex	Glide	Stress induced

that emanate from the seed. For InP growth, a similar technique [7.45] demonstrated that dislocation-free InP could be grown, but only up to 15 mm in diameter. When the diameter exceeds that, dislocations are generated at the periphery of the boule (not from the seed) due to the combination of tensile stresses and defects forming at the crystal surface. Reducing the density of dislocations in InP has been the subject of considerable effort during recent decades. The question has been raised of how best to reduce or minimize dislocation generation during InP bulk growth. One approach is the combination of magnetic field stabilization to reduce random thermal fluctuations together with Dash seeding to reduce dislocations from the seed.

7.4.1 Dislocation Reduction During Seeding

In principle, the formation of new dislocations within a crystal under normal growth conditions is nearly impossible. Stresses approaching the shear modulus (10–70 GPa) would be required to generate a new dislocation in the bulk. Thermoelastic stress during InP crystal growth rarely exceeds 10 MPa, and yet the dislocation density in pulled crystals is often greater than $5\times10^{4}\,\mathrm{cm}^{-3}$. This has led many to assume that all dislocations are generated from the seed. If this is true, a seeding technique that removes the grown-in dislocations could result in defect-free single crystals. A controlled seeding test for InP crystal growth was employed [7.46] to suppress the propagation of dislocations from the seed–melt interface. This study showed how the strategy of necking can be exploited to effectively eliminate dislocations in MLEC InP. Dislocations formed at the seed–melt interface are seen to glide out of the neck region. Understanding this mechanism of dislocation reduction in InP crystal growth may lead to better control of dislocation formation and migration. This study showed the origin of the dislocations to be induction by thermal stress rather than by replication of dislocations growing directly from the seed. The results of the experiments are tabulated in Table 7.1.

One source of dislocations is generated at the seed–crystal interface, and these dislocations may propagate through the crystal. The *Dash* seeding approach, discovered in 1958 [7.44], increases the pulling rate of the narrow neck until a defect-free condition appears. Dash speculated that dislocations leave the crystal, in the necked region, due to climb caused by vacancy supersaturation. One difference between compound semiconductors and silicon may be the mechanism of dislocation formation. In the case of Dash's silicon model, the dislocations are grown in at the solid–liquid interface and propagate along the growth direction until they move out of the narrow neck by the climb mechanism. On the other hand, for InP there is no evidence of straight-line axial dislocations; rather, it appears that segments of dislocations from the seed are threading into the newly grown crystal. The dislocations arise from stress in the solid, and they propagate

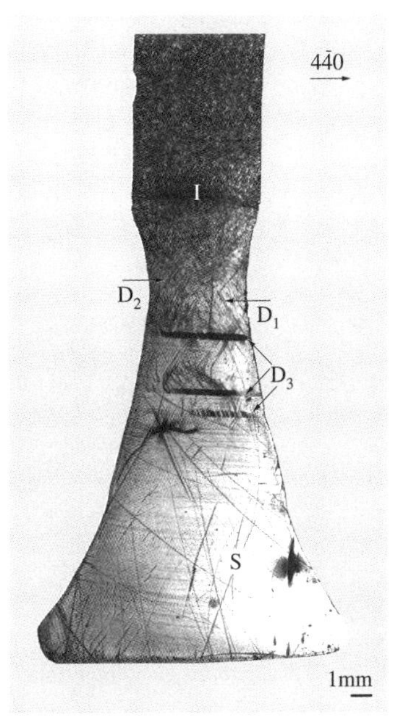

Fig. 7.15 SWBXT transmission topograph of sample A, an InP:S seed–neck region, showing dislocation clusters D_1, D_2, and D_3 and scratches S

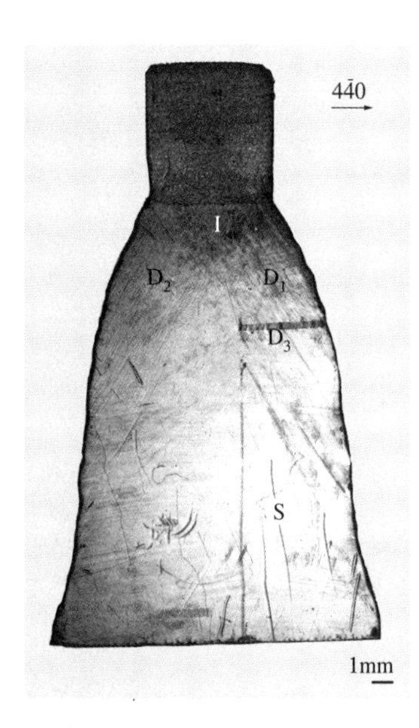

Fig. 7.16 SWBXT transmission topograph of sample B, an InP:S seed–neck region, showing dislocation clusters D_1, D_2, and D_3 and scratches S

by slip-induced glide along the closest-packed {111} planes.

Bliss et al. [7.46] demonstrated crystal growth with a controlled seeding technique that suppressed the propagation of dislocations from the seed–melt interface. Details of the mechanism leading to dislocation density reduction during necking were revealed, and the strategy of necking as an effective means to eliminate defects in InP was exploited. The effectiveness of the necking process as a strategy for reducing the dislocation density in MLEC-grown InP crystals was discussed.

Figures 7.15 and 7.16 show scanned transmission topographs recorded from two (110) InP slices. Figure 7.17 shows a geometric sketch of the dislocations in both crystals. Sample A (Fig. 7.15) has a narrow neck, whereas sample B (Fig. 7.16) expands in diameter from the seed. These crystals formed different shapes due to different growth conditions during initial seeding. For sample B, the initial temperature was slightly colder than normal – a condition called a *cold seed*. In contrast, the initial melt temperature for sample A was higher, and the meniscus diameter was smaller; this is called a *hot seed*. As the temperature was reduced, solidification proceeded to move down the narrow meniscus until the diameter started to increase. Comparing these two samples with different shapes, it

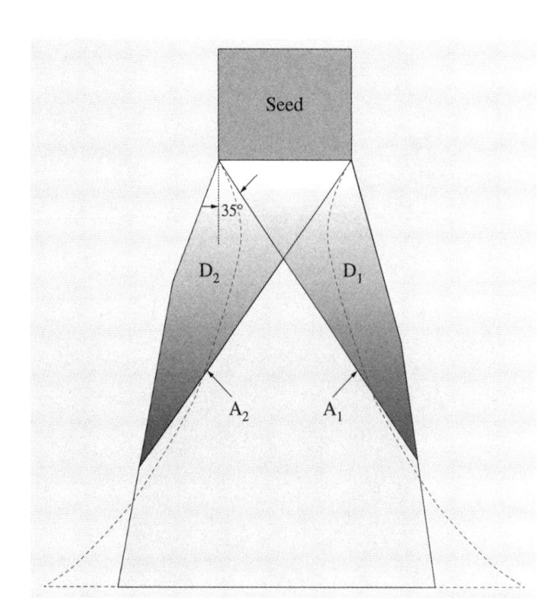

Fig. 7.17 Geometric sketch of the dislocations in both sample A and B

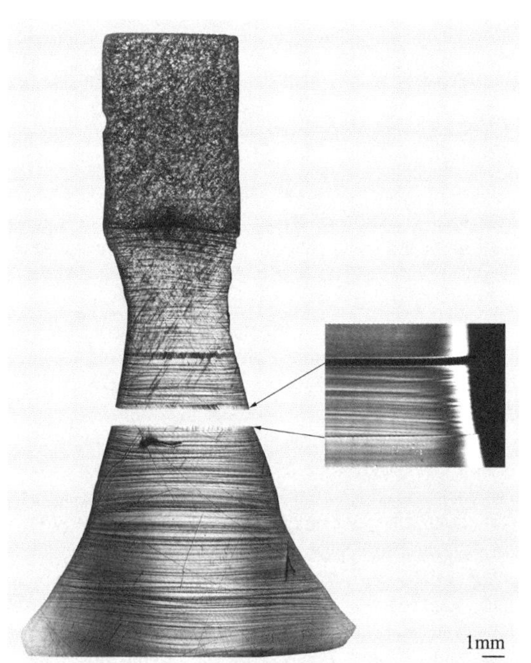

Fig. 7.18 SWBXT transmission topograph of sample A, a (110) slice of InP:S seed–neck region, with $g = 004$. *Inset*: Nomarski image of the etched surface around the twin lamella

is possible to evaluate the effectiveness of the necking process.

The diffraction vector for both topographs in Figs. 7.15 and 7.16 is $4\bar{4}0$, and the radiation wavelength is about 0.48 Å, slightly larger than the indium K-absorption edge (0.44 Å). In the figures the seed–crystal interface is represented; D_1 and D_2 are two groups of dislocations lying on $(1\bar{1}1)$ and $(\bar{1}11)$ planes, respectively. D_3 are dislocations lying on $(11\bar{1})$ planes, which are 35° from the (110) surface. A twin lamella intersects the surface in Fig. 7.15, bounded by two D_3 clusters, as can be seen more clearly in Fig. 7.18, the 004 topograph of sample A. By analyzing the projection lengths of the short, straight segments of D_3 dislocations, it was determined that they are parallel to the $(11\bar{1})_P$ twin plane. The dislocation density in the D_3 clusters is higher even than the dislocations in the seed (above $2\times10^4\,cm^{-2}$), so that individual dislocations cannot easily be resolved. The D_3 dislocation cluster indicates that there is plastic deformation along the twin plane. Thus, the dislocations are channeled by the twin lamella. The subsequent stress upon cooling is relaxed by the growth of D_3 clusters of dislocations adjacent to the twin lamella.

Most of the dislocations appear as intermittent short lines in both samples. The angles between the dislocation lines and the $[00\bar{1}]$ growth direction are about 35°. In Fig. 7.17 the dashed lines represent the cross-sectional shape of sample A while the solid lines correspond to sample B. A_1 and A_2 are two lines drawn from the bottom edges of the seed, at 35° to the growth axis. Most of the dislocations are distributed in the region of crystal above the lines A_1 and A_2. Below this region, very few dislocations can be found. Such dislocation configurations have been reproducibly observed in almost all MLEC-grown necked InP crystals. This demonstrates that these dislocations are closely related to the seed–crystal interface.

7.4.2 Analysis of Dislocations

In order to characterize the Burgers vector of the dislocations observed, transmission topographs with various diffraction $\boldsymbol{g}$ vectors were taken. Figures 7.19 and 7.20 show enlarged topographs near the seed–crystal interface of sample A, where the diffraction vectors are (a) $3\bar{3}3$ and (b) $33\bar{3}$. For the topographs in Figs. 7.19 and 7.20, some of the dislocations disappear because of the extinction conditions

$$\text{Screw:}\quad \boldsymbol{g}\cdot\boldsymbol{b}=0\,,$$

$$\text{Edge:}\quad \begin{cases}\boldsymbol{g}\cdot\boldsymbol{b}=0\,,\\ \boldsymbol{g}\cdot(\boldsymbol{b}\times\boldsymbol{l})=0\,,\end{cases} \tag{7.7}$$

where $\boldsymbol{b}$ is the dislocation Burgers vector and $\boldsymbol{l}$ is the dislocation line direction.

The extinction conditions for D_1 and D_2 dislocations are summarized in Table 7.2. From the extinction criteria of dislocation contrast on x-ray topographs [7.47], it can be determined that both D_1 and D_2 are segments of dislocations with mainly screw character. As shown in Table 7.2, D_1 can be explicitly determined as composing the dislocations with the line directions of [011] and $[10\bar{1}]$ and Burgers vectors $\frac{1}{2}[011]$ and $\frac{1}{2}[10\bar{1}]$, respectively. In the case of D_2, the

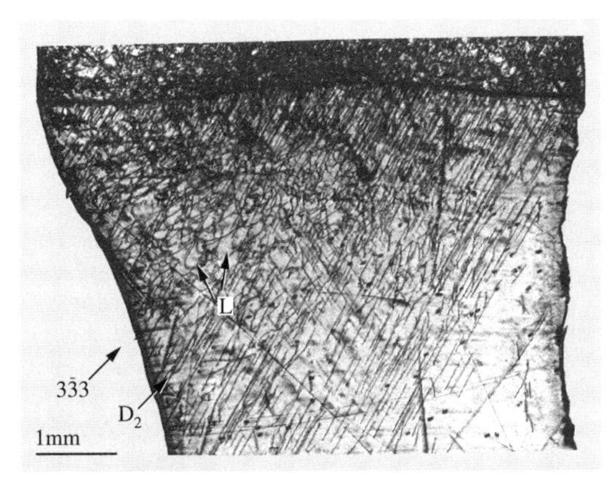

Fig. 7.19 Magnified x-ray topograph of neck region from sample A, showing D_2 defects

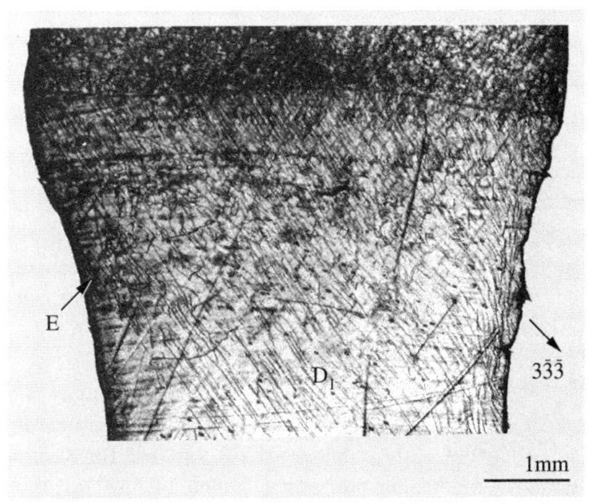

Fig. 7.20 Magnified x-ray topograph of neck region from sample A, showing D_1 defects

Table 7.2 Summary of x-ray data for allowed and extinct diffraction vectors for D_1 and D_2 dislocations

Dislocations	Visible	Invisible	Assigned slip system
D_1	$004, 4\bar{4}0, 2\bar{4}2$ etc.	$3\bar{3}3, 4\bar{2}2$	$(1\bar{1}1)\frac{1}{2}[011]$
	$004, 4\bar{4}0, 4\bar{2}2$ etc.	$3\bar{3}3, 2\bar{4}2$	$(1\bar{1}1)\frac{1}{2}[10\bar{1}]$
D_2	$004, 4\bar{4}0, 4\bar{2}\bar{2}$ etc.	$3\bar{3}\bar{3}, \bar{2}42$	$(\bar{1}11)\frac{1}{2}[101]$
	$004, 4\bar{4}0, \bar{2}42$ etc.	$3\bar{3}\bar{3}, 4\bar{2}\bar{2}$	$(\bar{1}11)\frac{1}{2}[01\bar{1}]$

dislocation lines are mainly along [101] and $[01\bar{1}]$ with Burgers vectors of $\frac{1}{2}[101]$ and $\frac{1}{2}[01\bar{1}]$, respectively.

Unlike Dash's straight axial growth dislocations, which originate in the seed, the topographs presented here show dislocations that consist of short segments lining up on three {111} planes. Careful observation of the spatial distribution and projected directions of these short dislocation segments, as determined above, confirms that they cannot be traced directly back to the seed, and therefore they cannot be growth dislocations. In addition, dislocation half-loops, indicating glide processes, can be observed in the topographs. These half-loops contain segments with line directions that are either parallel to the growth front or curve back against the growth front, which cannot occur if the dislocation is a growth dislocation [7.48]. It is thus apparent that most of the dislocations observed in these samples are not growth dislocations but are generated by postgrowth plastic deformation, i. e., by dislocation movement or interaction following crystal growth (behind the growth front). This is not surprising, considering the stresses generated due to the steep temperature gradients during MLEC growth. These straight segments of dislocations with mainly screw character are the remnants of dislocation loops generated via the stress-induced multiplication of dislocation segments from the seed, which thread into the newly grown region of crystal. As is common in crystals with the zincblende structure, the velocities of edge dislocations are greater than their screw counterparts. Consequently, rapidly moving edge segments, which in this case escape through the outer periphery of the necked region, may leave behind a remnant of trailing screw segments. The dislocation loops will expand in such a fashion that the screw segments will enlarge laterally as well as extending their lengths. Therefore, the screw segments may not be directly traced back to the seed interface. No clear evidence is found supporting the conjecture of *Dash* [7.44] that the reduced dislocation density resulting from the necking process is associated with the climb-induced motion of dislocations.

These experiments demonstrate that most of the dislocations in MLEC-grown InP crystals originate from the seed–crystal interface, propagate via slip processes on {111} planes, and eventually exit the crystal through the periphery of the necked region. It is apparent that one can effectively limit the dislocation regions by choosing proper growth conditions, i. e., by adopting the necking technique. At least for small-diameter crystals, a pyramid-shaped sector which is nearly dislocation free is formed below the dislocation regions. However, the defect-free region cannot be successfully expanded as the crystal grows to full diameter. New dislocations are observed at the periphery as the crystal diameter increases. These dislocations are clearly not generated from the seed, but rather they seem to be propagating from the outer edge where faults exist at the shoulder surface.

7.5 Magnetic Field Effects on Impurity Segregation

Incorporation of dopant impurities from the melt into the crystal is dependent on the applied magnetic field, as described in Sect. 7.2.2. The use of impurity doping requires control of uniformity on both the macroscale and microscale. The magnetic field has been shown to contribute to axial as well as radial uniformity. This is important because all of the commercial uses of InP material require impurity doping in the crystal. Unlike GaAs bulk material, where a native defect EL2 contributes the deep level responsible for semi-insulating electronic properties, in the case of InP the dopant impurity iron must be added to provide the deep-level defect. Iron is the only impurity species that can compensate the residual shallow donors of InP to produce semi-insulating material with resistivity at least $10^7\ \Omega\,\mathrm{cm}$. However, the distribution of iron in the crystal can have two possible negative consequences. Macrosegregation of impurities can lead to yield problems; if maximum solubility is reached, precipitate formation, or interface breakdown due to constitutional supercooling will occur. On the other hand, the microsegregation of impurities can lead to striations, especially pronounced in the case of sulfur doping. And for the case of iron, with a solubility limit of $3\times10^{17}\ \mathrm{cm^{-3}}$, at high doping concentrations the nucleation of precipitates becomes a problem.

7.5.1 Compensation Mechanism of InP:Fe

It is known that Fe substitutes for In in InP and primarily adopts the neutral Fe^{3+} state (so called to denote its oxidation state). However, some of the iron is compensated by shallow donors to form Fe^{2+}, and the ratio between the ionized and neutral concentrations, Fe^{2+}/Fe^{3+} was shown [7.49] to vary linearly with the free carrier concentration n. A linear dependence of n on Fe^{2+}/Fe^{3+} is expected since both quantities are proportional to the term in the Fermi distribution function $\exp[-E_F/(k_B T)]$, where E_F is the Fermi energy and k_B is the Boltzmann constant. The free carrier concentration can be expressed as

$$n \cong N_C \exp\left(-\frac{E_c - E_A}{k_B T}\right) \frac{[Fe^{2+}]}{[Fe^{3+}]} \frac{g(Fe^{3+})}{g(Fe^{2+})}, \quad (7.8)$$

where N_C is the density of states in the conduction band and g represents the degeneracies of the two iron defect states. The energy difference $E_c - E_A$ is simply the change in Gibbs free energy associated with the transfer of an electron from the occupied iron acceptor level to the conduction band. The Gibbs free energy can be expressed in terms of the enthalpy H and entropy S as $G = H - TS$. The above equation can then be rewritten as

$$n \cong N_C \exp\left(-\frac{H}{k_B T}\right) \exp\left(\frac{S}{k_B}\right) \frac{[Fe^{2+}]}{[Fe^{3+}]} \frac{g(Fe^{3+})}{g(Fe^{2+})}. \quad (7.9)$$

The slope of an Arrhenius plot of log n versus $1/T$ gives the change in enthalpy H or the activation energy extrapolated to temperature $T = 0$; this value of 0.63 eV is the same as that measured by temperature-dependent Hall effect or by deep level transient spectroscopy (DLTS). However, the entropy S can also be extracted to give the temperature shift of the iron acceptor level. *Zach* determined this value and found that the temperature shift was different from the temperature shift of the bandgap [7.49]. This is illustrated in Fig. 7.21, which shows the relative shift of the bandgap versus the deep defect level. The results indicate that the thermodynamic position of the Fe^{2+}/Fe^{3+} level at room temperature is 0.49 eV below the conduction band, and it is this energy that determines the carrier concentration. For example, referring to (7.9), the room-temperature carrier concentration for a half-occupied iron acceptor level, $Fe^{2+}/Fe^{3+} = \frac{1}{2}$, would be $n \cong 10^9\,cm^{-3}$. In this circumstance, with Fe-doping levels even twice as high as residual shallow donors, at room temperature InP material is still conducting.

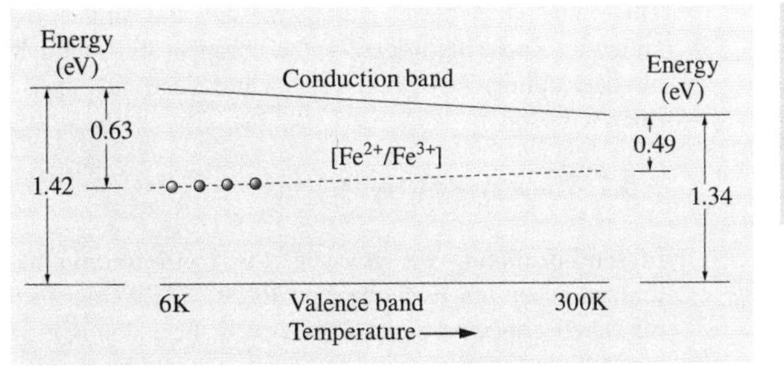

Fig. 7.21 Energy band diagram versus temperature for InP:Fe, showing the shift in bandgap energy (slope $= 3.8\times10^{-4}$ eV/K) and the shift in the Fe acceptor ionization energy (slope $= 4.7\times10^{-4}$ eV/K)

The Fermi energy can only be reduced to the midgap level as the occupancy of the iron acceptor level *decreases*, so that for $Fe^{2+}/Fe^{3+} = \frac{1}{10}$, the free carrier concentration will be $n \cong 10^7$. In other words, for typical semi-insulating InP with residual donor concentrations of $1-3\times10^{15}\,cm^{-3}$, a minimum Fe doping level of $1-3\times10^{16}\,cm^{-3}$ is required.

Zach [7.49] calculated the temperature shift of the iron level from the conduction band to be -4.7×10^{-4} eV/K. This value was later verified [7.50] by an optical experiment using two-wave mixing photorefractive and absorptive gain to evaluate the Fe^{2+}/Fe^{3+} ratio in InP:Fe. As they varied the temperature of the crystal they observed that the wavelength of absorption gain was strongly dependent on temperature. For example, when the wavelength is fixed at 973 nm a null absorption gain is obtained at a crystal temperature of 285 K, while for a wavelength fixed to 991 nm a null gain is obtained at a crystal temperature of 330 K. The null point corresponds to the disappearance of the photorefractive space-charge grating because the creation of holes exactly compensates the creation of electrons ($\sigma_p p_{T0} = \sigma_n n_{T0}$). In this terminology, p_{T0} refers to Fe^{3+} and n_{T0} refers to Fe^{2+}. These results were explained by the temperature dependence of the iron level position as a function of temperature. That is, a shift of the iron level introduces a change in the optical cross sections. Using the temperature for which the absorption gain is null at $\lambda = 991$ and 973 nm, the temperature shift of the iron level with respect to the conduction band was found to be

$$\frac{dE_{Fe\text{-}CB}}{dT} = \frac{E_{991\,nm} - E_{973\,nm}}{T^{\Gamma_\alpha=0}_{991\,nm} - T^{\Gamma_\alpha=0}_{973\,nm}} \approx -5.1\times10^{-4}\,eV/K. \quad (7.10)$$

This value is in close agreement with the value obtained by Zach. Significantly, the compensation model has been confirmed by both optical and electrical experiments.

7.5.2 The Role of Hydrogen

In semi-insulating InP material the concentration of ionized acceptors Fe^{2+} appears to be higher than the measured concentrations of Si and S. The donor impurities that compensate the iron acceptors in InP cannot all be accounted for through trace impurity analysis by glow-discharge mass spectroscopy (GDMS). Additional residual donors, not detected by impurity analysis, are present in these samples. It has been proposed that the other donor which contributes to the ionized Fe^{2+} concentration is hydrogen [7.52].

Several studies [7.53–55] of the absorption spectra of hydrogen in InP have established the structure of hydrogen-related complexes. The local vibrational mode (LVM) absorption at $2315.6\,cm^{-1}$ in LEC InP shown in Fig. 7.22 is due to the P–H stretching modes in an environment that has tetrahedral symmetry [7.55]. Isotopic co-doping with deuterium and hydrogen was used [7.52] to investigate the LVM in InP. The splitting in the D–H spectrum was consistent with a defect having tetrahedral symmetry, and the authors proposed that the absorption is due to a fully passivated indium vacancy [$V_{In}(PH_4)$]. Investigation of the electrical properties of the hydrogen defect in InP [7.56] showed that this defect has a characteristic donor behavior.

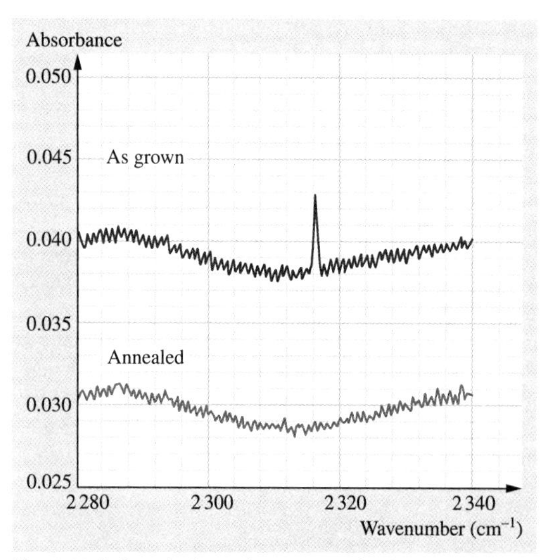

Fig. 7.22 Infrared absorption spectra in the range of the InP:H LVM line before and after annealing (after [7.50])

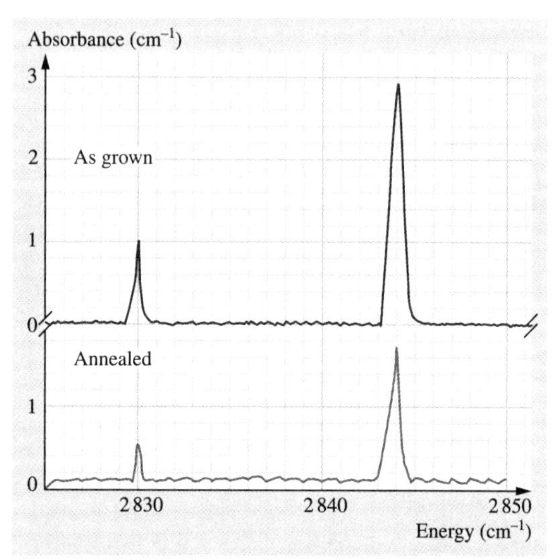

Fig. 7.23 Infrared absorption spectra in the range of the Fe^{2+} intracenter absorption before and after annealing (after [7.51])

The $V_{In}H_4$ defect disappears during annealing at $\approx 900\,°C$ under a phosphorus overpressure, as shown in Fig. 7.22. For undoped bulk InP a reduction in free carrier concentration and an increase in mobility after annealing is also observed, consistent with a decrease in the donor-related $V_{In}H_4$ defect. For the case of Fe-doped material, annealing experiments [7.51, 57] showed the role of hydrogen-related donors in controlling resistivity. In these experiments the Fe^{2+} absorption peak was measured quantitatively to determine the residual shallow donor concentration. For semi-insulating Fe-doped material, annealing reduces the Fe^{2+} absorption peak as seen in Fig. 7.23. The change occurs without any measurable increase in Fe^{3+} by diffusion or other mechanisms. How to quantify the effect of hydrogen in Fe-doped InP is the key to understanding the $V_{In}H_4$ defect – and crucial for producing semi-insulating InP with a minimum concentration of iron.

7.5.3 Annealing Experiments

It has become apparent in the last few years that post-growth thermal treatment can bring about modification of the intrinsic defect concentrations in InP crystals [7.58–63]. It is now understood that the conversion

Table 7.3 Effect of annealing and cool-down on Fe-doped InP samples

Sample cool-down procedure	$[Fe^{2+}]$ (cm^{-3}) as grown	$[Fe^{2+}]$ (cm^{-3}) annealed	$\Delta[Fe^{2+}]$ (cm^{-3})	VH_4 integrated absorption (cm^{-2})
Slow cool	3.3×10^{15}	1.5×10^{15}	1.8×10^{15}	0.089
Fast cool	3.3×10^{15}	1.7×10^{15}	1.6×10^{15}	0.097

from n-type to semi-insulating behavior by annealing at $\approx 950\,°C$ is a consequence of the escape of shallow donors (the VH_4 defect) and the presence of a residual concentration of iron. Some commercial vendors now offer undoped InP that is wafer-annealed under FeP_2 atmosphere to provide semi-insulating properties with high mobility [7.64]. Some questions still persist, however, concerning the precise mechanism of thermal conversion. Does high-temperature treatment annihilate the VH_4 defect complex, or does it merely stimulate the diffusion to free surfaces? Likewise, if iron atoms are *electrically active* only on substitutional In sites, what happens to the InP crystal when iron diffusion occurs through an interstitial mechanism? After annealing, are some iron atoms quenched into inactive interstitial sites? The annealing experiments discussed in [7.65] were designed to answer some of these questions.

In order to determine the activation behavior of iron, Fe-doped InP crystals were grown by the MLEC method, such that the total iron concentration ranged from 1×10^{16} to $1\times10^{17}\,cm^{-3}$ from seed to tail. The conductivity type of these crystals was semi-insulating (i.e., with free carrier concentration $n < 2\times10^{8}\,cm^{-3}$). The crystals were cut into rectangular $1\times1\times2\,cm^3$ polyhedra. Optical absorption spectra in the range $500-3500\,cm^{-1}$ (60–300 meV) were measured with a Digilab 80E-V vacuum Fourier-transform spectrometer. Samples were mounted in an exchange gas optical cryostat and cooled to temperatures between 6.5 and 12 K. The highest instrumental resolution used was $0.125\,cm^{-1}$, and standard measurements were performed at $0.25\,cm^{-1}$ resolution.

These crystals were annealed in quartz ampoules with sufficient phosphorus to provide an overpressure of 5 atm at 900 °C, well above the equilibrium partial pressure of P over InP. After being held at 900 °C for 36 h the samples were cooled either slowly or rapidly to room temperature. Slow cooling was performed at a rate of less than 1 K/min, whereas rapid cooling was in excess of 5 K/min.

The results of the annealing experiments can be summarized as follows: In the case of slowly cooled samples, measurements of the hydrogenated vacancy VH_4 from the integrated absorption at $2316\,cm^{-1}$ showed virtually complete annihilation of the defect after annealing. In these same samples, the Fe^{2+} absorption lines were measured before and after annealing. The results are tabulated in Table 7.3.

The reduction in the Fe^{2+} concentration after annealing is directly proportional to the reduction of integrated absorption due to the hydrogen defect. The following quantitative calibration correlates the hydrogen–vacancy complex with the change in Fe^{2+} (equivalent to the net donor concentration)

$$[V_{In}H_4] = 4.2\times10^{16}\,(cm^{-1}) \times \text{Absorbance}\,(cm^{-2})\,. \quad (7.11)$$

For as-grown InP crystals with net donor concentration less than $4\times10^{15}\,cm^{-3}$, the measured absorbance value for the $2316\,cm^{-1}$ feature is slightly less than $0.1\,cm^{-2}$. This explains why undoped InP crystals are always n-type with *intrinsic* free carrier concentrations of $2-4\times10^{15}\,cm^{-3}$. After annealing, the net donor concentration will be reduced so that it is possible to convert n-type InP with low iron doping to become semi-insulating.

For the case of the rapidly cooled sample, the results are somewhat different. It can be seen from the data in Table 7.3 that the annealing process is somewhat less efficient than for the slowly cooled sample. The change in Fe^{2+} concentration is not as large, indicating that VH_4 donor annihilation is either incomplete or that new compensating acceptors have been introduced during cool-down. Annihilation of VH_4 donors appears to be complete because of the total disappearance of the absorption peak at $2316\,cm^{-1}$. However, the mechanism of acceptors being quenched in during cool-down remains a possibility. Two observations concerning the quenched sample spectrum in Fig. 7.24 may supply clues to the answer. First, a sharp peak is found at $2204\,cm^{-1}$. This peak has been identified [7.66] as an LVM belonging to VH_2, an intermediary in the breakup of the VH_4 defect. As the number of H atoms in the vacancy decreases, the P–H bonds lengthen and the frequency of the vibrational mode decreases. As each hydrogen atom is removed from the vacancy, an electron is removed from the highest occupied state. The

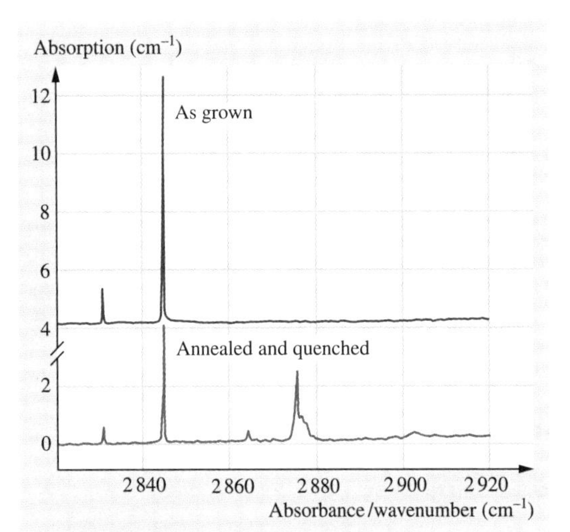

Fig. 7.24 A comparison of the absorption peaks in the vicinity of Fe^{2+} intracenter absorption, before and after annealing with rapid cooling

resulting defects, $V_{In}H_2$, $V_{In}H$, and V_{In}, have been identified as single, double, and triple acceptors, respectively, thereby compensating the residual donor concentration. The appearance of the VH_2 defect is likely contributing to the inefficient thermal conversion of quenched InP samples.

Another interesting feature in the IR absorption spectrum of rapidly cooled samples is seen in Fig. 7.24. Before annealing, in the upper plot, two peaks associated with the Fe^{2+} intracenter electronic transition are seen at 2830 and 2844 cm^{-1}. After annealing followed by rapid cooling, the integrated area of the two peaks is reduced, but not as much as for slowly cooled samples, and some new absorption lines emerge at 2864 cm^{-1} and a possible triplet centered around 2877 cm^{-1}. The new peaks are not identified in the literature, and it is not certain if they are caused by electronic transitions or by local vibrational modes. However, because of their similar spacing and proximity to the well-known iron peaks, one could suspect an iron-related defect. Possibly a high-temperature defect becomes quenched in the rapidly cooled samples. For example, if iron diffuses at high temperature by an interstitial mechanism, it is likely that rapid cooling could leave some iron atoms on inactive sites. Substitutional iron (Fe^{3+}) and a low Fe^{2+}/Fe^{3+} ratio are required for semi-insulating InP. If iron is trapped on interstitial sites, this will also contribute to the inefficiency of thermal conversion.

Fornari has observed [7.67] that the electrical activity of Fe in as-grown InP is dependent on the annealing cycle which occurs naturally during crystal growth. In other words, the concentration of active iron in a given cross section of the ingot depends on the thermal history and the concentration of In vacancies in that section. This observation led to the development of undoped semi-insulating InP wafers obtained by Fe diffusion [7.68]. In this new approach, unintentionally doped InP crystals were grown and sliced into wafers, which were then annealed at high temperature in an iron phosphide atmosphere. Subsequent electrical characterization showed that the wafers became semi-insulating with resistivities above $10^7\,\Omega\,\text{cm}$, and mobilities of 3000–4000 cm^2/(Vs). They also showed that Fe-diffused wafers are more uniform than Fe-doped crystals grown from the melt.

High-temperature annealing of InP has been shown to be an effective way to improve semi-insulating crystal properties. The anneal not only reduces the hydrogen-related donor complexes but also apparently causes Fe redistribution. The concentration of indium vacancies is critical to the annealing process. Substitutional iron requires an indium vacancy in order to be electrically active, just as hydrogen becomes a donor only in the configuration $V_{In}H_4$. With this understanding, it is possible to lower the critical threshold concentration of Fe in semi-insulating InP.

7.6 Optical Characterization of InP:Fe

In order to understand the spatial distribution of iron in InP, a mapping tool that can distinguish between the ionized state Fe^{2+} and the unoccupied Fe^{3+} is needed. Recent advances in optical characterization have developed scanning photocurrent (sPC) as a complement to scanning photoluminescence (PL) measurements [7.69]. Photocurrent mapping experiments are carried out at liquid-nitrogen temperature to improve the contrast [7.70]. The photocurrent is excited by extrinsic light (either $\lambda = 1.32$ or $1.06\,\mu\text{m}$ from a continuous wave (CW) Nd:YAG laser), which implies a probing depth equal to the wafer thickness; on the other hand, PL utilizes above-bandgap excitation wavelength, so the probed depth is on the order of

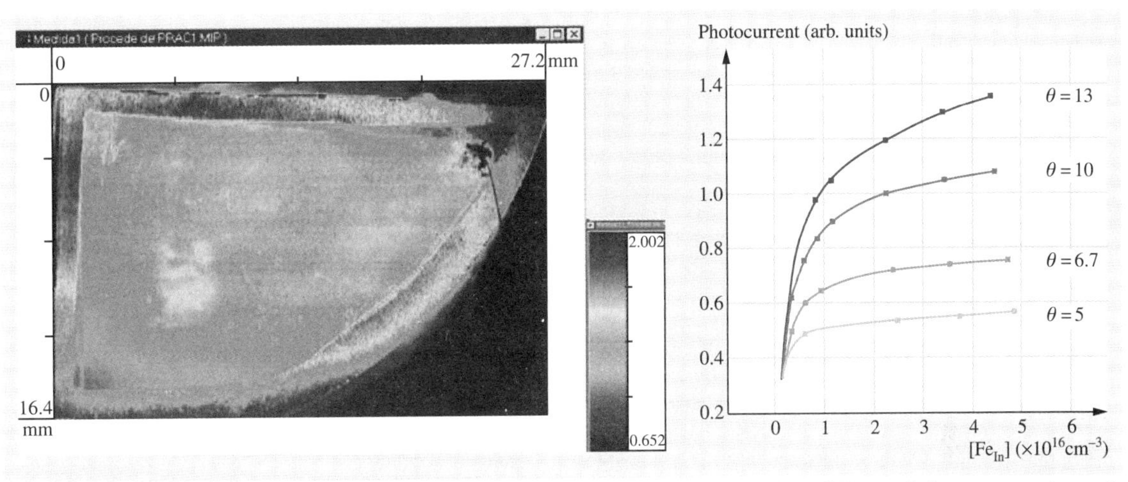

Fig. 7.25 Scanning PC measurement of InP:Fe with photocurrent intensity as a function of Fe_{In} and the compensation ratio $[Fe^{3+}]/[Fe^{2+}]$ (after [7.71])

the minority-carrier diffusion length, or about 1 μm. In order to explain the photoconductivity (PC) contrast, a model was developed [7.71] for Fe-doped InP and a direct correlation was found between the PC intensity and the compensation ratio $[Fe^{3+}]/[Fe^{2+}]$. This is just the opposite of the PL intensity. It would seem that the contrast in PC images will always be anticorrelated to that of PL images. In fact, the comparison is more subtle, as shown graphically in Figs. 7.25 and 7.26, where the calculated brightness varies with total Fe concentration, considering the compensation ratio as a slope parameter.

Now one can compare the PL and PC images of the same Fe-doped InP wafer to determine if the iron is active or inactive, i. e., substitutional or interstitial. If the two images are anticorrelated as expected, then the compensation ratio will be rather uniform across the wafer. On the other hand, if the both PL and PC images show bright areas at the same location, interesting new possibilities arise. If some areas are correlated and

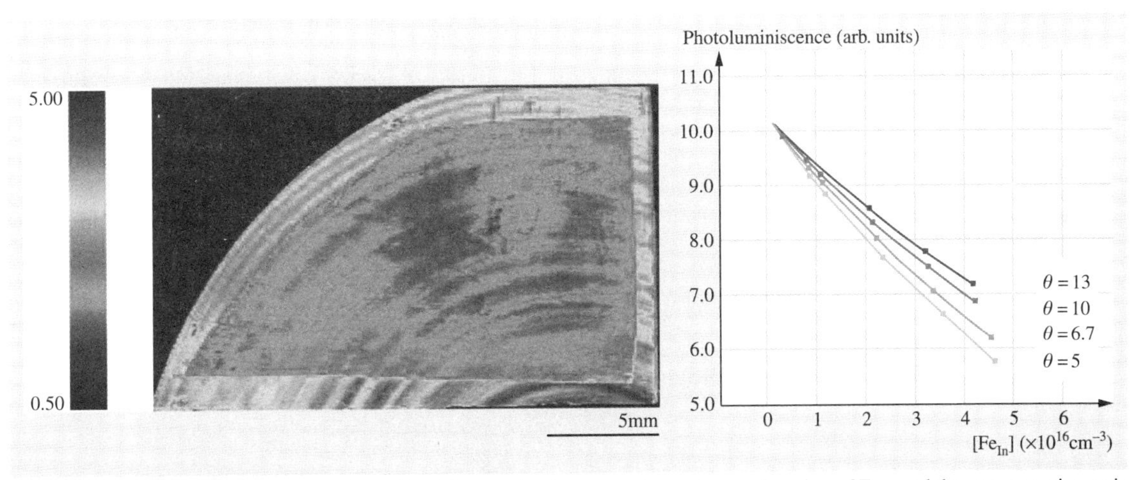

Fig. 7.26 Scanning PC measurement of InP:Fe with photoluminescence intensity as a function of Fe_{In} and the compensation ratio $[Fe^{3+}]/[Fe^{2+}]$ (after [7.71])

others anticorrelated, this indicates a nonuniform compensation ratio. Figures 7.25 and 7.26 contain PL and PC images of the same quarter wafer grown for this study of iron distribution. A visual comparison of the two images reveals a bright PC area, which coincides with a bright PL area. This correlated region is a region of low iron concentration but with high compensation. If this is true, it means that the Fe^{2+} concentration must be very low in this region. That means that both the iron atoms and the residual donors are reduced during the crystal growth and cooling process, possibly by a gettering mechanism.

High PL intensity is indicative of low Fe^{3+} concentration. The concentration may be reduced either because the total iron concentration here is low, or because some of the total iron atoms have migrated to inactive sites. On the other hand, high PC intensity indicates a high $[Fe^{3+}]/[Fe^{2+}]$ compensation ratio. Since the bright region must contain a low Fe^{3+} concentration, then the ionized Fe^{2+} must be even lower to account for the high PC intensity.

The PL and PC images present clear evidence that both iron and residual donors are diffusing in the solid during crystal growth. Furthermore, it appears from the annealing experiments that rapid cooling can increase wafer inhomogeneity. Two possible interpretations have been considered. First, the iron may diffuse interstitially, and remain at interstitial sites during rapid cool-down. Second, the iron and residual donors may be precipitating on dislocations by the mechanism of gettering. Either explanation could account for the inhomogeneity seen in the PL and PC images.

In summary, there is no concrete proof that interstitial iron atoms exist at room temperature in InP crystals. However, it can be inferred that the presence of interstitial iron contributes to the inhomogeneity of InP electrical and optical properties. This research has demonstrated that iron atoms as well as residual donors are diffusing rapidly through the crystal and forming precipitates during growth and cooling. Therefore, control over the annealing and cooling cycle is a prime factor in producing semi-insulating InP wafers.

7.7 Summary

The use of applied magnetic fields during liquid-encapsulated crystal growth has been explored as a means to control defect formation in InP crystals. Since magnetic field growth was first suggested as a means to control crystal defects, experimental work has proceeded at several laboratories to understand its effects on melt growth. However, its use has never been adopted for commercial production. The risk may be too high when there is so small a body of knowledge on the subject. This chapter gives a practical review of the advantages and disadvantages of magnetic fields for liquid-encapsulated growth of InP. With the aid of computer modeling, it is possible to design a system with the proper configuration of magnetic field that will optimize the growth properties of InP. Avoidance of twinning, crystal shape control, and control of dopant distribution are goals that should be of interest for commercial production.

This chapter has also contributed to a practical understanding of how dislocations are generated and the mechanism by which they propagate in LEC InP. Dislocations are generated after growth, several millimeters away from the solid–liquid interface. Experiments to control dislocation density by Dash seeding were made possible by magnetic stabilization, but except for small-diameter crystals, the dislocation density is not reduced by magnetic fields, either axial or cusped.

Finally, the activity of iron as a dopant species to control the semi-insulating behavior of InP has been discussed. The simple process of Fe melt-doping is shown to have practical limitations. Control of compensating species and consideration of where the Fe sits in the crystal structure of InP can now be understood in the light of annealing studies. The most uniform crystals may be those that are grown without intentional Fe dopant, and then sliced and annealed to allow iron diffusion after growth. New tools for mapping the charge state of iron in InP have contributed greatly to clarify our understanding of iron activation.

References

7.1 J.E. Bowers, H. Park, A.W. Fang, R. Jones, O. Cohen, M. Paniccia: A technology for integrating active photonic devices on SOI wafers, Proc. Int. Conf. InP Relat. Mater. (Princeton 2006) pp. 218–221

7.2 E.P.A. Metz, R. Miller, R. Mazelsky: A technique for pulling single crystals of volatile materials, J. Appl. Phys. **33**, 2016–2017 (1962)

7.3 B. Mullin, R. Heritage, C. Holiday, B. Straughan: Liquid encapsulation crystal pulling at high pressures, J. Cryst. Growth **3–4**, 284 (1968)

7.4 K.J. Bachmann, E. Buehler: The growth of InP crystals from the melt, J. Electron. Mater. **3**, 279 (1974)

7.5 L. Henry, E.M. Swiggard: InP growth and properties, J. Electron. Mater. **7**, 647–657 (1978)

7.6 D.F. Bliss: InP bulk crystal growth and characterization. In: *InP-Based Materials and Devices: Physics and Technology*, ed. by O. Wada, H. Hasegawa (Wiley, New York 1999), Chap. 5

7.7 I.R. Grant: Indium phosphide crystal growth. In: *Bulk Crystal Growth of Electronic, Optical and Optoelectronic Materials*, ed. by P. Capper (Wiley, Chichester 2005), Chap. 4

7.8 H. Utech, M. Flemings: Elimination of solute banding in indium antimonide crystals by growth in a magnetic field, J. Appl. Phys. **37**, 2021–2024 (1966)

7.9 H. Chedzey, D. Hurle: Avoidance of growth-striae in semiconductor and metal crystals grown by zone-melting techniques, Nature **210**, 933–934 (1966)

7.10 H. Miyairi, T. Inada, M. Eguchi, T. Fukuda: Growth and properties of InP single crystals grown by the magnetic field applied LEC method, J. Cryst. Growth **79**, 291–295 (1986)

7.11 S. Bachowski, D.F. Bliss, B. Ahern, R.M. Hilton: Magnetically stabilized Kyropoulos and Czochralski growth of InP, 2nd Int. Conf. InP Relat. Mater. (Denver, 1990) pp. 30–34

7.12 J. Burton, R. Prim, W. Slichter: The distribution of solute in crystals grown from the melt. Part I Theoretical, J. Chem. Phys. **21**, 1987–1991 (1953)

7.13 T. Hicks, N. Riley: Boundary layers in magnetic Czochralski crystal growth, J. Cryst. Growth **96**, 957–968 (1989)

7.14 D. Hurle, R. Series: Effective distribution coefficient in magnetic Czochralski growth, J. Cryst. Growth **73**, 1–9 (1985)

7.15 J. Czochralski: Ein neues Verfahren zur Messung der Kristallisationsgeschwindigkeit der Metalle, Z. Phys. Chem. **92**, 219 (1918), in German

7.16 S. Yoshida, S. Ozawa, T. Kijima, J. Suzuki, T. Kikuta: InP single crystal growth with controlled supercooling during the early stage by a modified LEC method, J. Cryst. Growth **113**, 221–226 (1991)

7.17 D. Bliss, R. Hilton, J. Adamski: MLEK crystal growth of large diameter (100) indium phosphide, J. Cryst. Growth **128**, 451–456 (1993)

7.18 D. Bliss, R. Hilton, S. Bachowski, J. Adamski: MLEK crystal growth of (100) indium phosphide, J. Electron. Mater. **20**, 967–971 (1991)

7.19 S. Ozawa, T. Kimura, J. Kobayashi, T. Fukuda: Programmed magnetic field applied liquid encapsulated Czochralski crystal growth, Appl. Phys. Lett. **50**, 329–331 (1987)

7.20 S. Kyropoulos: Ein Verfahren zur Herstellung grosser Kristalle, Z. Anorg. Allg. Chem. **154**, 308–311 (1926), in German

7.21 A.F. Wells: *Crystal Growth*, Annual Reports on the Progress of Chemistry (Chemical Society, London 1946) pp. 62–87

7.22 J.L. Morton, N. Ma, D. Bliss, G. Bryan: Diffusion-controlled dopant transport during magnetically-stabilized liquid-encapsulated Czochralski growth of compound semiconductor crystals, ASME J. Fluids Eng. **123**(4), 893–898 (2001)

7.23 D.T.J. Hurle, R.W. Series: Use of a magnetic field in melt growth. In: *Handbook of Crystal Growth*, Vol. 2A, ed. by D.T.J. Hurle (Elsevier, Amsterdam 1994) pp. 261–285

7.24 J.S. Walker: Models of melt motion, heat transfer, and mass transportduring crystal growth with strong magnetic fields. In: *Progress in Crystal Growth and Characterization of Materials*, Vol. 38, ed. by K.W. Benz (Elsevier, Amsterdam 1999) pp. 195–213

7.25 N. Ma, J. Walker, D. Bliss, G. Bryant: Forced convection during liquid encapsulated crystal growth with an axial magnetic field, J. Fluids Eng. **120**, 844–850 (1998)

7.26 J.L. Morton, N. Ma, D.F. Bliss, G.G. Bryant: Magnetic field effects during liquid-encapsulated Czochralski growth of doped photonic semiconductor crystals, J. Cryst. Growth **250**(1/2), 174–182 (2003)

7.27 Y.F. Zou, H. Zhang, V. Prasad: Dynamics of melt-crystal interface and coupled convection-stress predictions for Czochralski crystal growth processes, J. Cryst. Growth **166**, 476–482 (1996)

7.28 H. Zhang, V. Prasad, D.F. Bliss: Modeling of high pressure, liquid-encapsulated Czochralski growth of InP crystals, J. Cryst. Growth **169**, 250–260 (1996)

7.29 D. Hurle: A mechanism for twin formation during Czochralski and encapsulated vertical Bridgman growth of III–V compound semiconductors, J. Cryst. Growth **147**, 239–250 (1995)

7.30 J.C. Brice: Facet formation during crystal pulling, J. Cryst. Growth **6**, 205–206 (1970)

7.31 K.F. Hulme, J.B. Mullin: Indium antimonide: A review of its preparation, properties and device applications. In: *Solid State Electron*, Vol. 5 (Pergamon, London 1962) pp. 211–247

7.32 W. Bonner: Reproducible preparation of twin-free InP crystals using the LEC technique, Mater. Res. Bull. **15**, 63–72 (1980)

7.33 H. Chung, M. Dudley, D.J. Larson Jr., D.T.J. Hurle, D.F. Bliss, V. Prasad: The mechanism of growth-twin formation in zincblende crystals: new insights from a study of magnetic liquid encapsulated Czochralski-grown InP single crystals, J. Cryst. Growth **187**, 9–17 (1998)

7.34 G.G. Bryant, D.F. Bliss, D. Leahy, R. Lancto, N. Ma, J. Walker: Crystal growth of bulk InP from magnetically stabilized melts with a cusped field, Proc. Int. Conf. InP Relat. Mater. (Hyannis 1997) pp. 416–419

7.35 J.B. Mullin, A. Royle, B.W. Straughan: The preparation and electrical properties of InP crystals grown by liquid encapsulation, Int. Symp. GaAs Relat. Compd., Aachen (IOP, London, Bristol 1970) pp. 41–49

7.36 K.J. Bachmann, E. Buehler: The growth of InP crystals from the melt, J. Electron. Mater. **3**, 279–302 (1974)

7.37 M. Dudley, B. Raghothamachar, Y. Guo, X.R. Huang, H. Chung, D.T.J. Hurle, D.F. Bliss: The influence of polarity on twinning in zincblende structure crystals: new insights from a study of magnetic liquid encapsulated Czochralski grown InP single crystals, J. Cryst. Growth **192**, 1–10 (1998)

7.38 W.A. Bonner: InP synthesis and LEC growth of twin-free crystals, J. Cryst. Growth **54**, 21–31 (1981)

7.39 A. Steinemann, U. Zimmerli: Growth peculiarities of GaAs single crystals, Solid State Electron. **6**, 597–604 (1963)

7.40 B.N. Oshcherin: On surface energies of $A^N B^{8-N}$ semiconducting compounds, Phys. Status Solidi (a) **34**, K181–K186 (1976)

7.41 E. Beam, H. Temkin, S. Mahajan: Influence of dislocation density on *I–V* characteristics of InP photodiodes, Semicond. Sci. Technol. **7**, A229–A232 (1992)

7.42 R.K. Jain, D. Flood: Influence of the dislocation density on the performance of heteroepitaxial InP solar cells, IEEE Trans. Electron. Dev. **40**, 1928–1933 (1993)

7.43 T. Lee, C. Burrus: Dark current and breakdown characteristics of dislocation-free InP photodiodes, Appl. Phys. Lett. **36**, 587–589 (1980)

7.44 W.C. Dash: Single crystals free of dislocations, J. Appl. Phys. **29**, 736–737 (1958)

7.45 S. Shinoyama, C. Uemura, A. Yamamoto, S. Tohno: Growth of dislocation-free undoped InP crystals, Jpn. J. Appl. Phys. **19**, L331–L334 (1980)

7.46 D.F. Bliss, J.Y. Zhao, G. Bryant, R. Lancto, M. Dudley, V. Prasad: Dislocation generation and propogation near the seed-crystal interface during MLEC crystal growth of sulfur-doped InP, Proc. 11th Int. Conf. InP Relat. Mater. (IEEE, Davos 1998) p. 163

7.47 M. Dudley: X-ray topography. In: *Encyclopedia of Advanced Materials*, Vol. 4, ed. by D. Bloor, R.J. Brook, M.C. Flemings, S. Mahajan (Pergamon, Oxford 1994) pp. 2950–2956

7.48 H. Klapper: *Characterization of Crystal Growth Defects by X-ray Methods*, ed. by B.K. Tanner, D.K. Bowen (Plenum Press, New York London 1980) p. 133

7.49 F.X. Zach: New insights into the compensation mechanism of Fe-doped InP, J. Appl. Phys. **75**, 7894 (1994)

7.50 M. Chauvet, S.A. Hawkins, G.J. Salamo, D.F. Bliss, G. Bryant: Evaluation of InP:Fe parameters by measurement of two wave mixing photorefractive and absorptive gain, J. Electron. Mater. **27**, 883–890 (1998)

7.51 J. Wolk, G. Iseler, G. Bryant, E. Bourret-Courchesne, D. Bliss: Annealing behavior of the hydrogen-related defect in LEC indium phosphide, Proc. 9th Int. Conf. InP Relat. Mater. (Hyannis 1997) pp. 408–411

7.52 F.X. Zach, E.E. Haller, D. Gabbe, G. Iseler, G.G. Bryant, D.F. Bliss: Electrical properties of the hydrogen defect in InP and the microscopic structure of the 2316 cm^{-1} hydrogen related line, J. Electron. Mater. **25**, 331–335 (1996)

7.53 J. Pankove, N. Johnson: *Hydrogen in Semiconductors* (Academic, Orlando 1991)

7.54 B. Pajot, J. Chevallier, A. Jalil, B. Rose: Spectroscopic evidence for hydrogen-phosphorus pairing in zinc-doped InP containing hydrogen, Semicond. Sci. Technol. **4**, 91–93 (1989)

7.55 R. Darwich, B. Pajot, B. Rose, D. Robein, B. Theys, R. Rahbi, C. Porte, F. Gendron: Experimental study of the hydrogen complexes in indium phosphide, Phys. Rev. B **48**, 48 (1993)

7.56 C. Ewels, S. Oberg, R. Jones, B. Pajot, P. Briddon: Vacancy- and acceptor-H complexes in InP, Semicond. Sci. Technol. **11**, 502–507 (1996)

7.57 A. Zappettini, R. Fornari, R. Capelletti: Electrical and optical properties of semi-insulating InP obtained by wafer and ingot annealing, Mater. Sci. Eng. B **45**, 147–151 (1997)

7.58 R. Fornari, A. Brinciotti, E. Gombia, R. Mosca, A. Huber, C. Grattepain: Annealing-related compensation in bulk undoped InP, Proc. 8th Conf. Semi-insulating III–V Mater., ed. by M. Godlewski (World Scientific, Warsaw 1994) pp. 283–286

7.59 G. Hirt, D. Wolf, G. Müller: Quantitative study of the contribution of deep and shallow levels to the compensation mechanisms in annealed InP, J. Appl. Phys. **74**, 5538–5545 (1993)

7.60 P.B. Klein, R.L. Henry, T.A. Kennedy, N.D. Wilsey: Semi-insulating behavior in undoped LEC InP after annealing in phosphorus. In: *Defects in Semiconductors*, Vol. 10–12, ed. by H.J. von Bardeleben *Materials Science Forum* (Trans. Tech. Pubs. 1986) pp. 1259–1264

7.61 K. Kainosho, H. Shimakura, H. Yamamoto, O. Oda: Undoped semi-insulating InP by high pressure annealing, Appl. Phys. Lett. **59**, 932–934 (1991)

7.62 D. Wolf, G. Hirt, G. Müller: Control of low Fe content in the preparation of semi-insulating InP by wafer annealing, J. Electron. Mater. **24**, 93–97 (1995)

7.63 K. Kainosho, M. Ohta, M. Uchida, M. Nakamura, O. Oda: Effect of annealing conditions on the uniformity of undoped semi-insulating InP, J. Electron. Mater. **25**, 353–356 (1996)

7.64 K. Kuriyama, K. Ushiyama, T. Tsunoda, M. Uchida, K. Yokoyama: Uniformity of deep levels in semi-insulating InP obtained by multiple-step wafer annealing, J. Electron. Mater. **27**, 462–465 (1998)

7.65 Q. Ye, J.A. Wolk, E.D. Bourret-Courchesne, D.F. Bliss: Annealing behavior of the hydrogen-vacancy complex in bulk InP, MRS Symp. Proc. H, Vol. 513 (1998) pp. 241–246

7.66 C.P. Ewels, S. Öberg, R. Jones, B. Pajot, P.R. Briddon: Vacancy- and acceptor-H complexes in InP, Semicond. Sci. Technol. **11**, 502–507 (1996)

7.67 R. Fornari: On the electrical activity of Fe LEC indium phosphide, Semicond. Sci. Technol. **14**, 246–250 (1999)

7.68 R. Fornari, T. Görög, J. Jimenez, E. De la Puente, M. Avella, I. Grant, M. Brozel, M. Nicholis: Uniformity of semi-insulating InP wafers obtained by Fe diffusion, J. Appl. Phys. **88**, 5225–5229 (2000)

7.69 M. Avella, J. Jimenez, A. Alvarez, R. Fornari, E. Gigli-oli, A. Sentiri: Uniformity and physical properties of semi-insulating Fe-doped InP after wafer annealing, J. Appl. Phys. **82**, 3836–3845 (1997)

7.70 A. Alvarez, M. Avella, J. Jiménez, M.A. Gonzalez, R. Fornari: Photocurrent contrast in semi-insulating Fe-doped InP, Semicond. Sci. Technol. **11**, 941–946 (1996)

7.71 M. Avella, J. Jiménez, A. Alvarez, M.A. Gonzalez, L.F. Sanz: A photocurrent study of semiinsulating Fe-doped InP, Mater. Sci. Eng. B **28**, 111–114 (1994)

8. Czochralski Silicon Single Crystals for Semiconductor and Solar Cell Applications

Koichi Kakimoto

This chapter reviews growth and characterization of Czochralski silicon single crystals for semiconductor and solar cell applications. Magnetic-field-applied Czochralski growth systems and unidirectional solidification systems are the focus for large-scale integrated (LSI) circuits and solar applications, for which control of melt flow is a key issue to realize high-quality crystals.

Over the past 50years, single crystals of semiconductors such as silicon (Si), gallium arsenide (GaAs), and indium phosphide (InP) have become increasingly key materials in the fields of computer and information technology. Attempts to produce pure silicon (i. e., a defect-free single crystal of silicon) were motivated by the desire to obtain ultralarge-scale integrated circuits (ULSIs) in which microvoids of about 10 nm diameter [8.1] are formed during crystal growth. Research over the past decade on crystal growth of silicon has focused on analysis of the formation of such microvoids during crystal growth using mass-transfer and reaction equations and on the temperature field in the crystals, obtained from global modeling. Control of solid–liquid interface shapes of GaAs and InP has been extensively studied to find a way to prevent the formation of dislocations and polygonization during crystal growth [8.2–5].

Microvoids are formed by the agglomeration of vacancies introduced at the solid–liquid interface of silicon. In most past studies, it has been difficult to reduce the total number of such microvoids in a whole crystal because the vacancy flux in silicon crystals must be controlled to reduce the probability of agglomeration. One of the key points for controlling the vacancy flux in crystals, especially that near a solid–liquid interface, is control of the convection of the melt, through which the shape of the solid–liquid interface can be controlled. From this point of view, efforts have been made to control the periodic and/or turbulent flow of melt inside a crucible of large diameter. Crystal growth industries have mainly focused on quantitative prediction of a solid–liquid interface, point defect distribution, oxygen concentration, and dislocation-free growth. Steady (DC) and/or dynamic (AC) [8.6–44] magnetic fields, including electromagnetic fields, are opening up new fields to meet the increasing demand for large-diameter crystals.

A transverse magnetic field (i. e., TMCZ) is the type that has been utilized for commercial production, especially for large-diameter crystals. A lot of research [8.45–49] on the TMCZ method has been published. Numerical calculation [8.50–68] of these transverse magnetic fields is one of the key issues to predict temperature and impurity distributions in the system. Most numerical studies on CZ-Si growth in a transverse magnetic field have been limited to three-dimensional analysis of melt flow in a crucible by imposing a flat melt–crystal interface and external thermal boundary conditions in the models. However, since the real shape of the melt–crystal interface and the thermal field near it are of great interest commercially, three-dimensional (3-D) global modeling that takes into account the high degree of nonlinearity of the growth system, the inherent three-dimensionality of the melt flow, and the thermal field under the influence of a transverse magnetic field is necessary. However, there have been few studies using such modeling.

This chapter reviews crystal growth and characterization of CZ silicon. The effects of magnetic fields such as vertical and TMCZ methods on convection of the melt are also discussed. A means for solving the problem of convection computationally is also described.

8.1 Silicon Single Crystals for LSIs and Solar Applications

Figure 8.1 shows the structure of a typical furnace for CZ growth of silicon crystals. The heater, crucible heater, and thermal shields are made of carbon and/or carbon composite. Polycrystalline silicon as a raw material is placed in a quartz crucible. The growth furnace is evacuated, then Ar gas is introduced into the furnace to prevent oxidation of the silicon crystal and the melt. Subsequently, a seed crystal is attached to the top of the melt, then pulled upwards to grow silicon single crystals. The grown crystal is detached from the melt after the end of growth. Finally, the crystal is cooled to room temperature. The diameters of the crystal and crucible are currently 300 and 1200 mm, respectively. The diameter of the crystal is now increasing to 450 mm in research and development, where the requirement on defect density becomes critical. Therefore, precise control of growth conditions is of great importance to grow defect-free crystals.

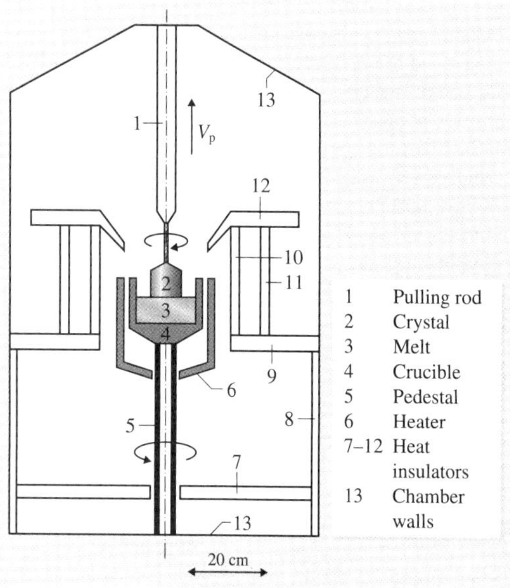

Fig. 8.1 Typical geometry of silicon Czochralski crystal growth furnace

8.1.1 Conventional Czochralski Silicon

A seed crystal is usually suspended by a wire, then touches the surface of the silicon melt. The crucible usually rotates to stabilize the flow of the melt through centrifugal forces. The pulling rate of the crystal is also controlled to keep the crystal diameter constant. Monitoring of the diameter is usually carried out by a camera that monitors the meniscus shape. The pulling rate of the crystal is kept at about 1 mm/min, which is important for production efficiency. Oxygen concentration in a crystal is usually controlled by adjusting the pressure and the flow rate of the Ar gas. The crystal and crucible are usually rotated to control impurity and oxygen concentrations as well as the shape of the interface between the crystal and the melt. Moreover, a cone is located near the melt–gas interface to rectify the gas flow just above the melt surface. This cone can control the evaporation rate of oxygen from the top of the melt, which enables control of the concentration of oxygen incorporated from the melt into the crystal.

Bulk crystalline silicon of high quality has become an essential material for today's information society. The distribution of temperature in a crystal during growth affects the distribution of vacancies and the formation of voids in the crystal. Therefore, it is important to control the temperature distribution in a crystal through control of the flow of the melt. To date, the temperature distribution in the furnace has been controlled by selecting a configuration of the thermal shields based on a global model including heat and mass transfer by radiation, convection, and conduction.

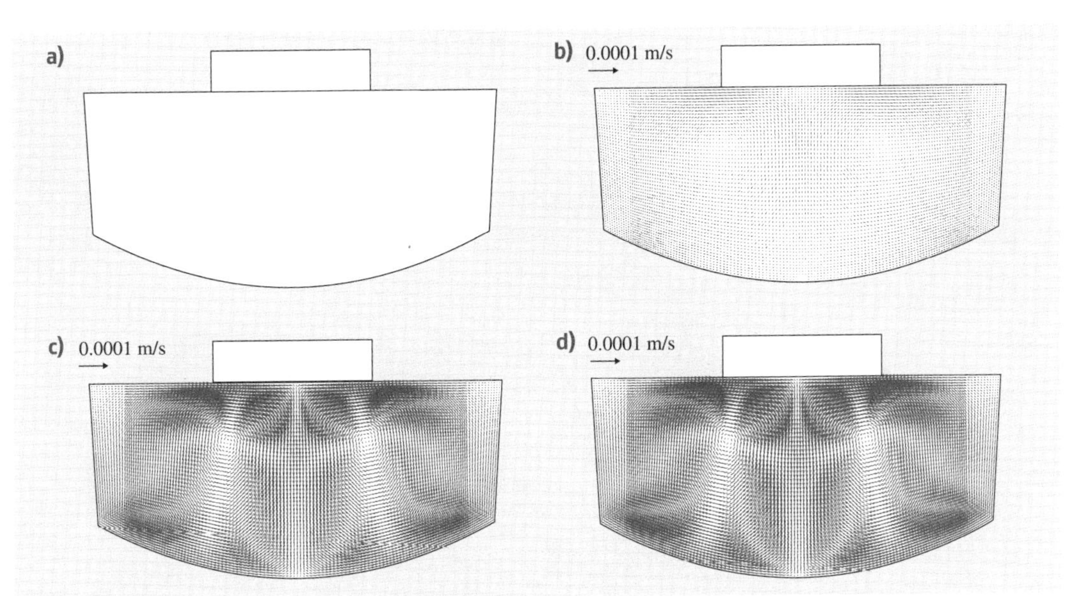

Fig. 8.2a–d Velocity profiles of Czochralski growth of silicon without temperature effect. (**a**) No forces, (**b**) with crystal rotation, (**c**) with crucible rotation, (**d**) with crystal and crucible rotations

Figure 8.2 shows velocity profiles of Czochralski growth of silicon without the effect of temperature, while Fig. 8.3 shows calculated temperature and velocity profiles of Czochralski growth of silicon including the effect of temperature. These figures clarify how crystal and/or crucible rotation and gravity affect the convection of the melt. This configuration contains melt and a crystal of silicon. Operating conditions in terms of the crystal (ω_s) and crucible (ω_c) rotation rates for Figs. 8.2 and 8.3 are listed in Table 8.1.

Figure 8.2a has no flow since there are no external or internal forces on the melt. Figure 8.2b shows the velocity profile with crystal rotation only, in which case the velocity is low. This is due to the low viscosity of the silicon melt, which cannot diffuse momentum effectively from the crystal to the melt. Figure 8.2c,d shows similar profiles with the crucible rotating, showing the large effect of crucible rotation on convection of the melt.

Figure 8.3 shows the temperature profiles under the conditions listed in Table 8.1. Figure 8.3a and b show almost the same profiles of temperature and velocity, similar to the relationship between Fig. 8.2a,b. This is because of the low viscosity of the melt.

Figure 8.3c,d shows a low velocity and a temperature profile similar to that for the conduction-dominated case. This small velocity is attributed to the law of conservation of angular momentum in the rotating melt [8.26]. The momentum (Navier–Stokes) equation for the rotating melt contains terms for the Coriolis and centrifugal forces, as shown in (8.1), for the rotating coordinate system

$$\frac{\partial \boldsymbol{u}}{\mathrm{d}t} = -\boldsymbol{u}\nabla\boldsymbol{u} - 2(\Omega\boldsymbol{k})\times\boldsymbol{u} + (\Omega\boldsymbol{k})\times(\Omega\boldsymbol{k})\times\boldsymbol{r} + \frac{1}{\rho}\nabla p + \mu/\rho\Delta\boldsymbol{u} + \boldsymbol{g}\beta(T - T_0)\,, \quad (8.1)$$

where $\boldsymbol{u}$ and $\boldsymbol{r}$ are the vectors of relative velocity on a rotational basis and position, respectively, Ω denotes the crucible rotation rate, p and μ represent the pressure and viscosity of the melt, and $\boldsymbol{g}$, β, and T_0 are the vectors of gravitational acceleration, the volume expansion coefficient, and the reference temperature

Table 8.1 Operating conditions of crystal (ω_s) and crucible (ω_c) rotation rates

Figure 8.2	a	b	c	d
ω_s	0	−3	0	−3
ω_c	0	0	10	10
Figure 8.3	**a**	**b**	**c**	**d**
ω_s	0	−3	0	−3
ω_c	0	0	10	10

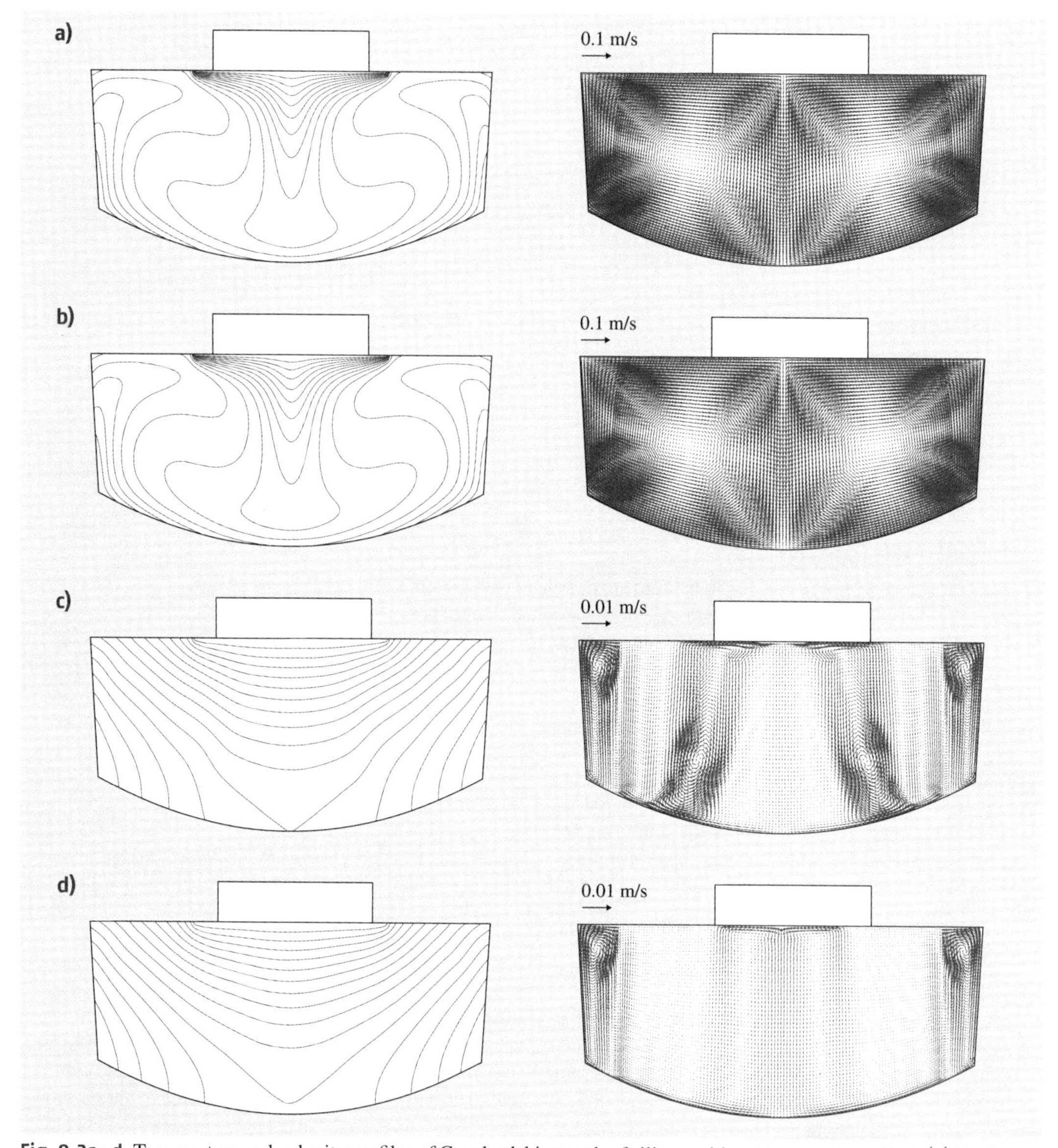

Fig. 8.3a–d Temperature and velocity profiles of Czochralski growth of silicon without temperature effect. (**a**) No forces, (**b**) with crystal rotation, (**c**) with crucible rotation, (**d**) with crystal and crucible rotations. Temperature difference between the contours is $\Delta T = 5\,\mathrm{K}$. Consequently, the centrifugal force always acts in the opposite direction. This means that the melt motion in the radial direction is suppressed by the crucible rotation

corresponding to specific mass, and the reference temperature of 1685 K, respectively. The second and third terms on the right-hand side of (8.1) express the Coriolis force and centrifugal acceleration, respectively. The vector $\boldsymbol{k}$ in (8.1) is a unit vector in the z-direction. The centrifugal acceleration vector ($\boldsymbol{a}_{\mathrm{cen}}$) can be expressed as (8.2).

$$a_{cen} = \Omega^2 r = \frac{L^2}{r^3} , \quad (8.2)$$

where L is the angular momentum of the melt.

When a small volume element moves instantaneously from position r to r' $(= r + \Delta r)$, the excess force shown in (8.3) is caused by the conservation of angular momentum

$$\Delta a_{cen} = L^2 \left(\frac{1}{r^3} - \frac{1}{r'^3} \right) . \quad (8.3)$$

8.1.2 Magnetic Czochralski (MCZ) Silicon

As the diameter of the crystal increases we need to stabilize the flow of the melt. There have been a lot of papers regarding the magnetic-field-applied Czochralski method, especially for large-diameter crystals [8.35–45]. Research on electromagnetic hydrodynamics has a long history in the field of steel and metal manufacturing processes. Since molten silicon, like molten steel or metal, has many free electrons, electromagnetic hydrodynamics can be used to control the convection in metallically conducting molten silicon through the application of magnetic and/or electric fields.

The electric current (J) in the melt and the Lorentz force (F) induced by the current in the case of a steady electromagnetic field are shown in (8.4) and (8.5), respectively, where σ_e, E, B, and v are the electric conductivity of the melt, electric field, magnetic flux density, and velocity of the melt, respectively.

$$J = \sigma_e (E + v \times B) , \quad (8.4)$$

$$F = J \times B . \quad (8.5)$$

Due to the continuity condition on the electric current in the melt, (8.6) (a Poisson-type equation) should be satisfied, since there is no source of charge in this case:

$$\nabla \cdot J = 0 . \quad (8.6)$$

The typical magnetic fields used in the Czochralski method are shown in Fig. 8.4. Figure 8.4a and b show schematic diagrams of the vertical magnetic-field-applied Czochralski (VMCZ) and the transverse magnetic-field-applied Czochralski (TMCZ) method, respectively. One or two coils are set parallel to the pulling axis in VMCZ, while two coils are set perpendicular to the axis in TMCZ. For the cusp-shaped magnetic-field-applied Czochralski method there are two coils, which are set parallel to the axis with opposite current directions. Therefore, inhomogeneous magnetic fields are applied to the melt.

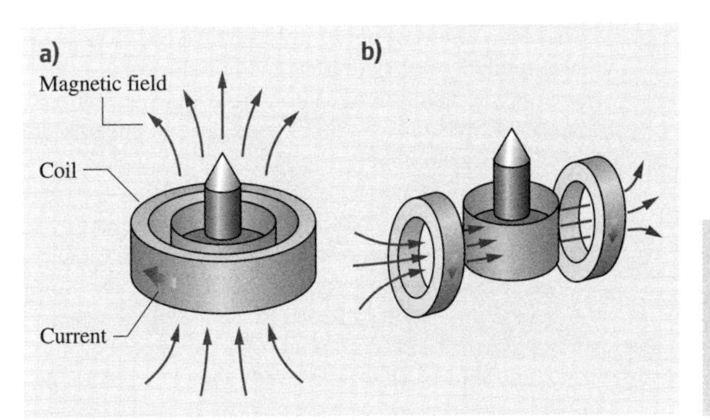

Fig. 8.4a,b Schematic diagram of a VMCZ (**a**) and TMCZ (**b**) system. Static current is applied to the cylindrical coil. Schematic diagram of electric currents and Lorentz forces under vertical magnetic fields

The vertical magnetic-field-applied Czochralski (VMCZ) method was one of the methods used in early works on magnetic-field-applied crystal growth. Magnetic fields are applied in the z-direction; therefore, motion of the melt in the radial and/or azimuthal directions reacts with the magnetic field, while melt motion in the vertical direction does not react with the field. Electric current is induced in the azimuthal direction by the radial motion of the melt under a vertical magnetic field. Therefore, the Lorenz force works in the opposite direction and suppresses the velocity of the melt in the radial direction.

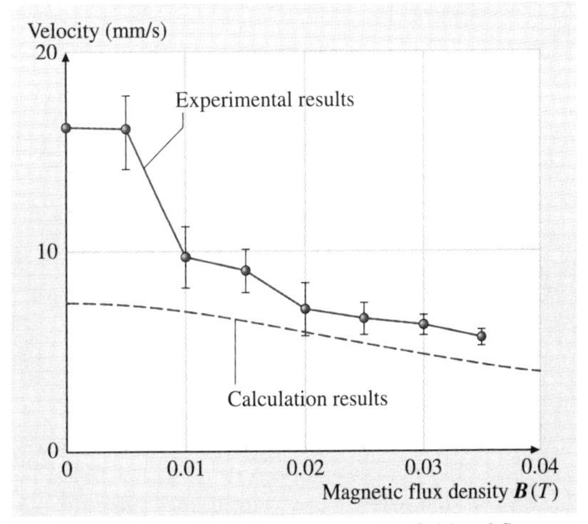

Fig. 8.5 Relationship between magnetic field and flow velocity in a meridional plane under vertical magnetic fields (after [8.35, 50])

Radial current, which is induced by a coupling with the azimuthal velocity and magnetic field, cannot flow through the crucible wall as it is electrically insulating. Thus, the Lorentz force cannot work, and the melt flows freely in the azimuthal direction.

Figure 8.5 shows experimental results for the melt velocity in a meridional plane obtained by a visualization technique using an x-ray radiography method [8.52]. The dots show experimental data while the lines show results of numerical calculation using a three-dimensional configuration of the melt. This figure clearly shows that there is a reduction in melt motion in the meridional plane. It was clarified from the visualization that the motion in the azimuthal direction was not suppressed.

The TMCZ system has a nonaxisymmetric configuration, and temperature and velocity fields therefore have twofold symmetry. Although this system has such a symmetry, it has been used for actual production of silicon for charge-coupled devices (CCDs), since the system enables crystals with low oxygen concentration to be produced. CCDs should have homogeneous and low oxygen concentration for reduction of inhomogeneity of image cells in the devices.

An elliptic temperature distribution due to inhomogeneous heat transfer in the melt can be seen in

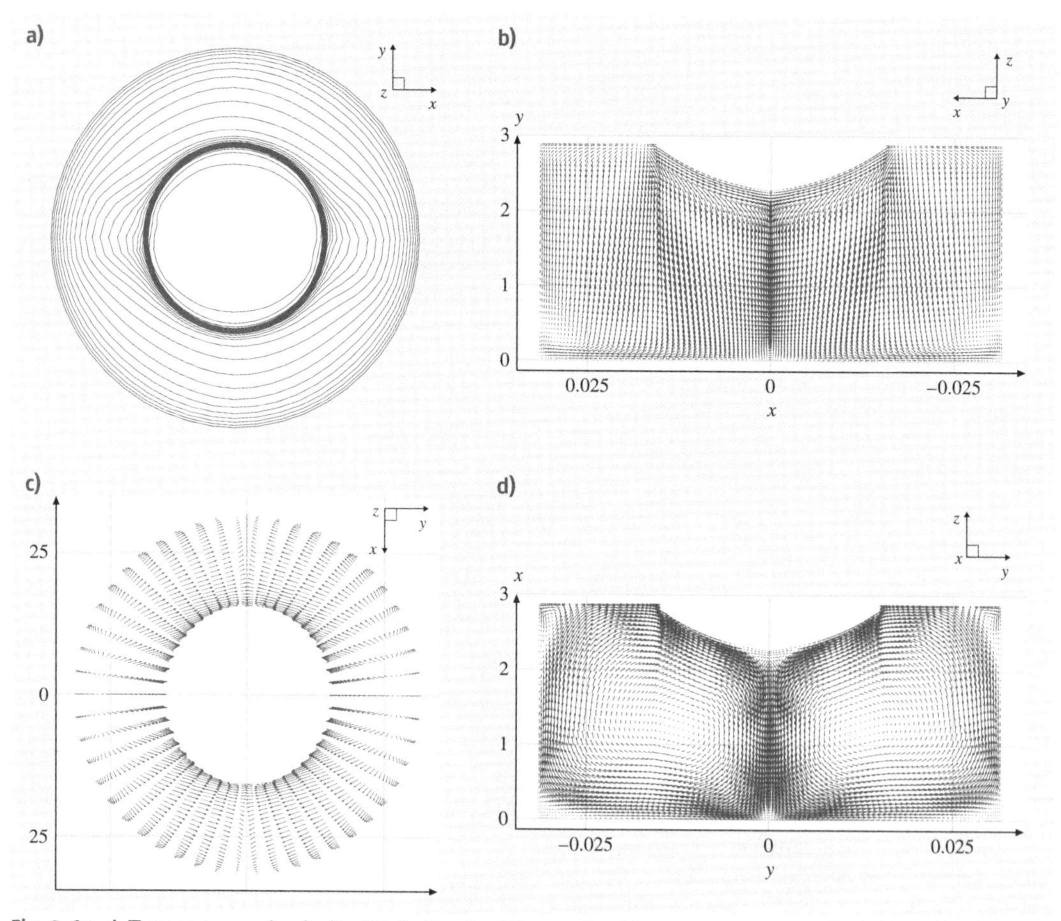

Fig. 8.6a–d Temperature and velocity distribution in silicon melt with a transverse magnetic field applied in the x-direction. Distributions of temperature **(a)** and velocity **(c)** at the top of the melt. Velocity distributions in planes perpendicular **(b)** and parallel **(d)** to the magnetic field. Temperature difference between the contours is 2 K. Average velocities of **(b)**, **(c)** and **(d)** are 0.2, 0.18 and 0.19 mm/s, respectively

Fig. 8.6a. As shown schematically in Figure 8.6a,b, only down-flow is formed in a plane parallel to the magnetic field, while two roll cells are formed in a plane perpendicular to the field. Consequently, thin boundary layers of velocity, temperature, and oxygen near the crucible wall are formed.

This phenomenon is a characteristic of a transverse magnetic field, which is static and nonaxisymmetric. If an axisymmetric magnetic field such as a vertical or cusp-shaped field is used, the melt rotates with the same angular velocity as the crucible.

8.2 Control of Crystal Defects in Czochralski Silicon

Microvoids are formed by agglomeration of vacancies that are introduced at a solid–liquid interface of silicon [8.1, 51–55]. In most past studies, it has been difficult to reduce the total number of such microvoids in a whole crystal because the vacancy flux in silicon crystals must be controlled to reduce the probability of agglomeration.

One of the key points for controlling the vacancy flux in crystals, especially that near a solid–liquid interface, is control of the convection of melt, by which the shape of the solid–liquid interface can be controlled. From this point of view, efforts have been made to control the periodic and/or turbulent flow of melt inside a crucible of large diameter. Crystal growth industries have mainly focused on quantitative prediction of a solid–liquid interface, point defect distribution, oxygen concentration, and dislocation-free growth.

8.2.1 Criterion for Characteristic Defect Formation

Point defects such as a vacancy and an interstitial atom form clusters which become microvoids and dislocation clusters. Such defects degrade characteristics of LSIs by formation of leakage current paths. The microvoids and dislocation clusters are formed by supersaturation of vacancy and interstitial atoms during the crystal cooling, respectively. If the total number of vacancy and interstitial atoms is the same, they can compensate, and perfect crystal will then be formed. However, such condition cannot be realized due to differences in their equilibrium concentrations at melting point and their diffusion constants. The interstitial atoms migrate fast near the melting point while vacancies move with the pulling speed since their diffusion constant is small.

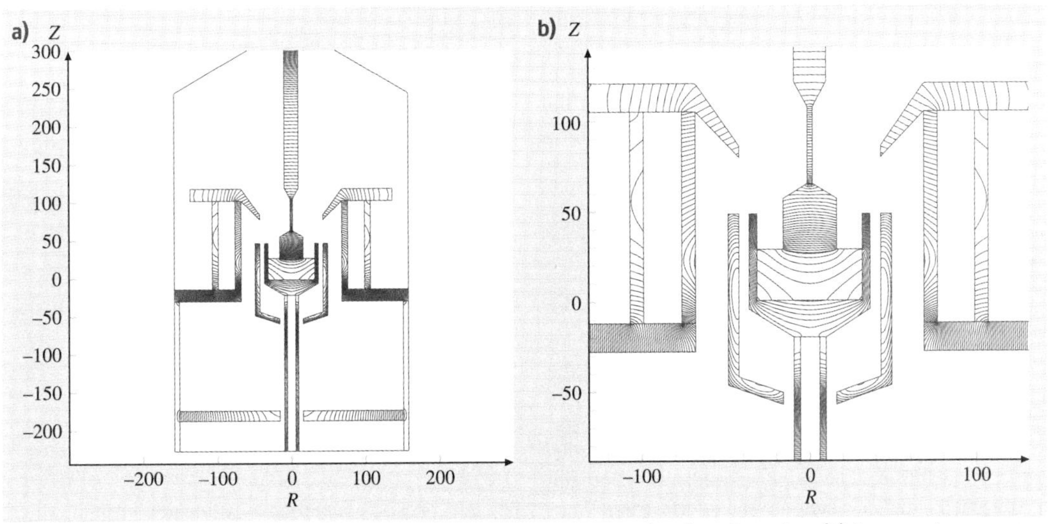

Fig. 8.7 **(a)** Temperature distribution in a global model with a two-dimensional configuration. **(b)** A zoomed temperature distribution in a global model with a two-dimensional configuration. Temperature difference between the contours is 5 K

Volonkov proposed (8.7) as the criterion to obtain perfect crystals [8.52]

$$\frac{V}{G} = 0.62 , \tag{8.7}$$

where V and G are the crystal pulling rate and the temperature gradient of the crystal, respectively.

The flux of vacancy transfer in a crystal is determined by the crystal pulling speed since vacancies, with a slow diffusion constant, move with the pulled crystals. However, the flux of interstitial atoms in a crystal is determined by the temperature gradient of the crystal since interstitial atoms diffuse based on the gradient of the equilibrium concentration. Such concentration gradients are defined by the temperature gradient in a crystal. Therefore, the dominant defect can be determined by the relationship expressed by (8.7).

To obtain a perfect crystal one has to control the flux of defects in a crystal. We need to control these fluxes by controlling the temperature distribution in the furnace and the shape of the solid–liquid interface, which affects the temperature gradient in the crystal. We usually use a global model that can predict the temperature distribution in a furnace, including the shape of the solid–liquid interface. The global model contains conduction, convection, and radiation of heat and mass transfer in the furnace.

Figure 8.7a,b [8.55] shows two-dimensional temperature distributions in a furnace at Kyushu University. We can recognize temperature distributions in each part of the crystal, crucible, melt, heater, and thermal shields. If the system is completely axisymmetric, a two-dimensional configuration can be imposed due to the almost axisymmetric configuration of a Czochralski furnace. However, many studies have shown that flow in the melt has a three-dimensional structure. Therefore, we have to change the configuration from two to three dimensional.

8.2.2 Effect of Pulling Rate and Temperature Gradient

We usually use magnetic fields to control temperature gradient in a crystal through control of the flow of the melt. The results of a series of computations with various intensities of magnetic field and various crystal pulling rates are summarized in Figs. 8.8 and 8.9. Figure 8.8 shows the axial temperature gradients in both the crystal and the melt at a melt–crystal interface as a function of crystal pulling rate. Solid lines show the results with and without a magnetic field, while dashed lines

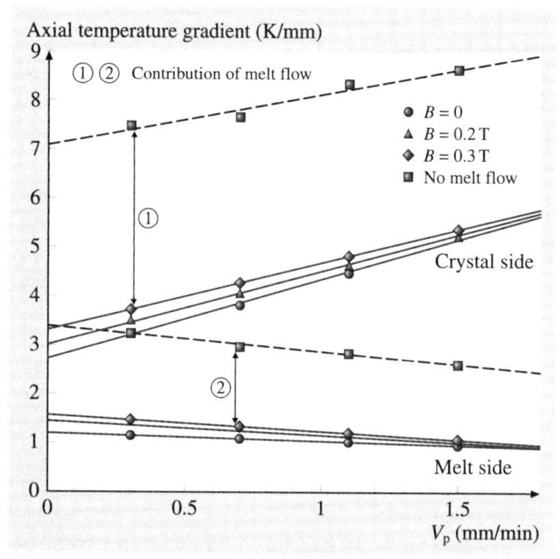

Fig. 8.8 Axial temperature gradients in the crystal (*upper part*) and melt (*lower part*) near the interface as a function of pulling rate of a crystal at different intensities of the applied magnetic field (after [8.55])

show the results without convection in the melt, approximately corresponding to the case with a magnetic field intensity of infinite value. The arrows in the figure show the contribution of the convection of the melt. The val-

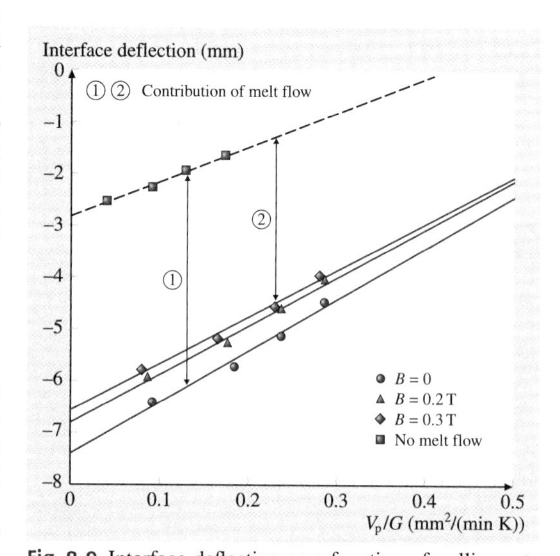

Fig. 8.9 Interface deflection as a function of pulling rate of a crystal at different intensities of the applied magnetic field (after [8.55])

ues of the axial temperature gradients in the melt and crystal were obtained by averaging in the central area of the interface. The values of the axial temperature gradients in the melt and crystal at the interface are not identical even when the crystal pulling rate is zero since the thermal conductivity of the crystal is not equal to that of the melt.

These results show that the axial temperature gradients in the melt and crystal near the interface increase with increasing magnetic field intensity. With increasing crystal pulling rate, the temperature gradient near the interface in the crystal increases, while that in the melt decreases. Meanwhile, the difference between the case with finite (including zero) magnetic field intensity and the case without melt convection becomes smaller. Since this difference is due to the melt convection, this result indicates that the contribution of melt flow becomes smaller with increasing crystal pulling rate.

These phenomena can be explained as follows. When we apply a magnetic field of large intensity natural convection of the melt is suppressed, resulting in a more inhomogeneous temperature distribution in the melt. Therefore, the temperature gradient in the melt increases with increasing magnetic field intensity. Meanwhile, because of heat balance between the liquid and solid at the interface, the temperature gradient in the crystal near the interface also increases accordingly. However, melt convection still remains, even if we apply a magnetic field with a relatively large intensity to the melt. As a result, even when a relatively large magnetic field of 0.3 T is applied to the system, the temperature gradients near the interface in both the crystal and the melt are far from those without melt convection, as shown in Fig. 8.8. On the other hand, since a larger crystal growth rate always results in lower heater power, the temperature on the sidewall of the melt becomes smaller due to the lower heater power. The temperature difference becomes smaller and the temperature distribution becomes less inhomogeneous in the melt. This in turn leads to weaker melt convection due to decrease in the thermal buoyancy force induced by the temperature gradient in the melt. Therefore, with increase in the crystal pulling rate, both the axial temperature gradient in the melt near the interface and the contribution of melt convection decrease. However, since the heat released due to solidification at the interface is proportional to the crystal growth rate and it is transported away from the interface through the crystal, a larger axial temperature gradient field is generated in the crystal near the interface when the crystal pulling rate is increased.

Figure 8.9 shows the interface deflection toward the melt as a function of the ratio between the crystal pulling rate (V_P) and the temperature gradient in the crystal near the interface (G). The values of the interface deflection and the parameter V_P/G were obtained by averaging in the central area of the interface. The arrows in the figure show the contribution of convection in the melt. The interface moves upward to the crystal side with increase in either the magnetic field intensity or the value of the parameter V_P/G. This tendency is consistent with that of the axial temperature gradient in the crystal near an interface shown in Fig. 8.8. This is because the interface shape is mainly determined by the temperature distribution in the crystal close to the interface and the melt convection in the crucible. As explained in the previous section, when a magnetic field of large intensity is applied to the system or a larger pulling rate is applied to the crystal, the melt convection is suppressed and the axial temperature gradient in the crystal near the interface increases. The melt–crystal interface then moves upward to the crystal side in order to accommodate the increased axial temperature gradient in the crystal near the interface and the contribution of melt convection in the crucible becomes smaller.

8.3 Growth and Characterization of Silicon Multicrystal for Solar Cell Applications

Casting is a key method for large-scale production of multicrystalline silicon for use in highly efficient solar cells [8.56, 69–71]. Since the efficiency of solar cells depends on the quality of the multicrystalline silicon, which is in turn determined by the crystallization process, it is important to investigate and optimize the casting process to control the distributions of temperature and iron in a silicon ingot during the solidification process. Moreover, dislocation plays an important role in the efficiency of solar cells. Such properties should be controlled carefully by using a large-scale calculation in order to obtain silicon crystals for solar cells.

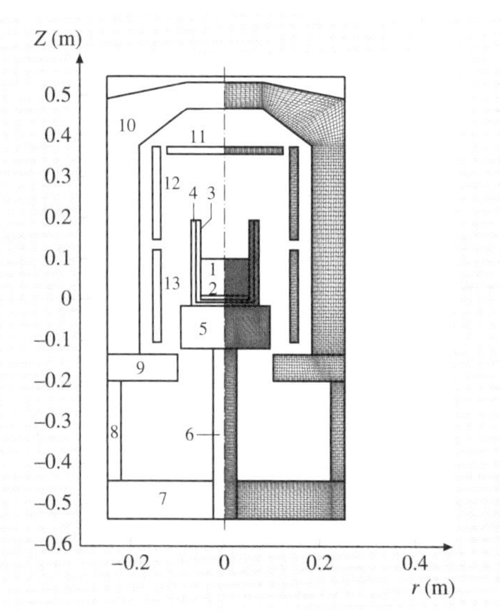

Fig. 8.10 Configuration and computation grid of a casting furnace. The melt, a crystal, a crucible, a crucible holder and pedestals are denoted as *1*, *2*, *3*, *4*, *5*, and *6*, respectively. Thermal shields are labeled *7–11*. The numbers *12* and *13* indicate multiple heater

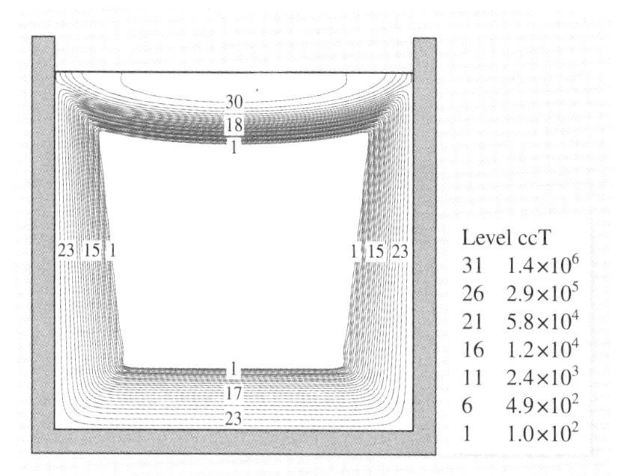

Fig. 8.11 Distribution of iron concentration in a solidified silicon ingot after the solidification process. The scale of iron concentration should be multiplied by $1 \times 10^{10}\,\mathrm{cm}^{-3}$. The periphery of the crystal contains on the order of $1 \times 10^{15}\,\mathrm{cm}^{-3}$

8.3.1 Recent Development of Crystalline Silicon for Solar Cells

Numerical simulation has become a powerful tool for investigation and optimization of the casting process and crystal growth process with the development of computer technology and computational techniques [8.56, 57]. Since a casting furnace has a highly nonlinear conjugated thermal system, transient simulation with global modeling is an essential tool for investigation and improvement of the casting process from melting to cooling through the solidification process. We developed a transient code with a global model for the casting process and carried out calculations to investigate the distributions of temperature and iron in a silicon ingot during the casting process. Time-dependent distributions of iron and temperature in a silicon ingot were investigated. Figure 8.10 shows a typical casting furnace for production of multicrystalline silicon for solar cells.

Figure 8.11 shows the distribution of iron concentration in a solidified silicon ingot that had been cooled for 1 h during the cooling process. The figure shows a vertical cross-section of the iron concentration in the crystal. The scale of the iron concentration should be multiplied by $1 \times 10^{10}\,\mathrm{cm}^{-3}$; therefore, the periphery of the crystal contains on the order of $1 \times 10^{15}\,\mathrm{cm}^{-3}$.

Areas with high iron concentration were formed at the top of the melt due to segregation of iron. Moreover, areas with high concentration of iron were formed close to the crucible walls. Such areas were formed by diffusion, which occurred during solidification and the cooling process. This is based on the small activation energy of iron diffusion in solid silicon. The central area of the ingot has a low concentration of iron after the solidification process, as shown in Fig. 8.11.

8.4 Summary

Crystal growth and characterization of Czochralski silicon single crystals for semiconductor and solar cell application is a key technology for information and renewable energy. Such activity enables us to open up a new world through technological breakthroughs. Both experimental and numerical studies on crystal growth help us to understand and derive new concepts for new crystal growth technology.

References

8.1 M. Itsumi, H. Akiya, T. Ueki: The composition of octahedron structures that act as an origin of defects in thermal SiO_2 on Chochralski silicon, J. Appl. Phys. **78**, 5984–5988 (1995)

8.2 K. Koai, A. Seidel, H.-J. Leister, G. Müller, A. Koehler: Modeling of thermal fluid-flow in the liquid encapsulated Czochralski process and comparison with experiments, J. Cryst. Growth **137**, 41–47 (1994)

8.3 H.-J. Leister, M. Peric: Numerical-simulation of a 3-D Czochralski flow by a finite volume multigrid-algorithm, J. Cryst. Growth **123**, 567–574 (1992)

8.4 H. Yamagishi, M. Kuramoto, Y. Shiraishi: CZ crystal growth development in super silicon crystal project, Solid State Phenom. **57–8**, 37–39 (1997)

8.5 Y.C. Won, K. Kakimoto, H. Ozoe: Transient three-dimensional flow characteristics of Si melt in a Czochralski configuration under a cusp-shaped magnetic field, Numer. Heat Transf. **A36**, 551–561 (1999)

8.6 K.-W. Yi, M. Watanabe, M. Eguchi, K. Kakimoto, T. Hibiya: Change in velocity in silicon melt of the Czochralski (CZ) process in a vertical magnetic field, Jpn. J. Appl. Phys. **33**, L487–L490 (1994)

8.7 M.G. Williams, J.S. Walker, W.E. Langlois: Melt motion in a Czochralski puller with a weak transverse magnetic-field, J. Cryst. Growth **100**, 233–253 (1990)

8.8 A.E. Organ, N. Riley: Oxygen-transport in magnetic Czochralski growth of silicon, J. Cryst. Growth **82**, 465–476 (1987)

8.9 J.S. Walker, M.G. Williams: Centrifugal pumping during Czochralski silicon growth with a strong transverse magnetic-field, J. Cryst. Growth **137**, 32–36 (1994)

8.10 J. Baumgartl, M. Gewald, R. Rupp, J. Stierlen, G. Müller: Studies of buoyancy driven convection in a vertical cylinder with parabolic temperature profile, Proc. 7th Eur. Symp. Mater. Fluid Sci. Microgravity, Oxford (1989) pp. 10–15

8.11 L.N. Hjellming, J.S. Walker: Melt motion in a Czochralski crystal puller with an axial magnetic-field-uncertainty in the thermal constants, J. Cryst. Growth **87**, 18–32 (1988)

8.12 S. Kobayashi: Numerical-analysis of oxygen-transport in magnetic Czochralski growth of silicon, J. Cryst. Growth **85**, 69–74 (1987)

8.13 M. Akamatsu, K. Kakimoto, H. Ozoe: Effect of crucible rotation on the melt convection and the structure in a Czochralski method, Transp. Phenom. Therm. Sci. Process Eng. **3**, 637–642 (1997)

8.14 K.-W. Yi, K. Kakimoto, M. Eguchi, M. Watanabe, T. Shyo, T. Hibiya: Spoke patterns on molten silicon in Czochralski system, J. Cryst. Growth **144**, 20–28 (1994)

8.15 K. Kakimoto, H. Ozoe: Oxygen distribution at a solid-liquid interface of silicon under transverse magnetic fields, J. Cryst. Growth **212**, 429–437 (2000)

8.16 R.A. Brown, T.A. Kinney, P.A. Sackinger, D.E. Bornside: Toward an integrated analysis of Czochralski growth, J. Cryst. Growth **97**, 99–115 (1989)

8.17 H. Hirata, N. Inoue: Study of thermal symmetry in Czochralski silicon melt under a vertical magnetic field, Jpn. J. Appl. Phys. **23**, L527–L530 (1984)

8.18 H. Hirata, K. Hoshikawa: Silicon crystal growth in a cusp magnetic-field, J. Cryst. Growth **96**, 747–755 (1989)

8.19 H. Hirata, K. Hoshikawa: Homogeneous increase in oxygen concentration in Czochralski silicon-crystals by a cusp magnetic-field, J. Cryst. Growth **98**, 777–781 (1989)

8.20 H. Hirata, K. Hoshikawa: Silicon crystal growth in a cusp magnetic field, J. Cryst. Growth **96**, 747–755 (1989)

8.21 K. Hoshi, T. Suzuki, Y. Okubo, N. Isawa: Extended Abstracts, Electrochem. Soc. Spring Meet., Vol. 80-1 (The Electrochem. Soc., Pennington 1980) p. 811

8.22 K. Hoshikawa: Czochralski silicon crystal growth in the vertical magnetic field, Jpn. J. Appl. Phys. **21**, L545–L547 (1982)

8.23 K. Hoshikawa, H. Kohda, H. Hirata: Homogeneity of vertical magnetic field applied LEC GaAs crystal, Jpn. J. Appl. Phys. **23**, L195–L197 (1984)

8.24 K. Kakimoto, L.J. Liu: Numerical study of the effects of cusp-shaped magnetic fields and thermal conductivity on the melt-crystal interface in CZ crystal growth, Cryst. Res. Technol. **38**, 716–725 (2003)

8.25 K. Kakimoto: Use of an inhomogeneous magnetic fields for silicon crystal growth, Proc. 2nd Workshop High Magn. Fields, ed. by H. Schneider-Muntau (World Scientific, New York 1997) pp. 21–24

8.26 K. Kakimoto: Flow instability during crystal growth from the melt, Prog. Cryst. Growth Charact. **30**, 191–215 (1995)

8.27 K. Kakimoto, Y.W. Yi, M. Eguchi: Oxygen transfer during single silicon crystal growth in Czochralski system with vertical magnetic fields, J. Cryst. Growth **163**, 238–242 (1996)

8.28 K. Kakimoto, Y.W. Yi: Use of magnetic fields in crystal growth from semiconductor melts, Physica B **216**, 406–408 (1996)

8.29 K.M. Kim, W.E. Langlois: Computer-simulation of boron transport in magnetic Czochralski growth of silicon, J. Electrochem. Soc. **133**, 2586–2590 (1986)

8.30 A.E. Organ, N. Riley: Oxygen-transport in magnetic Czochralski growth of silicon, J. Cryst. Growth **82**, 465–476 (1987)

8.31 Z.A. Salnick: Oxygen in Czochralski silicon crystals grown under an axial magnetic field, J. Cryst. Growth **121**, 775–780 (1992)

8.32 T. Suzuki, N. Isawa, Y. Okubo, K. Hoshi: Oxygen in Czochralski silicon crystals grown under a transverse magnetic field, Semiconductor Silicon 1981, ed. by H.R. Huff, R.J. Kriegler, Y. Takeishi (The Electrochem. Soc., Pennington 1981) pp. 90–94

8.33 R.N. Thomas, H.M. Hobgood, P.S. Ravishankar, T.T. Braggins: Oxygen distribution in silicon crystals grown by transverse magnetic fields, Solid State Technol. **April**, 163–170 (1990)

8.34 M. Watanabe, M. Eguchi, K. Kakimoto, T. Hibiya: The baroclinic flow instability in rotating silicon melt, J. Cryst. Growth **128**, 288–292 (1993)

8.35 M. Watanabe, M. Eguchi, K. Kakimoto, T. Hibiya: Flow mode transition and its effects on crystal-melt interface shape and oxygen distribution for Czochralski-grown Si single crystals, J. Cryst. Growth **151**, 285–290 (1995)

8.36 M.J. Wargo, A.F. Witt: Real-time thermal imaging for analysisi and control of crystal-growth by the Czochralski technique, J. Cryst. Growth **116**, 213–224 (1955)

8.37 K.-W. Yi, M. Watanabe, M. Eguchi, K. Kakimoto, T. Hibiya: Change in velocity in silicon melt of the Czochralski (CZ) process in a vertical magnetic field, Jpn. J. Appl. Phys. **33**, L487–L490 (1994)

8.38 Y. Gelfgat, J. Krumins, B.Q. Li: Effects of system parameters on MHD flows in rotating magnetic fields, J. Cryst. Growth **210**, 788–796 (2000)

8.39 Y. Gelfgat, E. Jpriede: The influence of combined electromagnetic fields on the heat and mass transfer in a cylindrical vessel with the melt, Magnetohydrodynamics **31**, 102–110 (1995)

8.40 R.U. Barz, G. Gerbeth, Y. Gelfgat: Numerical simulation of MHD rotator action on hydrodynamics and heat transfer in single crystal growth processes, J. Cryst. Growth **180**, 388–400 (1997)

8.41 T. Kaiser, K.W. Benz: Taylor vortex instabilities induced by a rotating magnetic field: A numerical approach, Phys. Fluids **10**, 1104–1110 (1998)

8.42 F.-U. Brucker, K. Schwerdtfeger: Single-crystal growth with Czochralski method involving rotational electromagnetic stirring of the melt, J. Cryst. Growth **139**, 351–356 (1994)

8.43 J. Virbulis, T. Wetzel, A. Muiznieks, B. Hanna, E. Dornberger, E. Tomzig, A. Muhlbauer, W. von Ammon: Stress-induced dislocation generation in large FZ- and CZ-silicon single crystals – Numerical model and qualitative considerations, Proc. 3rd Int. Workshop Model. Cryst. Growth (2000) pp. 31–33

8.44 L.J. Liu, T. Kitashima, K. Kakimoto: Three-dimantional calculation of Si-CZ growth, Proc. Int. Symp. Process. Technol. Market Dev. 300 mm Si Mater. (ISPM-300mm Si), Beijing (2003) pp. 2551–2555

8.45 O. Grabner, G. Mueller, E. Tomzig, W. von Ammon: Effects of various magnetic field configurations on temperature distributions in Czochralski silicon melts, Microelectron. Eng. **56**, 83–88 (2001)

8.46 K. Kakimoto, K.-W. Yi, M. Eguchi: Oxygen transfer during single silicon crystal growth in Czochralski system with vertical magnetic fields, J. Cryst. Growth **163**, 238–242 (1996)

8.47 A. Krauze, A. Muiznieks, A. Muhlbauer, T. Wetzel, L. Gorbunov, A. Pedchenko, J. Virbulis: Numerical 2-D modelling of turbulent melt flow in CZ system with dynamic magnetic fields, J. Cryst. Growth **266**, 40–47 (2004)

8.48 H. Ozoe, M. Iwamoto: Combined effects of crucible rotation and horizontal magnetic field on dopant concentration in a Czochralski melt, J. Cryst. Growth **142**, 236–244 (1994)

8.49 P. Sabhapathy, M.E. Salcudean: Numerical study of Czochralski growth of silicon in an axisymmetric magnetic field, J. Cryst. Growth **113**, 164–180 (1991)

8.50 K. Kakimoto, H. Watanabe, M. Eguchi, T. Hibiya: Direct observation by X-ray radiography of convection of molten silicon in the Czochralski growth method, J. Cryst. Growth **88**, 365–370 (1988)

8.51 K. Nakamura, S. Maeda, S. Togawa, T. Saisyoji, T. Tomioka: Effect of the shape of crystal-melt interface on point defect reaction in silicon crystals, ECS Proc. **17**, 31–33 (2000)

8.52 V. Voronkov: The mechanism of swirl defects formation in silicon, J. Cryst. Growth **59**, 625–643 (1982)

8.53 W. von Ammon, E. Dornberger, H. Oelkrug, H. Weider: The dependence of bulk defects on the axial temperature gradient od silicon crystals during Czochralski growth, J. Cryst. Growth **151**, 273–277 (1995)

8.54 K. Nakamura, T. Saisyoji, J. Tomioka: Grown-in defects in silicon crystals, J. Cryst. Growth **237**, 1678–1684 (2002)

8.55 L. Liu, S. Nakano, K. Kakimoto: An analysis of temperature distribution near the melt-crystal interface in silicon Czochralski growth with a transverse magnetic field, J. Cryst. Growth **282**, 49–59 (2005)

8.56 D. Franke, T. Rettelbach, C. Habler, W. Koch, A. Muller: Silicon ingot casting: process development by numerical simulations, Sol. Energy Mater. Sol. Cells **72**, 83–92 (2002)

8.57 M. Ghosh, J. Bahr, A. Muller: Silicon ingot casting: process development by numerical simulations, Proc. 19th Euro. Photovolt. Sol. Energy Conf., Paris (2004) pp. 560–563

8.58 D. Vizman, S. Eichler, J. Friedrich, G. Müller: Three-dimensional modeling of melt flow and interface shape in the industrial liquid-encapsulated Czochralski growth of GaAs, J. Cryst. Growth **266**, 396–403 (2004)

8.59 A. Krauze, A. Muiznieks, A. Muhlbauer, T. Wetzel, W. von Ammon: Numerical 3-D modelling of turbulent melt flow in a large CZ system with horizontal DC

magnetic field. II. Comparison with measurements, J. Cryst. Growth **265**, 14–257 (2004)
8.60 L.J. Liu, K. Kakimoto: D global analysis CZ-Si growth in transverse magnetic field with rotating crucible and crystal, Cryst. Res. Technol. **40**, 347–351 (2005)
8.61 K. Kakimoto, L.J. Liu: Numerical study of the effects of cusp-shaped magnetic fields and thermal conductivity on the melt-crystal interface in CZ crystal growth, Cryst. Res. Technol. **38**, 716–725 (2003)
8.62 J.J. Derby, R.A. Brown: Thermal-capillary analysis of Czochralski and liquid encapsulated Czochralski crystal growth, J. Cryst. Growth **75**, 227–240 (1986)
8.63 F. Dupret, P. Nicodeme, Y. Ryckmans, P. Wouters, M.J. Crochet: Global modeling of heat-transfer in crystal growth furnaces, Int. J. Heat Mass Transf. **33**, 1849–1871 (1990)
8.64 M. Li, Y. Li, N. Imaishi, T. Tsukada: Global simulation of a silicon Czochralski furnace, J. Cryst. Growth **234**, 32–46 (2002)
8.65 V.V. Kalaev, I.Y. Evstratov, N.Y. Makarov: Gas flow effect on global heat transport and melt convection in Czochralski silicon growth, J. Cryst. Growth **249**, 87–99 (2003)
8.66 L. Liu, K. Kakimoto: Partly three-dimensional global modeling of a silicon Czochralski furnace II. Model application: Analysis of a silicon Czochralski furnace in a transverse magnetic field, Int. J. Heat Mass Transf. **48**, 4492–4497 (2005)
8.67 L. Liu, S. Nakano, K. Kakimoto: An analysis of temperature distribution near the melt-crystal interface in silicon Czochralski growth with a transverse magnetic field, J. Cryst. Growth **282**, 49–59 (2005)
8.68 L. Liu, K. Kakimoto: Partly three-dimensional global modeling of a silicon Czochralski furnace. I. Principles, formulation and implementation of the model, Int. J. Heat Mass Transf. **48**, 4481–4491 (2005)
8.69 E.W. Weber: Transition-metal profiles in a silicon crystal, Appl. Phys. **A30**, 1–15 (1983)
8.70 W. Zuhlener, D. Huber: Czochralski crystal growth of silicon. In: *Crystal-Growth, Properties and Applications*, Vol. 8, ed. by J. Grabmaier (Springer, Berlin, Heidelberg 1988) pp. 1–12
8.71 D. Macdonald, A. Cuevas, A. Kinomura, Y. Nakano, J.J. Geerligs: Transition-metal profiles in a multicrystalline silicon ingot, J. Appl. Phys. **97**, 33523–33527 (2005)

9. Czochralski Growth of Oxide Photorefractive Crystals

Ernesto Diéguez, Jose Luis Plaza, Mohan D. Aggarwal, Ashok K. Batra

Czochralski crystal growth is one of the major methods of crystal growth from melt for bulk single crystals for commercial and technological applications. Most crystals, such as semiconductors and oxides, are grown from melt using this technique due to the much faster growth rates achievable. A detailed description of the process can only be given for specific materials; there is no universal crystal pulling system available commercially. The details of the basic principle and the design of automatic diameter control Czochralski crystal growth system elements are given in this chapter so as to enable any researcher to design and fabricate his/her own system. This chapter is devoted to the growth of bulk oxide photorefractive materials such as lithium niobate and sillenite crystals including the development in these materials during the last decade. A number of problems (and possible solutions) encountered by the authors during growth in their respective laboratories over the last two decades are discussed.

Section 9.2 provides the introduction to crystals and crystal growth mechanism and various methods of growing photorefractive crystals. Section 9.3 discusses in detail the Czochralski method of crystal growth, including selection of appropriate components for setting up a crystal growth system such as the heating system design, and raising, lowering, and rotation mechanisms. Section 9.4 discusses the growth and properties of lithium niobate crystals. A brief introduction to other photorefractive crystals is given in Sect. 9.5. The details of the growth and properties of sillenite crystals are given in Sect. 9.6. Section 9.7 summarizes the present state of these two important crystals in terms of growth and applications.

9.1 Background

With the rapid growth of the electronic and optoelectronic industries, the demand for crystalline materials has increased dramatically over the past few decades. The requirement for better, cheaper, and larger single crystals has driven extensive research and development in crystal growth. A major factor behind such growth is the advent of high-power and efficient solid-state lasers, in combination with the use of materials that exhibit large second- or third-order nonlinearities [9.1–3]. In such conditions, nonlinear optics is becoming an important technology in the design of new laser sources emitting in the visible or near infrared. Nonlinear optics also enables the attainment of new functionalities in laser systems and in optoelectronic signal transmission and processing. The class of nonlinear phenomena based on the photorefractive effects in electrooptic crystals will undoubtedly play a major role in these various applications of laser photonics. Since its discovery, photorefraction has stimulated much basic research, covering both fundamental materials studies and their applications in dynamic holography, laser beam control, and optical processing. For many years the challenge was only to understand the basic mechanisms of photorefraction giving rise to this effect in different crystals, which drove research and development in this area. More and more materials were found to be photorefractive. New electrooptic interactions were discovered. The fields of nonlinear optics, optical spectroscopy, electrooptics, ferroelectrics, electronic transport, and Fourier optics were brought together to develop a complete understanding of the complex microscopic mechanism involved. Another important aspect of photorefractive materials is their ability to perform efficient energy transfer between a signal and a reference beam interfering in the volume of the crystal. This property opens up a wide range of applications, including image amplification, optical phase conjugation with gain, and self-pumped optical cavities. To achieve these results and develop applications requires us to optimize nonlinear material properties, and therefore the choice of the best material is a critical issue. Nonlinear photorefractive optics is now well established and has reached scientific maturity. It contributes to stimulate basic research in solid-state physics and to investigate in detail the mechanisms of charge transport in different types of ferroelectric or semiconductor crystals such as $LiNbO_3$, $BaTiO_3$, $KNbO_3$, $Bi_{12}SiO_{20}$ or GaAs.

The Czochralski method is most popular and useful for the growth of large oxides crystals. This chapter is devoted to the growth of bulk photorefractive materials from the Czochralski method, summarizing the development in these materials during the last decade. We focus our attention on the two most important photorefractive materials, $LiNbO_3$ and sillenites, which are considered the most relevant, especially because in these two materials all the problems and circumstances inherent to bulk crystal growth of photorefractive oxide materials are concentrated. For this reason the chapter includes various sections devoted to each of these materials.

9.2 Crystal Growth

9.2.1 Czochralski Method of Crystal Growth

The CZ technique is named after J. Czochralski. Czochralski's invention was discovered essentially by accident. As the story goes, the young Czochralski, then chief of AEG's metal laboratory in Berlin, was studying the crystallization of metals. A crucible containing molten tin was left on his table for slow cooling and crystallization. Czochralski was preparing his notes on the experiments carried out during the day when at some point, lost in thought, he dipped his pen into this crucible instead of the inkwell placed near the crucible. He withdrew it quickly and saw a thin thread of solidified metal hanging at the tip of the nib. Thus, the discovery was made. He had discovered a phenomenon never observed before: crystallization by pulling from the surface of a melt. However, careful observation of this accidental process provided a discovery of great importance. Later, the nib slot, in which crystallization was initiated, was replaced by a special narrow capillary and later by a seed crystal of the material to be grown [9.4]. *Teal* and *Little* also grew germanium single crystals by similar technique [9.5]. In the recent past, *Brandle* has given a detailed description of Czochralski growth of oxides along with current status [9.6].

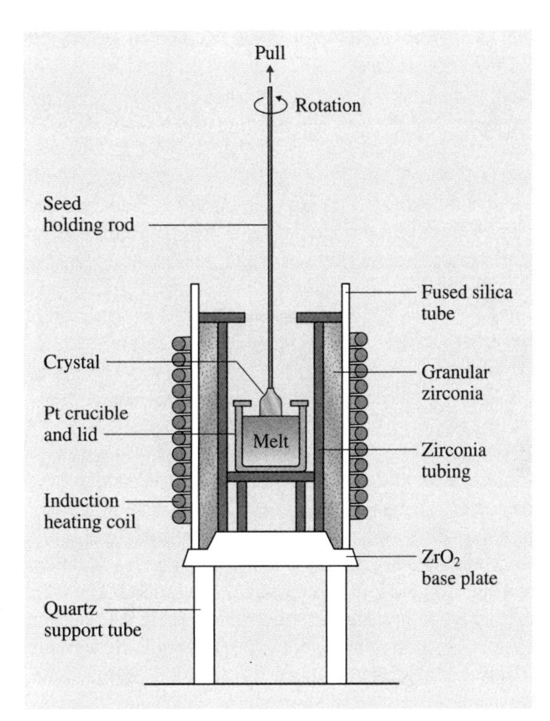

Fig. 9.1 A basic diagram of the Czochralski method ▶

A basic diagram of the CZ method is illustrated in Fig. 9.1 with various components to explain the process. To be considered as a possible candidate for crystal growth by the Czochralski technique, the material should have a relatively low vapor pressure. The crucible material should be nonreactive with the crystal growth material above its melting point. The Czochralski system is based on the following principle: the material is melted in a crucible and is kept for a certain time at a temperature above the melting point, then the temperature is reduced to a value slightly above the freezing point. The freezing point is judged by cooling the melt until crystals start to appear on the surface, then the temperature is slightly lowered and a seed (cut in the appropriate orientation) is inserted into the melt. It is kept at that position for a little while, then the pulling mechanism is started. The seed forms a crystallization center if the temperatures have been chosen correctly. If the crystal starts growing very fast and becomes visible to the naked eye, then the temperature of the melt needs to be increased by a degree or so; the melt must not be overheated as this would cause the crystal to dissolve and separate from the melt. The diameter of the pulled crystal is controlled by manipulating the temperature of the melt and the pulling rate. Suitable engineering of both axial and radial temperature gradients is needed to grow single crystals of desired dimensions reliably.

9.3 Design and Development of Czochralski Growth System

In this section, various components and instruments used for developing the automatic diameter control CZ technique are discussed so that the reader can set up his/her own system. The major components are: furnace design, heating methods viz. resistive or inductive, temperature control techniques, crucible selection, and crystal rotation and pulling arrangement [9.7].

9.3.1 Furnace Construction

The furnace used for crystal pulling can vary from the very simple, e.g., a resistance wound heating element, to one which is extremely complex because of thermal and chemical constraints placed upon it by the crystal. Two types of furnaces can be used; the first type is for oxide crystal growth and is generally composed of ceramic and noble-metal parts, whereas the second type is for semiconductor growth and is usually composed of graphite and fused-silica parts. For growing lithium niobate crystals, a resistance or inductive furnace can be utilized.

9.3.2 Heating Methods

Selection of the heating method depends on the operational temperature needed to grow single crystals of the desired material. This normally depends on the melting temperature, which can be determined from the phase diagram, which is essentially a first road map for the crystal grower. The basis for selection of the heating method is also determined by the following factors: the method of heat transfer (radiation, conduction or convection), the rate of heating or thermal transfer, the degree of uniformity of the temperatures, the shape of the temperature gradients, the precision needed in controlling the temperature, the furnace atmosphere, and

Table 9.1 Some useful properties of common resistance and heating elements

Material and trade name	Maximum operating temperature (°C)	Permissible atmosphere
Kanthal	1250	Oxidizing or reducing
Sintered SiC (Globar)	1600	Oxidizing
$MoSi_2$ (Kanthal Super)	1650	Oxidizing
Graphite	2500	Inert or vacuum
Tungsten	3000	Inert or vacuum
Nichrome	1200	Oxidizing or reducing
Platinum	1450	Oxidizing or reducing

the cost of the heating equipment. The variety of heating methods for crystal growth may be categorized mainly in two categories, viz. resistance heating or induction heating. Some useful properties of common resistance heating element materials are listed in Table 9.1. As can be seen, resistance-heated furnaces are normally limited to lower temperature ranges (up to 1600 °C) and can have one of the several types of elements listed in the table. A general advantage of resistance-heated furnaces over other types is their greater electrical efficiency and reduced operational costs.

High-frequency heating is of major importance in crystal growth because a large range of temperatures can be achieved with a reasonably high efficiency of energy transfer and it can be used in a variety of processes. Radiofrequency (RF) induction heating provides the cleanest and most readily available method of heating precious-metal crucibles, although in the interest of economy, resistance heating is sometimes used, especially for lower-melting-point materials. Induction heating normally occurs in a conducting material due to eddy currents induced in the conductor by the electromagnetic field from a high-frequency current-carrying coil (the RF work coil) that surrounds the charge. The useful RF band is typically 100 kHz to 10 MHz, although for relatively large metallic ingots, frequencies in the medium frequency band (0.5–10 kHz) are useful and advantageous. In fact, most of the heat is generated in the skin layer over the coupled flux volume of the conductor. If a nonconducting crucible is required, a graphite susceptor should be utilized, which is heated first and then conducts heat to the charge in the crucible.

This source of heating is generally used for higher-melting oxides such as sapphire (Al_2O_3) and garnets; however, it can be used for the growth of semiconductor materials. Usually RF generators operate in the above-mentioned frequency range and with a wide power range of 10–100 kW. For general laboratory use, a 10–30 kW RF generator is suitable and provides the capability to grow crystals up to about 30–50 mm in diameter.

9.3.3 Temperature Control Techniques

Thermocouples are routinely used for measuring the temperatures in these systems. There are a large number of thermocouple available including Chromel/Alumel, Pt-Pt10%Rh, and Pt-Pt13%Rh. Junctions between two thermocouples metals are produced by either fusing the two wires in a gas flame or by spot welding. Mechanical and brazed connections can also be used. Nowadays cold junction compensation is available for all types of thermocouples in any temperature measuring thermocouple thermometer. Basically these provide a zero suppression in accordance with the electromotive force (EMF) generated by the cold junction. Some of the most commonly used thermocouples are listed in Table 9.2.

Most conventional temperature controllers, whether analog or microprocessor based, are three-mode proportional–integral–differential (PID) controllers. This means that the control algorithm is based on a proportional gain, an integral action, and a derivative action. The proportional band simply amplifies the error between the set point and the measured value to establish a power level. The term *proportional band* (PB) expresses the gain of the controller as a percentage of the span of the instrument. A 25% PB equates to a gain of 4, whereas a 10% PB corresponds to a gain of 10. Given the case of a controller with a span of 1000 degrees, a PB of 10% defines a control range of 100 degrees around the set point. If the measured value is 25 degrees below the set point, the output level will be 25% heat. The proportional band determines the magnitude of the

Table 9.2 Most commonly used temperature sensors

Type and trade name	Thermocouple/ sensor elements	Range (°C)
K	Chromel/Alumel	0–999
J	Iron/Constantan	0–500
R	Pt /13%Rh-Pt	0–1760
S	Pt /10%Rh-Pt	17–1760
B	Pt / 6% Rh-Pt /30%Rh	24–1820
C	W/5%Re-W/26%Re	17–2320
Spot pyrometer	Optical pyrometer	600–3000

response to an error. If the proportional band is too small, meaning high gain, the system oscillates and is overresponsive. A wide proportional band, i.e., low gain, could lead to control *wander* due to a lack of responsiveness. The ideal situation is achieved when the proportional band is as narrow as possible without causing oscillation.

Integral action, or automatic reset, is probably the most important factor governing control at a set point. The integral term slowly shifts the output level as a result of an error between the set point and the measured value. If the measured value is below the set point the integral action will gradually increase the output power level in an attempt to correct this error. Expressed as a time constant, the longer the integral time constant, the more slowly the power level will be shifted (the fewer repeats/min, the slower the response). If the integral term is set to a fast value the power level will be shifted too quickly, thus causing oscillation since the controller is trying to work faster than the load can change. Conversely, an integral time constant which is too long will result in very sluggish control. The derivative action or rate provides a sudden change in output power level as a result of a quick change in measured value. If the measured value drops quickly the derivative term will provide a large change in output level in an attempt to correct the perturbation before it goes too far. Derivative action is probably the most misunderstood of the three. It is also most beneficial in recovering from small perturbations.

An optical pyrometer is another well-accepted noncontact temperature measuring device. Optical pyrometers work for temperatures up to 3000 °C.

9.3.4 Common Crucible Materials

The chemical and physical stability of the crucible in the processing environment are important factors that dictate the selection of the material used to fabricate it. The design and choice of the material for the crucible are critical. The crucible should not contaminate or stress the crystal. If possible it should have a thermal conductivity similar to that of the charge to ensure that planar isotherm profiles exist in the vicinity of the melting point of the charge, thus ensuring that the crystallizing interface is near planar. For crystal growth of metals, quartz and graphite are most often used. Some of the common materials used to make crucibles are listed in Table 9.3.

Table 9.3 Some commonly used crucible materials

Materials	Max. operating temperature (°C)	Melting point (°C)
Platinum	1500	1770
Silica	1550	1700
Alumina	1800	2040
Iridium	2100	2466
Molybdenum	2300	2620
Tungsten	2800	3410±20
Graphite	3000	3652

9.3.5 Crystal Rotation and Pulling Arrangement

A schematic of the crystal pulling arrangement used in author's laboratory is presented in Fig. 9.2. In this arrangement, circular motion of the motor is converted to linear motion by using a lead screw with a specific pitch. A rotation motor is attached to the cantilever arm. A load cell measures the weight of the growing crystal and also holds the seed rod. A gearbox arrangement is used to adjust the pulling rate, and the rotation rate is adjusted separately using a stepper motor and its controller.

9.3.6 The Czochralski Crystal Growth System

Using the above-described components, automatic diameter control for a CZ system has been designed and fabricated [9.8, 9]. The system consists of an electrical furnace with global heat elements that generates the required temperature gradient, seed rotation and lowering mechanisms, and an Hottinger Baldwin Messtechnik GmbH (HBM) electronic balance/load cell which can measure the weight of a growing crystal to the nearest 0.1 g. The temperature of the furnace goes up to 1500 °C, controlled using a Eurotherm 818 temperature controller. The data for the control thermocouple and from the balance are monitored by a personal computer, which through a central program, written in Visual Basic 6.0/BASIC (described below), sets the temperature controller to the required temperature set point. Two different stepper motors interfaced to a personal computer control the pulling and rotation of seed crystals. A photograph of the CZ crystal growth system used in the authors' laboratory using an induction heating system is shown in Fig. 9.3. Figure 9.4 shows a block diagram of the complete setup of the same Czochralski crystal growth system. An isometric view of a Czochralski crystal puller used in the laboratory at Alabama A&M University is shown in Fig. 9.5.

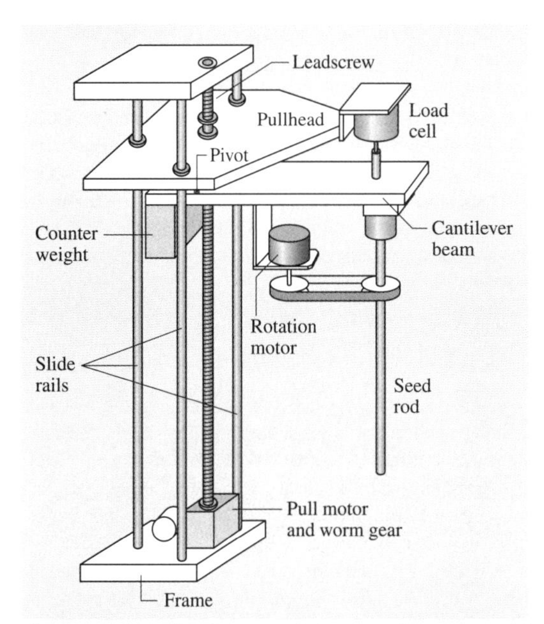

Fig. 9.2 Schematic diagram of the crystal rotation and pulling mechanisms with load cell for weighing the growing crystal

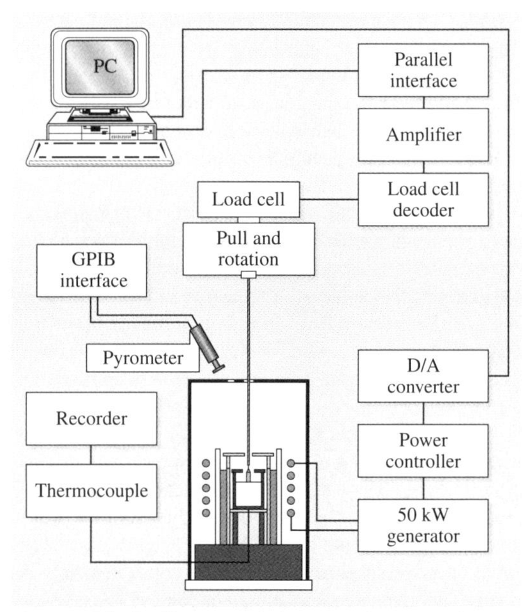

Fig. 9.4 Block diagram of the complete experimental set up for the Czochralski crystal growth technique (GPIB – general purpose interface bus, D/A – digital to analog)

Fig. 9.3 Growth chamber with crystal pulling and rotation system for Czochralski crystal growth

Fig. 9.5 An isometric view of Czochralski crystal puller at Alabama A&M University (M.S – mild steel) ▶

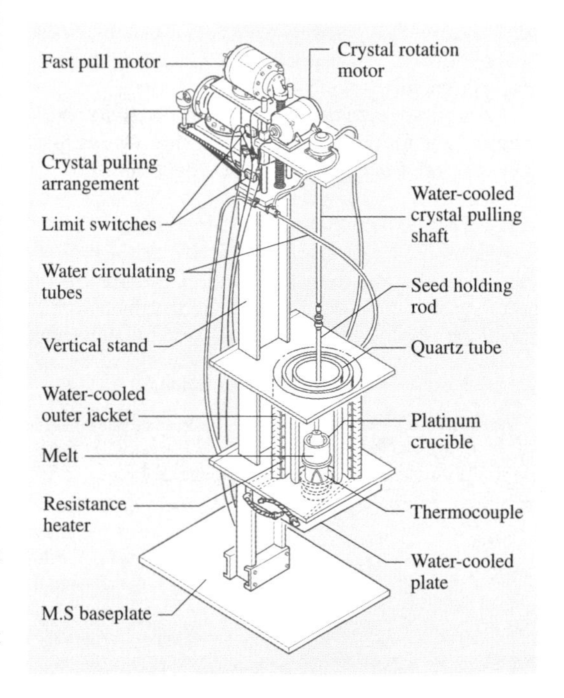

9.3.7 Automatic Diameter Control for Czochralski Crystal Growth Technique

In order to reproducibly grow high-quality single crystals the best control techniques need to be considered. In this subsection we will discuss the use of automatic diameter control systems for Czochralski crystal growth from melt [9.9, 10]. Various techniques for automatic diameter control are available, including optical methods (i. e., analysis of a video or x-ray image of the growing crystal) as well as weight monitoring, which consists of weighing the crystal and seed holder, as used in the authors' laboratory in the present system, or by weighing the crucible and its contents. The relative merits of some of these systems are reviewed in the literature [9.7]. Briefly, optical methods depend on the availability of a clear optical path to view the growth interface, which may be difficult to maintain over the time required for growth. Weighing the crucible gives a poorer ratio of signal (weight change per unit time) to total weight than weighing only the crystal and seed holder. In addition, if induction heating is used, the radiofrequency (RF) field exerts a vertical force on the crucible (proportional to the instantaneous power) which must be taken into account to obtain accurate weight change data. To control the crystal growth process and achieve/maintain a specific crystal diameter, a program in Visual Basic 6.0 was designed and developed. The Visual Basic program has near-real-time access to all data, control characteristics, and user-friendly interfaces for user interaction with the software and access to archival data. The algorithm, written in Visual Basic, provides control of the complete crystal profile during seed extension and growth of the crystal to the desired diameter, termination of the growth process, and cooling of the grown crystal to room temperature. The slope of crystal weight over time is used to compute the diameter of crystal. All the parameters required for growth runs need to be optimized for each material, and the desired sizes of crystal are stored in a parameters file. The parameters stored in the file are crystal density, melt density, crucible diameter, seed length, seed diameter, cone angle crystal diameter and growth rate, control loop parameters including proportional–integral–derivative (PID) values for different stages of growth, fitting factors for minimizing noise during data acquisition, and feedback values. The controlled parameter is the power to the furnace through the Eurotherm 818 process controller. In the current application, the weight reading is taken at 6 s intervals but can be varied. At start up, the program asks for a history file name and parameter file to be used. After selecting the desired file corresponding to the crystal to be grown, the user enters the main parameter screen, as shown in Fig. 9.6. The main parameter screen exhibits various choices/command buttons to activate functions such as:

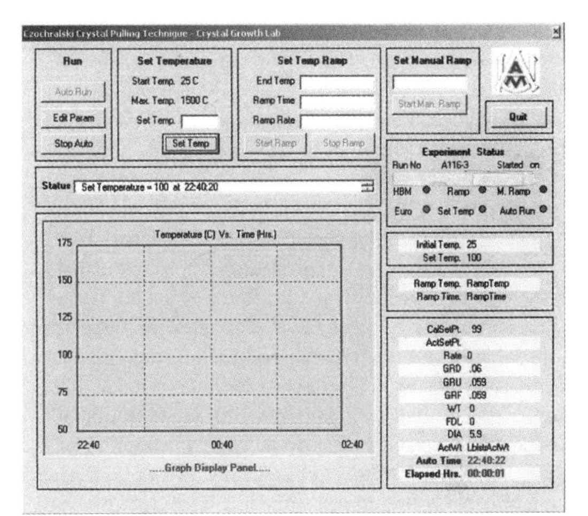

Fig. 9.6 Main parameter screen of the automatic diameter control software written in Visual Basic for the Czochralski crystal growth system

1. Set manual ramp
2. Ramp time
3. Ramp rate
4. End ramp
5. Edit parameter
6. Stop auto
7. Program end
8. Quit

When the *Auto* command button is pressed, it calls a subroutine to read crystal weight, calculates and displays relevant growth parameters, graphically displays temperature versus time, stores data in the history file, and sets the power level based on the calculated parameters.

In this program, one can set the desired growth profile of the crystal in terms of parameters such as seed length, cone profile, and desired crystal diameter. In the auto mode, this automatic diameter control program performs the desired functions and eventually a crystal with uniform diameter is obtained.

For a detailed description of crystal growth from melt, the text by *Hurle* is recommended [9.11].

9.4 Growth of Lithium Niobate Crystals and Its Characteristics

After more than 40 years of research into $LiNbO_3$ (LN), hundreds of publications have been devoted to this very important material, one of the most important photorefractive (PR) materials ever known, considering the huge number of applications it finds in our daily life.

LN is an excellent material, very attractive due to its nonlinear, electrooptic, piezoelectric, acoustical, and photorefractive properties, high electrooptic coefficient, transparency range, and the availability of large and good-quality single crystals. For these reasons, there are a huge number of applications of LN in a lot of fields such as linear and nonlinear optical devices, acoustooptic modulators, second-harmonic generation, integrated optics applications, and bulk and waveguide optoelectronic devices. In addition, LN occupies an important place in the field of laser materials, since *Johnson* and *Ballman* [9.12] reported for the first time pulsed stimulated emission by using Nd-doped congruent LN (Nd:CLN) crystals five decades ago.

The two principal topics that have been addressed in the last decade from the point of view of materials preparation can be summarized as the development of large stoichiometric bulk crystals and periodic structures. In the former case, CLN is commercially available, but stoichiometric LN (SLN) is still in its infancy, although there are high expectations for the Asiatic research area [9.13–15]. In the latter case, the interest emerges from LN's applications in electrooptical linear device fabrication.

CLN crystals are commercially grown on large scale by the Czochralski (CZ) method [9.16]. However it has been recognized lately that many properties of SLN crystals appear to be superior to those of CLN. The former shows larger electrooptic and nonlinear effects than those of CLN crystal [9.17, 18]. In addition, the SLN crystal coercive field has been reported to be much lower, approximately one-fifth that of CLN crystal at room temperature [9.19, 20].

Congruent lithium niobate is a very well-known material with congruent melting point and a Li/Nb ratio of about 48.6/51.4 mol %. This means that CLN is Li deficient and that Nb antisites are simultaneously present. Although these intrinsic defects represent a disadvantage of CLN which limits its technical applications, the possibility to fill these vacancies with a large number of dopants has been an extraordinary input for the applications of LN. One could say that the properties of CLN can be tailored with low concentration of a given dopant. A few examples support this statement: active rare-earth ions allow laser action in the visible–infrared (IR) region; transition metals such as Fe and Mn increase the PR sensitivity; Ti diffusion leads to the fabrication of excellent optical waveguides; and Mg increases the resistance against optical damage. On the other hand, the applications of CLN often normally require monodomain samples, which involves an extra process for poling the initially unpoled material by applying high electric fields. These two key issues, congruency and poling processes, will be of special interest in the following paragraphs.

Stoichiometric or nearly stoichiometric lithium niobate (nSLN) is extremely important compared with CLN for several reasons: low concentration of intrinsic defects, large electrooptic coefficient, shorter absorption edge, lower coercive electric fields required for ferroelectric domain switching, lower optical damage in PR experiments, higher nonlinear optical coefficient, and lower extraordinary refractive index with constant ordinary index. Bearing in mind these properties and the future and attractive applications of SLN crystals, many efforts are being conducted towards research into this important material.

In the last decade, extensive research has been carried out in order to solve the problems related to the preparation of bulk single crystals. Now, at the beginning of the 21st century, the major scientific problems related to the preparation of bulk single crystals have been solved and now it is time for industrial developments. Below, some of the advances reported in the literature in the last decade in the field of the preparation and growth defects of LN crystals are briefly reviewed. However, we would like to remark that it would be impractical to summarize fully the large effort applied by hundreds of researchers worldwide on this topic (see, for example, two of the most recent books devoted to this material [9.1, 2]).

9.4.1 Crystal Growth of Lithium Niobate

Considering the phase diagram [9.10] and the physical and chemical properties of LN, the conventional Czochralski (CZ) growth method is practically the only way for growing bulk single crystals, and there are no significant publications related to its growth by other methods. The phase diagram is illustrated in Fig. 9.7.

The experimental conditions of the CZ method for growth of LN are well established and can be summarized as follows:

1. There are a large number of authors who use the oxide Nb_2O_5 and the lithium carbonate Li_2CO_3 as starting materials. In this case it is necessary to carry out a two-step sintering process: (1) heating at 700–850 °C for 2–12 h according to the temperature used for drying and calcination of carbonate, and (2) heating at around 1150 °C for 2–4 h for sintering, followed by a grounding process at room temperature (RT) [9.21–27]. However, this process can be avoided using commercial LN of the highest chemical grade, which enables the growth of LN crystals with excellent physical properties. The same statements hold for growing other niobates or tantalates [9.27].
2. The standard growth parameters can be summarized as follows [9.23, 26, 28–34]: for CLN a pulling rate of 2–4 mm/h, with a rotation rate of 30–40 rpm, with axial temperature gradients in the solid–liquid interface (SLI) close to 10 K/cm. This value is easily attainable in resistance furnaces. However, active or passive afterheaters are recommended to be used with induction heaters. The growth process is generally terminated by a cooling process with cooling rates in the range 20–30 K/h followed by a fast cooling to RT. However some authors use an annealed step at 1000 °C for 24 h to remove strains [9.30].
   For SLN the pulling rate must be decreased to a value between 0.2 and 1 mm/h while the rotation rate is in the range of 5–15 rpm. The axial temperature gradient is kept at relatively lower values compared with for CLN, while the cooling rates employed are generally the same as for CLN. As an example, Fig. 9.8 shows a schematic diagram and a picture of standard CZ equipment currently used for growing bulk LN crystals [9.35, 36]. It is also

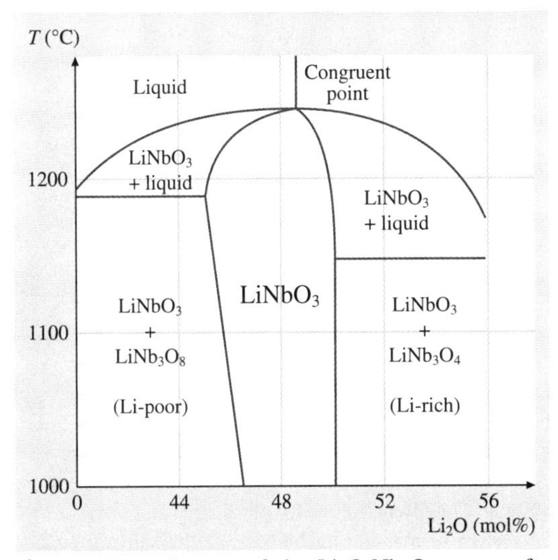

Fig. 9.7 Phase diagram of the Li_2O-Nb_2O_5 system for growing lithium niobate crystals

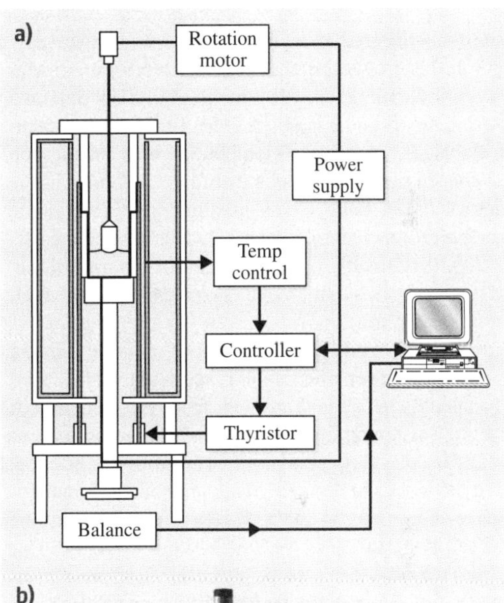

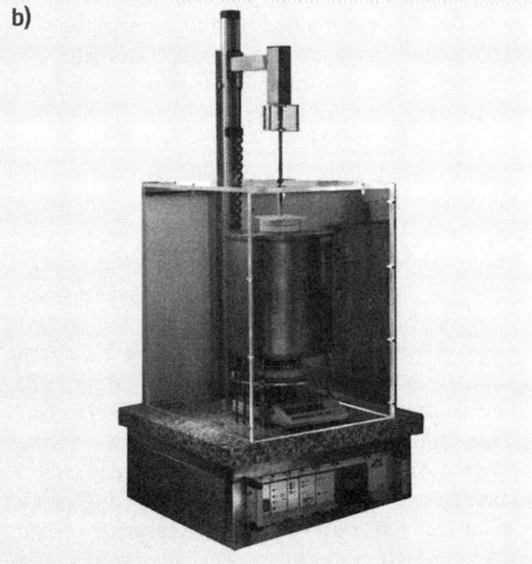

Fig. 9.8 (a) Schematic diagram and (b) conventional Czochralski crystal growth equipment with diameter control system

important to have excellent diameter control to obtain high-quality crystals. Figure 9.9 shows a CLN crystal grown by using a system like that shown in Fig. 9.8b.

3. However, the growth of bulk SLN crystals by the conventional CZ method is difficult and requires special conditions and/or technology, which are summarized in the following paragraphs.

Off-stoichiometric melts can be used with excess Li_2O in the growth process with a Li/Nb molar ratio of 1.38–1.41 [9.23–26, 30]. The yield of SLN obtained by using this process is about 10%. In this case a compositional inhomogeneity along the ingot is found, with a Li concentration difference from top to bottom in the range of 0.3–4 mol %. This inhomogeneity along the growth axis represents a great disadvantage of the off-stoichiometric melt growth process. This compositional variation increases as the Li_2O concentration in the melt increases with time.

SLN bulk crystals generally have a single-domain structure. However when an unpoled CLN seed is used in an stoichiometric melt growth, there is a diffusion of Li at high temperature from the stoichiometric melt to the CLN seed, giving rise to a compositional gradient along the length of the seed. This phenomenon induces a space-charge electric field, which in turn leads to the seed self-poling [9.37].

SLN crystals can also be successfully grown without any further experimental modification in the standard CZ technology by using K_2O-rich CLN melts. This can be considered as a mixture of oxides K_2O-Li_2O-NbO_5 where the ratio of Li/Nb is the same as that used for SLN melts [9.31, 32].

The derivative of the mass variation versus time obtained during the growth of this crystal is also shown in Fig. 9.9.

A concentration in the range 6–8 wt % of K_2O is necessary to obtain SLN crystals; other alkali oxides such as Na_2O, Rb_2O, and Cs_2O can also be used in lower concentrations, but in this case it is more difficult to obtain high-quality SLN [9.24]. The experimental conditions to be used in a K_2O-rich CLN melt growth of SLN crystals are the standard ones normally used in the CZ method. However, the rotation and pulling rates have to be low due to the higher viscosity of the melt, with values in the ranges 8–10 rpm and 0.1–0.3 m/h, respectively. The benefits of K_2O in growing SLN are related to the fact that K ions provide a similar chemical environment to that of Li ions. It has also been proven that K is not incorporated into the crystal, while other alkali ions such as Cs and Na are entrapped in microinclusions [9.24]. Another advantage of K_2O is the lowering of the crystallization temperature close to 20 °C. As a rule, all SLN crystals obtained from K_2O-rich melts have monodomain nature, with the advantage that, when using unpoled CLN seeds, they will be self-poled during the experiment [9.38].

The use of a double crucible in the CZ (DCCZ) method has proved to be an efficient approach for growing SLN, as described in [9.13, 27, 39] and references therein. The experimental setup of DCCZ is based on the use of a double crucible and/or a powder supply control in an extra upper crucible: although the experimental conditions are very strict and are generally difficult to attain in any standard crystal growth laboratory, crystals of up to 7.5–10 cm can be grown. Other benefits of this method include a constant surface melt level, which has two important consequences: constant convection throughout the growth process, and a fully stoichiometric crystal as a result of the composition adjustment of the melt supplied by the extra upper crucible. The DCCZ method is also very efficient for obtaining homogenously doped CLN and SLN bulk crystals. Another recently developed, modified version of the double-crucible method is described in the following subsection.

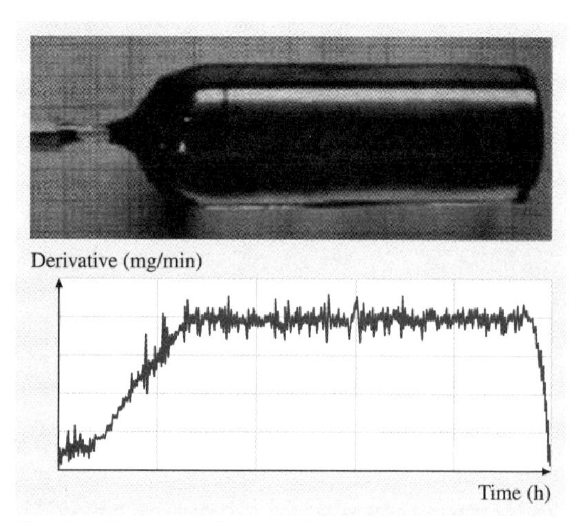

Fig. 9.9 Congruent lithium niobate crystal grown using the system shown in Fig. 9.8 and the derivative of mass variation with time

9.4.2 Mold-Pushing Melt-Supply Double-Crucible Czochralski Apparatus

The double-crucible Czochralski (DCCZ) growth apparatus consisted of three parts: the double crucible, the continuous supply system, and the Czochralski growth system. The double crucible consists of a growth crucible and a melt-supplying crucible. The two crucibles are connected by some holes. The continuous supply system consists of a mold made of Pt, which is placed in the melt-supply crucible, and a lowering mechanism. When pulling the crystal, the mold is lowered to push the melt from the melt-supply crucible into the growth crucible. The lowering rate of the mold is controlled independently by the lowering mechanism; the rate is controllable and can be adjusted in real time according to the rate of change of the crystal weight. A schematic drawing of the growth apparatus is shown in Fig. 9.10. The diameter of the as-grown crystal is controlled by an automatic weighing system, which controls the heating power. The composition of the as-grown crystal is influenced by the area ratio of the cross section of the crystal to that of the mold, the compositions of the melts in the two crucibles, and the velocity ratio of crystal pulling to mold lowering. Compared with the common DCCZ method, this technique avoids the problems related to the powder supply, inconsistency of melt composition, complexity, and the high price of the growth device. Researchers into this approach called it the mold-pushing melt-supplying (MPMS) technique [9.40].

Growth of near-stoichiometric $LiNbO_3$ crystals was illustrated by the authors [9.40].

In this work, the purity of the raw materials (Li_2CO_3 and Nb_2O_5) was 99.99%. The Li_2O content of the melt in the growth crucible was set to be 58 mol %, whereas in the melt supply crucible it was 50 mol %. The diameter of the as-grown crystal was kept unchanged throughout the whole run, while the rate of lowering was set to be equal to that of crystal pulling. The c-axis seed crystal was rotated at 5–15 rpm and both the pulling rate of the crystal and the lowering rate of the mold were 0.5–1.0 mm/h. The melt height in the growth crucible was kept unchanged during the whole process of crystal growth by controlling the melt supply from the melt-supply crucible. The diameter of the as-grown crystal was about 43 mm. The height of the crystal was 45 mm. The crystal was clear, without inclusions, free of cracks, and light yellow in color.

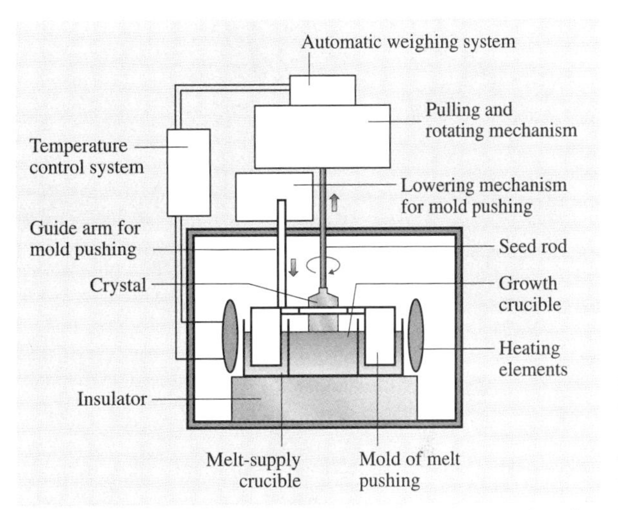

Fig. 9.10 Schematic diagram of mold pushing melt supply using DC Czochralski crystal growth system

9.4.3 Congruent Lithium Niobate Crystal Growth by Automatic Diameter Control Method

The apparatus described in Sect. 9.3.7 has been used for growing lithium niobate crystals [9.7–10]. A platinum crucible is filled with 300 g LN powder (purchased from Johnson Matthey, grade 99.9999% purity). The system is heated from room temperature to a temperature above

Fig. 9.11 A lithium niobate single crystal pulled from a Czochralski crystal growth system after cooling to room temperature

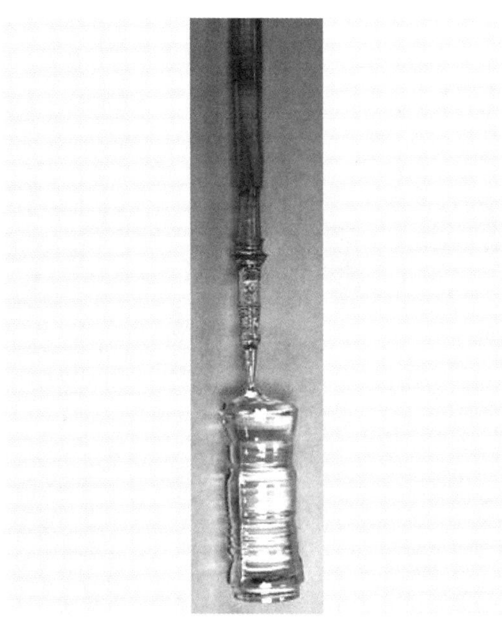

Fig. 9.12 A lithium niobate single crystal after necking of the seed crystal

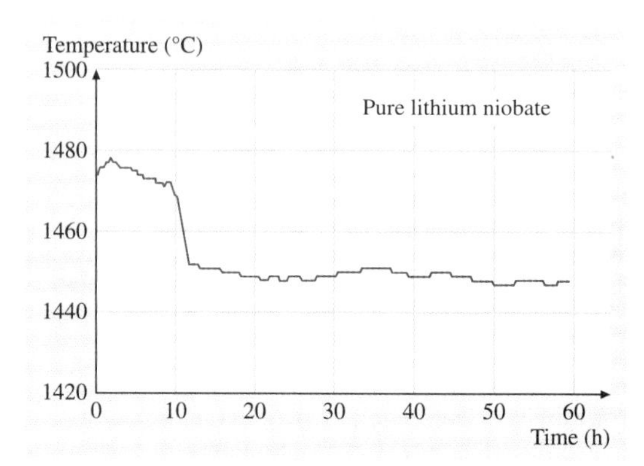

Fig. 9.14 Growth temperature versus time profile of the growth of lithium niobate crystal

1260 °C during a 24 h period using ramping mode. After this, the temperature is increased at a rate of 2 K every 30 min until all the charge is molten (complete molten state is in the vicinity of 1260 °C). A 5 mm-diameter seed oriented in the ⟨001⟩ direction is dipped into the melt. At this stage, the weight of the dipped seed is monitored so that the seed does not melt, and after 2–3 h of equilibration it is slowly pulled at a rate of 2 mm/h. The rotation of the seed is kept at around 20 rpm. At this time the necking process begins. When the seed crystal starts to grow initially, the system is put into *Auto* mode by pressing the *Auto Run* button. After the crystal has grown to the desired size, and the crystal is withdrawn from the melt, the auto mode is stopped and the cool-down ramp is started to bring it to room temperature in 48 h. Figure 9.11 shows a grown crystal of LN being pulled out of the furnace after growth. Figure 9.12 shows a photograph of another LN single crystal along with the seed that, grown in the authors' laboratory. Figure 9.13 shows a photograph of pure and doped lithium niobate crystals.

From the history file, the growth data are analyzed, and the relevant parameters are plotted and presented

Fig. 9.13 A number of pure and doped lithium niobate crystals grown at Alabama A&M University

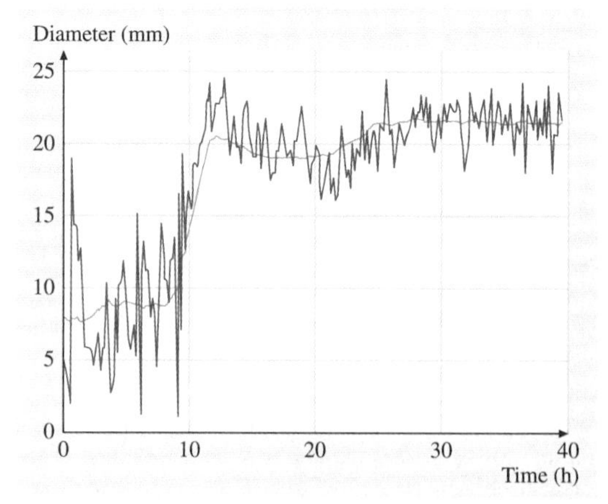

Fig. 9.15 Full diameter growth profile of lithium niobate crystal

in Figs. 9.14 and 9.15. After an initial adjustment, the power (temperature) rose during the 6 h period of growth at the diameter of the seed and then ramped down to allow the crystal to grow to its final desired diameter. The remainder of the growth shows an increase in power (temperature) due to increasing heat loss from the crystal surface. As the temperature fluctuates the automatic diameter program kicks in and controls the heat flow so that the curve of crystal diameter versus time (Fig. 9.15) is almost uniform. The irregularities in the curve represent the processing time during which the software calculates the diameter of the crystal.

9.4.4 Poling of Lithium Niobate

LN is a ferroelectric material at RT with a Curie temperature of around 1210 °C. In normal growth conditions, bulk CLN crystal has multidomain nature, while bulk SLN crystals have monodomain structure when any poling process is used. The top zone is a $+Z$ surface while the bottom one corresponds to a $-Z$ surface [9.30]. The origin of the internal field is related to nonstoichiometric point defects, such as lithium vacancies and niobium antisites, where hydrogen itself does not seem to play an active role [9.41]. When an external field reverses the polarization direction, the internal field tends to realign parallel to the new polarization direction with increasing temperature.

LN-based devices normally require monodomain crystals in order to eliminate undesirable multiplicity of ferroelectric domains. In order to produce poled LN crystals several techniques have been developed for use either during or after the growth process of bulk crystals. Let us summarize the most recent published results on this topic:

1. By far the most well-established poling technique is that based on the use of an external electric field applied to samples obtained from a bulk CLN ingot. For CLN, electric fields greater than the coercive field of 21 kV/mm are applied by using an appropriated mask with conductive electrodes. The process is carried out under controlled temperature of the order of 200 °C, followed by thermal annealing, and cooling of the sample while maintaining the electric field [9.25, 41–43]. The electrodes are based on a saturated aqueous solution of salts such as LiCl or KNO_3, with well-defined silicone rings. The poling process can be monitored in several ways, such as by measuring the charge delivered to the crystal, controlling the poling current under different experimental conditions, and analyzing the switching times, domain stabilization, back-switching phenomenon, etc. [9.43]. These results are important for the design of effective recipes which can be very useful for the fabrication of more complicated domain structures. In all these applications the nature of the LN plays a key role because the switching electric field dramatically decreases with increasing Li content, in such a way that for SLN this electric field has a zero value, according to its monodomain nature [9.42].
2. The poling process in CLN crystals can be carried out in situ during the growth experiment by applying current densities in the range 1–3 mA/cm^2 across the growing crystal. In these experiments the seed holder acts as the negative electrode and the bottom of the Pt crucible as the positive one [9.25, 42]. The current density value is reduced when the Li concentration increases during the growth experiment, being one order of magnitude lower in quasi-SLN crystals. Moreover, small current densities help to reduce any possible problems related with the melt electrolysis and/or seed decomposition, which result in catastrophic consequences for the crystal growth. The constant current density induces a transient voltage due to the domain polarization which follows an experimental law, with a profile similar to the external shape of the crystal, which can be related to the microscopic domain polarization [9.44]. As a consequence this method can be used for growing bulk monodomain CLN single crystals along the full length of the crystal.
3. A similar approach is the poling of the whole crystal after the growth process in the same experimental CZ equipment, at temperatures close to the melting point when the remaining melt is solidified [9.22]. This process gives good results, although the use of an appropriate ratio between the applied electric field and the switching time is essential in order to prevent formation of new defects due to the high ionic conductivity at high temperature.
4. Patterned domains can also be prepared by using special techniques, for example, electron-beam writing and pulsed laser at given wavelengths below and close to the absorption edge [9.18,45]. Both techniques have the advantage of creating ferroelectric surface domains with micro- and nanoscale dimensions, which can be analyzed in situ with appropriate microscopy facilities.

9.4.5 Periodically Poled Lithium Niobate Structures

In the last decade extensive research has been devoted to periodic LN structures. A variety of potential and promising applications have been found based on both periodic and aperiodic poled LN (PPLN and APPLN) and opposite domain LN (ODLN). Periodic structures considered as the organization of domains in a given way can be successfully applied to the fabrication of optoelectronic devices. Other structures such as ODLN, in which the polarization vector is aligned head to head and tail to tail, can be used in acoustic devices [9.46].

PPLN are structures where the polarization vector point is antiparallel, showing exceptional properties that are extremely useful for fabrication of electrooptic devices such as those conceived for the generation of light in the low-wavelength region through the quasi-phase-matched second-harmonic generation process (QPMSHG).

Although both structures, PPLN and ODLN, have extraordinary importance, we will focus our attention on PPLN structures considering the huge number of applications and possibilities for the future development of superior nonlinear optical media.

It is well known that birefringent phase-matching frequency requires single-domain crystals, while periodic-domain structures show the effect of QPM frequency conversion. These periodic modulations of the sign of the nonlinear coefficient can be obtained in ferroelectric crystals by periodically alternating the sign of the electric field of their ferroelectric domains. The benefits of PPLN structures over single-domain LN can be summarized as follows:

1. The highest value of the component of the second-order nonlinear tensor in the QPM wave-mixing process involved.
2. The reduction of photorefractive (PR) damage without the need for codoping of crystals with elements such as Mg as is the case for LN. Moreover, this SHG capability combined with LN's properties for hosting laser ions allows the development of optoelectronic devices such as self-frequency-doubling lasers, simultaneous multi-self-frequencies, and many other possibilities. Therefore, considering that the fabrication of new important devices requires microengineering of the domain structures, the first prerequisite is the manufacture of PPLN structures with controlled domain thickness.

The fabrication of PPLN structures can be performed sample by sample or just in the bulk crystal. In the former case, the most well-known and efficient technique for the preparation of these structures is the patterning of the surface by using standard optical lithography and suitable photoresist masks followed by the application of an external electric field [9.27, 47]. There are intrinsic drawbacks associated with this process: the high electric fields needed for inverting domains; domain widening due to intrinsic difficulties during the preparation of regular gratings with thickness of several μm; preprocessing of the sample in order to prevent it from breaking; limitations in the preparation of low-period structures; problems associated with the inverted domain depth; flip-back effects in which reversed domains can revert to their original orientation when the poling field is dropped to zero; control of the poling pulse length, which must be large enough to prevent flip-back; the presence of a leakage current superimposed on the poling current; post-bake and cooling processes needed to suppress the pyroelectrically induced surface charge on the crystal surface; the thermal history of the sample prior the poling process; etc.. In spite of these problems associated with the preparation of PPLN structures by the application of an external electric field, this is the most widely used method due to the possibility of defining standard conditions to be applied in a fixed way.

A more sophisticated process for PPLN fabrication is the engineering of microdomains by applying a voltage between the tip and back electrodes in a high voltage atomic force microscope (AFM) [9.48]. The evolution of the domain generation process is similar to that of the conventional approach based on the application of an electric field. However in this AFM-based technique the nucleation begins near the charged tip working as an active nucleus center for domain formation. This process has been successfully applied to LN and $TaLiO_3$ structures using thick single-crystal samples.

Another interesting method for preparation of PPLN structures is off-centered CZ, which enables generation of periodic domains during the crystal growth process without applying any external electric field [9.49–52]. In this case, the formation of domain structures is related to the ratio between the rotation and pulling rates in a CZ system where the seed holder is displaced from the central symmetry axis of the furnace. Furthermore, when this ratio between the rotation and pulling rates changes during a growth process, aperiodic domains (APPLN) are formed. The off-centered CZ method is based on the interplay between the forced convection existing in the melt and an appropriate ratio of rotation and pulling rates. Nevertheless one must also consider the effect of the SLI, which affects the formation of PPLN structure [9.53]. Furthermore, it is necessary to perform a quenching process after finishing the PPLN preparation, as slow cooling rates destroy the PPLN structures. Two advantages of this method can be mentioned. First the preparation of a large volume of PPLN structures, and second the ability to engineer PPLN-doped simultaneously with rare active laser ions and/or codoped with other interesting impurities such as Mg.

An issue which has been a matter of controversy is the domain composition in PPLN structures. It seems already well established that the positive domains contain higher Li concentration compared with the negative ones, which are Nb rich. In addition, the dopant concentration remains constant along the positive and negative domains, as has been proven for Nd, Er, and Y [9.54]. The influence of diffusion processes on ferroelectric domains has also been analyzed. For example the diffusion of fluorine ions along the positive domains has been ascertained by using secondary-ion mass spectrometry (SIMS) after a HF etching process [9.55]. Different etching rates between positive and negative domains have also been observed. This behavior can be understood by considering that Li outdiffusion occurs during the formation of periodic structures. This process, in turn, favors fluorine indiffusion preferentially in the positive domains, which are associated with positive charge regions as a consequence of Li outdiffusion.

Rare-earth (RE)-doped PPLN structures have additional physical properties which are advantageous for optical processes such as self-frequency doubling and self-frequency addition, as they combine nonlinear processes with host crystals containing RE elements [9.47, 51, 54, 56–58]. The possibility to fabricate PPLN-based optical devices has been proven in Er-, Nd-, and Yb-doped and Mg:Zn codoped PPLNs, the later two dopants being used in order to fully reduce the PR damage and obtain stable laser action at RT. However, it is also necessary to consider some drawbacks, such as the maximum limit in RE concentration which can be reached in the PPLN structures. This is the case of Er:PPLN, with a segregation coefficient of 1.2, where it is possible to create PPLN structures in the range of 0.5–1 mol %. However for lower Er concentration values irregular structures are obtained, while for concentrations higher than 1 mol %, disordered structures and clusters are formed. Similar behavior has been observed in Hf-doped PPLN structures, where the concentration induces polydomain structures for values higher than 0.5 mol %, or inhomogeneous structures for values lower than 0.25 mol %. This can be explained according to the Hf location in the LN lattice as has been previously described [9.53, and references therein]. It can be concluded that the behavior of Hf in PPLN structures can be considered as an exceptional case which induces a periodic modulation in Nb and Hf concentration as a function of the initial Hf concentration [9.53].

A critical issue in PPLN structures arises from the difficulty in visualizing them nondestructively using general techniques such as optical microscopy and low-voltage scanning electron microscopy. Nevertheless, there are other approaches which can also be used for this purpose [9.59]:

1. Considering the bulk of periodic structures as a volume diffraction grating, the use of a laser beam on a periodic structure supplies information about spatial frequency and orientation. In addition, the diffraction efficiency gives information about the magnitude of the periodic changes, where the number of diffracted spots and the spot size is the result of the properties of the domain. Light diffraction from periodic domains can be understood assuming that the refractive index changes at regions close to the domain walls, where the accumulated charge at the domain edges is probably due to the alternating Li/Nb ratio giving rise to a periodic change in the dielectric constant and hence in the refractive index of the sample.
2. Scanning force microscopy (SFM) has the advantage of simultaneous correlation of topographic and electrostatic images. Negative domains exhibit greater hardness, and positive ones show stronger attractive forces in the electrostatic image. Therefore, the etching rate in negative domains is faster than in positive ones. The contrast shown by the electrostatic images is explained according to the difference in the dielectric polarization, probably due to different Li/Nb ratio between the domains.

RE-doped PPLN and APPLN structures are promising materials for the development of optical devices based on nonlinear optical frequency conversion and the generation of visible laser radiation from efficient lasers operating in the near infrared. Some interesting examples of these kind of optical applications are: laser-diode pumped PPLN:Yb,Mg microchip laser sources; self-frequency mixing in APPLN:Nd [9.60]; multi-self-frequency conversion in APPLN:Nd [9.57], etc.

Part B | 9.4

9.4.6 Doped Lithium Niobate Crystals

The main motivation for doping crystals of inorganic materials is for study of the nature of the different compounds. Fundamental questions appear in the chemistry of solids regarding the behavior of dopants during crystal growth, the charge state of these dopants, and the positions they occupy in the crystalline lattice. The modification of various physical and chemical crystal properties is also another powerful motivation for doping.

Er, Pr, Yb [9.61–66], Nd, Fe, Zn, Mg [9.67, 68], Zn [9.69, 70] In [9.71, 72], and Sc [9.73] can be considered as the most widely used dopants in LN. Most of them (Mg, Fe, Zn, In, Sc, etc.) have been studied as they are considered optical-damage-resistant elements and therefore would offer the best solution for the optical damage which severely limits the holographic storage, one of the most important applications of LN. In particular Nd is of primary importance due to its applications in laser technology. However, one of the main problems associated with LN-based continuous-wave (CW) laser oscillation technology is the difficulty in achieving this effect in these crystals (Nd:CLN). This is due to the high photorefractive damage (optical damage) which induces severe changes in the birefringence when the laser intensity is high [9.74–76]. However, it was soon recognized that this difficulty could be easily overcome by codoping Nd:CLN crystals with Mg (Nd:Mg:CLN), which successfully enables laser oscillation [9.77]. In addition an increased thermal conductivity has been reported in this material, which is advantageous even if other laser properties are almost the same.

Many studies have been devoted to the study of Nd segregation in LN crystals [9.78], through which the Nd effective distribution coefficient k_{eff} has been calculated to be 0.12. However, at relatively low Nd concentrations, Nd-doped LN normally exhibits a uniform Nd concentration along the pulling direction, enabling the growth of high-quality Nd-doped LN crystals. However the growth difficulties increase for highly Nd-doped Nd:Mg:SLN crystals.

Regarding Mg-doped LN, chemical analysis has been developed and the k_{eff} value reported for Mg in Nd:Mg:SLN crystal is 1. In addition, this value does not seem to be altered when codoping with Nd.

Some other studies have also been conducted in order to establish if CLN and SLN require the same amount of Mg in order to reduce, to the same level, the photorefractive damage. It has been demonstrated that, while 5 mol % Mg is needed [9.79] in CLN, only 1 mol % Mg is enough in SLN crystals grown from Li-rich solutions (Li-58 mol % self-flux) in order to obtain the same photorefractive damage suppression effect [9.80,81]. Lower required Mg dopant concentration has been reported for different growth methods. For example, in SLN crystal grown from K_2O-Li_2O-Nb_2O_3 flux [9.82], 0.2 mol % Mg is already enough to suppress the photorefractive damage. Therefore, it can be stated that Mg-doped SLN (Mg:SLN) crystals offer a great deal of advantages over congruent crystals as the difficulty to obtain high-quality crystals increases when increasing Mg doping concentration.

Doping of LN aimed at the drastic reduction of optical damage results in a secondary effect affecting the photoconductivity. This optical property drastically increases for dopant concentrations exceeding a certain threshold value [9.83]. Some recent works devoted to study of the increasing photoconductivity in LN established a relationship between this effect and a reduced Fe^{3+} electron trapping cross section which, in turn, is related to the substitutional site occupied by the Fe^{3+} ions [9.84]. In fact, laser-induced refractive index variations have been correlated to the Fe site in the LN lattice [9.85]. Extensive research has recently been conducted on these lines regarding Fe dopant in LN. For example, *Zhen* et al. [9.86] studied Zn:Fe:$LiNbO_3$ with different ZnO concentrations and several Li/Nb ratios, concluding that the optical damage increased for certain ZnO concentrations above a threshold value.

Further research devoted to the reduction of the optical damage in high-power laser applications has recently been reported [9.87]. As previously stated, due to the lower intrinsic defects in SLN, a smaller amount of MgO is enough to enhance the photorefractive damage threshold up to 2 MW/cm^2 [9.82]. However this is not the only type of optical damage which can be found in LN. There is another kind of damage, the so-called dark trace, which still appears in MgO-doped LN crystals. Dark trace occurs when illuminating LN with a high-power laser beam, thus limiting the potential use of this material in high-power applications. ZnO doping in LN has been reported as the solution to this problem, demonstrating the effective removal of the dark trace [9.70,87]. These authors reported that ZnO doping of SLN could increase the optical resistance to the same level reported for CLN with 4–6 mol % ZnO doping (120 MW/cm^2).

Laser and holographic image recording applications have deeply motivated extensive research into Pr-doped LN [9.82–87]. However the crystal emission mechanisms of this important material are still not well understood. In the past the quenching of the $3P_0$ population was associated with multiphonon processes. Only very recently [9.88] have other mechanisms, including exciton trapping to Pr^{3+} ions, been proposed.

Ti has also been used as a dopant in LN waveguides in order to reduce one of the well-known intrinsic drawbacks of LN when applied to the fabrication of optical modulators: the so-called direct-current (DC) drift [9.89–92]. This effect consists of the time variation of the bias voltage in an optical modulator. DC-drift characteristics are normally measured by the detection of the optical modulator output intensity. Polarization-induced space charge and ion migrations are the two driving forces inducing DC drift. The dielectric nature of LN induces electrical relaxation related to this effect. Two approaches have been used in order to reduce this effect: the first one consists of using a buffer layer over the LN waveguide; the second is mainly related to the proton impurities reduction in the LN wafer [9.93–98].

9.4.7 Relevant Properties and Characteristics

Thermal conductivity is one of the most important parameters for laser materials. High values of this physical property are desirable for laser applications. There are various methods available to determine the parameters involved in the mathematical expression for thermal conductivity, i. e., the thermal diffusivity and specific heat. Amongst these, the laser-flash method [9.97] is currently used for measuring thermal diffusivity while differential scanning calorimetry (DSC) is normally used for measuring specific heat. Thermal conductivity of LN crystals is sensitive to nonstoichiometric defects (Nb antisites). In particular, for Nd:Mg:SLN crystals it was found to be about 1.3 times as large as that of the Nd,Mg codoped congruent LN (Nd:Mg:CLN) [9.78]. These authors showed that, while density and specific heat were almost insensitive to nonstoichiometric defects (Nb antisites) and dopants, the thermal conductivity strongly depends on these two parameters.

Elastic properties have also been extensively studied in LN crystals. New methods have recently been introduced to improve the accuracy compared with early data about the elastic constants of this crystal. As an example, we can mention the work carried out by *Hassel* et al. [9.79], who used a novel method consisting of laser Doppler acoustic spectroscopy in order to obtain the 16 independent piezoelectric coefficients, the elastic constants, and the anelastic coefficients of LN single crystals.

Regarding the optical properties of LN, many studies have been carried out in order to ascertain the absorption and emission spectra of pure and doped lithium niobate crystals. Among these we can mention, as an example, the optical absorption and emission spectra obtained from Nd:Mg:SLN crystals which have been analyzed and compared with those of Nd:Mg:CLN crystals. The two spectra were found to be very similar to one another [9.99].

Another important measurement which is often developed in LN crystals is analysis of the OH^- absorption band, whose peak position is generally determined in order to estimate whether the resistance to photorefractive damage in a crystal is weak or not. Fourier-transform infrared (FTIR) spectroscopy is commonly used to obtain the OH^- absorption peak. In the case of Mg:CLN crystals with photorefractive damage the OH^- band peak is located at 2.87 μm. On the other hand, in the case of Mg-doped crystals without damage, it is located at 2.83 μm [9.100, 101].

Photorefractivity is another advantageous physical property exhibited by LN crystals. This effect has been related to the hole polaron bound to Li vacancies and to three further, electronic polaronic structures, namely the free Nb_{Nb}^{5+} small (to intermediate) polaron, the Nb_{Li}^{4+} electron bound to the antisite defect, and the $Nb_{Li}^{4+}-Nb_{Nb}^{4+}$ bipolaron [9.98, 102].

Photorefractive properties in LN allow refractive index variations which can be electrooptically induced, in turn, by ionic distribution variations. These phenomena can all be used in order to permanently fix volume phase holograms recorded in LN for a long period of time, by using, for example, thermal fixing [9.103–107]. The use of the PR properties of LN for hologram thermal fixing involves three main stages. First the sample is submitted to a short heating process (a few minutes) at temperatures in the range of 120–160 °C in order to accommodate the holograms. During this process, ion migration effectively screens the initially recorded space-charge distribution. In a second stage, the ions are frozen in their new lattice positions by cooling the sample at room temperature. In the last stage, the charge distribution has to be developed by uniform illumination of the sample. Therefore fixed volume holograms can be contained by electrooptically induced variations of the refractive index generated by the above-mentioned charge distributions. Holograms lifetimes as long as 540 years have been reported recently in Fe-doped (0.1 mol %):Mg codoped (5.5 mol %) LN crystals.

The electrical conductivity of LN has also been studied for many years. It is well known that this physical property can be enhanced for concentrations of divalent ions above a certain threshold (4.5 mol %). Some researchers attributed this electrical conductivity enhancement to faster recovery or complete removal of optical damage [9.103]. Photoconductivity measurements have been developed in Cr-doped and Mg-codoped LN crystals [9.108]. In this case, the photocurrent has been explained by considering two main involved physical processes, namely, Cr^{3+} ionization and generation of Nb_{Nb}^{4+} small polarons acting as charge carriers. These experiments also showed that there exists a threshold temperature (140 K) below which photoconductivity is suppressed.

Other LN physical properties have also been shown to be strongly affected by the presence of divalent ions, such as changes in the lattice constants, and variations in the refractive indexes and in the location of the absorption edge [9.109–113].

9.5 Other Oxide Photorefractive Crystals

The list of other oxide photorefractive (PR) crystals [9.1] can be organized into two groups: pure niobates ($KNbO_3$, KN), tantalates ($LiTaO_3$, LTa; $KTaO_3$, KTa), and titanates ($BaTiO_3$, BTi), and their mixture compounds, e.g., niobates ($Sr_xBa_{1-x}Nb_2O_6$, SBN), titanates ($Ba_{1-x}Ca_xTiO_3$, BCTi), and tantalates ($KTa_{1-x}Nb_xO_3$, KTaN); and in the second group other PR oxides such as tellurites (Bi_2TeO_5) and germanates ($Pb_5Ge_3O_{11}$), although the importance of the former group is very well recognized due the huge number of established applications. Considering the impossibility of ordering these PR oxides according to importance, we include just a few words concerning the topics that have received the concentrated attention of researchers worldwide over the last decade.

The importance of LTa is well recognized due to their pyroelectric, piezoelectric, and optical properties. Just some brief comments about these applications:

1. Due to the unusually high electric field induced by external thermal gradients, large crystals up to 1 cm^3 have been used recently in order to induce nuclear fusion [9.114]. LTa offers the possibility to fabricate small, compact nuclear-fusion devices by using a small single crystal located in a chamber filled with deuterium gas, where a small furnace is used in order to supply heat to the crystal, creating a huge potential difference due to the pyroelectric effect. In turn, this potential is effectively used in order to accelerate the deuterium ions against a deuterated target, enabling the delivery of neutrons by nuclear fusion.
2. LTa crystals are also important for other applications such as multiplexing, demultiplexing, ultrafast optical amplifiers, and waveguide lasers [9.115–117]. However, LTa crystals are far less widespread despite the much higher optical damage threshold of LTa crystals compared with LN. Very recent works have been devoted to exploiting the PR properties of LTa crystals for optical waveguides fabrication [9.118]. In another field, poled LTa crystals have also recently been investigated due to their potential applications in optically integrated devices [9.114].
3. Like LN, LTa crystals also show an extremely large piezoelectric effect [9.119, 120]. These physical property together with its optical nonlinearity offer great potential for fabricating sensors, transducers, actuators, etc. In particular, for acoustic applications, large piezoelectric constants and electromechanical coupling factors are required. Several recent works have been devoted to ascertaining how to improve and enhance these parameters [9.19, 121, 122].

On the other hand, SBN crystals show ferroelectricity with Curie temperatures of between 320 and 470 K for $0.25 < x < 0.73$. The ferroelectric properties of SBN critically affect their application in optical storage and light-wave amplification devices. Some studies have been conducted in order to measure the modification of their ferroelectric properties with RE doping. Following this idea an effective lowering of the phase-transition temperature T_c has been achieved [9.123, 124]. SBN also exhibits outstanding piezoelectric, electrooptic, pyroelectric, and nonlinear optical properties. In this way, amongst the many recent works on SBN on the last topic, we can mention those based on the analysis of emission and excitation spectra obtained from RE-doped SBN crystals. These studies have shown that the intrinsic disorder of the host strongly affects the spectral bands through inhomogeneous broadening [9.125]. Broad emission bands in the near infrared have also been reported for Cr-doped SBN crystals [9.126].

There are other specific features concerning the physical and chemical characteristics and properties of the PR oxides: LTa is less sensitive to PR damage than LN, although it has the same structure with Li deficiency and a large number of Ta antisites, with a ferroelectric transition at 690 °C [9.15, 127–129]. BTi presents a transition at 120 °C from cubic to ferroelectric tetragonal phase, being stable at RT [9.1]. SBN, which has a lattice parameter that depends on the chemical composition of the solid solution [9.130] and a tetragonal tungsten bronze structure over a wide solid-solution range, presents a very small thermal conductivity with a value 1/60 of that of sapphire, which influences the form of the SLI during melt growth [9.131, 132]; KTaN, which shows similar valence state and ionic radius to Ta/Nb, presents difficulties for growing bulk single crystal with constant Ta/Nb ratio [9.133]; BCTi presents a great drawback due to the phase transition from tetragonal to orthorhombic at RT which deteriorates the optical quality [9.1].

Some of the preceding comments can affect the conditions of single-crystal growth, although there are a lot a similarities in the crystal growth method if one compare these PR oxides with the previous LN crystals. The CZ method, with the modifications discussed in the previous section on LN, is applicable to these PR oxides, although some features are specific:

- Top seed solution growth (TSSG) with optimized flux composition, generally in a solution rich in K_2O and in a process which requires extraordinary technical conditions, is used for KN [9.127].
- A variant of DCCZ with an automatic powder feeding system for obtaining crystals of 10 cm diameter [9.27, 129] or special parameters in the classical CZ method [9.128, 129] or TSSG with Li_2O excess [9.134] is used in the case of LTa.
- For SBN crystals a congruent composition is claimed for Sr composition of 0.61 mol % [9.135]; the double-crucible Stepanov technique with sophisticated dies is used, where the roll of the geometry of the die is fundamental to obtaining high-quality large (several cubic centimeters) crystals, although the vertical Bridgman method can exceptionally be used up to 25 mm crystal diameter [9.135], in both cases with free striations. As a consequence of the anomalous values of the thermal conductivity, a concave SLI is normally obtained due to the heat of solidification; a flat interface is very difficult to obtain [9.131].
- A curious method, named the step cooling technique by spontaneous nucleation, is used for KTaN with a gradient temperature accuracy of 0.1 K, and excess K_2CO_3 as a flux [9.133].

The growth defects that appear in LN crystals are still present in these PR oxides, although in a more quantitative way. In LTa crystals, multidomain ingots are obtained, which require a conventional poling process after growth, with a continuous core of inclusions along the crystal boule and growth defects such as twin planes, growth twins, and cracks which destroy the crystal quality [9.15, 27, 129, 136]; otherwise the importance of the seeding process and tail design seems to have a great importance in reducing the formation of mechanical twins and cracks [9.27, 129]. In SBN, although striations correlated with the temperature fluctuation appear, the authors claim that in general the crystals have very high optical homogeneity, are free of cracks, and have good transparency [9.20, 135, 137], although one must consider that growth by the CZ method is technically difficult, with highly facetted radial morphology [9.135].

The stoichiometry in PR oxide crystals is generally speaking a common characteristic considering that all of them are a mixture of more than two oxides. In LTa crystals there is a continuous change of Li_2O content, probably due to the Li outdiffusion from the crystal surface, an effect which is greater in near-stoichiometric crystals [9.15]. SBN has an open structure and the composition ratio of Sr/Ba can vary locally; considering the different ionic sizes of Sr and Ba, which gives rise to different preferred sites, its properties can be easily tailored [9.131, 132].

Doping of crystals with an appropriate impurity opens up the possible applications of PR oxides. Some singular results are the following: in LTa, doping with Nb helps to obtain crystals with highly homogeneous optical quality [9.136]; the PR response increases by two orders of magnitude by codoping of ZnO in LTa:Fe compared with LTa:Fe [9.128]. Some other dopants show interesting results: in SBN the effective coefficient of Nd is close to unity [9.137, 138]; different valence states have been analyzed by electron paramagnetic resonance (EPR) in BTi doped with Rh as well as Fe [9.1]; Bi_2TeO_5, being a competitor for data storage applications with LN:Fe, is an interesting host for dopants such as Cr, V, Mo [9.1] etc.

9.6 Growth of Sillenite Crystals and Its Characteristics

The term *sillenite compounds* is applied to the series of materials which show structural similarities to γ-Bi_2O_3. This family of compounds is very well known since the early 1970s when the first bulk crystals were grown by using CZ technique. With a structure containing Bi_2O_3 as the primary oxide compound, more than 60 single crystals have been obtained where the second oxide is a tetravalent oxide – general formula MO_2 with M = Ge, Si, Ti – such as GeO_2, SiO_2, and TiO_2 as the most common ones. The general ratios between the two oxides is 6 : 1 and 2 : 3, giving the formulas $Bi_{12}MO_{20}$ and $Bi_4M_3O_{12}$. However this ratio can largely vary according to the phase diagram. Other $Bi_xM_yO_z$ double oxides are also known, where the cation M can be Ga, Zn, Ba, etc., with ratio between the two oxides different from that previously mentioned [9.139].

Bismuth sillenites $Bi_{12}MO_{20}$ (BMO, with M = Si, Ge, Ti representing a tetravalent ion occupying a tetrahedral site in a body-centered cubic cell with space group I23 [9.135–137], play an important role as the main building blocks for the development of applications in the visible spectral range. These photorefractive crystals are of great importance in holographic applications, phase conjugation, and optical switching. In particular, sillenite crystals have recently found a prominent place in the field of metrology for the development of compact holographic interferometers [9.140–144]. As an example, we can mention holographic thermal fixing in $Bi_{20}SiO_{20}$ (BSO) single crystal [9.145, 146]. Extensive studies in terms of optimum dopants or thermal treatment have also been devoted to the enhancement of BMO photorefractive properties in the near infrared. This range is of great importance due to the recent development of laser diodes operating in this spectral region [9.147]. The early investigations around 1967 on the physical properties of sillenite crystals were mainly focused on congruently melting $Bi_{12}SiO_{20}$ and $Bi_{12}GeO_{20}$, and on the substitution of Si and Ge by other elements such as Ti, Ga, and Zn [9.148]. However in the last few years this research has also been directed towards the Bi_2O_3-B_2-O_3 system [9.149], where in addition to the well-known boron sillenite, there are two other compounds with relevant physical properties and applications: $Bi_4B_2O_9$ and BiB_3O_6. The former crystallizes in the monoclinic structure (space group $P2_l/c$) and has been reported to show an extreme indicatrix dispersion [9.145]. The new material BiB_3O_6, which crystallizes in the centrosymmetric monoclinic space group $C2$, presents extremely good properties for effective frequency conversion devices [9.150].

Furthermore the most important sillenites used in photorefractive PR devices are those with formula $Bi_{12}MO_{20}$ (M = Ge, Si, Ti), so-called BGO, BSO, and BTO, respectively; those compounds, with chemical formula $Bi_4M_3O_{12}$, are relevant for manufacturing scintillator devices. Attending to the goal of this chapter, we will focus our attention on PR sillenite compounds.

BGO, BSO, and BTO are well known as good piezoelectric, excellent acoustooptic, and extraordinary electrooptic materials. In particular their high optical activity and high-speed PR response in the visible range make these materials of prominent interest. Regarding these important properties, PR sillenites are used extensively in a large number of applications: light spatial modulators, imaging treatment, hologram recording phase conjugation, optical-fiber electric field sensors, etc. Although good and excellent reviews have been published in the last decade, the last book, published by *Gunter* and *Huignard* [9.1], provides an excellent summary of the most recent and important applications of PR materials.

For these reasons, high-quality crystals with appropriate dopant concentrations are required. This is not an easy task in these compounds due to at least two main reasons that arise from the crystal growth process: the presence of a central core and the tendency for facetting. These questions will be discussed below.

9.6.1 Growth of Bulk Sillenite Crystals

The growth of PR sillenites compounds has been developed since the early 1970s. BGO and BSO have congruent melting points, while BTO melts incongruently at 875 °C, where it decomposes into Bi_2O_3 and $Bi_4Ti_3O_{12}$ [9.151, 152]. In spite of this drawback in the BTO compounds growth process, very high-quality reproducible crystals have been obtained with TiO_2 concentration in the range of 8–10 mol %, with slightly different compositions amongst authors. A solid solution with a clear retrograde solidus curve has been demonstrated, and good reproducibility of the growth process can be achieved with alow standard deviation of the lattice constant [9.87, 152, 153]. Several works have shown that other sillenites with cations such as Zn, Ga, Ba, etc. present some substantial differences compared with the common ones such as narrow temperature and compositional ranges [9.139].

The starting materials for growing sillenites are commonly oxides such as Bi_2O_3, GeO_2, SiO_2, and TiO_2. As a general rule high-chemical-quality starting materials must be used in order to avoid several problems during the growth process, and to avoid the fatal consequences that impurities in ppm amounts present in the starting materials would have in the final bulk crystal. When oxide compounds are used as starting materials, a sintering process must be developed at around 800 °C for 24 h, followed by a grounding process at RT; this process is not mandatory as it could be done during the growth process, maintaining the starting products at the same temperature and for the same period of time as in the sintering process [9.154, 155]. Nevertheless, longer periods of time in the melted phase will increase the risk of evaporation of the constituent Bi_2O_3.

A variety of techniques have been proposed for growing bulk PR sillenites crystals such as Czochralski, Stepanov, Hydrothermal, and Bridgman methods. Among these, the CZ technique is without doubt the most widely used one, with very well-established growth parameters.

Typical growth parameters for bulk BGO and BSO crystals are the followings: pulling rate 0.3–1.8 mm/h; rotation rate 10–45 rpm; temperature gradient over the melt of 10–35 K/cm; and cooling rate 15–25 K/h. Furthermore for BTO crystals the growth parameters are stricter with narrower ranges, i. e.: pulling rate 0.1–0.3 mm/h; rotation rate 10–20 rpm; axial temperature gradient over the melt 10–20 K/cm; and cooling rate 10–20 K/h [9.134, 151, 153–159]. Proper selection of the growth parameters is of primary importance due to its influence on the solid–liquid interface (SLI).

Some modifications in the CZ growth method have been published in the last decade which could be summarized as:

1. Zone melting CZ, consisting of a modified CZ technique where two crucibles are used: an outer crucible which continuously feeds the inner crucible, where the standard CZ process is followed. There are two critical steps: the adjustment of the outer crucible at the exact position for the crystal growth to reach the adequate temperature profile, and the removal of both crucibles after the process is finished [9.154].
2. A pulling down method which continuously feeds the bulk growing crystal in a similar way as is done in the micro-pulling down method resembling the Verneuil process [9.160]; the melt is fed downwards by the action of gravity, and a molten zone is formed over the solidified material. The main difficulty in this technical approach is to obtain an adequate temperature gradient to reach a constant diameter and a controlled solid–liquid interface. Nevertheless, the great advantage is the removal of Bi_2O_3 evaporation by controlling the correct feeding.
3. The Bridgman technique is rarely used for the growth of PR sillenite compounds; nevertheless some approaches have been developed using Pt tubes encapsulated inside a ceramic tube. Al_2O_3 powder is also used between them in order to avoid thermal expansion problems which occur during the growth process [9.87]; the advantages of this modification are associated with benefits in the growth crystals claimed by the authors: lower dislocations density and striation free. Nevertheless one could say that the method is out of use compared with the standard CZ method.
4. Growth experiments in microgravity conditions [9.161] have been carried out with a similar technical Bridgman approach to the previous one and compared with the same experiments carried out on the ground; the results obtained support the ideas obtained in other recent microgravity growth experiments:
   a) In microgravity conditions the crystallinity is better according to the results of rocking curves, which demonstrate at least two times better quality of crystals grown under microgravity conditions.
   b) The phenomena of dewetting is observed, which is a great advantage of using the Bridgman method.
   c) When doped crystals are studied, there is a compositional homogeneity along the ingot, with nearly constant dopant concentration, while results on the ground follow the Sheil law with variation in the dopant composition along the ingot.
   d) Under the same experimental conditions, the solid–liquid interface is slightly convex under microgravity conditions compared with the concave one which appears on the ground.

In summary, the results of microgravity growth experiments show better perspectives than ground-based experiments.

9.6.2 Solid–Liquid Interface

The solid–liquid interface (SLI) in PR sillenites is a scientific topic which has been studied through the last decades due to its great influence on the quality of the bulk crystals. In fact, it is important to recognize that it is necessary to have complete control of the SLI in order to obtain good-quality PR sillenite crystals. Let us summarize the most important results obtained in the last decade:

1. The form of the SLI can be concave, flat or convex as seen from the melt. It is well recognized that an extremely large concave SLI will lead to dramatic consequence in the crystal such as increased stress, higher dislocations density, and microcracking; on the other hand, an extremely large convex interface will be the origin of core phenomena, which negatively affect the quality of PR sillenite bulk crystals [9.139].
2. The SLI is a consequence of the fluxes which exist in the melt, resulting from competition between free and forced convection, which are themselves governed principally by two growth parameters: the axial temperature gradient and the rotation rate; in fact, three zones must be considered: the shoulder zone, where the free convection is the dominant process and where a strong convex SLI is expected unless the rotation is drastically increased; the body zone, where the balance between forced convection and free convection must be obtained and a flat interface is normally obtained; and the last part of the growth process with a low value of the melt depth, where forced convection is dominant and a strong concave SLI is expected. All these situations have been analyzed and precisely identified by simulation tools using commercial codes [9.151].
3. The most important growth parameters that influence the SLI are: the axial temperature gradient, the pulling rate, and the rotation rate, which must be precisely controlled [9.139, 151, 153, 162, 163]:
   a) The axial temperature gradient for crystal growth of PR sillenites is on the order of 10–20 K/cm, which is easily obtainable with resistance furnaces but very difficult to obtain with RF furnaces; in both cases the use of passive or active afterheaters is highly recommended, although the use of active afterheaters where the temperature is modified during the growth process can be more efficient.
   b) The pulling rate must be kept within the limits discussed before, in which case its influence on the SLI is minimum.
   c) The rotation rate is a decisive growth parameter to achieve an adequate SLI, together with the axial temperature gradient [9.130].
4. For standard growth conditions, as is normally the case in a general CZ process, the SLI changes throughout the growth process as a consequence of:
   a) The change in crystal diameter from the shoulder to the body and from the last part to freezing
   b) The reduction of melt height when the crystal is pulled up.

   In an adequate, standard CZ process with constant rotation and pulling rates, in the initial stage of shoulder growth the SLI is convex and facetted; then this rounded shape changes from convex to flat, remaining so until a slightly concave shape is formed as the melt is reduced. These results are clearly visible when a crystal is withdrawn at different times during the growth process; furthermore simulations carried out considering specific growth parameters support these changes in the SLI [9.151, 162].
5. The consequence of a nonflat interface are thermal stresses, core phenomena, facetted interfaces, crystal cracking, impurity inhomogeneities, gas bubble entrapment, etc.
6. A flat or nearly convex interface is required to get high-quality PR sillenite bulk crystals.
7. A flat or nearly convex interface can be obtained with particular experimental conditions such as: variations in the rotation rate during the first step of the growth process during shoulder formation, a constant rotation rate during the growth of the body of the crystals, and with further variation in the rotation rate during the last part until freezing of the melt. When this situation occurs with a perfectly oriented seed crystal, we can say that we approach near-equilibrium conditions, and the result is the crystal shown in Fig. 9.16.

Furthermore, some other arrangements have been proposed such as metallic shields located on the seed holder to modify the radiation effect and as a consequence to influence the SLI [9.130, 139, 140]. Experimental results together with simulation data have shown the great influence that metallic or ceramic shields located on the seed holder have, flattening a deep interface [9.140].

From the above considerations on the SLI, two general conclusions can be drawn:

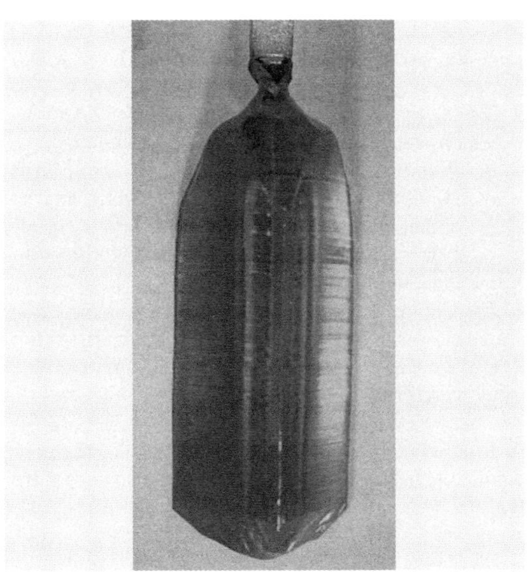

Fig. 9.16 Perfectly oriented BSO crystal obtained from nearly equilibrium conditions

1. The control of the SLI is extraordinary important to obtain high-quality crystals, and for this reason it is necessary to take special care of the growth parameters that modify the SLI, especially the axial temperature gradient and the rotation rate.
2. To obtain high-quality crystal a planar or slightly curved convex solid–liquid interface is necessary.

9.6.3 Core Effect

The core effect in PR sillenite materials has been one of the main scientific topics that have been studied for many years, due to its catastrophic consequences on the quality of crystals; see [9.139, 151, 153, 156–161] and references therein. One could say that the core effect is one of the main problems to solve in the preparation of bulk PR sillenite crystals. Let us comment on the most recent ideas which have been published in this field:

1. The core is a darker area that appears along the central part of the bulk crystal, normally full of defects and bubbles, which spreads from the shoulder part of the crystal and extends along the whole crystal, occupying a cross section of several millimeters, destroying its final crystal quality.
2. The core appears at the first stage of the growth process, just after the touching of the seed in the CZ method if no special conditions are used, a situation in the seeding process which also happens during growth by the Bridgman method in microgravity conditions, where similar defects in the core region were observed, as revealed by interferograms [9.161]. The core appears at this stage of the growth unless some conditions such us a high rotation rate are used in order to avoid the convex SLI. In fact, higher rotation rates at the shoulder position compared with the standard rotation value must be used to avoid the formation of the core region [9.156].
3. The shape of the core depends on the direction of the seed crystal. In fact when the growth direction is along the [100] and [110] directions, the core is favored and facets with several orientations are formed, while for the [111] growth direction it is easier to obtain core-free crystals for a nearly flat interface [9.157], although one must comment that there is not complete agreement between authors on this point [9.156]. Also, while for convex interfaces the core is central, for concave interfaces the core spreads out across the diameter of the crystal, forming a crown in the sample perpendicular to the growth direction; it is therefore clear that the shape of the core is related to SLI curvature [9.157, 162].
4. In general there is a consensus about the origin of the core and the relationship between the core and the facets: in some cases the core appears when there is a large convex interface at the liquid, where small facets are tangential to the interface, and in fact a facetted region is located in the core zone [9.147, 158]; in other cases, the core could be the result of small facets, and in these facets the concentration of impurities would be larger compared with the nonfacet region [9.139]. Nevertheless, it seems that in CZ growth the core effect can be avoided if one facet occupies the entire interface, which is not possible in the Bridgman method, where the core can occupy about half of the area of the grown crystal [9.139].
5. On the other hand, very recent results [9.164] show that specular reflection due to the shoulder side surface drastically changes the temperature fields and results in the appearance of a thin, cool area near the crystal axis which provokes large convexity of the SLI that may be responsible for the formation of the dark core in the center of the crystal. In this way, the angle and length of the shoulder will be of extraordinary importance for the formation of the core. In fact, experiments carried out in this regard have shown that, when a large shoulder of several

centimeters is used and the diameter of the crystal is limited to 1 cm, the core area disappears along the whole crystal.

6. There have been several studies devoted to impurity analysis of the core. It seems that deep studies on this subject have concluded that the concentration of coloring impurities such as Fe, Mn, Ni, and Co is higher in the core, while the concentration of bleaching impurities such as Ca, Al, Mg, P, and Cl is lower [9.139]. One must indicate at this point that, when the starting materials used for crystals growth are of highest commercial quality, impurities of Fe are ever present at concentrations of a few ppm, as has been detected in the core of PR sillenites by EPR measurements. In fact EPR studies show that Fe^{3+} ions occupy the positions of Si and Ge. Also, in the growth of BSO and BGO bulk crystals, analysis of the core shows a lower concentration of the cations Si and Ge, respectively, compared with noncore regions [9.156]. For this reason one must conclude that the core in PR sillenites is an impurity-getter region.
7. Moreover, the optical properties of the core region are different compared with those of noncore region, for example, the refractive index, which is at least five times lower in the core region [9.156]. On the other hand the cathodoluminescence emission band centered at 640 nm appears in both regions, being of higher intensity in the core region; one must therefore considered that, during electron radiation of the noncore region, the intensity of the 640 nm emission increases in intensity due to an ionized impurity such as Cr, Fe or Bi antisites as potential candidates, and as a consequence one could conclude that these emission centers are present in the core region [9.165]; for this reason one could say that in the original noncore region and under electron irradiation it is possible to create a *core* region with the same orange color and properties as the original core region.
8. There is agreement about the influence of the shoulder angle on the formation and extinction of the core. In fact, by increasing the cone angle, the dark core is reduced, which is explained as a consequence of the flattening of the SLI, and at the same time by smoothing of T irregularities [9.157]. The consequence of this effect is very important, because if one could relate the appearance of the core to the shoulder cone, it would be necessary to strictly control the shape and angle of the shoulder in the first stages of the growth process.
9. A general rule for the disappearance of the core is the presence of a nearly flat interface, which completely eliminates the core region [9.153, 163]. There are some other experimental conditions which reduce or eliminate the core formation:
   a) With temperature axial gradients lower than 10 °C in the vicinity of the SLI, which can be obtained either with passive afterheaters in resistance furnaces or with active afterheaters on induction heaters [9.156];
   b) The rotation rate is probably the most critical condition, because using a critical rotation rate in the shoulder part will eliminate the core [9.156].

In consequence, one must conclude that the core region can be avoided if some critical conditions are fulfilled.

9.6.4 Morphology and Faceting

PR sillenites can develop a polyhedral external morphology formed by the intersection of the narrow facets located at the outer part of the interface with the free surface of the melt. In fact, if special care is not taken during the growth process, such as the seed orientation, these facets are ever present and form a disordered external shape, due to this strong tendency for faceting and growth in typical habits [9.139]. Understanding facet formation during the growth process is rather complicated because there is interplay between the continuum transport phenomena and the interface growth kinetics.

The key question for the evolution of the external morphological is a balance between kinetic phenomena and thermal gradients. In fact the thermal conditions at the free surface around the crystals critically depend upon the radiant heat transfer from the melt surface and the structure of the flow field within the melt. These two conditions must be considered both in experimental conditions during the growth process and in simulation studies which can be carried out in order to understand the morphology of PR sillenite materials.

The shape of PR sillenite compounds is related to the radiative exchange that occurs during the growth process due to the growth regimes that occur throughout a complete growth experiment [9.166]. In fact, three regimes can be described:

1. During the shouldering step, radiative exchange occurs between the free melted surface and the upper surface of the growth chamber; in this case, taking

into account that the radiation losses are very high due to the low temperature of the upper surface, and at the same time considering the relatively uniformity of the radiation losses in the radial direction, both conditions will encouraged a facetted exterior shape.
2. During the growth of the crystal body, the thermal losses decrease due to heat exchange between the melt surface and the growing crystal; in this case the radial gradient becomes larger, and as a consequence the crystal has a circular cross section.
3. Nevertheless, during the last part of the growth process and due to the reduced melt height, the dominant mechanism of convection is forced convection with a decrease in natural convection, which reduces the radial gradient, resulting in a crystal which exhibits a polar habit.

All three of these regimes influence the formation of a facetted crystal, and for this reason one must control the growth conditions in order to understand the appearance of facets in the crystal. Also, for a given radial temperature gradient and a nonflat interface, an induced stress appears in the growing crystal which will produce a polarization in the PR sillenite crystals in a given direction, resulting in preferential growth of facets [9.159]. In this way, considering the similar magnitude of the piezoelectric coefficients of the PR sillenites, one could conclude that these materials will exhibit similar behavior in terms of facet formation when grown under the same conditions.

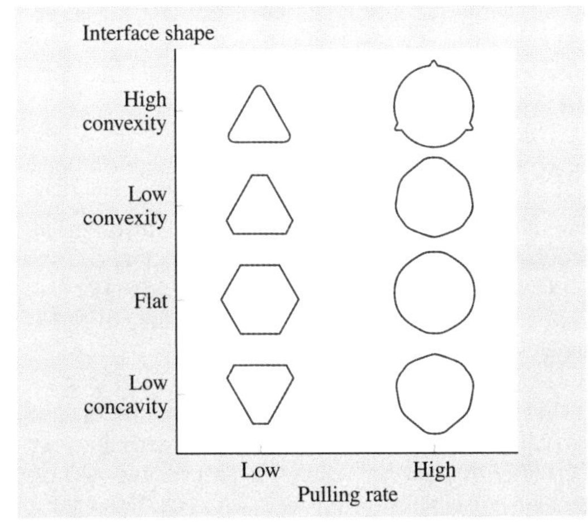

Fig. 9.17 Different morphologies in sillenite crystals for different experimental conditions

There are several factors which influence the morphology and facets in PR sillenite crystals, and all of them must be considered simultaneously. One must control the rotation rate, the pulling rate, and the axial and temperature gradients, as the most important factors. In fact all of these are growth parameters that control the SLI, and for this reason one must always consider the form of the SLI. Figure 9.17 shows an example of the morphology that appears in PR sillenite crystals for different combinations of experimental conditions for pulling rate and interface shape. For example, a constant pulling rate combined with a variable rotation rate with a low axial temperature gradient will be the most adequate set of growth parameters for obtaining near-equilibrium interface kinetics in order to eliminate inversion of the morphology in such a way that the seed direction will predict the facetted crystal and the orientation of the facets in the growing crystal [9.159]. Therefore, if one controls the seed orientation and applies adequate growth parameters, one can obtain experimental growth conditions that yield near-equilibrium conditions and thereby a crystal with controlled facetting.

9.6.5 Other Growth Defects

One could say that a perfect PR sillenite crystal without growth defects is in reality a dream, after considering the previous comments about the solid–liquid interface, the core effect, and the morphology of these materials. Furthermore, there are other growth defects that must be considered, about which a few comments are given below:

1. There are always native defects in PR sillenites crystals such as Bi antisites as a consequence of the phase diagram of these compounds. Nevertheless deeper studies have shown that a complex formed by a Bi antisite and an oxygen vacancy, or even more so the same Bi antisite with an oxygen divacancy, could be the most common native defect in these materials [9.1, 167]. In fact, one cannot avoid these defects because they are intrinsic and always present, although one can increase their concentration by going outside the congruent composition in the phase diagram.
2. The presence of growth striations is related to temperature fluctuations at the interface during the growth process [9.160]. In fact, these temperature fluctuations can result in deviations from stoichiometry and variation in impurity concentration when

a dopant is used, and in same cases the appearance of metastable phases [9.139]. Furthermore, larger temperature fluctuations will create problems associated with constitutional supercooling. For all of these reasons, growth striations are problems associated with a system with segregation coefficients different from unity, and they can be measured, for example, by resulting variations in the optical density [9.130]. Nevertheless there is not a clear relationship between temperature oscillations and growth striations, as happens in other materials.

3. The use of oxides such as Bi_2O_3 as starting materials could be a source of Pt inclusions due to the high reactivity of bismuth with the wall crucible [9.139]. In fact, in single crystals obtained with completely sintered BMO materials, these Pt inclusions can be totally avoided. Nevertheless deeper studies of BSO single crystals have shown the presence of Pt inclusions in dendrite form [9.139]. Some other inclusions and large precipitates have been found after uncontrolled remelting growth originating from a process of constitutional supercooling [9.132, 168].
4. The presence of other phases with different composition is another type of inclusion that must be considered [9.139]. The most common of these is γ-Bi_2O_3, followed by Bi_2SiO_5 and Bi_2GeO_5, although careful control of the melt temperature can eliminate these phases. It is recognized that superheating above the melting temperature on the order of 25 °C is required for the growth process, with the additional benefit of eliminating other structures different from the bulk BMO. Nevertheless lower or higher superheating temperatures must be clearly avoided.
5. Another growth defect in BMO crystals is the presence of bubbles, which often appear in various zones of the crystal such as the core. The bubbles are attributed to constitutional supercooling, which often occurs when the compositional difference between the melt and the growing crystal is large, probably as a consequence of Bi_2O_3 evaporation. This circumstance can be avoided in the zone-melting double-crucible Czochralski technique due to the thin melting zone used in the growth process [9.154, 160].
6. The first method to apply to reduce the dislocation density in a bulk crystal is the use of appropriate growth parameters. Nevertheless the second approach is the use of a seed of the highest quality; otherwise, at the seed–crystal junction the original dislocations present in the seed will propagate into the bulk crystal as individual or bundles of dislocations [9.137, 139]. For this reason, a neck is strictly necessary in PR sillenite bulk crystal growth by CZ technique.

9.6.6 Doping of Sillenites

A vast range of elements have been tried as dopants in sillenite crystals. For example *Tassev* et al. [9.169] reported the optical activity of $Bi_{12}Si_{12}O_{20}$ (BSO), $Bi_{12}GeO_{20}$ (BGO), and $Bi_{12}TiO_{20}$ (BTO) crystals doped with Al, P, and V and codoped with Al and P. They demonstrated important changes near the absorption shoulder in the absorption spectra of these crystals in the range of 480–590 nm. These dopants have been shown to have an important effect on the rotatory power of the crystals.

Chromium in a wide range of concentrations has also been studied as a dopant in sillenite crystals [9.170]. Extensive research has been devoted to explaining the optical absorption background [9.171] and optical activity [9.172] of Cr-doped BGO crystals. The relevance of local structure properties such as the location and local lattice distortion on the optical properties and applications has been extensively demonstrated in the literature [9.173, 174]. Strong variations in the optical and physical photorefractive properties of BTO single crystals has been reported by *Mokrushina* et al. [9.175] and *Mokrushina* [9.53], showing at least two charge states occupying probably different positions in the sillenite unit cell. This dopant shifts the absorption edge towards lower frequencies and a new absorption band appears in the near infrared when the chromium concentration in the crystals is increased.

Other rare-earth ions such as Nd^{3+} in sillenite crystals are also of primary importance for new optical devices. As a result of this promising future, Nd:BGO crystals have been widely investigated. Second-harmonic generation, optical rotation, and the fabrication of laser-diode-pumped microchip lasers are some of the challenging topics for deep study in Nd^{3+}-doped sillenites [9.173, 176–179].

Very recently some interesting studies have been devoted to the behavior of several dopants in the sillenite crystal structure, in particular when impurity ions are located at trigonal Bi^{3+} sites. As an example, we can mention recent work on dopants such as Cr^{3+} and Nd^{3+} ions in BGO crystals [9.180]. The local structures of these ions at the Bi^{3+} sites have been analyzed. It has been demonstrated that, contrary to the general assumption that these impurities are located exactly at Bi sites, they undergo center displacements related to these sites.

Other dopants have been used in order to induce photorefractive and photochromic properties in sillenite crystals. As an example we can mention some recent works on ruthenium-doped BSO and BTO crystals [9.140, 181, 182]. Ru ions can be readily incorporated into sillenite crystals such as BSO. This ion is normally located at Bi substitutional sites, enhancing the photorefractive properties of sillenite crystals at long wavelengths (red light) [9.183].

Electrical conductivity properties in sillenite crystals have also been recently investigated. In particular, electrical properties as a function of temperature have been analyzed in BTO crystals. The results show that the electrical resistance of this material can be modeled as a thermistor with negative temperature coefficient [9.184].

9.6.7 Relevant Properties

The mechanical properties of sillenites have been studied, and the elastic constants have been determined for some sillenite crystals [9.170], including the borate sillenites. The stronger Coulomb contribution of Si^{4+}, Ti^{4+}, and Ge^{4+} compared with B^{3+} seems to be the reason why the elastic constants corresponding to boron sillenites are slightly smaller than those corresponding to Si, Ti, and Ge.

Photorefractivity occurs in photoconductive crystals with a noncentrosymmetric structure when two coherent beams are used to create an interference pattern. In illuminated areas, electrons (holes) are ionized from a defect (intrinsic or extrinsic) to the conduction or valence band of the material. Once in this state, they can move (diffusion or drift regime) to dark areas and be efficiently trapped on the same or a different defect. A space-charge field is thus created, implying a modulation of the refractive index via the electrooptic (Pockels) effect. When trapping occurs at a level different from the original one, a modification of the absorption spectrum (photochromism) is frequently observed, especially at low temperature, due to the change in the concentration of intrinsic or extrinsic defects.

Recent studies have been devoted to analysis of the influence of holes bound to acceptor defects on the optical properties of sillenite crystals. Lattice distortion generally induces a strong localization of these systems at one of the oxygen legends related to the defect. Strong and wide absorption bands, usually in the visible region, have been reported to occur in these materials, and were related to polaron stabilization energies close to 1 eV. The formation of polarons in sillenite crystals can be characterized as follows: ideal MO_4 tetrahedra are considered as the building blocks of sillenite crystals, where M is regarded as the cation and can be Ti^{4+}, Si^{4+} or Ge^{4+}. In real samples, such as the well-known sillenite $Bi_{12}MO_{20}$ crystal, Bi antisite acceptors can be formed as a large number of the M ions can be replaced by Bi^{3+} ions. These kinds of ions present the very stable $6s^2$ configuration, which is why an electron will be taken from one of the four O^{2-} ions [9.185] instead of from the Bi^{3+}, thus forming a bound O^- polaron [9.186–188]; this is an example of where O^- is formed by direct electron ionization to the conduction band. Previous analyses [9.189, 190] have demonstrated that this defect corresponds to the localization of a hole at an oxygen site with a trigonal C_{3v} noncentrosymmetric nature in the paramagnetic state. Therefore space charges generated by inhomogeneous illumination are transposed into refractive index changes via the Pockels effect. This photorefractive effect shows quite a fast response to illumination variations, mainly related to the high mobility of the electrons in the conduction band, forming large polarons [9.191]

All of the statements above show that the main intrinsic absorption of the sillenites, causing their typical honey-yellow coloration, is due to charge transfer excitations of an O^- bound polaron next to $Bi^{3+}M$ antisite defects [9.60]. Very recently, the Bi-richest sillenite, $Bi_{24.5}BO_{38.25}$, has been successfully grown by the TSSG method [9.170]. The high values obtained for the refractive indexes of this boron sillenite compared with silicon sillenites were attributed to its high Bi content.

As previously mentioned, BSO [9.192] can be used for holographic applications. Hologram thermal fixing in BSO is induced by hydrogen impurities. This BSO material exhibits some differences compared with other crystals such as Fe-doped LN, including higher fixing temperatures (220 °C in BSO versus 140 °C in Fe:LN) and higher activation energy (1.44 eV) compared with Fe-doped LN (0.95 eV) [9.145, 146]. As a result, thermally fixed room-temperature lifetimes in BSO are much longer than those of holograms recorded in LN. Holographic fixing has also been achieved in other sillenites, such as BTO [9.193]. In this case, the fixing method consisted of the application of an alternating-current (AC) field under simultaneous heating at 90 °C. Two key material properties in holographic technology are: (1) high enough diffraction efficiencies and

(2) fast responses for processing of unfixed holograms. In this sense, sillenites are very promising materials for practical holographic applications, as long extrapolated hologram lifetimes of 10^4 years have been reported for particular materials such as BSO crystals [9.145].

9.6.8 Growth of Photorefractive Bismuth Silicon Oxide Crystals

For a typical $Bi_{12}SiO_{20}$ single-crystal growth run, 6 moles of Bi_2O_3 and 1 mole of SiO_2 (522.9 g of Bi_2O_3 and 11.237 g of SiO_2) are thoroughly mixed and loaded into a platinum crucible that is 5 cm in diameter and 5 cm high. The charge of 534.1 g is calculated to fill the crucible to 1.25 cm below the top with molten $Bi_{12}SiO_{20}$. From the phase diagram (Fig. 9.18), the melting point of $Bi_{12}SiO_{20}$ is about 910 °C [9.10]. The charge is then melted at around 920 °C in an induction furnace and the remaining material is loaded into the hot crucible. After all the material has been added, the charge is allowed to remain molten for 20 h. A seed crystal is slowly lowered to touch the melt surface in the center of the crucible. The seed is rotated at between 10–20 rpm and, for best results, pulled at between 2–2.5 mm/h.

If the interface is flat and the boule is a great deal smaller in diameter than the crucible, the interface will be elevated above the mean height of the melt and a column of liquid will be weighed along with the crystal. In the more common case of oxide crystals, the interface is not flat but rather convex toward the melt and the boule diameter is > 0.5 times the crucible diameter.

Thus, part of the interface extends below the level of the melt and that part of the solid is buoyed up by the liquid so that the measured weight is less than that of the crystal.

In both cases, however, the shape of the melt surface is not flat near the crystal and near the crucible wall. If the melt wets the solid crystal, the melt surface will rise to come into contact with the solid at a small angle, leaving a concave curved ring of liquid above the mean surface of the melt. If the melt does not wet the solid, the curvature is convex and the melt surface is lowered to a level near the surface of the solid.

From above, the surface of the melt appears to have a particular temperature due to the actual temperature of the melt and the additional influences of the radiation from the bulk of the melt and the inner surface of the crucible containing the melt. Both are modified by the emissivity of the liquid surface, the reflection of ambient light (if any), the reflectivity of the liquid surface, and the angle of incidence of the ambient light.

For the flat portion of the melt surface, the contribution of ambient reflections is quite small since the light would have to come from the (relatively cold) top of the furnace setup. However, if the surface is curved concavely, light is reflected from the inside of the crucible

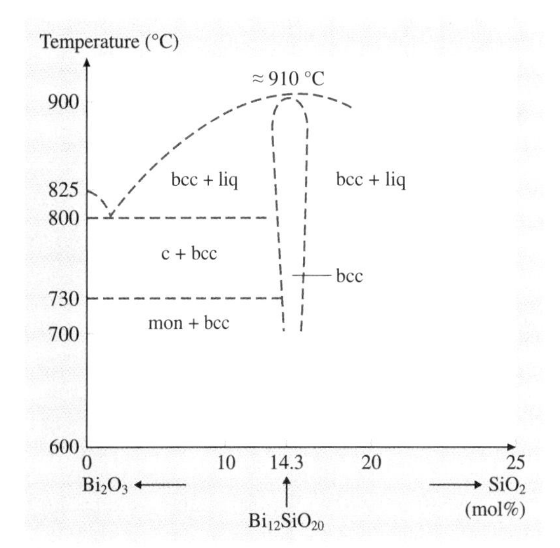

Fig. 9.18 Phase diagram the Bi_2O_3-$0SiO_2$ system for growth of $Bi_{12}SiO_{20}$ crystals

Fig. 9.19 BSO crystal grown using automatic control at Alabama A&M University

wall above the melt level, which is the hottest part of the growth environment. Thus, the meniscus wetting the growing crystal should appear hotter than the flat melt surface. If the melt does not wet the solid, the meniscus is convex and will reflect the crystal surface above it, which is colder than the melt and generally has a lower emissivity than the liquid. Thus the ring in this case should appear darker than the flat melt surface.

Radiative heat loss (proportional to the fourth power of the temperature) produces large temperature gradients in the crystals as they cool from their melting point, resulting in thermal stress that can result in shattering of the crystal some time after it has reached room temperature. These effects are minimized by the use of afterheaters.

Typical dimensions of successfully grown BSO crystals are 22 mm diameter and 75 mm long, and their typical weight is 250–300 g. A photograph of a BSO crystal grown using automatic diameter control is shown in Fig. 9.19.

9.7 Conclusions

The principal circumstances relating to the bulk growth of the photorefractive materials lithium niobate ($LiNbO_3$) and sillenites have been reviewed. Both of these kinds of crystals are grown by Czochralski method, with similar problems originating from similar effects. This implies that they have similar solid–liquid interface and other crucial growth parameters. In $LiNbO_3$ the two most important problems to solve are their stoichiometry and the growth conditions to obtain periodic structures, which have been practically solved through interesting research solutions. Lithium niobate has an extensive array of properties that have made it one of the most widely used ferroelectric materials, with one of its major applications being the manufacture of surface acoustic wave devices, employed in high-speed signal filtering in televisions and mobile phones. In fact, this material is often dubbed the silicon of nonlinear optics. It is widely sought after for applications for acoustic wave transducers, optical amplitude and phase modulators, second-harmonic generators, Q-switches, beam deflectors, phase conjugators, optical waveguides, holographic memory elements, and holographic data-processing devices. In sillenites, there is still a long way to go to reach the final goal of obtaining large high-quality bulk crystals, because problems inherent to their physicochemical properties such as the core effect and the solid–liquid interface are ever present. However, it can be seen that, in the last decade, there have been a large number of publications and an extraordinary high level of effort with the final result that most of these problems have been solved, as summarized in this chapter.

References

9.1 P. Günter, J.P. Huignard: *Photorefractive Materials and Their Applications 2* (Springer, Berlin 2007)

9.2 J. Frejlich: *Photorefractive Materials* (Wiley, New York 2007)

9.3 K.K. Wong: *Properties of Lithium Niobate* (Inspec, London 2002)

9.4 J. Czochralski: Ein neues Verfahren zur Messung der Kristallisationsgeschwindigkeit der Metalle, Z. Phys. Chem. **92**, 219–221 (1918), in Germany

9.5 G. Teal, J.B. Little: Growth of germanium single crystals, Phys. Rev. **78**, 647 (1950)

9.6 C.D. Brandle: Czochralski growth of oxides, J. Cryst. Growth **264**, 593–604 (2004)

9.7 M.D. Aggarwal, W.S. Wang, K. Bhat, P.G. Penn, D.O. Frazier: Photonic crystals: crystal growth processing and physical properties. In: *Handbook of Advanced Electronic and Photonic Materials and Devices*, Vol. 9, ed. by H.S. Nalwa (Academic Press, New York 2001) pp. 193–228

9.8 M.D. Aggarwal, T. Gebre, A.K. Batra, M.E. Edwards, R.B. Lal, B.G. Penn, D.O. Frazier: Growth of nonlinear optical materials at Alabama A and M University, Proc. SPIE, Vol. 4813 (2002) pp. 51–65

9.9 T. Gebre, D.E. Edwards, M.D. Aggarwal, A.K. Batra, M.E. Edwards, D. Patel, L. Huey, R.B. Lal: Electrooptic characterization and Czochralski growth technique of pure and doped lithium niobate crystals, Proc. SPIE, Vol. 5560, ed. by F.T. Yu, R. Guo, S. Yin (2004) pp. 17–25, Photorefractive Fiber and Crystal Devices: Materials, Optical Properties, and Applications IX

9.10 A.K. Batra, T. Gebre, J. Stephens, M.D. Aggarwal, R.B. Lal: Growth and characteristics of single crystals of lithium niobate, Proc. Mater. Res. Soc. Symp, Vol. 747 (2003) pp. 365–370

9.11 D.T.J. Hurle: *Crystal Pulling From the Melt* (Springer, Berlin 1993)

9.12 L.F. Johnson, A.A. Ballman: Coherent emission from rare earth ions in electro-optic crystals, J. Appl. Phys. **40**, 297–302 (1969)

9.13 Y. Furukawa, K. Kitamura, E. Suzike, K. Niwa: Stoichiometric $LiTaO_3$ single crystal growth by double crucible Czochralski method using automatic powder supply system, J. Cryst. Growth **197**, 889–895 (1999)
9.14 X. Li, K. Terabe, H. Hatano, K. Kitamura: Domain patterning in $LiNbO_3$ and $LiTaO_3$ by focused electron beam, J. Cryst. Growth **292**, 324–327 (2006)
9.15 D. Shumov, J. Rottenberg, S. Samuelson: Growth of 3-inch diameter near-stoichiometric $LiTaO_3$ by conventional Czochralski technique, J. Cryst. Growth **287**, 296–299 (2006)
9.16 P.F. Bordui, R.G. Norwood, C.D. Bird, G.D. Calvert: Compositional uniformity in growth and poling of large-diameter lithium niobate crystals, J. Cryst. Growth **113**, 61–68 (1991)
9.17 T. Fujiwara, M. Takahashi, M. Ohama, A.J. Ikushima, Y. Furukawa, K. Kitamura: Comparison of electro-optic effect between stoichiometric and congruent $LiNbO_3$, Electron. Lett. **35**, 499–501 (1999)
9.18 T. Fujiwara, A.J. Ikushima, Y. Furukawa, K. Kitamura: *New Aspects of Nonlinear Optical Materials and Devices* (Tech. Dig. Meet., Okazaki 1999)
9.19 M. Nakamura, M. Sekita, S. Takekawa, K. Kitamura: Crystal growth and characterization of Nd, Mg co-doped near-stoichiometric $LiNbO_3$, J. Cryst. Growth **290**, 144–148 (2006)
9.20 V. Gopalan, T.E. Mitchell, Y. Furukawa, K. Kitamura: The role of nonstoichiometry in 180° domain switching of $LiNbO_3$ crystals, Appl. Phys. Lett. **72**, 1981–1983 (1998)
9.21 L. Arizmendi, V. de Andrés, E.J. de Miguel-Sanz, M. Carrascosa: Determination of proton diffusion anisotropy by thermal decay of fixed holograms with K-vector perpendicular to the c-axis in $LiNbO_3$:Fe, Appl. Phys. **80**, 351–354 (2005)
9.22 V. Bermúdez, P.S. Dutta, M.D. Serrano, E. Diéguez: In situ poling of $LiNbO_3$ bulk crystals below the Curie temperature by application of electric field after growth, J. Cryst. Growth **169**, 409–412 (1996)
9.23 T. Zhang, B. Wang, S. Fand, D. Ma: Growth and photorefractive properties of an Fe-doped near-stoichiometric $LiNbO_3$ crystal, J. Phys. D Appl. Phys. **38**, 2013–2016 (2005)
9.24 G. Dravecz, Á. Péter, K. Polgár, L. Kovács: Alkali metal oxide solvents in the growth of stoichiometric $LiNbO_3$ single crystal, J. Cryst. Growth **286**, 334–337 (2006)
9.25 G. Bhagavannarayana, G.C. Budakoti, K.K. Maurya, B. Kumar: Enhancement of crystalline, piezoelectric and opticalquality of $LiNbO_3$ single crystals by post-growth annealing and poling, J. Cryst. Growth **282**, 394–401 (2005)
9.26 S.H. Yao, X.B. Hu, J.Y. Wang, H. Liu, L. Gao, X.F. Cheng, X. Yin, X.F. Chen: Growth and characterization of near stoichiometric $LiNbO_3$ single crystal, Cryst. Res. Technol. **42**, 114–118 (2007)
9.27 S. Kumaragurubran, S. Takekawa, M. Nakamura, K. Kitamura: Growth of 4-in diameter MgO-doped near-stoichiometric lithium tantalate single crystals and fabrication of periodically poled structures, J. Cryst. Growth **292**, 332–336 (2006)
9.28 V. Bermúdez, M.D. Serrano, J. Tornero, E. Diéguez: Er incorporation into congruent $LiNbO_3$ crystals, Solid State Commun. **112**, 699–703 (1999)
9.29 T.P.J. Han, F. Jaque, V. Bermúdez, E. Diéguez: Luminescence of the Cr^{+3} R-lines in pure and MgO co-doped near stoichiometric $LiNbO_3$:Cr crystals, Chem. Phys. Lett. **369**, 519–524 (2003)
9.30 M. Nakamura, M. Sekita, S. Takekawa, K. Kitamura: Crystal growth and characterization of Nd, Mg co-doped near-stoichiometric $LiNbO_3$, J. Cryst. Growth **290**, 144–148 (2006)
9.31 T.P. Han, F. Jaque, V. Bermúdez, E. Diéguez: The role of the Mg^{+2} ions in Cr^{+3} spectroscopy for near-stoichiometric $LiNbO_3$ crystals, J. Phys. Condens. Matter **15**, 281–290 (2003)
9.32 S. Kar, R. Bhatt, V. Shukla, R.K. Choubey, P. Sen, K.S. Bartwal: Optical behaviour of VTE treated near stoichiometric $LiNbO_3$ crystals, Solid State Commun. **137**, 283–287 (2006)
9.33 R.K. Choubey, P. Sen, S. Kar, G. Bhagavannarayana, K.S. Bartwal: Effect of codoping on crystalline perfection of Mg:Cr:$LiNbO_3$ crystals, Solid State Commun. **140**, 120–124 (2006)
9.34 Y. Furukawa, K. Kitamura, S. Takekawa, K. Niwa, Y. Yajima, H. Iyi, I. Mnushkina, P. Guggenheim, J.M. Martin: The correlation of MgO-doped near-stoichiometric $LiNbO_3$ composition to the defect structure, J. Cryst. Growth **211**, 230–236 (2000)
9.35 M. Mühlberg, M. Burianek, H. Edongue, C. Poetsch: $Bi_4B_2O_9$ – Crystal growth and some new attractive properties, J. Cryst. Growth **240**, 740–744 (2002)
9.36 J.C. Rojo, E. Diéguez, J. Jeffrey, J. Derby: A heat shield to control thermal gradients, melt convection and interface shape during shouldering in Czochralski oxide growth, J. Cryst. Growth **200**, 329–334 (1999)
9.37 V. Bermúdez, P.S. Dutta, M.D. Serrano, E. Diéguez: Effect of crystal composition on the domain structure of $LiNbO_3$ grown from Li-rich melts by the Czochralski technique, J. Cryst. Growth **172**, 269–273 (1997)
9.38 V. Bermúdez, P.S. Dutta, M.D. Serrano, E. Diéguez: On the single domain nature of stochiometric $LiNbO_3$ grown from melts containing K_2O, Appl. Phys. Lett. **70**, 729–731 (1997)
9.39 J. Sun, Y. Kong, L. Zhang, W. Yan, X. Wang, J. Xu, G. Zhang: Growth of large-diameter nearly stoichiometric lithium niobate crystals by continuous melt supplying system, J. Cryst. Growth **292**, 351–354 (2006)
9.40 Y. Zhen, E. Shi, Z. Lu, S. Chi, S. Wang, W. Zhang: A novel technique to grow stoichiometric lithium niobate single crystals, J. Cryst. Growth **211**, 895–898 (2005)
9.41 V. Gopalan, T.E. Mitchell: The role of nonstoichiometry in 180° domain swiching of $LiNbO_3$ crystals, Appl. Phys. Lett. **72**, 1981–1983 (1998)

9.42 V. Bermúdez, L. Huang, D. Hui, S. Field, E. Diéguez: Role of stoichiometric point defect in electric-field-poling lithium niobate, Appl. Phys. A **70**, 591–594 (2000)

9.43 F. Yazdani, M.L. Sundheimer, A.S.L. Gomes: Ferroelectric domain inversion in congruent lithium Niobate, Proc. SBMO/IEEE MTT-S IMOC (2003) pp. 453–458

9.44 V. Bermúdez, P.S. Dutta, M.D. Serrano, E. Diéguez: Transient electrical field characteristics due to polarization of domains in bulk $LiNbO_3$ during Czochralski growth, J. Appl. Phys. **81**, 862–864 (1997)

9.45 C.E. Valdivia, C.L. Sones, J.G. Scott, S. Mailis, R.W. Eason: Nanoscale surface domain formation on the $+z$ face of lithium niobate by pulsed ultraviolet laser illumination, Appl. Phys. Lett. **86**, 022906-3 (2005)

9.46 V. Bermúdez, F. Caccavale, E. Diéguez: Domain walls characterizatin of the opposite domain lithium niobate structures, J. Cryst. Growth **219**, 413–418 (2000)

9.47 K. Nakamura, J. Kurz, K. Parameswaran, M.M. Fejer: Periodic poling of magnesium-oxide doped lithium niobate, J. Appl. Phys. **91**, 4528–4534 (2002)

9.48 M. Molotskii, A. Agronin, P. Urenski, M. Shvebelman, G. Rosenman, Y. Rosenwaks: Ferroelectric domain breakdown, Phys. Rev. Lett. **90**, 107601–107604 (2003)

9.49 E. Kokanyan, E. Diéguez: New perspectives of lithium niobate crystals, J. Optoelectron. Adv. Mater. **2**, 205–214 (2000)

9.50 V. Bermúdez, D. Callejo, R. Vilaplana, J. Capmany, E. Diéguez: Engineering of lithium niobate domain structure through the off-centered Czochralski growth technique, J. Cryst. Growth **237**, 677–681 (2002)

9.51 V. Bermúdez, J. Capmany, J. García Solé, E. Diéguez: Growth and second harmonic generation characterization of Er^{+3} doped bulk periodically poled $LiNbO_3$, Appl. Phys. Lett. **73**, 593–595 (1998)

9.52 V. Bermúdez, M.D. Serrano, E. Diéguez: Bulk periodic poled lithium niobate crystals doped with Er and Yb, J. Cryst. Growth **200**, 185–190 (1999)

9.53 D. Callejo, V. Bermúdez, E. Diéguez: Influence of Hf ions in the formation of periodically poled lithium niobate structures, J. Phys. Condens. Matter **13**, 1337–1342 (2001)

9.54 V. Bermúdez, D. Callejo, F. Caccavale, F. Segato, F. Agulló-Rueda, E. Diéguez: On the compositional nature of bulk doped periodic poled lithium niobate crystals, Solid State Commun. **114**, 555–559 (2000)

9.55 V. Bermúdez, F. Caccavale, C. Sada, F. Segato, E. Diéguez: Etching effect on periodic domain structures of lithium niobate crystals, J. Cryst. Growth **191**, 589–593 (1998)

9.56 V. Bermúdez, M.D. Serrano, P.S. Dutta, E. Diéguez: Opposite domain formation in Er-doped $LiNbO_3$ bulk crystals grown by the off-centered Czochralski technique, J. Cryst. Growth **203**, 179–185 (1999)

9.57 J. Capmany, V. Bermúdez, D. Callejo, J. García Solé, E. Diéguez: Continuous wave simultaneous multi-self-frequency conversion in Nd^{+3} doped aperiodically poled bulk lithium niobate, Appl. Phys. Lett. **10**, 1225–1227 (2000)

9.58 J. Capmany, D. Callejo, V. Bermúdez, E. Diéguez, D. Artigas, L. Torner: Continuous-wave self-pumped optical parametric oscillator based on Yb^{+3} doped bulk periodically poled $LiNbO_3$ (MgO), Appl. Phys. Lett. **79**, 292–295 (2001)

9.59 V. Bermúdez, A. Gil, L. Arizmendi, J. Colchero, A.M. Baró, E. Diéguez: Techniques of observation and characterization of the domain structure in periodically poled lithium niobate, J. Mater. Res. **15**, 2814–2820 (2000)

9.60 J. Capmany, J.A. Pereda, V. Bermúdez, D. Callejo, E. Diéguez: Laser frequency converter for continuous-wave tunable Ti:sapphire laser based on aperiodically poled $LiNbO_3$:Nd^{+3}, Appl. Phys. Lett. **79**, 1751–1753 (2001)

9.61 A. Lorenzo, L.E. Bausá, J. García Solé: Optical spectroscopy of Pr^{3+} ions in $LiNbO_3$, Phys. Rev. B **51**, 16643–16650 (1995)

9.62 A. Lorenzo, H. Jarezic, B. Roux, G. Boulon, L.E. Bausá, J. García Solé: Lattice location of Pr^{3+} ions in $LiNbO_3$, Phys. Rev. B **52**, 6278–6284 (1995)

9.63 J.E. Muñoz-Santiuste, A. Lorenzo, L.E. Bausá, J. García Solé: Crystal field and energy levels of Pr^{3+} centres in $LiNbO_3$, J. Phys. Condens. Matter **10**, 7653–7664 (1998)

9.64 R. Piramidowicz, I. Pracka, W. Wolinski, M. Malinowski: Blue-green emission of Pr^{3+} ions in $LiNbO_3$, J. Phys. Condens. Matter **12**, 709 (2000)

9.65 W. Gryk, B. Kuklinski, M. Grinberg, M. Malinowski: High pressure spectroscopy of Pr^{3+} in $LiNbO_3$, J. Alloy. Compd. **380**, 230 (2004)

9.66 Y.S. Bai, R.R. Neugaonkar, R. Katchu: High-efficiency nonvolatile holographic storage with two-step recording in praseodymium-doped lithium niobate by use of continuous-wave lasers, Opt. Lett. **22**, 334–336 (1997)

9.67 G. Zhong, J. Jian, Z. Wu: Measurements of optically induced refractive index damage in lithium niobate, Proc. 11th Int. Conf. Quantum Electron. (IEEE, New York 1990) p. 631

9.68 M. Nakamura, M. Sekita, S. Takekawa, K. Kitamura: Crystal growth and characterization of Nd, Mg co-doped near-stoichiometric $LiNbO_3$, J. Cryst. Growth **290**, 144–148 (2006)

9.69 X.H. Zhen, Q. Li, Y.H. Xu: Influence of microscopic defects on optical damage resistance of Zn:Fe:$LiNbO_3$ crystals, Cryst. Res. Technol. **41**, 276–279 (2006)

9.70 T.R. Volk, V.I. Pryalkin, N.M. Rubinina: Optical-damage-resistant $LiNbO_3$:Zn crystal, Opt. Lett. **15**, 996 (1990)

9.71 T.R. Volk, N.M. Rubinina: A new optical damage resistant impurity in lithium niobate crystals: indium, Ferroelectr. Lett. **14**, 37–43 (1992)

9.72 Y. Kong, J. Wen, H. Wang: New doped lithium niobate crystal with high resistance to photorefraction – $LiNbO_3$:In, Appl. Phys. Lett. **66**, 280–281 (1995)

9.73 J.K. Yamanoto, K. Kitamura, N. Iyi, S. Kimura, Y. Furukawa, M. Sato: Increased optical damage resistance in Sc_2O_3-doped $LiNbO_3$, Appl. Phys. Lett. **61**, 2156–2158 (1992)

9.74 I.P. Kaminov, L.W. Stulz: Nd:$LiNbO_3$ laser, IEEE J. Quantum Electron. **QE-11**, 306–308 (1975)

9.75 Y. Furukawa, K. Kitamura, S. Takekawa, K. Niwa, Y. Yajima, N. Iyi, I. Mnushkina, P. Guggenheim, J.M. Martin: The correlation of MgO-doped near-stoichiometric $LiNbO_3$ composition to the defect structure, J. Cryst. Growth **211**, 230–236 (1990)

9.76 M.H. Li, Y.H. Xu, R. Wang, X.H. Zhen, C.Z. Zhao: Second harmonic generation in Zn-doped Li-rich $LiNbO_3$ crystals, Cryst. Res. Technol. **36**, 191–195 (2001)

9.77 A. Cordova-Plaza, T.Y. Fan, M.J.F. Digonnet, R.L. Byer, H.J. Shaw: Nd:MgO:$LiNbO_3$ continuous-wave laser pumped by a laser diode, Opt. Lett. **13**, 209 (1988)

9.78 M. Nakamura, M. Sekita, S. Takekawa, K. Kitamura: Crystal growth and characterization of Nd, Mg co-doped near-stoichiometric $LiNbO_3$, J. Cryst. Growth **290**, 144–148 (2006)

9.79 H. Ledbetter, H. Ogi, N. Nakamura: Elastic, anelastic, piezoelectric coefficients of monocrystal lithium niobate, Mech. Mater. **36**, 941–947 (2004)

9.80 D.A. Bryan, R. Gerson, H.E. Tomaschke: Increased optical damage resistance in lithium niobate, Appl. Phys. Lett. **44**, 847–849 (1984)

9.81 Y. Furukawa, K. Kitamura, S. Takakawa, A. Miyamoto, M. Terao, N. Suda: Photorefraction in $LiNbO_3$ as a function of [Li]/[Nb] and MgO concentrations, Appl. Phys. Lett. **77**, 2494–2496 (2000)

9.82 M. Nakamura, S. Higuchi, S. Takekawa, K. Terabe, Y. Furukawa, K. Kitamura: Optical damage resistance and refractive indices in near-stoichiometric MgO-doped $LiNbO_3$, Jpn. J. Appl. Phys. **41**, L49 (2002)

9.83 Á. Péter, K. Polgár, L. Kovács, K. Lengyel: Threshold concentration of MgO in near-stoichiometric $LiNbO_3$ crystals, J. Cryst. Growth **284**, 149–155 (2005)

9.84 T. Volk, N. Rubinina, M. Wöhlecke: Optical-damage-resistant impurities in lithium niobate, J. Opt. Soc. Am. B **11**, 1681 (1994)

9.85 R. Gerson, J.E. Firchhoff, H.E. Halliburton, D.A. Bryan: Microscopic mechanism of suppressing photorefraction in $LiNbO_3$:Mg,Fe crystals, J. Appl. Phys. **60**, 3553–3557 (1986)

9.86 J.J. Liu, W.L. Zhang, G.Y. Zhang: Microscopic mechanism of suppressing photorefraction in $LiNbO_3$:Mg,Fe crystals, Solid State Commun. **98**, 523–526 (1996)

9.87 C.B. Tsai, W.T. Hsu, M.D. Shih, Y.Y. Lin, Y.C. Huang, C.K. Hsieh, W.C. Hsu, R.T. Hsu, C.W. Lan: Growth and characterizations of ZnO-doped near-stoichiometric $LiNbO_3$ crystals by zone-leveling Czochralski method, J. Cryst. Growth **289**, 145–150 (2006)

9.88 C. Koepke, K. Wisniewski, D. Dyl, M. Grinberg, M. Malinowski: Evidence for existence of the trapped exciton states in Pr^{3+}-doped $LiNbO_3$ crystal, Opt. Mater. **28**, 137–142 (2006)

9.89 H. Nagata, J. Ichikawa: Progress and problems in reliability of Ti:$LiNbO_3$ optical intensity modulators, Opt. Eng. **34**(11), 3284–3293 (1995)

9.90 X. Liang, X. Xuewu, T.C. Chong, Y. Shaoning, Y. Fengliang, T.Y. Soon: Lithium in-diffusion treatment of thick $LiNbO_3$ crystals by the vapor transport equilibration method, J. Cryst. Growth **260**, 143 (2004)

9.91 V.V. Atuchin, T.I. Grigorieva, I.E. Kalabin, V.G. Kesler, L.D. Pokrovsky, D.I. Shevtsov: Comparative analysis of electronic structure of Ti:$LiNbO_3$ and $LiNbO_3$ surfaces, J. Cryst. Growth **275**(1/2), e1603–e1607 (2005)

9.92 H. Nagata, Y. Li, W.R. Bosenberg, G.L. Reiff: DC drift of x-cut $LiNbO_3$ modulators, IEEE Photon. Technol. Lett. **16**, 2233–2235 (2004)

9.93 S. Yamada, M. Minakata: DC drift phenomena in $LiNbO_3$ optical waveguide devices, Jpn. J. Appl. Phys. **20**, 733–737 (1981)

9.94 R.A. Beaker: Circuit effect in $LiNbO_3$ channel-waveguide modulators, Opt. Lett. **10**, 417–420 (1985)

9.95 H. Nagata, K. Kiuchi: Temperature dependence of DC drift of Ti:$LiNbO_3$ optical modulators with sputter deposited SiO_2 buffer layer, J. Appl. Phys. **73**(9), 4162–4164 (1993)

9.96 H. Nagata, K. Kiuchi, S. Shimotsu, J. Ogiwara, J. Minowa: Estimation of direct current bias and drift of Ti:$LiNbO_3$ optical modulators, J. Appl. Phys. **76**(3), 1405–1408 (1994)

9.97 X.H. Zhen, Q. Li, Y.H. Xu: Influence of microscopic defects on optical damage resistance of Zn:Fe:$LiNbO_3$ crystals, Cryst. Res. Technol. **41**(3), 276–279 (2006)

9.98 W.J. Parker, R.J. Jenkins, C.P. Butler, G.L. Abbott: Flash method of determining thermal diffusivity, heat capacity, and thermal conductivity, J. Appl. Phys. **32**, 1679–1684 (1961)

9.99 P. Herth, T. Granzow, D. Schaniel, T. Woike, M. Imlau, E. Krätzig: Evidence for light-induced hole polarons in $LiNbO_3$, Phys. Rev. Lett. **95**, 067404–067408 (2005)

9.100 O.F. Schirmer: O^- bound small polarons in oxide materials, J. Phys. Condens. Matter **18**, 667–704 (2006)

9.101 A.L. Shluger, A.M. Stoneham: Small polarons in real crystals, J. Phys. Condens. Matter **5**, 3049 (1993)

9.102 O.F. Schirmer, O. Thiemann, M. Wöhlecke: Defects in $LiNbO_3$ – I. Experimental aspects, J. Phys. Chem. Solids **52**, 185–200 (1991)

9.103 L. Arizmendi: Photonic applications of lithium niobate crystals, Phys. Status Solidi (a) **201**, 253–283 (2004)

9.104 B.K. Das, H. Suche, W. Sohler: Single-frequency Ti:Er:$LiNbO_3$ distributed Bragg reflector waveguide laser with thermally fixed photorefractive cavity, Appl. Phys. B **73**, 439–442 (2001)

9.105 J. Hukriede, D. Runde, D. Kip: Fabrication and application of holographic Bragg gratings in lithium niobate channel waveguides, J. Phys. D Appl. Phys. **36**, R1–R16 (2003)
9.106 I. Nee, O. Beyer, M. Müller, K. Buse: Circuit effect in $LiNbO_3$ channel-waveguide modulators, J. Opt. Soc. Am. B **20**, 1593–1602 (2003)
9.107 L. Arizmendi, F.J. López-Barberá: Lifetime of thermally fixed holograms in $LiNbO_3$ crystals doped with Mg and Fe, Appl. Phys. B **86**, 105–109 (2007)
9.108 L. Arizmendi, C. de las Heras, F. Jaque, A. Suchocki, S. Kobyakov, T.P.J. Han: Photoconductivity in $LiNbO_3$ crystals, codoped with MgO and Cr_2O_3, Appl. Phys. B **87**, 123–127 (2007)
9.109 L. Arizmendi, F. Agulló-López: $LiNbO_3$: A paradigm for photorefractive materials, MRS Bulletin **19**, 32–38 (1994)
9.110 J. Diaz-Caro, J. García-Solé, D. Bravo, J.A. Sanz-García, F.J. López, F. Jaque: MgO codoping-induced change in the site distribution of Cr^{3+} ions in $LiNbO_3$, Phys. Rev. B **54**, 13042–13046 (1996)
9.111 F. Jaque, T.P.J. Han, G. Lifante: Comparative study of the singularity in the optical properties of congruent doped $LiNbO_3$ crystals, J. Lumin. **248**, 102–103 (2003)
9.112 G.A. Torchia, J.A. Sanz-García, J. Díaz-Caro: Redistribution of Cr^{3+} ions from Li^{+} to Nb^{5+} sites in ZnO codoped $LiNbO_3$:Cr crystals, F. Jaque, T.P.J. Han, Chem. Phys. Lett. **288**, 65–70 (1998)
9.113 S.A. Basun, A.A. Kaplyanskii, A.B. Kutsenko, V. Dierolf, T. Troester, S.E. Kapphan, K. Polgár: Optical characterization of Cr^{3+} centers in $LiNbO_3$, Appl. Phys. B **73**, 453–461 (2001)
9.114 B. Naranjo, J.K. Gimzewski, S. Putterman: Observation of nuclear fusion driven by a pyroelectric crystal, Nature **434**, 1115–1117 (2005)
9.115 S. Tascu, P. Moretti, S. Kostritskii, B. Jacquier: Optical near-field measurements of guided modes in various processed $LiNbO_3$ and $LiTaO_3$ channel waveguides, Opt. Mater. **24**, 297–302 (2003)
9.116 Y.N. Korkishko, V.A. Fedorov, S.M. Kostritskii, A.N. Alkaev, E.I. Maslennikov, E.M. Paderin, D.V. Apraksin, F. Laurell: Proton exchanged $LiNbO_3$ and $LiTaO_3$ optical waveguides and integrated optic devices, Microelectron. Eng. **69**, 228–236 (2003)
9.117 P. Nekvindova, J. Špirková, J. Červená, M. Budnar, A. Razpet, B. Zorko, P. Pelicon: Annealed proton exchanged optical waveguides in lithium niobate: differences between the X- and Z-cuts, Opt. Mater. **19**, 245–253 (2002)
9.118 L. Salavcová, J. Špirková, F. Ondráček, A. Macková, J. Vacík, U. Kreissig, F. Eichhorn, R. Groetzschel: Study of anomalous behaviour of $LiTaO_3$ during the annealed proton exchange process of optical waveguide's formation – comparison with $LiNbO_3$, Opt. Mater. **29**, 913–918 (2007)
9.119 Z.D. Gao, Q.J. Wang, Y. Zhang, S.N. Zhu: Etching study of poled lithium tantalate crystal using wet etching technique with ultrasonic assistance, Opt. Mater. **30**(6), 847–850 (2008)
9.120 X.-H. Du, J. Zheng, U. Belegundu, K. Uchino: Crystal orientation dependence of piezoelectric properties of lead zirconate titanate near the morphotropic phase boundary, Appl. Phys. Lett. **72**, 2421–2423 (1998)
9.121 X.-H. Du, J. Zheng, K. Uchino: Crystal orientation dependence of piezoelectric properties in lead zirconate titanate: theoretical expectation for thin films, Jpn. J. Appl. Phys. **36**, 5580–5587 (1997)
9.122 W. Yue, J. Yi-jian: Crystal orientation dependence of piezoelectric properties in $LiNbO_3$ and $LiTaO_3$, Opt. Mater. **23**, 403–408 (2003)
9.123 T. Volk, V. Salobutin, L. Ivleva, N. Polzkov, R. Pankrath, M. Wöhlecke: Atomic structure of $Sr_{0.75}Ba_{0.25}Nb_2O_6$ single crystal and composition-structure-property relation in $(Sr,Ba)Nb_2O_6$ solid solutions, Sov. Phys. Solid State **42**, 2066–2071 (2000)
9.124 T. Volk, L. Ivleva, P. Lykov, N. Polozkov, V. Salobutin, R. Pankrath, M. Wöhlecke: Effects of rare-earth impurity doping on the ferroelectric and photorefractive properties of strontium-barium niobate crystals, Opt. Mater. **18**, 179–182 (2001)
9.125 M. Bettinelli, A. Speghini, A. Ródenas, P. Molina, M.O. Ramírez, B. Capote, D. Jaque, L.E. Bausá, J. García Solé: Luminescence of lanthanide ions in strontium barium niobate, J. Lumin. **122**, 307–310 (2007)
9.126 T.P.J. Han, F. Jaque, D. Jaque, J. García-Sole, L. Ivleva: Luminescence life time and time-resolved spectroscopy, of Cr^{3+} ions in strontium barium niobate, J. Lumin. **119**, 453–456 (2006)
9.127 T. Tagaki, T. Fujii, Y. Sakabe: Growth and characterization of $KNbO_3$ by vertical Bridgman method, J. Cryst. Growth **259**, 296–301 (2003)
9.128 R. Ilangovan, G. Ravi, C. Subramanian, P. Ramasamy, S. Sakai: Growth and characterization of potassium tantatale niobate single crystals by step-cooling technique, J. Cryst. Growth **237**, 694–699 (2002)
9.129 G. Ravi, R. Jayavel, S. Takekawa, M. Kanamura, K. Kitamura: Effect of niobium substitution in stoichiometric lithium tantalite (SLT) single crystals, J. Cryst. Growth **250**, 146–151 (2003)
9.130 M. Ulex, R. Pankrath, K. Betzler: Growth of strontium barium niobate: the liquidus-solidus phase diagram, J. Cryst. Growth **271**, 128–133 (2004)
9.131 S. Kuraragurubaran, S. Takekawa, M. Nakamura, K. Kitamura: Growth of 4-in diameter near-stoichiometric lithium tantalate single crystals, J. Cryst. Growth **285**, 88–95 (2005)
9.132 S. Fang, B. Wang, T. Zhans, F. Ling, R. Wang: Growth and photorefractive properties of Zn, Fe double-doped $LiTaO_3$ crystal, Opt. Mater. **28**, 207–211 (2006)
9.133 H.S. Lee, J.P. Wilde, R.S. Feigelson: Bridgman growth of strontium barium niobate single crystals, J. Cryst. Growth **187**, 89–101 (1998)

Part B | 9

9.134 J.B. Gruber, T.H. Allik, D.K. Sardar, R.J. Yow, M. Scripsick, B. Wechsler: Crystal growth and spectroscopic characterization of Yb^{+3}:$LiTaO_3$, J. Lumin. **117**, 233–238 (2006)

9.135 S. Takekawa, Y. Furukawa, M. Lee, K. Kitamura: Double crucible Stepanov technique for the growth of striation – free SBN single crystal, J. Cryst. Growth **229**, 238–242 (2001)

9.136 L.I. Ivleva, T.R. Volk, D.V. Isakov, V.V. Gladkii, N.M. Polozkov, P.A. Lykov: Growth and ferroelectric properties of Nd-doped strontium-barium niobate crystals, J. Cryst. Growth **237**, 700–702 (2002)

9.137 C. Nitash, M. Göbbels: Phase relations and lattice parameters in the system SrO-BaO-Nb_2O_5 focusing on SBN ($Sr_xBa_{1-x}Nb_2O_6$), J. Cryst. Growth **269**, 324–332 (2004)

9.138 S. Aravazhi, A. Tapponnier, D. Günther, P. Günther: Growth and characterization of barium-doped potassium tantalite crystals, J. Cryst. Growth **282**, 66–71 (2005)

9.139 H.A. Wang, C.H. Lee, F.A. Kröger, R.T. Cox: Point defects in α-Al_2O_3:Mg studied by electrical conductivity, optical absorption, and ESR, Phys. Rev. B **27**, 3821–3841 (1983)

9.140 F. Ramaz, L. Rakitina, M. Gospodinov, B. Briat: Photorefractive and photochromic properties of ruthenium-doped $Bi_{12}SiO_{20}$, Opt. Mater. **27**, 1547–1559 (2005)

9.141 M.P. Georges, V.S. Scauffaire, P.C. Lemaire: Compact and portable holographic camera using photorefractive crystals. Application in various metrological problems, Appl. Phys. B **72**, 761–765 (2001)

9.142 A.R. Lobato, S. Lanfredi, J.F. Carvalho, A.C. Hernandes: Synthesis, crystal growth and characterization of γ-phase bismuth titanium oxide with gallium, Mater. Res. **3**, 92–96 (2000)

9.143 M. Valant, D. Suvorov: Processing and dielectric properties of sillenite compounds Bi[12]MO[20-δ] (M = Si, Ge, Ti, Pb, Mn, $B_{1/2}P_{1/2}$), J. Am. Ceram. Soc. **84**(12), 2900–2904 (2001)

9.144 M. Valent, D. Suvorov: Synthesis and characterization of a new sillenite compound – $Bi_{12}(B_{0.5}P_{0.5})O_{20}$, J. Am. Ceram. Soc. **85**, 355–358 (2002)

9.145 L. Arizmendi, J.F. López-Barberá, M. Carrascosa: Twelve-fold increase of diffraction efficiency of thermally fixed holograms in $Bi_{12}SiO_{20}$, J. Appl. Phys. **97**, 073505–073507 (2005)

9.146 E.M. de Miguel-Sanz, M. Carrascosa, L. Arizmendi: Effect of the oxidation state and hydrogen concentration on the lifetime of thermally fixed holograms in $LiNbO_3$:Fe, Phys. Rev. B **65**, 165101 (2002)

9.147 B. Briat, V.G. Grachev, G.I. Malovichko, O.F. Schirmer, M. Wöhlecke: Defects in inorganic photorefractive materials and their investigations. In: *Photorefractive Materials*, Vol. 2, ed. by P. Günter, J.P. Huignard (Springer, Berlin 2007), Chap. 2

9.148 B. Briat, C.L. Boudy, J.C. Launay: Magnetic and natural circular dichroism of $Bi_{12}GeO_{20}$: Evidence for several paramagnetic centres, Ferroelectrics **125**, 467–469 (1992)

9.149 M. Burianek, S. Haussühl, M. Kugler, V. Wirth, M. Mühlberg: Some physical properties of boron sillenite: $Bi_{24.5}BO_{38.25}$, Cryst. Res. Technol. **41**, 375–378 (2006)

9.150 H. Hellwig, J. Liebertz, L. Bohatý: Exceptional large nonlinear optical coefficients in the monoclinic bismuth borate BiB_3O_6 (BIBO), Solid State Commun. **109**, 249–251 (1999)

9.151 J.F. Carvalho, A.C. Hernandes: Large $Bi_{12}TiO_{20}$ single crystals: A study of intrinsic defects and growth parameters, J. Cryst. Growth **205**, 185–190 (1999)

9.152 S. Miyazawa, T. Tabata: Bi_2O_3-TiO_2 binary phase diagram study for TSSG pulling of $Bi_{12}TiO_{20}$ single crystals, J. Cryst. Growth **191**, 512–516 (1998)

9.153 A. Majchrowski, M.T. Borowiec, J. Żmija, H. Szymczak, E. Michalski, M. Barański: Crystal growth of mixed titanium sillenites, Cryst. Res. Technol. **37**, 797–802 (2002)

9.154 C.W. Lan, H.J. Chen, C.B. Tsai: Zone-melting Czochralski pulling growth of $Bi_{12}SiO_20$ single crystals, J. Cryst. Growth **245**, 56–62 (2002)

9.155 M.T. Santos, L. Arizmendi, D. Bravo, E. Diéguez: Analysis of the core in $Bi_{12}SiO_{20}$ and $Bi_{12}GeO_{20}$ crystals grown by the Czochralski method, Mater. Res. Bull. **31**, 389–396 (1996)

9.156 J.C. Rojo, E. Diéguez: *Bismuth Germanate, Titanate and Silicate, Encyclopedia of Materials: Science and Technology* (Elsevier, Amsterdam 2001)

9.157 V. Bermúdez, O.N. Budenkova, V.S. Yuferev, M.G. Vasiliev, E.N. Bystrova, V.V. Kalaev, J.C. Rojo, E. Diéguez: Effect of the shouldering angle on the shape of the solid–liquid interface and temperature fields in sillenite-type crystals growth, J. Cryst. Growth **279**, 82–87 (2005)

9.158 S. Kumaragurubaran, S.M. Babu, K. Kitamur, S. Takegawa, C. Subramanian: Defect analysis in Czochralski grown $Bi_{12}SiO_{20}$ crystals, J. Cryst. Growth **239**, 233–237 (2001)

9.159 M.T. Santos, C. Marín, E. Diéguez: Morphology of $Bi_{12}GeO_{20}$ crystals grown along the $\langle 111 \rangle$ directions by the Czochralski method, J. Cryst. Growth **160**, 283–288 (1996)

9.160 S. Maida, M. Higuchi, K. Kodaira: Growth of $Bi_{12}SiO_{20}$ single crystals by the pulling-down method with continuous feeding, J. Cryst. Growth **205**, 317–322 (1999)

9.161 Y.F. Zhou, J.C. Wang, L.A. Tang, Z.L. Pan, N.F. Chen, W.C. Chen, Y.Y. Huang, W. He: Space growth studies of Ce-doped $Bi_{12}SiO_{20}$ single crystals, Mater. Sci. Eng. B **113**, 179–183 (2004)

9.162 M.T. Santos, J.C. Rojo, A. Cintas, L. Arizmendi, E. Diéguez: Changes in the solid–liquid interface during the growth of $Bi_{12}SiO_{20}$, $Bi_{12}GeO_{20}$ and $LiNbO_3$ crystals grown by the Czochralski method, J. Cryst. Growth **156**, 413–420 (1995)

9.163 J. Martínez-López, M. González-Mañas, J.C. Rojo, B. Capelle, M.A. Caballero, E. Diéguez: X-ray topographic characterization of growth defects in sillenite type crystals, Ann. Chim. Mater. **22**, 687–690 (1997)

9.164 O.N. Budenkova, M.G. Vasiliev, V.S. Yuferev, N. Bystrova, V.V. Kalaev, V. Bermúdez, E. Diéguez, Y.N. Makarov: Simulation of global heat transfer in the Czochralski process for BGO sillenite crystals, J. Cryst. Growth **266**, 103–108 (2004)

9.165 A. Cremades, M.T. Santos, A. Remón, J.A. García, E. Diéguez, J. Piqueras: Cathodoluminescence and photoluminescence in the core region of $Bi_{12}GeO_{20}$ and $Bi_{12}SiO_{20}$ crystals, J. Appl. Phys. **79**, 7186–7190 (1996)

9.166 J.C. Rojo, C. Marín, J.J. Derby, E. Diéguez: Heat transfer and the external morphology of Czochralski-grown sillenite compounds, J. Cryst. Growth **183**, 604–613 (1998)

9.167 N. de Diego, F. Plazaola, J. del Río, M.T. Santos, E. Diéguez: Positron annihilation in $Bi_{12}XO_{20}$ (X = Ge, Si, Ti) structures, J. Phys. Condens. Matter **8**, 1301–1306 (1996)

9.168 H.S. Horowitz, A.J. Jacobson, J.M. Newsam, J.T. Lewandowski, M.E. Leonowicz: Solution synthesis and characterization of sillenite phases: $Bi_{24}M_2O_{40}$ (M = Si, Ge, V, As, P), Solid State Ion. **32/33**, 678–690 (1989)

9.169 V. Tassev, G. Diankov, M. Gospodinov: Optical activity of doped sillenite crystals, Mater. Res. Bull. **30**, 1263–1267 (1995)

9.170 M. Burianek, S. Haussühl, M. Kugler, V. Wirth, M. Mühlberg: Some physical properties of boron sillenite: $Bi_{24.5}BO_{38.25}$, Cryst. Res. Technol. **41**(4), 375–378 (2006)

9.171 H. Marquet, M. Tapiero, J.C. Merle, J.P. Zielinger, J.C. Launa: Determination of the factors controlling the optical background absorption in nominally undoped and doped sillenites, Opt. Mater. **11**, 53–65 (1998)

9.172 V.L. Tassev, G.L. Diankov, M. Gospodinov: Measurement of optical activity and Faraday effect in pure and doped sillenite crystals, Proc. SPIE **2529**, 223–230 (1995)

9.173 D. Bravo, F.J. Lopez: The EPR technique as a tool for the understanding of laser systems, the case of Cr^{3+} and Cr^{4+} ions in $Bi_4Ge_3O_{12}$, Opt. Mater. **13**, 141–145 (1999)

9.174 D. Bravo, A. Martin, A.A. Kaminskii, F.J. Lopez: EPR spectra of Cr^{3+} ions in $LiNbO_3$:ZnO and $LiNbO_3$:CaO, Radiat. Eff. Defects Solids **135**, 261–264 (1995)

9.175 E.V. Mokrushina, A.A. Nechitailov, V.V. Prokofiev: A method for determining the charge state of chromium in Cr-doped $Bi_{12}SiO_{20}$ and $Bi_{12}TiO_{20}$, Opt. Commun. **123**, 592–596 (1996)

9.176 A.K. Jazmati, G. Vazquez, P.D. Townsend: Second harmonic generation from RE doped BGO waveguides, Nucl. Instrum. Methods B **166–167**, 592–596 (2000)

9.177 A.K. Jazmati, P.D. Townsend: Optical rotation in a $Bi_4Ge_3O_{12}$:RE surface modified by He-ion beam implantation, Nucl. Instrum. Methods B **148**, 698–703 (1999)

9.178 J.B. Shim, J.H. Lee, A. Yoshikawa, M. Nikl, D.H. Yoon, T. Fukuda: Growth of $Bi_4Ge_3O_{12}$ single crystal by the micro-pulling-down method from bismuth rich composition, J. Cryst. Growth **243**, 157–163 (2002)

9.179 X.Q. Feng, G.Q. Hu, Z.W. Yin, Y.P. Huang, S. Kapphan, C. Fisher, F.Z. Zhou, Y. Yang, D.Y. Fan: Growth, laser and magneto-optic properties of Nd-doped $Bi_4Ge_3O_{12}$ crystals, Mater. Sci. Eng. B **23**, 83–87 (1994)

9.180 S.Y. Wu, H.N. Dong: Investigations on the local structures of $Cr^{3+}(3d^3)$ and $Nd^{3+}(4f^3)$ ions at the trigonal Bi^{3+} sites in $Bi_4Ge_3O_{12}$, Opt. Mater. **28**, 1095–1100 (2006)

9.181 V. Marinova, M.L. Hsieh, S.H. Lin, K.Y. Hsu: Effect of ruthenium doping on the optical and photorefractive properties of $Bi_{12}TiO_{20}$ single crystals, Opt. Commun. **203**, 377–384 (2002)

9.182 V. Marinova, S.H. Lin, V. Sainov, M. Gospodinov, K.Y. Hsu: Light-induced properties of Ru-doped $Bi_{12}TiO_{20}$ crystals, J. Opt. A. Pure Appl. Opt. **5**, S500–S506 (2003)

9.183 K. Buse, H. Hesse, U. van Stevendaal, S. Loheide, D. Sabbert, E. Kratzig: Photorefractive properties of ruthenium-doped potassium niobate, Appl. Phys. A **59**, 563–567 (1994)

9.184 S. Lanfredi, M.A.L. Nobrea: Conductivity mechanism analysis at high temperature in bismuth titanate: A single crystal with sillenite-type structure, Appl. Phys. Lett. **86**, 081916 (2005)

9.185 B. Briat, V.G. Grachev, G.I. Malovichko, O.F. Schirmer, M. Wöhlecke: Defects in inorganic photorefractive materials and their investigations. In: *Photorefractive Materials and Their Applications*, Vol. 2, ed. by P. Gunter, J.P. Huignard (Springer, Berlin 2007)

9.186 B. Briat, A. Hamri, N.V. Romanov, F. Ramaz, J.C. Launay, O. Thiemann, H.J. Reyher: Magnetic circular dichroism and the optical detection of magnetic resonance for the Bi antisite defect in $Bi_{12}SiO_{20}$, J. Phys. Condens. Matter **7**, 6951–6959 (1995)

9.187 H.J. Reyher, U. Hellwig, O. Thiemann: Optically detected magnetic resonance of the bismuth-on-metal-site intrinsic defect in photorefractive sillenite crystals, Phys. Rev. B **47**, 5638–5645 (1993)

9.188 I. Biaggio, R.W. Hellwarth, J.P. Partanen: Band mobility of photoexcited electrons in $Bi_{12}SiO_{20}$, Phys. Rev. Lett. **78**, 891–894 (1997)

9.189 W. Rehwald, K. Frick, G.K. Lang, E. Meier: Doping effects upon the ultrasonic attenuation of $Bi_{12}SiO_{20}$, J. Appl. Phys. **47**, 1292–1294 (1976)

9.190 P.K. Grewal, M.J. Lea: Ultrasonic attenuation in pure and doped $Bi_{12}GeO_{20}$, J. Phys. C Solid State Phys. **16**, 247–257 (1983)

9.191 B.K. Meyer, H. Alves, D.M. Hofmann, W. Kriegseis, D. Forster, F. Bertram, J. Christen, A. Hoffmann, M. Strassburg, M. Dworzak, U. Haboeck, A.V. Rodina: Bound exciton and donor–acceptor pair recombinations in ZnO, Phys. Status Solidi (b) **241**, 231–260 (2004)
9.192 S.L. Lee, C.K. Lee, D.C. Sinclair: Synthesis and characterisation of bismuth phosphate-based sillenites, Solid State Ion. **176**, 393–400 (2005)
9.193 D.C.N. Swindells, J.L. Gonzalez: Absolute configuration and optical activity of laevorotatory $Bi_{12}TiO_{20}$, Acta Crystallogr. B **44**, 12–15 (1988)

10. Bulk Crystal Growth of Ternary III–V Semiconductors

Partha S. Dutta

Ternary semiconductor substrates with variable bandgaps and lattice constants are key enablers for next-generation advanced electronic, optoelectronic, and photovoltaic devices. This chapter presents a comprehensive review of the crystal growth challenges and methods to grow large-diameter, compositionally homogeneous, bulk ternary III–V semiconductors based on As, P, and Sb compounds such as GaInSb, GaInAs, InAsP, AlGaSb, etc. The Bridgman and gradient freezing techniques are the most successfully used methods for growing ternary crystals with a wide range of alloy compositions. Control of heat and mass transport during the growth of ternary compounds is crucial for achieving high-quality crystals. Melt mixing and melt replenishment methods are discussed. The scale-up issues for commercial viability of ternary substrates is also outlined.

10.1 III–V Ternary Semiconductors

Ternary compounds are synthesized by mixing three elements. These are also referred to as pseudobinary or tertiary compounds or alloys. From the periodic table, one can mix two group III elements and one group V element to form ternaries such as $Ga_{1-x}In_xAs$. Alternatively one could mix one group III element and two group V elements to form ternaries such as $InP_{1-y}As_y$. Here x and y are mole percentages. The values of x and y are between 0 (0 mol %) and 1 (100 mol %). *Pseudobinary* is an equivalent term since a ternary compound such as $Ga_{1-x}In_xAs$ can be viewed as comprising of x mol % of InAs and $(1-x)$ mol % of GaAs. Ternary compounds are attractive as substrates materials for electronic and optoelectronic applications since one can tune the lattice parameter or the bandgap energy of the ternary materials by choosing appropriate chemical compositions [10.1–7].

Figure 10.1 shows the lattice parameters and bandgap energies of the III–V semiconductors [10.1]. The values of bandgap and lattice parameter of the binary compounds such as GaAs, InP, GaSb, InSb, etc. are based on experimental data. Curves joining two binary compounds would represent different ternary compounds. For example, a curve between GaAs and InAs would represent the ternary $Ga_xIn_{1-x}As$ $(0 < x < 1)$ compound. The curves shown in the literature are primarily extrapolated based on experimental data points of few alloy compositions.

The lattice constants of ternary compounds (a_t) varies linearly with composition (x) between the two binary lattice constants (a_1 and a_2). This is known as *Vegard's law*

$$a_t = xa_1 + (1-x)a_2 \,. \tag{10.1}$$

For example

$$a(Ga_{1-x}In_xAs) = xa(InAs) + (1-x)a(GaAs) \,.$$

The lattice constants for binary compound semiconductors are listed in Table 10.1.

The bandgap of any ternary compound varies nonlinearly with the alloy composition (x or y). The

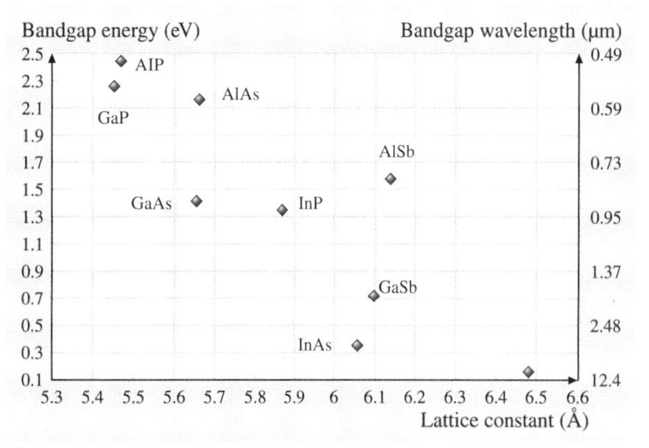

Fig. 10.1 Lattice constant, bandgap energy, and corresponding wavelength for As-, P-, and Sb-based III–V binary semiconductors

Table 10.1 Lattice constant, direct (D) and indirect (I) bandgap energy, and corresponding wavelength for As-, P-, and Sb-based III–V binary semiconductors

Material	Lattice constant (Å)	Bandgap energy (eV)	Bandgap wavelength (μm)
AlP	5.467	2.45 (I)	0.50
AlAs	5.660	2.14 (I)	0.58
AlSb	6.136	1.63 (I)	0.76
GaP	5.4512	2.268 (I)	0.55
GaAs	5.6532	1.424 (D)	0.87
GaSb	6.0959	0.725 (D)	1.70
InP	5.8687	1.34 (D)	0.92
InAs	6.0583	0.356 (D)	3.5
InSb	6.4794	0.18 (D)	6.88

Table 10.2 Dependence of bandgap on alloy composition for As-, P-, and Sb-based ternary semiconductors

Alloy	Direct energy gap (eV)
$Al_xIn_{1-x}P$	$1.34+2.23x$
$Al_xGa_{1-x}As$	$1.424+1.247x,\ 0<x<0.45$
	$1.424+1.087x+0.438x^2,\ x>0.45$
$Al_xIn_{1-x}As$	$0.356+2.35x+0.24x^2$
$Al_xGa_{1-x}Sb$	$0.726+1.10x+0.47x^2$
$Al_xIn_{1-x}Sb$	$0.18+1.621x+0.43x^2$
$Ga_xIn_{1-x}P$	$1.34+0.511x+0.6043x^2,\ 0<x<0.55$
$Ga_{1-x}In_xAs$	$0.356x+1.425(1-x)-0.436x(1-x)$
$Ga_{1-x}In_xSb$	$0.18x+0.726(1-x)-0.415x(1-x)$
GaP_xAs_{1-x}	$1.424+1.172x+0.186x^2$
InP_xAs_{1-x}	$0.356+0.675x+0.32x^2$
$InAs_ySb_{1-y}$	$0.356y+0.18(1-y)-0.58y(1-y)$
$GaAs_ySb_{1-y}$	$1.424y+0.726(1-y)-1.2y(1-y)$

experimentally measured bandgap (E_g) for a ternary alloy can be fitted to a quadratic equation [10.1]

$$E_g(x)[\mathrm{eV}] = xE_1 + (1-x)E_2 - cx(1-x) , \quad (10.2)$$

where E_1 is the bandgap at $x = 1$, E_2 is the bandgap at $x = 0$, and the *composition independent* constant c is the bowing parameter accounting for the nonlinearity.

The above equation (10.2) can be rearranged as

$$E_g(x) = a + bx + cx^2 ,$$

where

$$a = E_2 , \; b = E_1 - E_2 - c , \; c = c .$$

Table 10.2 summarizes the bandgap equations for various III–V ternary compounds. The bandgaps and lattice parameters shown in Fig. 10.1 as well as listed in Tables 10.1 and 10.2 are for compounds with the same crystal structure (zincblende or cubic). If the crystal structure changes, the lattice parameter and bandgap of the same material will be different. For example, the II–VI compound ZnS can exist in hexagonal or cubic phase.

10.2 Need for Ternary Substrates

For most semiconductor devices, multilayered thin-film structures comprising various ternary and quaternary compounds (also known as heterostructure) are necessary (Fig. 10.2). Ideally one would like to grow the entire thin-film structure on a substrate with same lattice constant (referred to as lattice-matched substrate). Unfortunately, device-grade single-crystal substrates of only binary compounds (such as GaAs, GaSb, InP, GaP, InAs, and InSb) with a few discrete lattice constants are commercially available (Fig. 10.1). Hence thin epitaxial layers are grown on binary substrates using liquid-phase epitaxy (LPE), metalorganic vapor-phase epitaxy (MOCVD or OMVPE) or molecular-beam epitaxy (MBE) techniques, as shown in Fig. 10.2. Due to the lattice mismatch between the epilayer and the substrate, misfit dislocations originate at the growth interface and propagate into the device layers [10.1, 8]. Typical misfit dislocation density is in the range of $10^6 - 10^9\,\mathrm{cm}^{-2}$ [10.8]. Such a high density of dislocations leads to degradation of the electrical and optical characteristics of the devices. To reduce this misfit dislocation density, a variety of buffer layers are grown between the substrate and the device layers. However, the buffer layer technology necessary to relieve misfit-related stresses is not optimized for all systems, and often devices exhibit poor characteristics due to interfacial defects. Hence substrates with variable lattice constants are highly desirable to enhance the performance of electronic and optoelectronic devices.

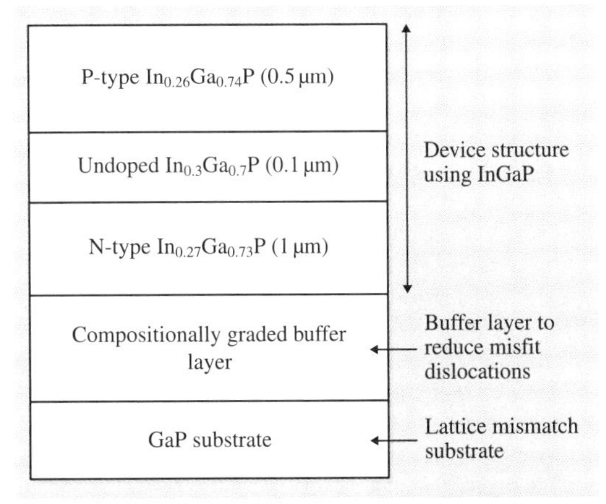

Fig. 10.2 A typical III–V ternary-based optoelectronic device structure grown on a lattice-mismatched binary substrate using a compositionally graded buffer layer

The availability of substrates with tunable bandgap and lattice constant would also open up numerous possibilities of interesting bandgap engineering in homo- and heteroepitaxial devices with improved performances, new features, reduced fabrication complexity, and better cost effective. Some appropriate examples are: GaInAs laser diodes on lattice-matched substrates with high characteristics temperature [10.9–11], low-cost, high-quantum-efficiency photodetector [10.12, 13], and thermophotovoltaic cells [10.14] with diffused p-n homojunction on GaInSb bulk substrates, high-efficiency tandem solar cells [10.15], and antimonide-based quantum well and superlattice structures lattice matched to GaSb for high-efficiency mid-infrared sources and detectors [10.16]. Future devices with improved performances are being sought for a variety of military and civilian applications. These include: infrared (IR) imaging sensors for missile and surveillance systems; IR sources for deceptive jamming system; monitoring and detecting environmental pollution, fire, greenhouse-gas fluxes, industrial gas purity, trace moisture in corrosive gases, and microleaks of toxic gases; high-performance and economical off-grid heat-electricity cogeneration systems; automobiles with ultralow emission (hybrid fuel–thermophotovoltaic); high sensitivity piezosensors and actuators; and quantum computing using electron spin coupling. Ternary substrates will provide a broad base of multicomponent alloy semiconductors with compatible heterostructure devices for these applications.

Bulk single-crystal substrates of ternary semiconductors are still not available from any commercial vendor in spite of four decades of research on ternary substrates. The reason is not due to the lack of market for these substrates, but due to inherent problems in growing them from melt [10.17–31]. The development of electronic, optoelectronic, and photovoltaic devices based on ternary lattice-matched substrates has not been successful due to the poor substrate quality of mixed alloys and/or low yield of wafers with the same alloy composition. This chapter lays out the challenges and potential solutions that could be adopted for future commercial development of ternary substrates.

10.3 Criteria for Device-Grade Ternary Substrates

For successful incorporation of ternary substrates into future commercial device technologies, the wafers need to meet several important criteria as discussed below:

(a) Single crystal wafers of 2 in (50 mm) or larger diameter must be available. Most commercial epitaxial growth and device fabrication equipment is geared towards handling wafers of 50 mm or larger. Hence the growth and fabrication conditions for any specific device structure need to be optimized using the geometries that will ultimately be used in large-scale production.
(b) The spatial compositional inhomogeneity across the entire wafer should be minimal (less than 0.5 mol %). Variations in composition will lead to differences in final device characteristics fabricated across the wafer.
(c) Wafers should be completely free from cracks, metallic inclusions, and multiphase regions. These are commonly observed defects in ternary crystals and hence a great deal of care is necessary to avoid them during growth.
(d) The dislocation density should be similar to that in existing high-quality commercial binary substrates (less than $1000\,cm^{-2}$). Since the dislocations in the substrate propagate into the epilayers, lower dislocation density in the starting wafer is always desirable for better yield of reliable devices [10.1].
(e) A wide range of doping should be possible in the substrate material in order to achieve desirable optical and electrical properties [10.32–38]. For electronic devices, semi-insulating substrates are necessary. For infrared photodetector applications with back illumination (radiation incident from the back of the substrate) or for light-emitting diodes, the substrate must be optically transparent for wavelengths that are being detected or emitted in the epilayers. The electrical resistivity or optical transparency can be altered by suitable impurity doping of the bulk crystal. Even if lattice matching is achieved using a certain substrate material, if they do not have the necessary electrical or optical characteristics, the substrates are of little use for end applications. For example: high-purity undoped substrates of GaSb are completely opaque for radiation with wavelengths less than its bandgap [10.35–37]. This is due to large concentration of native defects such as vacancies and antisites in the grown crystals. These defects act as p-type dopants, leading to optical absorption by free carrier mechanisms and low electrical resistivity. High-resistivity (semi-insulating) GaSb substrates are not commercially available. The optical transparency of GaSb (for below bandgap radiation) as well as its resistivity can be enhanced by impurity compensation [10.32, 37].
(f) Ternary crystals tend to be brittler than binary crystals and can be damaged during wafer slicing and polishing. This is due to high built-in strain in the crystals. Ensuring that the crystals can be processed into wafers is crucial. This requires proper thermal conditions during crystal growth and postgrowth annealing treatments.
(g) The wafer should have high-quality polished and chemically treated surface to enable epigrowth (also referred to as *epiready* surfaces) of high-quality layers. Certain materials such as AlSb, even if available as high-quality substrates, cannot be easily adopted for epigrowth or device fabrication due to the challenges with the surface oxidation. Similarly, antimonide-based wafer surfaces such as GaSb or

InSb need special chemical processing before growing device-quality epilayers. Hence developing the polishing recipe for each substrate material is necessary for their application.

(h) Finally, the cost of the final wafers should not be significantly higher than that of commercially available binary substrates. Though bulk substrates contribute to a small fraction of the cost for the entire device, high cost of the substrates poses a barrier for adoption unless it is clearly established that, by using lattice-matched substrates, significant improvements in device characteristics can be achieved.

These attributes can be translated into numerous constraints during crystal growth. As will be discussed in this chapter, many of the requirements are counteractive, and appropriate trade-offs in the growth conditions are necessary.

The growth conditions that are necessary for high-quality ternary crystals have been summarized below:

1. Crack-free and inclusion-free crystals can be grown by avoiding constitutional supercooling. That means that the crystal growth rate must be lower than the rate at which excess constituents rejected at the melt–crystal interface (due to segregation) are mixed back into the growth melt. This requires forced convective mixing in the growth melt and near the melt–crystal interface.
2. Low dislocation density and strain can be achieved if the growth takes place under a low temperature gradient and the crystals are cooled slowly after solidification.
3. High yield refers to obtaining wafers of the same composition from a single ingot. This is possible only when the crystal grown has the same composition along the growth direction. This requires replenishment of the melt with the constituents to maintain the same melt composition during the entire growth.
4. Rapid or uncontrolled melt replenishment leads to high level of supersaturation in the melt with the replenished constituents. This triggers polycrystalline growth due to random nucleation in the melt. To maintain single-crystallinity during growth, the rate at which the solute is fed to the melt must be precisely controlled at all times to match the crystal growth rate. This requires special solute feeding processes and forced convective mixing in the melt.
5. Low cost of final wafers directly relates to the growth rate of the crystal and the volume of starting melt versus the volume of final crystal. This would require: (a) consuming the entire melt during the growth and (b) rapidly transporting the dissolved constituents (replenishing elements or compounds) to the growth interface by forced convection. In addition, rapid dissolution of the replenished constituents (solute) is necessary. This would require that the dissolution of the solute occurs in a melt zone that is always undersaturated.
6. Compositionally homogeneous wafer would translate to uniform composition in the crystal perpendicular to the growth direction. This would require a planar melt–crystal interface during growth. This can be achieved by balancing the heat transfer at the melt–crystal interface by a combination of temperature gradient imposed by the furnace and the forced convective mixing in the growth melt. At the same time, temperature fluctuations due to forced convection in the growth melt (leading to composition fluctuation and local constitutional supercooling) must be eliminated. This can be achieved by having a low temperature gradient near the liquid–solid interface.
7. Large-diameter wafers are necessary for commercial applications. Hence any technology that is being developed for growth of alloy semiconductors must scale up to dimensions required for practical usage. This would require optimization of the heat and mass transport processes between various melt zones during growth for the length scales of interest. This is purely dictated by the design of the experimental setup and the process parameters. Bridgman and gradient freezing types of method are becoming more popular for large-diameter binary crystal growth with very low defect content. Hence the advancement made for binary growth could be adopted for future ternary crystal growth technology.

To achieve these goals, a lot of effort in engineering of the heat and mass transport processes during the growth of ternary alloys from melts is necessary. In this chapter, we discuss the interdependencies between various experimental parameters and their effects on the compositional homogeneity of ternary crystals. The focus will be on the Bridgman techniques as it offers a variety of beneficial attributes that are necessary for ternary crystal growth processes. In the next section, we briefly review the Bridgman and gradient freezing directional solidification processes.

10.4 Introduction to Bridgman Crystal Growth Techniques

10.4.1 Bridgman Techniques

The Bridgman technique (also referred to as the Bridgman–Stockbarger method) is one of the oldest techniques used for growing crystals [10.39]. The crystal growth can be implemented in either a vertical (vertical Bridgman technique) or horizontal system configuration (horizontal Bridgman technique). Schematics of the two configurations are shown in Figs. 10.3 and 10.4. The growth systems typically consist of a single- or multizone furnace. A single-zone furnace has a parabolic temperature profile with the highest temperature being at the center along the length of the furnace, as shown in Figs. 10.3 and 10.4. On both sides of the hottest section, a temperature gradient exists that is used during the crystal growth. For a multizone furnace, specific temperature gradients between different zones can be established as described later.

The principle of crystal growth using Bridgman technique is based on directional solidification by translating a molten charge (melt) from the hot to the cold zone of the furnace, as depicted in Figs. 10.3 and 10.4. The presence of a seed at the end of the crucible (container) ensures single-crystal growth along specific crystallographic orientation. The process for generating the *first* single crystalline seed of any new material has been described in Sect. 10.4.3. Using the single-crystal seed, the entire growth process takes place in the following sequence. At the beginning of the experiment, the crucible with the polycrystalline charge and seed is placed inside the growth chamber. Then the chamber is evacuated by a vacuum pump and refilled with inert gas. The temperature of the furnace is then raised. A proportional–integral–differential (PID) control mechanism controls the power to the heater elements, maintaining the desired temperatures. The PID controller also controls the power during the heating and cooling stages of the furnace. After the furnace is heated to a temperature above the melting point of the polycrystalline charge, the crucible is slowly translated into the hot zone to melt the polycrystalline charge completely and bring it into contact with the seed. Section 10.4.4 presents the intricacies of the seeding process. After the melt touches the seed, a portion of the seed is remelted to expose a fresh growth interface. The melt is thoroughly mixed using forced convection generated by rotating the crucible (using stepper motor control, for example). The homogenization of the melt can also occur by natural convection and diffusion in the melt without any forced convection. Hence melts can be homogenized by simply leaving the melt at temper-

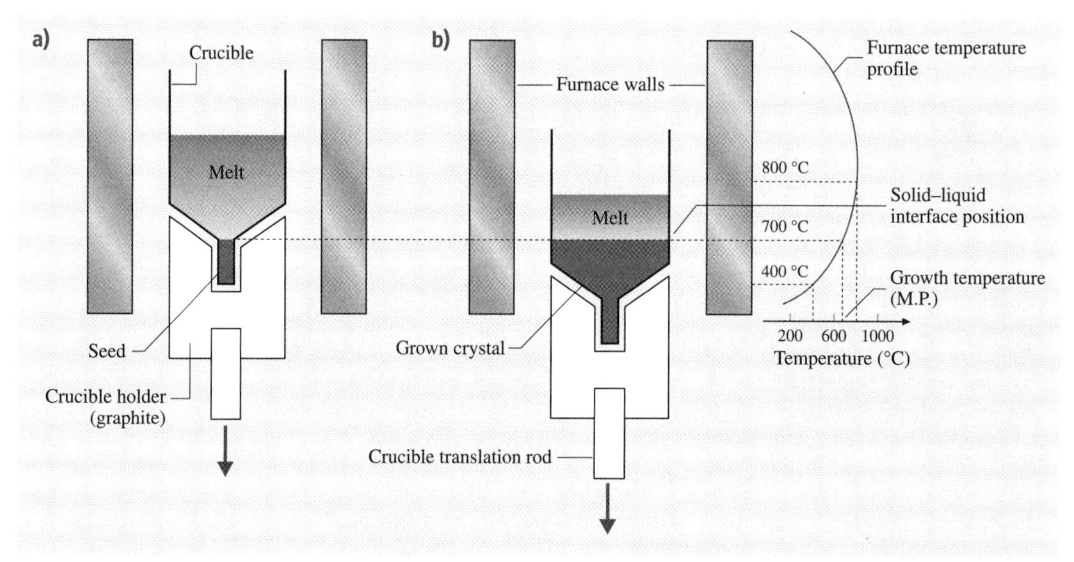

Fig. 10.3a,b Schematic diagram of a vertical Bridgman (VB) crystal growth process in a single-zone furnace: (**a**) at the beginning of the experiment and (**b**) with partially grown crystal

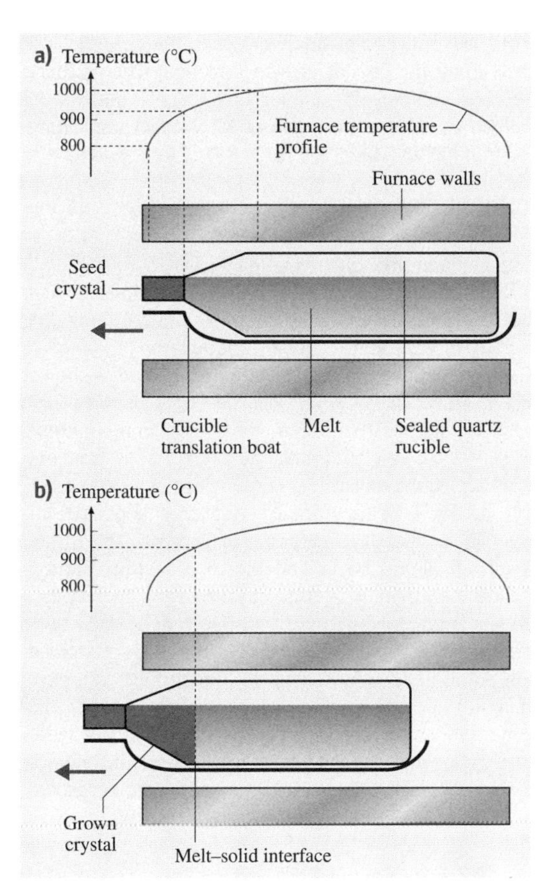

Fig. 10.4a,b Schematic diagram of a horizontal Bridgman (HB) crystal growth process in a single-zone furnace: (**a**) at the beginning of the experiment and (**b**) with partially grown crystal

atures higher than its melting point for a long duration. After the melt is completely mixed, crystal growth is initiated by cooling the melt from the seed end. This is done by translating the crucible slowly into the cooler section of the furnace. As the crucible is translated, the temperature at the bottom of crucible falls below the solidification temperature (the melting point of the material) and hence the melt starts to become a solid at the bottom of the crucible (at the seed–melt interface). After the melt has been completely translated below the melting point of the material, the entire molten charge converts to a solid ingot (also known as a boule). The crucible holder (typically made out of graphite) is connected to a stainless-steel shaft, which in turn is connected to a linear slide operated with the use of a computer-controlled stepper motor assembly. The computer program controls the lowering rates of the crucible, and hence the rate of crystallization (growth rate) can be accurately controlled over long periods of time.

The growth rates for different crystals need to be optimized in order to grow single crystals with high crystalline quality. For example, the typical growth rate for III–V binary semiconductors is in the range of 0.5–3 mm/h, while for ternary crystals it is in the range of 0.1–1 mm/h. The shaft supporting the crucible can also be rotated with the help of a high-torque stepper motor. As will be discussed later, this is essential for efficient mixing of the melt during growth. After the entire molten charge is directionally solidified, the temperature of the furnace is decreased slowly to reach room temperature. The postgrowth cooling rate must be controlled in order to avoid thermal shock (due to rapid cooling) to the solid ingot that could lead to mechanical cracks in the crystal. Typical cooling rates are in the range of 10–50 °C/h, depending on the material. At the end of the experiment, the crystal is removed from the crucible and sliced to prepare the substrates. Figure 10.5 shows a typical vertical Bridgman grown ingot of GaInSb from the author's laboratory.

Though the concept of vertical and horizontal Bridgman techniques is similar, there are certain advantages and disadvantages of the two methods. The wafers extracted from vertical Bridgman grown crystals are perfectly circular in shape, unlike the D-shaped wafers from horizontal Bridgman grown crystals. For

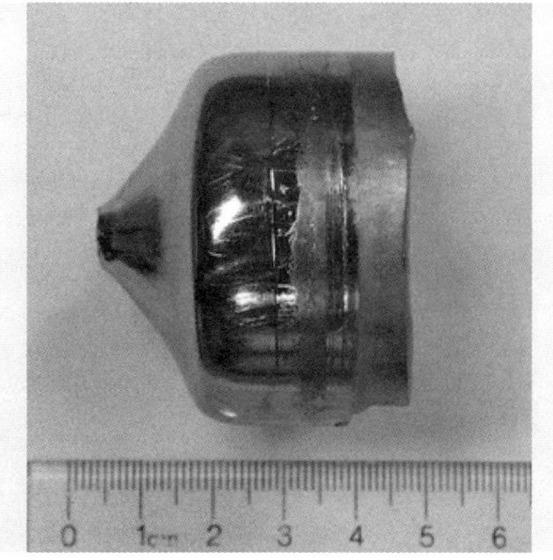

Fig. 10.5 Vertical Bridgman grown $Ga_{1-x}In_xSb$ single crystal (from the author's laboratory)

large-scale epitaxial growth on the substrates and device fabrication, circular wafers are certainly beneficial. The primary advantages of the horizontal Bridgman grown crystals are the high crystalline quality (such as low dislocation density) of the crystals and the stoichiometry control along the entire length of the crystal. In the horizontal Bridgman method (Fig. 10.4a,b), the crystal experiences lower stress due to the free surface on the top of the melt. During the growth, the solid does not touch the crucible on the top and hence is free to expand. Another advantage of the horizontal growth process is the enhanced mixing in the melt due to thermal convection (flows due to temperature gradient) at every location in the melt along the growth direction. As will be discussed later, this helps in ensuring stoichiometry (the composition of the crystal) along the length of the crystal by maintaining an overpressure of the volatile species such as the group V elements during III–V semiconductor growth.

Finally, it must be mentioned that, in the Bridgman techniques, the relative motion between the crucible and the furnace is all that matters. Instead of using crucible translation, the crucible can be kept stationary and the furnace translated to achieve directional solidification. In numerous cases, the crucible system design containing the melt, seed, and the overpressure elements is complex (as will be discussed below for GaAs and InP crystal growth) and hence translating the crucible poses problems. It becomes much easier to translate the furnace along the crystal growth direction. The gradient freezing techniques discussed below is a sophisticated method to achieve the same effect without translating the crucible or furnace.

10.4.2 Gradient Freezing Techniques

The gradient freezing technique is analogous to the Bridgman technique discussed above except for the fact that the temperature gradient is translated along the melt to implement directional solidification [10.40–44]. The schematic of the gradient freeze method is shown in Fig. 10.6. The principle can be implemented in either vertical (VGF) or horizontal (HGF) configurations, as in the case of Bridgman methods. In the gradient freezing techniques, the crucible with the seed and the melt as well as the furnace system is kept stationary. The temperature gradient, as shown in Fig. 10.6, is moved along the crystal growth direction starting from the seed–melt interface to the end of the crystal. This is accomplished by a multiple-zone furnace system wherein the power to each zone is programmed and controlled by individual PID controllers. The use of multiple heater zones is necessary to maintain the same temperature gradient at the melt–solid interface during the entire crystal growth experiment. If a single-zone furnace is used, the temperature gradient at the

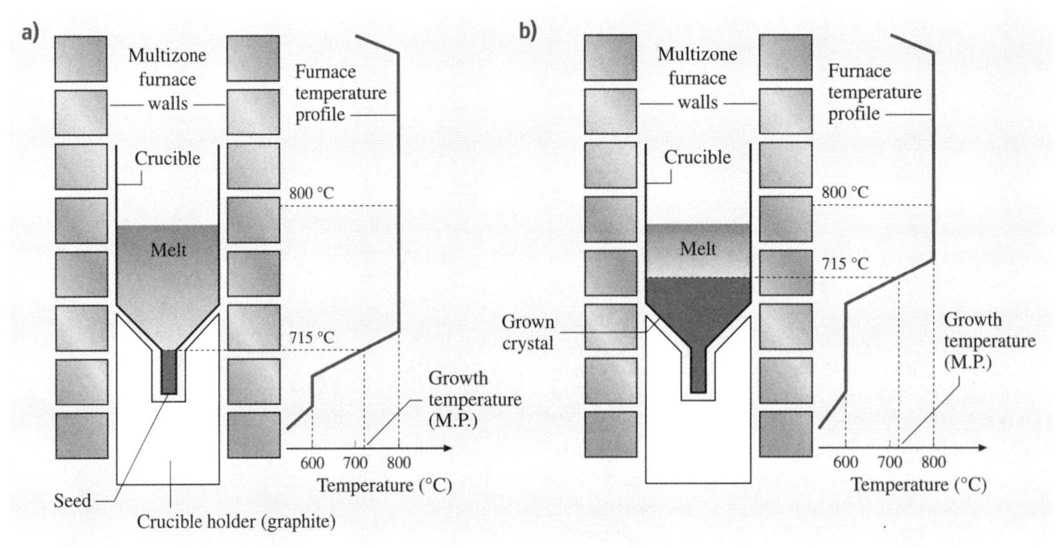

Fig. 10.6a,b Schematic diagram of a vertical gradient freeze (VGF) crystal growth process in a multizone (eight-zone) furnace: (**a**) at the beginning of the experiment and (**b**) with partially grown crystal

melt–solid interface changes at different locations in the furnace leading to variation in crystal growth rate with time. Since the heat flux through the growing crystal and melt changes with time, the multiple-zone system ensures continuously varying power to individual zones so as to maintain the same temperature gradient at the liquid–solid interface. The quality of the crystals grown by this method is far superior to that produced by traditional methods such as liquid encapsulated Czochralski (LEC). Since LEC growth takes place under a large temperature gradient (typically several tens of °C/cm), the dislocation density in grown crystals is of the order of $10\,000\,cm^{-2}$ [10.1, 42, 45]. Wafer breakage in the device production process due to large thermal stress in the wafers is a significant problem. On the other hand, the temperature gradient used in the Bridgman or gradient freezing methods is very small (1–10 °C/cm). Hence a dislocation density less than $500\,cm^{-2}$ has been achieved in VGF-grown crystals. As a result of this, the VGF and HGF techniques have received widespread commercial usage for both III–V and II–VI semiconductor crystal growth [10.40–44].

In a typical crystal growth experiment, the crucible with the charge is placed in the growth chamber. Then the chamber is evacuated by a vacuum pump and refilled with inert gas. The power to the individual element is increased with time to melt the polycrystalline charge completely and bring it into contact with the seed. A portion of the seed is also remelted to expose a fresh growth interface (Fig. 10.6a). Crystal growth is initiated by cooling the melt from the seed end. The temperature gradient at each and every point along the growth direction can be controlled very accurately due to the multiple zones (Fig. 10.6b). The sophisticated thermal controls also make gradient freezing systems more complex than typical Bridgman systems. Unlike simple resistance heater wires wound around tubes as in Bridgman-type systems, gradient freezing units have advanced heater designs. Some systems have water cooling tubes between the heater plates to ensure sharp and controllable temperature gradients along the length of the furnace.

For semiconductor crystal growth, oxygen in the melt must be eliminated completely. Hence crystal growth chambers are designed to be vacuum-tight as well as to withstand high internal gas pressures. After loading the crucible with charge, the growth chamber is evacuated to a high vacuum (1 mTorr or less pressure) and repeatedly flushed with an inert gas mixture such as argon/hydrogen or hydrogen/nitrogen to ensure the removal of moisture and oxygen from inside the chamber. During the charge synthesis and growth, the chamber is filled with an inert gas and a pressure is maintained beyond the vapor pressure of the melt in the chamber.

10.4.3 Seed Generation for New Materials

For single-crystal growth of any material, a single-crystalline seed of specific orientation is necessary. For a new material under development, single-crystal seeds are not available. Single-crystal seeds can be generated by vertical Bridgman growth using specially tipped crucibles [10.39] as shown in Fig. 10.7. There are two ways this can be made possible. In the first scenario (Fig. 10.7a), a natural seed selection process can occur at the tip region of the crucible. A specific nucleus can grow faster than the others and outgrow the rest of the nuclei. However this process is completely random and cannot be expected to repeat in every experiment. Hence a more predictive way of selecting an individual seed orientation is via the necking process (similar to what is being done in the Czochralski crystal growth process). In this process (Fig. 10.7b), randomly nucleated grains are subjected to a filtering process at the necking point in the crucible so that only one orientation reaches the main body of the crucible and grows into a bulk ingot. However, the production of a single-crystal grain cannot always be guaranteed by either of these processes. Random secondary nuclei from the crucible walls, such as shown in Fig. 10.7b, are often seen. Hence liquid encapsulation with low-melting-point liquids or crucible coating, e.g., with carbon, is necessary to avoid the melt touching the crucible wall during growth. This process,

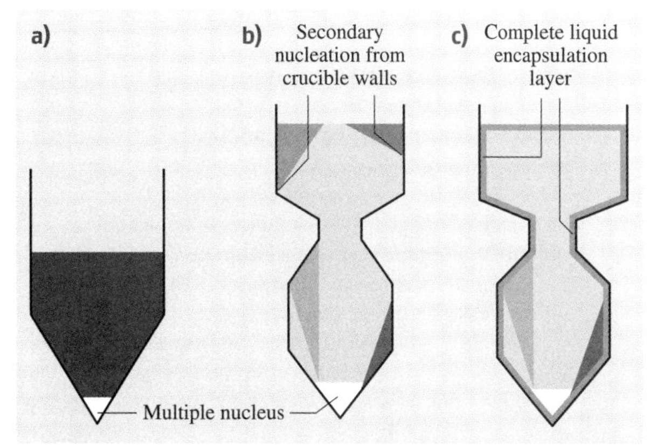

Fig. 10.7a–c Single-crystal seed generation processes: (**a**) by natural selection, (**b**) by necking, and (**c**) by necking and liquid encapsulation

depicted in Fig. 10.7c, is the most predictive way to ensure a single-crystal seed generation process. Due to the random nature of the initial seed orientation, the single-crystal seed is then oriented along a preferred crystallographic direction using the x-ray orientation method. The ingot is then mined to obtain a seed with a specific orientation for further crystal growth experiments. For a cylindrical seed, a diamond core drill could be used. For rectangular-shaped seeds, the ingot can be sliced using a blade or a wire saw.

10.4.4 The Seeding Process

The initial stage of the crystal growth experiments wherein a single-crystalline seed is contacted with the molten charge is crucial and requires adequate care and experience. High-quality single-crystal seeds are precious and used repeatedly for a large number of experiments. The seeds are smaller in diameter with respect to the final diameter of the grown crystals. For example, a typical seed diameter is 5 mm (with 5–6 cm in length), while the grown crystals could be 50–75 mm in diameter. The seed needs to be partially remelted (≈ 0.5 cm) to expose a fresh growth interface and then the crystal growth is started. Unlike in the Czochralski growth method, the seed–melt interface during Bridgman growth (in most cases) cannot be observed due to the opacity of the crucible, melt, and/or the growth chamber. Hence for precise remelting of the seed during crystal growth, one needs to conduct a priori thorough analysis of the thermal environment of the growth chamber. Translating the crucible into the hot zone even by 1 mm could sometimes remelt the entire length (and hence loss) of the seed. This is due to the heat transfer from the melt to the seed by conduction and convection in the hot melt. For Bridgman and gradient freezing types of crystal growth experiments, in order to ensure the exact seeding position in the furnace, prior experimentation needs to be done with dummy melts and seeds. Since the heat transfer between the furnace, crucible, melt, and growth chamber is quite complex, any optimization of the seeding position needs to be carried out using a specific set of crucible diameter, crucible holder design, crucible material (pyrolytic boron nitride (pBN), silica, graphite, alumina, etc.), melt depth, seed length, etc. If any of the above parameters is altered during the actual crystal growth experiment, the seeding position will change.

For the seeding position determination, two sets of experiments need to be carried out. In the first experiment, a polycrystalline charge (of the same material to be grown) is melted into a crucible with the exact shape as is to be used in the final crystal growth experiment. The melt is then directionally solidified as in the real crystal growth experiment. The polycrystalline ingot, which is shaped like a crystal with a seed at the end (Fig. 10.3b), is removed from the crucible and then chemically etched to make the surface shiny (as described in Sect. 10.5.4). In the second experiment, this polycrystalline ingot is placed inside an identical crucible (as in the first experiment). The crucible is then slowly translated into the hot zone of the furnace to remelt the ingot to a desired point and then directionally solidified by translating the crucible back into the cold zone. After the experiment, one can easily observe the interface between the remelted and unmelted portions of the ingot. By repeating this process, one can identify the crucible location in the furnace that will provide the desired seed–melt interface. Since this process is rigorous, once the crucible location has been identified for proper seeding, it is advisable to maintain the same conditions during actual crystal growth experiments. Slight changes such as an increase or decrease in melt height or thickness of crucible wall could lead to a large variation in the seeding interface location. Hence in large-scale crystal production, the design of the entire crystal growth systems is properly analyzed and crystal growth conditions are kept unchanged.

10.4.5 Growth Rate Determination Methods

The crystal growth rate is an important factor that dictates the crystalline quality and microstructure of the grown crystals. For crystals grown from melt in crucibles using Bridgman or gradient freezing techniques, precise determination of growth rates requires considerable experimental effort. There are a few techniques that have been developed and used for determining the growth rate of crystals, as briefly discussed below [10.46–65]. The choice of technique is decided by the crystal growth temperature, and the optical, electrical, and thermal properties of the melt, crystal, and crucible material.

Depending on the crystal to be grown, a proper choice of crucible material is necessary based on its chemical reactivity with the melt. There may be more than one option for the crucible material for any specific melt. The final selection of the crucible material is made based on several other factors such as heat transfer and thermal configurations, use of melt encapsulations, the thermal expansion coefficient of the grown crystal, application of external fields during growth, etc. Hence, based on the crucible material, a specific growth rate determination method is used. For example, crucible material such as silica (quartz) is optically transparent and hence the melt–crystal interface can be directly visualized during growth if a transparent furnace configuration is used. On the other hand crucibles made of pyrolytic boron nitride (pBN), alumina, graphite, etc. are opaque and hence indirect methods for melt–solid interface visualization are necessary. The thermal conductivity of the crucible material contributes to the heat transport from the hot zone to the cold zone of the furnace and thus has a significant effect on the melt–crystal interface shape and the growth rate of crystals. In the case of crystals grown under applied electric and magnetic fields, it is important to select a crucible material with appropriate electrical conductivity.

For crystals with low growth temperature (below 1000 °C) that can be grown in a transparent crucible such as silica, the determination of growth rate is relatively straightforward. Since furnaces with optically transparent walls can be used for low growth temperatures, one can directly observe the melt and crystal inside the crucible. In these furnaces, a resistive heating coil is enclosed inside a transparent or semitransparent silica tube. Usually the furnace wall is coated with a thin film of reflective metal such as gold to reduce radiative heat losses. By in situ observation of the length of solidified melt with time (due to directional cooling), one can easily calculate the growth rate. This method has been used for determining the growth rate of optically transparent crystals as well as metal and semiconductor crystals [10.28, 46]. In the case of optically transparent material, the transparencies of the melt and the solid need to be different in order to observe a clear demarcation at the melt–crystal interface. For metallic or semiconductor crystals, typically the optical reflectivities of the melt and the solidified crystal are different, thus providing a clear delineation at the melt–crystal interface. This method cannot be used when the crystals need to be grown in an opaque crucible such as pBN or when a furnace with opaque insulation wall is used. Typically, for growth temperature exceeding 1000 °C, the furnace wall is insulated with alumina or quartz wool and is opaque. In such scenarios, there are four methods for determining the growth rate, namely, melt quenching, Peltier interface demarcation (PD), periodic external field application, and real-time radioscopic x-ray/γ-ray visualization.

In the traditional melt quenching method [10.47–49], the crystal growth experiment is terminated by rapid cooling of the melt after a certain period. When the solidified ingot is sliced along the growth direction, one can easily observe two interfaces. The first interface is seen at the seeding point (at the first-to-freeze position on the seed crystal). The second interface is seen where the growth is terminated by rapid quenching. By measuring the distance between the two interface positions and the total growth duration, one can obtain the average growth rate. The sliced specimens generally requires metallographic preparation via coarse grinding and fine polishing (using abrasive powders or polishing slurries) followed by chemical etching (with a suitable etchant solution) to reveal the interface demarcation striations under Nomarski interference contrast microscopy. The quenching method is a universal technique that can be used for any material system irrespective of optical, electrical, and thermal properties. However it gives no information on the evolution of the solid–liquid interface position during growth. This is a major limitation since the crystal growth rate varies with time during the experiment as a result of varying heat flow through the melt and the crystal (due to changing axial temperature gradient in the melt and solid) as well as alloy segregation (as in the case of ternary and quaternary semiconductor alloys).

For crystals and melts with good electrical conductivity, the PD can be employed [10.50–56]. In a typical experiment, two electrical contacts are made: one at the top of the melt using an appropriate electrode material (that will not react with the melt) and one at the bottom of the seed crystal. In this method, thermoelectric effects induced by a current pulse passing through the melt and the directionally solidifying sample results in the creation of a rapid thermal perturbation at the solid–liquid interface. A series of perturbations result in concentration variations which, after being revealed by postgrowth metallography, delineate the instantaneous interface shape at successive times during growth and enable one to follow the time evolution of the solid–liquid interface. This method has been successfully used for metallic alloys and semiconductor materials. Since electric current needs to flow through the melt and the crystal, it is difficult to implement the PD process in materials with low electrical conductivity.

An elegant way of determining the crystal growth rate is by the application of periodic external fields during growth [10.57–60] such as by rotating magnetic fields or alternating electric fields. In the absence of an applied magnetic field, growth striations appear in the crystals due to natural convection. If the melt convection is suppressed by applied fields, the striations are not seen. By periodically switching on and off the external magnetic fields, one can create or suppress the growth striations. Postgrowth analysis of the specimens by selective chemical etching reveals the regions with striations and striation-free zones. By correlating the length of the individual zones with the field application durations, one can easily determine the growth rate. This technique cannot be used if the growth conditions have been optimized to obtain striation-free crystals without external fields. The application of alternating electric field has been found to alter the microstructure of the crystalline material. Hence, by switching on and off the field, one will obtain regions with different microstructures. The growth rate can be determined easily by postgrowth analysis of the cross-sectional view of the specimen after metallography.

A technique which is very attractive for real-time interface shape monitoring as well as growth rate determination is the use of x-rays and gamma-rays [10.61–65]. This technique is based on the differential transmission of the x-rays or gamma-rays through the solid and the liquid phases. This difference in transmittance could be due to: (a) density difference between the liquid and solid phases, (b) crystalline structure (long- or short-range ordering), (c) dopant (impurity) distribution in the two phases (due to impurity segregation), etc. In the case of dopant distribution, the absorption coefficient of the radiation increases with increasing impurity concentration. This results in image contrast between the solid and liquid phases due to differences in the amount of radiation absorbed (or transmitted) through the individual regions (solid or liquid). One advantage of the radioscopic system is the ability to melt a single sample, solidify it, remelt it, and then change experimental parameters for a different case study. In this manner one can produce a large range of data in a single experiment.

10.5 Overview of III–V Binary Crystal Growth Technologies

Since the focus of this chapter is on ternary crystal growth, the topic of binary III–V bulk crystal growth (excluding the nitrides) using Bridgman or gradient freezing types of processes will be briefly reviewed. This is done to highlight the fundamental differences between the growth conditions of binary versus ternary compounds and the additional advances necessary for ternary crystal growth.

10.5.1 Phase Equilibria for Binary Compounds

Binary compounds such as GaAs, InP, GaSb, etc. are synthesized by mixing the individual elements such as gallium, indium, arsenic, phosphorus, antimony, etc. beyond the melting points of the compounds. Table 10.3 lists the melting points of As-, P-, and Sb-based III–V compounds [10.66, 67]. For compound synthesis, one needs to review the thermodynamic phase diagram. Figure 10.8 shows a schematic phase diagram typical of any III–V binary compound (except the nitrides). This schematic diagram depicting GaAs is not the actual phase diagram, but rather a sketch to explain the important features of the phase formation during compound synthesis and crystal growth. Due to their very high melting temperatures and extreme vapor pressures (exceeding 40 000 atm), GaN, AlN, and InN cannot be grown from stoichiometric melts by Czochralski or Bridgman-type techniques [10.68].

The phase diagrams of various III–V binary compounds can be found in the literature [10.66, 67]. According to the phase diagram shown in Fig. 10.8, the most stable phase is the one where the ratio of group III to group V mole fraction is close to 1 [10.1, 69]. Due to the narrow stability region, the solidus is shown by a vertical line at a composition of 50 at.% (the stoichiometric composition). As shown in Fig. 10.8, the

Table 10.3 Melting points of elements and III–V binary compounds

Material	Melting point (°C)
Ga	29.8
In	156.6
Al	660.4
Sb	630.7
P (red)	416 (sublimes)
As	614 (sublimes)
AlP	> 2000
GaP	1480
InP	1062
AlAs	1740
GaAs	1238
InAs	942
AlSb	1065
GaSb	712
InSb	527

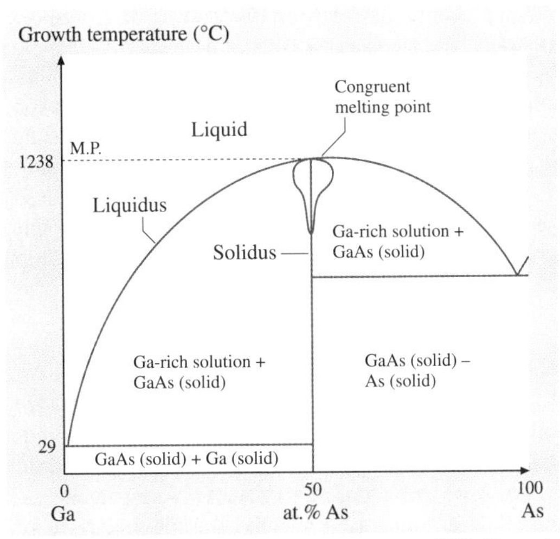

Fig. 10.8 Schematic binary phase diagram of III–V semiconductor (depicting GaAs)

stoichiometric melting point for GaAs is 1238 °C. The congruent melting point is typically slightly shifted from the stoichiometric composition. In the case of GaAs, the congruent melting point is on the arsenic-rich side of the phase diagram [10.69], and in the case of GaSb it is on the Ga-rich side [10.16]. The slight deviation from stoichiometry leads to point defects (native defects such as vacancies, interstitials, and antisites of group III and V elements) in the crystals, which affect the electrical and optical properties of the grown crystals [10.1, 16, 69].

Referring to Fig. 10.8, above the liquid curve (melting temperature), a uniform liquid region (melt) exists. Depending on the ratio of gallium to arsenic, the melting temperature varies. Irrespective of the growth temperature, the composition of the solid (solidus) that precipitate out of the liquid is always the stoichiometric compound GaAs. On the gallium-rich side, the melting point can be as low as 29 °C, which is the melting point of liquid gallium. However the solubility of arsenic in gallium decreases sharply as the temperature decreases. As the melt temperature is decreased, GaAs precipitates out first and then the excess gallium or arsenic solidifies. This phase formation property is used advantageously to grow binary compounds from group III- or group V-rich solutions at temperatures lower than the stoichiometric melting point. On both sides of the stoichiometric melting point, equilibrium regions with solid GaAs and Ga-rich or As-rich solution exist. Below a specific temperature, regions shown as solid GaAs plus solid arsenic or solid gallium exist, depending on the excess element in the starting melt.

Due to the high vapor pressure of the group V elements, III–V compounds generally dissociate above the melting points. However, the dissociation rate depends on the partial vapor pressures. Antimony is the least volatile species amongst the group V elements. The partial vapor pressure of antimony at the melting point of GaSb (712 °C) is 3×10^{-6} Torr. Typically, during 10 h of GaSb growth, 10^{-3} moles of Sb would be lost from the melt [10.16]. The rate of antimony loss from AlSb melt at 1300 °C is ≈ 0.25 g/h [10.70]. The next most volatile species is arsenic. The vapor pressure of arsenic in equilibrium with GaAs melt at its melting point (1238 °C) is 1 atm. The phosphorus vapor pressure is the highest. At the melting point of InP (1062 °C), the vapor pressure in equilibrium with InP melt is ≈ 27 atmosphere. For comparison, the partial pressure of Ga, In, and Al is less than 10^{-6} Torr at the melting points of various compounds. To avoid evaporation of the group V element from the melt, pressure above the partial vapor pressure needs to be applied to the melt. This requires specially made high-pressure stainless-steel vessels, inside which the furnace is assembled.

10.5.2 Binary Compound Synthesis

In this section, we describe Bridgman or gradient freezing types of process that are being used for synthesizing the binary compounds [10.1, 16, 39–44, 66, 71–73, and references therein]. In this process, the starting ma-

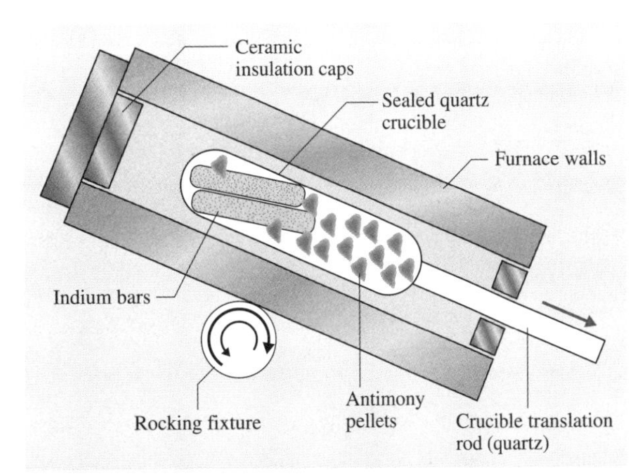

Fig. 10.9 InSb polycrystalline charge synthesis in a sealed crucible using a rocking furnace

terials (charge) consisting of equal mole fractions of the group III and group V elements are weighed and placed into a crucible. Depending on the compound to be synthesized, the placement of the elemental charges inside the crucible could be different, as described below. The crucible with the charge is then heated beyond the melting point of the binary compound and the compound synthesized in liquid form. Thereafter the melt is directionally solidified and slowly cooled to room temperature. Amongst the group V elements, arsenic and phosphorus tend to be extremely volatile. Hence an overpressure of the group V element is necessary during the synthesis. Different configurations are used for compound synthesis and crystal growth of InP, GaAs, GaSb, AlSb, etc. Schematic diagrams of the experimental setups used for synthesizing arsenide-, phosphide-, and antimonide-based compounds are shown in Figs. 10.9–10.15. In this section, we summarize the general schemes for the compound synthesis.

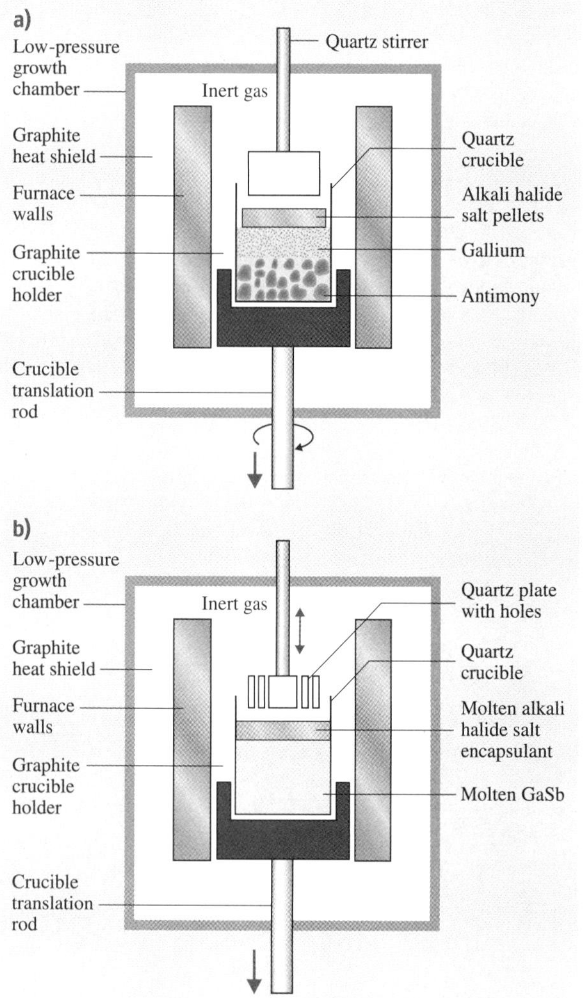

a)
Low-pressure growth chamber
Graphite heat shield
Furnace walls
Graphite crucible holder
Crucible translation rod
Inert gas
Quartz stirrer
Quartz crucible
Alkali halide salt pellets
Gallium
Antimony

b)
Low-pressure growth chamber
Graphite heat shield
Furnace walls
Graphite crucible holder
Crucible translation rod
Inert gas
Quartz plate with holes
Quartz crucible
Molten alkali halide salt encapsulant
Molten GaSb

Fig. 10.10a,b GaSb polycrystalline charge synthesis in an open crucible by (a) using a stirrer and (b) using a quartz baffle for melt mixing

Antimonide-Based Compounds

GaSb and InSb are the easiest compounds from the point of view of synthesis. The vapor pressure of antimony is very low at the melting points of the compounds. Hence both InSb and GaSb can be synthesized by simply melting and mixing the two constituents elements at a temperature beyond the melting point of the binary. The synthesis could be carried out either in a sealed (Fig. 10.9) or open crucible (Fig. 10.10). Inside the sealed crucibles, the vacuum level is typically 10^{-6} Torr or less. For open crucibles, inert-gas ambient is necessary. For synthesis of InSb, indium and antimony are mixed in equal mole fraction (1 : 1). Hence 114.82 g of indium and 121.75 g of antimony are necessary. The two elements are then heated to at least 20 °C more than the melting point of InSb (527 °C) and thoroughly mixed for several hours (10–12 h). Melt mixing could be implemented in sealed crucibles by rocked the crucible along with the furnace like a see-saw as shown in Fig. 10.9. At the end of the homogenization process, the furnace is held horizontal or vertical and the crucible is translated out of the furnace at a rate of 3–5 mm/h to directionally solidify the melt.

Figure 10.10a shows a schematic of GaSb synthesis in an open crucible configuration. The crucible is packed with antimony, gallium, and alkali halide salt (for melt encapsulation). The entire charge is heated to beyond the melting point of GaSb (712 °C) and the compound is synthesized by mixing the melt. For melt homogenization, the crucible can be rotated with a stationary stirrer in the melt, as shown in Fig. 10.10a. Another efficient approach for melt mixing is by using a baffle or a plate with holes made out of the same material as the crucible (Fig. 10.10b). The baffle or mixing plate can be moved back and forth from the top to the bottom of the melt, creating turbulent flow through the narrow regions across the moving object. This leads to very rapid mixing. Hence large melts can be thoroughly mixed in minutes as opposed to hours (as is necessary with crucible rotation or rocking). After the melt has been completely homogenized, it is directionally solidified by translating the crucible at a rate of 3–5 mm/h in a furnace with a temperature gradient of 10–15 °C/cm until the entire liquid turns solid. After that the furnace is slowly cooled to room temperature at a rate of 15–20 °C/h. The presence of alkali halide salt encapsulation on the top of the melt helps avoid volatilization of antimony from the melt surface. As discussed earlier, the usage of liquid encapsulation during single-crystal growth is crucial to avoid melts touching the crucible walls and polycrystalline growth. The synthesis process for InSb and GaSb is exactly the same except for the synthesis temperature.

For AlSb, the problem is the sticking of the aluminum with crucible materials such as silica, pBN or graphite. Recently, a new process for synthesis and growth of AlSb in silica crucibles using alkali halide salts as encapsulants has been demonstrated [10.71]. The adhesion of AlSb melt to silica crucible could be eliminated by employing a LiCl/KCl encapsulation [10.72] in conjunction with excess antimony in the melt. This process is shown schematically in Fig. 10.11a–c. The placement of the elements in the crucible (as shown in Fig. 10.11a) is crucial for successful synthesis of the compound without rupturing the crucible during the experiment. Care should be taken to avoid the elemental aluminum coming into contact with the silica crucible. The LiCl/KCl (58 : 42 mol %) eutectic salt mixture first melts and covers the crucible walls. Then the antimony melts and encapsulates the aluminum until it melts. The temperature of the melt is then increased to beyond the melting point of AlSb (1065 °C). The rest of the synthesis steps are similar to those for GaSb shown in Fig. 10.10.

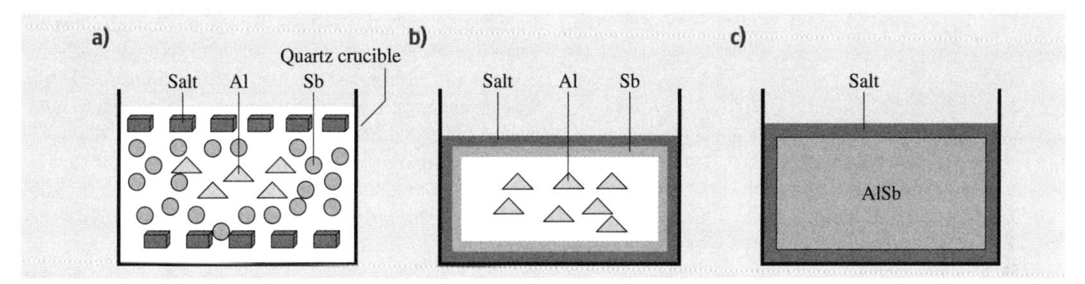

Fig. 10.11a–c Materials stacking scheme for AlSb synthesis in a quartz crucible

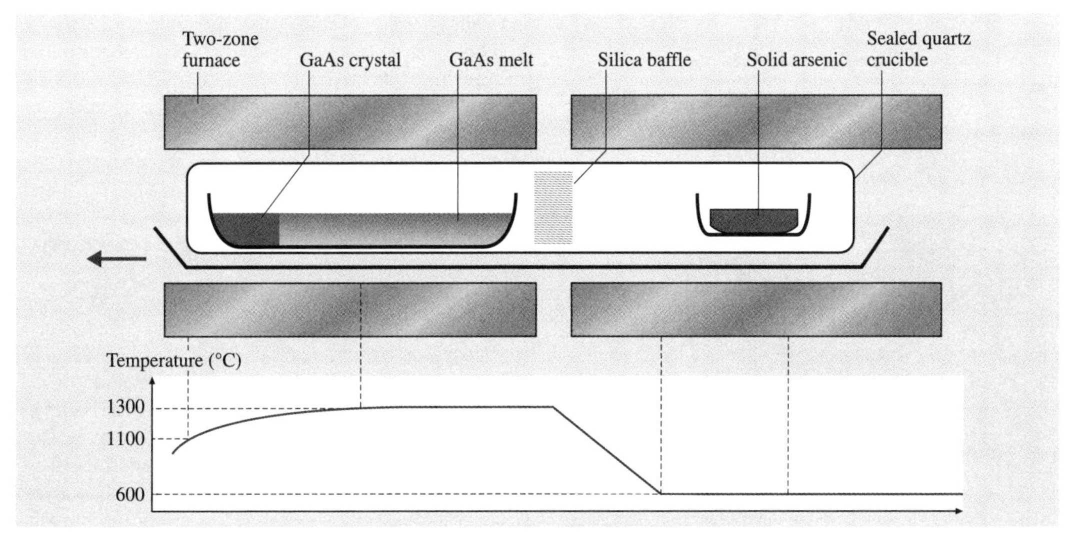

Fig. 10.12 GaAs synthesis inside sealed tube using a horizontal Bridgman configuration

Arsenide-Based Compounds

There are three general synthesis procedures for arsenic-based compounds. The configurations are shown in Figs. 10.12–10.14. Here we provide the specific examples of GaAs synthesis.

For GaAs, the synthesis is done under excess pressure of As. In the horizontal Bridgman method shown in Fig. 10.12, a sealed quartz tube containing 7 N pure gallium at one end separated and 6 N pure arsenic at the other end is used. The quartz tube is placed in a two-zone furnace. The arsenic is kept between 600–620 °C (which results in ≈ 1 atm of arsenic vapor pressure) in the lower-temperature zone. The quartz boat containing the gallium is kept in the higher-temperature zone of the furnace above the melting point of GaAs (1238 °C). The arsenic vapor reacts with gallium to form the GaAs melt. A quartz baffle with narrow constrictions (silica baffle) between the two zones controls the transport of the arsenic. After the melt is homogenized for 12–24 h, the melt is directionally solidified by translating the crucible in a temperature gradient (10–20 °C/cm) at a rate of 1–3 mm/h. After the entire melt solidifies, the ingot is cooled slowly to room temperature over a period of 48 h.

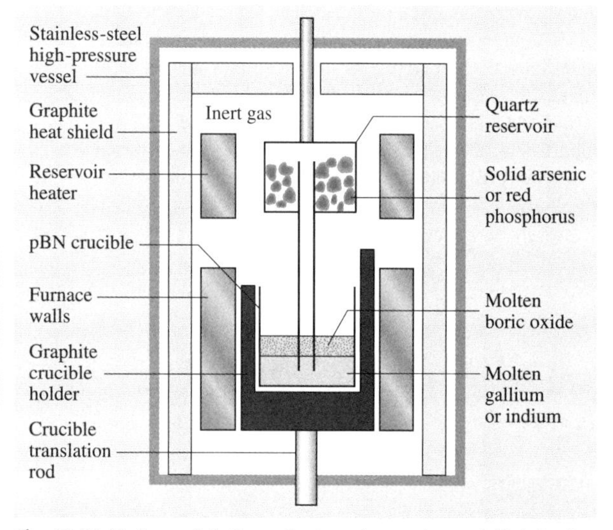

Fig. 10.13 GaAs and InP synthesis using a As and P injection method inside a high-pressure chamber

Another method for synthesizing GaAs is the injection method (Fig. 10.13). In this method, the group V volatile element such as arsenic is contained in a quartz ampoule and heated to form vapors. The vapors are transported into the B_2O_3-covered molten gallium to form the compound (GaAs). The melt is maintained at a temperature higher than the melting point of GaAs. The pressure on the top of the melt must be maintained in the range 1–2 MPa (≈ 10–20 atm). After the synthesis, the charge is directionally solidified and slowly cooled to room temperature. Since B_2O_3 reacts with quartz and crucible ruptures during cooling, pyrolytic boron nitride (pBN) crucibles are necessary for this process.

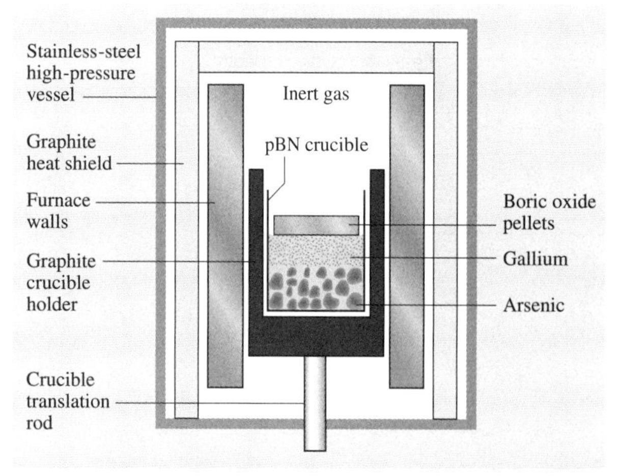

Fig. 10.14 Direct synthesis of GaAs in an open crucible inside a high-pressure chamber

Direct synthesis of GaAs can be implemented by using the configuration shown in Fig. 10.14. The arsenic and gallium are packed together as shown in the figure with boric oxide as an encapsulant. The chamber in the pressure must be maintained to around 60 atm during the synthesis using nitrogen or argon. The synthesis of GaAs occurs around 700 °C by an exothermic reaction. After the compound is formed, the temperature is increased above the melting point of GaAs (1238 °C) and the pressure of the chamber can be reduced to 5–10 atm. The melt is then directionally solidified as in the previous cases.

Phosphide-Based Compounds

Phosphide-based materials such as InP and GaP have much higher vapor pressures than arsenide-based compounds. Hence specific modifications in configurations are necessary for the synthesis and growth of phosphide-based compounds. A stainless-steel-lined pressure vessel designed for 150–170 atm and continuous operation at 60–70 atm is necessary for InP growth. The direct synthesis method, which is the preferred syn-

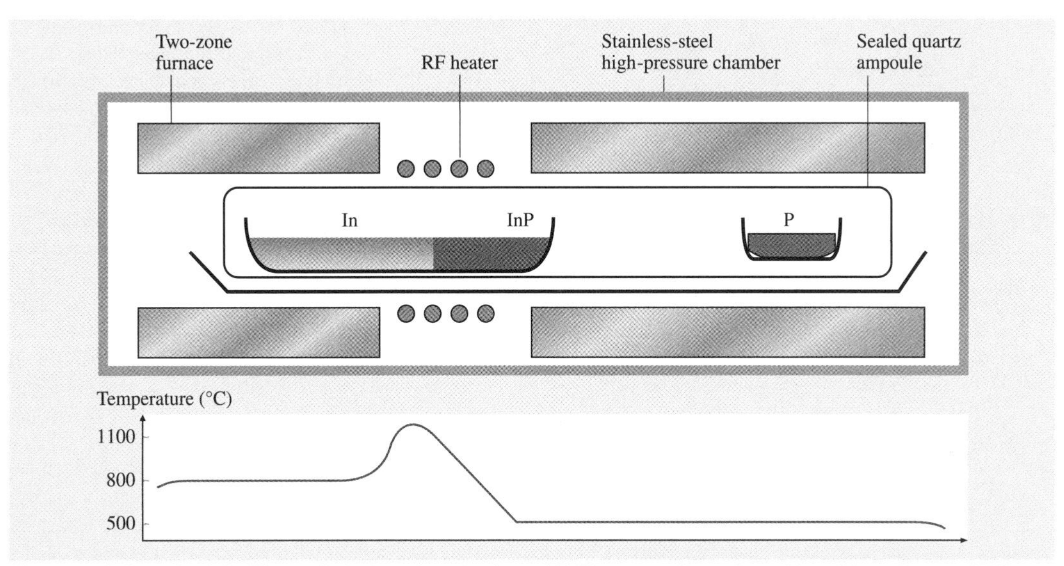

Fig. 10.15 InP synthesis inside a sealed tube using a horizontal Bridgman configuration with multizone heaters (RF heater refers to radio frequency heater coil)

thesis method for GaAs, is not possible for InP. Special handling procedures for phosphorus are necessary due to significant risk of ignition.

Two common InP synthesis techniques are the horizontal Bridgman (or gradient freezing) and the injection methods, as shown in Figs. 10.12 and 10.13. In the horizontal Bridgman process with a two-zone furnace (Fig. 10.12), indium is kept in a boat at one end and red phosphorus at the other end, and the silica tube sealed under 10^{-6} Torr. The sealed quartz crucible is kept inside a stainless-steel pressure vessel with preferably 45–50 atm of inert gas overpressure to avoid rupturing of the crucible. The red phosphorus is slowly heated to around 550 °C to maintain a phosphorus pressure of 27 atm. The indium boat is maintained at 1075–1080 °C. The phosphorus transports to the indium boat to form InP. After the homogenization of the melt, which could take 16–24 h, the melt is directionally solidified at a cooling rate of 0.5 °C/h until the entire melt solidifies and the solid reaches around 1000 °C. The crystal is then cooled to room temperature over a period of 30–40 h. Another common horizontal Bridgman-type configuration used for InP synthesis is shown in Fig. 10.15. The three-zone temperature profile resembles a zone-refining process and makes it easy to control the phosphorus vapor pressure on the melt and to avoid the supercooling effects seen in InP.

For the injection method shown in Fig. 10.13, the red phosphorus reservoir is maintained around 550 °C and the indium is heated to 1080 °C for the compound formation. The inert gas pressure in the chamber is maintained around 45–50 atm. The phosphorus vapors diffuse through the boric oxide to form InP. The melt is then homogenized and directionally solidified.

10.5.3 Single-Crystal Growth Processes

During the growth of binary compounds from melt, the macroscopic composition of the melt and the growing crystal remains constant (segregation coefficient equal to unity). Hence there is no need for solute replenishment processes (unlike in ternary crystal growth). Single-crystal growth is carried out by using a presynthesized polycrystalline charge and a seed. In addition, melt encapsulation is necessary for two reasons: (a) to avoid the ecsape of volatile species from the melt, and (b) to avoid contact between semiconductor melt and crucible walls during growth so that secondary nuclei, which lead to polycrystalline grains, do not originate. Details of crystal growth configurations and conditions can be found in the literature [10.1,16,40–44,66,71,73, and references therein]. In this section, we present the schematics of the growth setups and summarize the growth conditions.

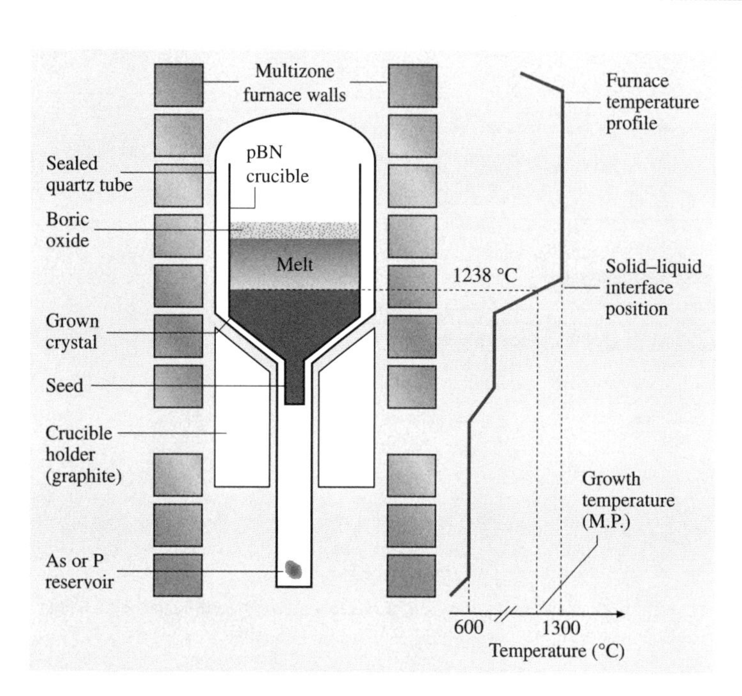

Fig. 10.16 Single-crystal growth of GaAs or InP inside a sealed tube using the vertical gradient freeze (VGF) method

Two typical VGF growth setups used for GaAs and InP are shown schematically in Figs. 10.16 and 10.17. In the sealed tube configuration (Fig. 10.16), a reservoir of a small amount of As or P is kept at a lower temperature (the same as that used during synthesis) to maintain an adequate vapor pressure over the melt surface. In the open tube configuration (Fig. 10.17), the entire melt is pressured with an inert gas at a level of 10–20 atm for GaAs and 45–50 atm for InP growth. pBN crucibles along with ultralow-water-content boric oxide (B_2O_3) encapsulation is used. The boric oxide melts at 450 °C and has a low viscosity at the growth temperatures of GaAs and InP. The typical furnace temperature gradient is in the range of 5–10 °C/cm and the crystal growth rate is in the range of 0.5–2 mm/h. Growth of low-dislocation-density large-diameter (typically 100–150 mm) GaAs and InP has been demonstrated using the VGF technique. Semi-insulating GaAs crystals with a diameter of 200 mm are being produced today by VGF.

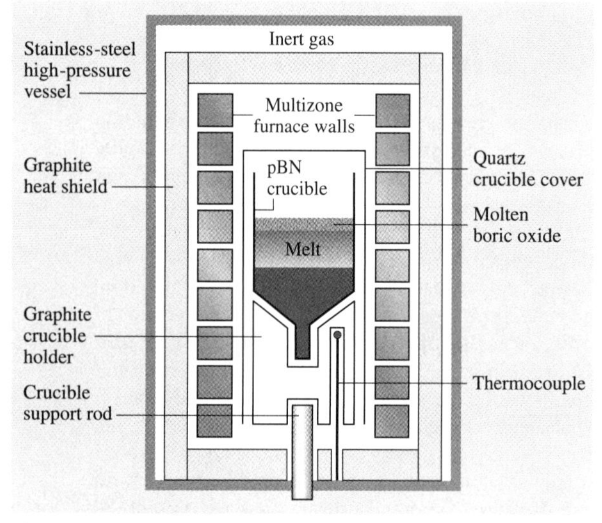

Fig. 10.17 Single-crystal growth of GaAs or InP in an open crucible using the vertical gradient freeze (VGF) method

For the growth of GaSb and InSb, B_2O_3 encapsulation is not suitable since it is very viscous at low temperatures [10.73]. Instead of B_2O_3, alkali halide eutectic salts such as LiCl/KCl (58 : 42 mol %) and NaCl/KCl (50 : 50 mol %) with low melting temperatures (between 350 and 600 °C) and significantly lower viscosity are used. These alkali halide salts do not react with the antimonide-based compounds and have no effect on the electrical and optical properties of the grown crystals [10.72]. Silica (quartz) crucibles are commonly used for GaSb and InSb growth. It should be noted that melt encapsulation for GaSb or InSb growth is not an absolute necessity. The vapor pressure of Sb is significantly lower compared with that of P or As. Nevertheless, alkali halide salt encapsulation does help in reducing the probability of polycrystalline growth from the crucible walls. Typically, thick oxide layers are present on GaSb and InSb polycrystalline surfaces (even after chemical etching). When the polycrystalline charge is melted, an oxide scum develops on the surface of the liquid. The scum in contact with the crucible walls enhances the sticking of the melt to the crucible and hence polycrystalline grains are formed. The salts help in dissolving the oxide scum and lead to a cleaner melt surface and avoid sticking of the melt with the crucible walls. The growth setup for GaSb and InSb is much simpler since there is no need for a high-pressure vessel. Typical vertical Bridgman or VGF can be used for single-crystal growth of low-dislocation-density GaSb and InSb. The typical furnace temperature gradient used for 50 mm diameter crystals is in the range of 5–15 °C/cm and the crystal growth rate is in the range of 1–3 mm/h.

10.5.4 Cleaning Procedures for Growth Chamber, Crucible, and Charge

Due to the high-purity nature of the semiconductor materials, the growth chambers, crucibles, and charge require special cleaning procedures. The growth chambers needs to be cleaned prior to every experiment. Stainless-steel chamber surfaces must be scrubbed and cleaned to remove deposits of elemental and compounded species such as arsenic, antimony, GaAs, etc. from previous experiments. Proper safety masks, handling, and disposal procedures must be followed. After scrubbing, the chamber surfaces could be rinsed and wiped with an organic solvent such as methanol. If silica tubing is used in the growth chamber, they need to be chemically cleaned following the same procedures as for the crucibles as described below. All graphite parts such as the crucible holder must be cleaned with organic solvents (successively in xylene, acetone, and methanol) and then baked at high temperature ($\approx$ 1200 °C) under high vacuum (less than 1 mTorr pressure) or under a flowing argon/hydrogen gas mixture.

Two most common crucible materials for III–V semiconductors are high-purity silica (quartz) and pyrolytic boron nitride (pBN). High-purity, low-porosity graphite is also a suitable material. However due to difficulty in handling graphite, it is not used for bulk crystal growth. Prior to crystal growth, the silica and pBN crucibles need to be degreased in warm xylene followed by acetone and methanol. Depending on the residue material left behind from the previous experiment, such as GaAs or GaSb, the crucibles are chemically treated in a suitable chemical etchant solution to remove the residues. Typical etchants consist of acids such as nitric acid (HNO_3), hydrofluoric acid (HF) and glacial acetic acid (CH_3COOH) mixed in various volume ratios [10.74]. The crucibles are finally rinsed in high-purity deionized (DI) water and methanol, and dried with nitrogen gas.

For polycrystalline charge synthesis, the elemental constituents are accurately weighed according to the melt composition required. Elements with purity level of 6 N (99.9999%) or 7 N (99.99999%) are commercially available and used these days. The elements are available in various forms; for example, gallium comes in squeeze bottles or in the form of solidified rods; indium comes in tear drops or bar forms; while antimony, phosphorus, and arsenic come in small pellets. If the elements are opened from packed containers, they can be used the experiments without any cleaning. However, oftentimes the elemental charge possesses a thick oxide layer on the surface. Hence it is necessary to clean the charge before synthesizing compounds. Common cleaning procedures involve acid rinsing the oxide layer from the surface of the elements followed by cleaning in high-purity water, methanol rinsing, and nitrogen drying. Typical etchants for elemental cleaning includes hydrochloric (HCl), nitric (HNO_3), and hydrofluoric (HF) acids [10.66]. When polycrystalline compounded materials such as GaAs, GaSb, InSb, etc. are used for single-crystal growth experiments, they are etched in a variety of acids that act as etchants for the specific compound [10.74]. One such etchant that is widely used is a mixture of HNO_3, CH_3COOH, and HF in the volume ratio 5 : 3 : 3, respectively. The polycrystalline charges are finally washed in high-purity water, rinsed with methanol, and dried by blowing nitrogen gas. Since there are numerous options for chemicals used in cleaning elemental and compound charges, no specific recommendations are made here. Every crystal grower adopts specific cleaning procedures based on the ease of handling certain chemicals, and experience.

10.6 Phase Equilibria for Ternary Compounds

There are three types of phase diagrams that can be used for the growth of ternary compounds, namely pseudobinary, ternary, and quaternary phase diagrams [10.17,29,30,66,75–77]. In this section, we will discuss the methods to use these phase diagram for determining melt or solution composition and the growth temperature for growing a crystal of specific composition.

10.6.1 Pseudobinary Phase Diagram

The most common phase diagrams used for melt growth of ternary crystals are the pseudobinary plots, as shown in Fig. 10.18 (depicting the GaInSb system [10.77]). These are known as pseudobinary diagrams because a ternary crystal such as $Ga_{0.6}In_{0.4}Sb$ can be thought of as a mixture of two binary compounds, namely, 60 mol % of GaSb and 40 mol % of InSb. In this diagram there are two curves: liquidus and solidus. The melting points of the two constituent binaries are shown on the y-axis at the two ends points. The melting points for pure InSb and pure GaSb are 527 and 712 °C, respectively. Any point on the liquidus curve represents the temperature above which the ternary of a specific composition is completely liquid. For example, $Ga_{0.6}In_{0.4}Sb$ or a mixture of 60 mol % GaSb and 40 mol % InSb will be completely liquid above 660 °C. When the individual binaries are mixed together, the InSb first melts at 525 °C and starts dissolving GaSb. As the temperature is increased, the amount of GaSb dissolved increases, resulting in a melt composition with increasing GaSb mole fraction or decreasing InSb mole fraction. The ternary compounds can also be prepared by mixing individual elements such as Ga, In, Sb, etc. For example, $Ga_{0.6}In_{0.4}Sb$ can be prepared by mixing 60 mol % Ga, 40 mol % In, and 100 mol % Sb. The liquidus and solidus temperatures are independent of the preparation methodologies.

The solidus represents the temperature below which a ternary of specific composition is completely solid. For example, $Ga_{0.6}In_{0.4}Sb$ will be completely solid below 570 °C. Between the liquidus temperature and the solidus temperature for a specific ternary composition, there exist a two-phase region, as shown in Fig. 10.18, where a portion of the material is in liquid state and the rest is in solid state. In pseudobinary phase diagrams, there exist a single growth temperature and a single melt composition for any solid composition. For example, for growing $Ga_{0.8}In_{0.2}Sb$ crystals, the melt composition must be $Ga_{0.30}In_{0.70}Sb$ and the growth temperature is 605 °C. A horizontal line connecting a point in the liquidus and solidus is known as a tie-line. This is different than the ternary and quaternary phase diagrams (to be discussed next), where a specific solid composition can be obtained at numerous liquid compositions and growth temperatures.

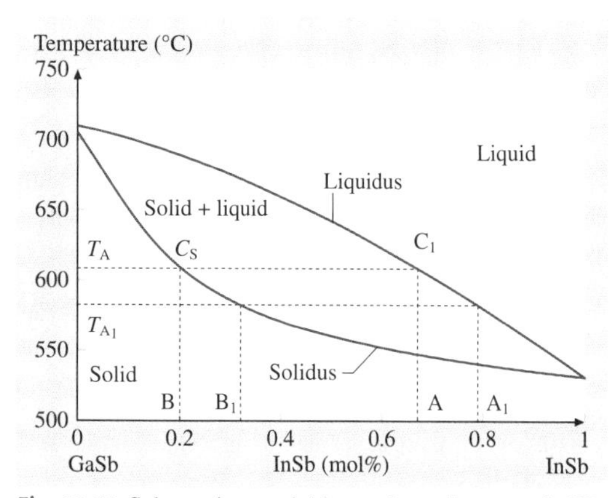

Fig. 10.18 Schematic pseudobinary phase diagram of a III–V ternary compound (depicting GaInSb)

10.6.2 Ternary Phase Diagram

Using the pseudobinary phase diagrams discussed above, the growth temperature of any ternary crystal is restricted to lie the melting points of the two constituent binaries. Growth of crystals from solution at lower temperatures has many advantages such as reduced native defects and lower dislocation densities that result in superior materials properties. Ternary phase diagrams can be used to grow crystals at low temperatures. Figure 10.19 shows a typical ternary phase diagram for the GaInSb system. This diagram is a schematic drawn to demonstrate the features of the ternary phase diagrams. The three vertices on the equilateral triangle represent the three constituent elements comprising the ternary compound. For GaInSb growth, the constituents are Ga, In, and Sb. The melting point of Ga, In, and Sb are 29, 156, and 630 °C, respectively. The low melting point of one or two constituent elements helps to decrease the growth temperature significantly. Each point within the triangular area represents a specific solution (liquid) composition (Ga : In : Sb in mol %). The lines drawn parallel to each face are used to determine the solution composition. For example, as one traverses the space from the Sb vertex of the triangle downwards, the Sb content changes from 100 mol % to 0 mol % at the opposite face (which connects In and Ga). The same methodology can be applied for calculating the In and Ga compositions. The compositions of the liquid at the four circular dots in Fig. 10.19 have been calculated by drawing three straight lines parallel to each triangle axis, intersecting at the circular dot. The compositions for In, Ga, and Sb are then estimated by measuring the distance between the point of interest and each vertex of the triangle.

Referring to Fig. 10.19, the liquidus are shown by solid lines and the solidus are shown by dashed lines. Each liquidus curve corresponds to a specific solution (liquid) temperature. For example, the two liquidus shown correspond to 575 and 625 °C. Each solidus curve corresponds to a specific ternary solid composition. For example, the two solidus curves shown correspond to gallium mole fractions of 0.7 and 0.8 in the crystal. For the growth of crystal of any desired composition from a solution, one needs to find a specific solution composition and the growth temperature. This is done by finding a point in the ternary phase plane where the solidus and liquidus curves cross each other such as the points (In,Ga,Sb: 50,15,35 and 35,35,32) in Fig. 10.19. Hence, to grow a crystal of $Ga_{0.7}In_{0.3}Sb$ at 575 °C, the solution composition should be $(In : Ga : Sb = 50 : 15 : 35$ by mol %). It should be pointed out that there is more than one option for the growth temperature and liquid composition for the same crystal composition. It is also interesting to note that, even though the liquid compositions are completely nonstoichiometric ($Ga_xIn_ySb_z$; $0 < x, y, z < 1$), the grown solids are stoichiometric, of the form $Ga_{1-x}In_xSb$ $(0 < x < 1)$.

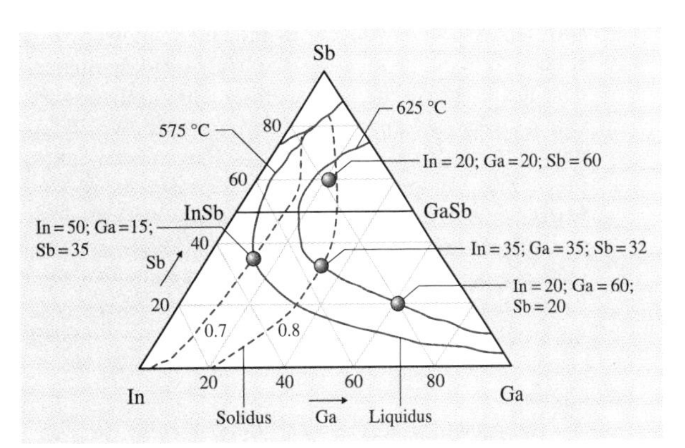

Fig. 10.19 Schematic ternary phase diagram of a III–V ternary compound (depicting GaInSb)

10.6.3 Quaternary Phase Diagram

Recently, it has been demonstrated that quaternary melts can be used to grow ternary crystals [10.29, 30, 78]. The growth of $Ga_{1-x}In_xAs$ from stoichiometric $Ga_{1-x}In_xAs_ySb_{1-y}$ melt has been clearly demonstrated [10.29, 78]. Other III–V ternary systems such as GaInP, GaAsSb, and AlGaP, can also be grown from GaInPSb, GaInAsSb, and AlGaPSb melts, respectively [10.30]. The use of a quaternary melt enables the growth of ternary crystals at lower temperature compared with that possible using pseudobinary phase diagrams. Figure 10.20 shows a schematic phase diagram of the Ga-In-As-Sb system at 950 °C. The four constituent binaries for Ga-In-As-Sb are GaAs, InAs, InSb, and GaSb. The quaternary phase diagram shows a liquidus curve and a solidus curve connected with tie-lines across the two-phase regions. The melt composition for growing a specific solid composition can be obtained by selecting a specific tie-line in the phase diagram. Like in the case of the ternary phase diagrams, depending on the growth temperature, the liquid composition for growing a particular solid composition varies. For example, the liquid composition of Ga-In-As-Sb to grow $Ga_{0.2}In_{0.8}As$ at 950 °C and 1000 °C are different. Using MTDATA, the National Physical Laboratory (NPL, UK) database for metallurgical thermochemistry, phase diagrams for the $In_xGa_{1-x}As_ySb_{1-y}$ system have been calculated for a wide temperature range [10.29, 30]. For temperatures below the melting point of InAs (942 °C),

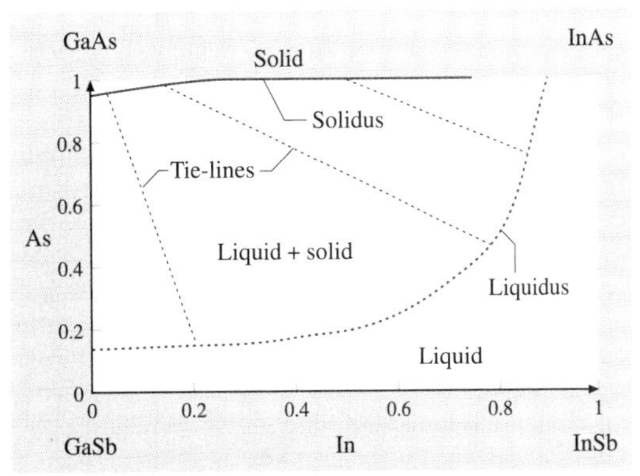

Fig. 10.20 Schematic quaternary phase diagram for III–V ternary compound growth (depicting GaInAs) from quaternary melt

$Ga_{1-x}In_xAs$ (with residual Sb) and $Ga_{1-x}In_xAs_ySb_{1-y}$ solids can be grown from the Ga-In-As-Sb quaternary melts. Above the InAs melting point, only ternary $Ga_{1-x}In_xAs$ is formed from the $Ga_{1-x}In_xAs_ySb_{1-y}$ melt. Figure 10.20 clearly shows that a wide range of GaInAs compositions could be grown using Ga-In-As-Sb melt at temperatures as low as 950 °C. Due to the fact that the first-to-freeze solid is ternary, the melt composition changes to a nonstoichiometric Ga-In-As-Sb from a stoichiometric starting melt (excess Sb in the solution). Hence replenishment of depleted species is necessary to keep the melt stoichiometric at all times.

The growth of a ternary from quaternary melt occurs at higher temperatures compared with growth from ternary solutions. Nevertheless, there are several technical advantages in using the quaternary melt. Since the quaternary melts can be prepared from presynthesized binary compounds, it enables the usage of liquid encapsulations during the growth. The metal-rich ternary solutions (such as Ga, In, Al) tend to stick to the crucible, making them difficult to mix and posing problems during Bridgman growth. Also the solubility of compounds in ternary solutions is significantly lower than in quaternary melts. This impacts the crystal growth rates. Another advantage of using the quaternary melts is seeding during a typical Bridgman growth. Due to the numerous technical benefits of the quaternary melts for ternary growth at low temperatures, this method will receive significant interest in the future, especially for crystals with high growth temperature (such as phosphides and arsenides).

10.7 Alloy Segregation in Ternary Semiconductors

All three types of phase diagrams discussed above for the growth of ternary crystals exhibit a common phenomenon. The solid composition precipitating out of the melt or solution is different from the liquid composition. This leads to alloy segregation, as discussed here. For the sake of simplicity, we will only discuss the pseudobinary phase diagrams since in this case one only needs to track the changes in concentration of one of the binary constituents. For example, during the growth of $Ga_{1-x}In_xSb$, either the change in InSb concentration or GaSb concentration needs to be analyzed. The InSb and GaSb concentrations are complementary to each other. On the other hand, while using ternary or quaternary phase diagrams, the individual elemental concentrations such as Ga, In, As, Sb, etc. need to be analyzed since the concentration of each elemental species varies independently during the growth.

As shown in Fig. 10.18, pseudobinary phase diagrams exhibit a separation between the solidus and liquidus curves. Starting from a melt of alloy composition (A), when the liquid is cooled below the liquidus temperature T_A, the composition of the solid precipitating out of the liquid will have a composition corresponding to the point on the solidus (B) at the temperature T_A. Since the solid composition (B) is different than the liquid composition (A), the melt composition will change as soon as the solid starts precipitating out of the liquid. During crystal growth, the rejected species accumulate in front of the interface and spread into the liquid phase by diffusion and mixing induced by convection. From the phase diagram, it is clear that the rejected species is the lower-melting binary constituent. For example, during the growth of $Ga_{1-x}In_xSb$, the precipitating solid is rich in GaSb compared with the liquid. Hence the excess InSb is rejected into the melt at the solidification interface and the melt composition changes to a point indicated by A_1 in Fig. 10.18. The next precipitation event occurs at a lower temperature shown by T_{A_1} with the precipitating solid of composition B_1. This continues and hence, during normal directional solidification process from melt as in Bridgman or Czochralski growth, the melt composition continuously changes with time, as in turn does the composition of the solidifying crystal.

The changing alloy composition in the solid can be theoretically analyzed using the fundamental equilibrium distribution (partition) coefficient or chemical segregation coefficient, k_0. This can be derived from the equilibrium phase diagram by the ratio of the solidus (C_s) and liquidus concentration (C_l) of a species at a given temperature (Fig. 10.18) as follows

$$k_0 = \frac{C_s}{C_l} . \tag{10.3}$$

In general, segregation depends on the growth rate and hence is described by an effective distribution coefficient, k_{eff}. If the melt is mixed continuously during a directional solidification crystal growth experiment, the alloy concentration in the solid is given by the well-known equation [10.79–82]

$$C_s = k_{eff} C_0 (1-g)^{k_{eff}-1} , \tag{10.4}$$

where g is the fraction of melt solidified and C_0 the initial concentration of the melt (at the beginning of the growth). In this equation k_{eff} does not vary with alloy concentration, an assumption which is invalid for ternary crystal growth, as discussed below.

In a typical crystal growth experiment, the rejected species at the solid–liquid interface takes a finite time to diffuse back into the melt, giving rise to a completely homogeneous melt. The mixing rate of the rejected species with the rest of the melt can be enhanced by increasing the temperature gradient across the melt–solid interface, decreasing the cooling rate of the melt (decreasing the crystal growth rate), and by using forced convection. There are limitations with each of these processes. Large temperature gradients lead to temperature fluctuations, almost inevitably producing striations. Slow cooling increases the duration for crystal growth, which is undesirable for a variety of reasons (increase in cost, reduced equipment lifetime, unexpected long-term instability in furnace power, enhanced degradation of grown crystals due to volatilization of elements from the surface of the crystals and melt, etc.). Forced mixing has its own limitations such as complicated fluid patterns and associated temperature and solutal fluctuations. Thus, under normal conditions, the growth process is dominated by a stagnant (or diffusion) boundary layer of thickness δ, beyond which stirring in the bulk liquid ensures a uniform bulk concentration. Within the diffusion boundary layer, the composition of the rejected species changes continuously starting from the crystal surface to the end of the boundary layer [10.81]. Burton, Prim, and Slichter (BPS) derived the following equation to correlate the boundary-layer thickness with the effective segregation coefficient [10.81]

$$k_{eff} = \frac{k_0}{k_0 + (1-k_0)\exp(-\Delta)} ,$$
$$\Delta = R\delta/D , \qquad (10.5)$$

where k_0 is the equilibrium segregation coefficient as estimated from the equilibrium phase diagram, D is the diffusion coefficient of the rejected species in the melt, R is the crystal growth rate, and δ is the thickness of the boundary layer. For crystal grown with forced convection, the boundary-layer thickness δ is given by

$$\delta = 1.6 D^{\frac{1}{3}} \nu^{\frac{1}{6}} \Omega^{-\frac{1}{2}} , \qquad (10.6)$$

where ν is the kinematic viscosity of the melt, Ω is the rotational rate of the crystal or seed (in the case of Czochralski growth), and D is the diffusion coefficient.

As can be seen from the above equation, when Δ approaches zero, k_{eff} equals k_0. In the above equation, the diffusion coefficient D is a fundamental material property and dependent on the alloy composition as well as the temperature. For example, the diffusion coefficient of InSb in GaInSb melt at 550 °C is different from that at 600 °C. Similarly, the diffusion coefficient of InSb in a Ga melt at 550 °C is different from that in GaInSb melt at the same temperature. The diffusion boundary-layer thickness can be altered by the hydrodynamic mixing in the melt as a result of natural and forced convection as well as diffusion. In the analysis shown above, k_0 is assumed to be constant with time. However during ternary crystal growth, the liquid concentration moves along the liquidus. For each corresponding liquidus and solidus points, the segregation coefficient is different. Hence the k_0 needs to be varied for analyzing the solid concentration along the crystal growth direction. Figure 10.21 shows the theoretically calculated gallium mole fraction along the length of $Ga_{1-x}In_xSb$ crystals for different starting melt compositions, assuming completely homogeneous melt at all times. In these plots, the difference in segregation coefficient with change in melt composition was taken into account. The segregation coefficient as a function of alloy composition was extracted from the phase diagram of GaInSb, as depicted in Fig. 10.18. As clearly seen in Fig. 10.21, the solid composition changes continuously in the crystal due to changes in the melt composition. The rate at which the alloy composition changes along the crystal is dependent on the extent of separation between the liquidus and solidus curves, which is a material-dependent

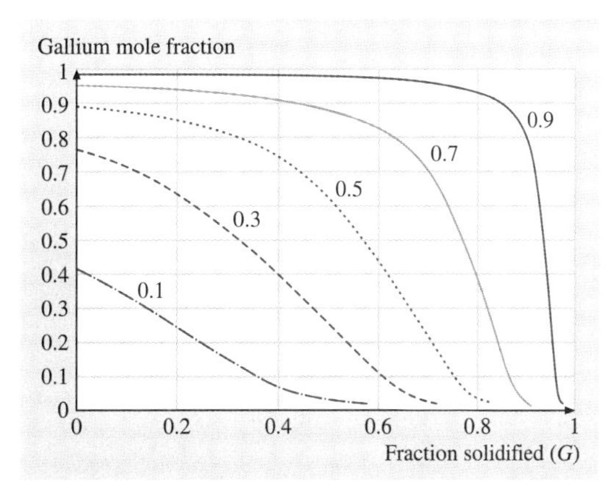

Fig. 10.21 Theoretical gallium concentration along the growth axis for $Ga_xIn_{1-x}Sb$ crystals grown from different starting melt compositions (under normal freezing conditions). The fractions next to each curve represent the gallium mole fraction in the starting melt

property. A narrow gap between the two curves is desirable for growing homogeneous crystals and avoiding defects such as cracks in the crystals, as discussed in the next section.

The other extreme for alloy segregation behavior is observed under steady-state diffusion in a convection-free melt. This is represented by the well-known equation for diffusion-controlled segregation as [10.80, 82]

$$\frac{C_s(x)}{C_0} = k_0 + (1-k_0)\left[1 - \exp\left(-\frac{k_0 R x}{D}\right)\right] , \quad (10.7)$$

where $C_s(x)$ is the alloy concentration in the solid at a distance x measured from the beginning of the growth. In this scenario, the alloy composition in the diffusion boundary layer increases with time since there is no convective flow in the melt to extract the species from the diffusion boundary layer. This leads to a rapid increase in the alloy composition in the solid (initial transient) followed by a uniform region in the crystal. Figure 10.22 shows a typical diffusion-controlled alloy composition profile in a solid (10.7). For the sake of comparison, a normal freezing curve (10.4) with the same segregation coefficient has been included. While plotting (10.7), the parameter x has been normalized to the solidified fraction g. The uniform region in the diffusion-controlled profile appears very attractive for growing uniform ternary crystals. However, the technical problem lies in the fact that it is very difficult to achieve pure diffusion-controlled conditions. This would require complete elimination of convection in the melt. Melt convections can be reduced by using magnetic fields [10.83–85], submerged baffles in the melt [10.86, 87], and under microgravity conditions [10.88–93]. However, these approaches have not been used in large-scale ternary crystal growth. There are other technical issues such as composition and length of the diffusion boundary layer necessary for growing ternary crystals with alloy compositions that are far from the binary compositions. Hence for the rest of this chapter, we will focus on the alloy composition changes that are typically seen in experiments with normal freezing conditions.

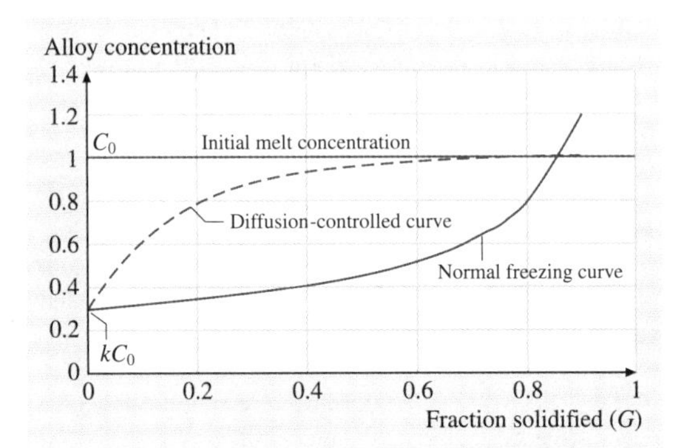

Fig. 10.22 Theoretical alloy composition profiles along the growth axis for crystals grown with complete melt mixing (normal freezing) and in the absence of melt convection (diffusion-controlled growth)

10.8 Crack Formation in Ternary Crystals

10.8.1 Phenomena of Crack Formation

One of the consequences of the alloy segregation is the cracking of the crystals during growth. If the growth rates are not low enough, constitutional supercooling occurs close to the liquid–solid interface, resulting in composition fluctuation and sudden transition from single to polycrystals. The local compositional inhomogeneity in the solid along with the wide difference in the lattice constants and the thermal expansion coefficients of constituent binary compounds introduces considerable strain, and invariably leads to cracking of the crystals. For the GaInSb system, the lattice parameter varies from 6.096 Å (for GaSb) to 6.479 Å (for InSb) and the thermal expansion coefficient from 7.75×10^{-6} to $5.37\times10^{-6}\,°C^{-1}$ [10.1]. Figure 10.23

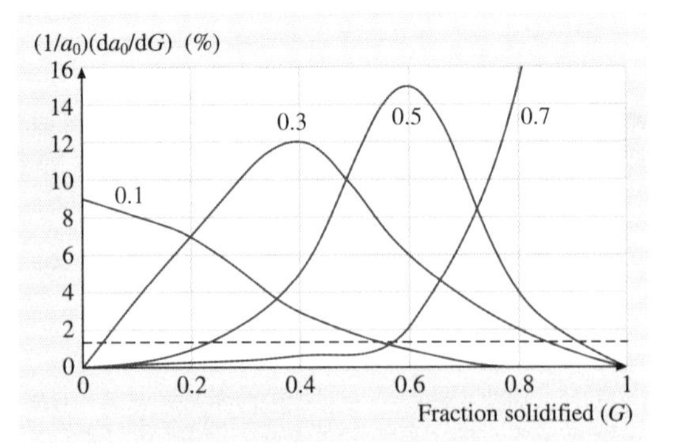

Fig. 10.23 Theoretical misfit strain gradient versus fraction of solidified melt calculated from the theoretical concentration curves shown in Fig. 10.21 for $Ga_xIn_{1-x}Sb$ crystals

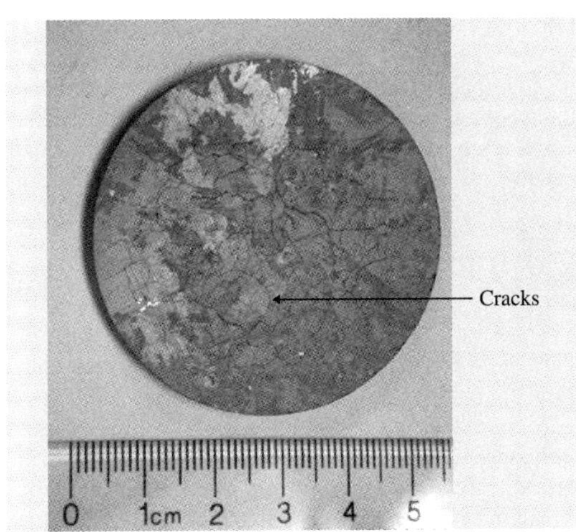

Fig. 10.24 Typical microcracks seen in ternary III–V crystals. The composition of the wafer shown here is $Ga_{0.7}In_{0.3}Sb$

shows the theoretical misfit strain gradient curves calculated using the theoretical alloy composition profiles shown in Fig. 10.21 and the lattice parameter a_0 for each composition using Vegard's law (10.1). It has been empirically observed that crystals inevitably crack if the misfit strain gradient exceeds 1–2% [10.17]. It is obvious that an alloy system with larger separation between the solidus and liquidus line has higher strain gradient than those with smaller separation. Moreover, cracking may occur even below 1% misfit strain gradient if there is a thermal strain caused by the large temperature gradient in the solid or during rapid cooling for the crystals. Figure 10.24 shows microcracks in a wafer sliced from a $Ga_{0.7}In_{0.3}Sb$ polycrystal grown using the vertical Bridgman method. The growth conditions used for this crystal were not optimized for avoiding cracks (as discussed below).

Cracks in crystals make the wafers completely unusable for any basic research or device applications. Hence eliminating cracks in the crystals by optimizing growth conditions is the first task for any ternary crystal growth. The cracks can be somewhat eliminated by zone leveling, slow cooling or prolonged annealing of the solidified ingot [10.17]. However, the concentration gradient does not smoothen completely by thermal treatment due to the low solid-state diffusion coefficients of III–V compounds. Hence microscopic cracks and stresses remain. Cracks can be avoided if and only if smooth change in alloy composition is ensured in the

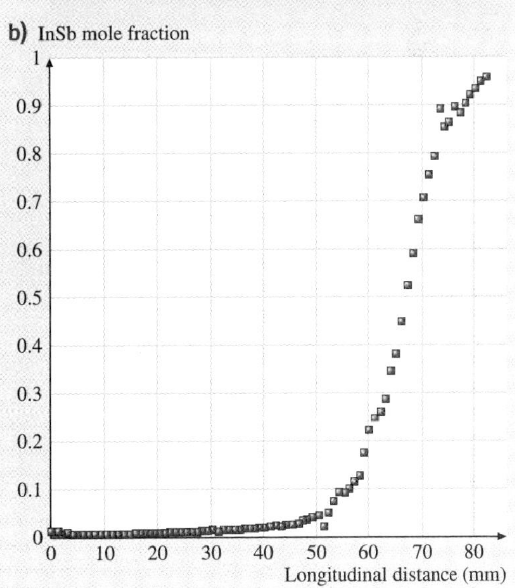

Fig. 10.25 (a) Cross section of a GaInSb crystal showing strain-gradient-related cracks and (b) the longitudinal composition profile measured along the center of the crystal

crystals during growth. Referring to Fig. 10.23, strain-related cracks originate in the crystal where the strain gradient exceeds 1–2%, as shown by the horizontal dashed line. For a crystal grown with starting gallium mole fraction of 0.7 in the melt, cracks should appear when the fraction solidified (G) is close to 60%. This has been experimentally confirmed during growth of GaInSb. Figure 10.25a shows a cross section of a crystal with a visible crack around 6 cm from the bottom of the crystal. This coincides with the longitudinal composition profile measured in the same crystal as shown in Fig. 10.25b. The region where the composition starts changing rapidly (between 60 and 70 mm) is exactly where the crack in the crystal appears. It is interesting to observe that the cracks propagate across multiple grains (seen in both Figs. 10.24 and 10.25a) and the crack network cannot be correlated with the polycrystalline grain structure. Strain-related cracks seem to occur after the solidification front (melt–solid interface) had crossed the respective position. This clearly indicates that these cracks are not due to constitutional supercooling during growth, but are rather due to strain in the crystal. Rapid change in composition such as in Fig. 10.25b can also lead to constitutional supercooling and interface breakdown with sudden change in the grain structure and the appearance of cracks. However these cracks are easily distinguishable from strain-related cracks. Since there are two origins of cracks, different crystal growth conditions are necessary to eliminate them.

10.8.2 Elimination of Cracks

Misfit strain-related cracks are related to the rate at which the alloy composition and more importantly the lattice parameter and other physical properties such as thermal expansion coefficient change along the length of the crystal. Hence these cracks can be eliminated by ensuring gradual changes in alloy composition that will result in a strain gradient lower than that which initiates cracking. This is especially important during the seed formation process, as described in the next section. Since the materials properties for binary compounds vary significantly, the density of cracks varies for each ternary compound. Theoretical analysis of strain gradient as in Fig. 10.23 coupled with experimental trials is necessary to determine the compositional gradient in the crystal that will not lead to cracking.

Cracks originating from the constitutional supercooling process can be eliminated by avoiding supercooling during growth. This requires thinning the diffusion boundary layer at the melt–solid interface. Supersaturation in the boundary layer results in spurious nucleation and the formation of polycrystals [10.94,95]. A boundary layer with a diffusion-controlled (exponential) concentration distribution ahead of the growth interface causes constitutional instability. A region of constitutional supercooling emerges if the equilibrium liquidus temperature gradient exceeds the actual temperature gradient in the melt at the solid–liquid interface [10.82]. The composition of the diffusion boundary layer (enriched with the lower-melting-point binary compound such as InSb during the growth of GaInSb) depends on the melt volume, the growth rate, and the degree of melt stirring. Theoretical analysis [10.80] has shown that, for preventing constitutional supercooling, the ratio of the temperature gradient in the melt near the growth interface (G) to the growth rate (R) should exceed a critical value given by

$$\frac{G}{R} \geq \frac{mC_0(1-k_0)}{k_0 D} , \tag{10.8}$$

where m is the slope of the liquidus in the phase diagram.

Since avoiding constitutional supercooling is absolutely necessary during ternary growth, the role of melt stirring is crucial. Melt stirring can also enhance the growth rates by thinning the diffusion boundary layer at the growth interface. Melt stirring and homogenization can be achieved by natural convection, diffusion, and forced convection. Mass transport by natural convection takes place in most growth experiments due to the thermal convection caused by the temperature differences as well as the solutal convection caused by the density difference arising from the variation in solute concentration [10.96]. The natural convection and diffusion phenomena are weak. As a result of this, growth rates that are necessary for crack-free crystals are very low. Figure 10.26 shows the experimental crystal growth rates that were found to result in crack-free GaInSb crystals of different composition [10.18]. It is clear from the figure that, by using forced convection in the melt,

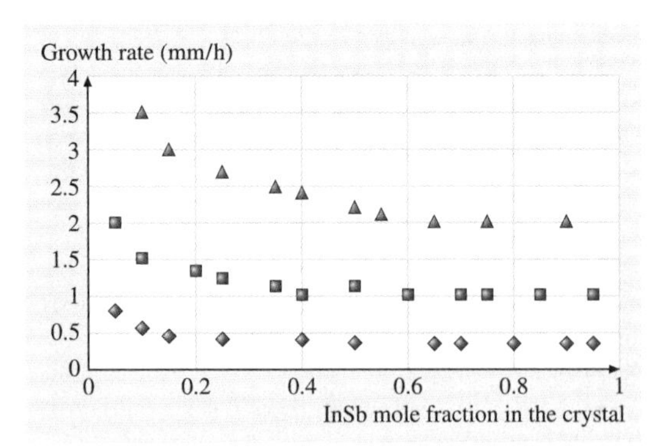

Fig. 10.26 Growth rate necessary for crack-free GaInSb crystals as a function of InSb mole fraction in the solid: crystals grown using temperature gradient of (◆) 10 °C/cm and (□) 20 °C/cm without forced convection; (▲) crystals grown using 10 °C/cm temperature gradient with forced convection in the melt

the growth rates can be increased. For melts without any forced convection, a larger axial temperature gradient (20 °C/cm) helps to increase growth rates for crack-free crystals. This is expected from *Tiller*'s analysis [10.80], as discussed earlier. While increasing the axial temperature gradient helps to avoid constitutional supercooling, it increases the melt–solid interface curvature, due to which the radial composition profile in the wafers varies (as discussed later). Hence a lower temperature gradient is necessary for the growth of ternary compounds, which also helps in eliminating strain-related cracks in the crystal. To avoid constitutional supercooling under a lower temperature gradient, either the crystal growth rate needs to be decreased or melt stirring (by forced convection) needs to be enhanced. There are no thermal and solutal convection cells in the homogenized solution due to stirring and thus the supersaturation is not likely to go beyond the critical values for spontaneous nucleation. The second effect of homogenization is that the growing crystal always faces a homogeneous diffusion field, and thus inhomogeneities in crystals such as striations are avoided or decreased to a minimum level. Forced convection leads to a decrease in the metastable region and supersaturation. This decrease in supersaturation corresponds to a narrowing of the metastability range, so that no enhanced spontaneous nucleation is seen. Stirring decreases the diffusion boundary layer and reduces the inhomogeneous supersaturation across the growing crystal, which allows a faster stable growth rate than without stirring [10.97].

Fluid mixing patterns used for forced convection can be very complicated and significantly impact on crystalline quality. Melt mixing schemes reported in the literature [10.18, 25, 98, 99] during ternary crystal growth include: unidirectional rotation of the crucible, the accelerated crucible rotation technique (ACRT), and mixing the melt using baffles and stirrers. Unidirectional rotation, even at high rotation rates (100–200 rpm), does not result in effective mixing for large-diameter crystals (50 mm and higher). In ACRT, the crucible is periodically accelerated and decelerated (around the growth axis) to promote efficient mixing of the melt, as shown in Fig. 10.27. However, there can be dead (unmixed) zones even in a thoroughly mixed melt, especially at the center of the crystal [10.99]. These types of unmixed zones lead to compositional fluctuations in the grown crystals. Hence it is essential to optimize the rotation schemes for each crucible diameter, melt size, and fluid viscosity in the ACRT scheme. In the baffle mixing scheme [10.25], the melt is homogenized very efficiently either by rotating the baffle in the melt or by translating the baffle perpendicular to the growth interface, similar to shown in Fig. 10.10b (for polycrystalline charge synthesis). In the case where the baffle is translated perpendicular to the growth interface, obtaining single crystals is quite difficult due to thermal fluctuations at the melt–solid interface. Very small-sized grains are observed by using this process. In the case of stirrer mixing, a stationary stirrer is held inside the melt with a uniformly rotating crucible. The stirrer also helps in efficient solute transport during solute feeding processes (discussed later). The stirrer mixing scheme is simple and can be easily adopted for large melt mixing. In the future, it will be worthwhile investigating the effect of magnetic field stirring on the crack elimination process.

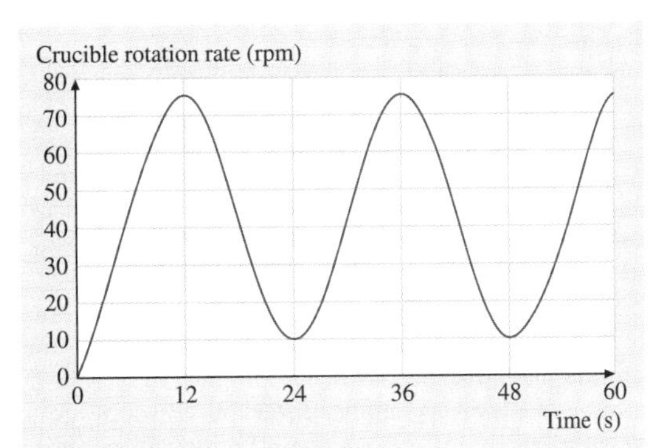

Fig. 10.27 Accelerated crucible rotation rates used for melt mixing and solute transport during vertical Bridgman growth of ternary crystals

Fig. 10.28 Crack-free ternary wafer of $Ga_{0.76}In_{0.24}Sb$

10.8.3 Crystal Growth Rate for Crack-Free Ternary Crystals

As is evident from Fig. 10.26, crystal growth rates can be enhanced by a factor of four by efficient melt mixing. Typical growth rates for the GaInSb system are in the range of 0.3–0.5 mm/h (with 10 °C/cm thermal gradient). By using the optimized ACRT scheme, as shown in Fig. 10.27, the growth rate for crack-free crystals can be increased to 2 mm/h under the same thermal gradient (10 °C/cm). Figure 10.28 shows a crack-free polycrystalline wafer of $Ga_{0.76}In_{0.24}Sb$ grown using the ACRT scheme. The growth rate was $\approx$ 1 mm/h, similar to that used for the cracked wafer shown in Fig. 10.24. In general, for III–V ternary crystals, with an axial temperature gradient of $\approx$ 10 °C/cm, the maximum growth rate for crack-free single crystals is in the range of 0.02–0.5 mm/h, depending on the specific alloy system and its composition [10.18]. These rates are significantly lower than the binary growth rates (1–3 mm/h) under similar thermal gradient. A delicate balance between heat and mass transport strategies is necessary for ternary single-crystal growth without cracks. By using efficient melt stirring, the crystal growth rates for ternary materials can approach those for binary compounds grown under the same temperature gradient and group V overpressure.

10.9 Single-Crystalline Ternary Seed Generation Processes

For III–V ternary alloys, single-crystal growth poses a major challenge. Unlike for binary compounds, growth of spontaneously nucleated single-crystal ternary seeds such as from the tip of a crucible in the Bridgman method has not been possible. The preferred method for generating a seed of any ternary composition is by starting from a binary seed. However there is a major criterion that needs to be satisfied. It has been empirically observed that, for alloy systems with wide lattice mismatch between the end binaries (GaAs–InAs, GaSb–InSb, InAs–InSb, etc.), the first-to-freeze section of the crystal must have a composition less than or equal to 5 mol % with respect to the seed in order to maintain single-crystallinity [10.19–21, 100]. For example, during the growth of single-crystal $Ga_{1-x}In_xAs$ on a GaAs seed, the first-to-freeze $Ga_{1-x}In_xAs$ should have $x < 0.05$ [10.100]. With $x > 0.05$, polycrystalline growth occurs on a single-crystal seed [10.100]. Hence the compositional grading during ternary seed generation needs to be carefully controlled to obtain single-crystal seeds.

There are three common ways of generating a compositionally graded ternary seed starting from a seed of a binary compound, as discussed below.

10.9.1 Bootstrapping Method

The most common method is the *bootstrapping* process, wherein seeding is initiated from a binary seed and successive experiments are performed with increasing solute concentration to reach the target alloy composition in the seed. Figure 10.29 schematically depicts the axial composition in a single-crystal GaInAs seed generated by the bootstrapping method. For example: to generate a $Ga_{1-x}In_xAs$ seed, one would start from a GaAs or InAs seed and grow successive ingots of step-graded GaInAs. The maximum allowable step-like change in composition between the last-to-freeze of any ingot with the first-to-freeze of the next ingot must be equal or less than 5 mol %, as shown in Fig. 10.29. Due to change in the segregation coefficients with alloy composition (dictated by the pseudobinary phase diagrams), there is an increase in the InAs concentration during growth beyond each step. *Bonner* and coworkers [10.101, 102] employed this method to grow ternary seeds of $Ga_{1-x}In_xAs$ up to $x = 0.12$ by using the Czochralski technique. *Tanaka* et al. [10.23] employed multistep pulling solute feeding Czochralski

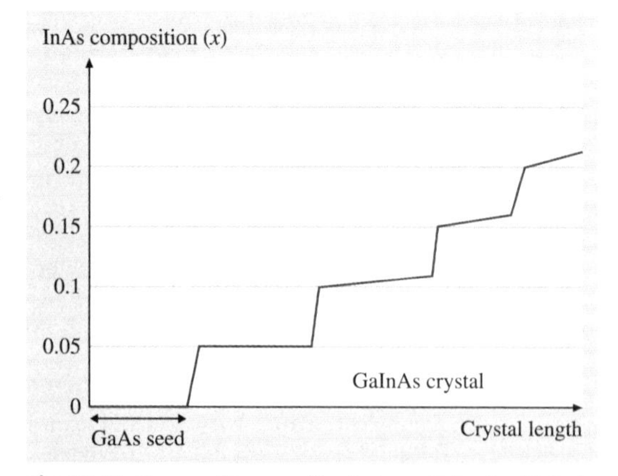

Fig. 10.29 Composition profile (schematic) of a GaInAs single-crystal seed generated by the bootstrapping process

method to grow $Ga_{1-x}In_xSb$ single crystals using composition steps of $x = 0.04, 0.08, 0.12$, and 0.18. Though the bootstrapping technique is the most straightforward way of generating single-crystal ternary seeds, it has several limitations. First of all, seed generation with concentrated alloy composition (e.g., $x = 0.5$) can be time consuming since the step in alloy composition in each growth run can only be 5 mol %. For vertical Bridgman growth, it is very difficult to use this method. The melt composition for successive growth in contact with the previous seed needs to be carefully monitored to avoid excessive dissolution of the seed crystal.

10.9.2 Directional Solidification by Normal Freezing

This method is based on the intrinsically changing axial composition profile in a crystal that has been directionally solidified by normal freezing, as depicted in Fig. 10.21 [10.100, 103, 104]. The process is shown schematically in Fig. 10.30a–d. In this process, a binary seed such as GaAs is used. A GaInAs melt is prepared by dissolving InAs and GaAs with the appropriate melt composition that would lead to a first-to-freeze ternary crystal region with 5 mol % compositional step as required for the ternary seed generation process. The melt is directionally solidified by Bridgman or gradient freezing method to obtain a ternary graded crystal. Figure 10.31 schematically depicts the axial InAs composition profile in a vertical Bridgman grown GaInAs crystal. Depending on the composition of the seed necessary for the actual crystal growth experiment, the crystal could be sliced at an appropriate axial position along the length of the crystal. The main problem with this approach is when the composition starts changing rapidly in the crystal; microscopic fluctuation of the composition and polycrystallinity occur due to constitutional supercooling unless the growth rate is significantly reduced. Hence only a part of crystal is sin-

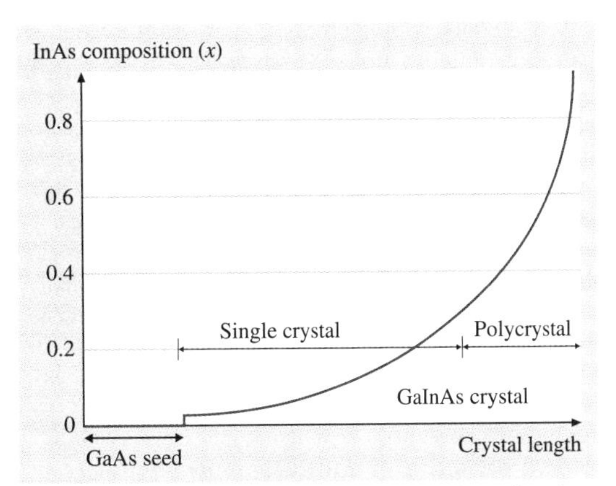

Fig. 10.31 Composition profile (schematic) in a directionally solidified ternary single-crystal seed

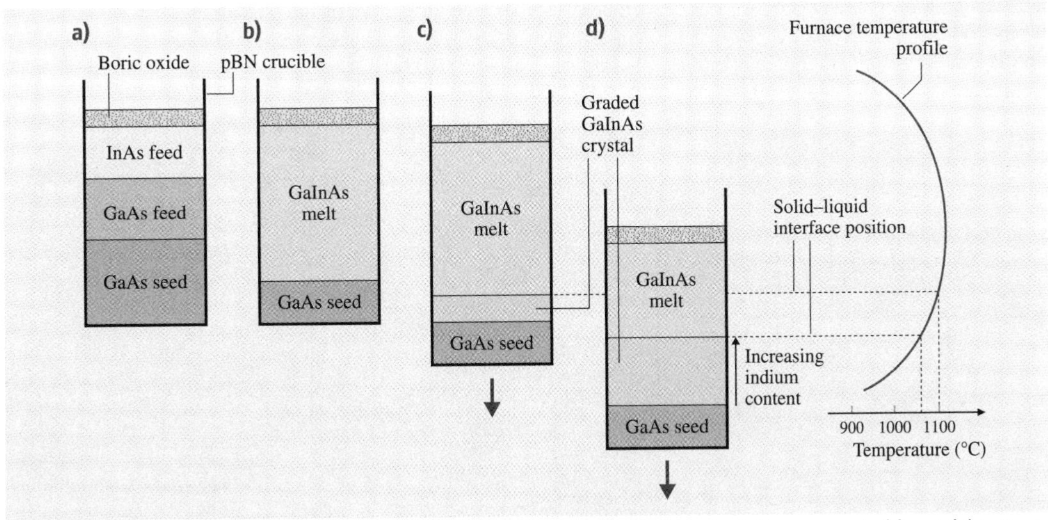

Fig. 10.30a–d GaInAs single-crystal seed generation by vertical Bridgman method (the *arrows* in (c) and (d) indicate the lowering of the crucible to the cold zone of the furnace during crystal growth)

gle and a bootstrapping process is necessary to change the composition of the seed beyond the point where the single- to polycrystal transition occurs. Even if one manages to grow a complete single crystal by reducing the growth rate, using a seed with a steep change in composition still remains a problem. During the subsequent crystal growth experiment, the seed needs to be dissolved slightly by contact with the ternary melt. However, any dissolution in the steep compositionally graded region will change the composition of the seed at the growth interface. Hence using the compositionally graded seed for a lattice-matched crystal growth experiment is not practical.

10.9.3 Directional Solidification by Solute Diffusion and Precipitation

This process is based on solute diffusion in the melt due to a concentration gradient [10.18, 98, 105, 106]. A schematic diagram of the approach is shown in Fig. 10.32a–d for growing GaInAs seeds. For GaInAs graded seed growth, the lower melting binary (InAs) is used as a seed. The charges in the crucible are stacked as shown in Fig. 10.32a. The crucible is placed in a temperature gradient in such a way that the InAs feed and a portion of the InAs seed remelts to dissolve a portion of the GaAs feed, forming a GaInAs melt. While selecting the temperature profile in the furnace, care should be taken to avoid melting of the GaAs feed. The dissolved GaAs from the feed diffuses towards the seed due to the concentration gradient. Once the solute content in the melt at any point reaches the solidus composition (corresponding to the solidus temperature), solidification occurs. The growth rate is thus limited by the solute diffusion rate. During the growth, neither the crucible nor the furnace is translated. The composition profile in the graded crystal will depend on the temperature profile imposed by the furnace and the solidus curve in the pseudobinary phase diagram. As the GaAs in the melt precipitates and the crystal grows, the melt gets richer in InAs and hence more GaAs is dissolved from the feed. Ultimately, a crystal is obtained with compositional grading from InAs towards GaAs. Theoretically, if the furnace temperature profile matches the solidus temperature curve of the phase diagram, the compositional grading in the crystal will be linear, as depicted in Fig. 10.33.

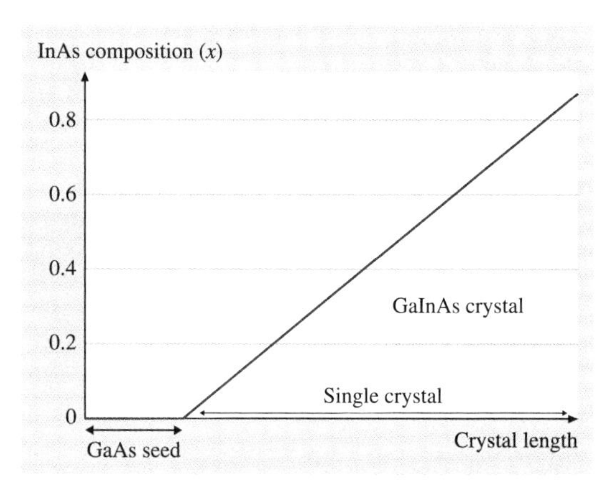

Fig. 10.33 Composition profile (schematic) of a GaInAs single-crystal seed generated by solute dissolution and diffusion

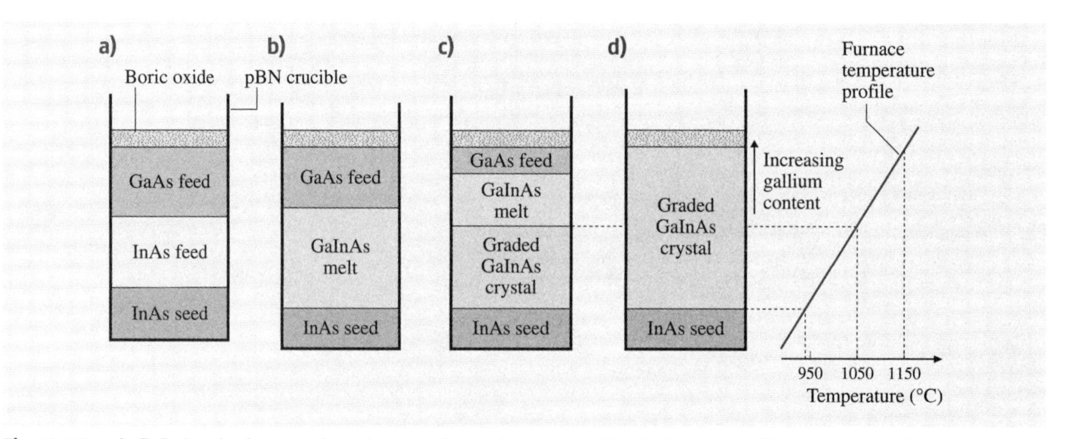

Fig. 10.32a–d GaInAs single-crystal seed generation using solute dissolution and diffusion method (with stationary crucible and furnace)

10.9.4 Growth of Lattice-Mismatched Ternary on Binary Using Quaternary Grading

One of the interesting aspects of the thermochemistry shown in Fig. 10.20 is the growth of ternary compounds from quaternary melts [10.29, 30, 78]. The concept is shown schematically in Fig. 10.34a. By using a GaAs seed and a GaInAsSb melt of a specific composition at a specific growth temperature, one can grow a GaInAs ternary crystal of uniform composition. The composition of the uniform ternary crystal is dependent on the quaternary melt composition and growth temperature. By carefully analyzing the region between the ternary crystal and the binary seed, a quaternary transition region has been observed where the composition changes from the binary to the ternary. The thickness of this graded quaternary region can be in the range of 10–100 μm, depending on the melt composition. In this process, no efforts are made to control the growth conditions to grow the graded quaternary intermediate layer. The compositional grading occurs spontaneously during the growth process. Figure 10.34b shows a compositionally homogenous $Ga_{0.86}In_{0.14}As$ grown on GaAs substrate using a quaternary $Ga_{0.07}In_{0.93}As_{0.06}Sb_{0.94}$ melt at 700 °C [10.78]. The GaInAs ternary region does contain residual Sb (less than 1 mol %). However, for all practical purposes, it can be considered a ternary compound. The effect of residual antimony on the lattice parameter and bandgap is negligible. This process could be used in the future to grow ternary seeds of a desired composition starting from a binary seed and quaternary melt by simple Bridgman or gradient freezing method without the need for any sophisticated grading procedures.

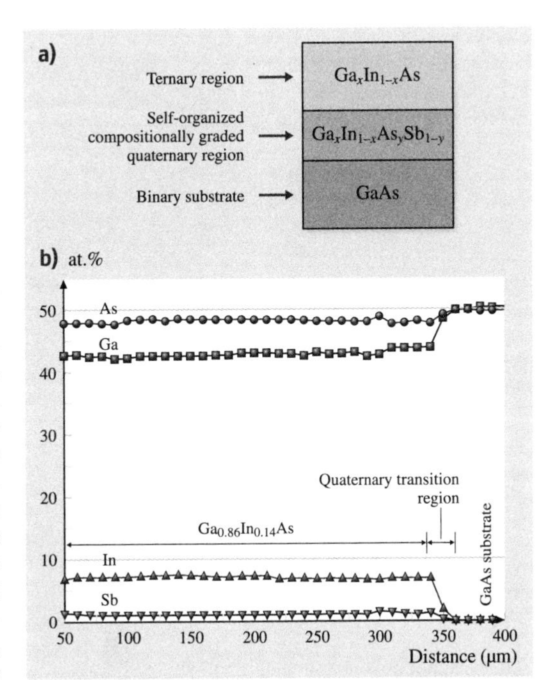

Fig. 10.34 (a) GaInAs seed generation using a self-organized quaternary grading method and (b) experimental composition profile for $Ga_{0.86}In_{0.14}As$ grown on GaAs

10.10 Solute Feeding Processes for Homogeneous Alloy Growth

Having grown a ternary single-crystal seed of the desired composition, the next step is to grow a ternary crystal of homogeneous composition from which a large number of wafers (of the same composition) can be sliced. We know that the melt composition as well as the ternary crystal composition changes continuously during the experiment (due to alloy segregation). Hence to grow homogeneous crystals, it is necessary to develop melt replenishment processes that can maintain the composition of the melt constant over the entire duration of the crystal growth experiment. In this section, we discuss the various melt replenishment methods and their limitations.

10.10.1 Growth from Large-Volume Melts

One simple idea would be to grow a small crystal from a huge volume of melt [10.107]. Referring to Fig. 10.21, few wafers of constant composition could be extracted from the first-to-freeze section of the crystal. The melt could then be reused for the next experiment by replenishing with fresh material to account for the extracted wafers. However, this is not a practical solution. Contamination of the melt by impurities from repeated use, the large crystal growth systems required, and the possibility of damaging the crucible during cooling of the melt make this process unattractive for ternary substrate technology. For large-scale crystal growth production, it is desirable to consume the entire melt during a single experiment. Hence, growth from large-volume melts is not practiced for commercial applications.

10.10.2 Solute Feeding Using Double-Crucible Configuration

Melt replenishment during growth is also known as solute feeding. Many interesting solute feeding concepts have been tried by different researchers. The simplest of these processes is the solute feeding in the Czochralski (CZ) configuration [10.108–112]. Figure 10.35a shows a schematic diagram of this process for GaInAs growth. As the melt becomes depleted in GaAs, the feed rod is inserted to dissolve and maintain the composition of the melt at the liquidus composition. The problem with this process is the uncontrolled dissolution of the feed rod. If the radial temperature in the growth melt is uniform, the feed rod dissolves slower than the crystal pulling rate. This is due to the long duration it takes for the melt to uniformly develop the undersaturation level after the GaAs is depleted by the growing crystal. Hence the feed rod tends to dissolve at a slower rate than the rate at which the GaAs is depleted in the melt. This leads to fluctuation in the composition of the growing crystal. On the other hand, if the radial temperature at the edge of the crystal is increased to enhance the dissolution of the feed rod, rapid transport of GaAs to the growth interface often leads to rapid polycrystalline growth. Hence the feed dissolution rate and GaAs depletion rate need to be matched. This is quite challenging given the fact that the melt height changes during growth in the CZ technique, and thus so does the radial temperature gradient. To restrict the rapid transport of solute from the feed to the growth interface, a double-crucible CZ method as shown in Fig. 10.35b can be employed. In this configuration, the depleting component can be added to the outer crucible instead of the inner crucible. The temperature of the outer crucible is higher than that of the inner crucible, due to which the solubility of the GaAs feed in the GaInAs melt is higher. The narrow gap between the two crucibles controls the feeding rate of GaAs from the outer to inner crucible. This method also shares one of the limitations of the single-crucible CZ technique: once the outer melt is supersaturated with GaAs, the dissolution of the feed rod stops. The outer melt supersaturation level decreases only when the GaAs has been transported to the inner melt and the inner and outer melt have homogenized. Hence the feed rod does not dissolve continuously during the growth, which leads to fluctuations in the alloy composition in the crystal.

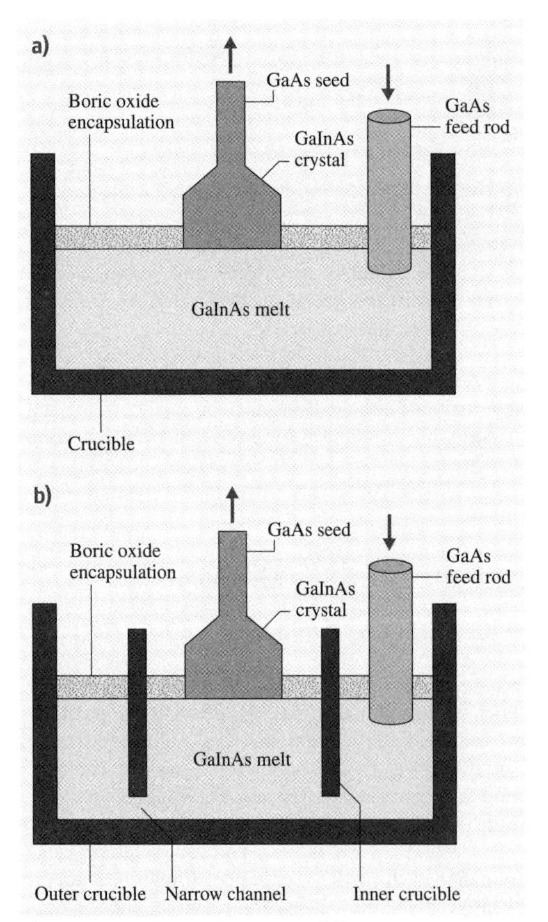

Fig. 10.35a,b Solute feeding in (**a**) single-crucible and (**b**) double-crucible Czochralski (CZ) growth configurations

Variations on the double-crucible CZ configuration (refer to Fig. 10.36) have been attempted to control the solute transport rate between the inner and outer crucibles [10.23, 106, 113–116]. In both of these configurations, a piston attached below the outer crucible is used to keep the volume of the melt constant in the inner crucible. The diameter (Fig. 10.36a) or the length (Fig. 10.36b) of the channel is varied to control the solute feeding rate. The limitation with these processes is the blocking of the channels by the transporting solute. A change in liquidus temperature occurs along the

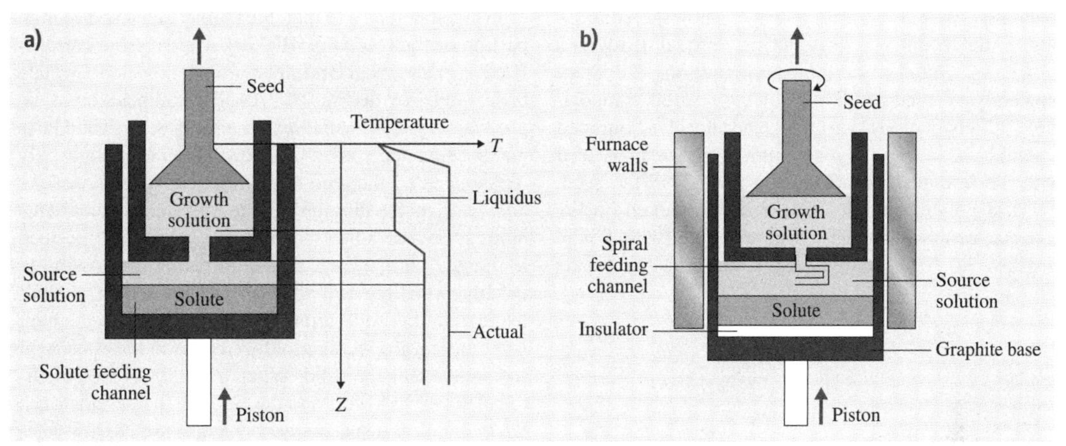

Fig. 10.36a,b Solute feed rate control in the double-crucible CZ method by optimizing (**a**) feeding channel diameter and (**b**) feeding channel length

length of the supply channel, as shown in Fig. 10.36a. This leads to precipitation of the solute in the channel towards the lower-temperature side. If the channel width is made larger to solve this problem, back-diffusion of species from the inner to the outer crucible takes place, which leads to variations in the composition of the grown crystal. Another problem is the optimization of the length of the spiral feeding channel (Fig. 10.36b). The diffusion rate of species in the channel dictates the growth rate of the crystal. A shorter channel leads to back-diffusion of species from the inner to the outer melt, changing the composition of the growth solution. On the other hand, a longer channel leads to slow diffusion of species from the outer to the inner melt.

10.10.3 Solute Feeding in the Vertical Bridgman Method

The double-crucible CZ concept for solute feeding can be applied to the Bridgman or gradient freeze type

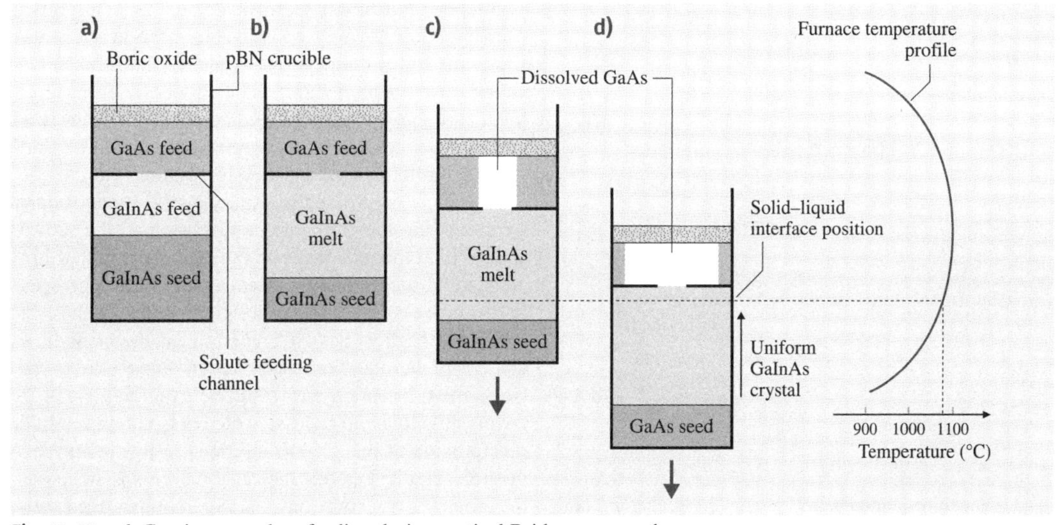

Fig. 10.37a–d Continuous solute feeding during vertical Bridgman growth

methods [10.21, 104]. Figure 10.37 shows a continuous solute feeding process during vertical Bridgman growth of GaInAs. In this configuration, a ternary single-crystal seed of GaInAs of the desired composition (solidus composition) is placed in a flat-bottomed crucible. A GaInAs polycrystalline homogeneous feed with the liquidus composition is placed on the seed. A GaAs polycrystalline solute feed is stacked on the GaInAs feed, separated by a quartz plate with a hole (Fig. 10.37a). The entire stack is then heated to the growth temperature to melt the GaInAs feed and a small portion of the feed (Fig. 10.37b). The crucible is translated as in a vertical Bridgman growth process to initiate the crystal growth (Fig. 10.37c). The composition of the growing crystal is same as the ternary seed. The GaAs that is being depleted in the growth melt is continuously replenished by the feed at the top. The diameter of the hole restricts the dissolution rate of the GaAs feed. This helps to avoid rapid transport of GaAs from the feed to the growth interface, thus eliminating the possibility of polycrystalline growth [10.104]. The technical challenge with this configuration is the variation in the melt level during growth due to differences in the density of the melt versus the solid. When the GaInAs feed melts, the melt height lowers and the GaAs feed does not touch the melt unless a special mechanism is used to lower the feed. When the crystal grows, the solid expands, due to which the melt height increases. Hence there should be proper mechanism for moving the GaAs feed up into the crucible.

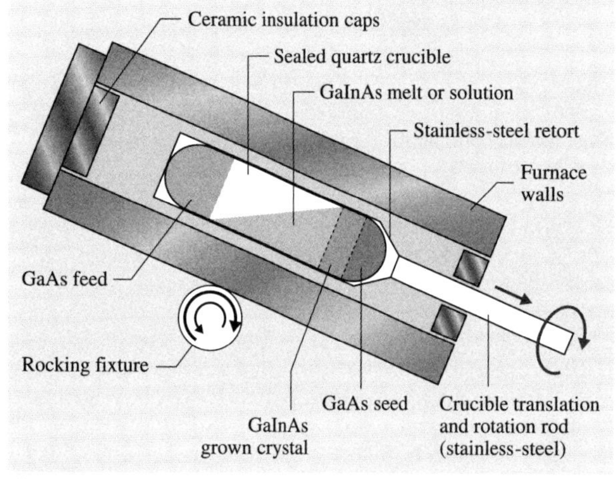

Fig. 10.38 Solute dissolution and feeding in a rocking furnace assembly

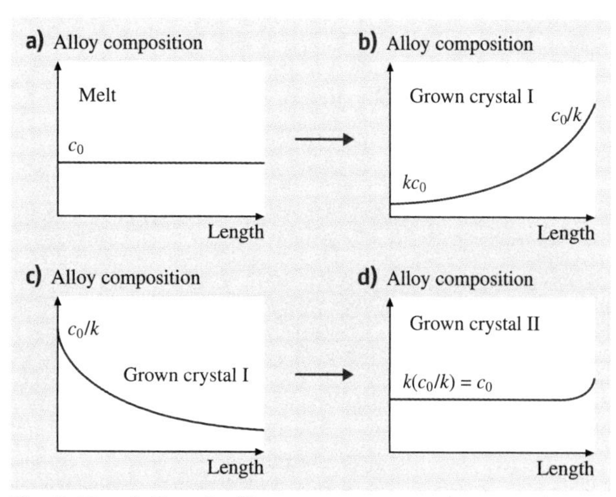

Fig. 10.39a–d Growth of homogeneous ternary by sequential normal freezing and zone growth methods

10.10.4 Solute Feeding by Crucible Oscillation

The method for synthesizing GaSb and InSb in sealed crucible as depicted in Fig. 10.9 can be adopted to perform solute feeding process during ternary growth [10.117]. Figure 10.38 shows the rocking furnace configuration for GaInAs growth. At one end of the crucible, the GaAs seed is placed, onto which a graded-composition GaInAs crystal is first grown from the GaInAs melt by translating the crucible towards the lower-temperature region of the furnace (as in the Bridgman method). Thereafter the furnace is tilted periodically such that the GaInAs melt touches the GaAs feed to replenish the depleted species. The crucible is continuously translated to continue the growth. Since the GaAs level in the ternary melt is kept constant by dissolving the feed, the composition of the grown crystal is homogeneous. For radial compositional homogeneity, the crucible can be rotated during growth. In this process, the temperature gradient across the crucible length plays a significant role in the crystalline quality. If thermal fluctuation occurs due to the oscillating melt, the grain size could be extremely small in the crystal. To avoid excessive feed dissolution or rapid precipitation at the growth interface, a low axial temperature gradient is desirable. However this increases the probability of constitutional supercooling. Hence efficient melt mixing by forced convection (continuous oscillation) in conjunction with a slow growth rate is necessary for this process.

10.10.5 Growth Using Compositionally Graded Feed

The graded concentration solute feeding process is depicted in Fig. 10.39 [10.118]. In this method, starting from a homogenous melt of composition C_0, a crystal is first grown using the Bridgman or gradient freezing method. The axial alloy composition in the grown crystal is shown in Fig. 10.39b. According to segregation phenomenon, the composition in the crystal starts from kC_0 (k being the segregation coefficient) and continuously builds up beyond C_0/k. The crystal is then reversed and a traveling zone heater is traversed across the crystal starting from the C_0/k side. The final composition in the crystal is uniform and equal to C_0. A traveling heater method (THM) can also be used to practice this concept [10.119]. This method needs a purely diffusive growth condition. Such conditions can only be found in microgravity situations. Convective flow in the liquid zone will change the solute concentration and result in nonuniform growth. Magnetic fields can be used to dampen convective flows. However for large-scale ternary substrate production, this method is not suited.

10.10.6 Periodic Solute Feeding Process

From the processes described above, it is clear that any large-scale ternary crystal growth process will require mechanisms for precise control of melt composition. Another technical problem is the preparation of the growth melt during the experiment. Ensuring that the melt composition is in equilibrium with the seed composition is crucial for avoiding excessive seed dissolution. The rate of solute dissolution and transport to the growth interface also needs to be controlled to achieve high crystalline quality and growth rates suitable for commercial applications. To address all of these technical issues collectively, the process shown in Fig. 10.40 has been developed and successfully employed for large-diameter ternary crystal growth. In this process, starting from a binary seed, the graded ternary seed and the homogeneous ternary crystal is grown in a single experiment. The schematic of the experimental growth configuration and crystal growth sequence for GaInSb is shown in Fig. 10.40a–d [10.28, 98, 120, 121].

To start with, a InSb single-crystalline seed is placed at the bottom of the crucible along with InSb polycrystalline charge. A GaSb polycrystal feed is suspended from the top of the growth chamber. After heating and stabilizing the furnace to obtain a specific temperature gradient, the crucible is raised into the furnace.

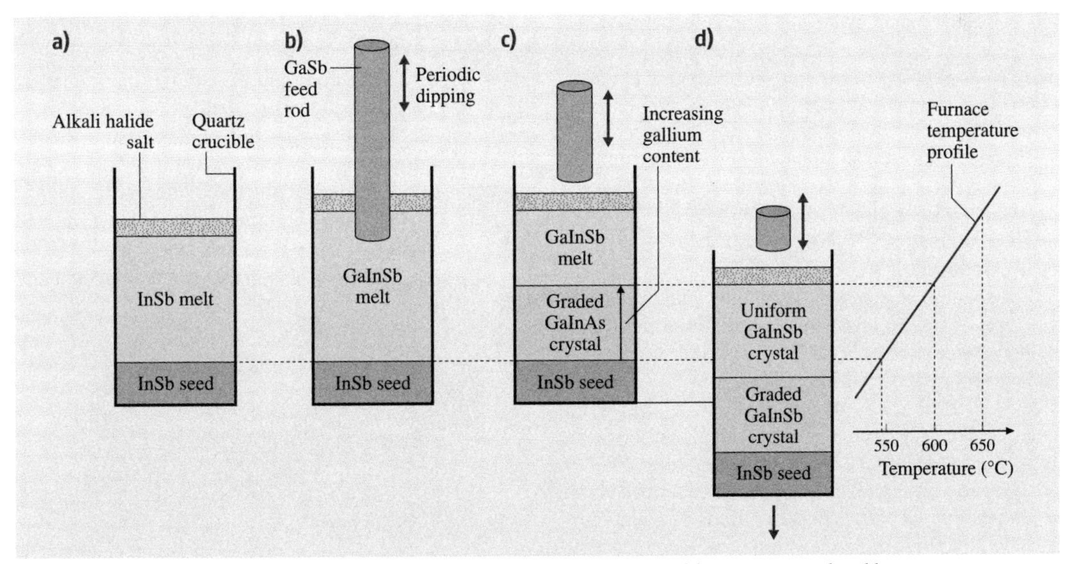

Fig. 10.40a–d Periodic solute feeding process for growth of graded-composition ternary seed and homogeneous ternary crystal in a single experiment by sequentially using solute diffusion and normal freezing processes

The InSb feed melts along with part of the InSb seed to obtain a melt–solid interface (Fig. 10.40a). At this point, the GaSb feed is lowered and allowed to touch the top of the InSb melt for a few seconds. The crucible is set to perform accelerated crucible rotation (ACRT), as shown in Fig. 10.27, and a homogeneous GaInSb melt is prepared (Fig. 10.40b). The GaSb feed is lowered periodically to touch the GaInSb melt. The dissolved species is transported rapidly to the InSb seed interface as a result of ACRT melt mixing. Crystal growth is initiated in the crucible as a result of the increasing level of the solute concentration at the solid–liquid interface with time. When the solute concentration in the melt near the seed interface reaches the liquidus composition, precipitation takes place and GaInSb starts growing on InSb (Fig. 10.40c). The crucible is kept at the same position in the furnace while the ACRT continues. As the GaSb feed is periodically lowered to touch the melt for few seconds, more solute is dissolved and transported to the growth interface and the crystal growth continues. Crystal grown by this method is compositionally graded along the growth axis with increasing gallium concentration in $Ga_{1-x}In_xSb$ (decreasing indium concentration). The rate of compositional grading is decided by the solidus temperature in the pseudobinary phase diagram and the axial furnace temperature gradient. The axial composition in the crystal is graded until a desirable alloy composition is achieved and then a homogeneous composition crystal length is grown. During the compositionally graded crystal growth, the melt–solid interface automatically rises from the cooler to the hotter zone in the furnace. Thus the melt–solid interface shape changes continuously during the growth. The effect of temperature gradient on the melt–solid interface shape and the radial compositional variation will be discussed in the next section. To grow crystal with axially uniform composition, the crucible is translated into the lower-temperature zone of the furnace while the GaSb feed dissolution is continued by the periodic dipping method. While the GaSb at the melt–solid interface is depleted by preferential incorporation in the crystal, it is being replenished by the feed dissolution. Hence the melt–solid interface remains at the same position in the furnace until the entire melt has solidified (Fig. 10.40d).

The periodic solute feeding process described above in conjunction with efficient melt mixing provides complete control over the composition profile in the graded seed and the homogeneous crystal [10.120]. The entire experimental process could be automated using programmable stepper motors for the periodic solute feeding, ACRT, and crucible translation processes. The main advantage of this process is the single experiment in which a ternary homogeneous crystal of any alloy composition can be obtained starting from a binary single-crystal seed and polycrystalline binary feed materials. For high-quality crystal growth, the periodic solute feeding process parameters need to be fine-tuned. This is necessary to precisely control the rate at which solute depletes and replenishes in the melt. The major process parameters that affect the crystalline quality include: the solute dipping time and frequency, the mixing strategy of the melt for solute transport, and the thermal gradient in the melt. Dipping time denotes the actual time that the melt is in contact with the feed. This determines the amount of feed introduced in the melt over one period of the dipping cycle. Due to the temperature gradient in the melt, the top of the melt (higher-temperature zone) is usually undersaturated. Hence there exists a danger of uncontrolled dissolution if the solute feeding rod is dipped for longer than the required time. Thus precise control of dipping time is required. Excess dissolution of the solute can promote random nucleation in the entire melt volume. Additionally, this can lead to an oversupply of the solute to the growth interface leading to small grains or causing a change in grain structure, as shown in Fig. 10.41. Dipping frequency defines the period of the dipping cycle and should depend on the growth rate. It takes finite time for the solute to reach the growth surface. Further, the growth kinetics determines the actual growth rate. The thermal gradient in the melt dictates the melt homogenization efficiency. For lower thermal gradients, the time for melt homogenization is longer than for high thermal gradient. Hence the dipping frequency needs to be lower for lower thermal gradients in order to prevent constitutional supercooling and so-

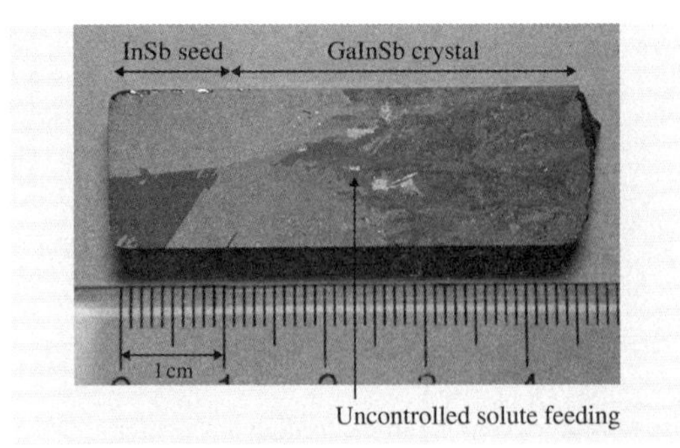

Fig. 10.41 Transition from large to small grains due to uncontrolled (rapid) solute feeding process during growth

lute precipitation in the melt. Mixing of the melt greatly affects the crystal growth. Efficient mixing can remove the excess solute rejected at the growth interface in less time than without mixing, thereby decreasing the possibility of constitutional supercooling. This suppresses the chances of interface breakdown, leading to conditions conducive for single-crystal growth. Based on experimental crystal growth results, a set of optimized periodic solute feeding parameters for GaInSb growth include: solute contact area of 20 mm, solute dipping time of 2–3 s, solute dipping frequency of 3–4 times per hour, temperature gradient near the melt–solid interface of 15 °C/cm, and ACRT melt mixing process. Figure 10.42 shows elemental composition profiles of Ga, In, and Sb for the graded GaInSb seed (shown in Fig. 10.41) grown using the periodic solute feeding process with the optimized parameters. The superior control of the composition profile shown in this diagram clearly demonstrates the effectiveness of the periodic solute feeding process. Beyond 47 mm, the solute feeding process was discontinued and the crystal was directionally solidified, giving rise to a composition profile as expected from the normal freezing process. This process has been demonstrated for the growth of homogeneous $Ga_{0.3}In_{0.7}Sb$ crystals [10.120].

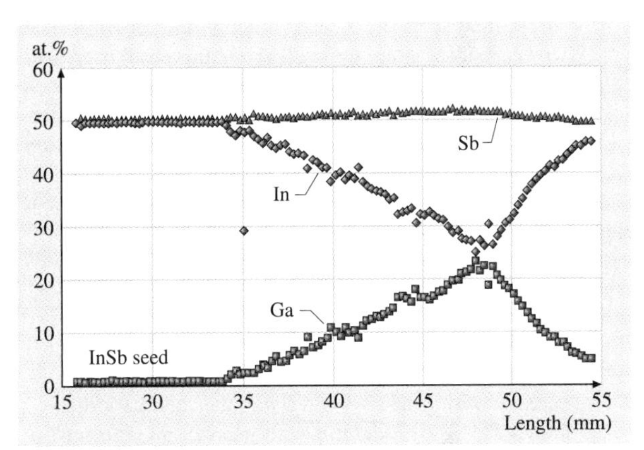

Fig. 10.42 Elemental composition profile along the growth direction of the GaInSb graded seed crystal grown using periodic solute feeding process shown in Fig. 10.40

Another method for periodic solute feeding process is shown in Fig. 10.43. In this process, the higher-melting binary is used as the starting seed material for the ternary crystal growth. In Fig. 10.43, the process for GaInSb growth starting from GaSb seed is illustrated. The experiment starts with a GaSb single-

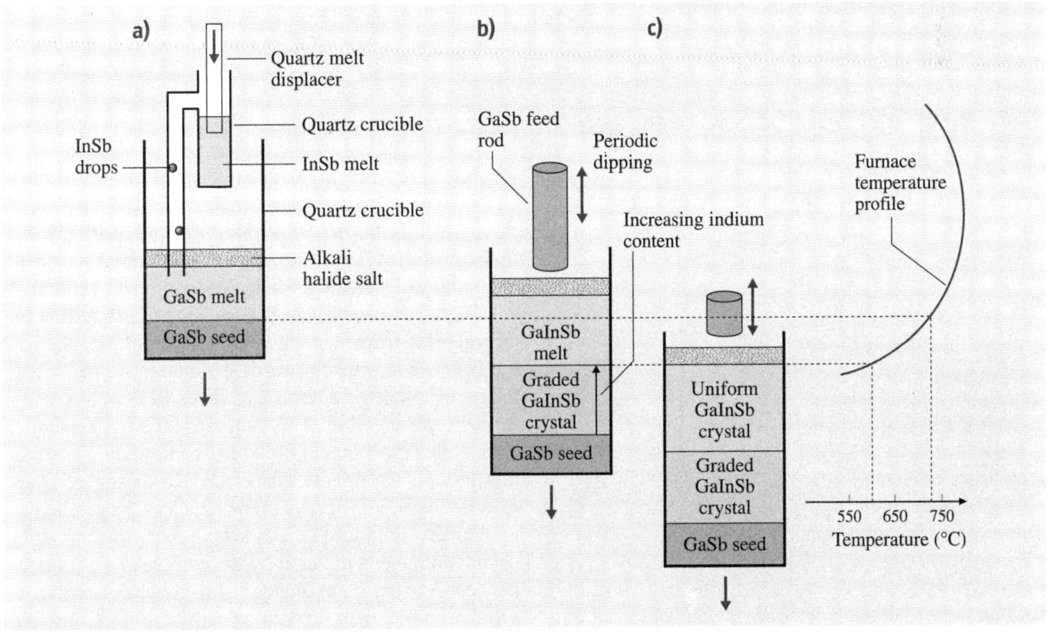

Fig. 10.43a–c Periodic solute feeding process for growth of graded-composition ternary seed and homogeneous ternary crystal in a single experiment by vertical Bridgman method

crystalline seed and a GaSb melt in the crucible. The lower-melting-point binary (InSb in this case) is present in a quartz melt displacer. After the furnace has been heated, the GaSb feed melts to form a melt–solid interface. The crucible is set for accelerated crucible rotation (ACRT) as in the previous case. By pressing the piston, the InSb can be periodically fed into the GaSb melt. After the initial InSb melt is dispensed into the growth melt to form GaInSb, the crucible is translated towards the lower-temperature zone of the furnace as in Bridgman growth. The graded-composition GaInSb ternary starts from the GaSb seed with increasing InSb content in the crystal. The InSb concentration in the melt increases with time as it is fed into the melt periodically. After the graded GaInSb seed growth, a GaSb feed is introduced into the melt periodically as in the previous case. The crucible translation is continued while the GaSb feed is periodically fed into the melt (Fig. 10.43c). This step results in a compositionally uniform GaInSb region as before (Fig. 10.40d). Care must be taken to ensure that the GaSb feed shown in Fig. 10.43b,c is introduced into the lower-temperature zone of the furnace without melting. The process illustrated in Fig. 10.43 can also be implemented by the gradient freezing process. In that case, the furnace temperature is decreased below the melting point of GaSb before it is introduced into the GaInSb melt for the homogeneous growth region. Figure 10.44 shows the axial InSb mole fraction in a GaInSb crystal grown by the periodic InSb feeding process. The composition of the crystal was continuously varied from GaSb to InSb, showing the flexibility of the process for controlling the alloy concentration in the crystal.

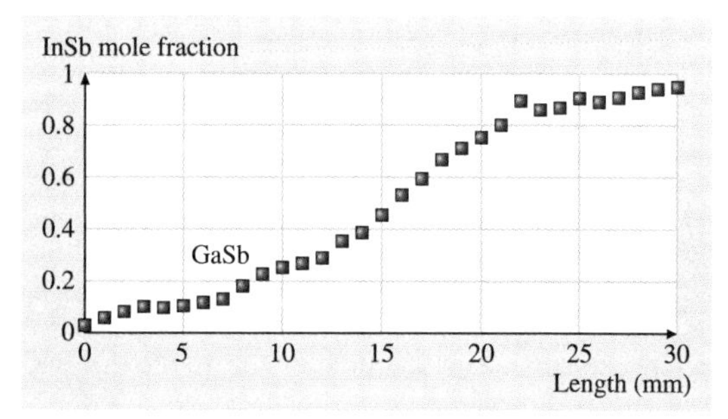

Fig. 10.44 InSb mole fraction along the growth direction of GaInSb crystal grown using the first step of the process (feeding InSb) shown in Fig. 10.43

The periodic solute feeding process described in this section can be universally applied for the growth of any ternary alloy system. The same experimental configurations that are being used for the respective binary growth can be used with the addition of the periodic solute feeding mechanism.

10.11 Role of Melt–Solid Interface Shapes

The melt–solid interface shape during crystal growth plays a significant role in the quality of the grown crystals [10.100, 122, 123]. As shown in Fig. 10.45, with respect to the solid, the melt–solid interface shape could be (a) convex, (b) planar or (c) concave. The shape represents the isotherm corresponding to the melting point of the material [10.124]. A convex interface with respect to the solid is created when the temperature gradient at the growth interface is low since the heat extraction through the grown crystal becomes dominant, whereas the concave interface towards the melt is formed when the temperature gradient is large because heat is lost out through the crucible. The concave interface as shown means that the temperature at the center is higher than that around the edges. The arrows in the figure indicate the direction of heat flow, which is perpendicular to the melt–solid interface at every point. The curvature of the interface is dictated by the complex heat transfer between various components in the growth system (furnace, crucible, melt, crystal, encapsulant, gas, crucible holders, etc.). In general, a concave interface shape is the least desirable since it encourages polycrystalline

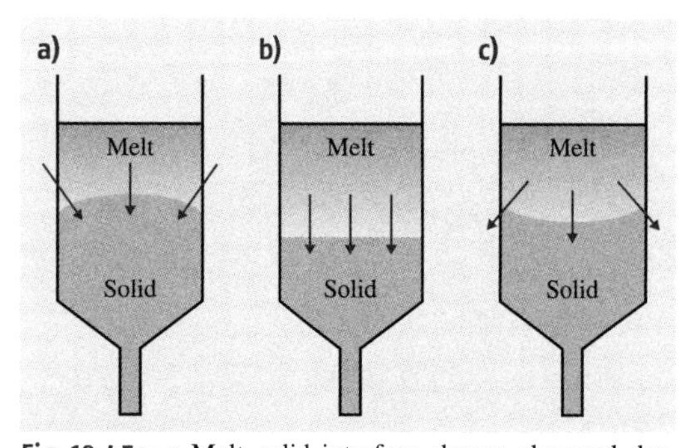

Fig. 10.45a–c Melt–solid interface shapes observed during vertical Bridgman growth: (**a**) convex, (**b**) planar, and (**c**) concave interface shapes

grains to originate from the walls of the crucible and grow into the crystal. The convex interface shape is desirable from the point of view of grain elimination (any polycrystalline grain will grow outwards to the edge of the crucible). However a convex interface tends to produce twinning in the crystal. Both concave and convex interfaces will produce nonuniform stress along the radial direction, thus producing a high dislocation density in localized regions across the wafer. The planar interface is the ideal for crystal growth. A significant amount of computational modeling and experimental research is being conducted to study the interface shape in real crystal growth configurations [10.125–127]. This is being done to identify thermal and growth conditions under which a planar melt–solid interface could be obtained.

The shape of the melt–solid interface determines the radial compositional profile in ternary crystals [10.100]. A planar interface is absolutely necessary for obtaining wafers with uniform alloy composition. Hence it is very important to understand the heat transfer processes during ternary crystal growth and identify suitable conditions that could lead to planar interface during growth. In general, for a Bridgman-type configuration, the primary heat flow mechanisms are: radiative heat transfer from the hot zone of furnace to the melt, convection and conduction in the melt, conduction in the solid (crystal) and radiative heat transfer from the crystal to the cold zone of the furnace [10.128]. Other effects such as latent heat release at the melt–solid interface, convective flow, etc. make secondary contributions. Convection in the liquid can be driven by several forces. In the presence of gravitational forces, the most importance source of convective flow is due to the density differences caused by temperature and composition differences in the liquid. Additional flow can be driven by the volume change accompanying the phase change. If free surfaces (i. e., liquid–vapor, liquid–liquid) are present, Marangoni convections due to the surface tension gradient originating from the difference of temperature and solute concentration occur. The density-driven convective flow due to the temperature difference can be prevented if the temperature increases with height (the stabilizing condition, as in the case of vertical Bridgman configuration) and no temperature gradient is present in the radial direction even in the presence of gravitational forces [10.96]. If a solute is present, the growth process will result in compositional differences ahead of the growth interface because the rejected solute from the crystal is lower or higher in density than the solvent; these composition differences may result in a thermosolutal convective instability [10.96]. A global model incorporating the materials properties of all the components present in the growth chamber is necessary to predict the melt–solid interface shape to a high degree of accuracy. In this section, we will discuss the experimental parameters that alter the melt–solid interface curvature during ternary crystal growth using Bridgman or gradient freeze type methods.

During Bridgman growth of ternary crystals, the temperature gradient in the melt, at the melt–solid interface, and in the crystal plays a significant role in determining the curvature of the interface [10.121].

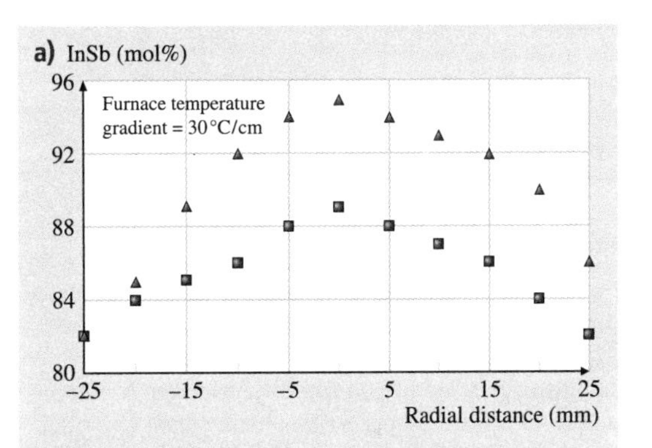

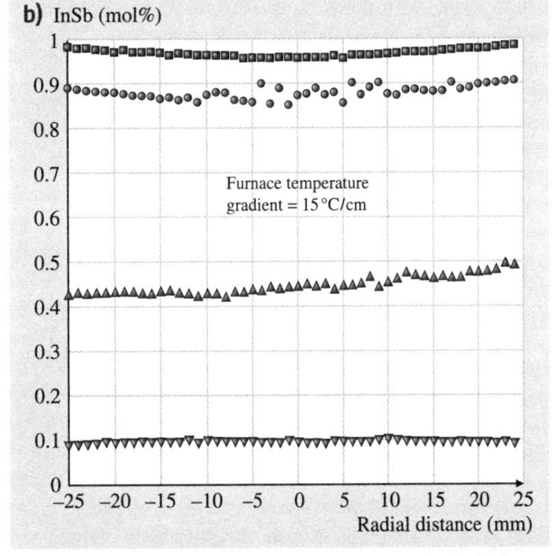

Fig. 10.46a,b Correlation of melt–solid (MS]M–S) interface shape with radial InSb composition profile in GaInSb crystals: (**a**) concave M–S interface resulting in convex InSb profile (▲: without melt stirring, □: with melt stirring) and (**b**) planar M–S interfaces resulting in uniform InSb profiles (with melt stirring)

Part B | 10.11

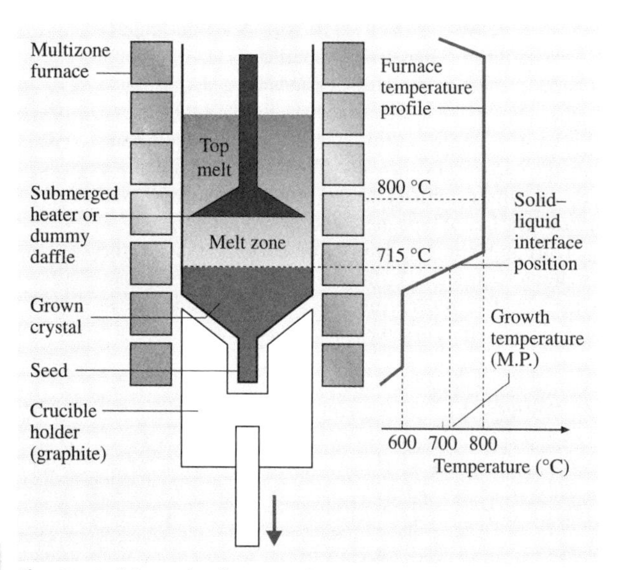

Fig. 10.47 Schematic diagram of vertical Bridgman growth configuration with submerged heater or baffle in the melt used for flattening the melt–solid interface shape

Interestingly, the effect of forced convection on the interface curvature seems to be weaker than the effect of temperature gradient. Figure 10.46 shows the radial InSb composition profiles for vertical Bridgman grown GaInSb crystals with furnace temperature gradients of 30 and 15 °C/cm. Under an axial temperature gradient of 30 °C/cm, the growth interface is highly concave with respect to the solid (leading to a convex InSb radial profile). Figure 10.46a also shows the radial InSb profile for a crystal grown using melt stirring (stationary stirrer in the melt with constant crucible rotation of 100 rpm, as in Fig. 10.10a). The curved compositional profile clearly indicates that high axial gradients lead to curved interfaces which are very difficult to flatten even with strong fluid stirring. The convex radial profile is also seen in the crystal grown with ACRT (Fig. 10.27). This clearly demonstrates the strong influence of temperature gradient on the interface curvature. Higher temperature gradients are known to produce concave melt–solid interface shapes. Lowering the gradient would certainly flatten the interface. However very low gradients must be avoided since the probability of constitutional supercooling will be enhanced with low gradients. For ternary crystals with diameter of 50 mm, the optimum temperature gradient is in the range of 10–15 °C/cm, wherein constitutional supercooling can be avoided by suitable melt stirring and a flat interface could be achieved. Figure 10.46b shows the radial InSb composition profiles in a GaInSb crystal grown with a temperature gradient of 15 °C/cm and with melt stirring using ACRT (Fig. 10.27). The interface curvature is extremely flat, except towards the end of growth where the interface curvature varies slightly. If melt stirring is not used under the same gradient of 15 °C/cm, constitutional supercooling was observed for growth rate of 0.7 mm/h or higher. Hence a combination of low temperature gradient and low growth rate or melt stirring is necessary to obtain a flat interface without constitutional supercooling [10.121].

A sophisticated method for achieving planar melt–solid interface shape during crystal growth is shown in Fig. 10.47 [10.86, 87]. In this configuration (the submerged heater method, SHM), a heater is enclosed inside a quartz disc-shaped hollow enclosure (baffle). A shallow growth melt is contained between the baffle and the growth interface. Heat is axially supplied to the melt by the disc-shaped heater submerged in the melt, which is held at a temperature higher than the melting point of the materials and is extracted downward through the crystal during the growth. This results in flattening of the melt–solid interface. It is also possible to use the submerged baffle without power. The thermal conductivity of the baffle on its own could flatten the interface without the addition of extra heat. This scheme enables the crystal to grow under steady-stable conditions due to the low level of convection in the small melt zone and the continuous and stable replenishment of melt and impurities. While the crucible is lowered, the large melt above the baffle provides a liquid feed to the small melt enclosed between the crystal and the baffle. The downward flow in the gap between the crucible and the baffle inhibits back-diffusion, and buoyancy-driven convection is not expected since the top is hotter than the bottom in the small melt zone. The submerged baffle can also be rotated to create forced convection in the growth melt. This has been employed to eliminate cracks in GaInSb ternary crystals [10.25]. Advanced fluid flow patterns [10.129] will be necessary to achieve a high degree of compositional homogeneity in ternary crystals. In the future, a combination of ACRT [10.130–132], magnetic fields [10.133], and SHM [10.87] could enable the necessary heat and mass transport for growing compositionally tailored device-grade ternary crystals.

10.12 Conclusion

Ternary single-crystal substrates are important for advanced photonic, electronic, and photovoltaic-based power generation device structures. Large-scale crystal growth technology for growing reliable device-grade ternary substrates is still in its infancy and requires substantial engineering development. The Bridgman or gradient freezing process is the most suitable method for growing large-diameter III–V ternary bulk crystals. Ternary crystal growth requires low temperature gradient and low growth rate. Typical temperature gradient and growth rate are in the range of 5–15 °C/cm and 0.02–0.5 mm/h, respectively. A low temperature gradient helps to ensure a planar melt–solid interface shape, which is necessary for a uniform radial composition in the crystal as well as reducing the strain level in the crystal to avoid cracking. Low growth rates are necessary to avoid constitutional supercooling and interface breakdown, which lead to transition from single- to polycrystal growth and the generation of microcracks. Melt stirring using forced convection is crucial to avoid constitutional supercooling during growth. Stirring also helps in enhancing crystal growth rate. Special conditions are necessary for ternary single-crystal seed generation. The rate of compositional grading along the crystal during the seed generation process should be maintained below a specific level in order to avoid strain-gradient-related cracks in the crystal. Precise melt composition control using solute replenishment is absolutely essential for the growth of compositionally homogeneous crystals. In this chapter, an advanced crystal growth process has been described that could lead to growth of high-quality device-grade and commercially viable ternary substrates. The proof of concept for the process has been successfully demonstrated by laboratory-scale large-diameter crystal growth experiments. It is anticipated that the single-step process for seed generation and homogeneous crystal growth could be universally employed for all III–V ternary compounds.

References

10.1 V. Swaminathan, A.T. Macrander: *Materials Aspects of GaAs and InP Based Structures* (Prentice Hall, New Jersey 1991)

10.2 M. Neuberger: III–V ternary semiconducting compounds-data tables. In: *Handbook of Electronic Materials*, Vol. 7 (IFI/Plenum, New York 1972)

10.3 O. Madelung, M. Schulz (eds): Landolt–Börnstein, numerical data and functional relationships. In: *Science and Technology*, Semiconductors, Vol. 22(A) (Springer, New York 1987)

10.4 W.B. Pearson: *A Handbook of Lattice Spacings and Structures of Metals and Alloys*, Vol. 1 (Pergamon, New York 1956), Vol. 2 (1967)

10.5 M.B. Panish, M. Ilegems: Phase equilibria in ternary III–V systems, Prog. Solid State Chem. **7**, 39–83 (1972)

10.6 H.C. Casey Jr., M.B. Panish: *Heterojunction Lasers, Part B – Materials and Operating Characteristics*, Quantum Electronics Series (Academic, New York 1978)

10.7 I. Vurgaftman, J.R. Meyer, L.R. Ram-Mohan: Band parameters for III–V compound semiconductors and their alloys, J. Appl. Phys. **89**(11), 5815–5875 (2001)

10.8 Special issue on: Compliant and alternative substrate technology, J. Electron. Mater. **29** (2000)

10.9 K. Otsubo, H. Shoji, T. Kusunoki, T. Suzuki, T. Uchida, Y. Nishijina, K. Nakajima, H. Ishikawa: High T_0 (140 K) and low-threshold long-wavelength strained quantum well lasers on InGaAs ternary substances, Electron. Lett. **33**, 1795–1797 (1997)

10.10 H. Ishikawa: Theoretical gain of strained quantum well grown on an InGaAs ternary substrate, Appl. Phys. Lett. **63**, 712–714 (1993)

10.11 K. Otsubo, Y. Nishijima, T. Uchida, H. Shoji, K. Nakajima, H. Ishikawa: 1.3 μm InGaAs/InAlGaAs strained quantum well laser on InGaAs ternary substrates, Jpn. J. Appl. Phys. **38**, L312–L314 (1999)

10.12 H.X. Yuan, D. Grubisic, T.T.S. Wong: GaInSb photodetectors developed from single crystal bulk grown materials, J. Electron. Mater. **28**, 39–42 (1999)

10.13 T. Refaat, M.N. Abedin, V. Bhagwat, I.B. Bhat, P.S. Dutta, U.N. Singh: InGaSb photodetectors using InGaSb substrate for 2 μm applications, Appl. Phys. Lett. **85**(11), 1874–1876 (2004)

10.14 P.S. Dutta, J.M. Borrego, H. Ehsani, G. Rajagopalan, I.B. Bhat, R.J. Gutmann, G. Nichols, P.F. Baldasaro: GaSb and GaInSb thermophotovoltaic cells using diffused junction technology in bulk substrates,, AIP Conf. Proc. **653**, 392–401 (2002)

10.15 J. Merrill, D.C. Senft: Directions and materials challenges in high performance photovoltaics, J. Miner. Met. Mater. Soc. (JOM) **59**(12), 26–30 (2007)

10.16 P.S. Dutta, H.L. Bhat, V. Kumar: The physics and technology of gallium antimonide: An emerging optoelectronic material, J. Appl. Phys. **81**, 5821–5870 (1997)

10.17 K.J. Bachmann, F.A. Thiel, H. Schreiber Jr.: Melt and solution growth of bulk single crystals of quaternary III–V alloys, Prog. Cryst. Growth Charact. **2**, 171–206 (1979)

10.18 P.S. Dutta: III–V ternary bulk substrate growth technology: a review, J. Cryst. Growth **275**, 106–112 (2005)

10.19 W.A. Bonner, B.J. Skromme, E. Berry, H.L. Gilchrist, R.E. Nahory: Bulk single crystal GaInAs: LEC growth and characterization, Proc. 15th Int. Symp. GaAs Relat. Compd., Vol. 96, ed. by J.S. Harris (Institute of Physics, Bristol 1989) pp. 337–342

10.20 W.A. Bonner, B. Lent, D.J. Freschi, W. Hoke: Substrate quality of III–V single crystals for II–VI device applications: Growth and characterization, Proc. SPIE **2228**, 33–43 (1994)

10.21 S. Kodama, Y. Furumura, K. Kinoshita, H. Kato, S. Yoda: Single crystalline bulk growth of $In_{0.3}Ga_{0.7}As$ on GaAs seed using the multicomponent zone melting method, J. Cryst. Growth **208**, 165–170 (2000)

10.22 A. Mitric, T. Duffar, C. Diaz-Guerra, V. Corregidor, L.C. Alves, C. Garnier, G. Vian: Growth of GaInSb alloys by vertical Bridgman technique under alternating magnetic field, J. Cryst. Growth **287**(2), 224–229 (2006)

10.23 A. Tanaka, J. Shintani, M. Kimura, T. Sukegawa: Multi-step pulling of GaInSb bulk crystal from ternary solution, J. Cryst. Growth **209**, 625–629 (2000)

10.24 J.P. Garandet, T. Duffar, J.J. Favier: Vertical gradient freeze growth of ternary GaSb-InSb crystals, J. Cryst. Growth **106**, 426–436 (1990)

10.25 P.S. Dutta, A.G. Ostrogorsky: Suppression of cracks in $In_xGa_{1-x}Sb$ crystals through forced convection in the melt, J. Cryst. Growth **194**, 1–7 (1998)

10.26 P.S. Dutta, A.G. Ostrogorsky: Melt growth of quasi-binary $(GaSb)_{1-x}(InAs)_x$ crystals, J. Cryst. Growth **198/199**, 384–389 (1999)

10.27 P.S. Dutta, A.G. Ostrogorsky: Strong band gap narrowing in quasi-binary $(GaSb)_{1-x}(InAs)_x$ crystals grown from melt, J. Cryst. Growth **197**, 1–6 (1999)

10.28 P.S. Dutta, H.J. Kim, A. Chandola: Controlling heat and mass transport during the vertical Bridgman growth of homogeneous ternary III–V semiconductor alloys, Trans. Indian Inst. Met. **60**(2–3), 155–160 (2007)

10.29 P.S. Dutta, T.R. Miller: Engineering phase formation thermo-chemistry for crystal growth of homogeneous ternary and quaternary III–V compound semiconductors from melts, J. Electron. Mater. **29**, 956–963 (2000)

10.30 P.S. Dutta, T.R. Miller: Multicomponent homogeneous alloys and method for making same, US Patent 6613162 B1 (2003)

10.31 P.S. Dutta, A.G. Ostrogorsky: Alloys and methods for their preparation, US Patent 6273969 B1 (2001)

10.32 R. Pino, Y. Ko, P.S. Dutta: High-resistivity GaSb bulk crystals grown by the vertical Bridgman method, J. Electron. Mater. **33**(9), 1012–1015 (2004)

10.33 A. Chandola, H.J. Kim, S. Guha, L. Gonzalez, V. Kumar, P.S. Dutta: Below band-gap optical absorption in $Ga_xIn_{1-x}Sb$ alloys, J. Appl. Phys. **98**, 093103–093109 (2005)

10.34 H.J. Kim, A. Chandola, S. Guha, L. Gonzalez, V. Kumar, P.S. Dutta: Influence of native defects on the infrared transmission of undoped $Ga_{1-x}In_xSb$ bulk crystals, J. Electron. Mater. **34**(11), 1391–1398 (2005)

10.35 A. Chandola, R. Pino, P.S. Dutta: Below bandgap optical absorption in tellurium-doped GaSb, Semicond. Sci. Technol. **20**, 886–893 (2005)

10.36 R. Pino, Y. Ko, P.S. Dutta: Native defect compensation in III–V antimonide bulk substrates, Int. J. High-Speed Electron. Syst. **14**(3), 658–663 (2004)

10.37 R. Pino, Y. Ko, P.S. Dutta: Enhancement of infrared transmission in GaSb bulk crystals by carrier compensation, J. Appl. Phys. **96**(2), 1064–1067 (2004)

10.38 R. Pino, Y. Ko, P.S. Dutta, S. Guha, L. Gonzalez: Burstein–Moss shift in impurity-compensated bulk $Ga_{1-x}In_xSb$ substrates, J. Appl. Phys. **96**(9), 5349–5352 (2004)

10.39 W.D. Lawson, S. Nielsen: *Preparation of Single Crystals* (Butterworths Scientific Publications, London 1958)

10.40 W.A. Gault, E.M. Monberg, J.E. Clemens: A novel application of the vertical gradient freeze method to the growth of high quality III–V crystals, J. Cryst. Growth **74**, 491–506 (1986)

10.41 I.R. Grant: InP crystal growth. In: *Bulk Crystal Growth of Electronic, Optical and Optoelectronic Materials*, ed. by P. Capper (Wiley, England 2005), Chap. 4

10.42 M.R. Brozel, I.R. Grant: Growth of GaAs. In: *Bulk Crystal Growth of Electronic, Optical and Optoelectronic Materials*, ed. by P. Capper (Wiley, England 2005), Chap. 2

10.43 T. Asahi, K. Kainosho, K. Kohiro, A. Noda, K. Sato, O. Oda: Growth of III–V and II–VI single crystals by the vertical gradient freeze method. In: *Crystal Growth Technology*, ed. by H.J. Scheel, T. Fukuda (Wiley, England 2003), Chap. 15

10.44 T. Kawase, M. Tatsumi, Y. Nishida: Growth technology of III–V single crystals for production. In: *Crystal Growth Technology*, ed. by H.J. Scheel, T. Fukuda (Wiley, England 2003), Chap. 16

10.45 P. Rudolph, M. Jurisch: Fundamental and technological aspects of Czochralski growth of high quality semi-insulating GaAs crystals. In: *Crystal Growth Technology*, ed. by H.J. Scheel, T. Fukuda (Wiley, England 2003), Chap. 14

10.46 N.B. Singh, S.S. Mani, J.D. Adam, S.R. Coriell, M.E. Glicksman, W.M.B. Duval, G.J. Santoro, R. De-

Witt: Direct observations of interface instabilities, J. Cryst. Growth **166**, 364–369 (1996)

10.47 P. Capper, J.J.G. Gosney, C.L. Jones, M.J.T. Quelch: Quenching studies in Bridgman-grown $Cd_xHg_{1-x}Te$, J. Cryst. Growth **63**, 154–164 (1983)

10.48 P.S. Dutta, K.S. Sangunni, H.L. Bhat, V. Kumar: Experimental determination of melt–solid interface shapes and actual growth rates of gallium antimonide grown by vertical Bridgman technique, J. Cryst. Growth **141**, 476–478 (1994)

10.49 R.K. Route, M. Wolf, R.S. Feigelson: Interface studies during vertical Bridgman CdTe crystal growth, J. Cryst. Growth **70**, 379–385 (1984)

10.50 R. Singh, A.F. Witt, H.C. Gatos: Application of the Peltier effect for the determination of crystal growth rates, J. Electrochem. Soc. **115**, 112–113 (1968)

10.51 Y. Dabo, H. Nguyen Thi, S.R. Coriell, G.B. McFadden, Q. Li, B. Billia: Microsegregation in Peltier interface demarcation, J. Cryst. Growth **216**, 483–494 (2000)

10.52 L.L. Zheng, D.J. Larson Jr.: Thermoelectric effects on interface demarcation and directional solidification of bismuth, J. Cryst. Growth **180**, 293–304 (1997)

10.53 N. Duhanian, T. Duffar, C. Marin, E. Dieguez, J.P. Garandet, P. Dantan, G. Guiffant: Experimental study of the solid–liquid interface dynamics and chemical segregation in concentrated semiconductor alloy Bridgman growth, J. Cryst. Growth **275**, 422–432 (2005)

10.54 J.M. Bly, M.L. Kaforey, D.H. Matthiesen, A. Chait: Interface shape and growth rate analysis of Se/GaAs bulk crystals grown in the NASA crystal growth furnace (CGF), J. Cryst. Growth **174**, 220–225 (1997)

10.55 C.A. Wang, J.R. Carruthers, A.F. Witt: Growth rate dependence of the interface distribution coefficient in the system Ge–Ga, J. Cryst. Growth **60**, 144–146 (1982)

10.56 D.H. Matthiesen, M.E.K. Wiegel: Determination of the Peltier coeffcient of germanium in a vertical Bridgman–Stockbarger furnace, J. Cryst. Growth **174**, 194–201 (1997)

10.57 B. Fischer, J. Friedrich, H. Weimann, G. Muller: The use of time-dependent magnetic fields for control of convective flows in melt growth configurations, J. Cryst. Growth **198/199**, 170–175 (1999)

10.58 M.P. Volz, J.S. Walker, M. Schweizer, S.D. Cobb, F.R. Szofran: Bridgman growth of germanium crystals in a rotating magnetic field, J. Cryst. Growth **282**, 305–312 (2005)

10.59 Y. Ma, L.L. Zheng, D.J. Larson Jr.: Microstructure formation during BiMn/Bi eutectic growth with applied alternating electric fields, J. Cryst. Growth **262**, 620–630 (2004)

10.60 L.N. Brush, B.T. Murray: Crystal growth with applied current, J. Cryst. Growth **250**, 170–173 (2003)

10.61 T.A. Campbell, J.N. Koster: In situ visualization of constitutional supercooling within a Bridgman–Stockbarger system, J. Cryst. Growth **171**, 1–11 (1997)

10.62 T.A. Campbell, J.N. Koster: Visualization of liquid–solid interface morphologies in gallium subject to natural convection, J. Cryst. Growth **140**, 414–425 (1994)

10.63 T.A. Campbell, J.N. Koster: Radioscopic visualization of indium antimonide growth by the vertical Bridgman–Stockbarger technique, J. Cryst. Growth **147**, 408–410 (1995)

10.64 T. Schenk, H. Nguyen Thi, J. Gastaldi, G. Reinhart, V. Cristiglio, N. Mangelinck-Noel, H. Klein, J. Hartwig, B. Grushko, B. Billia, J. Baruchel: Application of synchrotron X-ray imaging to the study of directional solidification of aluminium-based alloys, J. Cryst. Growth **275**, 201–208 (2005)

10.65 P.G. Barber, R.K. Crouch, A.L. Fripp, W.J. Debnam, R.F. Berry, R. Simchick: A procedure to visualize the melt–solid interface in Bridgman grown germanium and lead tin telluride, J. Cryst. Growth **74**, 228–230 (1986)

10.66 R. K. Willardson, H.L. Goering: Preparation of III–V compounds, *Compound Semiconductors*, Vol. 1 (Reinhold Publishing Corporation, New York 1962)

10.67 M. Hansen (Ed.): *Constitution of Binary Alloys* (McGraw–Hill, New York 1958)

10.68 I. Grzegory, M. Bockowski, S. Porowski: GaN bulk substrates grown under pressure from solution in gallium. In: *Bulk Crystal Growth of Electronic, Optical and Optoelectronic Materials*, ed. by P. Capper (Wiley, England 2005), Chap. 6

10.69 P. Rudolph: Thermodynamic fundamentals of phase transitions applied to crystal growth processes. In: *Crystal Growth Technology*, ed. by H.J. Scheel, T. Fukuda (Wiley, England 2003), Chap. 2

10.70 C.T. Lin, E. Schonherr, H. Bender: Growth and characterization of doped and undoped AlSb single crystals, J. Cryst. Growth **104**, 653–660 (1990)

10.71 R. Pino, Y. Ko, P.S. Dutta: Adhesion-free growth of AlSb bulk crystals in silica crucibles, J. Cryst. Growth **290**, 29–34 (2006)

10.72 T. Duffar, J.M. Gourbil, P. Boiton, P. Dusserre, N. Eustathopoulos: Full encapsulation by molten salts during the Bridgman growth process, J. Cryst. Growth **179**, 356–362 (1997)

10.73 P.S. Dutta, A.G. Ostrogorsky, R.J. Gutmann: Bulk growth of GaSb and $Ga_xIn_{1-x}Sb$, Proc. 3rd NREL Conf. Thermophotovolt. Gener. Electr., AIP Conf. Proc., Vol. 401 (1997) pp. 157–166

10.74 A.R. Clawson: Guide to References on III–V semiconductor chemical etching, Mater. Sci. Eng. R **31**, 1–438 (2001)

10.75 K. Ishida, H. Tokunaga, H. Ohtani, T. Nishizawa: Data base for calculating phase diagrams of III–V alloy semiconductors, J. Cryst. Growth **98**, 140–147 (1989)

10.76 G.B. Stringfellow: Calculation of ternary and quaternary III–V phase diagrams, J. Cryst. Growth **27**, 21–34 (1974)

10.77 T.C. Yu, R.F. Brebrick: Thermodynamic analysis of the In-Ga-Sb System, Metall. Mater. Trans. A **25**, 2331–2340 (1994)

10.78 A. Kumar: Growth of Thick Lattice Mis-matched Layers of GaInAsSb on GaAs Substrates from Quaternary Melts. Ph.D. Thesis (Rensselaer Polytechnic Institute, Troy, New York 2006)

10.79 W.G. Pfann: *Zone-melting* (Wiley, New York 1959)

10.80 W.A. Tiller: *The Science of Crystallization: Macroscopic Phenomena and Defect Generation* (Cambridge Univ. Press, New York 1991)

10.81 J.A. Burton, R.C. Prim, W.P. Slichter: The distribution of solute in crystals grown from the melt. Part I. Theoretical, J. Chem. Phys. **21**, 1987–1991 (1953)

10.82 W.A. Tiller, K.A. Jackson, J.W. Rutter, B. Chalmers: The redistribution of solute atoms during the solidification of metals, Acta Metall. **1**, 428–437 (1953)

10.83 S. Sen, R.A. Lefever: Influence of magnetic field on vertical Bridgman–Stockbarger growth of InGaSb, J. Cryst. Growth **43**, 526–530 (1978)

10.84 H.P. Utech, M.C. Flemings: Elimination of solute banding in indium antimonide crystals by growth in a magnetic field, J. Appl. Phys. **37**, 2021–2024 (1966)

10.85 J. Kang, T. Fukuda: Growth exploration of compositionally uniform bulk semiconductors under a high magnetic field of 80 000 Gauss, Mater. Sci. Eng. B **75**, 149–152 (2000)

10.86 A.G. Ostrogorsky: Numerical simulation of single crystal growth by submerged heater method, J. Cryst. Growth **104**, 233–238 (1990)

10.87 A.G. Ostrogorsky, G. Müller: Normal and zone solidification using the submerged heater method, J. Cryst. Growth **137**, 64–71 (1994)

10.88 A.F. Witt, H.C. Gatos, M. Lichtensteiger, M.C. Lavine, C.J. Herman: Crystal growth and steady-state segregation under zero gravity: InSb, J. Electrochem. Soc. **122**, 276 (1975)

10.89 J.F. Yee, M.-C. Lin, K. Sarma, W.R. Wilcox: The influence of gravity on crystal defect formation in InSb-GaSb alloys, J. Cryst. Growth **30**, 185–192 (1975)

10.90 K. Okitsu, Y. Hayakawa, T. Yamaguchi, A. Hirata, S. Fujiwara, Y. Okano, N. Imaishi, S. Yoda, T. Oida, M. Kumagawa: Melt mixing of the In/GaSb/Sb solid combination by diffusion under microgravity, Jpn. J. Appl. Phys. **36**, 3613–3619 (1997)

10.91 Y. Hayakawa, K. Balakrishnan, H. Komatsu, N. Murakami, T. Nakamura, T. Koyama, T. Ozawa, Y. Okano, M. Miyazawa, S. Dost, L.H. Dao, M. Kumagawa: Drop experiments on crystallization of InGaSb semiconductor, J. Cryst. Growth **237–239**, 1831–1834 (2002)

10.92 A. Eyer, H. Leister, R. Nitsche: Floating zone growth of silicon under microgravity in a sounding rocket, J. Cryst. Growth **71**, 173–182 (1985)

10.93 C.H. Su, Y.G. Sha, S.L. Lehoczky, F.R. Szofran, C.C. Gillies, R.N. Scripa, S.D. Cobb, J.C. Wang: Crystal growth of HgZnTe alloy by directional solidification in low gravity environment, J. Cryst. Growth **234**, 487–497 (2002)

10.94 K. Hashio, M. Tatsumi, H. Kato, K. Kinoshita: Directional solidification of $In_xGa_{1-x}As$, J. Cryst. Growth **210**, 471–477 (2000)

10.95 W.W. Mullins, R.F. Sekerka: Stability of a planar interface during solidification of a dilute binary alloy, J. Appl. Phys. **35**, 444–451 (1964)

10.96 G.B. McFadden, S.R. Coriell: Thermosolutal convection during directional solidification. II. Flow Transitions, Phys. Fluids **30**(3), 659–671 (1987)

10.97 D. Elwell, H.J. Scheel: *Crystal Growth from High-Temperature Solutions* (Academic, London 1975)

10.98 H.J. Kim: Bulk Crystal Growth Process for Compositionally Homogeneous GaInSb Substrates. Ph.D. Thesis (Rensselaer Polytechnic Institute, Troy, New York 2005)

10.99 K.J. Vogel: Solute Redistribution and Constitutional Supercooling Effects in Vertical Bridgman Grown InGaSb by Accelerated Crucible Rotation Technique. Ph.D. Thesis (Rensselaer Polytechnic Institute, Troy, New York 2004)

10.100 Y. Nishijima, K. Nakajima, K. Otsubo, H. Ishikawa: InGaAs single crystal using a GaAs seed grown with the vertical gradient freeze technique, J. Cryst. Growth **197**, 769–776 (1999)

10.101 D. Reid, B. Lent, T. Bryskiewicz, P. Singer, E. Mortimer, W.A. Bonner: Cellular structure in LEC ternary $Ga_{1-x}In_xAs$ crystals, J. Cryst. Growth **174**, 250–255 (1997)

10.102 W.A. Bonner, R.E. Nahory, H.L. Glichrist, E. Berry: Semi-insulating single crystal GaInAs: LEC growth and Characterization, *Semi-Insulating III–V Materials* (1990) pp. 199–204

10.103 K. Nakajima, T. Kusunoki, K. Otsubo: Bridgman growth of compositionally graded $In_xGa_{1-x}As$ ($x = 0.05 - 0.30$) single crystals for use as seeds for $In_{0.25}Ga_{0.75}As$ crystal growth, J. Cryst. Growth **173**, 42–50 (1997)

10.104 Y. Nishijima, K. Nakajima, K. Otsubo, H. Ishikawa: InGaAs single crystal with a uniform composition in the growth direction grown on an InGaAs seed using the multicomponent zone growth method, J. Cryst. Growth **208**, 171–178 (2000)

10.105 T. Suzuki, K. Nakajima, T. Kusunoki, T. Katoh: Multicomponent zone melting growth of ternary InGaAs bulk crystal, J. Electron. Mater. **25**(3), 357–361 (1996)

10.106 A. Watanabe, A. Tanaka, T. Sukegawa: Pulling technique of a homogeneous GaInSb alloy under solute-feeding conditions, Jpn. J. Appl. Phys. **32**, L793–L795 (1993)

10.107 H.-J. Sell: Growth of GaInAs bulk mixed crystals as a substrate with a tailored lattice parameter, J. Cryst. Growth **107**, 396–402 (1991)
10.108 W.F. Leverton: Floating crucible technique for growing uniformly doped crystals, J. Appl. Phys. **29**, 1241–1244 (1958)
10.109 T. Kusunoki, K. Nakajima, K. Kuramata: Constant Temperature LEC growth of uniform composition InGaAs bulk crystals through continuous supply of GaAs, Inst. Phys. Conf. Ser. **129**, 37–42 (1992)
10.110 K. Nakajima, T. Kusunoki: Constant temperature growth of uniform composition InGaAs bulk crystals by supplying GaAs, Inst. Phys. Conf. Ser. **120**, 67–71 (1991)
10.111 T. Kusunoki, C. Takenaka, K. Nakajima: Growth of ternary $In_{0.14}Ga_{0.86}As$ bulk crystal with uniform composition at constant temperature through GaAs supply, J. Cryst. Growth **115**, 723–727 (1991)
10.112 T. Ashley, J.A. Beswick, B. Cockayne, C.T. Elliott: The growth of ternary substrates of indium gallium antimonide by the double crucible Czochralski technique, Inst. Phys. Conf. Ser. **144**, 209–213 (1995)
10.113 A. Tanaka, A. Watanabe, M. Kimura, T. Sukegawa: The solute-feeding Czochralski method for homogeneous GaInSb bulk alloy pulling, J. Cryst. Growth **135**, 269–272 (1994)
10.114 A. Tanaka, T. Yoneyama, M. Kimura, T. Sukegawa: Control of GaInSb alloy composition grown from ternary solution, J. Cryst. Growth **186**, 305–308 (1998)
10.115 M.H. Lin, S. Kou: Czochralski pulling of InSb single crystals from a molten zone on a solid feed, J. Cryst. Growth **193**, 443–445 (1998)
10.116 M.H. Lin, S. Kou: Dopant segregation control in Czochralski crystal growth with a wetted float, J. Cryst. Growth **132**, 461–466 (1993)
10.117 T. Ozawa, Y. Hayakawa, M. Kumagawa: Growth of III–V ternary and quaternary mixed crystals by the rotationary Bridgman method, J. Cryst. Growth **109**, 212–217 (1991), see also [10.134]
10.118 K. Kinoshita, H. Kato, S. Matsumoto, S. Yoda: Growth of homogeneous $In_{1-x}Ga_xSb$ crystals by the graded solute concentration method, J. Cryst. Growth **216**, 37–43 (2000)
10.119 K. Kinoshita, H. Kato, M. Iwai, T. Tsuru, M. Muramatsu, S. Yoda: Homogeneous $In_{0.3}Ga_{0.7}As$ crystal growth by the traveling liquidus-zone method, J. Cryst. Growth **225**, 59–66 (2001)
10.120 A. Chandola: Bulk Crystal Growth and Infrared Absorption Studies of GaInSb. Ph.D. Thesis (Rensselaer Polytechnic Institute, Troy, New York 2005)
10.121 H. Kim, A. Chandola, R. Bhat, P.S. Dutta: Forced convection induced thermal fluctuations at the solid–liquid interface and its effect on the radial alloy distribution in vertical Bridgman grown $Ga_{1-x}In_xSb$ bulk crystals, J. Cryst. Growth **289**, 450–457 (2006)
10.122 J.C. Brice: *The Growth of Crystals from Liquids* (North-Holland, Amsterdam 1973)
10.123 P.S. Dutta, K.S. Sangunni, H.L. Bhat, V. Kumar: Growth of gallium antimonide by vertical Bridgman technique with planar crystal-melt interface, J. Cryst. Growth **141**, 44–50 (1994)
10.124 C.E. Chang, W.R. Wilcox: Control of interface shape in the vertical Bridgman–Stockbarger technique, J. Cryst. Growth **21**, 135–140 (1974)
10.125 A. Yeckel, J.J. Derby: Computer modeling of bulk crystal growth. In: *Bulk Crystal Growth of Electronic, Optical and Optoelectronic Materials*, ed. by P. Capper (Wiley, England 2005), Chap. 3
10.126 A. Yeckel, J.J. Derby: Computational simulations of the growth of crystals from liquids. In: *Crystal Growth Technology*, ed. by H.J. Scheel, T. Fukuda (Wiley, England 2003), Chap. 6
10.127 V.I. Polezhaev: Modeling of technologically important hydrodynamics and heat/mass transfer processes during crystal growth. In: *Crystal Growth Technology*, ed. by H.J. Scheel, T. Fukuda (Wiley, England 2003), Chap. 8
10.128 C.L. Jones, P. Capper, J.J.G. Gosney: Thermal modeling of Bridgman crystal growth, J. Cryst. Growth **56**, 581–590 (1982)
10.129 H.P. Greenspan: *The Theory of Rotating Fluids* (Cambridge Univ. Press, London 1968)
10.130 H.J. Scheel, R.H. Swendsen: Evaluation of experimental parameters for growth of homogeneous solid solutions, J. Cryst. Growth **233**, 609–617 (2001)
10.131 H.J. Scheel, E.O. Schulz-Dubois: Flux growth of large crystals by accelerated crucible-rotation technique, J. Cryst. Growth **8**, 304–306 (1971)
10.132 H.J. Scheel: Accelerated crucible rotation: A novel stirring technique in high-temperature solution growth, J. Cryst. Growth **13/14**, 560–565 (1972)
10.133 J.B. Mullin: The Role of Magnetic Fields in Crystal Growth, Special Issue of Prog. Cryst. Growth Charact. Mater. **38**, 1–6 (1999), see whole issue
10.134 T. Ozawa, Y. Hayakawa, K. Balakrishna, F. Ohonishi, T. Koyama, M. Kumagawa: Growth of $In_xGa_{1-x}As$ bulk mixed crystals with a uniform composition by the rational Bridgman method, J. Cryst. Growth **229**, 124–129 (2001)

11. Growth and Characterization of Antimony-Based Narrow-Bandgap III–V Semiconductor Crystals for Infrared Detector Applications

Vijay K. Dixit, Handady L. Bhat

Materials for the generation and detection of 7–12 μm wavelength radiation continue to be of considerable interest for many applications such as night vision, medical imaging, sensitive pollution gas monitoring, etc. For such applications HgCdTe has been the main material of choice in the past. However, HgCdTe lacks stability and uniformity over a large area, and only works under cryogenic conditions. Because of these problems, antimony-based III–V materials have been considered as alternatives. Consequently, there has been a tremendous growth in research activity on InSb-based systems. In fact, InSb-based compounds have proved to be interesting materials for both basic and applied research. This chapter presents a comprehensive account of research carried out so far. It explores the materials aspects of indium antimonide (InSb), indium bismuth antimonide ($InBi_xSb_{1-x}$), indium arsenic antimonide ($InAs_xSb_{1-x}$), and indium bismuth arsenic antimonide ($InBi_xAs_ySb_{1-x-y}$) in terms of crystal growth in bulk and epitaxial forms and interesting device feasibility. The limiting single-phase composition of $InAs_xSb_{1-x}$ and $InBi_xSb_{1-x}$ using near-equilibrium technique has been also addressed. An overview of the structural, transport, optical, and device-related properties is presented. Some of the current areas of research and development have been critically reviewed and their significance for both understanding the basic physics as well as device applications are discussed. These include the role of defects and impurity on structural, optical, and electrical properties of the materials.

Tremendous efforts have been applied in the development of infrared (IR) detectors and sensor materials. Much of the research on IR optoelectronic has been focused mostly on military needs, with particular emphasis on the development of high-performance detectors operating in the 3–5 and 7–12 μm wavelength bands. In recent years civilian needs have become more dominant due to the development of IR light-emitting diodes (LED) and lasers, which provide low-cost sensitive pollution monitoring systems that detect trace gases by their fundamental vibrational–rotational absorption bonds [11.1]. Other applications include fire-fighting, environmental monitoring, fiber-optic and free-space optical communication systems, landfill gas monitoring, fuel-gas analysis, personal safety, sports, medicine, and a variety of horticultural uses that include total organic carbon dioxide measurement. For all these applications HgCdTe (MCT) has been the main material of choice in the past, but it has proved to be a difficult compound to prepare due to the high vapor pressure of Hg and weak Hg bond. Also its material parameters change with time, resulting in poor mechanical and thermal properties [11.2]. Because of these problems associated with MCT, antimony (Sb)-based III–V materials have been considered as more attractive alternatives. Also they have stronger covalent bonds between indium (In) and Sb, which makes them stable compounds. In particular, InSb-based materials such as $InAs_xSb_{1-x}$, $InBi_xSb_{1-x}$, and $InBi_xAs_ySb_{1-x-y}$ have been extensively explored for many years [11.2–12]. In this chapter, we will review mostly the investigations carried out on these materials for the past two decades. The review highlights the current status of understanding of the crystal growth process and various physical properties of InSb, $InBi_xSb_{1-x}$, $InAs_xSb_{1-x}$, and $InBi_xAs_ySb_{1-x-y}$. While InSb continues to be grown commercially, very few reports are available on the growth of large-sized crystals and wafers with high quality. Some recent studies have surely led to better understanding of the problems associated with the enhancement of the size and quality of the grown crystals.

Substantial bowing in the energy gap to values below those of InSb (0.17 eV) and InAs (0.38 eV) occurs in indium arsenic antimonide. The energy gap of $InAs_xSb_{1-x}$ continuously decreases with increasing x and attains a minimum value of 0.1 eV for $x = 0.4$, at room temperature [11.13–19]. Furthermore, the very low effective mass of $InAs_xSb_{1-x}$ across the compositional range raises the prospect of using this material extensively for detector applications. Although the benefits of alloying As in InSb were discovered more than 34 years ago and have been studied since then, all studies have been carried out on bulk samples made from crudely prepared polycrystals [11.19, 20]. The main growth limitations in $InAs_xSb_{1-x}$ arise from the wide separation between the liquidus and solidus curves in the temperature–composition phase diagram and the very low diffusion rates in the solid phase. Hence most bulk crystal growth (near equilibrium) has been carried out using gradient freeze and zone recrystallization and with long annealing duration, usually resulting in polycrystals.

$InBi_xSb_{1-x}$ is a very interesting material because it is composed of the semimetal InBi and the semiconductor InSb and hence the bandgap energy and lattice constant can be varied over a wide range. Consequently, these crystals are also useful for the applications mentioned above. However, this alloy is even more difficult to grow in single-crystal form because of the wide separation between the liquidus and solidus lines in its phase diagram. This leads to constitutional supercooling in the solution below the growth interface. Hence, the composition ratio in the grown crystals will not be the same as that in the solution. Recently *Dixit* et al. [11.9, 12] reported successful growth of $InAs_xSb_{1-x}$ and $InBi_xSb_{1-x}$ single crystals using the rotatory Bridgman method (RBM). Efforts are also being made for possible integration of these materials onto suitable semi-insulating and IR-transparent substrates. CdTe is the only semi-insulating and lattice-matched substrate available for this purpose, but it is very difficult to prevent In_2Te_3 precipitate formation at the interface during growth [11.21, 22]. Hence, Si, GaAs, and InP have been used as substrates for epitaxial growth of these materials [11.23–25]. In spite of the large lattice mismatch between GaAs and these materials, there are many reports on the growth of these heterostructures using molecular-beam epitaxy (MBE) [11.26, 27], metalorganic vapor-phase epitaxy (MOVPE) [11.28–32], liquid-phase epitaxy (LPE) [11.5, 10, 33–35], and melt epitaxy (ME) [11.36–38]. The lattice-mismatched heteroepitaxy affects the structural, optical, and electrical properties of these materials and these issues have been addressed in this review.

11.1 Importance of Antimony-Based Semiconductors

After silicon, germanium, and gallium arsenide, indium antimonide is perhaps the most studied semiconductor. This is because it has the smallest energy gap (0.17 eV) among the III–V binary semiconductors. InSb has a strong band nonparabolicity, a very large g factor and is intrinsic at room temperature [11.40–50]. Due to these intrinsic properties, its physics is qualitatively different from that of other common semiconductors. The ease with which quantum phenomena can be seen and cleanly modeled has for a long time made it a favorite of semiconductor researchers. Two material-specific parameters that define the utility of a possible device are the energy gap and the effective mass. These can be tuned by strained epitaxy to some extent. However, alloying of two or more semiconductors drastically widens the available parameter space of various physical properties. Most physical properties (energy gap, effective mass, and lattice constant) of the alloy are continuous functions of the alloy composition and interpolate between the end members as a polynomial of low order. The energy gap of InSb, although the smallest among the III–V binaries, is not small enough to be used in practical long-wavelength infrared detectors. Hence the material has to be engineered so that its gap is within one of the wavelength windows where the atmospheric gases are transparent, i.e., 3–5 and 7–12 μm. The desired gap tailoring can be accomplished by alloying with low-bandgap semiconductors or semimetal.

InSb–InAs is one of the interesting alloy of this class because the substitution of a fraction of antimony sites in InSb with isovalent arsenic reduces the energy gap to a value lower than the energy gap of either of the parent compounds; it consequently has the lowest energy gap among the III–V semiconductors. This system also has one of the largest bowing parameters among the semiconductor alloys (Fig. 11.1a) [11.13]. Therefore the effects of alloy disorder in determining the physical properties may be expected to be more significant. By alloying with a suitable fraction of InAs, a room-temperature energy gap in both atmospheric wavelength windows can be achieved.

The desired energy gap tailoring can be accomplished by alloying with InBi also (Fig. 11.1b). Since Bi is much larger than Sb it produces rapid reduction of the bandgap of InSb at the rate of 36 meV/%Bi [11.51–54]. Incorporation of Bi in InAsSb produces an even larger reduction in bandgap (55 meV/%Bi) as determined by absorption and photoluminescence studies. *Ma* et al. [11.54] reported Bi incorporation in $InAs_xSb_{1-x}$ lattice and decrease in the bandgap energy of MOVPE-grown epilayer. The high quality of these layers was evidenced by the production of photoluminescence. *Huang* et al. [11.51] calculated the expected bandgap energies for the $InBi_xAs_ySb_{1-x-y}$ alloys having the optimum Bi concentration by linearly interpolating the values of dE_g/dx between InAs and InSb, yielding $dE_g/dx = -55 \pm 19x$ (meV/%Bi). The variation of the energy gap with alloy composition manifests in the optical and electrical properties of respective materials.

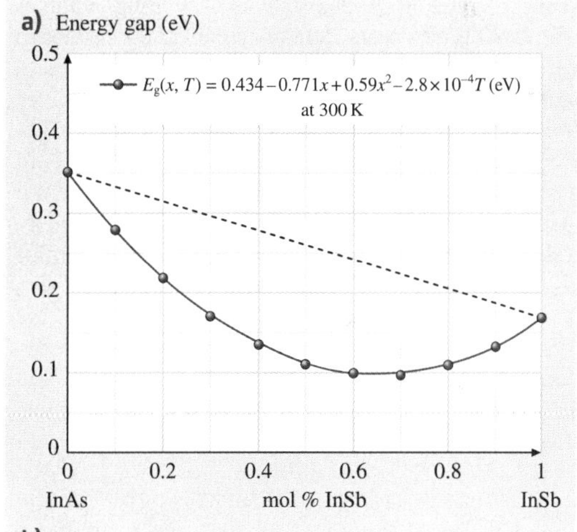

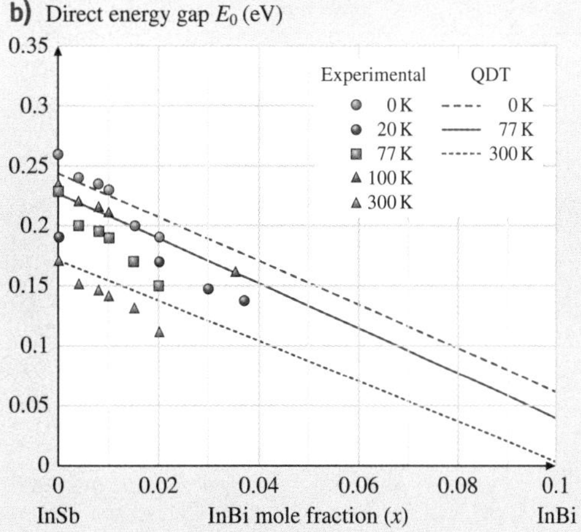

Fig. 11.1a,b Energy gap bowing for (a) $InAs_xSb_{1-x}$, and (b) $InBi_xSb_{1-x}$ (after [11.13, 39])

11.2 Phase Diagrams

In order to grow homogeneous and stoichiometric bulk and epitaxial single crystals of binaries and ternaries it is essential to understand the phase diagrams of these materials. The phase diagram of InSb, $InAs_xSb_{1-x}$, and $InBi_xSb_{1-x}$ are briefly described below.

11.2.1 InSb

The phase diagram of InSb was determined by *Liu* and *Peretti* [11.56] and is reproduced in Fig. 11.2a. It is characterized by the presence of two eutectics occurring at 0.8 and 68.2 at. % Sb. From the diagram it is clear that at the extreme left there exists a phase consisting of pure In (α-phase) with a melting point of 156 °C. At the extreme right is elemental Sb (γ-phase) with a melting point of 630 °C. InSb has a congruent melting temperature below the melting point of one of its constituents. This leads to certain differences between solution growth and congruent melt growth in terms of composition of In and Sb. Of interest in the phase diagram is the β-phase in the indium–antimony system, which has a congruent melting temperature of 525 °C. The transition from solid to liquid phase occurs at composition of 50 at. %. This is the point where the crystal is grown stoichiometrically. Alloys with a deviation as small as 0.5% from the stoichiometric ratio show phase separation, making the phase very sensitive to composition. The β-phase divides the diagram into two subsystems, namely the In–InSb (the $\alpha+\beta$-phase) and the InSb–Sb (the $\beta+\gamma$-phase) alloys. Also represented are the $L+\alpha$- and the $L+\gamma$-phases. Above the curve, InSb is in the liquid phase. For the growth of InSb epilayers using LPE, In–InSb region is preferred over other region, because In has a lower vapor pressure compared with Sb.

11.2.2 $InAs_xSb_{1-x}$

Shih and *Peretti* [11.55] investigated the phase diagram of the InAs–InSb system and obtained a degenerate eutectic diagram with the terminal solid solution with 2% InSb. *Goryunova* and *Fedorova* [11.58] also reported

Part B | 11.2

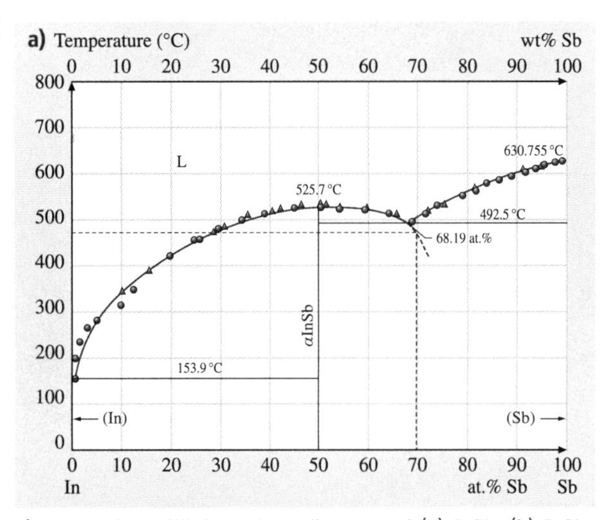

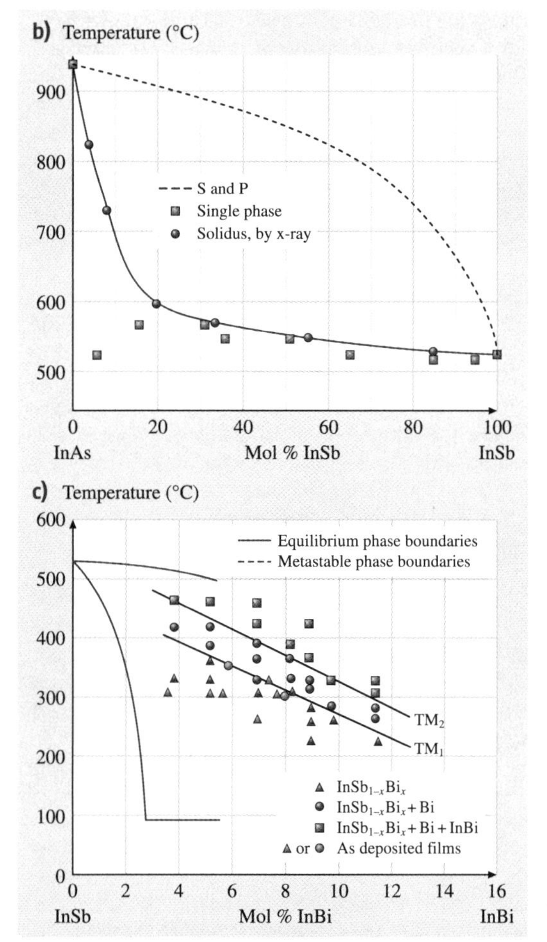

Fig. 11.2a–d Equilibrium phase diagrams of **(a)** InSb, **(b)** InSb–InAs, and **(c)** InSb–InBi (after [11.55–57])

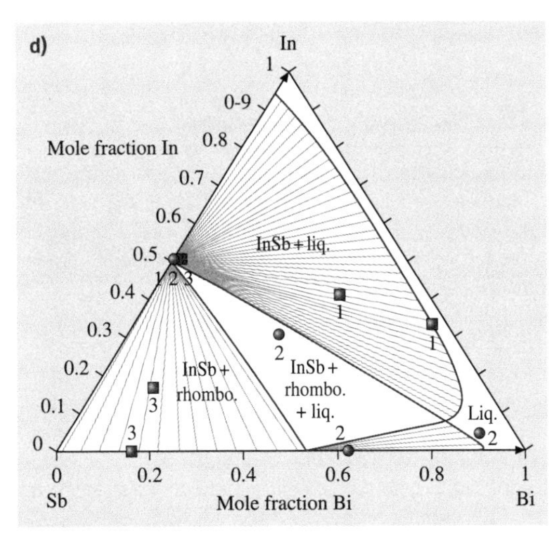

Fig. 11.2 (d) Equilibrium phase diagrams of $InBi_xSb_{1-x}$ at 300 K (after [11.62])

investigation on the same system with a maximum terminal solid solution with ≈ 3% InSb, which was in reasonable agreement with the result of *Shih* and *Peretti* [11.55]. Later, *Woolley* et al. [11.59] showed the complete solid solution for all compositions. *Stringfellow* and *Greene* [11.60] also reported the full phase diagram of $InAs_xSb_{1-x}$. The analysis was carried out using x-ray and differential thermal analysis. They made $InAs_xSb_{1-x}$ polycrystalline samples for the complete range of composition and annealed them at 525 °C for 3 months. Alloys over a considerable range of composition were approaching a single-phase state, but alloys in the approximate range of 90–40 mol % InAs showed an apparent two-phase equilibrium condition when investigated by x-ray methods. When alloys of the same composition were first annealed to single phase at 550 °C followed by annealing at 525 °C for approximately 3 months, no indication of any splitting into a two-phase condition could be observed by x-ray investigation. The complete phase diagram for all composition is shown in Fig. 11.2b. It would thus appear that at these compositions the diffusion rates in the solid are so low that, even at temperatures only ≈ 50 °C below the solidus, equilibrium cannot be attained under normal practical conditions, and the form of the equilibrium diagram at temperature below 550 °C cannot be determined. *Dutta* and *Miller* [11.61] also reported studies on the phase diagram of this material.

11.2.3 $InBi_xSb_{1-x}$

The phase diagram of $InBi_xSb_{1-x}$ was determined by *Joukoff* and *Jean-Louis* [11.63] using differential thermal analysis (DTA) only up to 6 at. % Bi. They found that the maximum equilibrium solubility limit of Bi was 2.6 at. % in InSb, but *Zilko* and *Greene* [11.57] reported that in metastable conditions the solubility of Bi in InSb is 12 at. %. As can be seen from Fig. 11.2c wide separation exists between the liquidus and solidus lines in the phase diagram, which leads to constitutional supercooling in the solution below the growth interface. Hence, the composition ratio in the grown crystals will not be same as that in the solution. Two metastable phase boundaries T_{M1} and T_{M2} are also shown in Fig. 11.2c. Below the T_{M1} phase boundary InBi–InSb forms one single stable $InBi_xSb_{1-x}$ phase, while in the region between the two-phase boundaries (T_{M1} and T_{M2}) the solution forms stable $InBi_xSb_{1-x}$ and Bi phase. In the region above T_{M2} the solution forms stable $InBi_xSb_{1-x}$, InBi, and Bi phases. Very recently *Minic* et al. [11.62] reported thermodynamic predictions for the equilibria of the InSbBi system. They estimated the phase diagram at 300 °C and compared it with experimental results obtained by scanning electron microscopy (SEM) analysis (Fig. 11.2d). They experimentally determined that liquid and rhombohedral phases of $In_{0.31}Sb_{0.38}Sn_{0.31}$ sample contained higher contents of bismuth compared with the prediction, which suggests the need for the introduction of ternary interaction parameters for the calculation.

11.3 Crystal Structure and Bonding

For understanding the optical and electrical properties of InSb, $InAs_xSb_{1-x}$, and $InBi_xSb_{1-x}$ crystals it is essential to understand the crystal structures and bonding of these materials, which are briefly described below.

11.3.1 Crystal Structure and Bonding of InSb

The structure of InSb was first determined by *Goldschmidt* [11.43] and later in detail by *Iandelli* [11.64] using conventional x-ray structure determination tech-

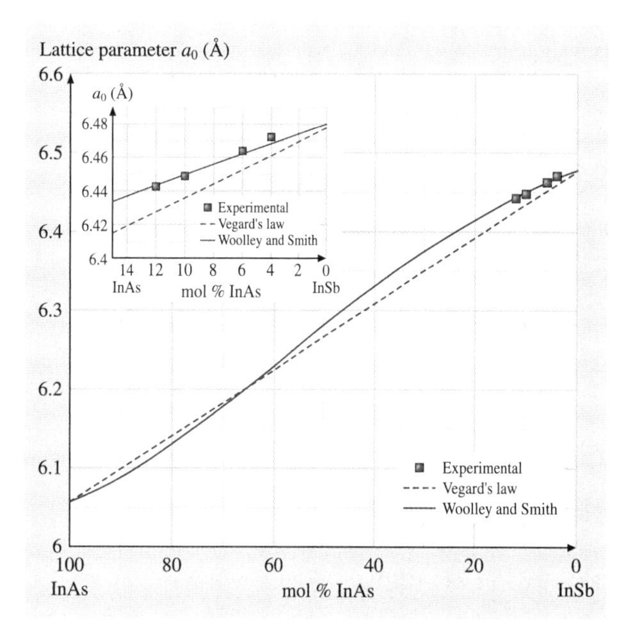

Fig. 11.3 Lattice parameter variation of InAsSb with composition (after [11.5, 20])

niques and found to be of zincblende type. The space group of InSb is $F\bar{4}3m$ [11.43] and hence it belongs to the point group $\bar{4}3m$. In the structure, each In atom is coordinated to four Sb atoms and vice versa. The presence of dissimilar atoms in the lattice imparts a directional character to the lattice. Hence, the direction from the In to the Sb is not equivalent to that from Sb to In. Thus, due to the lack of center of symmetry, the 111 direction forms the polar axis and there is a distinction between the [111] and the $[\bar{1}\bar{1}\bar{1}]$ directions. Conventionally, the direction from group III to V atom is $[\bar{1}\bar{1}\bar{1}]$ and from group V to III atom is [111]. In the zincblende structure, the crystal viewed perpendicular to [111] direction appears as sheets of alternate group III and V atoms stacked over one another. *Dewald* [11.65], while studying the growth of oxide layers on InSb, noticed a difference in the growth kinetics of (111) and $(\bar{1}\bar{1}\bar{1})$ surfaces. This difference in the behavior of the two surfaces was also noticed by other workers [11.66] during the observation of etch pits on InSb using CP4 ($HF : HNO_3 : CH_3COOH : H_2O = 3 : 5 : 3 : 20$) etchant. For example, it has been shown that dislocation etch pits appear on $(\bar{1}\bar{1}\bar{1})$ and not on (111) face. Also, growth of InSb crystals is easier on $(\bar{1}\bar{1}\bar{1})$ face than on (111) face. The bonding between In and Sb is a combination of two idealized states (purely ionic and purely covalent states) as represented by the equation below

$$\Phi = a_{\text{ionic}}\Phi_{\text{ionic}} + a_{\text{covalent}}\Phi_{\text{covalent}} . \qquad (11.1)$$

This means that the bonding is of mixed nature and is characterized by the ionicity, which is the ratio of a_{ionic} to a_{covalent} [11.68]. This lends a mixed character to the lattice, giving rise to a net polarity along certain directions.

11.3.2 Structural Properties of $InAs_xSb_{1-x}$

There are not many reports on the structural properties of $InAs_xSb_{1-x}$ except for a study on the variation of lattice parameter with composition [11.20]. The variation of lattice parameter with composition is shown in Fig. 11.3. The experimental curve (solid line) obtained by *Woolley* and *Warner* [11.20] for the whole compositional range (in polycrystalline samples) crosses Vegard's line at approximately 67 mol % InAs, indicating that Vegard's law is not satisfied except at one point ($x = 0.67$). Recently *Dixit* et al. [11.5, 69] also reported similar observation up to 12% As. Their results are shown in the inset of Fig. 11.3. The maximum difference in the lattice parameters (6.330 Å) obtained from experimental curve (from Woolley and Warner's work) and the Vegard line is 0.030 Å at 33% InSb.

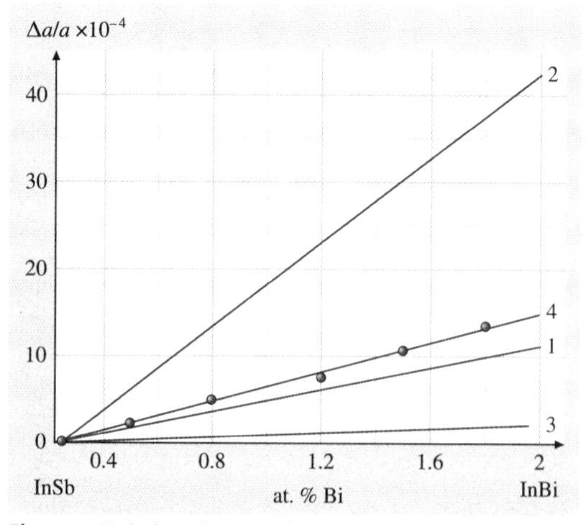

Fig. 11.4 Relative change of lattice parameter of InBiSb for Bi in (1) antimony position, (2) interstitial position, and (3) indium position, and (4) experimental curve (after [11.67])

11.3.3 Crystal Chemical Aspect of Bi Substitution in InSb

A large region of the phase diagram of $InBi_xSb_{1-x}$ system corresponds to the region of primary crystallization of solid solutions based on InSb. Since In forms two stable compounds with Bi, depending on the composition and temperature of the molten solution, Bi can occur in three basic forms, Bi, InBi, and In_2Bi, in the melt [11.66]. By varying the crystallization conditions, one can follow how various forms of Bi in the melt influence its solubility in InSb as well as structural properties of $InBi_xSb_{1-x}$. The character of Bi in the crystal lattice of InSb was determined by comparing the experimental composition dependence of the lattice parameter a with that calculated theoretically for substitutional and interstitial solid solutions and also by studying the spectra of diffuse scattering of x-rays. The values of a were measured with a precision of $\pm 5\times 10^{-5}$ Å [11.70]. By comparing the theoretical and experimental composition dependences of the relative change in the lattice parameters of $InBi_xSb_{1-x}$, one can determine the position of Bi in InSb, as shown in Fig. 11.4. The results indicate that, when the relative change in the lattice parameter with Bi composition is large, Bi goes into interstitial position, whereas when it is small Bi goes into In position. For moderate change in the lattice parameter with Bi composition, Bi goes into Sb position.

11.4 Material Synthesis and Purification

Since the discovery of InSb material, considerable attention has been devoted to its growth in single-crystal form [11.71–75]. Growth of good-quality InSb single crystal depends on the purity of the source materials and homogeneous synthesis of the starting materials. InSb is synthesized by taking stoichiometric amounts of In and Sb in a quartz ampoule, kept in an atmosphere of pure argon or hydrogen. The charge is melted and mixed in the molten state inside the quartz ampoule and stirred either mechanically or electromagnetically. Synthesized InSb carries impurities either from the source materials or from growth environment. Use of high-purity source materials helps to avoid severe contamination of the synthesized compound. For purification of InSb two important techniques have been used, viz. volatilization and zone refining.

11.4.1 Volatilization

In volatilization, the melt is kept at an elevated temperature for very long periods, so that impurities that have higher volatility evaporate out of the melt. In InSb, cadmium (Cd), an elemental acceptor impurity, is effectively removed by volatilization. In the case of InSb, antimony has a higher vapor pressure compared with In. Hence, Sb escapes from the charge during volatilization. Therefore, whenever this process is adopted, Sb is taken in excess to compensate for its escape from the melt during volatilization.

11.4.2 Zone Refining

In zone refining (a technique developed by *Pfann* [11.76]), purification takes place because of the impurity atom's preference to stay either in the molten zone or in the solid zone (depending on the value of a segregation coefficient) as it moves along the ingot. This preference for the impurity is measured by defining an equilibrium distribution coefficient, k_{eq}, given by C_s/C_l, the ratio of the concentration of the impurity in the solid and liquid phases. The larger the deviation of k from 1, the easier it is to remove the impurity through zone refining. *Harman* [11.77] carried out extensive experiments on zone refining of InSb and found two main impurities, viz. Zn and Te, that segregated from the ingot. While Zn had an acceptor-like characteristic in InSb, Te showed donor-like properties. *Strauss* [11.78] determined the distribution coefficients of various elements such as Cd, Zn, Se, and Te in InSb. Zinc has a k_{eff} of 2.3–3.5 in InSb, which makes it fairly easy to be zone refined, whereas Te has a k_{eff} of 1 in InSb, which makes it almost impossible to remove from InSb. One can use the relation $c(x) = k_{eff}c_0(1-g)^{k_{eff}} - 1$ to find the concentration at any location (x) of the growth front in different conditions, where g is the fraction of crystallized material and c_0 is the initial concentration. Here it is assumed that k_{eff} does not vary with concentration or temperature. Recent studies show that k_{eff} depends on growth rate also. Hence, k^* may in principle differ from k_{eff} and may depend on the growth rate, interface orientation, and solute concentration. Hence it is more appropriate to use the following equation

$$k_{\text{eff}} = \frac{k^*}{k^* + (1-k^*)\exp\left(\frac{-f\delta}{D}\right)}, \quad (11.2)$$

where f is the growth rate, δ is the diffusion thickness ($\delta = 1.6D^{1/3}\nu^{1/6}\omega^{-1/2}$), which is a function of rotation rate ω, the viscosity of the liquid ν, and the diffusion coefficient D of the impurity in the liquid. Under conditions of stirring at the interface, k^* is close to k_{eff}, but the two are quite different in real experimental situations. In fact, for certain impurities, very strong dependence of k^* on the crystallographic direction of ingot to be refined has been noticed, as is the case for Te in InSb. This anisotropic segregation created many difficulties for Te removal. It was thus difficult to remove Te effectively, as it always depended on ingot orientation and the ingots used were usually polycrystalline. However, the observation of the dependence of k^* on ingot orientation by *Mullin* and *Hulme* [11.79] showed that k^* is about 0.5 in the off-[111] direction and this can be effectively used to segregate Te from InSb ingots whose axis is crystallographically inclined away from [111] direction. In fact, k^* was not found to vary much on facets other than (111). Thus, researchers began to use off-[111] ingots for the refining process. Hence, the anisotropic segregation of Te in InSb, which posed problems earlier, was cleverly used to advantage in the refining process. However, this necessitated the use of oriented InSb ingots for the refining process rather than polycrystalline ingots as previously sought.

11.5 Bulk Growth of InSb

Since the discovery of InSb a large number of workers have performed growth experiments on InSb using a variety of techniques. Prominent examples are briefly described here.

11.5.1 Zone Melting

In the early stages, the growth technique used in the case of InSb was predominantly *zone melting* so that the refining and the growth could be carried out together. It was often found that multiple refining passes led to single-crystal growth. *Mueller* and *Jacobson* [11.80] used seeds of orientations [110], [111], and [211] to grow single crystals. Coherent twins were seen to occur, with the twin boundaries parallel to {111} planes. It was also observed that twinning occurred more readily when an [111] A surface was exposed to the melt. Superior growth characteristics of [111] B direction as compared with [111] A were demonstrated [11.81].

11.5.2 Vertical and Horizontal Bridgman Methods

Parker et al. [11.82] reported the growth of InSb single crystals by the horizontal Bridgman technique. Single crystals grown after refining had an etch pit density (EPD) of $10^5\,\text{cm}^{-2}$ with carrier mobility of $919\,000\,\text{cm}^2/(\text{V s})$ (at 80 K). InSb was also grown by vertical Bridgman technique in quartz crucibles to yield single crystals with mobilities on the order of $10^4\,\text{cm}^2/(\text{V s})$ [11.71]. *Bagai* et al. [11.72] used semicircular heaters to refine and grow single crystals with high mobility by horizontal Bridgman technique. Although the growth of InSb has been tried by a variety of methods, and single crystals have been obtained, various problems continue to plague its production. The origin of most defects is traced to the growth environment, particularly the temperature. Impurity striation which was a common problem in Czochralski (CZ) grown InSb was also encountered in the Bridgman configuration. *Zhou* et al. [11.83] applied accelerated crucible rotation technique (ACRT) to the directional solidification of Te:InSb in a Bridgman system with a view to reduce rotational striation. Alternate methods such as the application of magnetic field during growth have also been attempted to tackle this problem [11.84, 85]. The magnetic field is known to reduce temperature fluctuations by damping convection in the melt. Better axial homogeneity in Te:InSb was observed by growth in a magnetic field. In the last decade great strides have been made in rheology and temperature modeling during growth to map the exact thermal environment in the growth chamber. Such studies have yielded a variety of information that provides greater insight into the growth phenomena. In situ observation of the growth of InSb by Bridgman method [11.86] has shown considerable supercooling during growth. This raises serious doubts as to whether the growth of the crystal under such a situation is from a homogeneous melt, a condition essential for growth. There has been a lot of work to understand the role that the crystal–melt interface plays during growth. Growth constrained by the crucible, such as in Bridgman technique, has been modeled [11.87–89], and the role that the thermal properties of the melt and solid play in growth and interface shapes has been studied. A variety of quenching and etching techniques [11.90] have been developed to mark the growth history. Such experiments help researchers to study the origin of defects such as striation and coring, which lead to crystal inhomogeneity. Feedback from such studies helps to improve growth conditions. *Campbell* and *Koster* et al. [11.91] studied interface visualization through radiographic technique and calibrated interface temperature measurements to study growth of InSb from off-stoichiometric melts. Recently sequential-freeze technology has further been used for the study of segregation [11.92, 93]. This system has been used to test the effects of microgravity [11.94, 95] on growth. Large-sized InSb wafers are being produced (up to 100 mm diameter) in cutting-edge commercial product by Firebird Technology [11.96].

Czochralski Growth

Parker et al. also reported the growth of InSb by the CZ technique [11.82]. Polycrystalline zone-refined feed material was used as the starting charge for growth. In fact, crystals grown by CZ from zone-refined ingots as the starting material gave the best results, yielding an EPD as low as 500 cm^{-2}. *Allred* and *Bate* [11.97] grew single-crystal InSb by CZ technique along ⟨111⟩ and ⟨110⟩ directions. They used radioactive Se and Te as dopants to trace dopant homogeneity in the pulled crystals. It was observed through contact radiography that the dopant formed a central core in the crystal where it was incorporated in large amounts, termed the coring effect. This was seen particularly for crystals grown along ⟨111⟩ directions. When grown along directions such as ⟨110⟩, it was seen that the dopant was preferentially incorporated in the {111} facets that occurred at the edges of ⟨110⟩ boules. *Terashima* [11.98] introduced InN in InSb melt during the CZ growth to improve the crystals in terms of EPD. This was proposed to be the effect of nitrogen doping of the melt via InN. *Hurle* et al. [11.99] performed growth experiments of InSb from off-stoichiometric solutions which were either In or Sb rich. There exist many reports on the growth of InSb crystals using Czochralski method and the formation of various types of defects [11.97–102].

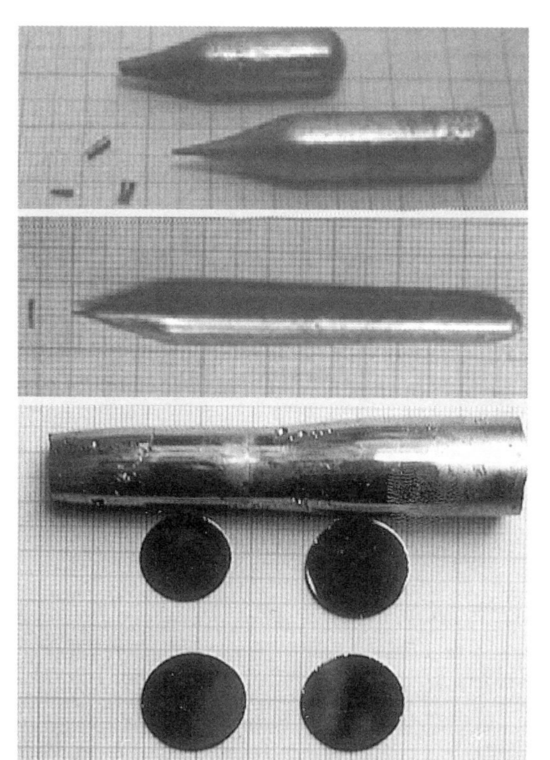

Fig. 11.5 InSb crystals and wafers of different sizes

Traveling Heater Method (THM)

Benz and *Müller* [11.103] conducted various experiments on growth of InSb by the traveling heater method. The THM is a solvent method, and inclusion of the solvent from the feed is common in crystals grown by this method. Both vertical and horizontal configurations have been used and have obtained single crystals with no inclusion. The defect densities in the grown crystals were evaluated to be about 1000 cm^{-2}. The optimized growth rate for the growth of inclusion-free crystals was about 2.5 mm/day. It was found that higher growth rates led to solvent inclusions. The results were similar in the horizontal case. *Hamaker* and *White* [11.104] used the temperature gradient zone melting (TGZM) technique to study the growth kinetics in InSb. Thus InSb bulk crystals can be grown by many techniques, but at present the economical and large-sized crystals are produced by Bridgman methods. Also at present there is no incentive to use more sophisticated technique for growth of InSb. Some crystals grown in the authors' laboratory are shown in Fig. 11.5 [11.11, 69, 105, 106].

11.5.3 Bulk Growth of $InAs_xSb_{1-x}$

Woolley and *Warner* [11.20] have reported two methods for the preparation of $InAs_xSb_{1-x}$ alloys, viz. slow directional freezing and slow zone recrystallization of suitable ingots. All the ingots were produced by melting together appropriate amounts of high-purity InAs and InSb. In slow directional freezing, ingots were prepared using equimolar proportions of InSb and InAs. Initially a furnace with a temperature gradient of 10 K/cm was used and temperature was controlled by an on–off controller, so that the time required to freeze the ingot completely was ≈2–3 months. It was found that each end of the ingot was in good single-phase condition and the composition varied from 3 mol % to 12 mol % of InSb at one end and 88 mol % to 98 mol % at the other end. In between these regions however there was some 3 cm of the ingot where there was a very rapid change in composition with position and where the material was not in equilibrium condition. An alternate method for preparing ingots of $InAs_xSb_{1-x}$ alloys is the zone recrystallization technique. Here larger temperature gradients are used so that constitutional supercooling could be less of a problem, and the rate of movement of the freezing surface is directly controlled by the movement of the heater. *Glazov* and *Poyarkov* [11.107] reported homogeneous growth of this material in thin-foil form using rapid quenching. Recently *Dixit* et al. [11.9] reported growth of bulk single crystals of $InAs_xSb_{1-x}$ for $0 \le x \le 0.05$ using RBM (Fig. 11.6). The phase diagram of In–As–Sb, which has been well studied both theoretically and experimentally [11.55, 61], was used to determine the

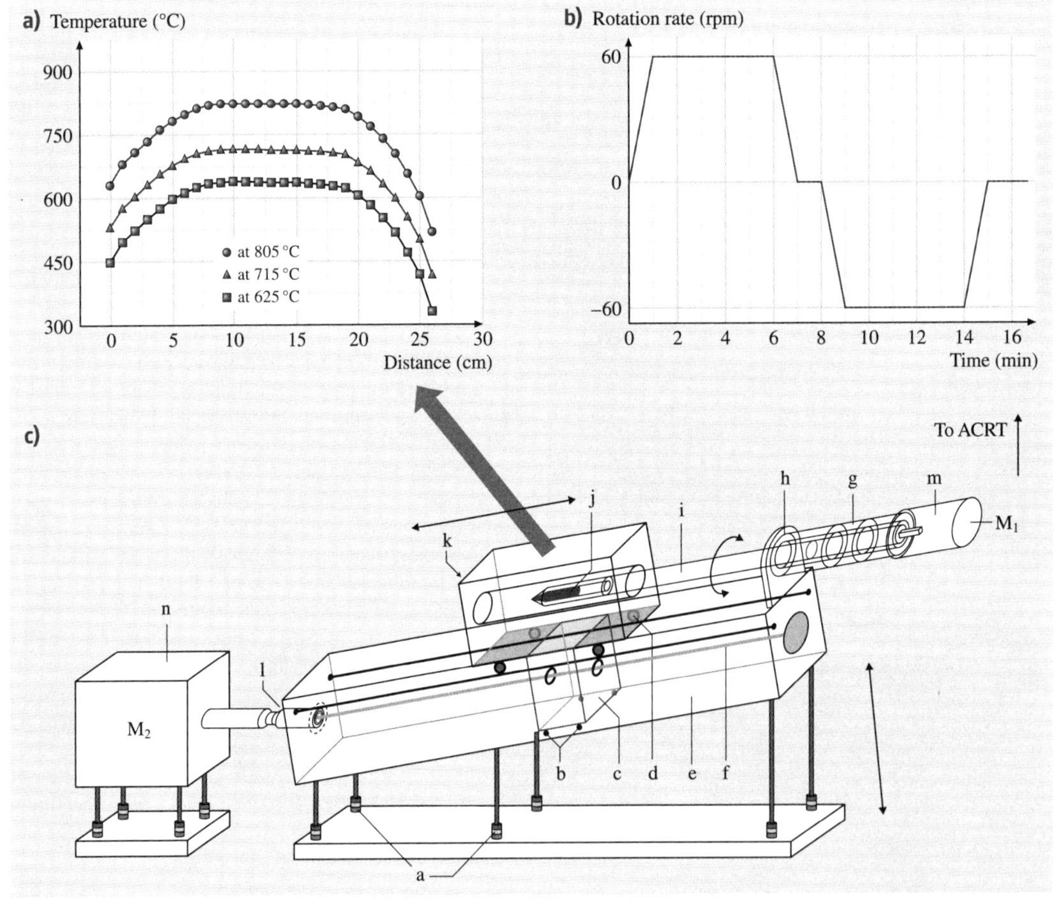

Fig. 11.6 **(c)** Schematic of RBM growth apparatus: (a – Antivibration mount, b – wheels, c – guided coupler, d – platform, e – aluminum frame, f – threaded lead screw, g – ball bearing, h – Wilson seal, i – quartz tube, j – ampoule, k – furnace, l – universal coupler, m – motor M_1, and n – geared motor M_2). **(a)** Temperature profile of the furnace and **(b)** ACR sequence

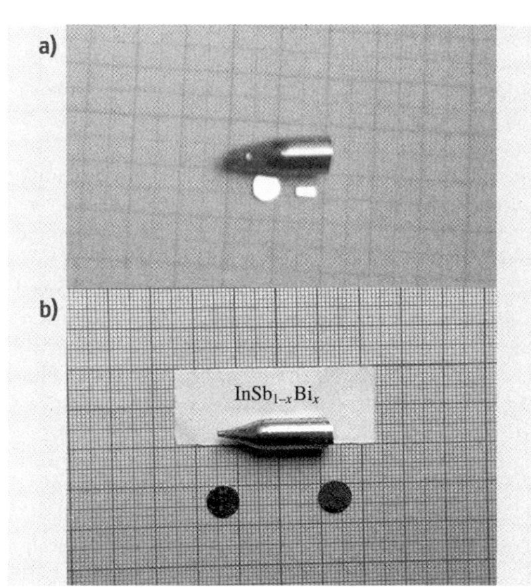

Fig. 11.7a,b Ingot and wafers of (**a**) $InAs_xSb_{1-x}$ and (**b**) $InBi_xSb_{1-x}$

initial reaction temperatures; these were in the range of 600–700 °C for various compositions in their experiments. $InAs_xSb_{1-x}$ was synthesized by diffusing arsenic into high-purity InSb. Although various molar proportions, viz. InSb : *As* = 95 : 5, 90 : 10, 85 : 15, 80 : 20, and 75 : 25, were tried, it was found that homogeneous single crystals of $InAs_xSb_{1-x}$ could be grown only up to starting proportion of 90 : 10, beyond which the grown crystals were inhomogeneous and often phase separated. It was observed experimentally that the liquid–solid interface was slightly concave and symmetric with respect to ampoule axis. Typical size of the grown crystals was 8 mm diameter and about 45 mm length. The ingots were single crystalline up to ≈35 mm, after which they were multigrain. Wafers were made from the 10–20 mm region from the tip of the ingot. Figure 11.7a shows a typical crystal and wafers. If extensive work is carried out in these directions, bulk growth of $InAs_xSb_{1-x}$ leading to large-size device-quality wafers can be achieved.

11.5.4 Bulk Growth of $InBi_xSb_{1-x}$

Compared with $InAs_xSb_{1-x}$ it is even more difficult to synthesize and grow single crystals of $InBi_xSb_{1-x}$. The first work on synthesis and growth of $InBi_xSb_{1-x}$ was reported by *Jean-Louis* et al. [11.108], followed by *Joukoff* and *Jean-Louis* [11.63]. The growth was carried out using Czochralski technique, using ⟨111⟩-oriented InSb seed rods of 3 mm diameter. Single crystal of $InBi_xSb_{1-x}$ could be obtained from a melt up to 30% InBi, which yielded $x = 0.001$, insufficient for development of detectors in the 8–12 μm atmospheric window. Hence they attempted growth with 70% InBi melt, which did not give single crystal. In the intermediate range of 50–60 % they could obtain single crystal up to $x = 0.026$. Later, in 1991, *Ozawa* et al. [11.109] reported growth of $InBi_xSb_{1-x}$ using RBM. They could grow $InBi_xSb_{1-x}$ single crystals with x varying from 0.016 to 0.03 from the seed region to the end of the ingot. This indicated that the segregation coefficient of Bi in InSb was less than unity. They could enhance the Bi content up to ≈ 5 at. % by codoping with gallium [11.110]. *Dixit* et al. [11.12] also reported growth of bulk single crystals of $InBi_xSb_{1-x}$ with $x = 0.067$ using RBM. $InBi_xSb_{1-x}$ was synthesized using various molar proportions (1 : 1, 1 : 3, 1 : 4, and 1 : 5) of InSb and InBi. It was found that, compared with 1 : 1 and 1 : 3, 1 : 4 yielded higher Bi concentration in $InBi_xSb_{1-x}$ crystal as well as better homogeneity. However further increase of InBi molar ratio to 1 : 5 led to formation of InBi and In_2Bi precipitates. Similar compositional ratio (1 : 4) was also employed by *Kumagawa* et al. [11.110]. Typical size of the grown crystals are 8 mm diameter and about 25 mm length. Figure 11.7b shows the grown crystal and wafers. Microgravity experiments on the growth yielded crystals with reduced segregation of InBi [11.94]. This is one avenue that could be explored for higher incorporation of Bi in the crystal, thereby improving suitability for device applications.

11.5.5 Growth of Thick Layers of InSb, $InAs_xSb_{1-x}$, and $InBi_xSb_{1-x}$, by Liquid-Phase Epitaxy

LPE, although a widely used general laboratory preparation technique, has not found much favor with these materials, mainly due to the low temperatures of growth, limited by the melting point of InSb. However, recently there has been renewed activity on LPE growth of these materials. Here we present a comprehensive report on the growth of InSb-based epitaxial layers grown using LPE. These layers mimic the bulk-like properties and can also be used as virtual epiready wafers for further growth of multilayered structures.

InSb Epilayers

Various efforts have been made to adapt LPE for growth of InSb. *Kumagawa* et al. [11.111] reported dopant

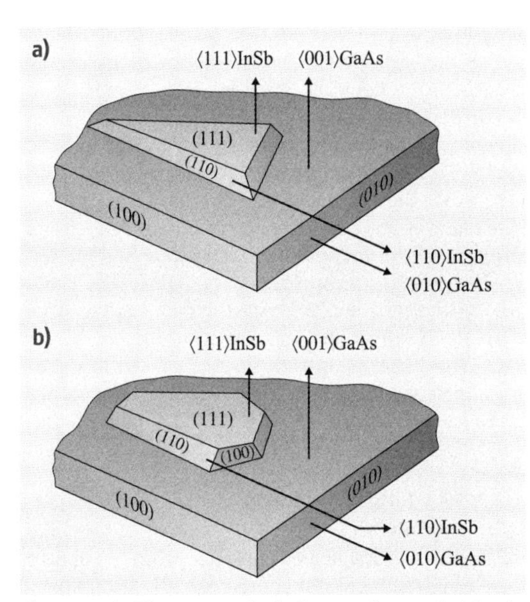

Fig. 11.8a,b The two typical shapes of growth islands: (**a**) triangular (60° angle) and (**b**) truncated triangular

modulation in InSb layers grown by liquid-phase electroepitaxy (LPEE). *Melngailis* and *Calawa* [11.112] reported growth of planar InSb laser structures by LPE. They used a stearic acid encapsulant for growth to avoid oxide layer formation during growth. *Holmes* and *Kamath* [11.113] have used infinite solution epitaxy to grow binary InSb and discussed the growth of continuous layers with constant ternary composition. However, other than a few early papers on homoepitaxy [11.111, 113], there are no reports on successful growth InSb/GaAs by LPE. Recently, *Dixit* et al. [11.10] reported successful epitaxial growth of this material on SI–GaAs substrate by LPE with extremely low ramp cooling rates. Growth was carried out on (001) semi-insulating GaAs substrate in a boat-slider-type LPE unit [11.114, 115]. The optimized III/V mass ratio used for the growth was ≈ 3.45. A ramp cooling routine was adopted with the intention of providing a driving force $\beta \propto \Delta T/T_g$ [11.116] (where ΔT denotes the supercooling of the solution and T_g the initial growth temperature) at every point of growth and induce orderly epitaxial growth. The sliding was achieved by an improvised electromechanical system. Films were grown under varied conditions of supersaturation by changing the ramp rate R and the cooling temperature step ΔT. During the course of this study, it was noticed that higher cooling rates led to formation of interesting island morphology. The two typical shapes were (a) triangular (60° angle) and (b) truncated triangular islands (Fig. 11.8). The results from various growth experiments are presented in Table 11.1. It is seen that the growth rate r (μm/h) and the ramp rate R have a direct correspondence as expected. Furthermore, as can be seen from Table 11.1, oriented films were grown only at $R = 0.2$ K/h, with ΔT of 2 K. Although oriented films were obtained by this way, coverage was low and morphology was poor. By increasing the growth temperature while keeping other conditions the same the growth rate could be decreased (#48A to #33B in Table 11.1), which is in agreement with *Elwenspoek*'s work [11.116]. By initiating the growth at 423 °C and decreasing the ramp rate to 0.2 K/h, oriented films of 3 μm thickness could be grown.

$InAs_xSb_{1-x}$ Epilayers

Until the 1990s growth of $InAs_xSb_{1-x}$ by LPE was achieved only for very high arsenic concentrations (for 3–5 μm gap) and on nearly lattice-matched InAs and

Table 11.1 Summary of various InSb/GaAs thin-film growth experiments

#	R (K/h)	T_g (°C)	ΔT (K)	r (μm/h)	hkl_{max}	Other peaks	$I_{004}/\sum I_{hkl}$	Thickness (μm)
43A	3	410	5	240	(111)	(311) (220) (004)	0.1	400
40A	2.5	410	10	85	(331)	(220)	0	330
50A	1.8	410	5	45	(004)	(220) (311)	0.63	120
45A	0.5	410	5	30	(004)	(311) (220)	0.62	300
48A	0.2	410	2	27	(004)	–	1	270
13B	3	423	11	14.7	(004)	–	1	54
20B	1.5	423	5	3	(004)	–	1	10
30B	0.8	423	3	1.6	(004)	–	1	6
33B	0.2	423	3	0.4	(004)	–	1	4
5C	0.2	423	2	0.7	(004)	–	1	7
7C	0.2	423	2	0.3	(004)	–	1	3

Table 11.2 Summary of various $InAs_xSb_{1-x}$/GaAs thin-film growth experiments

#	R (K/h)	T_g (°C)	ΔT (K)	r (μm/h)	hkl_{max}	Other peaks	$I_{004}/\sum I_{hkl}$	t (μm)	Composition (x)	
									EDAX	X-ray
11D	2.5	453	3	16.67	(004)	(220)	0.85	20	0.12±0.01	0.12±0.004
21D	1.5	450	3	6.5	(004)	(220)	0.85	13	0.10±0.01	0.12±0.004
39D	0.8	448	3	2.67	(004)	–	1	10	0.06±0.01	0.45±0.004
43D	0.6	446	3	2.10	(004)	–	1	8	0.06±0.01	0.42±0.004

Table 11.3 Summary of various $InBi_xSb_{1-x}$/GaAs thin-film growth experiments

#	R (K/h)	T_g (°C)	ΔT (K)	r (μm/h)	hkl_{max}	Other peaks	$I_{004}/\sum I_{hkl}$	Thickness (μm)
13C	2.5	355	5	7.5	(004)	(220)	0.73	15
20C	1.5	353	3	5.0	(004)	(220)	0.89	10
30C	0.6	351	2	1.79	(004)	–	1	6
33C	0.2	351	2	0.2	(004)	–	1	2

GaSb substrates [11.117, 118]. However, there does not seem to be any work on LPE growth of this system for low values of x. This is primarily due to problems associated with the low growth temperature and large lattice mismatch. Recently a few more results have been published on this growth by authors such as *Dixit* et al. [11.5], *Peng* et al. [11.34], and *Gao* et al. [11.37], employing LPE or melt epitaxy (ME). *Dixit* et al. [11.5] grew $InAs_xSb_{1-x}$ layers in the same boat-slider-type LPE unit used to grow InSb. Prior to growth, indium-rich InAsSb solution was prepared with indium and previously grown $InAs_xSb_{1-x}$ crystals [11.9]. The optimizcd In/(Sb + As) mass ratio used for the growth was ≈ 2.93. The solution temperature was increased to 5 °C above the growth temperature of 446 °C, and baking for 6 h was used to allow proper homogenization. Here also a ramp cooling routine was used. The results obtained under various growth conditions are summarized in Table 11.2. Large growth rates force nucleation with off-⟨100⟩ orientation, while oriented growth takes place on (100) GaAs at lower growth rates. It is to be noted that, when growth temperature increases above 446 °C, the arsenic incorporation into InSb increases due to the diffusion of arsenic from the GaAs substrate itself [11.119, 120]. Hence growth temperature was restricted to 446 °C. This led to the limitation of As incorporation to only 6 at. % in $InAs_xSb_{1-x}$ epitaxial layers. Also it was observed that the quality of the interface deteriorated with increasing arsenic composition. This is due to a miscibility gap in the phase diagram. Recently, relatively thick films (100–200 μm) were grown by *Gao* et al. [11.37] using ME at 600 °C. Very recently *Peng* et al. [11.34] also reported predominantly oriented growth of $InAs_xSb_{1-x}$ films with up to 30% As incorporation using conventional LPE. In this work, layers were grown at constant growth temperatures rather than employing a ramp cooling routine. Constant growth temperature helps the compositional uniformity of the $InAs_xSb_{1-x}$ films in the growth direction, similar to the observation of *Dixit* et al. [11.5] (by very slow growth rate condition).

$InBi_xSb_{1-x}$ Epilayers

Literature on LPE growth of $InBi_xSb_{1-x}$ is very limited. There are only a few reports on homoepitaxy of $InBi_xSb_{1-x}$. *Ufimtsev* et al. [11.121] reported the growth for $x = 0.01$–0.02. *Gao* and *Yamaguchi* [11.122] also reported homoepitaxy of $InBi_xSb_{1-x}$ using LPE, wherein they found that surface morphology was strongly dependent on cooling rate. Also Bi metal particles appeared on the surface of the epilayer. The cutoff wavelength reported for this film was 14 μm. Many reports can however be found in the literature on the heteroepitaxy of $InBi_xSb_{1-x}$ using a variety of other techniques. *Dixit* et al. [11.123] also reported growth of $InBi_xSb_{1-x}$ on GaAs using the boat-slider-type LPE unit. The optimized In/(Sb + Bi) mass ratio used for the growth was ≈ 1.86. The solution temperature was increased to ≈ 50 °C above the growth temperature (351 °C), and baking for a period of 6 h in ultra-pure hydrogen ambient was used for homogenization. Table 11.3 gives the correlation between imposed supersaturation ΔT (K) and the thickness of the grown films.

Work on the growth of $InBi_xAs_ySb_{1-x-y}$ by a variety of techniques has also been reported, including LPE [11.4, 8, 53, 124–126]. However, extensive work is needed for the development of the Sb-based stable detectors for the long-wavelength region.

11.6 Structural Properties of InSb, $InAs_xSb_{1-x}$, and $InBi_xSb_{1-x}$

Structural information is crucial for developing any technology. Particularly it is very important to understand the role of alloying and mismatch epitaxy in these systems. Structural information about these materials is obtained using various characterization techniques. The most commonly used techniques include high-resolution x-ray diffraction (HRXRD), chemical etching, x-ray topography, atomic force microscopy (AFM), scanning electron microscopy (SEM), and transmission electron microscopy (TEM).

11.6.1 InSb

Composition

Although InSb growth is very well studied, care must be taken to obtain crystals in stoichiometric form. Hence the starting material for all growth runs should be synthesized in one batch to avoid any batch-to-batch variation in composition, and care must be taken in transferring the material and to maintain accuracy to the level of μg/g of charge. Also, due to the vapor pressure difference, composition differences of the starting material have to be taken into account. The typical results of energy-dispersive x-ray analysis (EDAX) for samples made from grown InSb crystals showed that the crystals were radially and axially homogeneous. The typical radial composition profile for 20 mm InSb wafer is given in Table 11.4.

Orientation Determination

The crystallographic orientation of the grown crystals can be determined from x-ray diffraction experiments. The Laue pattern shows fourfold symmetry for ⟨001⟩ and threefold symmetry for ⟨111⟩ InSb wafer. From powder x-ray diffraction scans of the wafer one can determine the growth direction of the crystals. For example, a single peak corresponding to (111) and (110) planes is obtained for wafers made from crystals grown by horizontal and vertical Bridgman techniques.

Chemical Etching

InSb is the most perfect material among the III–V semiconductors available to date. For chemical etching, usually CP4 etchant has been found to be very effective (Fig. 11.9). Chemical etching producing pits has been used to quantify dislocations. In InSb, a one-to-one correspondence has been established between dislocations and etch pits. The pit count, EPD, gives an order of magnitude estimate of the dislocation content in the crystals. Typically 10–100 cm^{-2} etch pit density is specified in commercially available InSb. However, the best results are only 1 etch pit in 50 cm^2. T-shaped and *star*-type arrays of etch pits are more prominent in InSb due to an inherent difference between the (111) and $(\bar{1}\bar{1}\bar{1})$

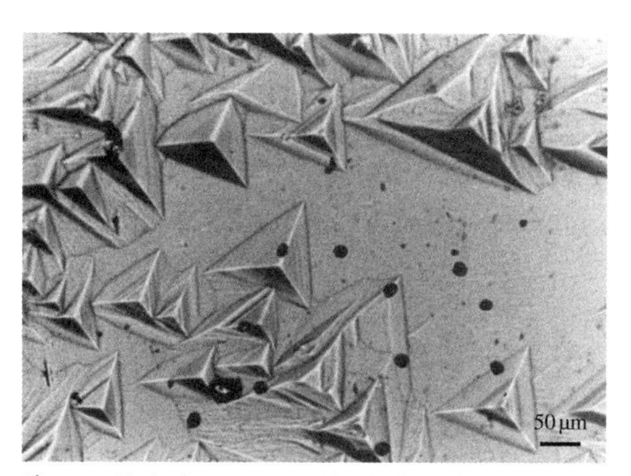

Fig. 11.9 Etch pit pattern on (111) InSb wafer

Table 11.4 Compositional profile of InSb wafer; *a*, *b*, *c*, *d* are marked in the schematic of the wafer *shown at the top of the table*

Compositional profile					
From *a* to *b*	**Atomic percentage**		**From *c* to *d***	**Atomic percentage**	
	In	**Sb**		**In**	**Sb**
1. (1 mm)	50.10	49.85	1. (1 mm)	50.09	49.91
2. (6 mm)	50.07	49.93	2. (7 mm)	50.11	49.89
3. (12 mm)	50.19	49.81	3. (13 mm)	50.19	49.81
4. (19 mm)	50.10	49.9	4. (20 mm)	50.23	49.77

planes [11.96]. Also this polarity influences many other properties; for example, damage depth and chemical attack rates are different for 111B (Sb planes) and 111A (In planes).

Slip System

{111} are the slip planes for InSb. The slip direction is one of the ⟨110⟩ directions. If InSb crystals are compressed or bent by applying stress above a critical value, arrays of slip dislocations can be produced, which can be viewed as etch pit arrays defining the traces of the {111} planes.

Twinning

During growth, a 60° lattice rotation about any III–V bond in a ⟨111⟩ direction preserves the tetrahedron of opposite species but in a lattice mirrored across the (111) plane, since both the nearest and second nearest neighbors remain at their normal distance; the energy required for this twinned growth is exceptionally low. For this reason twinning across the set of {111} planes is the principal complication in growing single crystals of InSb. Various other types of twinning are also reported along {211} and {111} planes for III–V as well as IV semiconductors but not in InSb [11.96, 127]. *Terashima* [11.98] reported Y-shaped defects in ⟨111⟩ pulled InSb crystals and these defects are eliminated by growing the crystal using a seed inclined at 5–10° from ⟨111⟩ towards ⟨110⟩.

High-Resolution X-Ray Diffraction Study

An ideal crystal on interaction with a x-ray beam diffracts a beam of extremely narrow width at the Bragg angle. However, in practice one sees a width of diffracted beam or rocking curve width of several hundreds of arc seconds in angular spread. This was first explained by *Darwin* [11.128], considering the crystal to be a macromosaic of crystals of slightly differing orientations. These macromosaics receive the x-ray at different incident angles and hence there is a finite spread in the resultant diffracted beam. Also, defects in the lattice such as vacancies, interstitial, dislocations, twins, and small-angle grain boundaries distort the lattice and contribute to broadening of the rocking curve. The relative widths of diffraction curves have been found useful for comparing the crystalline perfection of many semiconductors. Such studies have been made on InSb and are described by *Auleytner* [11.129]. High-resolution x-ray diffraction study is very effective for determining the crystalline quality [11.130] of InSb crystals. *Gartstein* and *Cowley* [11.131] and

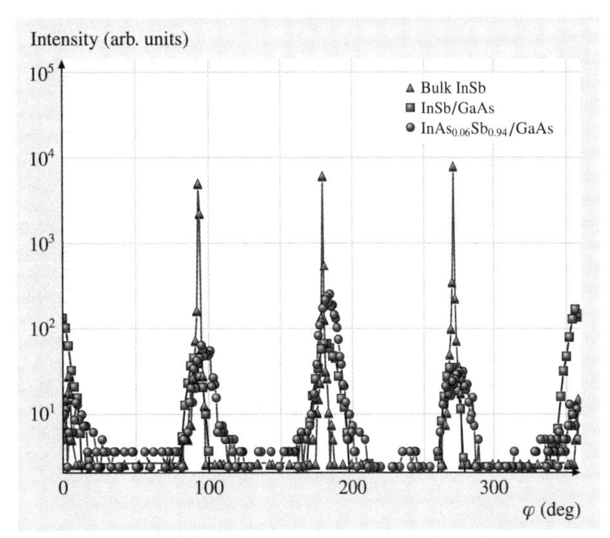

Fig. 11.10 X-ray diffraction patterns (ϕ scan) of single crystal and epilayers in asymmetric (115) reflection over 360° azimuths

Gartstein et al. [11.132] discuss the x-ray diffraction study of perfect and In^+-ion-implanted InSb single crystals. The rocking curves of these crystals are relatively large; however substantial reduction in radiation damage by permanent magnetic fields is noticed. Recently, *Dixit* [11.11, 69] carried out HRXRD studies on bulk InSb and Li:InSb using a Philips X'PERT high-resolution x-ray diffractometer. A ϕ scan of an asymmetric reflection was performed to determine the single-crystal nature of the bulk InSb crystal. As seen in Fig. 11.10 the ϕ scan over 360° azimuths gave four distinct peaks separated by 90° intervals with no scattering between them, confirming the single-crystalline nature of InSb. Subsequently several symmetric and asymmetric rocking curves were recorded. The rocking curve width (full-width at half-maximum, FWHM) was found to be 18″ (Fig. 11.11), which is approaching the theoretical limit of the rocking curve width for 004 reflection of InSb. Employing MOCVD *Gaskill* et al. [11.29] have grown layers of InSb on InSb substrates with rocking curve width of 34″, approaching the theoretical limit of 11″ for InSb. Interesting studies on the structural dynamics of InSb using time-resolved x-ray diffraction were reported by *Chin* et al. [11.133]

Synchrotron Radiation Topographic Study

Surowiec and *Tanner* [11.134] reported transmission x-ray topography studies on the dislocation configurations around microindentations on {111} surfaces

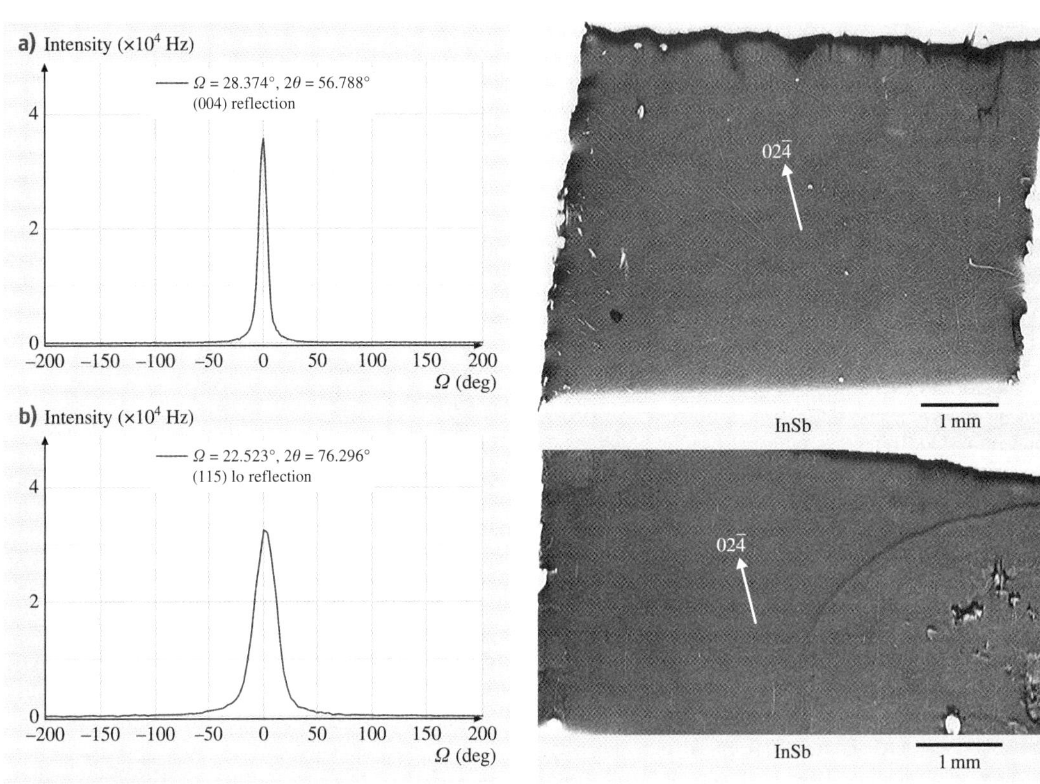

Fig. 11.11a,b Representative $\omega/2\theta$ scans of InSb for (**a**) (004) reflection and (**b**) (115) lo reflection

Fig. 11.12 Synchrotron reflection topographs of InSb wafers

of InSb. Glide occurs on only B-type {111} planes, and the most extended dislocation loops occur around A-surface indents. Glide occurs on inclined {111} planes and the loops have extended screw segments parallel to the surface. The mobility of these dislocations is about two or three times less than the former type. Direct evidence for the formation of edge dislocation barriers from reactions between dislocations gliding in the (111) plane is reported. *Dixit* et al. [11.69] also carried out synchrotron x-ray topographic study on high-quality InSb crystals. Reflection topographs of polished surfaces were taken with the normal tilted 5° from 024 reflection. A diffraction wavelength of 1.30 Å was used, and the sample-to-film distance was 20 cm. Results are shown in Fig. 11.12a,b. As can be seen from the figure, the topographic images are free from asterism. Surfaces show some polishing effects but they are free from any dislocation. Figure 11.12b shows some white circular features which are due to metal inclusions.

11.6.2 $InAs_xSb_{1-x}$

Dixit et al. [11.9, 69] estimated arsenic content in grown $InAs_xSb_{1-x}$ crystal by measuring the melting point using differential scanning calorimetry and inferred that it must be more than 3 at.% (Fig. 11.13). The composition x in $InAs_xSb_{1-x}$ crystals grown from starting proportions of 95 : 5 and 90 : 10 was 0.02 and 0.05, respectively, as determined from EDAX. Further confirmation of the composition was made using x-ray photoelectron spectroscopy (XPS). The binding energies were measured with respect to the 1s peak at 285 eV with a precision of ±0.1 eV. XPS of core-level regions of In(3d), Sb(3d), and As(3d) are shown in Fig. 11.14. Spin–orbit doublet peaks of $In(3d_{5/2,3/2})$, $Sb(3d_{5/2,3/2})$, and $As(3d_{5/2,3/2})$ show shifts compared with the respective metals [11.135]. The surface concentration ratio In(3d)/Sb(3d) has been estimated using the following equation

$$\frac{C_{\text{In}}}{C_{\text{Sb}}} = \frac{I_{\text{In}}\sigma_{\text{Sb}}\lambda_{\text{Sb}}D_{\text{E}}(\text{Sb})}{I_{\text{Sb}}\sigma_{\text{In}}\lambda_{\text{In}}D_{\text{E}}(\text{In})} , \tag{11.3}$$

where C, I, σ, λ, and D_{E} are the concentration, intensity, photoionization cross section, mean escape depth [11.135–137], and geometric factor respectively. Integrated intensities of metal peaks have been taken into account for calculating the concentrations. Surface concentration ratios of In(3d)/Sb(3d) and As(3d)/Sb(3d) can also be calculated using similar equations. The composition has been estimated to be $InAs_{0.055}Sb_{0.945}$, which is close to the bulk composition obtained from EDAX. From EDAX and XPS results

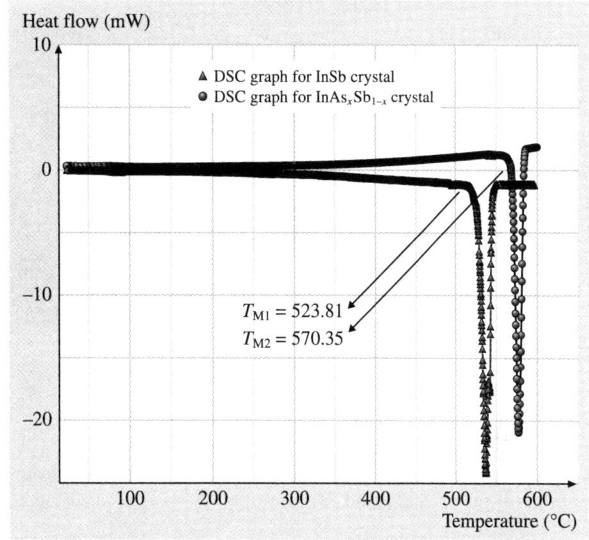

Fig. 11.13 Differential scanning calorimetry curves of $InAs_xSb_{1-x}$ for $x = 0$ and 0.05

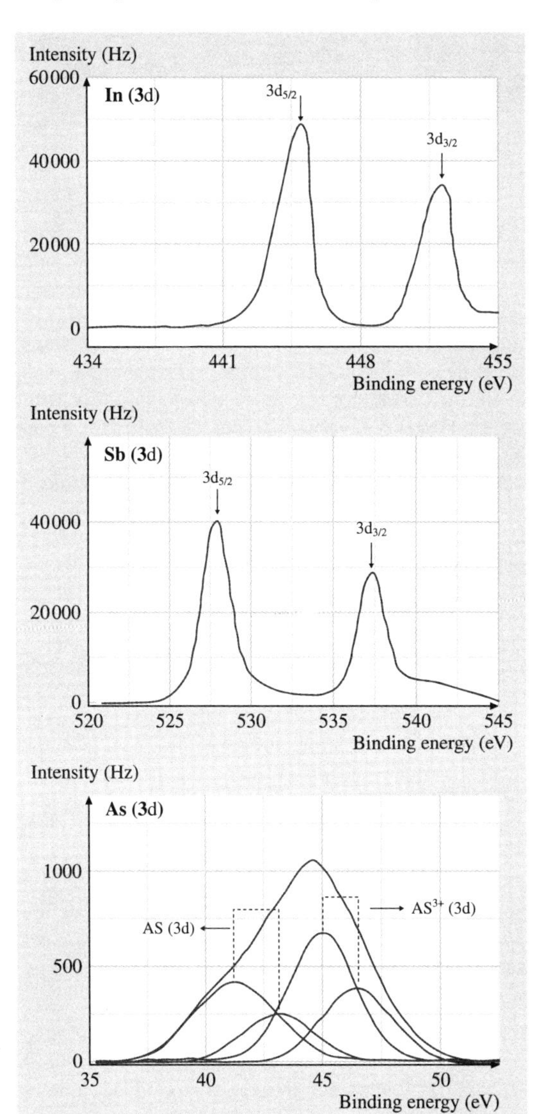

Fig. 11.14 X-ray photoelectron spectra of $InAs_{0.05}Sb_{0.95}$ for In, Sb, and As

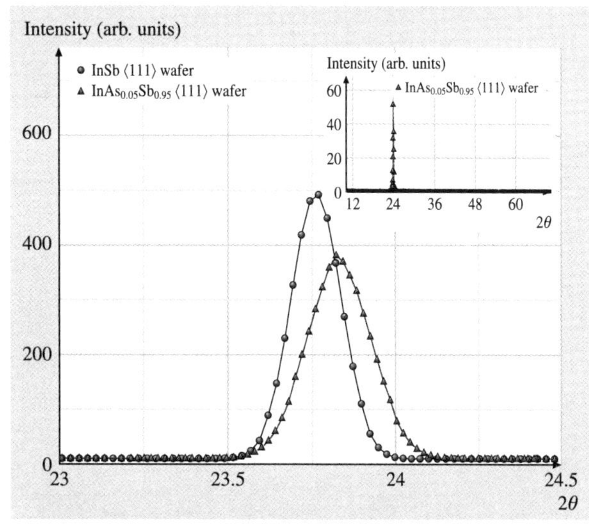

Fig. 11.15 X-ray diffraction peaks of single-crystal wafers of InSb and $InAs_{0.05}Sb_{0.95}$ for ⟨111⟩ reflection

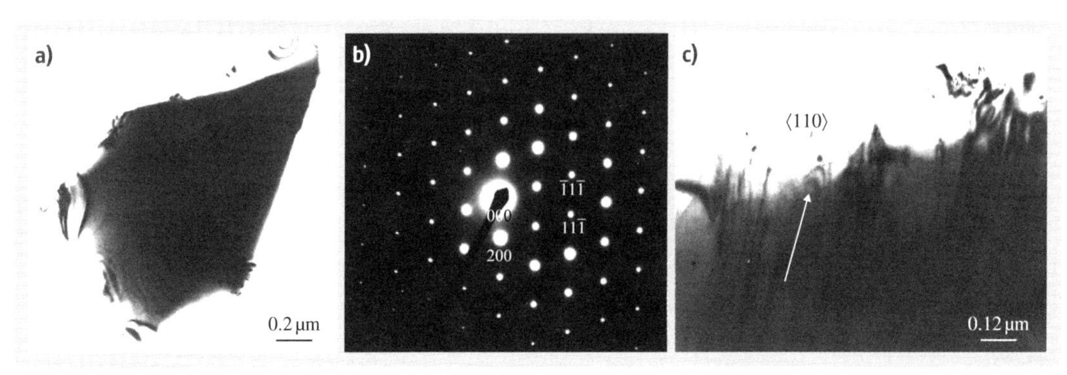

Fig. 11.16 (**a**) Bright-field image of $InAs_{0.05}Sb_{0.95}$, (**b**) selected-area diffraction pattern of (**a**), and (**c**) bright-field image showing dislocations along the ⟨110⟩ direction

it was concluded that the grown crystals were radially homogeneous but the As composition decreased from tip to the other end of the ingot. The lattice parameter of $InAs_{0.05}Sb_{0.95}$ as calculated from XRD powder patterns is 6.460 ± 0.006 Å and matches reasonably with that obtained from Vegard's law. $InAs_{0.05}Sb_{0.95}$ showed lattice contraction relative to that of InSb, as confirmed by XRD (Fig. 11.15). *Dixit* et al. [11.9] also obtained bright-field TEM images of $InAs_{0.05}Sb_{0.95}$ along with its selected-area electron diffraction pattern, confirming the absence of twinning (Fig. 11.16). However in the bright-field image, dislocations were observed primarily along ⟨110⟩ direction (Fig. 11.16). The estimated dislocation density was $\approx 10^8\,cm^{-2}$. This is comparable to those found for $InAs_xSb_{1-x}$ alloy grown by other techniques [11.138].

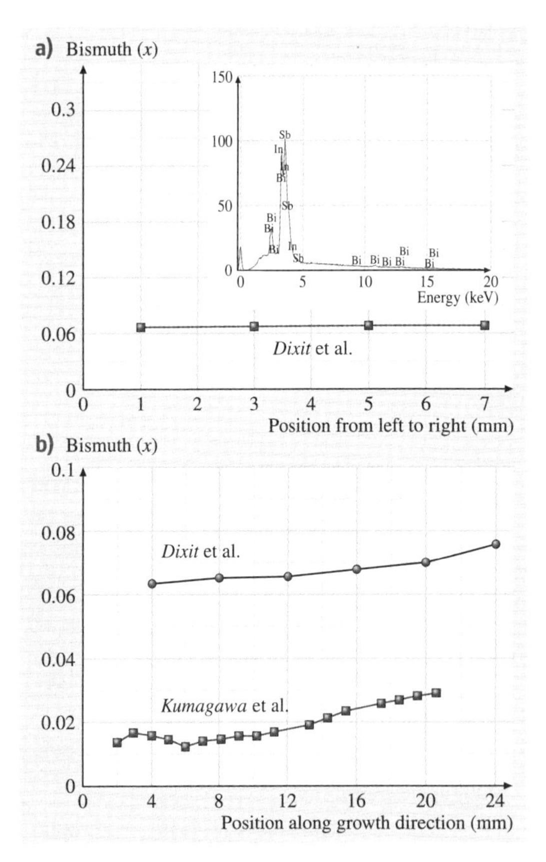

11.6.3 $InBi_xSb_{1-x}$

So far, the highest value of x in $InBi_xSb_{1-x}$ bulk single crystals is 0.064 ± 0.01, as determined from EDAX (inset in Fig. 11.17a). *Dixit* et al. [11.12] found that the Bi content in crystal grown by RBM increases from the tip to the upper end of the ingot, confirming that the equilibrium segregation coefficient of Bi in InSb is less than unity. Compared with *Kumagawa* et al. [11.110], they could achieve higher Bi incorporation into the crystals with better radial and axial homogeneity, as can be seen from Fig. 11.17. From powder x-ray diffraction patterns recorded for $InBi_xSb_{1-x}$ the lattice parameter was calculated and found to be 6.489 ± 0.006 Å. The composition of Bi as evaluated by Vegard's law

$$a_{InBi_xSb_{1-x}} = x a_{InBi} + (1-x) a_{InSb} , \qquad (11.4)$$

Fig. 11.17 (**a**) Radial and (**b**) axial compositional profiles of $InBi_xSb_{1-x}$ crystal; *inset* in (**a**) shows a typical EDAX spectrum ◀

is 6.79 at. %, which is in reasonable agreement with the EDAX result, where

$$a_{\mathrm{InBi}} = 6.640\,\text{Å}\ [11.39] \quad \text{and}$$
$$a_{\mathrm{InSb}} = 6.478\,\text{Å}\,,$$
$$a_{\mathrm{InBi}_x\mathrm{Sb}_{1-x}} = 6.490 \pm 0.006\,\text{Å}\,.$$

11.6.4 InSb, $InAs_xSb_{1-x}$, and $InBi_xSb_{1-x}$ Grown on GaAs

Dixit et al. [11.10, 115] carried out extensive HRXRD studies on these heterostructures grown using LPE. A ϕ scan of an asymmetric reflection recorded on InSb/GaAs gave four distinct layer peaks separated by 90° intervals with negligible scattering between them (Fig. 11.10). Layers show structural coherence with substrates as confirmed from various reflections (Fig. 11.18). The tilt angle $T_{\mathrm{l,s}}$ of the epilayer to the substrate was estimated by using the relation

$$T_{\mathrm{l,s}} = \frac{1}{2}\left[(\Delta\omega_1 - \Delta\omega_3)^2 + (\Delta\omega_2 - \Delta\omega_4)^2\right]^{1/2}, \quad (11.5)$$

where the $\Delta\omega_i$ are the peak separations at 90° intervals. The tilt was found to be quite small and was about 0.01°. Strain parameters were evaluated by a least-squares routine for the following relationship [11.139]

$$\Delta\omega = k_1\epsilon_\perp + k_2\epsilon_\parallel\,, \quad (11.6)$$

where $k_1 = \cos^2\phi\tan\theta_{\mathrm{B}} + 1/2\sin 2\phi$ and $k_2 = \sin^2\phi \times \tan\theta_{\mathrm{B}} - 1/2\sin 2\phi$, in which θ_{B} is the substrate Bragg angle and ϕ is the angle between the reflecting plane (*hkl*) and the sample surface (001) with ϕ being positive (negative) for low (high) glancing incidence. The perpendicular and in-plane x-ray strain parameters, $\epsilon_\perp$ and $\epsilon_\parallel$, in the above relationship are defined [11.140]

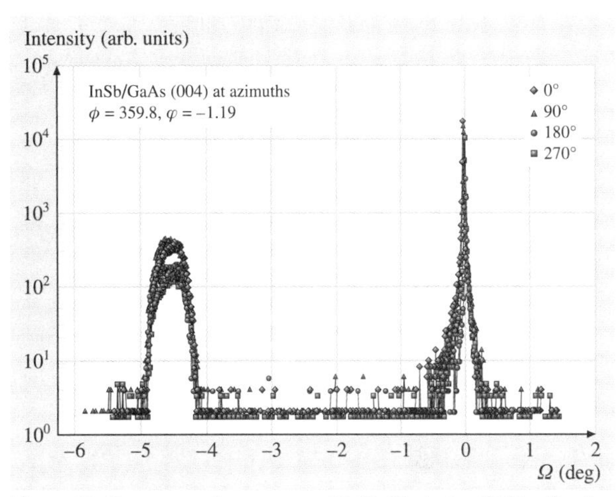

Fig. 11.18 Representative ω scans of InSb/GaAs for (004) reflection at four azimuths

with respect to the substrate lattice parameter as

$$\epsilon_\parallel = \frac{a_\parallel - a_{\mathrm{s}}}{a_{\mathrm{s}}}\,, \quad \epsilon_\perp = \frac{a_\perp - a_{\mathrm{s}}}{a_{\mathrm{s}}}\,, \quad (11.7)$$

where $a_\parallel$ and $a_\perp$ are the in-plane and out-of-plane lattice constants, respectively. The results of the least-square analysis are $\epsilon_\parallel = (99 \pm 10) \times 10^{-3}$ and $\epsilon_\perp = (108 \pm 3) \times 10^{-3}$, indicating considerable in-plane strain relaxation expected for the large mismatch. From these results they determined $a_\parallel = 6.213\,\text{Å}$ and $a_\perp = 6.264\,\text{Å}$, both approaching the standard lattice constant of InSb. The in-plane and out-of-plane residual strains as evaluated from $a_\parallel$ and $a_\perp$ are 0.0409 and 0.033, respectively, which are quite small. All the films shows almost complete relaxation, hence extensive dislocation network is to be expected, which was estimated using the FWHM of symmetric (004) reflec-

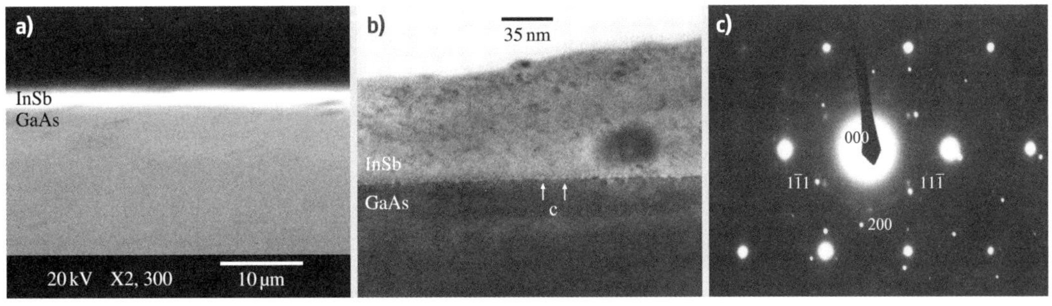

Fig. 11.19 (a) Scanning electron micrograph of InSb/GaAs interface, (b) bright-field image of the InSb/GaAs cross-section, and (c) selected-area diffraction pattern of ((b))

tion. With the magnitude of the Burger's vector taken as $1/(2\sqrt{2})a$ (which is the case for 60° misfit dislocation), where a is $(a_{InSb}+a_{GaAs})/2$ [11.141], the estimated average dislocation density is in the range 1.3×10^9–$4.7\times10^{10}\,cm^{-2}$, which compares favorably with the values reported for epitaxial layers grown by MBE and MOCVD [11.28, 29].

The cross section of the film (cleaved sample) observed under a scanning electron microscope (SEM), shown in Fig. 11.19a, clearly reflects reasonable sharpness of the interface between the InSb layer and GaAs substrate. The bright-field cross-sectional TEM image, taken with a JEM-200CX transmission electron microscope, indicates a sharp interface with very small coalescing islands (marked "C" in Fig. 11.19b). The selected-area electron diffraction pattern of an InSb–GaAs interface is shown in Fig. 11.19c. The diffraction spots occur in pairs, corresponding to InSb and GaAs, respectively, indicating that the layer and the substrate are epitaxially oriented. Very similar results were obtained for $InAs_xSb_{1-x}$/GaAs and $InBi_xSb_{1-x}$/GaAs heterostructures, confirming that heteroepitaxy is more dominant than alloying effects for these structures for small composition.

11.7 Physical Properties of InSb, $InAs_xSb_{1-x}$, and $InBi_xSb_{1-x}$

It is known that III–V compounds have band structures nearly similar to those of group IV semiconductors, although there are some differences which arise from the lack of inversion symmetry in their crystal structures. The effective mass of electrons in the conduction band varies from compound to compound and is the smallest for InSb-based materials amongst all III–V compounds. The effective masses of heavy holes, however, do not seem to very much in these compounds. The strong interaction between the valence and conduction bands results in nonparabolicity at the bottom of the conduction band in these materials. Hence their electrical and optical properties are different from various other compound semiconductors.

11.7.1 Band Structure of InSb, $InAs_xSb_{1-x}$, and $InBi_xSb_{1-x}$

The band structure of InSb was calculated using the $\boldsymbol{k}\cdot\boldsymbol{p}$ perturbation approach [11.142]. Various authors have postulated the band structure of InSb and experiments have verified the same. Experiments show that the minimum of the conduction band in indium antimonide lies at the center of the zone and that the band is spherically symmetrical. The most direct evidence for the spherical symmetry of the band has come from microwave cyclotron resonance experiments [11.143]. For a band with spherical symmetry, the longitudinal magnetoresistance should be zero, and it has been found experimentally in n-type InSb that it is indeed an order of magnitude smaller than the transverse magnetoresistance [11.144]. The effective mass of electrons at the bottom of the conduction band in InSb is only $0.013(\pm0.001)m_0$ [11.143]. This very small effective mass means that the band has very high curvature and very low density of states at its minimum. As a result a small number of electrons fill the band to high energy levels [11.143]. Hence material becomes degenerate at relatively low electron densities and the height of the Fermi level above the bottom of the conduction band increases rapidly with electron concentration. At a temperature T the conduction band is filled up to a level which is about $4k_BT$ below the Fermi level, so that in an impure sample the transitions are to a level well above the bottom of the conduction band. In this case the optical energy gap increases with electron concentration as

$$\Delta E=\Delta E_0+\left(1+\frac{m_e}{m_h}\right)(\zeta-4k_BT)\,, \qquad (11.8)$$

where ζ is the height of the Fermi level above the bottom of the band and ΔE_0 is the difference between the conduction-band minimum and the valence-band maximum. A striking property of electrons in the conduction band of indium antimonide is their very large magnetic moment, which is a consequence of interaction between the conduction and valence bands [11.145]. The g factor can be evaluated from

$$g=2\left[1-\left(\frac{m_0}{m_e}-1\right)\left(\frac{\Delta}{3E_g+2\Delta}\right)\right], \qquad (11.9)$$

where Δ is the spin–orbit interaction and E_g is the energy gap. The measured g values decrease from -50.7 for $2\times10^{14}\,cm^{-3}$ to -48.8 for $3\times10^{15}\,cm^{-3}$ electron concentration. The theoretical work on InSb suggests the existence of three valance bands: a *heavy-hole* band V_1, a *light-hole* band V_2, degenerate with V_1 at $\boldsymbol{k}=0$, and a band V_3 due to spin–orbit coupling (Fig. 11.20a).

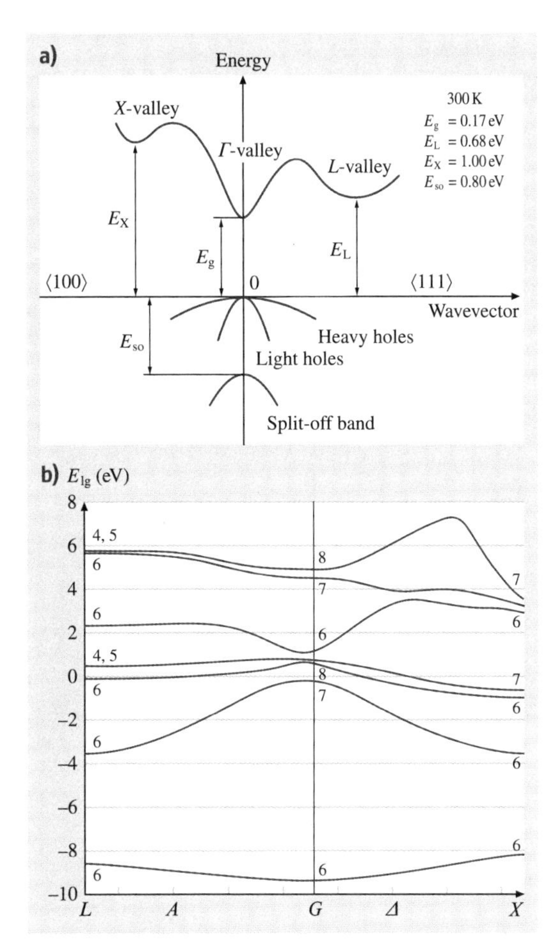

Fig. 11.20a,b Band structures of (**a**) InSb and (**b**) InBiSb (after [11.149])

The effective mass of the heavy hole lies in the range $0.55m_0 < m_h < 0.72m_0$ and that of the light hole is $0.015m_0$. The spin–orbit splitting of the valence band is estimated to be about 0.8 eV. The maximum of the valence band is not exactly at $\boldsymbol{k} = 0$ but is displaced in the ⟨111⟩ direction, although it is difficult to resolve this displaced maxima experimentally because of the small energy (10^{-4} eV). Since the momentum offset is so small in InSb it behaves as a direct-gap semiconductor. Magnetoresistance measurements have been used to measure the shape of the valence band [11.146]. The band structure for $InAs_xSb_{1-x}$ has also been studied within the framework of $\boldsymbol{k} \cdot \boldsymbol{p}$ theory and using this the variations of effective masses and g factor with composition were reported. By modifying the $\boldsymbol{k} \cdot \boldsymbol{p}$ theory and taking into account a multiband approximation as well as modified matrix elements the precise values of m and the g factor were obtained by *Hermann* and *Weisbuch* [11.147]. Furthermore, spin–orbit splitting of $InAs_xSb_{1-x}$ has been studied by several authors [11.147, 148]. They reported values quite different from those calculated by a virtual-crystal approximation in which fluctuations in the crystal potential had been neglected. However, for very small values of x ($0 < x < 0.07$) both give similar values.

The band structure of $InBi_xSb_{1-x}$ has been studied extensively by *Vyklyuk* et al. [11.149]. They calculated the electronic band structure of $InBi_xSb_{1-x}$ for $x = 0.05$ using a local empirical pseudopotential with spin–orbit interaction in the virtual-crystal approximation (Fig. 11.20b). Utilizing this band structure they calculated the absorption coefficient, which closely matches experimental values.

11.7.2 Transport Properties of InSb, $InAs_xSb_{1-x}$, and $InBi_xSb_{1-x}$

InSb

Intrinsic carrier concentration is of fundamental importance for semiconductors and their application in devices. Knowledge of the intrinsic carrier concentration should help in understanding the performance of detectors. At moderate carrier densities, the degeneracy in InSb varies considerably in the temperature range of measurements [11.150]. This has a significant effect in determining the temperature dependence of transport properties. At room temperature, the intrinsic carrier density in InSb is $\approx 2.02 \times 10^{16}\,\mathrm{cm^{-3}}$, which is often larger than the background doping. Therefore, even in n-type doped samples, transport experiments must be considered in the light of ambipolar conduction [11.151]. Figure 11.21 shows a plot of the intrinsic carrier concentration n_i as a function of temperature. Since the heavy holes are ≈ 50 times heavier than the light holes (which have almost the same effective mass as the conduction electrons) they have a correspondingly larger relative population density at a given temperature. The purest InSb crystals reported contained less than 10^{13} donors/cm^3 [11.152], but most of the information available on electrical properties refers to crystals with impurity concentration greater than $10^{14}\,\mathrm{cm^{-3}}$. The Hall coefficient in n-type samples varies little with temperature below 100 K. Above 150 K the purest sample is intrinsic. In n-type InSb the donor levels are merged with the conduction band, and at temperatures below the intrinsic range,

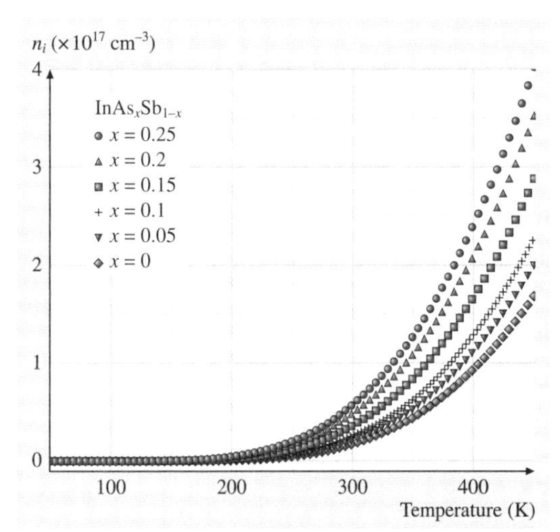

Fig. 11.21 Calculated intrinsic carrier concentration in InSb and $InAs_xSb_{1-x}$ as a function temperature

the concentration of conduction electrons does not decrease with temperature. Thus at this point material behaves like a metal, with few charge carriers of high mobility.

The temperature dependence of conductivity and Hall data were reported by many groups [11.153–155]. *Vinogradova* and coworkers [11.156, 157] have reported room-temperature mobilities higher than 100 000 $cm^2/(V\,s)$ in specimens with donor concentration of $10^{13}\,cm^{-3}$. A more likely value for samples with donor concentration lower than $5\times10^{14}\,cm^{-3}$ is 78 000 or 65 000, where the Hall factor (r_H) is taken as 1 or $\frac{3}{8}\pi$, respectively. The highest value reported for electron mobility at 77 K is $1.1\times10^6\,cm^2/(V\,s)$ for samples with donor concentration of $8\times10^{12}\,cm^{-3}$ [11.152]. The highest reliable hole mobility value at room temperature is $750\,cm^2/(V\,s)$ and that at 77 K is $10\,000\,cm^2/(V\,s)$. The mobility ratio (b) is also temperature dependent and for pure specimens is $\approx 6.3\,T^{1/2}$ above 250 K. At room temperature b is just 100. In impure specimens b is rather lower because the mobility of electrons is affected more than that of holes by impurity scattering. *Madelung* and *Weiss* [11.158] using resistivity data measured the thermal energy gap in InSb at 0 K to be 0.265 eV. *Hrostowski* et al. [11.153] also estimated it to be between 0.26 and 0.29 eV. Recently *Bansal* and *Venkataraman* [11.151] probed the effect of Landau level formation on the population of intrinsic electrons in InSb near room temperature in magnetic fields up to 16 T. Although the measured magnetic field dependence of the Hall coefficient is qualitatively similar to the published results, it is shown that the data may also be explained by simply including ambipolar conduction. Thus, the inference on band depopulation drawn from previous measurements on InSb is inconclusive unless both the Hall and magnetoresistive components of the resistivity tensor are simultaneously measured and modeled. When the model includes both depopulation and ambipolar conduction, reasonable agreement with theory can be achieved. *Drachenko* et al. [11.159] experimentally reevaluated the effective mass of electron in InSb for two different g-factor values. It was found that the effective mass m_e was equal to $0.0127m_0$ at 80 K for $g = 54$. On the other hand, taking the experimental value of g-factor reported by *Miura* et al. [11.160] as 70, m_e was evaluated as

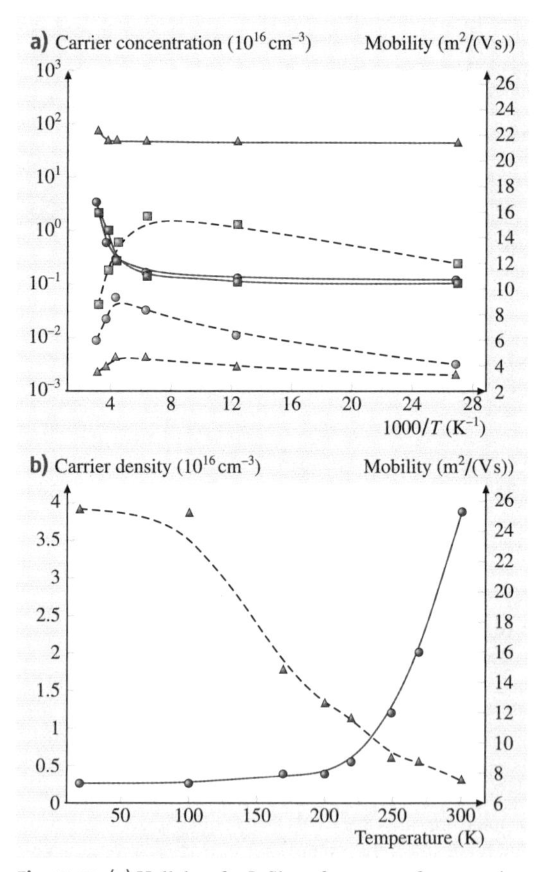

Fig. 11.22 (**a**) Hall data for InSb wafers grown from starting materials of different purity and (**b**) temperature-dependent Hall data of high-quality InSb wafer

$0.0133\,m_0$, which is very close to the recommended value ($0.0135\,m_0$) for InSb at helium temperatures. A very comprehensive account of band structure effects and transport calculations in narrow-gap semiconductors can be found in the book by *Nag* [11.161] and review articles by *Zawadzki* [11.162]. Recently *Bansal* and *Venkataraman* [11.151] and *Dixit* et al. [11.10, 115] have reported transport properties of these materials in both the bulk and epitaxy forms. Strong effects of impurities in the starting materials on carrier density variation as a function of temperature have been noticed (Fig. 11.22a) [11.115]. The bulk InSb sample may be taken as a reference against which the effects of epitaxy and alloying can be compared. The temperature dependence of the measured carrier concentration and mobility for a bulk InSb sample is shown in Fig. 11.22b. The background doping of bulk InSb is a reasonable indicator of its properties. The low-temperature mobility and carrier concentration of bulk sample were at least an order of magnitude better than those of InSb epitaxial films grown from starting materials of the same purity (5 N).

$InAs_xSb_{1-x}$

Apart from disorder contributions, transport in InSb–InAs alloys may also be treated in a similar way to that of InSb. Only the band parameters, viz. the energy gap, the spin–orbit splitting, the effective masses, and phonon energies, need to be suitably determined at a given alloy concentration. For this the simplest method has been suggested by *van Vechten* [11.148] in a disordered virtual-crystal model for his quantum dielectric theory [11.163]. The first calculation of the intrinsic carrier concentration in $InAs_xSb_{1-x}$ was carried out by *Rogalski* and *Jozwikowski* [11.18], who used a three-band approximation of $\boldsymbol{k}\cdot\boldsymbol{p}$ theory. The intrinsic carrier concentration and the reduced Fermi energy were calculated for $InAs_xSb_{1-x}$ with $0 \le x \le 1$ and $50\,K \le T \le 300\,K$. By fitting the calculated nonparabolic n_i values to the expression for parabolic bands, the following equation for the intrinsic carrier concentration has been obtained

$$n_i = (1.35 + 8.50x + 4.22\times10^{-3}T - 1.53\times10^{-3}xT - 6.73x^2)T^{3/2}E_g^{3/4} \times \exp\left(\frac{-E_g}{2k_BT}\right)10^{14}\,. \tag{11.10}$$

The calculated intrinsic carrier concentration in $InAs_xSb_{1-x}$ (for lower values of As) as a function of temperatures has already been shown in Fig. 11.21. As can be seen from Fig. 11.21, with increasing arsenic content the intrinsic carrier density increases. The conduction-band effective mass m_e was determined by *Berolo* et al. [11.164] for the whole range of compositions. The resultant conduction band effective mass m_e was given as

$$\frac{1}{m_e(x)} = \frac{1}{m_{ce}} + \frac{\delta E}{3}\left[\frac{1}{m_{hh}E_{gv}} + \frac{1}{m_{lh}E_{gv}} + \frac{1}{m_s(E_{gv}+\Delta_v)} - \frac{1}{m_{ce}}\left(\frac{2}{E_{gv}} + \frac{1}{E_{gv}+\Delta_v}\right)\right]. \tag{11.11}$$

In order to determine $m_e(x)$, all the parameters of above equation should be known. Here the difference $\delta E = E_{gv} - E_g$ is determined by the effects of the aperiodic potential due to disorder. E_{gv} and Δ_v are obtained

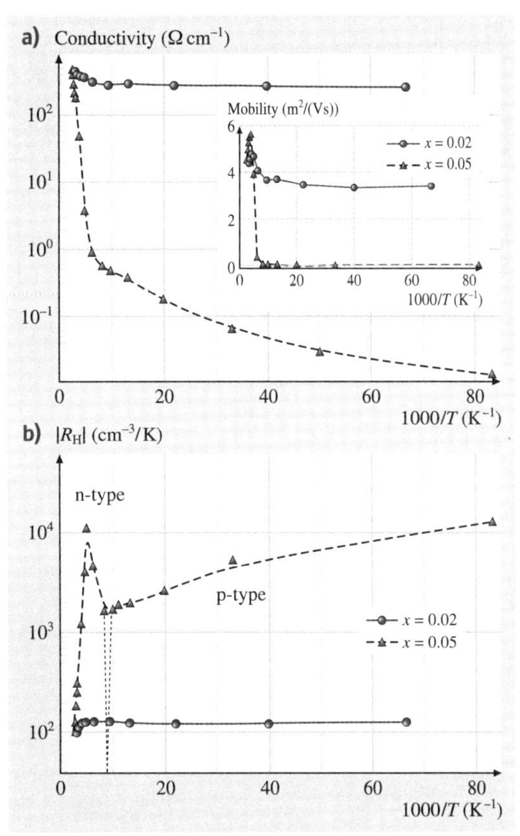

Fig. 11.23a,b Temperature dependence of (**a**) conductivity and mobility (*inset*) of $InAs_xSb_{1-x}$ and (**b**) Hall coefficient

using the virtual-crystal approximation (VCA), which can be approximated by the expressions $E_{gv} = 0.351 - 0.0176x$ and $\Delta_v = 0.39 - 0.42x$ at 300 K. m_{lh} and m_{hh} are the light- and heavy-hole effective masses and m_{ce} is the effective mass for the conduction band in the absence of conduction–valence band mixing.

Resistivity and low-field (< 0.2 T) Hall coefficient for $InAs_xSb_{1-x}$ samples with $x = 0.02$ and 0.05 were reported between 12 and 300 K [11.9]. The room-temperature (RT) mobility in the samples for $x = 0.02$ was $4.5 \times 10^4\,cm^2/(V\,s)$ and that for $x = 0.05$ was $5.6 \times 10^4\,cm^2/(V\,s)$. All samples with $x = 0.02$ were n-type and showed intrinsic behavior above ≈ 250 K. The representative results for $x = 0.02$ samples shown in Fig. 11.23a and b give a RT carrier concentration of $6.4 \times 10^{16}\,cm^{-3}$ and a background doping of $5 \times 10^{16}\,cm^{-3}$. On the other hand $InAs_{0.05}Sb_{0.95}$ shows a type conversion from n to p at 110 K. Below 77 K, the Hall coefficient and resistivity did not saturate and the conductivity below 30 K was strongly activated, indicating the presence of trap states. Further evidence of their presence is given by the pronounced tail in the absorption edge for this sample (as discussed in the next section). The hole mobility continuously dropped with decreasing temperature to $175\,cm^2/(V\,s)$ at 15 K.

$InBi_xSb_{1-x}$

Akchurin et al. [11.67] carried out experimental investigation on the behavior of Bi in $InBi_xSb_{1-x}$ solid solutions. It was found that bismuth doping in InSb strongly increases the electron density, which is due to the formation of Bi donor level [11.67]. An x-ray structure investigation indicated that this level is due to a complex state of Bi in the InSb lattice, representing a simultaneous combination of the substitutional and interstitial components. In the entire range of composition $InBi_xSb_{1-x}$ shows n-type behavior. The electron mobility of $InBi_xSb_{1-x}$ films decreases as the Bi concentration increases. This decrease of mobility is attributed to random alloy scattering [11.67]. The temperature dependence of resistivity also reveals the donor nature of Bi in $InBi_xSb_{1-x}$ [11.67]. *Dixit* et al. [11.12] also showed the donor nature of Bi in $InBi_xSb_{1-x}$ system.

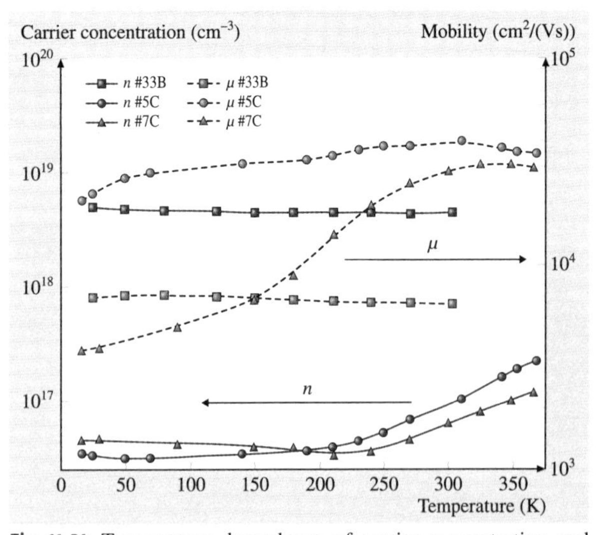

Fig. 11.24 Temperature dependence of carrier concentration and mobility of InSb/GaAs

InSb, $InAs_xSb_{1-x}$, and $InBi_xSb_{1-x}$ Grown on GaAs

For these materials, the unintentional background doping levels for the epitaxial layers were all similar, at around $\approx 10^{16}\,cm^{-3}$, almost an order of magnitude higher than that in pure bulk InSb prepared under similar conditions. The mechanism for this unintentional n-type doping is not clear, although material contamination during growth (or use of the starting material of lower purity [11.115]) was seen to increase the background doping levels and reduce the mobility (Fig. 11.24). The room-temperature mobility is $30\,000–50\,000\,cm^2/(V\,s)$ for InSb, $InAs_xSb_{1-x}$, and $InBi_xSb_{1-x}$ epilayers when grown from starting materials of the same purity (5 N). At around 275 K, the number of thermally generated electron–hole pairs become comparable to this background doping level and therefore the carrier concentration shows an activated behavior above this temperature. The activated region, when fitted to a relation $n_i = AT^{3/2}\exp(-E_g/2k_BT)$, yielded zero-temperature energy gap values of 0.23, 0.20, and 0.19 eV for InSb, $InAs_{0.06}Sb_{0.94}$, and $InBi_{0.04}Sb_{0.96}$, respectively, in reasonable agreement with more precise estimation from optical measurements (described in the next section). A comparison with published results indicates that both the carrier density and the mobility match with the results published on InSb, $InAs_xSb_{1-x}$, and $InBi_xSb_{1-x}$ epitaxial layers on GaAs, grown by MBE or MOCVD. Since between 10 and 250 K, an anomaly was observed in the carrier concentration for all the epitaxial layers, the measured Hall coefficient (R_H) showed a maximum at some intermediate temperature. In the past this has been attributed to multicarrier conduction, with contributions from interfacial, bulk, and possibly a surface layer, each with a different mobility, separately contributing to the total conductivity. However, LPE-grown epilayers are relatively thick and the observed properties did not correlate with film thickness. An alternative explanation in terms of variation of Hall factor (r_H) with the sample temperature and hence its degeneracy is more appropriate in this case. The Hall factor is exactly unity for a strongly degenerate sample but can take a larger value in nondegenerate samples, especially when long-range scattering potentials are active in limiting the sample mobility (e.g., $r_H = 1.93$ for ionized impurity scattering). The Hall factor for a partially degenerate gas, as in this case, can increase the measured Hall coefficient by $\approx 10\%$, which could explain the anomaly seen at most temperatures. The calculated temperature dependence of mobility due to various scattering mechanisms is shown in Fig. 11.25.

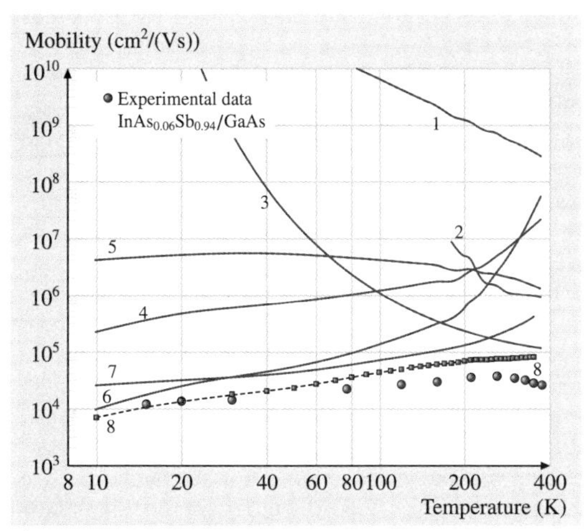

Fig. 11.25 Calculated temperature dependence of mobility due to various scattering mechanisms for $InAs_{0.06}Sb_{0.94}$/GaAs. (1) Acoustic phonon, (2) electron–hole, (3) polar optical, (4) ionized impurity, (5) alloy, (6) charged dislocation, (7) strain in dislocations, and (8) experimental data points

Magnetoresistance measurements were performed on $InAs_xSb_{1-x}$/GaAs by *Bansal* et al. [11.6]. The magnetoresistance signal starts as quadratic initially and becomes linear at higher magnetic fields (B), which is the expected behavior in the extreme quantum regime (Fig. 11.26). There was no magnetic-field-induced freeze-out observed at 4.2 K since, at background carrier densities of $\approx 10^{16}\,cm^{-3}$, the impurity-band wavefunctions, even at 6 T magnetic field, have a spatial extent larger than the interimpurity separation because of the very small effective mass. There was no appreciable change in resistance or magnetoresistance between 4.2 and 30 K, again indicating that parallel conduction effects are not important since these are known to give an appreciable temperature dependence. Unlike in metals, the oscillations are not periodic in $1/B$ because of a considerable magnetic field dependence of the Fermi energy and unequal Landau level spacing due to nonparabolic energy dispersion. The amplitude of oscillations is extremely sensitive to sample homogeneity and therefore, although the $\mu B > 1$ condition was easily met, the oscillation amplitudes were small. This was the first time a Shubnikov signal in InAsSb has been observed. These results were further investigated in detail by *Drachenko* et al. [11.159], who reported the first experimental data for the calculation of the effective mass for $InAs_xSb_{1-x}$. Using the expression for $E_g(T, x)$ in $InAs_xSb_{1-x}$ alloys, they obtained m_e of $0.0122m_0$ and $0.0117m_0$ for $InAs_{0.04}Sb_{0.96}$, and $InAs_{0.06}Sb_{0.94}$ respectively.

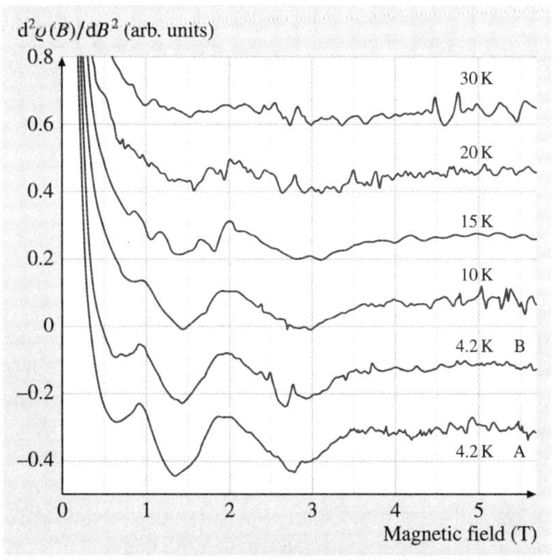

Fig. 11.26 Second derivative of the magnetoresistance signal of $InAs_{0.06}Sb_{0.94}$ measured at different temperatures. The electric field in B is twice that in A

$InBi_xSb_{1-x}$ and $InBi_xAs_ySb_{1-x-y}$ also show n-type behavior over the entire temperature range. The maximum electron mobility was $3.54\times10^4\,cm^2/(V\,s)$ and carrier concentration was $9.2\times10^{16}\,cm^{-3}$ at 300 K for $InBi_xSb_{1-x}$. Similarly $InBi_xAs_ySb_{1-x-y}$ has maximum electron mobility of $3.1\times10^4\,cm^2/(V\,s)$ and carrier concentration of $8.07\times10^{16}\,cm^{-3}$ at 300 K [11.8, 123].

11.7.3 Optical Properties of InSb, $InAs_xSb_{1-x}$, and $InBi_xSb_{1-x}$

The interest in antimonides is (to a large extent) because their energy gap is in the infrared. Optical properties of InSb have, of course, been long studied and are well understood [11.165, 166]. This section extends the studies to optical properties to determine the effect of alloying and heteroepitaxy on them.

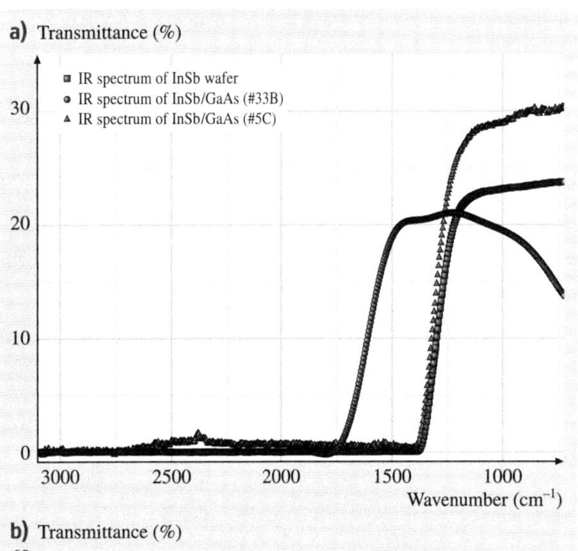

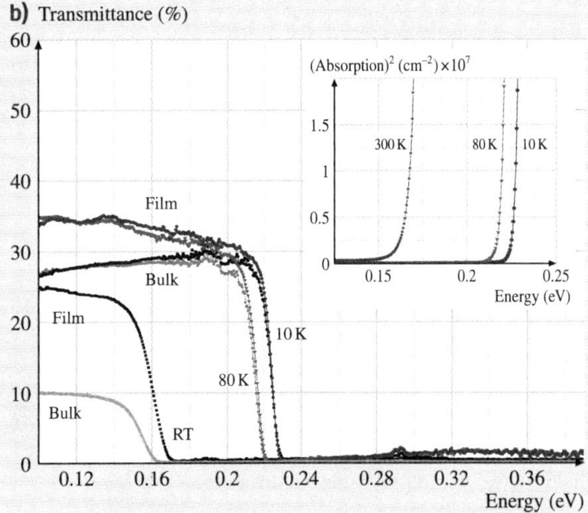

Fig. 11.27a,b Infrared spectra of InSb wafer and InSb/GaAs at (**a**) room temperature (**b**) different temperatures. The inset shows absorption squared versus energy plots at different temperatures

InSb

In InSb, as the photon energy decreases past the value corresponding to the energy gap (0.17 eV), transmission shows a sudden increase until saturation, after which it remains constant or oscillatory (depending on the scattering process). From the position and shape of this *absorption edge* we obtain information on the nature of transition from valence band to conduction band. With further lowering of the photon energy (0.14 eV) the optical properties are affected by interaction between the photons and free carriers, and resulting transition takes place within the valence or the conduction bands. At still lower energy (0.09 eV) there can be interaction between photons and the lattice. The optical properties in this region are influenced both by the lattice and the free carriers and are being studied using Raman spectroscopy. The absorption edge of InSb is extremely steep and the absorption coefficients change by two orders of magnitude within a range of 0.01–0.04 eV. The position of the absorption edge will be dependent on pressure since compression of the lattice also changes the energy gap. The dimensions of the lattice also change when temperature is altered, but here there is an additional effect on the optical properties because the change in lattice vibration affects the width of the energy levels.

The wavelength dependence of free carrier absorption is therefore dependent on the scattering mechanism. At temperatures at which $k_BT \gg h\nu$, the square-law dependence is applicable for both acoustic and nonpolar optical scattering. When photon energy is greater than k_BT, the absorption is proportional to $\lambda^{3/2}$ for acoustic scattering but for optical scattering the dependence is more complicated. For impurity scattering, the absorption coefficient (α) is proportional to λ^3. *Kurnick* and *Powell* [11.167] found that in InSb the free electron absorption had a λ^2 dependence, but the free-hole absorption was independent of wavelength. At 9 μm, the free-electron optical cross section ($=\alpha/n$) is 2.3×10^{-17} cm^2 but the hole cross section is 40 times larger. The electron cross section is in good agreement with the theoretical value [11.167]. Measurements by *Spitzer* and *Fan* [11.168] on n-type InSb were extended to 35 μm and their results show that the square-law dependence is not obeyed at wavelengths longer than 12 μm. However, *Moss* [11.169] pointed out that the refractive index of InSb is not constant in this wavelength region. This theory gives the wavelength dependence of the product $\eta\alpha$ (where η is refractive index), and if the results of Spitzer and Fan are used in conjunction with refractive index values, it is found that the product $\eta\alpha$ is proportional to λ^2.

Recently *Dixit* et al. [11.115] observed that the absorption edge is shifted from the expected value towards higher energy to a different extent depending on the purity of the starting materials (Fig. 11.27a) similar to other report in literature [11.170]. Hence it is clear that the energy gap of InSb is very sensitive to impurities. The temperature dependence of the absorption edge is not significant for heavily doped materials because the Fermi level lies inside the conduction band. On the other hand the absorption edge for moderately doped or undoped InSb is very sensitive to temperature, as shown in Fig. 11.27b. The absorption coefficient was calculated from the transmission T, using the well-known expression that accounts for multiple reflections within the sample, viz. $T=((1-R)^2\mathrm{e}^{-\alpha d})/((1-R)^2\mathrm{e}^{-2\alpha d})$. Here d is the thickness of the sample and $R=0.4$ is the reflection coefficient, assumed to be constant in the spectral range of measurement. This equation is easily inverted by substituting $z=\mathrm{e}^{-\alpha d}$ and then solving the quadratic equation in z. The energy gap, E_g, was obtained by fitting $\alpha=A(h\nu-E_g)^{1/2}$, to the experimental data around the absorption edge. Here α is the absorption coefficient in cm^{-1}, ν is the incident photon frequency, and A is a constant depending on the electron and hole effective masses and the optical transition matrix elements. The onset of absorption fits very well to the above relation, indicating a direct energy gap (inset of Fig. 11.27b). The energy gap was measured to be 0.172, 0.225, and 0.235 eV for RT, 80, and 10 K, respectively. The analysis of the temperature dependence of E_g is usually done by a three-parameter fit to Varshni's empirical relation [11.171], $E_g(T)=E_{gv}(0)-(\alpha T^2)/(\beta+T)$, where α and β are constants and $E_{gv}(0)$ is the energy gap at zero temperature. The energy gap of a semiconductor varies with temperature due to three distinct effects attributable to phonons [11.172]. These, all of similar importance, are the anharmonic (thermal expansion) and harmonic (Debye–Waller factor) phonon effects, and the temperature-dependent renormalization of the gap due to the electron–phonon self-energy correction (Fan's term). The value of the energy gap at different temperatures and the least-square fits to the above equation show good agreement. The fits yield $E_{gv}(0)$, α, and β as 0.235 ± 0.003 eV, $3.1\pm1.1\times10^{-4}$ eV/K, and 452 ± 190 K, respectively. These values and the temperature dependence of the energy gap are very similar to previous results [11.170], although these measurements were on wafers made from large-size crystals.

$InAs_xSb_{1-x}$

Woolley and *Warner* [11.13] carried out optical studies on $InAs_xSb_{1-x}$ samples made by directional freeze and zone recrystallization methods. Although the energy values are determined on crudely prepared single crystal and ploycrystals, the variation of energy gap with composition and temperature is reliable. The room-temperature values of E_g have been determined over the whole composition range. It is found that E_g falls as one compound is added to the other, the measured value of E_g reaching a minimum of 0.10 eV at approximately 60 mol% InSb. The variation of optical energy gap with alloy composition at room temperature has already been shown in Fig. 11.1a. For alloys near the center of the composition range the energy gap reduces. Due to the smaller energy gap and larger carrier concentration these alloys become degenerate at room temperature and hence the Fermi level goes into the conduction band. Thus the actual energy gap for the compounds in the intermediate range may in fact be smaller than the measured value because of the Moss–Burstein effect. The bandgap bowing in alloy systems has been explained by virtual-crystal analysis [11.173]. The energy gap variation in $InAs_xSb_{1-x}$ as a function of x is given as

$$E_g(x,T)=0.434-0.771x+0.59x^2 -2.8x10^{-4}T\ [\mathrm{eV}]\,. \quad (11.12)$$

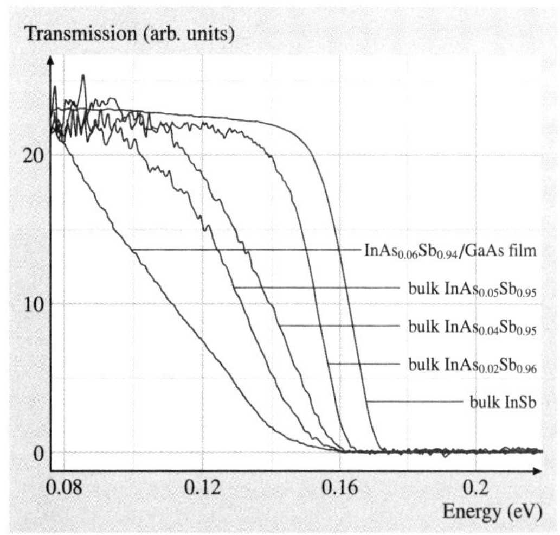

Fig. 11.28 Room-temperature transmission spectra of $InAs_xSb_{1-x}$ for various x values up to 0.6

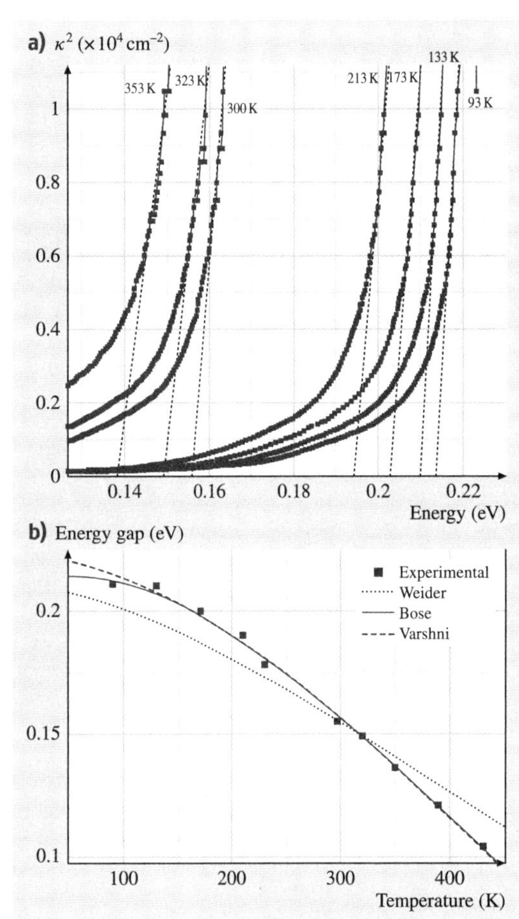

Fig. 11.29 (**a**) Plots of absorption squared versus energy for $InAs_xSb_{1-x}$. *Straight lines* imply a direct bandgap. (**b**) Temperature dependence of energy gap. Data fitted to equations of Bose and Varshni. Wieder and Clawson's formula is also plotted (after [11.176])

The absorption edge shifts to lower energy for $InAs_xSb_{1-x}$ as x increases. The room-temperature transmission spectra for $InAs_xSb_{1-x}$ as a function of x are shown in Fig. 11.28. The value of the energy gap was evaluated to be 0.17, 0.16, and 0.15 eV for $x = 0$, 0.02, and 0.05, respectively. The temperature dependence of the energy gap and free carrier absorption in a high-quality $InAs_{0.05}Sb_{0.95}$ single crystal were also studied between 90 K and 430 K through the absorption spectra [11.7] (Fig. 11.29a). The value of the energy gap at different temperatures along with the least-square fits to the above equation as well as Bose ($E_g(T) = E_{gBE}(0) - 2a_B/(\exp^{(\theta_{BE}/T)} - 1)$ [11.174, 175]) and Varshni ($E_g(T) = E_{gv}(0) - (\alpha T^2)/(\beta + T)$) equations are shown in Fig. 11.29b. The fits yield $E_{gv(0)}$, α, and β as 0.223 ± 0.003 eV, $7.1 \pm 2.3 \times 10^{-4}$ eV/K, and 675 ± 370 K, respectively. For $E_{gBE(0)}$, $2a_B$, and θ_{BE} the values are 0.214 ± 0.002 eV, 0.107 ± 0.01 eV, and 466 ± 46 K, respectively. It can be seen that the Bose and Varshni equations give nearly identical fits for $T > 125$ K (Fig. 11.29b). Below 125 K the Bose fit is better, as is usually the case. The value of $E_{gBE(0)}$ agrees with the relation $E_{g(0)} = 0.4324 - 0.8831(1-x) + 0.6853(1-x)^2$, which gives $E_{g(0)} = 0.212$ eV for 5 at.% arsenic. Through measurements on flash-evaporated films and using data from other groups, *Wieder* and *Clawson* [11.176] fitted the energy gap to the above equation. Owing to its inherent appeal and widespread use, this equation is also plotted in Fig. 11.29b, although the agreement is not good except at room temperature. *Bansal* et al. [11.7] showed that on the low energy side (< 100 meV) free electrons become the dominant source of absorption for $InAs_{0.05}Sb_{0.95}$. Due to the vast difference in strength, the contributions of band-to-band transitions and free carrier absorption (FCA) can be separated by a minimum in absorption which occurs at an intermediate energy. The temperature dependence of the absorption coefficient as a function of wavelength (14–25 μm) is shown in Fig. 11.30a. The studied sample was p-type at low temperatures and hence the FCA increased around RT when the sample became intrinsic and the Hall coefficient changed sign. Absorption below room temperature due to holes was too weak to be resolved from the higher-order interband absorption background because of their much larger effective mass. Using the value of the carrier concentration of this sample at 300 K, the FCA cross section was measured to be 7.35×10^{-16} cm² at 15 μm. The curves in the figure can be fitted to a power law of the form $\alpha = k\lambda^p$. The experimentally determined value of the exponent p is ≈ 1.5 near room temperature and is the one theoretically expected for acoustic-phonon-assisted FCA. The FCA exponent shows a steady drop with increasing temperature above 300 K; several reasons have been given for this. Firstly, above 350 K, there is an enhanced probability for second-order interband transitions accompanying the abrupt increase in the optical phonon occupancy (optical phonon energy 300 K). These broaden the absorption edge by ≈ 25 meV and overlap with FCA, making the total absorption curve flatter and extraction of the exponent unreliable. Secondly, beyond 400 K, where the pho-

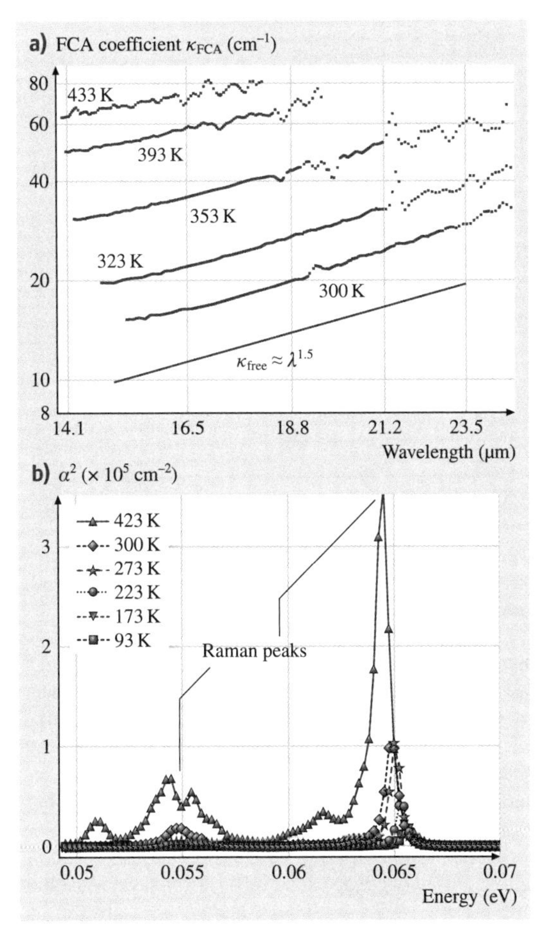

Fig. 11.30 (a) FCA spectra of $InAs_{0.06}Sb_{0.94}$ at different temperatures. Also $k_{FCA} \approx (\lambda)^{1.5}$ is plotted for comparison (after [11.7]) (b) Raman peaks of $InAs_{0.06}Sb_{0.94}$ at different temperatures (after [11.5])

ton energy is $\approx 2k_BT$, classical analysis may become applicable. This, when accompanied by an increase in the optical phonon occupancy, could also lead to a sharp drop in the value of the relaxation time, τ, due to enhanced electron–optical phonon scattering. As τ reaches close to 10^{-14} s, deviation from the quadratic to a non-power-law behavior is expected, which could also lead to the observed reduction in the value of p. At high temperatures, multiphoton absorption by IR-active lattice modes is significantly enhanced. This is seen as noise in the high-temperature absorption spectra appearing around the second harmonic of the Raman peaks [11.177] (≈ 50 meV) (Fig. 11.30b).

$InBi_xSb_{1-x}$

Quantum dielectric theory (QDT) has been used to predict the direct energy gap E_g of $InBi_xSb_{1-x}$ semiconductor alloy up to 10 mol% InBi [11.39]. The calculated composition dependence of E_g for $InBi_xSb_{1-x}$ is in good agreement with experimental results with a predicted semiconductor–semimetal transition (77 K) at $x = 0.124$. Figure 11.1b shows the E_g variation with InBi predicted using QDT for $InBi_xSb_{1-x}$ at 0, 77, and 300 K, along with experimental results. E_g(77 K) values corresponding to 8–12 μm are obtained at $x = 0.043$–0.070. The first-order calculation of E_g versus x follows the simple relation $E_0\,(\mathrm{eV}) = 0.23-1.85x$ at 77 K. *Vyklyuk* and coworkers [11.149] have obtained the absorption coefficient of $InBi_xSb_{1-x}$ in 0–10 eV range. The calculation of the absorption coefficient was based on the electron energy of $InBi_xSb_{1-x}$ using a local empirical pseudopotential method with a virtual-crystal approximation including spin–orbital interaction. The energy dependence of the absorption coefficient of $InBi_xSb_{1-x}$ showed an increase with Bi content and a shift in the absorption curve to lower energies [11.149]. The absorption edge of $InBi_xSb_{1-x}$ at RT showed a shift towards lower energy [11.12]. Bandgap as estimated from the IR absorption edge is 0.113 ± 0.009 eV, which is very close to the value obtained from Fig. 11.1b [11.39]. They also noted that the free-carrier absorption was more in $InBi_xSb_{1-x}$ crystal compared with InSb.

InSb, $InAs_xSb_{1-x}$, and $InBi_xSb_{1-x}$ Epilayers Grown on GaAs

Figure 11.27a,b also shows the absorption edge for InSb epitaxial layer at room temperature. The transmission spectra for the bulk and thin film are almost coincident, indicating that disorder effects are not strong enough to affect the optical absorption properties. Therefore, the effect of mismatched epitaxy is not evident in the optical absorption near the energy gap. This is in contrast to transport measurements, where heteroepitaxy was seen to be the most important factor affecting carrier mobility. Temperature-dependent optical absorption edge measurements on heteroepitaxial InSb are also shown in Fig. 11.27b. The difference in the cutoff energies between the spectra at 10, 80, and 300 K clearly shows the nonlinearity of the temperature dependence of the energy gap at low temperatures.

Figure 11.28 shows the transmission spectra measured for $InAs_xSb_{1-x}$ with different alloy concentrations. As expected, the absorption edge clearly shifts to lower photon energies as the composition of arsenic is increased. There is also an enhanced band tailing as the alloy fraction increases. Such band tails have been previously observed in similar anion-substituted mismatched alloy InP_xSb_{1-x} [11.178], and have been attributed to the formation of localized states as a result of composition fluctuations. It has been argued that composition and strain fluctuations effectively act as quantum wells which can trap electrons. A Gaussian distribution of their widths results in an exponentially decreasing density of localized states below the mobility edge. Recently, similar observations were reported by *Gao* et al. [11.179] on $InAs_xSb_{1-x}/GaAs$ ($x < 0.06$) grown using ME. They also explain their results through composition fluctuations. Very recently *Bansal* et al. [11.180] further studied this and distinguished three absorption region: band to band, Urbach edge, and free-carrier absorption regions. They modeled the Urbach region and determined a structural disorder energy of 30 meV for $InAs_xSb_{1-x}$. The RT energy gap for $InAs_xSb_{1-x}$, calculated by assuming a cutoff at the mid-transmittance wavelength [11.181], is as low as 0.1 eV. However a more reliable estimation, i. e., fitting the absorption coefficient to the relation $\alpha = A(E_g - h\nu)^{1/2}$, gave a value of 0.133 eV for $InAs_{0.06}Sb_{0.94}/GaAs$. *Wieder* and *Clawson*'s relation [11.176] gives the expected energy gap to be 0.146 eV for $x = 0.06$ at 300 K. Results for the bulk $InAs_{0.05}Sb_{0.95}$ sample showed perfect agreement with the above relation. Therefore, the 13 meV discrepancy is either due to error associated with calculating the energy gap (due to band tailing) or to a decrease in the gap due to residual strain. The residual strain, as evaluated from x-ray measurements, corresponds to a shift in the gap, $\Delta E = 2b(C_{11} + 2C_{12}/C_{11})e_{xx}$, where b is the deformation potential, C_{11} and C_{12} are the stress components, and e_{xx} is the in-plane residual strain, which is 0.0184 as evaluated from HRXRD. This yields a splitting energy (ΔE) of 15 meV, which may also explain the difference between the values for bulk and epitaxial $InAs_xSb_{1-x}$. The change in E_g value with temperature for a $InAs_{0.06}Sb_{0.94}/GaAs$ sample along with fits to Varshni's and Bose–Einstein-type relations are also shown in Fig. 11.30. Varshni parameters for this curve are $E_g(0) = 0.193 \pm 0.007$ eV, $\alpha = 3.01 \pm 3.1 \times 10^{-4}$ eV/K, and $\beta = 341 \pm 60$ K. The Bose–Einstein fit yields $E_{gBE(0)} = 0.19$ eV, $2a_B = 0.051$, and $\theta_{BE} = 395$ K. The set of Varshni parameters are different from those obtained for bulk crystal. *Marciniak* et al. [11.182] have suggested a straight-line relationship between α and β, where $\alpha/(271 + \beta) = 6.5 \times 10^{-7}$ eV/K^2. Both the parameters

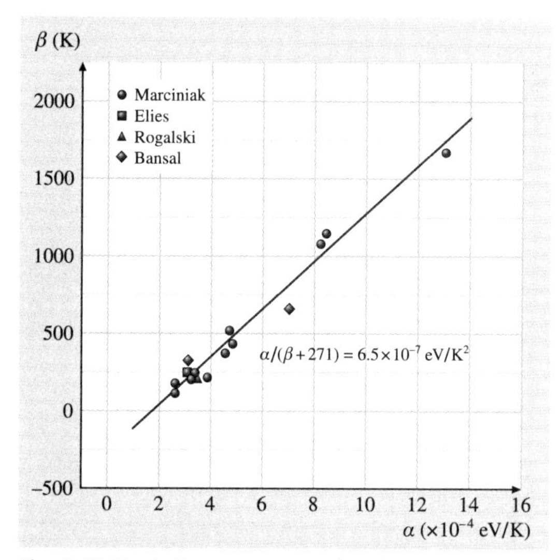

Fig. 11.31 Varshni's parameters for InAs–InSb alloys measured at different concentrations by different research groups (after [11.183])

obtained for $InAs_{0.05}Sb_{0.95}$ and $InAs_{0.06}Sb_{0.94}/GaAs$ fall close to this straight line [11.183] (Fig. 11.31). It should be emphasized that Fig. 11.31 is completely empirical. The parameters obtained for the bulk crystals should nevertheless be taken as more accurate because of the absence of strain and higher mobility.

Dixit et al. [11.8, 123] reported the room-temperature bandgap for $InBi_{0.04}Sb_{0.96}/GaAs$ and $InBi_{0.025}As_{0.105}Sb_{0.870}/GaAs$ using the absorption equation. The bandgaps were 0.134 and 0.113 eV, respectively. The spectra obtained for $InBi_{0.04}Sb_{0.96}/GaAs$ and $InBi_{0.025}As_{0.105}Sb_{0.870}$ layers also show a low energy tail in absorption. Although the bandgaps of $InAs_xSb_{1-x}/GaAs$ and $InBi_xSb_{1-x}/GaAs$ show a reduction compared with the InSb bandgap (0.17 eV), this reduction is lower than the reported theoretical values. The possible cause could be that not all As and Bi have substituted antimony sites, but could have taken interstitial positions.

11.7.4 Thermal Properties of InSb and Its Alloys

The linear expansion coefficient of InSb is 6.5×10^{-6} K^{-1} and 5.04×10^{-6} K^{-1} at 80 and 300 K, respectively [11.184]. The thermal conductivity of the melt and solid InSb, which are important parameters in controlling the heat flow during growth, are 0.123 and $0.0474\,W\,cm^{-1}K^{-1}$, respectively [11.185]. The specific heat of the melt and solid InSb are found to be 0.234 and $0.242\,J/(g\,K)$, respectively, and the latent heat of fusion of InSb is $108\,J/g$. Significant thermal parameters required for the growth are reported by *Duffer* et al. [11.94]. *Dixon* and *Furdyna* [11.186] reported the static dielectric constant of InSb lattice as 16.8 ± 0.2. There do not seem to be detailed reports on the thermal properties of other considered alloys.

11.8 Applications

InSb and related crystals are used in many applications. Here we emphasize only infrared detector and galvanomagnetic sensor applications.

Infrared Photodetectors

The properties of InSb as a material for infrared detectors have been extensively discussed for more than 46 years now. At room temperature the intrinsic carrier density in narrow-gap materials is in the range $10^{15}-10^{17}\,cm^{-3}$ and the thermal generation rate is $10^{23}-10^{24}\,cm^{-3}\,s^{-1}$. This results in a high noise level in photodetectors at room temperature. *Ashley* and *Elliot* [11.187] used *nonequilibrium* operation, which resulted in better detector performance at room temperature. The structure designed to achieve nonequilibrium (higher operating temperature) conditions is $p^+\underline{p}^+\pi n^+$ or $p^+\underline{p}^+\nu n^+$, where π and ν refer to intrinsic materials (either p- or n-type), which form an active region, and $\underline{p}^+$ refers to wider-bandgap materials. The active region has a low doping level and therefore is intrinsic at RT. The n- and p-type contacts are made to the active region via regions with larger energy gap or higher doping level or both, so that under appropriate bias conditions minimal transport of minority carriers through the active region is ensured. At zero bias, the band structure of the devices ensures little transport of the minority carriers from the contact regions so that additional noise is minimized. The detectivity (D^*) of these devices is $2.5 \times 10^9\,cm\,Hz^{1/2}\,W^{-1}$. For certain applications, particularly those requiring low temperature, it is often desirable to narrow the spectral responsivity, thereby increasing the detectivity by reducing the influence of background radiation. This can be achieved either by an external filter or by embodying the filter in the photodetector structure. *Djuric* et al. [11.188] made use of a remarkable technique which involves self-filtering based on the Moss–Burstein effect. The structure is an n^+-p-InSb photodiode (p-type InSb wafer and a heavily doped InSb layer using LPE). The quantum efficiency of the Moss–Burstein effect decreases almost linearly with wavelength. This allows approximately constant sensitivity over a wide range of wavelengths when choosing appropriate material parameters. Later *Bloom* and *Nemirovsky* [11.189] concentrated on the fabrication of these detectors by reducing the surface recombination rates. *Michel* et al. [11.190] also developed InSb photovoltaic structures on GaAs using MBE and demonstrated a near-bulk value for the carrier lifetime in spite of large dislocation densities.

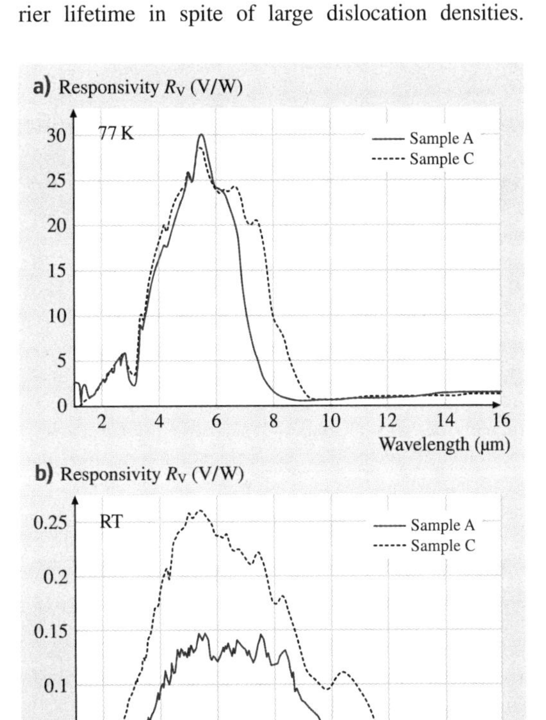

Fig. 11.32a,b Spectral responsivity of the photoconductors fabricated on InAsSb at **(a)** 77 K and **(b)** room temperature (after [11.34])

The highest carrier lifetime of 240 ns was reported for an InSb heteroepitaxial structure by these authors. $InSb/InAs_xSb_{1-x}/InSb$ ($p^+\pi\ n^+$) photovoltaic devices grown by LP-MOVPE and operated at room temperature in 8–13 μm was reported by *Kim* et al. [11.3]. Also, better performance was reported for photoconductors developed on $p\text{-}InAs_xSb_{1-x}/p\text{-}InSb/GaAs$ [11.3]. Very recently *Peng* et al. [11.34] developed infrared photodetectors in the 8–13 μm range using LPE. They fabricated photoconductors from LPE-grown InAsSb/GaAs, and a notable photoresponse beyond 8 μm was observed at RT. In particular, for $InAs_{0.3}Sb_{0.7}/GaAs$, a photoresponse up to 13 mm with maximum responsivity of 0.26 V/W was obtained at RT (Fig. 11.32). Hence, InAsSb/GaAs heterostructures grown using LPE demonstrate attractive properties suitable for room-temperature, long-wavelength infrared radiation. Details of the detectors fabricated employing MBE and MOVPE can be found in [11.191, 192].

Galvanomagnetic Applications

The most common, semiconductor-based magnetic field sensors are silicon (Si)-based Hall sensors. In general, the higher the mobility of the semiconductor and the thinner the active region, the better the galvanomagnetic device. The room-temperature mobility of undoped InSb is ≈ 55 times higher than that of Si. Hence InSb should be preferred over Si for use in Hall sensors. In fact, bulk InSb wafers have been used for many years in the fabrication of magnetic field sensors, such as magnetoresistors and Hall sensors. Magnetic field sensors are, in turn, used in conjunction with permanent magnets to make contactless potentiometers and rotary encoders. This sensing technology offers the most reliable way to convert a mechanical movement into an electrical signal, and is widespread in automotive applications. Recent developments in the growth of thin epitaxial layers of InSb on semi-insulating GaAs substrates have resulted in the development of magnetoresistors with excellent sensitivity and operating temperatures up to 285 °C which are also cost effective. Hall sensors and magnetotransistors of thin n-InSb films outperform their Si-based counterparts even with integrated amplification. *Oszwaldowski* [11.194] also suggested very specific Hall sensors that can be made from heavily doped ($1\text{–}2\times10^{18}\ cm^{-3}$) n-type InSb films. with these sensors they could achieve magnetic field sensitivity ≥ 0.05 V/T and temperature coefficient of the output voltage $\leq 0.01\%K^{-1}$, which are very difficult to achieve by any other Hall sensor. *Heremans* et al. [11.195] and *Heremans* [11.196] have described

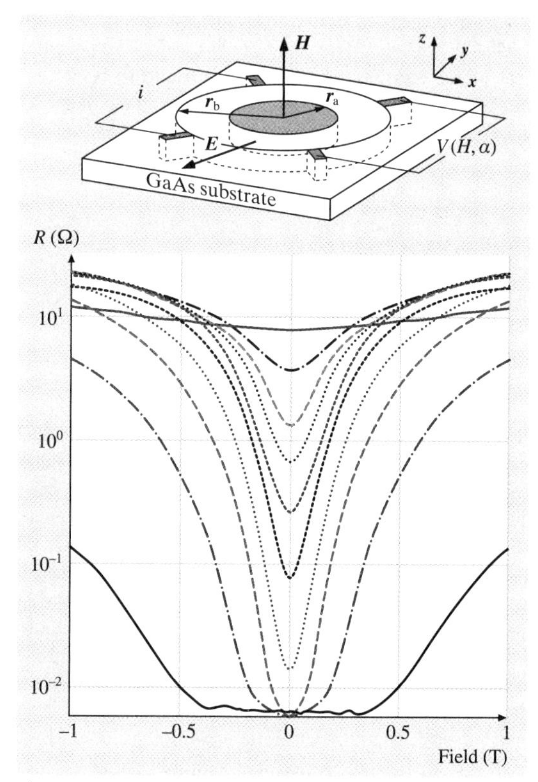

Fig. 11.33 RT resistance up to 1 T of a composite van der Pauw disk (vdP) of InSb and Au for a number of values of $a = 5r_a/r_b$. *Inset* schematic diagram of a cylindrical Au inhomogeneity (radius r_a) embedded in a homogeneous InSb vdP disk (radius r_b) (after [11.193])

device design criteria, materials requirements, and a direct comparison of the three types of galvanomagnetic devices, and have given the following guidelines for selecting sensors in different applications. According to them, NiFe-based magnetoresistors are preferred for sensing magnetic fields below 0.005 T, typical in magnetic read-out applications; Si Hall sensors with integrated amplification are useful in low-frequency applications in the field range of 0.005–0.03 T; and InSb-based magnetoresistors are most suitable in field ranges above 0.03 T but are not sensitive to the polarity of the field.

Recently *Solin* et al. [11.193] reported enhanced room-temperature geometric magnetoresistance in homogeneous nonmagnetic indium antimonide with an embedded concentric gold inhomogeneity (Fig. 11.33). The room-temperature geometric magnetoresistance is as high as 100%, 9100%, and 750 000% at magnetic fields of 0.05, 0.25, and 4.0 T, respectively. They found that for inhomogeneities of sufficiently large diameter relative to that of the surrounding disk, the resistance is field-independent up to an onset field ≈ 0.4 T, above which it increases rapidly. These results have been understood in terms of field-dependent deflection of current around the inhomogeneity.

11.9 Concluding Remarks and Future Outlook

In this chapter we have described and discussed certain specific characteristics of InSb, InSb-based ternary and quaternary crystals, and their thin films on highly mismatched semi-insulating GaAs substrates, with reference to infrared detection and galvanomagnetic applications. The use of these material systems as an alternative for such applications relies on the production of high-quality materials with low background doping level and defect density. While basic material quality is dictated by crystal growth conditions, the physical properties of the material are profoundly influenced by the process cycles and the conditions under which the devices are operated. Hence in making a good device, it is important to understand the material issues that are related to device performance and to achieve synergies between material preparation and device fabrication. Some of the issues which have been address in this review are:

1. How does mismatch heteroepitaxy affect the structural properties when the films are far beyond their critical thickness?
2. What do the interfaces look like?
3. Does strain modify the optical gap usefully?
4. How does the electron mobility depend on composition given that alloy scattering is negligible?
5. What role do dislocations play?

for which reasonable understanding has been provided.

It was found that scattering from dislocations, introduced as a result of GaAs heteroepitaxy, degrades mobility at low temperatures. However, room-temperature transport properties are comparatively unaffected. Bright-field cross-sectional TEM imaging shows a sharp interface with very small coalescing islands and the selected-area diffraction pattern indicates that the layer and the substrate are epitaxially oriented. Very similar results were obtained on $InAs_xSb_{1-x}$/GaAs and $InBi_xSb_{1-x}$/GaAs heterostructures. The energy gap of 0.133 eV for $InAs_{0.06}Sb_{0.94}$/GaAs shows a 13 meV reduction due to residual strain. This reduction is very useful to shift the energy gap to well within the 8–12 μm range. The interesting observation of band tailing as a result of alloying has been modeled through Urbach tail characteristics. Shubnikov–de Haas oscillations are observed in $InAs_{0.04}Sb_{0.96}$ and the effective mass of $InAs_xSb_{1-x}$ for low values of x is reported. The temperature dependence of the energy gap and the behavior of free-carrier absorption have been reported for InAsSb alloy for a composition for which the energy gap is 15% lower than that of InSb. The values of fundamental material parameters, such as the zero-temperature gap and its temperature coefficients, the effective mass, etc., were evaluated and analyzed within the existing theoretical models. From the FCA spectra, it was concluded that, for 5% arsenic, alloy scattering is not the mobility-limiting mechanism near RT, which is useful for reliably modeling and evaluating the performance of optical and electrical devices made from these alloys.

Although successful reports on the growth of InSb, $InAs_xSb_{1-x}$, and $InBi_xSb_{1-x}$ crystals and their epitaxial films on semi insulating GaAs using liquid-phase epitaxy are presented in this chapter, a number of issues still remain to be addressed. In growth utilizing nonequilibrium techniques, whether one can increase the composition of As and Bi in single crystals of $InAs_xSb_{1-x}$ and $InBi_xSb_{1-x}$ remains to be explored. Reduction in the film thickness of these materials using LPE is another challenge which further opens the scope for improving heteroepitaxy of these materials. From structural, optical, and transport studies, it is clear that InSb–InAs and InSb–InBi alloying cannot be understood by simply using the conventional theories of alloy disorder.

A large electronegativity and size difference between the host and substituted anions also lead to considerable structural disorder in the form of local strains, defects, and at times compositional fluctuations. Thermodynamic analysis using empirical theories of bond energies can help to produce quantitative estimates of these effects. Such a calculation would be useful in determining the practical limits on the quality of samples grown under equilibrium conditions. $InAs_xSb_{1-x}$,

Part B | 11.9

$InBi_xSb_{1-x}$, and $InBi_xAs_ySb_{1-x-y}$ are some of the many systems in the general class of highly mismatched alloys. A comparative analysis of other systems such as InPAs, InPSb, InNAs, and GaPSb over a wider compositional range would be useful in drawing generic conclusions regarding the properties of mismatched alloys. Experimental data on these systems are still limited. *Shan* et al. [11.197] have recently proposed a band anticrossing model to explain the large bowing in dilute nitride systems. Since it is the only theory that can be rigorously compared against experiment, it would be interesting to look for the anticrossing gap in $InAs_xSb_{1-x}$, $InBi_xSb_{1-x}$, and $InBi_xAs_ySb_{1-x-y}$ if they exist. Such and other interesting issues, when properly addressed, will fix these materials firmly in the realm of modern technology.

References

11.1 C.H. Wang, J.G. Crowder, V. Mannheim, T. Ashley, D.T. Dutton, A.D. Johnson, G.J. Pryce, S.D. Smith: Detection of nitrogen dioxide using a room temperature operationmid-infrared InSb light emitting diode, Electron. Lett. **34**, 300–3001 (1998)

11.2 J.J. Lee, J.D. Kim, M. Razeghi: Room temperature operation of 8–12 μm InSbBi infrared photodetectors on GaAs substrates, Appl. Phys. Lett. **73**, 602–604 (1998)

11.3 J.D. Kim, S. Kim, D. Wu, J. Wojkowski, J. Xu, J. Piotrowski, E. Bigan, M. Razeghi: 8–13 μm InAsSb heterojunction photodiode operating at near room temperature, Appl. Phys. Lett. **67**, 2645–2647 (1995)

11.4 K.Y. Ma, Z.M. Fang, D.H. Jaw, R.M. Cohen, G.B. Stringfellow, W.P. Kosar, D.W. Brown: Organometallic vapor phase epitaxial growth and characterization of InAsBi and InAsSbBi, Appl. Phys. Lett. **55**, 2420–2422 (1989)

11.5 V.K. Dixit, B. Bansal, V. Venkataraman, H.L. Bhat, K.S. Chandrasekharan, B.M. Arora: Studies on high resolution x-ray diffraction, optical and transport properties of $InAs_xSb_{1-x}$/GaAs ($x \leq 0.06$) heterostructure grown using liquid phase epitaxy, J. Appl. Phys. **96**, 4989–4995 (2004)

11.6 B. Bansal, V.K. Dixit, V. Venkataraman, H.L. Bhat: Transport, optical and magnetotransport properties of hetero-epitaxial $InAs_xSb_{1-x}$/GaAs ($x \leq 0.06$) and bulk $InAs_xSb_{1-x}$ ($x \leq 0.05$) crystals: experiment and theoretical analysis, Physica E **20**, 272–277 (2004)

11.7 B. Bansal, V.K. Dixit, V. Venkataraman, H.L. Bhat: Temperature dependence of the energy gap and free carrier absorption in bulk $InAs_{0.05}Sb_{0.95}$ single crystals, Appl. Phys. Lett. **82**, 4720–4722 (2003)

11.8 V.K. Dixit, K.S. Keerthi, P. Bera, M.S. Hegde, H.L. Bhat: Structural and compositional analysis of $InBi_xAs_ySb_{1-x-y}$ films grown on GaAs(001) substrates by liquid phase epitaxy, Appl. Surf. Sci. **220**, 321–326 (2003)

11.9 V.K. Dixit, B. Bansal, V. Venkataraman, H.L. Bhat: Structural, optical, and electrical properties of bulk single crystals of $InAs_xSb_{1-x}$ grown by rotatory Bridgman method, Appl. Phys. Lett. **81**, 1630–1632 (2002)

11.10 V.K. Dixit, B. Bansal, V. Venkataraman, G.N. Subbanna, K.S. Chandrasekharan, B.M. Arora, H.L. Bhat: High-mobility InSb epitaxial films grown on a GaAs(001) substrate using liquid-phase epitaxy, Appl. Phys. Lett. **80**, 2102–2104 (2002)

11.11 V.K. Dixit, B.V. Rodrigues, R. Venkataraghavan, K.S. Chandrasekharan, B.M. Arora, H.L. Bhat: Effect of lithium ion irradiation on transport and optical properties of Bridgman grown n-InSb single crystals, J. Appl. Phys. **90**, 1750–1755 (2001)

11.12 V.K. Dixit, B.V. Rodrigues, H.L. Bhat: Growth of $InBi_xSb_{1-x}$ crystals by rotatory Bridgman method and their characterization, J. Cryst. Growth **217**, 40–47 (2000)

11.13 J.C. Woolley, J. Warner: Optical energy-gap variaton in InAs–InSb alloys, Can. J. Phys. **42**, 1879–1885 (1964)

11.14 W.M. Coderre, J.C. Woolley: Electrical properties of $InAs_xSb_{1-x}$ alloys, Can. J. Phys. **46**, 1207–1214 (1968)

11.15 C.E.A. Grigorescu, R.A. Stradling: Semiconductor optical and electro-optical devices. In: *Handbook of Thin film Devices*, Vol. 2, ed. by M.H. Francombe (Academic, New York 2000) pp. 27–62

11.16 M.Y. Yen, B.F. Levine, C.G. Bethea, K.K. Choi, A.Y. Cho: Molecular beam epitaxial growth and optical properties of $InAs_xSb_{1-x}$ in 8–12 μm wavelength range, Appl. Phys. Lett. **50**, 927–929 (1987)

11.17 J.D. Kim, D. Wu, J. Wojkowski, J. Piotrowski, J. Xu, M. Razeghi: Long-wavelength InAsSb photoconductors operated at near room temperatures (200–300 K), Appl. Phys. Lett. **68**, 99 (1996)

11.18 A. Rogalski, K. Jozwikowski: Intrinsic carrier concentration and effective masses in $InAs_xSb_{1-x}$, Infrared Phys. **29**, 35–42 (1989)

11.19 I. Kudman, L. Ekstrom: Semiconducting properties of InSb-InAs alloys, J. Appl. Phys. **39**, 3385–3388 (1968)

11.20 J.C. Woolley, J. Warner: Preparation of InAs-InSb alloys, J. Electrochem. Soc. **111**, 1142–1145 (1964)

11.21 K. Sugiyama: Molecular beam epitaxy of InSb films on CdTe, J. Cryst. Growth **60**, 450–452 (1982)

11.22 R. Venkataraghavan, K.S.R.K. Rao, M.S. Hegde, H.L. Bhat: Influence of growth parameters on the surface and interface quality of laser deposited

InSb/CdTe heterostructures, Phys. Status Solidi (a) **163**, 93–100 (1997)
11.23 B.V. Rao, T. Okamoto, A. Shinmura, D. Gruznev, M. Mori, T. Tambo, C. Tatsuyama: Growth temperature effect on the heteroepitaxy of InSb on Si(111), Appl. Surf. Sci. **159/160**, 335–340 (2000)
11.24 S.D. Parker, R.L. Williams, R. Droopad, R.A. Stradling, K.W.J. Barnham, S.N. Holmes, J. Laverty, C.C. Phillips, E. Skuras, R. Thomas, X. Zhang, A. Staton-Beven, D.W. Pashley: Observation and control of the amphoteric behaviour of Si doped InSb grown on GaAs by MBE, Semicond. Sci. Technol. **4**, 663–676 (1989)
11.25 M. Mori, Y. Nizawa, Y. Nishi, K. Mae, T. Tambo, C. Tatsuyama: Effect of current flow direction on the heteroepitaxial growth of InSb films on Ge/Si(001) substrate heated by direct current, Appl. Surf. Sci. **159/160**, 328–334 (2000)
11.26 S.V. Ivanov, A.A. Boudza, R.N. Kutt, N.N. Ledentsov, B.Y. Meltser, S.S. Ruvimov, S.V. Shaposhnikov, P.S. Kopev: Molecular-Beam epitaxial growth of InSb/GaAs(100) and InSb/Si(100) heteroepitaxial layers (Thermodynamic analysis and characterization), J. Cryst. Growth **156**, 191–205 (1995)
11.27 P.E. Thompson, J.L. Davis, J. Waterman, R.J. Wagner, D. Gammon, D.K. Gaskill, R. Stahlbush: Use of atomic layer epitaxy buffer for the growth of InSb on GaAs by molecular beam epitaxy, J. Appl. Phys. **69**, 7166–7172 (1991)
11.28 B.S. Yoo, M.A. McKee, S.G. Kim, E.H. Lee: Structural and electrical properties of InSb epitaxial films grown on GaAs by low pressure MOCVD, Solid Stat. Commun. **88**, 447–450 (1993)
11.29 D.K. Gaskill, G.T. Stauf, N. Bottka: High mobility InSb grown by organometallic vapor-phase-epitaxy, Appl. Phys. Lett. **58**, 1905–1907 (1991)
11.30 M.C. Debnath, T. Zhang, C. Roberts, L.F. Cohen, R.A. Stradling: High-mobility InSb thin films on GaAs(001) substrate grown by the two-step growth process, J. Cryst. Growth **267**, 17–21 (2004)
11.31 T.R. Yang, Y. Cheng, J.B. Wang, Z.C. Feng: Optical and transport properties of InSb thin films grown on GaAs by metal organic chemical vapor deposition, Thin Solid Films **498**, 158–162 (2006)
11.32 T.R. Yang, Z.C. Feng, W. Lu, W.E. Collins: Far infrared reflectance spectroscopy of InSb thin films grown on GaAs by metal-organic vapor deposition, Proc. XIXth Int. Conf. Raman Spectrosc., ed. by P.M. Fredericks, R.L. Frost, L. Rintoul (2004) pp. 629–630
11.33 A. Kumar, P.S. Dutta: Growth of long wavelength $In_xGa_{1-x}As_ySb_{1-y}$ layers on GaAs from liquid phase, Appl. Phys. Lett. **89**, 162101–162103 (2006)
11.34 C. Peng, N.F. Chen, F. Gao, X. Zhang, C. Chen, J. Wu, Y. Yu: Liquid-phase-epitaxy-grown $InAs_xSb_{1-x}$/GaAs for room-temperature 8–12 μm infrared detectors, Appl. Phys. Lett. **88**, 242108–242110 (2006)
11.35 C.M. Ruiz, J.L. Plaza, V. Bermúdez, E. Diéguez: Study of induced structural defects on GaSb films grown on different substrates by the liquid phase epitaxy technique, J. Phys.: Condens. Matter **14**, 12755–12759 (2002)
11.36 Y.Z. Gao, H. Kan, M. Aoyama, T. Yamaguchi: Germanium and zinc-doped p-type InAsSb single crystals with a cutoff wavelength of 12.5 μm, Jpn. J. Appl. Phys. **39**, 2520–2522 (2000)
11.37 Y.Z. Gao, H. Kan, F.S. Gao, X.Y. Gong, T. Yamaguchi: Improved purity of long-wavelength InAsSb epilayers grown by melt epitaxy in fused silica boats, J. Cryst. Growth **234**, 85–90 (2002)
11.38 Y.Z. Gao, X.Y. Gong, Y.S. Gui, T. Yamaguchi, N. Dai: Electrical properties of melt-epitaxy-grown $InAs_{0.04}Sb_{0.96}$ layers with cutoff wavelength of 12 μm, Jpn. J. Appl. Phys. **43**, 1051–1054 (2004)
11.39 S.A. Barnett: Direct E_0 energy gaps of bismuth-containing III–V alloys predicted using quantum dielectric theory, J. Vac. Sci. Technol. A **5**, 2845–2848 (1987)
11.40 Y. Amemiya, H. Tareo, Y. Sakai: Electrical properties of InSb-based mixed crystals, J. Appl. Phys. **44**, 1625–1630 (1973)
11.41 W. Zawadzki: Electron transport in small-gap semiconductors, Adv. Phys. **23**, 435 (1974), and In: *Handbook on Semiconductors* ed. by T.S. Moss (North-Holland, Amsterdam 1982) p. 713
11.42 A. Thiel, H. Koelsch: Studies on indium, Z. Anorg. Chem. **66**, 288–321 (1910), in German
11.43 V.M. Goldschmidt: Crystal structure and chemical constitution, Trans. Faraday Soc. **25**(253), 253–282 (1929)
11.44 H. Welker: Über neue halbleitende Verbindungen, Z. Naturforsch. A **7**, 744–749 (1952)
11.45 L. Pincherle, J.M. Radcliffe: Semiconducting intermetallic compounds, Adv. Phys, Philos. Mag. Suppl. **5**, 271–322 (1956)
11.46 H. Welker, H. Wiess: *Solid State Physics*, Vol. 3 (Academic, New York 1956)
11.47 F.A. Cunnell, E.W. Saker: *Progress in Semiconductors*, Vol. 2 (Heywood, London 1959)
11.48 R.A. Smith: *Semiconductors* (Cambridge Univ. Press, Cambridge 1959)
11.49 M.J. Whelan: Properties of some covalent semiconductors. In: *Semiconductors*, ed. by J.B. Hannay (Rheinhold, New York 1959)
11.50 H.J. Hrostowski: Infrared absorption of semiconductors. In: *Semiconductors*, ed. by J.B. Hannay (Rheinhold, New York 1959)
11.51 K.T. Huang, C.T. Chiu, R.M. Cohen, G.B. Stringfellow: InAsSbBi alloys grown by organometallic vapor phase epitaxy, J. Appl. Phys. **75**, 2857–2863 (1994)
11.52 Q. Du, J. Alperin, W.T. Wang: Molecular beam epitaxial growth of GaInSbBi for infrared detector applications, J. Cryst. Growth **175**, 849–852 (1997)
11.53 T.P. Humphreys, P.K. Chiang, S.M. Bedair, N.R. Parikh: Metalorganic chemical vapor depo-

sition and characterization of the In-As-Sb-Bi material system for infrared detection, Appl. Phys. Lett. **53**, 142–144 (1988)

11.54 K.Y. Ma, Z.M. Fang, D.H. Jaw, R.M. Cohen, G.B. Stringfellow, W.P. Kosar, D.W. Brown: Organometallic vapor phase epitaxial growth and characterization of InAsBi and InAsSbBi, Appl. Phys. Lett. **55**, 2420–2422 (1989)

11.55 C.H. Shih, E.A. Peretti: The phase diagram of the system InAs-Sb, Trans. Am. Soc. Met. **46**, 389–396 (1954)

11.56 T.S. Liu, E.A. Peretti: The indium-antimony system, Trans. Am. Soc. Met. **44**, 539–548 (1951)

11.57 J.L. Zilko, J.E. Greene: Growth and phase stability of epitaxial metastable $InBi_xSb_{1-x}$ films on GaAs, J. Appl. Phys. **51**, 1549–1564 (1980)

11.58 N.A. Goryunova, N.N. Fedorova: On the question of the isomorphism of compounds of AIII–BV type, J. Tech. Phys. Moscow **24**, 1339–1341 (1955)

11.59 J.C. Woolley, B.A. Smith, D.G. Lee: Solid solution in the GaSb-InSb system, Proc. Phys. Soc. B **69**, 1339–1343 (1956)

11.60 G.B. Stringfellow, P.E. Greene: Calculation of III-V ternary phase diagram In-Ga-As and In-As-Sb, J. Phys. Chem. Solids **30**, 1779–1780 (1969)

11.61 P.S. Dutta, T.R. Miller: Engineering phase formation thermo-chemistry for crystal growth of homogeneous ternary and quaternary III-V compound semiconductors from melts, J. Electron. Mater. **29**, 956–963 (2000)

11.62 D. Minic, D. Manasijevic, D. Zivkovic, Z. Zivkovic: Phase equilibria in the In-Sb-Bi system at 300 °C, J. Serb. Chem. Soc. **71**, 843–847 (2006)

11.63 B. Joukoff, A.M. Jean-Louis: Growth of $InBi_xSb_{1-x}$ single crystals by Czochralski method, J. Cryst. Growth **12**, 169–172 (1972)

11.64 A. Iandelli: MX2-Verbindungen der Erdalkali- und Seltenen Erdmetalle mit Gallium, Indium und Thallium, Z. Anorg. Allg. Chem. **330**(3), 221–232 (1941), , in German

11.65 J.F. Dewald: The kinetics and mechanism of formation of anode films on single crystal InSb, J. Electrochem. Soc. **104**, 244–251 (1957)

11.66 G.N. Kozhemyakin: Influence of ultrasonic vibration on the growth of InSb crystals, J. Cryst. Growth **149**, 266–268 (1995)

11.67 R.K. Akchurin, V.G. Zinov'ev, V.B. Ufimtsev, V.T. Bublik, A.N. Morozov: Donor nature of bismuth in indium antimonide, Sov. Phys. Semicond. **16**, 126–129 (1982)

11.68 L. Pauling: *The Nature of Chemical Bond* (Oxford Univ. Press, London 1940)

11.69 V.K. Dixit: Bulk and Thin Film Growth of Pure and Substituted Indium Antimonide for Infrared Detector Applications. Ph.D. Thesis (Indian Institute of Science, Bangalore 2004)

11.70 W.L. Bond: Precision lattice constant determination, Acta. Crystallogr. **13**, 814–818 (1960)

11.71 R. Krishnaswamy: Compounding, zone refining and crystal growing of Indium Antimonide, J. Indian Chem. Soc. **LII**, 60–63 (1975)

11.72 R.K. Bagai, G.L. Seth, W.N. Borle: Growth of high purity indium antimony crystals for infrared detectors, Indian J. Pure Appl. Phys. **21**, 441–444 (1983)

11.73 T.A. Campbell, J.N. Koster: In situ visualization of constitutional supercooling within a Bridgman-Stockbarger system, J. Cryst. Growth **171**, 1–11 (1997)

11.74 D.B. Gadkari, K.B. Lal, A.P. Shah, B.M. Arora: Growth of high mobility InSb crystals, J. Cryst. Growth **173**, 585–588 (1997)

11.75 M.H. Lin, S. Kou: Czochralski pulling of InSb single crystals from a molten zone on a solid feed, J. Cryst. Growth **193**, 443–445 (1998)

11.76 W.G. Pfann: Principles of zone melting, J. Met. **4**, 747–753 (1952)

11.77 T.C. Harman: Effect of zone refining variables on the segregation of impurities in indium antimonide, J. Electrochem. Soc. **103**, 128–132 (1956)

11.78 A.J. Strauss: Distribution coefficients and carrier mobilities in InSb, J. Appl. Phys. **30**, 559–563 (1959)

11.79 J.B. Mullin, K.F. Hulme: Orientation dependent distribution coefficients in melt grown InSb crystals, J. Phys. Chem. Solids **17**, 1–6 (1960)

11.80 R.K. Mueller, R.L. Jacobson: Growth twins in indium antimonide, J. Appl. Phys. **32**, 550–551 (1961)

11.81 A.R. Murray, J.A. Baldrey, J.B. Mullin, O. Jones: A systematic study of zone refining of single crystal indium antimonide, J. Mater. Sci. **1**, 14–28 (1966)

11.82 S.G. Parker, O.W. Wilson, B.H. Barbee: Indium antimonide of high perfection, J. Electrochem. Soc. **112**, 80–81 (1965)

11.83 J. Zhou, M. Larrousse, W.R. Wilcox, L.L. Regel: Directional solidification with ACRT, J. Cryst. Growth **128**, 173–177 (1993)

11.84 K.M. Kim: Suppression of thermal convection by transverse magnetic field, J. Electrochem. Soc. **129**, 427–429 (1982)

11.85 J. Kang, Y. Okano, K. Hoshikawa, T. Fukuda: Influence of a high vertical magnetic field on Te dopant segregation in InSb grown by the vertical gradient freeze method, J. Cryst. Growth **140**, 435–438 (1994)

11.86 A.G. Ostrogorsky, H.J. Sell, S. Scharl, G. Müller: Convection and segregation during growth of Ge and InSb crystals by the submerged heater method, J. Cryst. Growth **128**, 201 (1993)

11.87 P.S. Dutta, H.L. Bhat, V. Kumar: Numerical analysis of melt-solid interface shapes and growth rates of gallium antimonide in a single-zone vertical Bridgman furnace, J. Cryst. Growth **154**, 213–222 (1995)

11.88 R. Venkataraghavan, K.S.R.K. Rao, H.L. Bhat: The effect of growth parameters on the position of the melt-solid interface in Bridgman growth of indium antimonide, J. Phys. D Appl. Phys. **30**, L61–L63 (1997)

11.89 N.K. Udayshankar, K. Gopalakrishna Naik, H.L. Bhat: The influence of temperature gradient and lowering speed on the melt-solid interface shape of $Ga_xIn_{1-x}Sb$ alloy crystals grown by vertical Bridgman technique, J. Cryst. Growth **203**, 333–339 (1999)

11.90 P.G. Barber, R.K. Crouch, A.L. Fripp, W.J. Debnam, R.F. Berry, R. Simchick: Modelling melt-solid interfaces in Bridgman growth, J. Cryst. Growth **97**, 672–674 (1989)

11.91 T.A. Campbell, J.N. Koster: Growth rate effects during indium-antimony crystal growth, Crystal. Res. Technol. **34**, 275–283 (1999)

11.92 M.J. Hui, K. Beatty, K. Blackmore, K. Jackson: Impurity distribution in InSb single crystals, J. Cryst. Growth **174**, 245–249 (1997)

11.93 T.A. Campbell, J.N. Koster: Compositional effects on solidification of congruently melting InSb, Crystal. Res. Technol. **33**, 717–731 (1998)

11.94 T. Duffar, C. Potard, P. Dusserre: Growth analysis of the InSb compound by a calorimetric method in microgravity results of the Spacelab-D1 experiment, J. Cryst. Growth **92**, 467–478 (1988)

11.95 R. F. Redden, W. F. H. Micklethwait: Final Report to the Canadian Space Agency, MiM/QUELD Increment II (1998)

11.96 W.F.H. Micklethwaite: Bulk growth of InSb and related ternary alloys. In: *Bulk Growth of Electronic, Optical and Optoelectronic Materials*, ed. by P. Capper (Wiley, Chichester 2005)

11.97 W.P. Allred, R.T. Bate: Anisotropic segregation in InSb, J. Electrochem. Soc. **108**, 258–261 (1961)

11.98 K. Terashima: Growth of highly homogeneous InSb single crystals, J. Cryst. Growth **60**, 363–368 (1982)

11.99 D.T.J. Hurle, O. Jones, J.B. Mullin: Growth of semiconducting compounds from non-stoichiometric melts, Solid Stat. Electron. **3**, 317–320 (1961)

11.100 J.W. Faust Jr., H.F. John: The growth of semiconductor crystals from solution using the twin-plane reentrant-edge mechanism, J. Phys. Chem. Solids **23**, 1407–1415 (1962)

11.101 K. Morizane, A.F. Witt, H.C. Gatos: Impurity distributions in single crystals. I. Impurity striations in nonrotated InSb crystals, J. Electrochem. Soc. **114**, 51–52 (1966)

11.102 H.C. Gatos, A.J. Strauss, M.C. Lavine, T.C. Harmon: Impurity striations in unrotated crystals of InSb, J. Appl. Phys. **32**, 2057–2058 (1961)

11.103 K.W. Benz, G. Müller: GaSb and InSb crystals grown by vertical and horizontal travelling heater method, J. Cryst. Growth **46**, 35–42 (1979)

11.104 R.W. Hamaker, W.B. White: Mechanism of single-crystal growth in InSb using temperature-gradient zone melting, J. Appl. Phys. **39**, 1758–1765 (1968)

11.105 N.K. Udayashankar, H.L. Bhat: Growth and characterization of indium antimonide and gallium antimonide crystals, Bull. Mater. Sci. **24**, 445–453 (2001)

11.106 R. Venkataraghavan, K.S.R.K. Rao, H.L. Bhat: The effect of temperature gradient and ampoule velocity on the composition and other properties of Bridgman-grown indium antimonide, J. Cryst. Growth **186**, 322–328 (1998)

11.107 V.M. Glazov, K.B. Poyarkov: InSb-InAs alloys prepared by rapid quenching (10^6–10^8 K/s), Inorg. Mater. **36**, 991–996 (2000)

11.108 A.M. Jean-Louis, B. Ayrault, J. Vargas: Properties of $InSb_{1-x}Bi_x$ alloys. 2. Optical absorption, Phys. Status Solidi (b) **34**, 341–342 (1969)

11.109 T. Ozawa, Y. Hayakawa, M. Kumagawa: Growth of III-V ternary and quaternary mixed crystals by the rotationary Bridgman method, J. Cryst. Growth **109**, 212–217 (1991)

11.110 M. Kumagawa, T. Ozawa, Y. Hayakawa: A new technique for the growth of III-V mixed crystal layers, Appl. Surf. Sci. **33/34**, 611–618 (1988)

11.111 M. Kumagawa, A.F. Witt, M. Lichtensteiger, H.C. Gatos: Current-controlled growth and dopant modulation in liquid phase epitaxy, J. Electrochem. Soc. **120**, 583–584 (1973)

11.112 I. Melngailis, A.R. Calawa: Solution regrowth of planar InSb laser structures, J. Electrochem. Soc. **113**, 58–59 (1966)

11.113 D.E. Holmes, G.S. Kamath: Growth-characteristics of LPE InSb and InGaSb, J. Electron. Mater. **9**, 95–110 (1980)

11.114 R. Venkataraghavan, N.K. Udayashankar, B.V. Rodrigues, K.S.R.K. Rao, H.L. Bhat: Design and fabrication of liquid phase epitaxy system, Bull. Mater. Sci. **22**, 133–137 (1999)

11.115 V.K. Dixit, B.V. Rodrigues, R. Venkataraghavan, K.S. Chandrasekharan, B.M. Arora, H.L. Bhat: Growth of InSb epitaxial layers on GaAs(001) substrates by LPE and their characterizations, J. Cryst. Growth **235**, 154–160 (2002)

11.116 M. Elwenspoek: On the estimate of the supersaturation of nonelectrolyte solutions from solubility data, J. Cryst. Growth **76**, 514–516 (1986)

11.117 A.S. Popov, A.M. Koinova, S.L. Tzeneva: The In-As-Sb phase diagram and LPE growth of InAsSb layers on InAs at extremely low temperatures, J. Cryst. Growth **186**, 338–343 (1998)

11.118 L.O. Bubulac, A.M. Andrews, E.R. Gertner, D.T. Cheung: Backside illuminated InAsSb/GaSb broadband detectors, Appl. Phys. Lett. **36**, 734 (1980)

11.119 M.C. Wagener, J.R. Botha, A.W.R. Leitch: Substitutional incorporation of arsenic from GaAs substrates into MOVPE grown InSbBi thin films, Physica B **308–310**, 866–869 (2001)

11.120 M.C. Wagener, J.R. Botha, A.W.R. Leitch: Characterization of secondary phases formed during MOVPE growth of InSbBi mixed crystals, J. Cryst. Growth **213**, 51–56 (2000)

11.121 V.B. Ufimtsev, V.G. Zinovev, M.R. Raukhman: Heterogeneous equilibria in the system In-Sb-Bi and liquid phase epitaxy of InSb based solid solution, Inorg. Mater. **15**, 1371–1374 (1979)
11.122 Y.Z. Gao, T. Yamaguchi: Liquid phase epitaxial growth and properties of InSbBi films grown from In, Bi and Sn solutions, Cryst. Res. Technol. **34**, 285–292 (1999)
11.123 V.K. Dixit, K.S. Keerthi, P. Bera, H.L. Bhat: Growth of $InBi_xSb_{1-x}$ films on GaAs(001) substrates using liquid phase epitaxy and their characterization, J. Cryst. Growth **241**, 171–176 (2002)
11.124 K.T. Huang, C.T. Chiu, R.M. Cohen, G.B. Stringfellow: $InBi_xAs_ySb_{1-x-y}$ alloys grown by organometallic vapor-phase epitaxy, J. Appl. Phys. **75**, 2857–2862 (1994)
11.125 Q. Du, J. Alperin, W.T. Wang: Molecular beam epitaxial growth of GaInSbBi for infrared detector applications, J. Cryst. Growth **175/176**, 849–852 (1997)
11.126 M. Oszwaldowski, T. Berus, J. Szade, K. Jόzwiak, I. Olejniczak, P. Konarski: Structural properties of InSbBi and InSbAsBi thin films prepared by the flash-evaporation method, Cryst. Res. Technol. **36**, 1155–1171 (2001)
11.127 P. Haasen: Twinning in indium antimonide, J. Met. **209**, 30–32 (1957)
11.128 C.G. Darwin: The reflexion of x-rays from imperfect crystals, Philos. Mag. **43**, 800–829 (1922)
11.129 J. Auleytner: *X-Ray Methods in the Study of Defects in Single Crystals* (Pergamon, Oxford 1967)
11.130 P.F. Fewster: *X-Ray Scattering from Semiconductors* (Imperial College Press, London 2000)
11.131 E. Gartstein, R.A. Cowley: The intensity patterns with a multicrystal diffractometer observed at a synchrotron source, Z. Naturforsch. A **48**, 519–522 (1992)
11.132 E. Gartstein, Y. Khait, V. Richter: An x-ray diffraction study of implantation damage in InSb reduced by a magnetic field, J. Phys. D Appl. Phys. **28**, A291–A294 (1995)
11.133 A.H. Chin, R.W. Schoenlein, T.E. Glover, P. Balling, W.P. Leemans, C.V. Shank: Ultrafast structural dynamics in InSb probed by time-resolved x-ray diffraction, Phys. Rev. Lett. **83**, 336–339 (1999)
11.134 M.R. Surowiec, B.K. Tanner: X-ray topography study of dislocations around indents on {111} surfaces of indium-antimonide, J. Appl. Cryst. **20**, 499–504 (1987)
11.135 D. Briggs, M.P. Seah: *Practical Surface Analysis by Auger and x-ray Photoelectron Spectroscopy* (Wiley, New York 1984), , Appendix 4
11.136 J.H. Scofield: Hartree–Slater subshell photoionization cross sections at 1254 and 1487 eV, J. Electron Spectrosc. Relat. Phenom. **8**, 129–137 (1976)
11.137 D.R. Penn: Quantitative chemical analysis by ESCA, J. Electron. Spectrosc. Relat. Phenom. **9**, 29–40 (1976)
11.138 R.J. Egan, V.W.L. Chin, T.L. Tansley: Dislocation scattering effects on electron mobility in InAsSb, J. Appl. Phys. **75**, 2473–2476 (1994)
11.139 C. Bocchi, C. Ferrari, P. Franzosi, A. Bosacchi, S. Franchi: Accurate determination of lattice mismatch in the epitaxial AlAs/GaAs system by high-resolution x-ray diffraction, J. Cryst. Growth **132**, 427–434 (1993)
11.140 C.R. Wie: High resolution x-ray diffraction characterization of semiconductor structures, Mater. Sci. Eng. R **13**, 1–56 (1994)
11.141 X. Weng, R.S. Goldman, D.L. Partin, J.P. Heremans: Evolution of structural and electronic properties of highly mismatched InSb films, J. Appl. Phys. **88**, 6276–6286 (2000)
11.142 E.O. Kane: Band structure of indium antimonide, J. Phys. Chem. Solids **1**, 249–261 (1957)
11.143 G. Dresselhaus, A.F. Kip, C. Kittel, G. Wagoner: Cyclotron and spin resonance in indium antimonide, Phys. Rev. **98**, 556–557 (1955)
11.144 H.P.R. Frederikse, W.R. Hosler: Galvanomagnetic effects in n-type indium antimonide, Phys. Rev. **108**, 1136 (1957)
11.145 L.M. Roth, B. Lax, S. Zwerdling: Theory of optical magneto-absorption effects in semiconductors, Phys. Rev. **114**, 90–103 (1959)
11.146 H.P.R. Frederikse, W.R. Hosler: Galvanomagnetic effects in p-type indium antimonide, Phys. Rev. **108**, 1146–1151 (1957)
11.147 C. Hermann, C. Weisbuch: $\boldsymbol{k} \cdot \boldsymbol{p}$ perturbation theory in III–V compounds and alloys reexamination, Phys. Rev. B **15**, 823–833 (1977)
11.148 J.A. van Vechten, O. Berolo, J.C. Woolley: Spin-orbit splitting in compositionally disordered semiconductors, Phys. Rev. Lett. **29**, 1400–1403 (1972)
11.149 J.I. Vyklyuk, V.G. Deibuk, I.M. Rarenko: Calculation of absorption coefficients of $InBi_xSb_{1-x}$ solid solutions, Semicond. Phys. Quantum Electron. Optoelectron. **3**, 174–177 (2000)
11.150 S.D. Smith, T.S. Moss, K.W. Taylor: The energy-dependence of electron mass in indium antimonide determined from measurements of the infrared Faraday effect, J. Phys. Chem. Solids **11**, 131–139 (1959)
11.151 B. Bansal, V. Venkataraman: Magnetic field induced band depopulation in intrinsic InSb: a revisit, J. Phys.: Condens. Matter **17**, 7053–7060 (2005)
11.152 C. Hilsum, A.C. Rose-Innes: *Semiconducting III–V Compounds* (Pergamon, New York 1961) pp. 128–
11.153 H.J. Hrostowski, F.J. Morin, T.H. Geballe, G.H. Wheatley: Hall effect and conductivity of InSb, Phys. Rev. **100**, 1672–1672 (1955)
11.154 H. Fritsche, K. Lark-Horovitz: Electrical properties of p-type indium antimonide at low temperatures, Phys. Rev. **99**, 400–405 (1955)
11.155 R. Barrie, J.T. Edmond: A study of the conduction band of InSb, J. Electron. **1**, 161–170 (1955)

11.156 K. Vinogradova, V. Galavanov, D. Nasledov: Production of high purity indium antimonide by zone fusion, Sov. Phys. Tech. Phys. **2**, 1832–1839 (1957)
11.157 K. Vinogradova, V. Galavanov, D. Nasledov, L. Soloveva: Production of high purity single crystals of InSb by zone melting, Sov. Phys. Solid. Stat. **1**, 364–367 (1959)
11.158 O. Madelung, H. Weiss: Die elektrischen Eigenschaften von Indiumantimoniden, Z. Naturforsch. **9a**, 527–534 (1954)
11.159 O. Drachenko, B. Bansal, V.V. Rylkov, J. Galibert, V.K. Dixit, J. Leotin: InAsSb/GaAs hetero-epitaxial crystals studied by cyclotron resonance measurements, 12th Int. Conf. Narrow Gap Semicond. (Toulouse, 2005)
11.160 N. Miura, G. Kido, S. Chikazumi: Infrared cyclotron resonance in InSb, GaAs and Ge in very high magnetic fields, Solid State Commun. **18**, 885–888 (1976)
11.161 B.R. Nag: *Electron Transport in Compound Semiconductors* (Springer, Berlin 1980)
11.162 W. Zawadzki: Electron transport in small gap semiconductors, Adv. Phys. **23**, 435–455 (1974)
11.163 J.A. van Vechten, T.K. Bergstresser: Electronic structures of semiconductor alloys, Phys. Rev. B **1**, 3351–3358 (1970)
11.164 O. Berolo, J.C. Woolley, J.A. van Vechten: Effect of disorder on conduction band effective mass, valence band spin orbit splitting and direct band gap in III–V alloys, Phys. Rev. B **8**, 3794–3798 (1973)
11.165 E.J. Johnson: Optical properties of III–V compounds. In: *Semiconductors and Semimetals*, Vol. 3, ed. by R.K. Willardson, A.C. Beer (Academic, New York 1967) pp. 154–
11.166 C.E.A. Grigorescu, R.A. Stradling: Antimony-based infrared materials and devices. In: *Handbook of Thin Film Devices*, Vol. 2, ed. by M.H. Francombe (Academic, New York 2000)
11.167 S.W. Kurnick, J.M. Powell: Optical absorption in pure single crystal InSb at 298 K and 78 K, Phys. Rev. **116**, 597–604 (1959)
11.168 W.G. Spitzer, H.Y. Fan: Infrared absorption in indium antimonide, Phys. Rev. **106**, 1893–1894 (1955)
11.169 T.S. Moss: *Optical Properties of Semiconductors* (Butterworths, London 1959)
11.170 O. Madelung (Ed.): *Semiconductors – Basic Data* (Springer, Berlin 1996)
11.171 Y.P. Varshni: Temperature dependence of energy gap in semiconductors, Physica **34**, 149–150 (1967)
11.172 M. Cardona: Renormalization of the optical response of semiconductors by electron-phonon interaction, Phys. Status Solidi (a) **188**, 1209–1232 (2001)
11.173 R.H. Parmenter: Energy levels of a disordered alloy, Phys. Rev. **97**, 587–598 (1955)
11.174 L. Malikova, W. Krystek, F.H. Pollak, N. Dai, A. Cavus, M.C. Tamargo: Temperature dependence of the direct gaps of ZnSe and $Zn_{0.56}Cd_{0.44}Se$, Phys. Rev. B **54**, 1819–1824 (1996)
11.175 L. Vina, S. Logothetidis, M. Cardona: Temperature dependance of the dielectric function of germanium, Phys. Rev. B **30**, 1979–1991 (1984)
11.176 H.H. Wieder, A.R. Clawson: Photo-electronic properties of $InAs_{0.07}Sb_{0.93}$ films, Thin Solid Films **15**, 217–221 (1973)
11.177 Y.B. Li, S.S. Dosanjh, I.T. Ferguson, A.G. Norman, A.G. de Oliveira, R.A. Stradling, R. Zallen: Raman scattering in $InAs_xSb_{1-x}$ alloys grown on GaAs by molecular beam epitaxy, Semicond. Sci. Technol. **7**, 567–570 (1992)
11.178 E.H. Reihlen, M.J. Jou, Z.M. Fang, G.B. Stringfellow: Optical absorption and emission of $InP_{1-x}Sb_x$ alloys, J. Appl. Phys. **68**, 4604–4609 (1990)
11.179 Y.S. Gao, X.D. Gong, T. Yamaguchi: Optical properties of InAsSb single crystals with cutoff wavelengths of 8–12 μm grown by melt-epitaxy, J. Appl. Phys. **45**, 5732–5734 (2006)
11.180 B. Bansal, V. K. Dixit, V. Venkataraman, H. L. Bhat: Alloying induced degradation of the absorption edge of $InAs_xSb_{1-x}$, Appl. Phys. Lett. **90**, 101905(1–3) (2007)
11.181 Y.Z. Gao, H. Kan, F.S. Gao, X.Y. Gong, T. Yamaguchi: Improved purity of long-wavelength InAsSb epilayers grown by melt epitaxy in fused silica boats, J. Cryst. Growth **234**, 85–90 (2002)
11.182 M.A. Marciniak, R.L. Hengehold, Y.K. Yeo, G.W. Turner: Optical characterization of molecular beam epitaxially grown InAsSb nearly lattice matched to GaSb, J. Appl. Phys. **84**, 480–488 (1998)
11.183 B. Bansal: Construction of a 17 Tesla Pulsed Magnet and Effect of Arsenic Alloying and Heteroepitaxy on Transport and Optical Properties of Indium Antimonide. Ph.D. Thesis (Indian Institute of Science, Bangalore 2004)
11.184 L. Bernstein, R.J. Beals: Thermal expansion and related bonding problems of some III–V compound semiconductors, J. Appl. Phys. **32**, 122–123 (1961)
11.185 A. Jordan: Estimated thermal diffusivity, Prandtl number and Grashof number of molten GaAs, InP GaSb J. Cryst. Growth **71**, 551–558 (1985)
11.186 J.R. Dixon, J.K. Furdyna: Measurement of the static dielectric constant of the InSb lattice via gyrotropic sphere resonances, Solid State Commun. **35**, 195–198 (1980)
11.187 T. Ashley, C.T. Elliott: Operation and properties of narrow gap semiconductor devices near room temperature using nonequilbrium techniques, Semicond. Sci. Technol. **6**, C99–C105 (1991)
11.188 Z. Djuric, V. Jovic, M. Matic, Z. Jaksic: IR photodetector with exclusion effect and self-filtering n^+ layer, Electron. Lett. **26**, 929–931 (1990)
11.189 I. Bloom, Y. Nemirovsky: Surface passivation of backside-illuminated indium antimonide focal plane array, IEEE Trans. Electron. Dev. **40**, 309–314 (1993)

11.190 E. Michel, J. Xu, J.D. Kim, I. Ferguson, M. Razeghi: InSb infrared photodetectors on Si substrates grown by molecular beam epitaxy, IEEE Photon. Technol. Lett. **8**, 673–675 (1996)

11.191 M. Razeghi: Overview of antimonide based III–V semiconductor epitaxial layers and their applications at the center for quantum devices, Eur. Phys. J. PA **23**, 149–205 (2003)

11.192 A. Rogalski: Heterostructure infrared photodiodes, Semicond. Phys. Quantum Electron. Optoelectron. **3**, 111–120 (2000)

11.193 S.A. Solin, T. Thio, D.R. Hines, J.J. Heremans: Enhanced room-temperature geometric magnetoresistance in inhomogeneous narrow-gap semiconductors, Science **289**, 1530–1532 (2000)

11.194 M. Oszwaldowski: Hall sensors based on heavily doped n-InSb thin films, Sens. Actuators A **68**, 234–237 (1998)

11.195 J. Heremans, D.L. Partin, C.M. Thrush, L. Green: Narrow gap semiconductor magnetic field sensors and applications, Semicond. Sci. Technol. **8**, S424–S430 (1993)

11.196 J. Heremans: Solid state magnetic field sensors and applications, J. Phys. D Appl. Phys. **26**, 1149 (1993)

11.197 W. Shan, W. Walukiewicz, J.W. Ager III, E.E. Haller, J.F. Geisz, D.J. Friedman, J.M. Olson, S.R. Kurtz: Band anticrossing in GaInNAs alloys, Phys. Rev. Lett. **82**, 1221–1224 (1999)

12. Crystal Growth of Oxides by Optical Floating Zone Technique

Hanna A. Dabkowska, Antoni B. Dabkowski

Single crystals of various congruently and incongruently melting oxides have been recently grown by the floating zone (FZ) and traveling solvent floating zone (TSFZ) techniques. For the incongruently melting materials, the use of solvent with an experimentally determined composition allows the establishment of the *practical* steady state much faster, leading to better, more stable growth. Growth conditions for different oxides are compared. Important problems in crystal characterization and assessment of micro- and macrodefects are briefly presented.

12.1 Historical Notes

W. G. Pfann, a metallurgist at Bell Laboratories, developed zone refining, a precursor of floating zone crystal growth, in 1951. *Pfann*'s process [12.1] involved placing the material to be crystallized in a crucible rather than suspending it in space as is the case in today's floating zone process. In 1952 Theurer, also at Bell Laboratories, created the floating zone process. Slightly later and independently, *von Emeis* at Siemens [12.2] as well as *Keck* and *Golay* [12.3] of the US Army Signal Corps developed a similar technique. Keck and Golay used *incandescent heating* with a short cylindrical tantalum heater rather than an optical system and published their results in 1952. As *Theurer*'s diary showed his priority in the invention, Bell Laboratories was eventually assigned a US patent [12.4]. According to *Mühlbauer* [12.5] the first firm to commercialize the float zone technique was Siemens in the early 1950s. They used the process in the course of manufacturing silicon, and patent for the technique known as the floating zone method was granted to them in 1953. They became the first company to produce ultrapure silicon single crystals for semiconductor devices by applying induction (radiofrequency, RF) heating to melt a narrow part of a cylindrical rod of polycrystalline material. Passing this molten zone along the rod transformed the ceramics into single crystals. In this case surface tension together with the electromagnetic force (*levitation*) supported the melt. The first company to manufacture a floating zone apparatus commercially was Ecco Corporation, North Bergen, NJ, in 1953.

The further development and production of float zone (FZ) silicon crystals took place in many countries. Detailed historical notes about the development of FZ technique can be found in [12.5]. Until the end of the 1950s, crystal diameter size did not exceed 25 mm, whereas during the 1960s and 1970s a dramatic increase from 33 to 50 and 75 mm was achieved. This diameter enlargement was a prerequisite for the development and fabrication of high-power discrete devices such as thyristors and others. The 100 mm crystal appears at the end of the 1970s, the 125 mm one in 1986, the worldwide first 200 mm (100) dislocation-free silicon crystal was grown at Wacker-Siltronic Company in September 2000.

Ferrites became important materials for microwave applications in the late 1950s and the floating zone technique was an obvious choice for crystal growth, but direct RF heating cannot be easily apply to oxides. Optical heating (*arc image furnaces*) utilizing carbon arc and elliptical mirrors was introduced for crystal growth of magnesium ferrite and sapphire by *De La Rue* and *Halden* [12.6] as a replacement for flame melting in the Verneuil method. His apparatus was based on earlier concepts and experiments with *image furnaces*, see *Null* and *Lozier* [12.7]. The concept of the image furnace was adopted to zone melting (in the geometry later call *pedestal growth*) for silicon by *Poplawsky* and *Thomas* [12.8] and soon *Poplawsky* applied this technique to ferrites [12.9]. The carbon arc heating is not an easy method to control, so at the same time the floating zone growth of yttrium iron garnet, $Y_3Fe_5O_{12}$ (YIG) was performed with RF heated $MoSi_2$ susceptor – which is also a kind of infrared (IR) optical heating – by *Abernethy* et al. [12.10]. As interest in ferrites was growing, soon more convenient halogen lamps and even more powerful xenon arc lamps were introduced and the apparatus for crystal growth of ferrites was developed by the groups of *Akashi* [12.11] and *Shindo* [12.12] in Japan.

12.2 Optical Floating Zone Technique – Application for Oxides

More recently the floating zone technique using halogen lamps and ellipsoidal mirrors (often called optical floating zone, OFZ) has been employed for crystal growth of a wide range of materials, including metals, oxides, and semiconductors [12.13–16]. With a growing number of optical systems around the world this technique is gaining popularity as the method of choice for the growth of various nonconventional oxides including high-temperature superconductors and new magnetic materials [12.17].

Crystals grown by the optical floating zone technique are of high quality but relatively small (usually not larger than a few millimeters in diameter and a few centimeters in length) so the majority of work is concentrated on new materials grown mainly for research purposes.

Recently high-quality crystals of β-Ga_2O_3 as large as 1 inch in diameter have been reported [12.18] but the only oxides grown by this technique for industrial applications are still $Y_3Fe_5O_{12}$ [12.12, 14, 19] and TiO_2 [12.20].

In this chapter the advantages and disadvantages of the FZ technique for the growth of crystals of congruently and incongruently melting oxides and their solid solutions are discussed. For incongruently melting materials a variation of the FZ technique, called the traveling solvent floating zone (TSFZ), technique is used. Both methods are suitable for the growth of solid solutions.

The list of oxide materials grown by FZ and TSFZ methods includes simple oxides such as β-Ga_2O_3 or TiO_2, as well as, complex oxides such as the spin Peierls material $CuGeO_3$ [12.17], high-temperature superconductors such as $Bi_2Sr_2CaCu_2O_n$ (BiSCCO) [12.17, 21], and frustrated antiferromagnets such as $RE_2Ti_2O_7$ (RE = rare earth) [12.22, 23].

The origin of most common macro- and microdefects – often seriously influencing crystal properties – is also briefly presented.

12.3 Optical Floating Zone and Traveling Solvent Crystal Growth Techniques

The idea of the optical floating zone is presented in Fig. 12.1. Two ceramic rods are mounted in such a way that their tips meet at the focal point of ellipsoidal mirrors. Halogen or xenon lamps of appropriate power sit at the other focal points of these mirrors. Any crystal growth process performed by the optical floating zone starts by melting the tips of polycrystalline rods, bringing them together and establishing a liquid called the *floating zone* between the bottom (seed) rod and a top (feed) rod (Fig. 12.2).

After the zone is created it starts moving upwards (either by moving the mirrors up or by moving the seed-and-feed setup down), the liquid cools and the material eventually crystallizes on the seed rod. During growth the rods rotate either in the same or in opposite directions with experimentally established rates. The rate of rotation is important as it is responsible for a pattern of forced convection flows within the zone and – as a result – for mixing of material, for the shape of the crystallization front (solid–liquid interface), and for the defects resulting from it. The rotation speed is optimized experimentally for each material and varies from 0 to 50 rpm. Successful growth requires a very stable zone. Stability of this zone depends on the quality of the starting rods as well as the alignment of both the feed and the seed rods. This is achieved by rigidly fixing the seed rod to the lower shaft. The feed rod can either be attached equally rigidly to the top holder or can hang loosely from a hook. Rigid mounting requires a very high-quality (straight and dense) rod and a very precise alignment. Another important factor is the alignment of the lamp and mirrors. Mirrors are usually factory preadjusted, but alignment of lamps critically influences the temperature distribution within the zone.

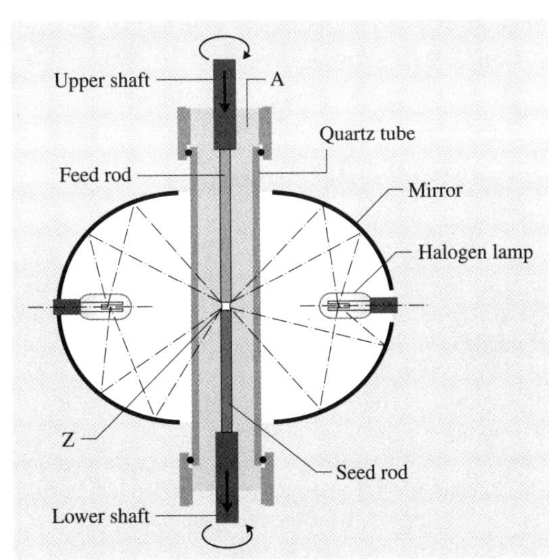

Fig. 12.1 Schematic diagram of optical floating zone apparatus (A – atmosphere, Z – floating zone)

It is advisable to start the growth on a crystalline, oriented seed, as this facilitates the beginning of crystallization and controls the appropriate orientation. It also prevents the soaking of the molten zone into the porous seed rod.

During the growth process only few parameters can be controlled. The pulling rate can be adjusted and this controls average growth rate. Due to the heat of crystallization this parameter also influence temperature near the crystallization front. The lamp power setting directly controls molten zone temperature, temperature gradients and also influences size and shape of the floating zone. The rate of rotation of both rods controls stirring of molten material which influence both, temperature distribution and composition within molten zone. Finally, adjusting the feeding rate controls the size of the zone.

There has been an intensive effort to model the silicon process [12.24, 25] but only a few modeling attempts have been made to understand what is happening inside the oxide molten zone [12.26–28]. The majority of modeling works deal with the floating zone method itself, investigating temperature oscillation in the zone [12.29], detailed lamp irradiation and thermal flows analysis [12.30, 31] or assessing the interface

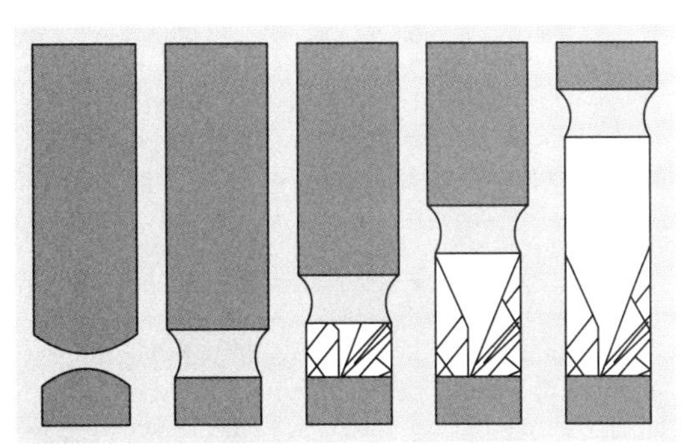

Fig. 12.2 Stages for nucleation on ceramic rods (see also Fig. 12.17)

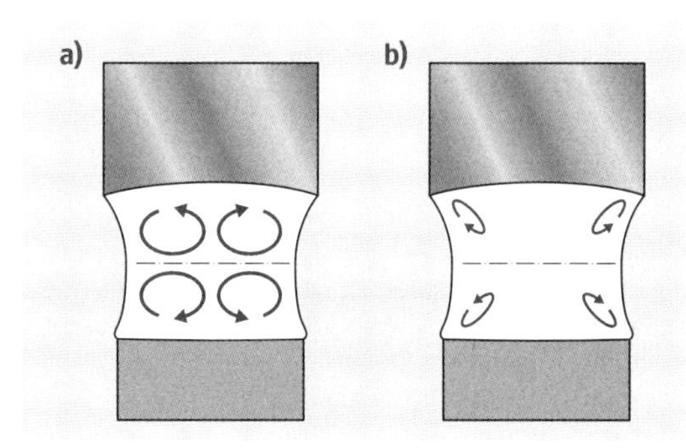

Fig. 12.3a,b Convection flows in the molten zone. (**a**) Combined buoyancy and forced convection for counter-rotation. (**b**) Marangoni convection (after *Brice* [12.32])

shape [12.27]. The recent and most likely the most advanced three-dimensional (3-D) modeling, preformed by *Lan* [12.28] for optical heating with two elliptical mirrors in the horizontal configuration, suggests that the temperature distribution is less symmetrical and that the pattern of convection cells is more complex then the one presented by *Brice* [12.32] in an older, simpler description of Si growth with RF heating. External heating (either optical or RF) causes significant overheating of the liquid zone surface. High temperature gradients in the molten zone lead to strong buoyancy as well as Marangoni convection [12.29]. If the feed and seed rods counter-rotate the convection patterns become even more complicated. To explain these patterns the existence of upper and lower convection cells has been suggested. A schematic explanation of convection flows in the molten zone is shown on Fig. 12.3. Because the flows on the cells' interface should be collinear, one can expect only limited liquid exchange and – for incongruently melting materials – upper and lower cells having distinctively different chemical compositions.

Using growth of $Y_3Al_5O_{12}$ as an example of a congruently melting material (melting point, mp = 1970 °C) grown in a double elliptical mirror system, *Lan* [12.28] concluded that the overheating (above mp of the material) is higher than 600 °C, with azimuthal differences of more than 400 °C (in the plane of lamps) and a temperature gradient near liquid–crystal interface in the range of 1500 °C/cm. As this model assumes no rotation, the realistic values of the temperature gradients as well as overheating with rotation will most likely be lower.

Very high thermal gradients, even with a small size of the liquid zone (typically 0.4 cm^3 for 0.7 cm rod diameter), cause fast convection flows. Indeed, the results of Lan's modeling suggest that the liquid velocity in the zone is in the range of up to a few cm/s.

In order to do the modelling one needs experimental values of material properties such as the surface tension, wetting angle, melt viscosity, and density as well as the optical properties, emissivity factor and thermal conductivity of the melt and the crystal. The change of these properties with temperature should be known as well. Although this information is available for $Y_3Al_5O_{12}$ [12.33], it is not yet available for the majority of oxides.

Due to experimental challenges and the high temperatures involved unfortunately it is not yet feasible to directly compare this modeling with experimental results.

12.4 Advantages and Limitations of the Floating Zone Techniques

The greatest advantages of the OFZ technique come from the fact that no crucible is necessary and that both congruently and incongruently melting materials can be grown. This allows for growth of large-sized crystals that was not possible before. The relatively high thermal gradient on the crystallization front characteristic of this method decreases the chance of constitutional supercooling and allows for faster growth of incongruently crystallizing materials (see the discussion in Sect. 12.7). It is also important to note that oxides with the highest melting temperatures can be grown using xenon lamps. The growth can be conducted at high pressure (up to 70 atm, depending on the furnace model) and in a controlled gas atmosphere. Solid solutions with controlled and uniform chemical composition can be prepared because – in contrast to crucible methods – the steady state, in principle, can be achieved. This is beneficial for crystallization of incongruently melting materials and for doped materials (with distribution coefficient different than 1), as well as materials in which, due to cation substitution, the congruently melting composition is not stoichiometric. The floating zone technique, when supported by characterization methods such as differential thermal analysis (DTA) or/and x-ray diffraction, is also an effective approach for the construction and investigation of phase diagrams [12.15].

There are some limitations to the growth of crystals by the OFZ method. As a rule this method is not suitable for materials with high vapor pressure, low surface tension or high viscosity as well as for materials that undergo a phase transition during cooling (because such crystals usually crack after growth) [12.34, 35].

The small volume of liquid in the zone makes the stability of this crystal growth method susceptible to fluctuations of power and/or short-time oscillations of gas pressure. Together with very high thermal gradients at the liquid–solid interface this leads to difficulties in achieving and maintaining a flat crystallization front and stable growth rate, which can result in many defects and growth instabilities [12.36]. Significant thermal and mechanical stresses limit the size and quality of crystals obtained. This problem is reduced if an afterheater is applied [12.14, 37, 38].

12.5 Optical Floating Zone Furnaces

Several types of optical floating zone furnaces are commercially available on the market, with two [12.14, 39, 41] or four [12.15, 40] ellipsoid mirrors. The idea of the furnace using only one mirror was also tested in 1969 [12.11, 13]. All these furnaces employ halogen or xenon arc lamps of different power as an energy source and – as already mentioned before – the growth can be carried in a controlled gas atmosphere and/or at high pressure. Using high pressure is advantageous in the case of crystallizing materials with high vapor pressure at crystallization temperature. As the diffusion coefficient is inversely proportional to pressure, higher pressure will slow the vapor transport from the source (the molten zone or its hottest regions) to the coolest part of the system (quartz tube). Furthermore, using gas with higher molecular mass (for example Ar instead of N_2) reduces the rate of evaporation. Additionally, the capability of FZ systems to apply high pressure is useful when high partial pressure of oxygen (p_{O_2}) is required to stabilize higher oxidation state of cation(s) and – as a result – crystallization of the appropriate phase. Oxygen pressure as high as 70 atm was used to growth of single crystals of ferrites [12.14] whereas for growth of layered cuprates application of O_2 stabilizes Cu^{2+} [12.17, 42, 43].

The main difference between the commercially available furnaces is that in one system the lamps

Fig. 12.4 Two (metallic) mirrors optical system of Nippon Electric Co. (now Canon). Mirrors are moved from their operational positions and the quartz tube is not installed for clarity of the view. M – mirrors, F – feed rod, S – seed rod, US – upper shaft, LS – lower shaft, TV – camera (after [12.39])

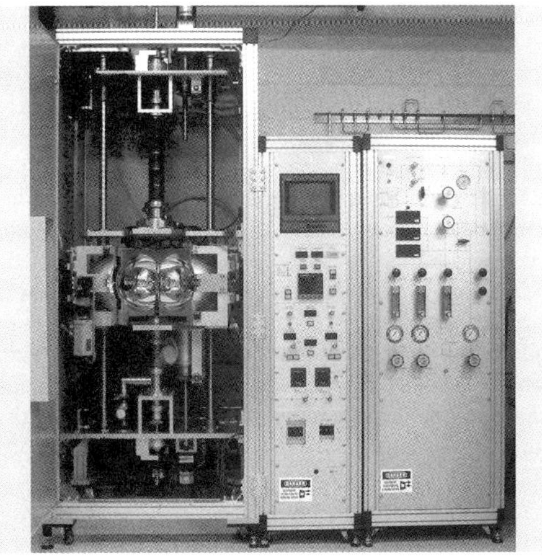

Fig. 12.5 General view of a crystal system machine (after [12.40]). The door to the four mirror optical system is open, and also the two front mirrors are open on their hinges to improve the view. The pressure control panel is on the *right side*

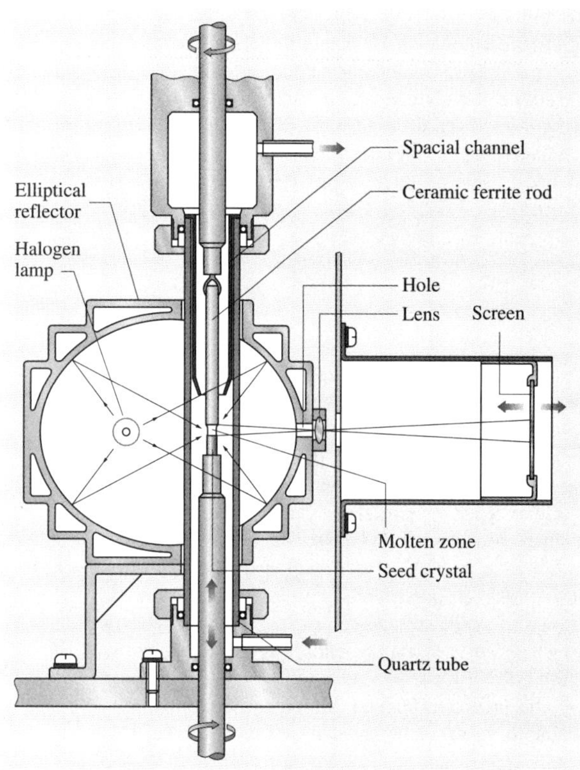

Fig. 12.6 Schematic drawing of a single elliptical mirror furnace (after *Akashi* et al. [12.11])

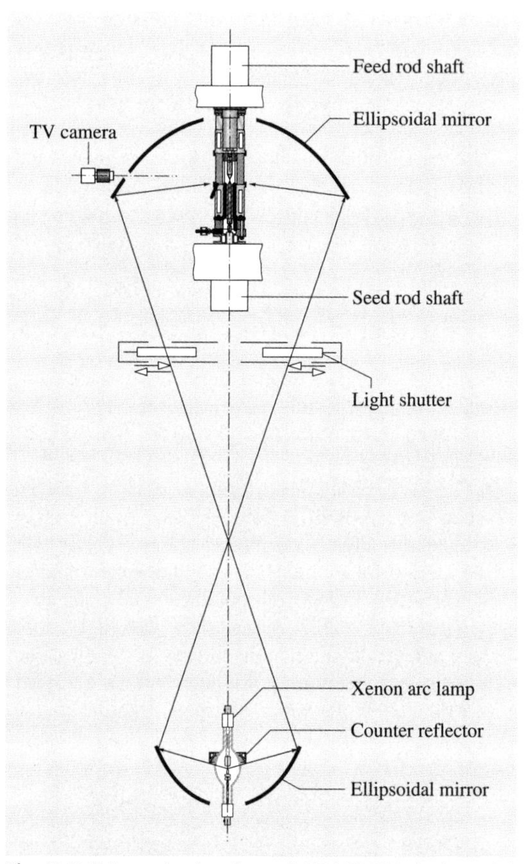

Fig. 12.7 Schematic drawing of URM2-ZN design with vertical optical axis (after [12.41])

move (Crystal System [12.40]), whereas in the others (NEC [12.39] and URN2-ZN [12.41]) the rods move while the lamps stay in one position. In addition in the URN2-ZN apparatus the elliptical mirrors are configured in such a way that the optical axis of the system is vertical and the mechanical axis (of the rods) overlap. The single xenon lamp is at the focus of the lower mirror and the FZ at the focus of the upper one. This optical system is more complex and access to the growth chamber seems to be more complicated, but one gains a uniform azimuthal temperature distribution and less overheating of liquid.

The practical, commercial realizations of OFZ are presented in Figs. 12.4 and 12.5 and drawings of other designs are presented on Figs. 12.6 and 12.7.

All of the furnaces are usually equipped with video cameras, allowing in situ observation of the crystal growth process during experiments. There are also options that allow for remote control of the process via the Internet. This is helpful for adjusting growth conditions during lengthy experiments.

12.6 Experimental Details of Ceramics and Rod Preparation for OFZT

Stability of the growth process – and the quality of the obtained crystal – depends strongly on the stability of the zone. This depends on the stability of the power and of the feed and seed rod shaft translations. Modern equipment is capable of stabilizing the power supplied to the halogen lamp with relative accuracy better than 10^{-4}, and the rate of shaft translation is realized with similar accuracy. Stability of the gas pres-

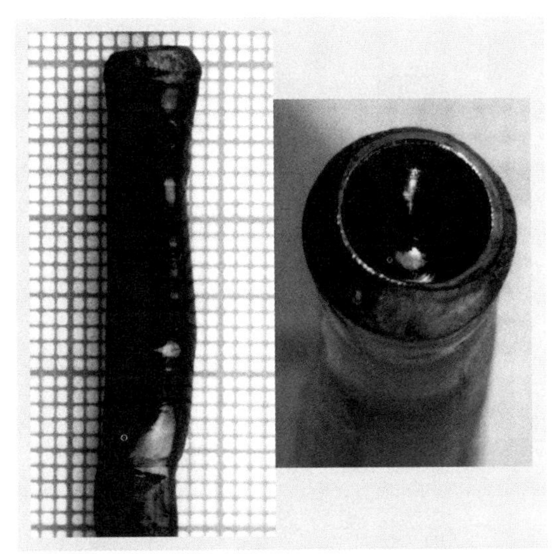

Fig. 12.8 $CoTiO_3$ crystal (growth rate 1 mm/h) showing internal cavity caused by a stable bubble seriously interrupting growth

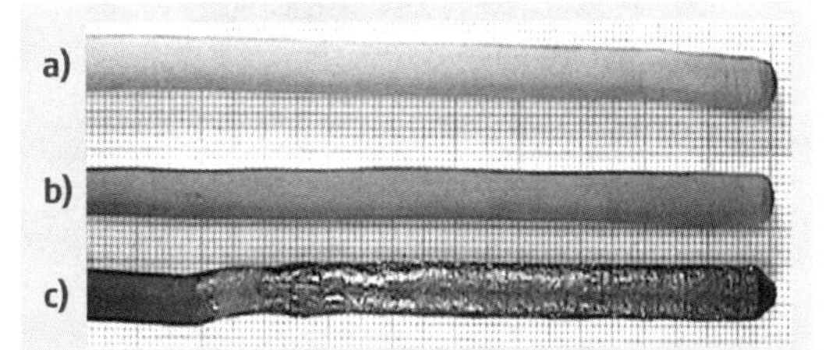

Fig. 12.9a–c Steps for preparation of ceramic feed rods of $SrCu_2(BO_3)_2$: (**a**) sintered in air, (**b**) sintered in O_2, (**c**) premelted in O_2

sure (and flow) is important for a constant temperature distribution as well, as even small fluctuations have to be avoided. Furthermore, achieving a stable liquid zone requires a homogeneous, uniformly dense ceramic rod acting as the source of material for crystallization. Gas bubbles incorporated into the feed rod can seriously influence the zone, interrupting the crystal growth process (Fig. 12.8).

Successful preparation of starting rods for crystal growth requires: muffle furnaces for preliminary ceramics preparation, furnaces with controlled atmosphere appropriate for rod sintering, a hydrostatic press, as well grinders and/or mortars, as the particle size after grinding is crucial for obtaining good-quality rods.

The first step in rod preparation is a typical ceramics synthesis. The batch of powders is weighed accordingly to the chemical reaction and is prepared by ball-mixing and then by manual or automatic grinding in the mortar. The powders are pelletized and annealed at an appropriate temperature and specific time. The quality of the prepared ceramics is assessed by x-ray diffraction and – if found to be acceptable – the material is reground and formed as a rod (typically 8–10 mm in diameter and 120–150 mm long) by either cold or hot pressing. The pressure has to be experimentally selected to avoid *overpressing*, with the typical range being 800–2500 atm. Polyvinyl alcohol, glycerol or other common additives are often used to reduce internal friction during pressing, allowing the production of denser, less porous ceramic rods. The pressed rod is later sintered at an optimized temperature and appropriate atmosphere.

The density of as-obtained ceramic rods (Fig. 12.9) should be measured and compared with the crystallographic density. This ratio depends as much on the material as on the quality of preparation and varies dramatically from under 60 to above 90%. In many cases, when evaporation is not an issue it is suggested to premelt (grow very fast) a less dense rod before performing the final crystal growth. This was found to be especially important for slower-grown, incongruently melting compounds (e.g., high temperature superconductors (HTSC) or $SrCu_2(BO_3)_2$). In the furnaces mentioned above, crystals up to 100–150 mm long and 10 mm in diameter can be grown, but the typical size of a good-quality grown crystal is usually smaller.

12.7 Stable Growth of Congruently and Incongruently Melting Oxides

Differential thermal analysis (DTA) should be routinely performed for all (attempted) new materials to establish their melting properties. For many new materials this is often not possible as DTA apparatus rated above 1500 °C are not common and analysis at elevated temperatures is often limited by a lack of appropriate crucible materials. In this case the melting properties are determined during preliminary growth. Depending on those properties either the direct crystallization or the traveling solvent zone (TSZ) approach is applied.

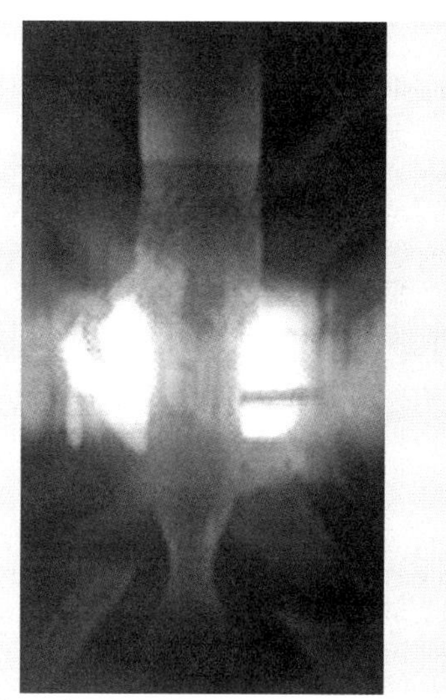

Fig. 12.10 Growth of congruently melting Al_2O_3 in an NEC furnace. Note *necking* on the *lower part* of the crystal

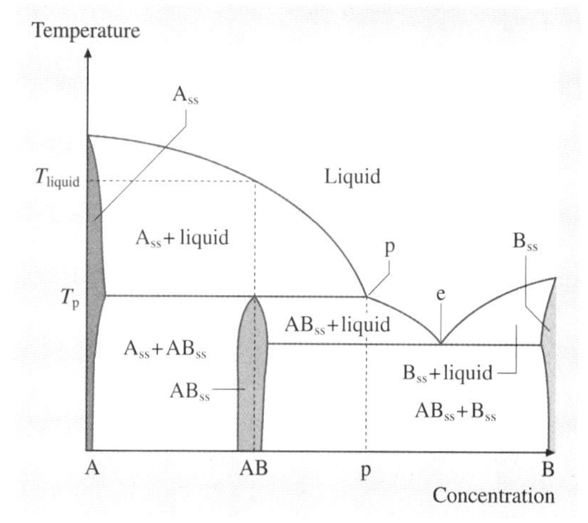

Fig. 12.11 Generic phase diagram of compound AB melting with peritectic decomposition

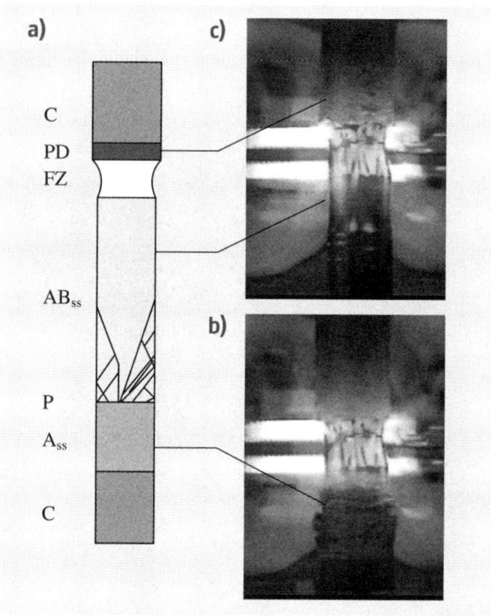

Fig. 12.12 (a) Simplified description of the creation of self-flux and growth of an incongruently melting oxide; C – ceramic rods with composition AB, A_{SS} – precipitation of phase A_{SS}. P – composition of liquid is passing peritectic point P, AB_{SS} – crystallization of AB_{SS} begins. FZ – floating zone, PD – peritectic decomposition of ceramic AB. On the beginning of growth primary phase A_{SS} crystallizes from FZ and liquid is continuously enriched in compound B until it reach peritectic composition P where AB_{SS} becomes primary phase solidifying with multiple nuclei. Grains with favoured orientation are growing faster and a single crystal of AB_{SS} can be grown. **(b,c)** growth of incongruently melting $Ba_xLa_{1-x}CuO_4$ in a Crystal System furnace; image **(c)** recorded 23 h after image **(b)**

For congruently melting oxides the composition of ceramic rods, growing crystal, and the melt is the same. The crystal growth process in this case is relatively fast, and the growth rates vary from 1 to as high as 50 mm/h. Rods rotate in opposite directions so that temperature uniformity as well as mixing of material inside the molten zone is achieved (Fig. 12.10).

When an incongruently melting material (Fig. 12.11) is being grown one has to use a solvent (flux) to make crystallization possible. The composition of the flux can be suggested on the basis of information from the appropriate phase diagram [12.44]. If this is not feasible it is suggested to use the self-flux approach. In such a case the zone is created by melting the ceramic rod (with composition similar to the composition of the required crystal) and carefully adjusting the temperature and growth speed until the evolution of the composition of liquid zone stabilizes growth with the required composition. In the case of peritectic transformation, the crystal growth starts with the precipitation of high-temperature primary phase on the seed rod (Fig. 12.12).

The zone composition then changes towards and beyond the peritectic liquid composition until a nearly

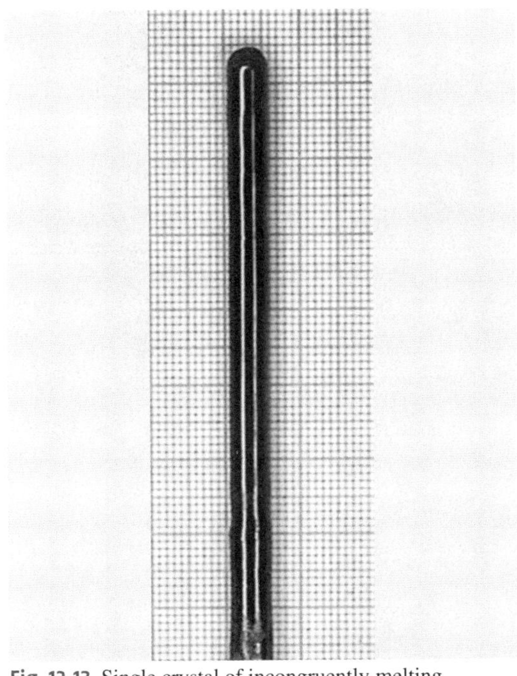

Fig. 12.13 Single crystal of incongruently melting $La_{2-x}Ba_xCuO_4$ grown by TSFZ technique (1 mm/h, 180 kPa O_2)

steady state is established and crystallization of the required phase begins. If the zone is quenched at this point it is possible to analyze the composition of the self-flux [12.45, 46].

In subsequent growths, to speed-up the process of achieving the nearly steady state, it is recommended to use a flux pellet with appropriate composition and size to create the molten zone. The flux pellet is synthesized and mounted between the feed and seed rods. As the temperature increases it melts and the rods become joined. At this point, the temperature has to be carefully adjusted again to allow for the establishment of a steady state, and then the growth starts. Dissolving the feed rod into the liquid in the zone continually restores the amount of material solidifying from the zone on the seed rod. The growth should be slow, due to the slow mass transport through the diffusion layer at the solid–liquid interface. For complex oxides such as $Bi_2Sr_2Ca_2Cu_3O_{10}$ [12.47] or $SrCu_2(BO_3)_2$ [12.48] it can be as slow as 0.1–0.2 mm/h, allowing crystallization of appropriate phase from a melt of significantly different composition.

Some materials decompose during growth and lose one of the components due to evaporation. In this case a small addition of the evaporating constituent to the feed rod proved to be successful (e.g., 1–1.5% CuO in $Ba_xLa_{1-x}CuO_4$ [12.42, 49], see Fig. 12.13).

12.8 Constitutional Supercooling and Crystallization Front Stability

For complex oxides the congruently melting composition is quite often not exactly stoichiometric. This effect is well known for $LiNbO_3$ and was also observed for RE garnets (where the smaller RE ions usually occupying dodecahedral position also substitute smaller cations at octahedral positions). In such a case the evolution of the molten zone composition helps to produce crystals with composition close to stoichiometric. In the case of peritectic-type melting of binary (or more complex) compounds with solubility in the solid phase (i.e., with distribution coefficient $\neq 1$) the composition of liquid and solid in equilibrium are different (Figs. 12.11 and 12.12). Convection and/or mechanical mixing stir the majority of the volume of the liquid phase and this volume has a relatively uniform composition, but the layer in the vicinity of the crystallization front is nearly stagnant. This layer is depleted of the species that are incorporated into the crystal and enriched in those species rejected from the solid state. In contrast to the volume of the well-stirred convection cell, mass transport in this layer is mostly driven by diffusion (*diffusion layer*) and a significant gradient of concentration is observed. For a distribution coefficient $k \ll 1$ this effect is more pronounced, and still more pronounced for the peritectic-type transition. As a result of this concentration gradient the solidus temperature decreases towards the solid–liquid interface. For lower thermal gradients in the diffusion layer a part of this layer can be supercooled more than liquid in close vicinity to the solid–liquid interface (Fig. 12.14), an effect referred to as *constitutional supercooling* [12.32, 50, 51].

As the growth rate is proportional to the degree of supercooling any positive fluctuation in the growth rate will lead to a further increase of growth rate (as it forces solid–liquid interface deeper into more supercooled liquid). The regions of the adjoining liquid will be even more depleted from crystal constituency and locally the solidus temperature will be even lower.

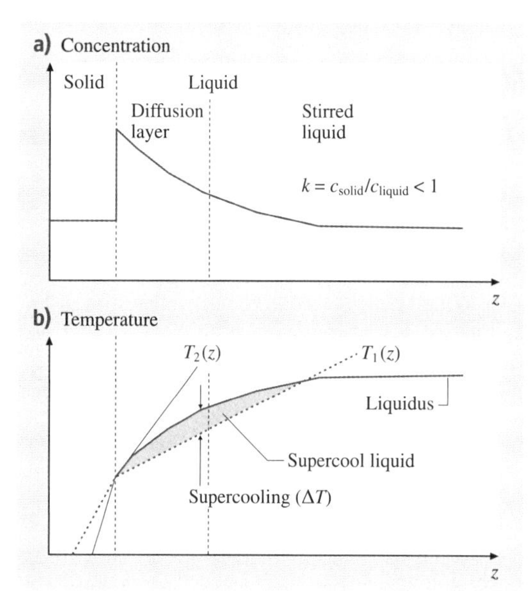

Fig. 12.14a,b Constitutional supercooling. **(a)** Concentration of dopant with distribution coefficient < 1 near the solid–liquid interface. **(b)** Conditions for constitutional supercooling near solid–liquid interface in the case of lower thermal gradient for temperature distribution $T_1(z)$

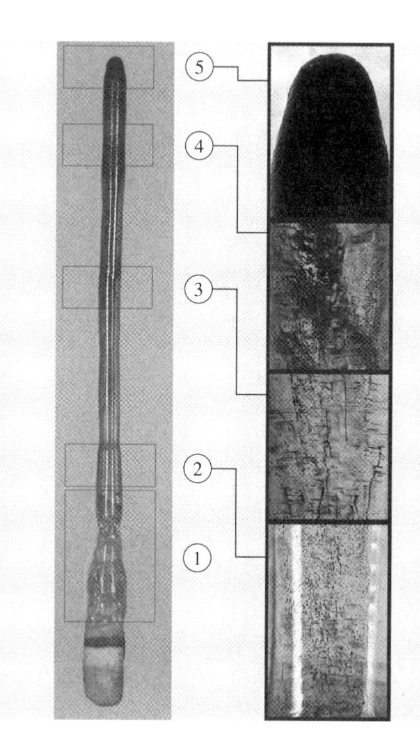

This causes crystallization front instability. Solute-rich channels form with the general direction parallel to the crystallization direction, which is not necessarily the pulling direction (note that both convex and concave crystallization fronts were observed in OFZ experiments) [12.52]. *Microfaceting* and *cellular growth* are often observed. For materials with high growth-rate anisotropy (especially for crystals with layered crystallographic structures such as $CuGeO_3$ [12.43] or $Bi_2Sr_2CaCu_2O_n$ [12.53]) this can result in platelike growth. Other defects such as flux tubes, precipitations, and dendrite growth can also be expected as a result of crystallization front instability (Figs. 12.15 and 12.16).

To avoid constitutional supercooling the growth rate should be decreased. This measure reduces concentration gradients and the possibility of supercooling. Higher thermal gradients (on the solid–liquid interface) can also be suggested as a remedy. Unfortunately this action will cause even larger temperature gradients in the growing crystal, which can result in cracking during cooling (Fig. 12.15).

Control of the thermal gradient in the liquid near the crystallization front is a challenging problem. One

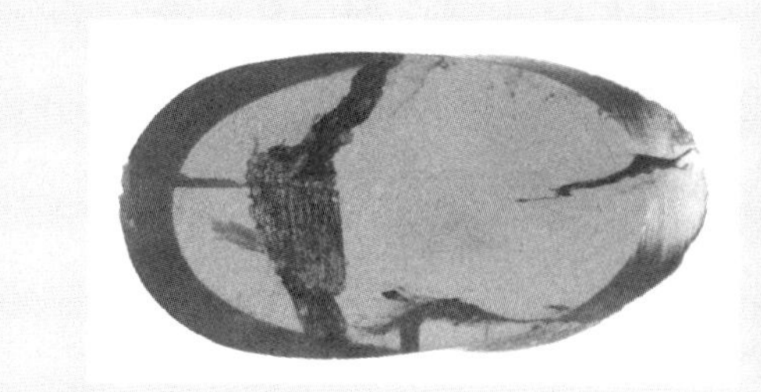
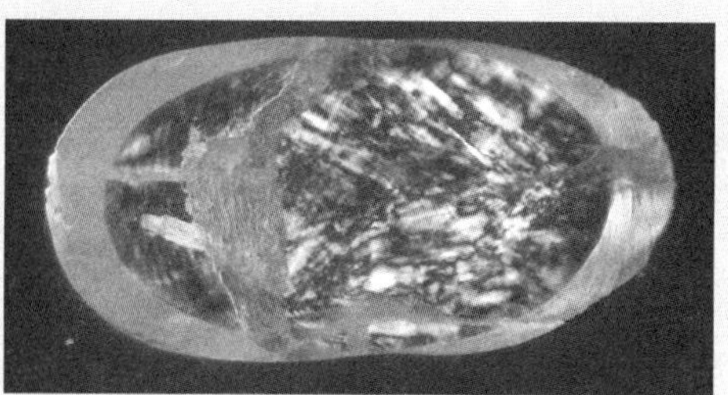

Fig. 12.15 Cut and polished slice of $Dy_2Ti_2O_7$ for defects observation. Transmission and dark field (crossed polarizers) macrophotographs

Fig. 12.16 Influence of evolution of composition of the molten zone on the crystallization process and defects in a *green sapphire* crystal grown by optical FZ. Al_2O_3 doped with $< 2\%$ of transition-metal oxides. Growth rate 8 mm/h in air, 25 rpm (counter-rotation). Transmission macrophotograph. (1) Initial part of growth. Nucleation begins on a ceramic seed rod with composition identical to the feed rod. The composition of molten zone is similar to that of the ceramic, resulting in low concentration of dopant in the crystal (*low coloration*). A multigrain crystal evolves into a single crystal after necking. Thermal stress during cooling causes cracks in the multigrain region. (2) Beginning of single-crystal growth. The concentration of dopant in the molten zone increases, as does the concentration of dopant in the crystal (*green coloration*). Isolated precipitations are seen as *dark spots* elongated in the growth direction (*white vertical strips* are caused by light reflection inside crystal). (3) Central part of the growth. The concentration of dopant in the solid (*greener coloration*) increases as a result of significantly higher accumulation of dopant in the molten zone. The density of precipitations increases due to more severe constitutional supercooling conditions. Note the striations, related to growth rate instability. These striations visualize a nearly flat crystallization front. (4) Advanced part of the growth. The concentration of dopant in the solid (*more green*) is higher due to the accumulation of dopant in the molten zone. This relative increase of dopants is lesser than observed in the lower part of crystal showing a tendency to stabilize. The density of precipitations is high. (5) The final part of the growth crystal growth termination. The crystal grows faster for a brief period of time and then the molten zone solidifies quickly (paraboloid tip). The fast growing part has a high density of precipitates and higher dopant concentration (*very green*). Very high concentration of dopant (and defects) is present in the solidified tip; this part of boule crystal is opaque ◀

of the possible approaches – anisotropic heating – was successfully applied by *Watauchi* et al. [12.54] for growth of congruently melting $CuGeO_3$ and incongruently melting $Sr_{14}Cu_{24}O_{41}$.

These effects are also depicted in Fig. 12.16. Single crystals of *green sapphire* (Al_2O_3 with < 2 mol % of transition-metal oxides added, dopant segregation coefficient < 1) have been grown with constant pulling rate and dopant accumulate in the FZ as growth progress. As the concentration of dopant increases, so does the degree of constitutional supercooling, creating increasing amount of defects. As the growth was not long enough the *steady state* could not be achieved (as it is an asymptotic process), but coloration of the upper part of the crystal indicates that as the crystal grows the dopant concentration apparently stabilizes – in contrast to the crucible methods.

12.9 Crystal Growth Termination and Cooling

As most of the heat is supplied to the crystal via the liquid zone, growth termination generates difficult-to-avoid *thermal shock* to the crystal. To lower the thermal shock the tip of as-grown crystal should be left in the hot zone during the lamp cooling process. The rate of lamp cooling should be related to the growth rate as switching off the lamps rapidly results in cracks in the crystal.

In some cases, it is advisable to relax the stresses created during the growth and cooling process by annealing the as-grown crystal at elevated temperature, then cooling it slowly to room temperature. After growth, annealing in a specific atmosphere often helps to reduce the number of defects present [12.51].

12.10 Characterization of Crystals Grown by the OFZ Technique

A good crystal – as defined by the *end user* – is in fact a crystal sample optimized for measurements or analysis by a particular method. Such a sample is usually oriented and cut to the required size. Special preparation of the surface of the crystal may also be necessary.

Large single crystals of silicon grown by the FZ method are of extremely good quality and can be dislocation free, but even they are not perfect as they have striations, oxygen incorporation, and other minute defects that influence or limit some demanding applica-

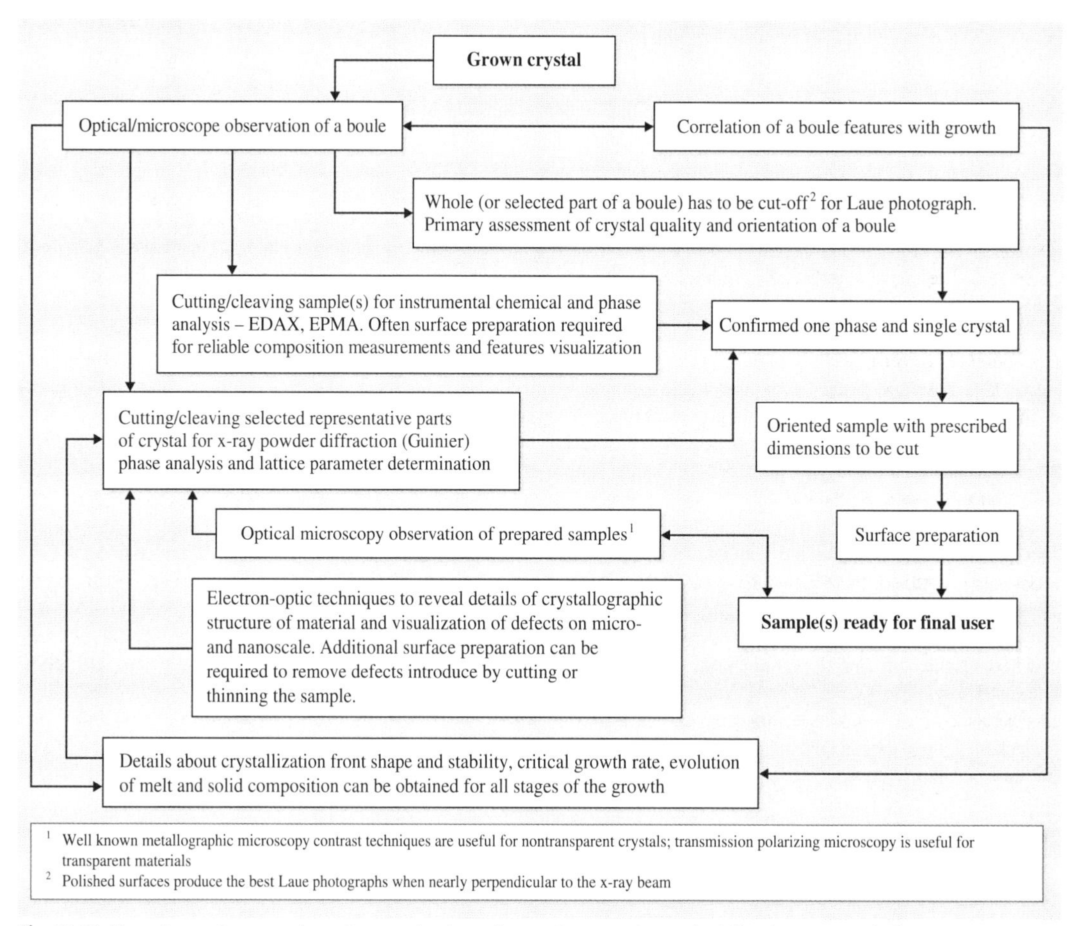

Fig. 12.16 Experimental approach to characterization of crystals grown by optical floating zone techniques

tions [12.55]. Improving the quality of a substantial size crystal in real life is a challenging and time-consuming task. It is mostly driven by specific applications or user demands – the silicon with its steady improvement is the best example.

Oxide crystals grown by OFZ method are not an exception to this rule and a lot of time and effort has been spent perfecting them. The more complicated the chemical formula of the attempted material, the more time is required to grow high-quality (although still not perfect) single crystals.

Careful characterization of both preprepared ceramics and the resulting crystals is essential for producing high-quality materials for further applications. This characterization depends as much on the material itself as on the purpose of its production. *Full characterization* of a crystal involves a lot of time and manpower and is rarely performed. As growers using floating zone technique are interested mostly in investigating specific effects (e.g., superconducting properties) they often neglect detailed characterization of the crystals obtained. Even more unsettling is the fact that reviewers of renowned journals accept works done on crystals with virtually no characterization mentioned. This creates a vicious circle, as a generation of students believes that anything grown is a single crystal. Figure 12.16 suggests the basic experimental approach to characterization and quality improvement of crystals grown by the OFZ technique.

The existence of the required phase is usually confirmed by x-ray powder diffraction. Powder diffraction

Fig. 12.17 Cross section of the initial part of $Sr_{14}Cu_{24}O_{41}$ crystal grown at 1 mm/h in 8 atm O_2. Note the visible grain boundaries on the axial cross section and a core on the cross section perpendicular to the boule axis

also allows the assessment of the uniformity of the material and addresses the problem of the existence of more then one phase. Accurate values of crystallographic lattice parameters can be measured by a Guiner camera with either Si or KCl as the internal standard, using a minute amount of material cut or cleaved from selected, characteristic parts of a boule. This information is important when crystals of solid solutions are obtained.

The observation of an as-grown transparent boule under an optical and/or polarizing microscope can reveal different defects in the crystal and their evolution during the growth process. It can also monitor the continuity of the growth and facilitate the observation of how changes of growth conditions (applied power, growth, and feed rate as well as speed of rotation) influence the quality of material grown (Fig. 12.17).

For nontransparent materials the surface features often mask the volume ones. This makes orientation of an as-grown FZ crystal very challenging. The majority of crystals grown by OFZ technique do not show distinctive facets (due to the high thermal gradients); usually they are round or oblong in cross section (Figs. 12.17 and 12.18). In this case sectioning of a grown boule and special surface preparation such as polishing and chemical etching are necessary before orientation by x-ray methods.

If as-grown crystals show distinctive facets then the facet quality may indicate overall crystal quality, and their shape and the angles between them as well as optical properties of materials can be used to orient the crystal. The existence of facets suggests a convex interface and can also be connected to growth striations. X-ray topography, discussed in detail in [12.56], con-

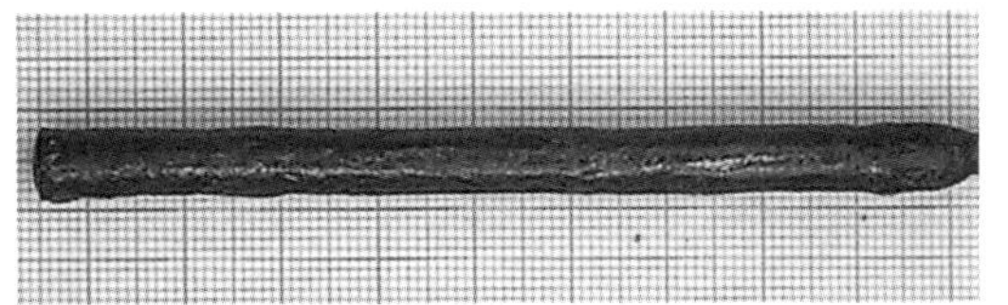

Fig. 12.18 Single crystal of $SrCu_2(^{11}BO_3)_2$ grown at 0.25 mm/h in 280 kPa (abs.) O_2

firmed a compositional difference between faceted and nonfaceted regions for YIG crystals [12.13, 57].

The Laue photography technique is a versatile method for crystal orientation (especially with available software [12.58]) and can also provide a preliminary assessment of the as-grown rod quality, answering the basic question of whether the material is a single crystal, multigrain sample (*blocks*) or polycrystalline boule with some degree of texture, etc. As the penetration of x-rays is in the micrometer range, the confirmation of crystallinity of a whole boule requires multiple Laue photographs (Fig. 12.19). If the Laue technique is used to assess local crystal quality, which relies on visual analysis of the shape of diffraction spots, its sensitivity is considerably limited and produces qualitative results only. For crystals of high quality collecting a *rocking curve* (ω-scan, which allows us to determine the full-width at half-maximum, FWHM) or pole figures using multiaxis diffractometers (preferably on polished and chemically treated

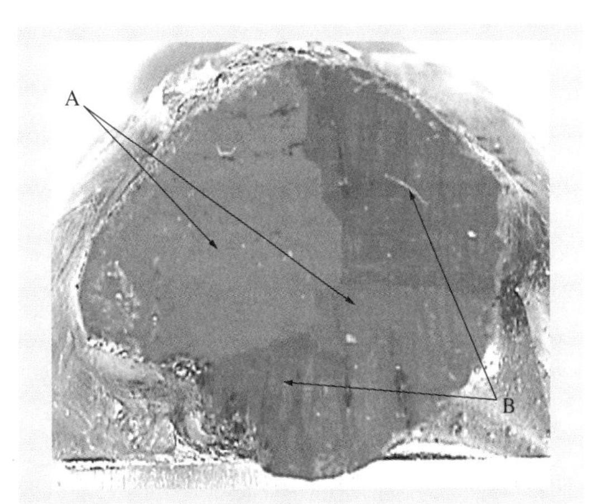

Fig. 12.19 Two grains in $SrCu_2(BO_3)_2$:Mg crystal. *Regions* marked A have nearly identical crystallographic orientation, but different from that in *regions* marked B, which are also co-oriented (characterization by Laue photography, courtesy of S. Dunsiger)

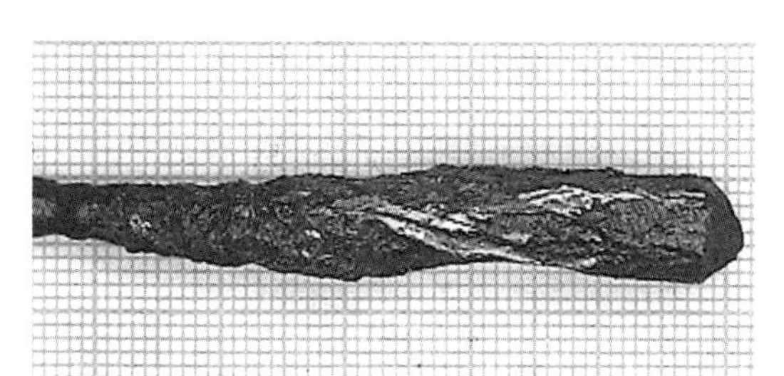

Fig. 12.20 Single crystal of $Sr_{1.95}Na_{0.05}Cu_2(^{11}BO_3)_2$ grown at 0.25 mm/h in 280 kPa (abs.) O_2

surfaces) addresses problems related to mosaic spread, grains, and twinning in a more quantitative manner. The rocking curve method can also be used for neutron diffraction [12.59].

Solving crystallographic structure provides the ultimate description of the crystal. Single-crystal x-ray diffraction is preformed on a selected, small (and usually perfect) piece of as-grown material and consequently does not provide information about the entire grown crystal.

Both x-ray and electron diffraction provide very valuable surface information. Neutrons, on the other hand, provide information about the entire volume of the crystal due to their low absorption by the majority of materials. Neutron diffraction techniques – if available – provide an excellent confirmation of the total crystal quality because all the substantial grains present in the as-grown crystal can be detected. This is especially important in the case of nontransparent crystals where the assessment of the quality of the whole boule can be performed without labor-intensive and destructive sectioning, surface preparation, etc. For example, high-resolution, elastic neutron scattering measurements on $SrCu_2(^{11}BO_3)_2$ confirmed that slowly (less than 0.3 mm/h) grown crystals (Fig. 12.18) are a single domain each, with a mosaic spread of the (110) Bragg peak of 0.3°. Note, that in this case, to avoid high thermal neutrons absorption in ^{10}B ($\approx 20\%$ of natural abundance) ^{11}B isotope enriched to 99.6% has been used [12.60].

Similar analysis preformed on $Sr_{1.95}Na_{0.05}Cu_2(^{11}BO_3)_2$ (Fig. 12.20) revealed the presence of 11 grains in similarly sized crystal grown in the same conditions.

Energy-dispersive x-ray analysis (EDAX) and electron microprobe analysis (EPMA) can confirm the exact chemical formula of crystals and solid solutions grown in different conditions. They also allow advanced phase analysis of obtained materials. Good oxide standards are essential for quantitative analysis.

Single crystals of oxides are generally investigated by the crystal users, according to physical properties and predicted applications. Investigation of these properties and their changes with different dopants and growth conditions are now the main purpose of growing crystals by OFZ and TSFZ methods. The majority of crystals grown for research purpose have unique magnetic, electrical, and crystallographic properties, which are investigated and reported in specific journals.

12.11 Determination of Defects in Crystals – The Experimental Approach

All crystals contain defects and impurities that influence their physical properties. To observe and assess the growth features an oriented crystal has to be cut and polished. Which defects are expected determines the way in which the sample is prepared for observation (Figs. 12.15–12.19). It should be specified what defects are expected in crystals grown by a given method, as many defects interfere with measurements and applications.

An excellent discussion of the defects present in oxide crystals grown from low-temperature solutions is given by *Rudolf* [12.61], from high-temperature solutions by *Elwell* and *Scheel* [12.50], and for oxides grown by Czochralski method by *Hurle* [12.51].

The defects present in oxide crystals grown by the optical FZ technique are similar to those found in crystals grown by other methods, but some result from the specific growth conditions in the OFZ, such as the very high temperature gradients (for both FZ and TSFZ methods) and growth from flux (for TSFZ method). In the next paragraph those defects and the reasons for their appearance are discussed.

Increasingly available and advanced electron-optics techniques with easily accessible instrumentation such as scanning electron microscopy (SEM), transmission electron microscopy (TEM), high-resolution transmission electron microscopy (HRTEM), electron energy-loss spectroscopy (EELS), and high-angle annular dark field in scanning transmission electron microscope (HAADF-STEM) answer many questions related to the real structure of crystal and the presence of defects on the scale from millimeters down to atomic

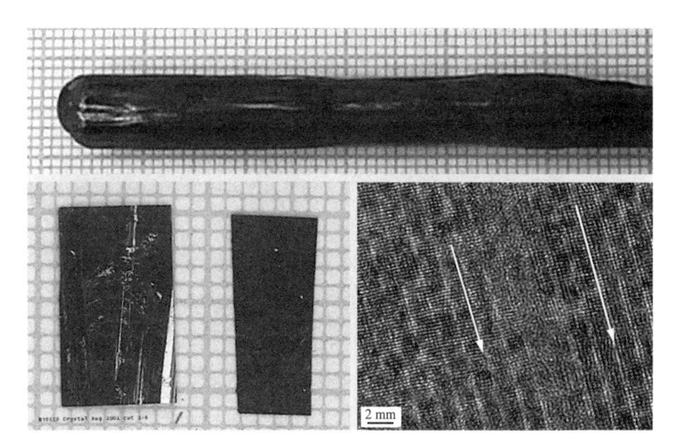

Fig. 12.21 *Top*: Single crystal of BiSCCO 2212 grown by TSFZ technique. *Left*: plates cleaved from this crystal. *Right*: HRTEM image of the (100) plane with grain boundary at the center. The *arrows* in the image indicate the direction of the *b*-axis in each grain (courtesy Y. Zhu) ▶

resolution. Additional sample preparation is necessary to fully utilize the potential of these techniques. A leading experimental work on defect characterization, using STEM, EELS, and TEM techniques on cleaved $Bi_2Sr_2CaCu_2O_n$ crystals grown slowly (0.21 mm/h) by TSFZ method (Fig. 12.21) has recently been published by *Zhu* et al. [12.62].

Figure 12.22 presents a comparison between known crystallographic structure and a HAADF-STEM image. This very sophisticated method of characterization of crystal grown by TSFZ method visualizes the displacement of Bi atoms in crystallographic lattice by intensity modulation; the results are consistent with (but much clearer than) those obtained with conventional high-resolution TEM on the same crystal sample. Figure 12.23 shows a detailed analysis of grain boundary performed on a HAADF image. The line defect has a lattice parameter suggesting that it is an intergrowth of Bi-2201 phase in Bi-2213 crystal. Stacking defects are characterisitc for BiSCCO superconductors and other layerd compounds, but it is common to find few hundred by few hundred nm area of the perfect atoms arrangements – see Fig. 12.24 in the crystal of $(La, Ba)_2CuO_4$ grown by TSFZ.

Local macrodefects (already mentioned in the section about instabilities in crystal growth), i.e., *microfaceting* (Fig. 12.20) and *cellular (grain) growth*, as well as *grain boundaries* (Figs. 12.12, 12.17, and 12.25), *flux tubes, precipitations, and dendrite growth* [12.62, 63] are all present as a result of crystallization front instability. They are often observed under

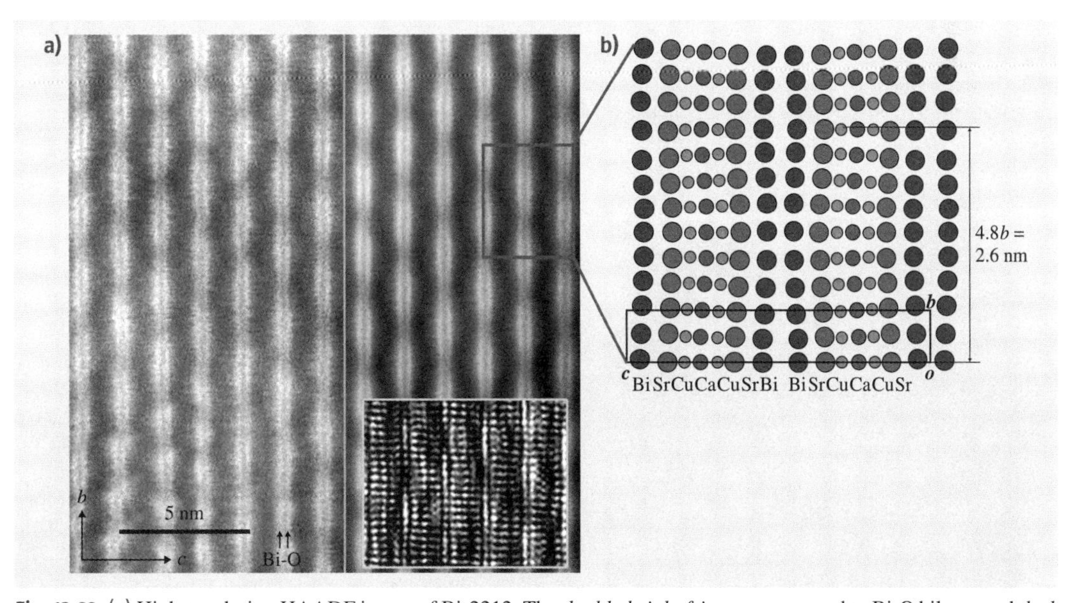

Fig. 12.22 **(a)** High-resolution HAADF image of Bi-2212. The *double bright fringes* correspond to Bi-O bilayer and *dark ribbons* to atomic layers of lighter elements. Compared with model of Bi-2212 incommensurate modulated structure **(b)** (after [12.62], courtesy Elsevier 2006)

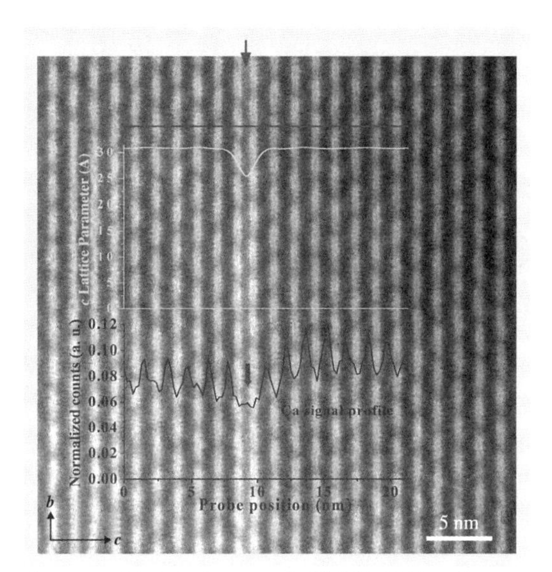

Fig. 12.23 HAADF images of BISCCO-2212 with intergrowth of 2201 phase. *Yellow profile*: calculated lattice parameter c; over the intergrowth $c = 25.5$ Å, indicating that this defect is an intergrowth of 2201. *Blue profile*: normalized Ca signal related to the local concentration of Ca, indicating absence of Ca in this layer (after [12.62], courtesy Elsevier 2006)

Fig. 12.25 Macrophotograph of polished cross section of La_2CuO_4 crystal grown 1 mm/h on seed, with visible large grains (polarized light, false colors)

polarizing microscope and/or by x-ray topography. These defects are more pronounced at the beginning of crystallization and can also be eliminated with the use of a crystalline (preferably oriented) seed and/or *necking* procedure (Fig. 12.10). Crystal quality can be further improved by adjusting the rotation so as to maintain a flat crystallization front and slower growth rate.

Impurities, inclusions, and precipitates (e.g., in $(La, Sr)_2CuO_4$ [12.63] or YVO_4 [12.64, 65]) are particles of the same or different phases embedded into otherwise good crystals. They can be generated by instabilities in the growth conditions leading to a noticeable amount of undissolved material in the molten zone. A slight change in stoichiometry of the starting rod can sometimes result in inclusion-free crystals as was proved for crystals of Mg_2TiO_4 grown from ceramic rod with Mg : Ti ratio of 2 : 1.01 [12.46]; a small excess of CuO is also used for various cuprates [12.21, 66, 67]. Dense feed rods and slower growths have been suggested for reducing the number of inclusions. Inclusions – on their own – are also the main cause of dislocations. The number of *precipitates* (such as the presence of Al_2O_3 in $NiAl_2O_4$ [12.52] or carbonates and BiO_x phase detected in $Bi_2Sr_2CaCu_2O_n$ [12.62, 68]) can sometimes be reduced with a slower growth rate or change of growth atmosphere.

Cracks, caused by mechanical stress due to cooling (e.g., in $Ca_2Al_2SiO_7$ [12.69] or $SrZrO_3$ [12.70]), can be diminished with the use of an afterheater [12.70, 71]. Sometimes, severe cracks are the result of phase transitions during cooling (e.g., $CaTiSiO_5$, Fig. 12.26). Change of the growth atmosphere (from air to N_2) proved to be helpful in the above case [12.35].

Fig. 12.24 HRTEM of perfect crystal of $La_{1.9}Ba_{0.1}CuO_4$ grown by TSFZ method (image courtesy of G. Botton and C. Maunders)

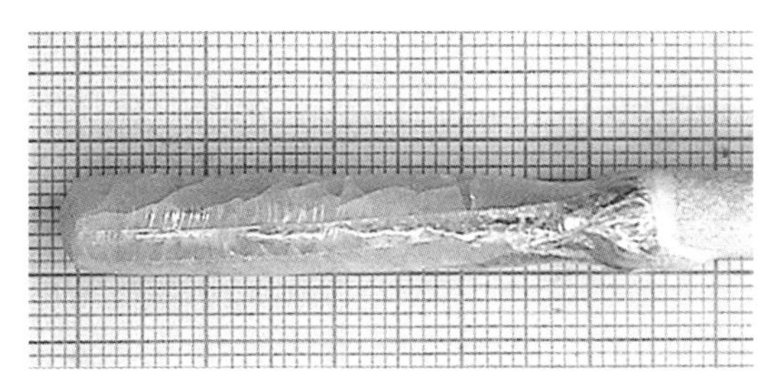

Fig. 12.26 $CaTiSiO_5$ undergoes phase transition during the final stage of cooling (about 235 °C). Initially transparent crystal cracked only when cooled to near room temperature; H. Dabkowska, A. Dabkowski, unpublished results

Gas bubbles are trapped inside crystals grown by the FZ method because the shape of the solid–liquid interface causes the bubbles to concentrate in the core region of the crystals (Fig. 12.8). This effect is often caused by either solubility of O_2 in the liquid phase or partial decomposition of oxides. It can be controlled to some extent by atmosphere composition, as described for Al_2O_3 [12.72, 73] and Li_3VO_4 [12.74].

Striations are caused by fluctuations of the temperature of the liquid near the crystallization front, originating from convention flows. Temperature fluctuations cause fluctuations of the growth rate, and as a result there are changes in chemical composition (as the effective segregation coefficient depends on the growth rate). Striations may be used as indicators of the shape of the solid–liquid interface [12.13, 65]. Optical observations on polished and etched YIG revealed the internal structure of a crystal, providing valuable information about the growth mechanism [12.12]. This proved that too fast a growth rate leads to *cellular growth*, which can often be corrected by lowering the growth speed.

Sheet defects [12.15, 17, 54] and *eutectic solidification* can be revealed by careful observation of the cross section of as-grown crystal when the coexistence of two or more phases is noticed. *Mao* et al. [12.75] combined microscope observation of polished Sr_2RuO_4 crystal with x-ray Laue photography and detected the presence of layered intergrowth of different phases (Sr and $SrRuO_3$) in the crystal. On the basis of this analysis it was possible to optimize the starting composition of the feed rod and the growth conditions, reducing the level of impurities and defects and increasing the T_c of ruthenate. Sometimes it is possible to understand and modify the physical properties of material by identifying defects resulting from the crystal growth. For example, eutectic solidification in crystals grown by OFZ was discussed by *Troileux* et al. in $La_{2-x}Sr_xCuO_4$ [12.76] and by *Fittipaldi* et al. [12.77] in the Sr-Ru-O system.

Twin boundaries analyzed by x-ray diffraction depend mostly on the crystallographic structure of the material (e.g., $Mg_3V_2O_8$ [12.78] and $La_{1-x}Ca_xMnO_3$ [12.79])

The influence of convection on dopant segregation is discussed in [12.36, 80, 81].

12.12 Details of Conditions for Growth of Selected Oxide Single Crystals by OFZ and TSFZ Methods

In Table 12.1 examples of oxides grown by FZ and TSFZ techniques and the growth conditions employed are listed. Crystals grown with the application of a solvent are marked with a dagger, and references referring to defects in each oxide crystal are marked with an asterisk.

Table 12.1 Examples of oxide crystals grown by optical floating zone technique and traveling solvent optical floating zone technique (marked by †). References discussing defects in obtained crystals are marked by *

Material	Growth			References and comments
	rate (mm/h)	atmosphere	rotation (rpm)	
Al_2O_3	10–1500	Air	0–200	[12.72, 73]*
$BaCo_2Si_2O_7$	1	Air	20–30	*H. Dabkowska*, unpublished
$Ba_3Cr_2O_8$	10	2 atm Ar	–	[12.82]
$BaFe_{12}O_{19}$†	6	70 atm O_2	–	[12.14]*
$BaTiO_3$†	1	O_2/Ar 1 : 1	20	[12.83]
$Ba_{1-x}Sr_xTiO_3$	1–2	O_2	–	[12.38]
$Bi_2Sr_2CaCu_2O_n$†	0.2–0.35	O_2	20–30	[12.21, 34, 53, 61, 62, 68, 84–88]

Part B | 12.12

Table 12.1 (cont.)

Material	Growth rate (mm/h)	atmosphere	rotation (rpm)	References and comments
$Bi_2Sr_2CaCu_2O_n$:Y† $Bi_2Sr_2CaCu_2O_n$:Li†	0.5	–	–	[12.89,90]
$Bi_2Sr_2Ca_2Cu_3O_{10}$†	0.05	O_2/Ar 1 : 4	10	[12.47]
$Bi_2Sr_2CuO_6$†	1.5	–	–	[12.91,92]
$Bi_{12}TiO_{20}$†	0.25	O_2 flow	35–45	[12.93]
$Ca_{12}Al_{14}O_{33}$	0.2	–	–	[12.94]*
$CaAl_2O_4$	4	$Ar-O_2$, Ar, $Ar-H_2$	21	[12.95]
$Ca_2Al_2SiO_7$	3	Low pressure	25	[12.69]*, [12.96]
Ca_2CuO_3†	1	1 atm O_2	20–30	[12.97]
$CaCu_3Ti_4O_{12}$†	6	O_2	30	[12.98,99]
Ca_2FeMoO_6	60	0.25–0.5 atm N_2	–	[12.100]*
$Ca_{0.5}La_{0.5}MnO_3$	2.5–10	Air, Ar/O_2	–	[12.101]
$Ca_2MgSi_2O_7$	2–3	Air, O_2	30–45	[12.96]
$CaTiSiO_5$	0.5–12	Air	15–30	[12.34,35,61,84]* phase transition
Ca_2RuO_4 $Ca_{2-x}La_xRuO_4$	45	Ar/O_2 9 : 1 10 atm	–	[12.102]*
$Ca_{2-x}Sr_xRuO_4$	20–50	9Ar : 1O_2 10 atm	30–50	[12.103]*
$CaYAlO_4$	3–5	Ar, O_2, N_2, air, O_2/N_2	15–20	[12.104]* color centers
$Ca_{2+x}Y_{2-x}Cu_5O_{10}$†	0.5	O_2	15	[12.105, 106]*
$CdCu_3Ti_4O_{12}$†	6	O_2	30	[12.99]
$CoTiO_3$	5	O_2	20	HAD, not published
$Co_3(VO_4)_2$†	0.5–3	Air	20–30	[12.107]
$CuGeO_3$	1–10	1 atm O_2	30	[12.17]*, [12.43,54, 108]
Fe_3O_4†	6	CO_2	–	[12.13]*
Ga_2O_3	5–10	Air, Ar	15–30	[12.18]* volatile
$GeCo_2O_4$	15–30	Air	–	[12.109] evaporation, low viscosity
$GeNi_2O_4$	40	10 atm, O_2	–	[12.109] evaporation, low viscosity
$Gd_3Fe_5O_{12}$†	1.5	70 atm O_2	–	[12.14]
La_2CuO_4†	0.5–1	O_2	25–30	[12.43, 110]
$La_{2-x}Sr_xCuO_4$†	1–15	O_2, air	40	[12.63]*, [12.76, 111–114]
$La_{2-x}Ba_xCuO_4$†	0.5–0.7	10^{-2} atm O_2	30	[12.49]*, [12.63, 106, 110, 114–118]
$(La_{1-x}Ca_x)_2CaCu_2O_{6+\delta}$†	0.35–1	10 atm O_2	–	[12.45]*
$(La_{1-x}Ca_x)_2CaSrCu_2O_6$†	0.5	10 atm O_2	–	[12.119]*
$La_{14-x}Ca_xCu_{24}O_{41}$†	1–1.5	13 atm O_2	40	[12.120]
$LaCoO_3$	20	O_2	–	[12.121, 122]
$LaFeO_3$, $YFeO_3$	20	O_2	–	[12.123]
$LaMnO_3$	10	Air	–	[12.123]
$La_{0.8}Sr_{0.2}MnO_3$	3–10	Air, Ar/O_2	15–50	[12.101, 124–127]
La_2NiO_4	20	O_2	30–50	[12.71]*
$La_{2-x}Sr_xNiO_{4+\delta}$	3–5	5–7 atm Ar : O_2	40	[12.59]
$LaTiO_3$	50	30% H_2 in Ar	–	[12.123]
$LiNbO_3$	9	–	–	[12.26, 27, 36]* laser heated
Li_3VO_4, β(II)-Li_3VO_4†	0.5–1	Ar, O_2, air	–	[12.37, 74]*
$LuFe_2O_4$†	1	Air	–	[12.128]*
$LuFeCoO_4$†	1	CO_2/CO	–	[12.128]*

Table 12.1 (cont.)

Material	Growth rate (mm/h)	atmosphere	rotation (rpm)	References and comments
$MgFe_2O_4$†	5	50 atm O_2	–	[12.14]
Mg_2SiO_4:Cr	–	Ar/O_2	–	[12.129]
$MgTi_2O_4$†	1–5	O_2	30	[12.46]* inclusions, precipitates
$Mg_3(VO_4)_2$†	0.5	O_2/N_2, O_2	16–18	[12.78]*
Na_xCoO_2	2	O_2	–	[12.130–132]
Nd_2CuO_4†, $Nd_{1.85}Ce_{0.15}Cu_{4-\delta}$†	0.5–0.7	O_2 flow, O_2/Ar, 8 atm O_2	30	[12.113]*, [12.133]
$Nd_{0.7}Sr_{0.3}MnO_3$	2.5–10	Air/O_2	–	[12.101]
$Nd_{1+x}Sr_{2-x}Mn_2O_7$	2–8	Air	25–30	[12.134]
$NdTiO_3$, $Nd_{1-x}TiO_3$	25	Ar/H_2 95 : 5	15	[12.135]
$NiAl_2O_4$	2.5–10	1–3 atm O_2	5–40	[12.52]*
$Ni_3(VO_4)_2$†	0.5–3	Air	20–30	[12.121]
$Pr_{1+x}Sr_{2-x}Mn_2O_7$	2–8	Air	25–30	[12.134]
$RE_3Ga_5O_{12}$†	10	Air	–	[12.14]
$REBa_2Cu_3O_{7-y}$† (RE = Y, La, Pr, Nd, Sm)	–	10^{-2} atm O_2	–	[12.136]*
$REFeO_3$†	3–10	70 atm O_2	–	[12.14]
$RE_2Ti_2O_7$ (RE = rare earth)	5–20	Ar, O_2	10–20	[12.22, 34, 61, 84]*, [12.23]
$SrAl_2O_4$	4	Ar-O_2, Ar, Ar-H_2	21	[12.95]
$Sr_3Cr_2O_8$	10/20/08	2 atm Ar	–	*H. Dabkowska*, unpublished
$Sr_2CuO_2Cl_2$	5	Ar, Ar/O_2	30	[12.137]*
$SrCuO_2$†	0.5–2	1 atm O_2	–	[12.43, 138, 139]
Sr_2CuO_3†	1	1 atm O_2	–	[12.43, 138]
$SrCu_2(BO_3)_2$†	0.2–0.5	O_2	10–20	[12.42, 48, 60, 140]
$SrCu_2(BO_3)_2$:Mg, La, Na†				low surface tension
$Sr_{14}Cu_{24}O_{41}$†	1	5 atm	–	[12.34, 42, 43, 54, 61, 84, 138]
$Sr_{14-x}Ca_xCu_{24}O_{41}$†	1	10 atm O_2	–	[12.141]
$SrFe_{12}O_{19}$†	6	50 atm O_2	–	[12.14]
$Sr_3Fe_2O_{7-x}$†	1–5	0.2–3 atm O_2	10–15	[12.142]
Sr_2RuO_4, Sr_2RuO_4:Ti	25–45	Ar/O_2 2 : 1	–	[12.34, 61, 75, 84]*, [12.143] volatile
$Sr_3Ru_2O_7$	15–20	Ar/O_2 1 : 9	–	[12.82, 144, 145]* volatile
$SrTiO_3$, $SrTiO_3$:La	15	15 atm	20–30	[12.14, 146]
$SrZrO_3$	5–65	Air	3–25	[12.70]* evaporation
TiO_2	5	Air, O_2	30	[12.14], [12.20]* ZrO_2 added
$Y_3Al_5O_{12}$	5	Air	–	[12.14, 28], [12.147]*
$YCrO_3$	20	Ar	–	[12.123]
$Y_3Fe_5O_{12}$†	1.5–3	15–30 atm O_2	–	[12.10, 14], [12.17, 19, 57]*
$Y_3Fe_{5-x}Al_xO_{12}$†	6	O_2	–	[12.12, 13]*, [12.19]
$Y_3Fe_{5-x}Ga_xO_{12}$†	1.5–3	20–30 atm O_2	–	[12.14]
YFe_2O_4†	2–10	CO_2/H_2	30	[12.148]*
$YTiO_3$	50	30% H_2 in Ar	–	[12.123]
YVO_4:Er, Ho, Tm	5–20	7% H_2 in Ar, O_2	20–30	[12.64]*, [12.65, 123]
$YbCoGaO_4$	1.7	Air	15–30	[12.34, 61, 84, 149]
$YbFeMgO_4$†	1	Air	–	[12.128]*
ZnO†	0.5–1	Air	20	[12.150]*

12.13 Conclusions

The field of crystal growth by the optical floating zone and traveling solvent floating zone techniques is definitely expanding. With a growing number of highly computerized furnaces the quality and amount of works reported on new and already known materials is rising. The main advantage of both discussed floating zone techniques lies in the high purity of the obtained crystals resulting from the absence of a container. This feature also allows us to melt and grow high-quality, relatively large and uniform crystals of oxides for which there is no container (crucible) – either because of very high melting point or the melt being very aggressive – and which cannot be obtained by other methods. Both of these methods also make it possible to prepare solid solutions of oxides, modifying their properties according to the future applications.

Numerical modeling of the floating zone silicon process has turned out to be an excellent tool for modern crystal growth practice. Chains of models that cover the complete floating-zone process have been developed over the years [12.24, 25]. Now it is possible to simulate numerically the growth situations precisely enough to improve the final crystal quality even for very large diameters. Further modeling efforts, especially for other materials for which it is possible to compare modeling predictions with experimental observations, are necessary to improve the understanding of transport processes in the molten zone and of heat transport in the whole system. Only then will it be possible to explain the changes observed in growth features as growth conditions change.

Apart from crystallizing new materials, future work in this area should also include:

- Investigation of phase diagrams
- Understanding/assessing high-temperature properties of molten oxides and salts
- Creating a user-friendly theoretical approach connecting crystal growth conditions with obtained results.

There is also the crucial but nearly untouched problem of automation of crystal growth. The automatic diameter control systems have proven to be useful in the Czochralski crystal growth of oxides [12.51] and one can expect that this approach can be successfully applied to grow them by the optical floating zone process as well.

It took about 50 years of research and financing to grow very high-quality, large single crystals of single element Si by FZ technique. Now the challenge is to follow this with growth of other technologically important materials. To achieve this continuous cooperation between a crystal grower and crystal user is necessary. Without good characterization of the crystals grown no progress in crystal growth can be achieved.

References

12.1 W.G. Pfann: Principles of zone-melting, J. Met. Trans. AIME **4**, 747 (1952)

12.2 R. Von Emeis: Tiegelfreies Ziehen von Silicium-Einkristallen, Z. Naturforsch. **9A**, 67 (1954), in German

12.3 H. Keck, M.J.E. Golay: Crystallization of silicon from a floating liquid zone, Phys. Rev. **89**, 1297 (1953)

12.4 H.C. Theuerer: Method of processing semiconductive materials, US Patent 3060123 (1962)

12.5 A. Mühlbauer: Innovative induction melting technologies: A historical review, Int. Scientific Colloq. Modell. Mater. Process. (Riga 2006)

12.6 R.E. De La Rue, F.A. Halden: Arc-image furnace for growth of single crystals, Rev. Sci. Instrum. **31**, 35–38 (1960)

12.7 M.R. Null, W.W. Lozier: Carbon arc image furnaces, Rev. Sci. Instrum. **29**, 163–170 (1958)

12.8 R.P. Poplawsky, J.E. Thomas Jr.: Floating zone crystals using arc image furance, Rev. Sci. Instrum. **31**, 1303–1308 (1960)

12.9 R.P. Poplawsky: Ferrite crystals using arc image furance, J. Appl. Phys. **33**, 1616–1617 (1961)

12.10 L.L. Abernethy, T.H. Ramsey Jr., J.W. Ross: Growth of yttrium iron garnet single crystals by the floating zone technique, J. Appl. Phys. **32**, 376S (1961)

12.11 T. Akashi, K. Matumi, T. Okada, T. Mizutani: Preparation of ferrite single crystals by new floating zone technique, IEEE Trans. Magn. **5**, 285–289 (1969)

12.12 I. Shindo, N. Ii, K. Kitamura, S. Kimura: Single crystal growth of substituted yttrium iron garnets $Y_3Fe_{5-x}(Ga,Al)_xO_{12}$ by the floating zone method, J. Cryst. Growth **46**, 307–313 (1979)

12.13 S. Kimura, K. Kitamura: Floating zone crystal growth and phase equilibria: A review, J. Am. Ceram. Soc. **75**(6), 1140–1146 (1992)

12.14 A.M. Balbashov, S.K. Egorov: Apparatus for growth of single crystals of oxide compounds by floating zone melting with radiation heating, J. Cryst. Growth **52**, 498–504 (1981)

12.15 I. Shindo: Determination of the phase diagram by the slow cooling float zone method: The system $MgO-TiO_2$, J. Cryst. Growth **50**, 839–851 (1980)
12.16 B. Moest, V.G. Glebovsky, H.H. Brongersma, R.H. Bergmans, A.W. Denier van der Gon, V.N. Semenov: Study of Pd single crystals grown by crucibleless zone melting, J. Cryst. Growth **192**, 410–416 (1998)
12.17 A. Revcolevschi, J. Jegoudez: Growth of large high-Tc single crystals by the floating zone method: A review, Progr. Mater. Sci. **42**, 321–339 (1997)
12.18 E.G. Villora, K. Shimamura, Y. Yoshikawa, K. Aoki, N. Ichinose: Large-size β-Ga_2O_3 single crystals and wafers, J. Cryst. Growth **270**, 420–426 (2004)
12.19 A.M. Balbashov, A.A. Tsvetkova, A.Y. Chervonenkis: Imperfections in crystals of yttrium-iron garnet grown from nonstoichiometric melts, Neorg. Mater. **11**, 108–111 (1975)
12.20 M. Higuchi, K. Kodaira: Effect of ZrO_2 addition on FZ growth of rutile single crystals, J. Cryst. Growth **123**, 495–499 (1992)
12.21 G.D. Gu, T. Egi, N. Koshizuka, P.A. Miles, G.J. Russell, S.J. Kennedy: Effect of growth conditions on crystal morphology and superconductivity of Bi-2212 oxide, Physica C **263**, 180–184 (1996)
12.22 J.S. Gardner, B.D. Gaulin, D.M. Paul: Single crystal growth by the floating zone method of a geometrically frustrated pyrochlore antiferromagnet $Tb_2Ti_2O_7$, J. Cryst. Growth **191**, 740–745 (1998)
12.23 J.C.P. Ruff, B.D. Gaulin, J.P. Castellan, K.C. Rule, J.P. Clancy, J. Rodriguez, H.A. Dabkowska: Structural fluctuations in the spin-liquid state of $Tb_7Ti_7O_7$, Phys. Rev. Lett. **99**, 237202 (2007)
12.24 A. Mühlbauer, A. Muiznieks, J. Virbulis, A. Lüdge, H. Riemann: Interface shape, heat-transfer and fluid-flow in the floating-zone growth of large silicon-crystals with the needle-eye technique, J. of Crystal Growth **151**, 66 (1995)
12.25 A. Rudevics, A. Muiznieks, G. Radnieks: Transient modeling of FZ crystal growth process and automatic adjusting of the HF inductor current and feed rod velocity, Proc. Joint 15th Riga and 6th Int. Conf. Fundam. Appl. MHD, Vol. 2 (2005) p. 229
12.26 J.C. Chen, H.-K. Wu: Numerical computation of heat flow, fluid flow and interface shapes in the float zone of lithium niobate during a melting process, J. Heat Mass Transf. **39**, 3707–3716 (1996)
12.27 C.J. Chen, H. Chieh: Measurement of the float-zone interface shape for lithium niobate, J. Cryst. Growth **149**, 87–95 (1995)
12.28 C.W. Lan: Three-dimensional simulation of floating-zone crystal growth of oxide crystals, J. Cryst. Growth **247**, 597–612 (2003)
12.29 Y.K. Yang, S. Kou: Temperature oscillation in a tin liquid bridge and critical Marangoni number dependency on Prandtl number, J. Cryst. Growth **222**, 135–143 (2001)
12.30 D. Rivas, C. Vazquez-Espi: An analysis of lamp irradiation in ellipsoidal mirror furnaces, J. Cryst. Growth **223**, 433–445 (2001)
12.31 S. Otani, T. Tanaka, Y. Ishizawa: Control of heat-flow to feed rod in floating zone system, J. Cryst. Growth **87**, 175–179 (1988)
12.32 J.C. Brice: *Crystal Growth Processes* (Blackie, Glasgow, London 1986)
12.33 V.J. Fratello, C.D. Brandle: Physical-properties of a $Y_3Al_5O_{12}$ melt, J. Cryst. Growth **128**, 1006–1010 (1993)
12.34 H.A. Dabkowska, B. D. Gaulin: *Crystal Growth of Technologically Important Electronic Materials*, ed. by K. Byrappa, T. Ochachi, M. Klapper, R. Fornari (Allied Publishers PVT, New Delhi 2003) pp. 341–354
12.35 I. Tanaka, T. Obuchi, H. Kojima: Growth and characterization of titanite ($CaTiSiO_5$) single crystals by the floating zone method, J. Cryst. Growth **87**, 169–174 (1988)
12.36 C.J. Chen, Y.-C. Lee, C. Hu: A simple method of examining the propagation of defects in the floating-zone solidification process of lithium niobate, J. Cryst. Growth **166**, 151–155 (1996)
12.37 W. Itoyama, K. Iishi, S. Sakata: Growth of β(II)-Li_3VO_4 single crystals by the floating zone technique with the aid of a heat reservoir, J. Cryst. Growth **158**, 534–539 (1996)
12.38 H. Kojima, M. Watanabe, I. Tanaka: Crystal growth of strontium substituted barium-titanate ($Ba_{1-x}Sr_xTiO_3$) by the floating-zone method, J. Cryst. Growth **155**, 70–74 (1995)
12.39 NEC now: CANON MACHINERY Inc. 85 Minami Yamada-cho Kusatsu-city Shiga pref. 525-8511 Japan
12.40 I. Shindo: Crystal System Inc., 9633 Kobuchisawa, Yamanashi 408 Japan
12.41 A. Balbashov: Private communications, Moscow Power Engeneering Institute Technical University, 14 Krasnokazarmennaya, Moscow, 111250, Russia
12.42 H.A. Dabkowska, B.D. Gaulin: Growth of single crystals of selected cuprates by the optical floating zone technique, J. Optoelectron. Adv. Mater. **9**, 1215–1220 (2007)
12.43 A. Revcolevschi, U. Ammerahl, G. Dhalenne: Crystal growth of pure and substituted low-dimentionality cuprates $CuGeO_3$, La_2CuO_4, $SrCuO_2$, Sr_2CuO_3 and $Sr_{14}Cu_{24}O_{41}$ by the floating zone and travelling solvent zone methods, J. Cryst. Growth **198/199**, 593–599 (1999)
12.44 E.M. Levin, C.R. Robins, H.F. McMurdie: *Phase Diagrams for Ceramists* (Am. Ceram. Soc, Columbus 1964), and the next volumes; recently edited and published by NIST & ACS
12.45 G.D. Gu, M. Hucker, Y.J. Kim, J.M. Tranquada, H. Dabkowska, G.M. Luke, T. Timusk, B.D. Gaulin, Q. Li, A.R. Moodenbaugh: Crystal growth and superconductivity of $(La_{1-x}Ca_x)_2CaCu_2O_{6+\delta}$, J. Phys. Chem. Solids **67**, 431–434 (2006)

12.46 I. Shindo, S. Kimura, K. Kitamura: Growth of Mg_2TiO_4 by the floating zone method, J. Mater. Sci. **14**, 1901–1906 (1979)

12.47 T. Fujii, T. Watanabe, A. Matsuda: Single-crystal growth of $Bi_2Sr_2Ca_2Cu_3O_{10+\delta}$ (Bi-2223) by TSFZ method, J. Cryst. Growth **223**, 175–180 (2001)

12.48 H.A. Dabkowska, A.B. Dabkowski, G.M. Luke, S.R. Dunsiger, S. Haravifard, M. Cecchinel, B.D. Gaulin: Crystal growth and magnetic behaviour of pure and doped $SrCu_2(^{11}BO_3)_2$, J. Cryst. Growth **306**, 123–128 (2007)

12.49 T. Adachi, T. Noji, Y. Koike: Crystal growth, transport properties and crystal structure of the single-crystal $La_{(2-x)}Ba_xCuO_4$ ($x = 0.11$), Phys. Rev. B **64**, 144524-1-6 (2001)

12.50 D. Elwell, H.J. Scheel: *Crystal Growth from High-Temperature Solutions* (Academic, London 1975)

12.51 D.T.J. Hurle: *Crystal Pulling from the Melt* (Springer, Berlin, New York 1993)

12.52 R. Subramanian, M. Higuchi, R. Dieckman: Growth of nickel aluminate single crystals by the floating-zone method, J. Cryst. Growth **143**, 311–316 (1994)

12.53 G.D. Gu, K. Takamuku, N. Koshizuka, S. Tanaka: Growth and superconductivity of $Bi_{2.1}CuSr_{1.9}Ca_{1.0}(Cu_{1-y}Fe_y)_2O_x$ single-crystal, J. Cryst. Growth **137**, 472–478 (1994)

12.54 S. Watauchi, M. Wakihara, I. Tanaka: Control of the anisotropic growth rates of oxide single crystals in floating zone growth, J. Cryst. Growth **229**, 423–427 (2001)

12.55 A. Mühlbauer, A. Muiznieks, J. Virbulis: Analysis of the dopant segregation effects at the floating zone growth of large silicon crystals, J. Cryst. Growth **180**, 372–380 (1997)

12.56 O.B. Raghothamachar, G. Dhanaraj, J. Bai, M. Dudley: Defect analysis in crystals using x-ray topography, Microsc. Res. Tech. **69**, 343–358 (2006)

12.57 K. Kitamura, S. Kimura, Y. Miyazawa, Y. Mori, O. Kamada: Stress-birefringence associated with facets of rare-earth garnets grown from the melt – a model and measurement of stress-birefringence observed in thin-sections, J. Cryst. Growth **62**, 351–359 (1983)

12.58 OrientExpress is a part of LMGP suite for Windows by Jean Laugier and Bernard Bochu (Laboratoire des Materiaux et du Génie Physique de l'Ecole Supérieure de Physique de Grenoble http://www.inpg.fr/LMGP/). Program can be downloaded from the LMGP Crystallography software suite website at: http://www.ccp14.ac.uk/ccp/web-mirrors/lmgp-laugier-bochu/ (last accessed August 6, 2009)

12.59 D. Prabhakaran, P. Isla, A.T. Boothroyd: Growth of large $La_{2-x}Sr_xNiO_{4+delta}$ single crystals by the floating-zone technique, J. Cryst. Growth **237**, 815–819 (2002), Part 1

12.60 S. Haravifard, S.R. Dunsiger, S. El Shawiish, B.D. Gaulin, H.A. Dabkowska, M.T.F. Telling, J. Bonca: In-gap excitation and finite triplet lifetimes in the dilute singlet ground state system $SrCu_{2-x}Mg_x(BO_3)_2$, Phys. Rev. Lett. **97**, 247206 (2006)

12.61 P. Rudolf: *Crystal Growth of Technologically Important Electronic Materials*, ed. by K. Byrappa, T. Ochachi, M. Klapper, R. Fornari (Allied Publishers PVT Ltd., New Delhi 2003) pp. 407–417

12.62 Y. Zhu, Y.M. Niewczas, M. Couillard, G.A. Botton: Single atomic layer detection of Ca and defect characterization of Bi-2212 with EELS in HA-ADF STEM, Ultramicroscopy **106**, 1076–1081 (2006)

12.63 K. Zhang, R. Mogilevsky, D.G. Hinks, J. Mitchell, A.J. Schultz, Y. Wang, V. Dravid: Crystal Growth of $(La,Sr)_2CuO_4$ by float zone melting, J. Cryst. Growth **169**, 73–78 (1996)

12.64 X.-L. Yan, X. Wu, J.-F. Zhou, Z.-G. Zhang, X.-M. Wang: Growth of laser single crystals Er:YVO_4 by floating zone method, J. Cryst. Growth **220**, 543–547 (2000)

12.65 X.-L. Yan, X. Wu, J.-F. Zhou, Z.-G. Zhang, X.-M. Wang, X.-M. Fu, P.-M. Jiang, Y.-D. Hu, J.-D. Hu, J.-L. Qiu: Growth of Tm:Ho:YVO4 laser single crystals by the floating zone method, J. Cryst. Growth **212**, 204–210 (2000)

12.66 R.S. Dusinger, Y. Zao, Z. Yamani, W.J.L. Buyers, H.A. Dabkowska, B.D. Gaulin: Incommensurate spin ordering and fluctuations in underdoped $La_{2-x}Ba_xCuO_4$, Phys. Rev. B **77**, 224410 (2008)

12.67 Y. Zhao, B.D. Gaulin, J.P. Castellan, J.P.C. Ruff, S.R. Dunsiger, G.D. Gu, H.A. Dabkowska: High-resolution x-ray scattering studies of structural phase transitions in underdoped $La_{2-x}Ba_xCuO_4$, Phys. Rev. B **76**, 184121 (2007)

12.68 A. Maljuk, B. Liang, C.T. Lin, G.A. Emelchenko: On the growth of overdoped Bi-2212 single crystals under high oxygen pressure, Physica C **335**, 140–146 (2001)

12.69 N. Britos, A.-M. Lejus, B. Viana, D. Vivien: Crystal growth and spectroscopy of Tm^{3+} doped $Ca_2Al_2SiO_7$, Eur. J. Solid State Inorg. Chem. **32**, 415–428 (1995)

12.70 D. Souptel, G. Behr, A.M. Balbashov: $SrZrO_3$ Single crystal growth by floating zone technique with radiation heating, J. Cryst. Growth **236**, 583–588 (2002)

12.71 K. Dembinski, J.M. Bassat, J.P. Coutures, P. Odier: Crystal growth of La_2NiO_4 by the floating zone method with a CW CO_2 laser – Preliminary characterizations, J. Mater. Sci. Lett. **6**, 1365–1367 (1987)

12.72 M. Saito: Growth process of gas bubble in ruby single crystals by floating zone method, J. Cryst. Growth **74**, 385–390 (1986)

12.73 M. Saito: Gas-bubble formation of ruby single-crystals by floating zone method with an infrared radiation convergence type heater, J. Cryst. Growth **71**, 664–672 (1985)

12.74 S. Sakata, W. Itoyama, I. Fujii, K. Iishi: Preparation of low temperature Li_3VO_4 single crystal by floating zone technique, J. Cryst. Growth **135**, 555–560 (1994)

12.75 Z.Q. Mao, Y. Maeno, H. Fukazawa: Crystal growth of Sr_2RuO_4, Mater. Res. Bull. **35**, 1813–1824 (2000)

12.76 L. Trouilleux, G. Dhalenne, A. Revcolevschi, P. Monod: Growth and anisotropic magnetic behavior of aligned eutectic-type structures in the system $La_{2-x}Sr_xCuO_4$ copper oxide, J. Cryst. Growth **91**, 268–273 (1988)

12.77 R. Fittipaldi, A. Vecchione, D.S. Sisti, S. Pace, S. Kittaka, Y. Maeno: Micro-crystallorgraphic structure of $Sr_2RuO_4/Sr_3Ru_2O_7$ eutectic crystals grown by floating zone method, IUCr XXI Congr. (Osaka 2008), MS13.5 C34

12.78 J.D. Pless, N. Erdman, D. Ko, L.D. Marks, P.C. Stair, K.R. Pöppelmeier: Single-crystal growth of magnesium orthovanadate, $Mg_3(VO_4)_2$ by the optical floating zone technique, Cryst. Growth Des. **3**, 615–619 (2003)

12.79 B.I. Belevtsev, D.G. Naugle, K.D.D. Rathnayaka, A. Parasiris, J. Fink-Finowicki: Extrinsic inhomogeneity effects in magnetic, transport and magnetoresistive properties of $La_{1-x}Ca_xMnO_3$ ($x \approx 0.33$) crystal prepared by the floating-zone method, Physica B: Cond. Matter **355**, 341–351 (2005)

12.80 C. Winkler, G. Amberg, T. Carlberg: Radial segregation due to weak convection in a floating Zone, J. Cryst. Growth **210**, 573–586 (2000)

12.81 J.-C. Chen, G.-H. Chin: Linear stability analysis of thermocapillary convection in the floating zone, J. Cryst. Growth **154**, 98–107 (1995)

12.82 A.A. Aczel, H.A. Dabkowska, P.R. Provencher, G.M. Luke: Crystal growths and characterization of the new spin dimmer system $Ba_3Cr_2O_8$, J. Cryst. Growth **310**, 870–873 (2008)

12.83 J. Furukawa, T. Tsukamoto: $BaTiO_3$ single crystal growth by traveling solvent floating zone technique, Jpn. J. Appl. Phys. **30**, 2391–2393 (1991)

12.84 H. Klapper: *Crystal Growth of Technologically Important Electronic Materials*, ed. by K. Byrappa, T. Ochachi, M. Klapper, R. Fornari (Allied Publishers PVT Ltd., New Delhi 2003) pp. 603–615

12.85 S. Takekawa, H. Nozaki: Single crystal growth of the superconductor $Bi_{2.0}(Bi_{0.2}Sr_{1.8}Ca_{1.0})Cu_{2.0}O_8$, J. Cryst. Growth **92**, 687–690 (1988)

12.86 G. Balakrishnan, D.McK. Paul, M.R. Lees: Superconducting properties of doped and off-stoichiometric $Bi_2Sr_2CaCu_2O_8$ single crystals, Physica B **194–196**, 2197–2198 (1994)

12.87 P. Murugakoothan, R. Jayavel, C.R.V. Rao, C. Subramanian, P. Ramasamy: Growth and characterization of $Bi_2Sr_2Ca_1Cu_2O_y$ by the floating zone method, Mater. Chem. Phys. **31**, 281–284 (1992)

12.88 Y. Huang, B.-L. Wang, M.-Y. Hong, M.-K. Wu: Single crystal preparation of $Bi_2Sr_2CaCu_2O_x$ superconductor by the travelling solvent floating zone method, Physica C **235–240**, 525–526 (1994)

12.89 K. Takamuku, K. Ikeda, T. Takata, T. Miyatake, I. Tomeno, S. Gotoh, N. Koshizuka: Single crystal growth and characterization of $Bi_2Sr_2Ca_{1-x}Y_xCu_2O_y$ by TSFZ method, Physica C **185–189**, 451–452 (1991)

12.90 T. Horiuchi, K. Kitahama, T. Kawai, S. Kawai, S. Hontsu, K. Ogura, I. Shiogaki, Y. Kawate: Li substitution to Bi-Sr-Ca-Cu-O superconductor, Physica C **185–189**, 629–630 (1991)

12.91 B. Liang, A. Maljuk, C.T. Lin: Growth of large superconducting $Bi_{2+x}Sr_{2-y}CuO_{6+\delta}$ single crystals by traveling solvent floating zone method, Physica C: Superconduct. Appl. **361**, 156–164 (2001)

12.92 M. Matsumoto, J. Shirafuji, K. Kitahama, S. Kawai, I. Shigaki, Y. Kawate: Preparation of $Bi_2Sr_2CuO_6$ single crystals by the travelling solvent floating zone method, Physica C **185–189**, 455–456 (1991)

12.93 S. Miyazawa, T. Tabata: Bi_2O_3-TiO_2 binary phase diagram study for TSSG pulling of $Bi_{12}TiO_{20}$ single crystals, J. Cryst. Growth **191**, 512–516 (1998)

12.94 S. Watauchi, I. Tanaka, K. Hayashi, M. Hirano, H. Hosono: Crystal growth of $Ca_{12}Al_{14}O_{33}$ by the floating zone method, J. Cryst. Growth **237**, 801–805 (2002)

12.95 T. Katsumata, T. Nabae, K. Sasajima, T. Matsuzawa: Growth and characteristic of long persistent $SrAl_2O_4$- and $CaAl_2O_4$-based phosphors crystals by floating zone technique, J. Cryst. Growth **83**, 361–365 (1998)

12.96 N.I. Shindo: Single crystal growth of akermanite ($Ca_2MgSi_2O_7$) and gehlenite ($Ca_2Al_2SiO_7$) by the floating zone method, J. Cryst. Growth **46**, 569–574 (1979)

12.97 J. Wada, S. Wakimoto, S. Hosoya, K. Yamada, Y. Endoh: Preparation of single crystal of Ca_2CuO_3 by TSFZ method, Physica C **244**, 193–195 (1997)

12.98 Y.J. Kim, S. Wakimoto, S.M. Shapiro, P.M. Gehring, A.P. Ramirez: Neutron scattering study of antiferromagnetic order in $CaCu_3Ti_4O_{12}$, Solid State Commun. **121**, 625 (2002)

12.99 C.C. Homes, T. Vogt, S.M. Shapiro, S. Wakimoto, M.A. Subramanian, A.P. Ramirez: Charge transfer in the high dielectric constant materials $CaCu_3Ti_4O_{12}$ and $CdCu_3Ti_4O_{12}$, Phys. Rev. B **67**, 092106 (2003)

12.100 L.B. Barbosa, D.R. Ardila, J.P. Andreeta: Growth of double perovskite Ca_2FeMoO_6 crystals by a floating zone technique, J. Cryst. Growth **254**, 378–383 (2003)

12.101 C. Kloc, S.-W. Cheong, P. Matl: Floating-zone crystal growth of perovskite manganites with colossal magnetoresistance, J. Cryst. Growth **191**, 294–297 (1998)

12.102 H. Fukazawa, S. Nakatsuji, Y. Maeno: Intrinsic properties of the mott insulator $Ca_2RuO_{4+\delta}$ ($\delta = 0$) studied with single crystals, Physica B **281**, 613–614 (2000)

12.103 S. Nakatsuji, Y. Maeno: Synthesis and single crystal growth of $Ca_{2-x}Sr_xRuO_4$, J. Solid State Chem. **156**, 26–31 (2001)

12.104 W. Wanyan, X. Yan, X. Wu, Z. Zhang, B. Hu, J. Zhou: Study of single-crystal growth of Tm^{3+}:$CaYAlO_4$ by the floating-zone method, J. Cryst. Growth **219**, 56–60 (2000)

12.105 K. Oka, H. Yamaguchi, T. Ito: Crystal growth of $Ca_{2+x}Y_{2-x}Cu_5O_{10}$ with edge sharing CuO_2 chains by the traveling-solvent floating-zone method, J. Cryst. Growth **229**, 419–422 (2001)

12.106 T. Ito, K. Oka: Growth and transport properties of single crystalline $La_{2-x}Ba_xCuO_4$, Part 1, Physica C **235**, 549–550 (1994)

12.107 G. Balakrishnan, O.A. Petrenko, M.R. Lees, D.M.K. Paul: Single crystals of the anisotropic Kagome staircase compounds $Ni_3V_2O_8$ and $Co_3V_2O_8$, J. Phys. Cond. Matter **16**, L347–L350 (2004)

12.108 G. Dhalenne, A. Revcolevschi, J.C. Rouchaud, M. Fedoroff: Floating zone crystal growth of pure and Si- or Zn-substituted copper germanate $CuGeO_3$, Mater. Res. Bull. **32**, 939–946 (1997)

12.109 S. Hara, Y. Yoshida, S.I. Ikeda, N. Shirakawa, M.K. Crawford, K. Takase, K. Tanako, K. Sekizawa: Crystal growth of germanium-based oxide spinels by the float zone method, J. Cryst. Growth **283**, 185–192 (2005)

12.110 X.L. Yan, J.F. Zhou, X.J. Niu, X.L. Chen, Q.Y. Tu, X. Wu: Crystal growth of La_2CuO_4 and $La_{2-x}Ba_xCuO_4$ by the tavelling solvent floating zone method, J. Cryst. Growth **242**, 161–166 (2002)

12.111 I. Tanaka, H. Kojima: Superconducting single crystals, Nature **337**, 21 (1985)

12.112 I. Tanaka, K. Yamane, H. Kojima: Single crystal growth of superconducting $La_{2-x}Sr_xCuO_4$ by the TSFZ method, J. Cryst. Growth **96**, 711–715 (1989)

12.113 A.M. Balbashev, D.A. Shulyatev, G.K. Panova, M.N. Khlopkin, N.A. Chernoplekov, A.A. Shikov, A.V. Suetin: The floating zone growth and superconductive properties of $La_{1.85}Sr_{0.15}CuO_4$ and $Nd_{1.85}Ce_{0.15}CuO_4$ single crystals, Physica C **256**, 371–377 (1996)

12.114 H. Kojima, Y. Yamamoto, Y. Mori, M.K.R. Khan, H. Tanabe, I. Tanaka: Single crystal growth of superconducting $La_{2-x}M_xCuO_4$ (M = Ca, Sr, Ba) by the TSFZ method, Physica C **293**, 14–19 (1997)

12.115 T. Ito, K. Oka: New technique for the crystal growth of $La_{2-x}Ba_xCuO_4$ ($x \leq 0.5$), Physica C **231**, 305–310 (1994)

12.116 J.D. Yu, J.D. Yanagida, H. Takashima, Y. Inaguma, M. Itoh, T. Nakamura: Single crystal growth of superconducting $La_{2-x}Ba_xCuO_4$ by TSFZ method, Physica C **209**, 442–448 (1993)

12.117 L.S. Jia, X.L. Yan, J.F. Zhou, X.L. Chen: Effect of pulling rates on quality of $La_{2-x}Ba_xCuO_4$ single crystal, Physica C: Supercond. Appl. **385**, 483–487 (2003)

12.118 I. Tanaka, H. Kojima: Single crystal growth of 2-1-4 system superconductors by the TSFZ method, Int. Workshop Superconduct. (Honolulu 1992) pp. 146–149

12.119 G.D. Gu, M. Hucker, Y.J. Kim, J.M. Tranquada, Q. Li, A.R. Moodenbaugh: Single crystal growth and superconductivity of $(La_{1-x}Sr_x)_2CaCu_2O_{6+\delta}$, J. Cryst. Growth **287**, 318–322 (2006)

12.120 U. Ammerahl, A. Revcolevschi: Crystal growth of the spin-ladder compound $(Ca,La)_{14}Cu_{24}O_{41}$ and observation of one-dimensional disorder, J. Cryst. Growth **197**, 825–832 (1999)

12.121 C. Zobel, M. Kriener, D. Bruns, J. Baier, M. Grüninger, T. Lorenz, P. Reutler, A. Revcolevschi: Evidence for a low-spin to intermediate-spin state transition in $LaCoO_3$, Phys. Rev. B **66**, 020402 (2002)

12.122 P. Aleshkevych, P.M. Baran, S.N. Barilo, J. Fink-Finowicki, H. Szymczak: Resonance and non-resonance microwave absorption in cobaltites, J. Phys. Cond. Matter **16**, L179–L186 (2004)

12.123 T.-H. Arima, Y. Tokura: Optical study of electronic structure in Perovskite-type RMO_3 (R = La, Y; M = Sc, Ti, V, Cr, Mn, Fe, Co, Ni, Cu), J. Phys. Soc. Jpn. **64**, 2488–2501 (1995)

12.124 Y. Tomioka, A. Asamitsu, Y. Tokura: Magnetotransport properties and magnetostructural phenomenon in single crystals of $La_{0.7}(Ca_{1-y}Sr_y)_{0.3}MnO_3$, Phys. Rev. B **63**, 024421 (2000)

12.125 A.M. De Leon-Guevara, P. Berthet, J. Berthon, F. Millot, A. Revcolevschi: Controlled reduction and oxidation of $La_{0.85}Sr_{0.15}MnO_3$ single crystals, J. Alloys Compd. **262/263**, 163–168 (1997)

12.126 A. Urushibara, Y. Moritomo, T. Arima, A. Asamitsu, G. Kido, Y. Tokura: Insulator-metal transition and giant magnetoresistance in $La_{1-x}Sr_xMnO_3$, Phys. Rev. B **51**, 14103–14109 (1995)

12.127 D. Prabhakaran, A.I. Coldea, A.T. Boothroyd, S. Blundell: Growth of large $La_{1-x}Sr_xMnO_3$ single crystals under argon pressure by the floating zone technique, J. Cryst. Growth **237**, 806–809 (2002)

12.128 J. Iida, S. Takekawa, N. Kimizuka: Single-crystal growth of $LuFe_2O_4$, $LuFeCoO_4$ and $YbFeMgO_4$ by the floating zone method, J. Cryst. Growth **102**, 398–400 (1990)

12.129 J.L. Mass, J.M. Burlitch, S.A. Markgraf, M. Higughi, R. Dieckmann, D.B. Barber, C.R. Pollock: Oxygen activity depedence of the chromium (IV) population in chromium-doped forsterite crystals grown by the floating zone technique, J. Cryst. Growth **165**, 250–257 (1996)

12.130 D.P. Chen, H.C. Chen, A. Maljuk, A. Kulakov, H. Zhang, P. Lemmens, P.C.T. Lin: Single-crystal growth and investigation of Na_xCoO_2 and $Na_xCoO_2 \cdot yH_2O$, Phys. Rev. B **70**, 024506 (2004)

12.131 C.T. Lin, D.P. Chen, P. Lemmens, X.N. Zhang, A. Maljuk, P.X. Zhang: Study of intercalation/deintercalation of Na_xCoO_2 single crystals, J. Cryst. Growth **275**, 606–616 (2005)

12.132 D. Prabhakaran, A.T. Boothroyd, R. Coldea, N.R. Charnley: Crystal growth of Na_xCoO_2 under different atmospheres, J. Cryst. Growth **271**, 74–80 (2004)

12.133 L.S. Jia, X.L. Yan, X.L. Chen: Growth of semiconducting Nd_2CuO_4 and as-grown superconductive $Nd_{1.85}Ce_{0.15}Cu_{4-z}$ single crystals, J. Cryst. Growth **254**, 437–442 (2003)

12.134 G. Balakrishnan, M.R. Lees, D.M. Paul: Single-crystal growth and properties of the double-layered manganese oxides, J. Phys. Cond. Matter **9**, L471–L474 (1997)

12.135 A.S. Sefat, J.E. Greedan, G.M. Luke, M. Niewczas, J.D. Garrett, H. Dabkowska, A. Dabkowski: Anderson-Mott transition induced by hole doping in $Nd_{1-x}TiO_3$, Phys. Rev. B **74**, 104419 (2006)

12.136 T. Ito, K. Oka: Crystal growth of $REBa_2Cu_3O_{7-y}$ and ambient atmosphere, Physica C **235**, 355–356 (1994), Part 1

12.137 N.T. Hien, J.J.M. Franse, J.J.M. Pothuizen, A.A. Menovsky: Growth and characterization of bulk $Sr_2CuO_2Cl_2$ single crystals, J. Cryst. Growth **171**, 102–108 (1997)

12.138 A. Revcolevschi, A. Vietkine, H. Moudden: Crystal growth and characterization of chain cuprates $SrCuO_2$, Sr_2CuO_3 and spin-ladder $Sr_{14}Cu_{24}O_{41}$, Physica C **282-287**, 493–494 (1997)

12.139 N. Ohashi, K. Fujiwara, T. Tsurumi, O. Fukunaga: Growth of orthorhombic $SrCuO_2$ by a travelling solvent floating zone method and its phase transformation under high pressure, J. Cryst. Growth **186**, 128–132 (1998)

12.140 H. Kageyama, K. Onizuka, T. Yamauchi, Y. Ueda: Crystal growth of the two-dimensional spin gap system $SrCu_2(BO_3)_2$, J. Cryst. Growth **206**, 65–67 (1999)

12.141 U. Ammerahl, G. Dhalenne, A. Revcolevschi, J. Berthon, H. Moudden: Crystal growth and characterization of the spin-ladder compound $(Sr,Ca)_{14}Cu_{24}O_{41}$, J. Cryst. Growth **193**, 55–60 (1998)

12.142 A. Maljuk, J. Strempfer, C. Ulrich, M. Sofin, L. Capogna, C.T. Lin, B. Keimer: Growth of $Sr_3Fe_2O_{7-x}$ single crystals by the floating zone method, J. Cryst. Growth **273**, 207–212 (2004)

12.143 Y. Maeno, H. Hashimoto, K. Yoshida, S. Nishizaki, T. Fujita, J.G. Bednorz, F. Lichtenberg: Superconductivity in a layered Perovskite without copper, Nature **372**, 532–534 (1994)

12.144 R.S. Perry, Y. Maeno: Systematic approach to the growth of high-quality single crystals of $Sr_3Ru_2O_7$, J. Cryst. Growth **271**, 134–141 (2006)

12.145 R. Fittipaldi, A. Vecchione, S. Fusanobori, K. Takizawa, H. Yaguchi, J. Hooper, R.S. Perry, Y. Maeno: Crystal growth of the new Sr_2RuO_4-$Sr_3Ru_2O_7$ eutectic system by a floating-zone method, J. Cryst. Growth **282**, 152–159 (2006)

12.146 K. Uematsu, O. Sakurai, N. Mizutani, M. Kato: Electrical properties of La-doped $SrTiO_3$ (La: 0.1 to 2.0 at %) single crystals grown by xenon-arc image floating zone method, J. Mater. Sci. **19**, 3671–3679 (1984)

12.147 A. Sugimoto, Y. Nob, K. Yamagishi: Crystal-growth and optical characterization of Cr,Ca-$Y_3AL_5O_{12}$, J. Cryst. Growth **140**, 349–354 (1994)

12.148 I. Shindo, N. Kimizuka, S. Kimura: Growth of YFe_2O_4 single-crystals by floating zone method, Mater. Res. Bull. **11**, 637–643 (1976)

12.149 H.A. Dabkowska, A. Dabkowski, G.M. Luke, B.D. Gaulin: Crystal growth structure and magnetic behaviour of ytterbium cobalt gallium oxide $YbCoGaO_4$, J. Cryst. Growth **234**, 411–414 (2002)

12.150 K. Oka, H. Shibata, S. Kashiwaya: Crystal growth of ZnO, J. Cryst. Growth **237**, 509–513 (2002)

13. Laser-Heated Pedestal Growth of Oxide Fibers

Marcello R.B. Andreeta, Antonio Carlos Hernandes

The laser-heated pedestal growth (LHPG) technique, when compared with conventional growth methods, presents many advantages, such as high pulling rates, a crucible-free process, and growth of high and low melting point materials. These special features make the LHPG technique a powerful material research tool. We describe the background history, theoretical fundamentals, and how the features of LHPG affect the growth of oxide fibers. We also present a list of materials processed by laser heating in recent decades, such as $LiNbO_3$, Sr–Ba–Nb–O, $Bi_{12}TiO_{20}$, Sr_2RuO_4, Bi–Sr–Ca–Cu–O, ZrO_2:Y_2O_3, $LaAlO_3$, and also the eutectic fibers of Al_2O_3:$GdAlO_3$, Al_2O_3:Y_2O_3, and ZrO_2:Al_2O_3.

Laser technology has been applied for many years to process a wide range of materials and devices. The characteristics that make laser radiation very attractive for this purpose are its coherence and collimation, which allow the adjustment of the energy distribution with conventional optical elements as well as location of the heat source outside the preparation chamber. The main applications of laser materials processing technology are: perforation, cutting and welding, sintering, surface thermal treatment, unidirectional solidification applied to fiber pulling, material texturing, and production of aligned eutectic materials.

Among all the laser-heated crystalline fiber-pulling techniques that have appeared over the last 40 years, laser-heated pedestal growth (LHPG) has become one of the most powerful tools in new and conventional materials research. The main advantages of this technique are its high pulling rates (about 60 times faster than conventional Czochralski technique) and the possibility to produce very high melting point materials. Besides that, it is a crucible-free technique, which allows the pulling of high-purity single crystals and composite fibers, avoiding mechanical stress and contamination due to the crucible material during the solidification process.

In addition to all of these advantages of the LHPG technique, the crystal geometric shape, its flexibility for small diameters, and its low cost make single-crystal fibers (SCF) produced by LHPG even more suitable to substitute for many of today's bulk devices, especially with respect to high melting point materials. However, for this purpose the SCF must have equal or superior optical and structural qualities compared with bulk devices.

In this chapter we describe the laser-heated pedestal growth technique as a powerful materials research tool and its features, such as the high axial thermal gradients at the solid–liquid interface. In order to produce this chapter, the authors analyzed more than 300 original scientific papers dealing with laser-heated pedestal growth. The publication dynamics on laser-heated fiber growth using the laser-heated floating-zone-like technique are presented in Fig. 13.1. Those are the results of the last two decades of research. From these data we can verify the great increase in publication rate over the last two decades, showing the importance of this technique in materials science. We also present the historical evolution, theoretical fundamentals, and how the unique features of the LHPG technique affect the quality of the pulled fibers and make it possible to pull incongruently melting and evaporating materials, as well as its influence on the microstructures of eutectic oxide composites.

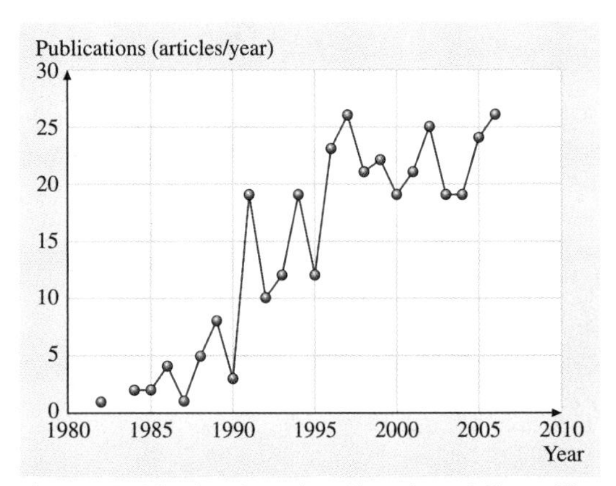

Fig. 13.1 Publication dynamics of laser-heated fiber pulling research

13.1 Fiber-Pulling Research

The first publications on fiber preparation from the melt date back to the beginning of the 20th century and were restricted to research on metal wires. The origin of this research started with the physicists *Baker* [13.1] and *da Costa Andrade* [13.2], who wrote in his personal reminiscences: "I continued my work on the creep of metals and accidentally made the first single crystal metal wires" [13.3]. However, in the same period, there was also the need to understand one of the topical areas of physical chemistry that had lasted for 50 years: the measurement of crystallization velocities [13.4,5]. It was with this idea in mind that *Czochralski* presented in 1918 a new *method for measurement of the crystallization velocity of metals* where he actually pulled fibers of low melting point metals from their melts [13.6]. Soon after this, *Gomperz* [13.7] used floating dies to control the fiber diameter [an early Stepanov or edge-defined film-fed growth (EFG) process] [13.5].

In the 1950s, single-crystal fiber attracted the attention of researchers due to its high crystalline perfection and outstanding mechanical properties. Those fibers were produced mainly by vapor-phase reactions [13.8]. Research into fiber pulling from the melt, especially of nonconducting materials, remained somewhat latent for approximately 20 years until the end of World War II (1945), the transistor revolution (1947), and the creation of the laser (1960). The works of *Labelle* and *Mlavsky* [13.9,10] on the pulling of sapphire fibers from the melt opened a new perspective in fiber research. In their work of 1971 they grew sapphire fibers by the now famous die-shapes growth method known as edge-defined film-fed growth (EFG) [13.8]. Due to its outstanding mechanical strength, these sapphire fibers were considered good candidates to be used as structural composites.

It was only in 1972 that fiber pulling from the melt received a major contribution. Due to the interest of industry in creating a technique for producing fibers with small diameter, high mechanical strength, and high purity, the floating zone technique was considered. It was *Haggerty* [13.11] who developed the method of four focused laser beams to create a laser-heated floating zone for fiber production. However, this was not the first time that lasers had been used in the growth of bulk samples, as demonstrated by the work of *Cockayne* and *Gasson* [13.12]. Haggerty grew fibers of Al_2O_3, Y_2O_3, TiC, and TiB_2 and investigated their mechanical properties. He concluded that, even without optimization of the growth process, the room-temperature strength of $Cr:Al_2O_3$ fibers could reach $9.65\times10^9\ N/m^2$, which had previously been observed only with small, *whisker* single crystals.

In the late 1970s, the development in fiber-optic communication and the need for miniaturization of optical devices, such as miniature solid-state lasers, provided a major impulse for the laser-heated fiber growth technique. During this period, we can highlight the works of *Burrus*, *Stone*, and *Coldren* on the production of high melting point oxide fibers (Nd:YAG, Nd:Y_2O_3, and Al_2O_3:Cr) and also the room-temperature operation of miniature lasers using Nd:YAG and Nd:Y_2O_3 fibers grown by the two focused laser beam heated pedestal growth method [13.13–15].

Until 1980, the laser-heated technique applied to crystal growth used only two or four laser beams focused over the source material. This condition usually leads to the creation of large radial thermal gradients in the molten zone that could be deleterious for growing crystal [13.8]. The main contribution on the subject of radial thermal flow in the laser-heated crystal growth technique was made by *Fejer* and *Feigelson* [13.16, 17]. They incorporated into the focusing system a special optical component known as a *reflaxicon* [13.18], consisting of an inner cone surrounded by a larger coaxial cone section, both with a reflective surface, which converts the cylindrical laser beam into a larger-diameter hollow cylinder surface. This optical component allowed the energy to be distributed radially over the molten zone, reducing the radial thermal gradients [13.19]. This new version of the laser floating zone technique became known as laser-heated pedestal growth, or simply LHPG.

In the setup based on the Stanford version of the LHPG technique, a CO_2 laser (continuous wave (CW), $\lambda = 10.6\,\mu m$) or Nd:YAG laser (CW, $\lambda = 1.06\,\mu m$) can be used with the laser cavity cooled by water or air flow, depending on the nominal laser power. It is quite common to use a He–Ne laser, propagating parallel to CO_2 laser beam, to act as a guide for the eye in the optical alignment. The laser beams are guided into a closed chamber through a ZnSe window and hit the reflaxicon [13.18], where they are converted as mentioned above into a cylindrical shell and guided to a spherical or parabolic mirror, and focused over the source material as shown in Fig. 13.2. The focusing mirror has a hole in its center to allow the seed holder to be placed in its optical axis and also to avoid vapor material deposition onto the optical mirrors during fiber pulling. The use of a growth chamber allows for a controlled growth atmosphere (special gases or vacuum). There are normally two windows in the growth chamber for visualization of the growth process: it is possible to see the molten zone by using magnifying lenses and with a video system mounted 90° off-axis from the first one. This video system has a microscope attached to it, allowing the viewing of the molten zone and the growing fiber (Fig. 13.3).

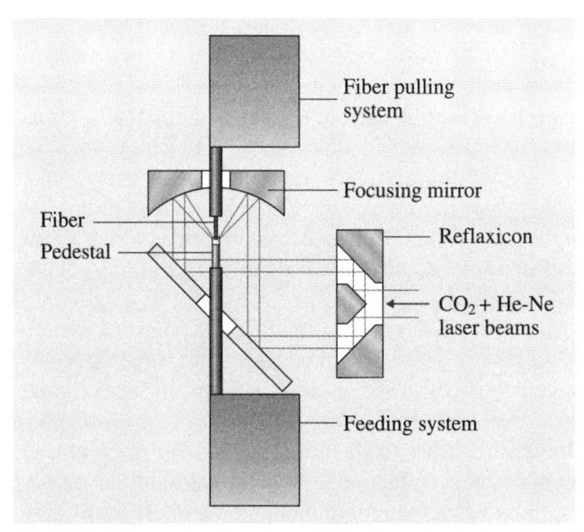

Fig. 13.2 Schematic drawing of the laser-heated pedestal growth (LHPG) technique (after [13.16–19])

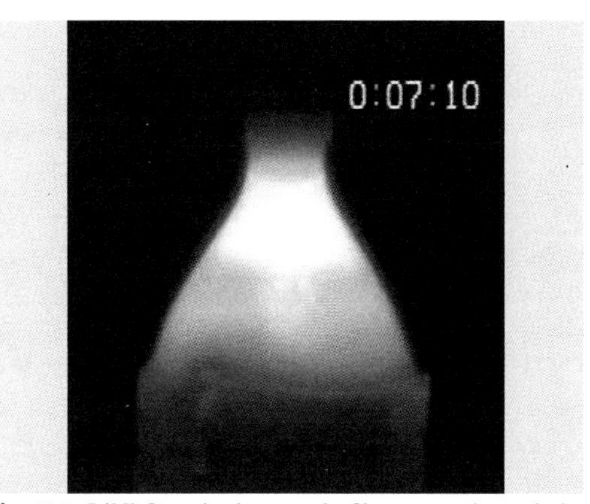

Fig. 13.3 $LiNbO_3$ single-crystal fiber growth and its molten zone by the LHPG technique

Since the implantation of the LHPG system at Stanford University laboratories, several modifications have been proposed to improve it. Table 13.1 lists some of these modifications in the LHPG technique over the last decades. The first modification to the LHPG system was made by *Fejer* himself (1985) when he implemented high-speed diameter control of the fiber during the pulling process [13.20]. Diameter control proved to be essential to obtain high-quality single-crystal fibers and also to improve the efficiency of resulting fiber-optical devices (such as waveguides and solid-state lasers) by decreasing the light scattering at the surface of the fiber.

Due to the fact that the LHPG technique has very localized heating, it generates a very large thermal gradient at the growth interface (on the order of

Table 13.1 Laser-heated fiber-pulling technique: evolution/modifications over the past decades

Year	Authors	Improvement	Reference
1971	Haggerty	Four laser beams focused at the molten zone	[13.11]
1982	Fejer et al.	Reflaxicon – LHPG	[13.16]
1985	Fejer et al.	Fiber diameter control	[13.20]
1992	Uda and Tiller	Resistive afterheater	[13.21]
1993	Sugiyama et al.	Optical afterheater	[13.22]
1994	Phomsakha et al.	Gaussian reflector	[13.23]
1995	Yokoo et al.	Indirect laser heating (organic compounds) – ILHPG	[13.24]
1995	Imai et al.	Gas flux to increase axial thermal gradient	[13.25]
1996	Brueck et al.	Ultrahigh-vacuum LHPG	[13.26]
1997	Brenier et al.	Ferroelectric domain inversion by in situ electric field poling	[13.27]
1997	Nubling and Harrington	New reflaxicon without shadows in the molten zone	[13.28]
1999	Ardila et al.	LHPG in a high isostatic pressure environment	[13.29]
2001	Ardila et al.	Bifocal spherical mirror	[13.30]
2001	Laversenne et al.	Gradient crystals (pedestal preparation)	[13.31]
2002	Andreeta et al.	Thermal gradient control by the modification of the focal spot	[13.32]
2003	Andreeta et al.	Automatic diameter control based on artificial vision	[13.33]
2004	Carrasco et al.	Electrical-assisted LHPG – EALHPG	[13.34]
2005	Lo et al.	Double-clad fiber crystal	[13.35]

10^3–10^4 °C/cm). *Uda* and *Tiller* [13.21] used a Pt–Rh(10%) resistive afterheater in order to control the axial thermal gradient at the solid–liquid growth interface [13.21] (Fig. 13.4a). They showed that Cr^{3+} doping concentration in $LiNbO_3$ single-crystal fiber is affected by an electric field generated at the growing interface, which is proportional to the axial temperature gradient. They could also calculate the diffusion coefficient for Cr in the liquid phase (3.8×10^{-6} cm^2/s). *Sugiyama* et al., in 1993 [13.22], proposed the use of an optical afterheater in order to create a more homogenous temperature distribution around the growing fiber, to allow the pulling of larger-diameter (0.5–1.0 mm) high-quality *a*-axis strontium barium niobate (SBN) fibers for holographic applications (Fig. 13.4b). Larger-diameter fibers presented optical inhomogeneity resulting from internal stress induced during solidification and from anisotropic growth kinetics in the radial direction. Sugiyama et al. reported the improvement in the fiber morphology and also in the fiber composition homogeneity with the use of the afterheater. In the Fejer version of the LHPG technique, the arrangement results in very tight focusing on the shoulder of the molten zone. *Phomsakha* et al. proposed the use of a Gaussian reflector [13.23] to substitute the reflaxicon (Fig. 13.4c). In this way, they achieved a condition for sapphire fiber pulling where not only is the laser radiation spread more evenly over the molten zone but it also impinges on a short length of the newly grown fiber. In this configuration the fiber-pulling limit for high-quality sapphire fibers was extended from 2–3 mm/min to as high as 20 mm/min using He atmosphere. The authors believe that the appropriate distribution of the laser beam in the vicinity of the molten zone is important in minimizing the magnitude of convection currents, which could improve the sapphire fiber quality and allow greater pulling speed.

In 1995 *Yokoo* et al. modified the LHPG system in order to grow low melting point organic compounds [13.24]. The LHPG system in its original configuration cannot be applied directly to grow organic single fibers due to their very low melting point and poor thermal conductivity. Even though sublimation occurs more easily due to direct and localized heating by the laser in those compounds, this does not allow for stabilization of the molten zone. The new configura-

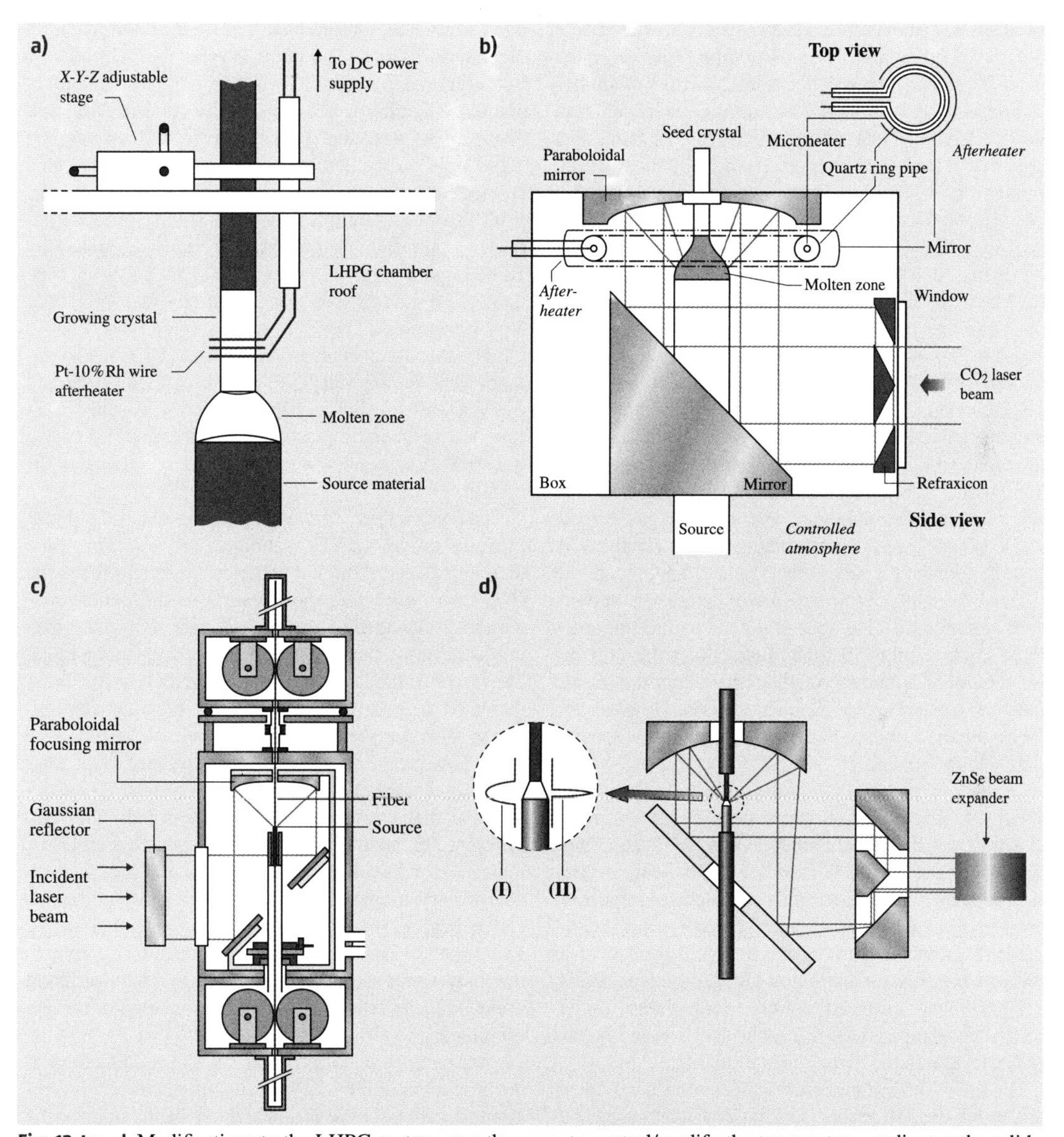

Fig. 13.4a–d Modifications to the LHPG system over the years to control/modify the temperature gradient at the solid–liquid interface. (**a**) Resistive afterheater (after [13.21]), (**b**) optical afterheater (after [13.22]), (**c**) Gaussian reflector replacing the reflaxicon device (after [13.23]) and (**d**) Creation of a controlled optical aberration at the focusing point; (I) divergent laser beam and (II) parallel laser beam (after [13.32])

tion is based on the insertion of a glass tube to stop the laser beam before it reaches the material to be melted, and the introduction of a gas flow. The authors called this new method indirect laser-heated pedestal growth (ILHPG). The source rod and seed crystal are inserted into the glass tube from opposite sides of the tube. The CO_2 laser beam is focused onto the glass tube circularly with an axially symmetric irradiance pattern. The laser beam is absorbed by the glass and retransmitted by black-body radiation emission from its inner surface.

As a result, direct and localized heating of the source rod is avoided. With this configuration they were able to produce single-crystal fibers of 2-adamantylamino-5-nitropyridine (AANP). The authors were able to produce AANP crystal fibers with lengths of more than 20 mm and diameters ranging from 300 μm to 2 mm. In addition, it has been experimentally confirmed that the efficiency of second-harmonic generation (SHG) for rod-like AANP crystal grown by the ILHPG method is higher than that for conventional AANP crystal grown by Bridgman–Stockbarger method.

While many researchers were making efforts to decrease or homogenize the axial and radial temperature gradient at the solid–liquid interface, respectively, *Imai* and coworkers (1995), for the growth of potassium niobium tantalate (KTN) fibers, went in the opposite direction [13.25], observing that the thermal axial gradient was not high enough to prevent constitutional supercooling. The authors constructed a gas blower whose nozzle encircled the fiber symmetrically. The flow surrounds the fiber, the molten zone, and the pedestal-like tube. With this modification the authors were able to pull KTN fiber of 400 μm at pulling rates up to 0.4 mm/min, 50 times faster than the conventional top seed solution growth method, suppressing the constitutional supercooling phenomenon. However the fibers still had cracks and showed striations, located mainly at the center.

Brueck et al., in 1996, used an ultrahigh-vacuum (UHV) modified LHPG to grow metal fibers, returning, in a way, to the birth of fiber research [13.26]. They grew UPt_3 and UNi_2Al_3 in order to study their electric property, which is strongly affected by impurities and oxidation. The fibers were about 600 μm in diameter and showed superconductivity behavior down to 400 and 600 mK for UPt_3 and UNi_2Al_3, respectively. UNi_2Al_3 fiber presented a higher temperature transition when compared with Czochralski single crystals (300 mK) due to the higher crystal quality.

In 1997 *Nubling* and *Harrington* introduced a new version of the reflaxicon [13.28]. The inner cone was made from a ZnSe window with an Ag-coated diamond-turned cone in the center. By using a window and reflecting cone, they eliminated the need for supporting spokes for the cone and thus the laser beam was not obscured. The entire optical system could produce a calculated spot size as small as 22 μm. For larger-diameter fiber they were able to adjust the spacing between the reflaxicon elements so that it was possible to create a larger focused spot that led to more stable growth. In the same year, *Brenier* et al. developed a method for producing periodic poling lithium niobate (PPLN) [13.27] single-crystal fiber. They added two identical tungsten electrodes, 250 or 750 μm in diameter, parallel to the growth direction (a-axis) and 6 mm apart, near the growth interface. They applied a 200 V periodic alternating voltage signal and created an electric field parallel and antiparallel to the c-axis. With this configuration they were able to produce periodic alternating ferroelectric domains separated by 1.68 μm with a typical period of 204 ms for the electric field and a pulling speed of 120 mm/h. In 1999 *Ardila* et al. developed a system to grow single-crystal fiber in a high isostatic pressure environment [13.29]. The authors have shown that fiber growth of $LiNbO_3$:Er and $Ca_3V_2O_8$:Er are favored by a high O_2 isostatic pressure during the pulling process. In the case of $Ca_3V_2O_8$ fiber the best results were obtained by a 10 bar O_2 pressure.

Two years later, *Ardila* et al. proposed another modification to the LHPG technique [13.30]. This time they introduced a bifocal spherical mirror to focus the CO_2 laser. With this modification to the optical system, they managed to create an optical afterheater with the same laser beam as used to melt the compound. They were able to grow c-axis $LiNbO_3$ with 2 mm diameter without cracks, doubling the diameter obtained with the previous focusing mirror. In the same year, *Laversenne* et al. [13.31] developed a process to produce concentration-gradient crystals. They describe a methodology to produce pedestals with two different composition (or concentration) regions. During the growth process the liquid phase continuously changes its composition, thus allowing crystallization at the beginning with one composition and finishing with another. This means that, in the same experiment, one may create a *spectroscopic library* (with different doping concentrations) for the determination of the optimum desired properties.

Andreeta et al. proposed in 2002 a modification to the alignment of the LHPG system that allowed the creation of an optical aberration at the focusing point of the CO_2 laser beam [13.32]. In this way it was possible to decrease the axial gradient while maintaining the annular radial distribution in the fiber growth. The authors have shown that the introduction of a 1° deviation from the original parallel laser beam entering the reflaxicon could expand the width of the laser beam focused at the molten zone by a factor of 10 (Fig. 13.4d). This leads to a reduction of about 1000 °C/cm in the axial temperature gradient at the growing interface (from 3800 °C/cm to 2700 °C/cm) for the Bi–Sr–Ca–Cu–O (BSCCO) system used as a model material. The following year *Andreeta* et al. published an article describing another approach for automatic diameter control in the LHPG technique [13.33]. In this new approach an optical image of the molten zone is digitized and its height and the fiber diameter are monitored by a microcomputer, in an attempt to effectively control the shape of the molten zone and not only the fiber diameter itself.

In 2004, *Carrasco* et al., using a Nd:YAG-LHPG technique, introduced an electric current flow through the crystallization interface in BSCCO fiber pulling [13.34]. They showed that the direction of current flux modifies the fiber quality and also that the choice of appropriate current direction and intensity can improve the texture of the BSCCO fibers produced.

Finally, in 2005, *Lo* et al. [13.35] developed a method to produce YAG fibers with a clad structure by introducing a previously grown fiber with a diameter of 68 μm into a fused-silica capillary tube with 76 and 320 μm inner and outer diameter, respectively. They then applied the LHPG technique again in the tube with the fiber and pulled in the downward direction. The result was a core (YAG fiber with 25 μm in diameter) and two more layers. The outer layer proved to be of SiO_2 whereas the inner layer was a mixture of YAG and SiO_2.

The history of fiber research is very rich and, as mentioned earlier, started before many of the traditional crystal growth techniques, such as Czochralski. The introduction of laser heating for the production of fiber-shaped materials expanded the previous physical limitations for new materials development. High melting point materials could be developed with minimal material lost. Due to its versatility and to achieve the fiber quality needed for some applications, almost once a year a new version/modification appears in the literature for the LHPG technique. Some of the changes and proposals for modification to the LHPG system described earlier are restricted to the developers' laboratory, mainly due to difficulties in their implementation. However, all of them slightly extended the potential of the LHPG technique to create new features for the development of new materials and also to improve the quality of known materials.

13.2 The Laser-Heated Pedestal Growth Technique

The LHPG technique, as described in Sect. 13.1, is basically a miniature floating-zone-like method where the heating element is substituted by a focused laser ring to generate the molten zone. The pulling process starts with basically four steps inside the growth chamber, as illustrated in Fig. 13.5. The first step is to align the seed and the pedestal mechanically, both centralized in the optical axis of the laser beam. The next step is to create a small molten zone on top of the pedestal, by turning on the laser and slowly increasing the laser power. In the following step the seed is introduced into the liquid and a molten zone is formed. Finally, the motors start the fiber pulling and rotation, as indicated in step IV of Fig. 13.5. In the following sections the experimental aspects of this technique are presented in terms of source and

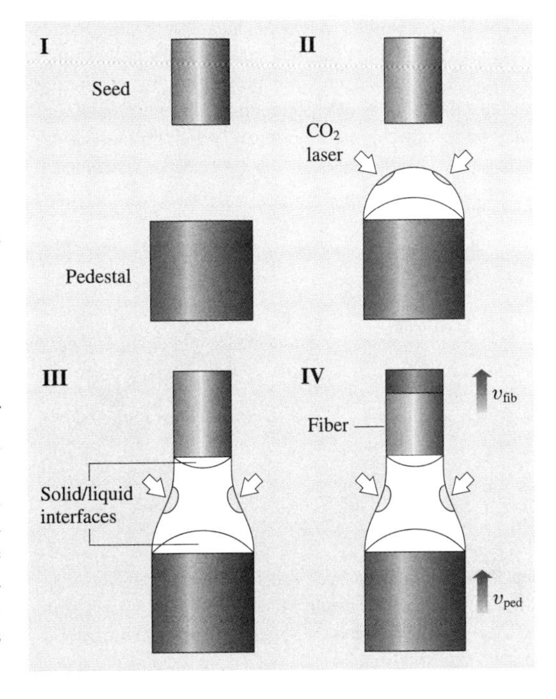

Fig. 13.5 Schematic drawing of the four steps performed in the LHPG technique: (I) mechanical alignment of the seed and the pedestal and (II) creation of a small molten zone on the top of the pedestal. (III) The seed is then introduced in the liquid phase and there is creation of a molten zone. (IV) The motors are started and the fiber-pulling process begins ▶

seed preparation and automatic diameter control (ADC) systems.

13.2.1 Source Preparation and Seeding

The source material (pedestals) used to grow single-crystal fibers can be prepared by cutting a previous bulk-grown crystal or dense ceramic or by the cold extrusion process. Pedestals obtained from bulk crystals or dense ceramics usually have a square section, which is not the best geometry to fit the cylindrical focusing laser beam. In this way one more step is needed before the fiber growth process, if one wants to avoid the corners of the square section pedestal going inside the liquid phase during the pulling process, destabilizing the contact angle. To avoid this effect it is necessary to round the pedestal.

In the cold extrusion process, an organic binder (such as polyvinyl alcohol) is added to the material of reacted or unreacted powder until it achieves a plastic-like consistency, and then it is forced, with a manual press, through a 5 mm long 1 mm diameter stainless-steel cylindrical tube [13.36]. This source material does not need to be cut, as it already has a convenient geometric shape. The pedestals are then allowed to dry at ambient temperature for at least 3 h, and can then be used as a source for crystal growth. This method for producing source materials has many advantages over conventional approaches, such as the small amount of materials needed and the possibility of producing source rods from reagents (green-rods), which also enables better stoichiometric control. When green rods are used in the LHPG system, the three steps in material preparation are reduced to just one (synthesis, sintering, and crystal growth). In previous works, we have shown that the fiber quality of high melting point materials is related to the pedestal heat treatment previous to the growth process [13.37, 38]. The longer the pedestal heat treatment prior to the growth process, the darker the resulting fiber. The origin of the dark color of the fiber may be related to the creation of oxygen vacancies when the pedestal is processed at high temperatures, since changes in the stoichiometry were within experimental error. Crystals grown from an unreacted source (cold extrusion process) are colorless and show good transparency, as shown in Fig. 13.6.

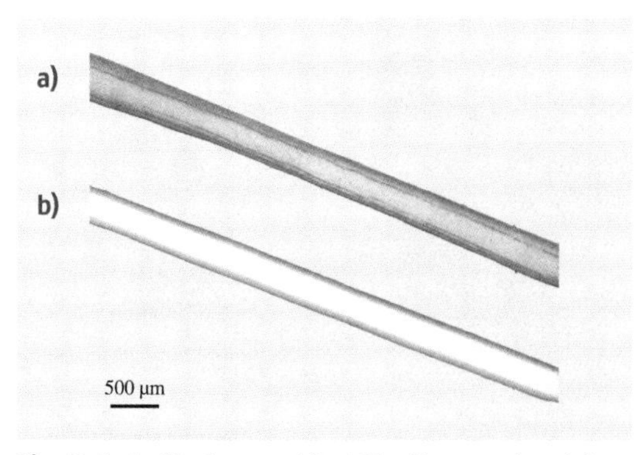

Fig. 13.6a,b Single-crystal $LaAlO_3$ fibers produced from different types of pedestals: (**a**) $LaAlO_3$ fiber grown from prereacted pedestal, and (**b**) $LaAlO_3$ fiber grown from unreacted pedestal (*green-rod*)

Another feature of the LHPG technique is the possibility to prepare pedestals with compositional variation, such as in the method developed by *Laversenne* et al. [13.31] to create a different doping concentration along the fiber axis. This method was also used by *Barbosa* et al. [13.39] by conveniently joining two cold-extruded sections of $CaMoO_4$ and $SrMoO_4$. The obtained fiber was pure $CaMoO_4$ at one end and pure $SrMoO_4$ at the other, both single crystals, with a continuous lattice parameter variation between the two pure sections (a gradient crystal fiber). This kind of crystal can be useful for x-ray optics applications.

The seeds used in the LHPG system must have the same characteristics as for any other crystal growth technique. However, since LHPG is a very powerful tool for new materials development, it is usual to use wires (Pt, Au, etc.) or small pieces of the pedestal (polycrystalline material) as a startup seed for the pulling process due to the impossibility of obtaining previous oriented crystals by any other growth technique. As would be expected, it has been shown that the seed plays a major role in the quality of the grown fiber. *Lu* et al. [13.40] studied seeding effects in the Bi–Sr–Ca–Cu–O system. They used Pt and Au wires to start the pulling process. The use of wires of Pt creates a very high thermal gradient at the wire–melt interface, thus nucleating grains that are aligned with the preferential growth direction perpendicular to the wire surface. While most of the grains grow out of the fiber, some of them grow parallel to the fiber axis and later dominate the grain structure. Grain selection usually takes place in the first 1–4 mm of the grown fiber, and depends on the diameter. For the same materials Au wires did not show the same results. The difference between Pt and Au wire seeding, according to the authors, is due to the reduced wettability or higher thermal conductivity of Au. Due to the considerably larger thermal conductivity of gold wires, the laser power (melt temperature) had to be increased significantly to cause the melt to adhere to the gold wire, and then decreased abruptly to create a normal-sized molten zone. In this way, higher-temperature phases dominate the beginning of the pulling process (CaO + liquid and $CaSrCu_2O_4$ + liquid). *Chen* and coworkers [13.41] have shown that it is possible to pull stoichiometric $LiNbO_3$ single-crystal fibers from pedestals with 50 mol % Li_2O, if a seed with 58 mol % Li_2O is used. Using this strategy, in the beginning of the growth process an excess of Li_2O is inserted into the molten zone due to a small fraction of the seed that melts. In this way, since the early stages of the growth process, the crystal is allowed to solidify to 50 mol % Li_2O. In other words, when the solidified crystal composition reaches 50 mol % Li_2O, the fiber will have the same composition as the pedestal (source).

Another interesting work was performed by *Ishibashi* et al. [13.42], who studied facet suppression in the growth of YAG:Ca:Cr fiber by the LHPG technique. This is a major issue for optimization of solid-state laser efficiency and holographic recording. In order to improve the laser operation efficiency, the authors studied the best crystallographic seed orientation in the growth of YAG:Ca:Cr fibers. They found that facet suppression occurs for a crystallographic oriented seed at 15° from $\langle 100\rangle$ to $\langle 110\rangle$. For SBN single-crystal fibers, *Sugiyama* et al. [13.22] reported that the only way to obtain ridge fiber (facet suppression) is to use a ridge-fiber crystal as a seed.

13.2.2 Automatic Diameter Control Applied to LHPG

Despite the precautions intended to ensure stable growth conditions, as will be described in the following sections, it is found that excessive variations in the fiber diameter occur during open-loop growth. It is thus necessary to design closed-loop systems to regulate fiber diameter during the growth process. These systems contain basically two major components: a real-time dimension measurement apparatus and a proportional controller feeding back to the laser and/or motor speed ratio. Two main approaches have been developed to control the fiber diameter in the LHPG system. The first is based on quasisinusoidal interference fringes as the measurement technique observed in the far field of a fiber side-illuminated by a beam of coherent light [13.20]. The second method is based on artificial vision of the molten zone of the growing fiber [13.33].

The theoretical diameter variation resolution $(\Delta\phi/\phi)_{\min}$ of the system based on the interference fringes can be obtained from a ray-tracing analysis, as shown in (13.1)

$$\left(\frac{\Delta\phi}{\phi}\right)_{\min} = \left(\frac{s}{f}\right) g(n,\theta)\,, \tag{13.1}$$

where s is the distance between two adjacent elements in a photodiode array, f is the focal length of the lens that projects the interference pattern onto the photodiode array, and $g(n,\theta)$ is a geometrical factor dependent on the refraction index of the fiber n and the angle between the laser beam and the detector θ. This factor is obtained from ray-tracing analysis and is typically between 0.4 and 1.0. The quantity s/f can be described as the angular resolution of the optical system. The monotonic improvement in the resolution with increasing focal length predicted by (13.1) is eventually limited by signal-to-noise considerations. *Feyer* et al. [13.20] have shown that this ADC system allows the production of fibers with diameter fluctuations as small as 0.02%.

Another way to introduce an automatic diameter control system is to use the measurements of a charge-coupled device (CCD) camera image of the growing fiber as a feedback for the CO_2 laser and the fiber-pulling motor [13.33]. The video hardware captures the image, which is recorded and processed by a personal computer (PC) in gray scale mode. The black-body radiation from the molten zone and the high temperature gradient at the melting (pedestal) and crystallization (fiber) interfaces (10^3–10^4 °C/cm) make it easy to identify those interfaces positions, due to the good contrast in the image at those points in the LHPG technique. The software uses those positions to measure the fiber diameter and the height of the molten zone. In this way, two predefined areas in the image are measured. The first area is a vertical rectangle (3 × 354 pixels) and the

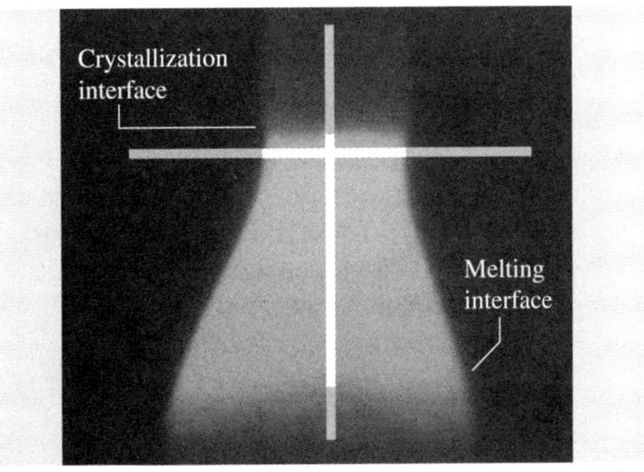

Fig. 13.7 Measurements on a captured image of a growing $LaAlO_3$ fiber (after [13.33])

other one is a horizontal rectangle (354×3 pixels), as shown in Fig. 13.7. The software recognizes the value of all the pixels (from 0 to 255) inside these areas. The next step is to count the number of white pixels inside the predefined area (molten material). The number of these pixels is proportional to the height of the molten zone (vertical) and to the fiber diameter (horizontal). Based on the vertical pixel count, the software recognizes the position of the crystallization interface, using this information to take the horizontal measurement just a few pixels below the crystallization interface (because of the better contrast on the molten zone). After selecting the desired diameter and molten zone height, the software modifies the laser power and pulling rate to match the number of pixels established by the crystal grower (the set point). One advantage of this system is that it is independent of the material used (opaque or transparent). Diameter control of better than 2% was obtained for growth of $LaAlO_3$ single-crystal fibers, using unreacted pedestals with an appropriate mixture of La_2O_3–Al_2O_3, with diameter fluctuation intentionally introduced in the pedestal up to 20% (Fig. 13.8). It has been shown that this ADC system can reduce the overall fiber diameter to less than 1% for dense pedestals, with regions on the order of 1 cm with near-zero variations.

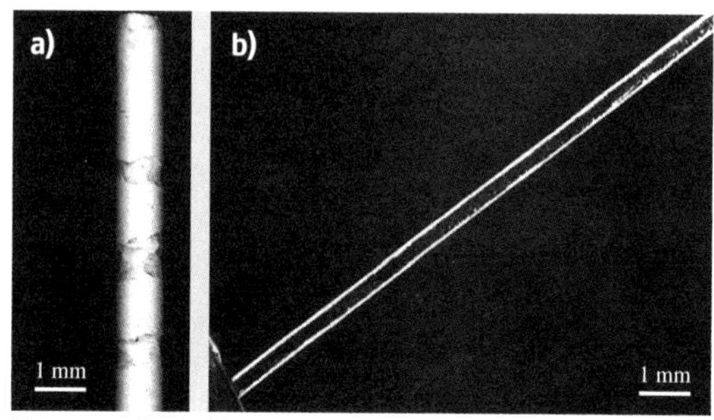

Fig. 13.8a,b A fiber pulled from a pedestal prepared with diameter fluctuations of up to 20%. (**a**) Unreacted pedestal (*green rod*) with the introduced diameter variation and (**b**) fiber pulled from the same pedestal showing that the diameter fluctuation was corrected by the automatic diameter control system (ADC)

Both systems have their advantages and limitations. The interference system showed better resolution results, however it is dependent on the fiber transparency, which affects the resolution of the apparatus: $g(n,\theta) = g(\theta) \approx 1/\theta$. On the other hand, the artificial vision system, although more versatile, is not adequate for low melting point materials (< 800 °C) due to their poor black-body emission.

13.3 Fundamentals

Many of the unique features of the LHPG technique are consequences of the very small liquid-phase volume necessary to grow crystals and in the following sections we shall discuss both experimental and theoretical aspects of the pulling process.

13.3.1 Conservation of Mass

It is possible to correlate the pedestal and fiber cross-section areas with their respective pulling rates if we consider that all molten materials from the pedestal are crystallized as a single-crystal fiber. If we are allowed to consider that the pedestal used has the same density of the growing crystal fiber, we can further simplify this relation as shown in (13.2)

$$\rho_{fib} r_{fib}^2 v_{fib} = \rho_{ped} r_{ped}^2 v_{ped}$$
$$\xrightarrow{\text{if: } \rho_{fib} = \rho_{ped}} r_{fib} = r_{ped}\sqrt{\frac{v_{ped}}{v_{fib}}}\,, \quad (13.2)$$

where ρ_{fib}, r_{fib}, v_{fib}, and ρ_{ped}, r_{ped}, v_{ped} are the density, fiber radius, and pulling rates of the fiber and pedestal, respectively.

In laser-heated pedestal growth, the laser power and pulling rates are parameters that we can modify during the crystal growth process. It is evident that the crystal diameter will be constant if all the physical parameters remain unchanged. However, it is reasonable to assume that small imperfections in the pedestals may occur. In this way, if some variation of the source cross-sections remains, it is necessary to change one of the growth parameters in order to keep the fiber diameter constant. The simplest solution is to modify the pulling rate, assuming that the pedestal upward pulling rate remains unchanged.

That is why a fast and effective diameter control system is important, allowing for only very small changes in the pulling rate to correct for possible nonuniformity of the geometry of the pedestals used.

13.3.2 Balance of Heat Transfer

A schematic representation of the most important heat fluxes during fiber pulling is illustrated in Fig. 13.9. The energy delivered by the CO_2 laser that is used to keep the molten zone height is: conducted through the fiber and pedestal, lost to the growth chamber by irradiation, due to the black-body emission, and convection to the fluid inside, when present. There is also the latent heat of solidification and melting generated at the fiber–liquid and pedestal–melt interfaces.

In the LHPG system it is reasonable to assume heat transfer in one dimension, considering the heat fluxes in a semi-infinite cylinder, moving with velocity v in the z-direction [13.43]. The equation for the heat transfer can be written as

$$\frac{\mathrm{d}^2Y}{\mathrm{d}Z^2} - 2(\mathrm{Bi}_{\mathrm{total}})Y - 2\mathrm{Pe}\frac{\mathrm{d}Y}{\mathrm{d}Z} = 2\mathrm{Pe}\frac{\mathrm{d}Y}{\mathrm{d}\tau}\,, \tag{13.3}$$

where $Y=(T-T_{\mathrm{o}})/(T_{\mathrm{i}}-T_{\mathrm{o}})$, $Z=z/a$, and $\tau = v_{\mathrm{fib}}t/a$; T_{o} is the ambient temperature in the surroundings of the growing fiber, T_{i} is the growing temperature interface, a is the fiber radius, Bi $=ha$ is the total Biot number, h is the effective cooling constant (convective and radiative), and Pe is the Péclet number defined as $\rho c_{\mathrm{p}} v_{\mathrm{fib}} a/2K$, where ρ is the solid density, c_{p} is the thermal capacity, and K is the solid thermal conductivity. From the solution of this equation it is possible to obtain an analytical solution for the thermal gradient at the crystal–liquid interface ($G_{\mathrm{s,i}}$) as shown in (13.4)

$$G_{\mathrm{s,i}} = \left(\frac{\mathrm{d}T}{\mathrm{d}z}\right)_{z=0} = -(T_{\mathrm{i}}-T_{\mathrm{o}})\left\{\left[\mathrm{Pe}^2+2(\mathrm{Bi})\right]^{\frac{1}{2}} - \mathrm{Pe}\right\}\left(\frac{1}{a}\right)\,. \tag{13.4}$$

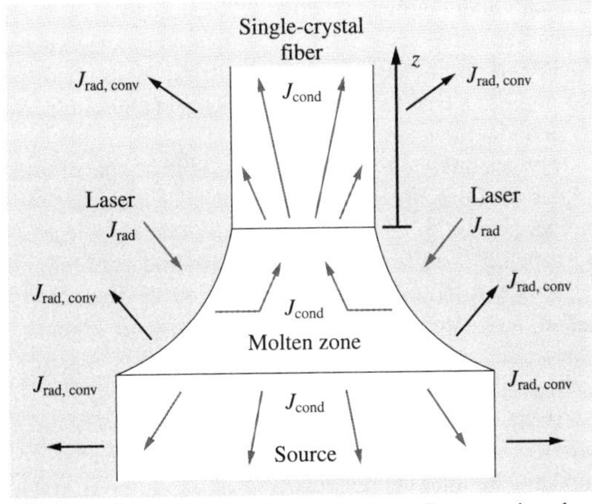

Fig. 13.9 Schematic drawing of the heat fluxes acting during the single-crystal fiber growth process

The laser-heated pedestal growth method produces steeper thermal gradients at the growing interface than any other crystal growth method (ranging from 10^3–10^4 °C/cm). This high temperature gradient at the growing interface is responsible for the high pulling rate in this technique, allows the pulling rate to be mm/min instead of mm/h as is typical for conventional crystal growth methods. However, these same elevated thermal gradients also have some disadvantages, one of which is that it produces greater thermal stresses which restrict the maximum crystal diameter that can be grown before cracks develop. *Brice* [13.44] has shown that this maximum diameter can be determined based on the mechanical and thermal properties of the compound (13.5)

$$\left|\frac{\mathrm{d}T}{\mathrm{d}z}\right|_{\mathrm{max}} \approx \frac{4\varepsilon_{\mathrm{b}}}{\alpha a^{\frac{3}{2}}}\left(\frac{1}{h}\right)^{\frac{1}{2}}\,, \tag{13.5}$$

where $|\mathrm{d}T/\mathrm{d}z|_{\mathrm{max}}$ is the absolute value of the maximum allowed thermal gradient at the growth interface, ε_{b} is the breaking strain, α is the coefficient of linear thermal expansion, a is the radius of the fiber, and h is the cooling constant.

The temperature gradient at the crystallization interface, in a conventional LHPG system, has a strong dependence on the fiber diameter. *Andreeta* [13.45] used lithium niobate as a material model due to its technological importance, for a study on the effects of the temperature gradients at the solid–liquid interface on the maximum attainable fiber diameter without cracks. For temperature measurements a very small Pt/Pt-Rh thermocouple (60 μm in diameter) attached to the pulling system was used. The thermocouple was inserted into the liquid phase and pulled through the solid–liquid interface. Single-crystal fibers of 40 mm in length and ranging from 300 to 1200 μm in diameter were pulled from single-crystal pedestals, along the c-axis. With this system it was possible to measure the high axial thermal gradient at the solid–liquid interface as a function of fiber radius (Fig. 13.10). From this data and the maximum diameter, above which the fiber cracks (13.5), we determined for lithium niobate a critical radius of 470 μm (Fig. 13.10).

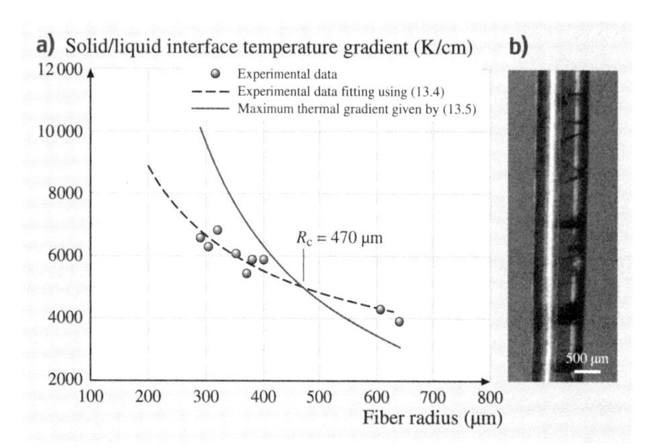

Fig. 13.10a,b Temperature gradient at the solid–liquid interface during $LiNbO_3$ single-crystal fiber growth. **(a)** The *continuous line* represents the maximum thermal gradient that the material can handle before it cracks (13.5) and the *dashed line* is the fitting of the temperature gradient measurements based on (13.4). $R_c = 470\,\mu m$ is the critical radius above which fiber cracking results during the growth process. **(b)** $LiNbO_3$ single-crystal fiber with 1 mm diameter, showing cracks produced by generated thermal stress

The high axial temperature gradient at the solid–liquid interface results in a wide number of advantages, including higher growth rates, ferroelectric domain alignment during growth ($LiNbO_3$), and the possibility of obtaining high-temperature metastable phases of certain compounds. Another important effect of the high axial temperature gradient is to prevent constitutional supercooling. The condition for constitutional supercooling can be written as

$$\frac{G}{v} \leq \frac{mC_s}{D_l}\left(\frac{1-k_0}{k_0}\right), \tag{13.6}$$

where G is the absolute value of the temperature gradient at the solid–liquid interface, v is the pulling rate, m is the absolute value of the liquidus slope, C_s is the concentration in the solid phase, D_l is the diffusion coefficient of the solute in the liquid, and k_0 is the equilibrium distribution coefficient (the ratio between the solid and liquid concentration $-C_s/C_l$). The smaller the fiber radius, the greater the pulling rate allowed before constitutional supercooling occurs. *Lu* et al. [13.40] systematically studied the influence of constitutional supercooling on the growth of $Bi_2Sr_2Ca_1Cu_2O_8$ crystal fibers. They found a critical value for the constitutional supercooling ($1.0\times10^{12}\,K\,s/m^2$) below which the pulled fiber became polycrystalline. They also estimated that, for a fiber 30 μm in diameter, the maximum speed allowed should be around 26 mm/h.

Unfortunately, the same axial temperature gradient is also responsible for a high level of stress in the grown fiber. The thermal stresses induced by temperature gradients are the main cause of the formation of various crystallographic defects in the fiber. The expression for approximate calculation of the dislocation density (N_d) in terms of the axial temperature gradient was performed by *Prokofiev* et al. for the $Bi_{12}SiO_{20}$ (BSO) compound [13.46], yielding

$$N_d = \frac{\alpha}{b}\frac{dT}{dz} - \frac{2\varepsilon_b}{ba}, \tag{13.7}$$

where b is the Burger's vector and dT/dz is the axial temperature gradient at the crystallization interface. Using the calculated values of the axial temperature gradient, it is possible to evaluate the dislocation density in the fiber. The calculated value of N_d for BSO fibers was $5\times10^5\,pits/cm^2$ for fibers with 1 mm diameter, which is in good agreement with the value estimated by the chemical etching method ($1\times10^5\,pits/cm^2$). Prokofiev et al. also estimated that a BSO fiber with diameter below 100 μm should be free-dislocation.

13.3.3 Mechanical Stability

The shape of the molten zone is a very important growth parameter, especially in a floating-zone-like technique such as LHPG. The variation of the contact angle is responsible for the diameter fluctuation of grown fibers. The stability of a liquid trapped between two cylindrical plates has been studied for more than a century. Interest in this topic starts mainly from the 1950s due to the importance of the new (at that time) technology of semiconductors crystals.

For an analysis of the molten zone shape of the miniature floating zone let us consider an isotropic molten zone with length L, volume V, and surface area S, suspended between two axial solid cylinders of radius r_{fib} and r_{ped}. *Kim* et al. [13.47] studied the maximum stable zone length for sapphire and silicon crystals. The value found was $L_{max} = \pi\phi$, where ϕ is the fiber diameter.

If we consider also that the liquid–fiber and liquid–pedestal contact angles are ψ_{fib} and ψ_{ped}, respectively, and that the total energy involved in the liquid is composed of potential gravitational and surface energy, the molten zone will acquire the shape that allows for total energy minimization as shown in (13.8)

$$E_T = E_g + E_S = \pi \rho g \int_0^L r^2(z) z \, dz + 2\pi\sigma \int_0^L \left(\sqrt{1+\left(\frac{dr}{dz}\right)^2}\right) r(z) \, dz \,, \quad (13.8)$$

where g is the acceleration due to gravity, ρ is the density of the liquid, and σ is the surface tension. Since the effect of gravity on LHPG is very small, it can be neglected in comparison with the surface energy [13.19]. In this way, the problem is reduced to the minimization of the surface energy. Figure 13.11 illustrates several profiles of molten zones obtained for different fiber–pedestal diameter reductions, based on the numerical solution of the minimization of the surface energy of the liquid phase. *Saitou* et al. [13.48] developed an analytical solution for the liquid-phase profile in the floating zone technique for a pedestal–crystal diameter ratio of 1. They showed that minimization of the surface free energy (13.8) followed by application of the principle of variation led to the dimensionless expression

$$\frac{dx}{dy} = \frac{\frac{y^2}{2} + \lambda y + C_0}{\sqrt{1 - \left(\frac{y^2}{2} + \lambda y + C_0\right)^2}} \,. \quad (13.9)$$

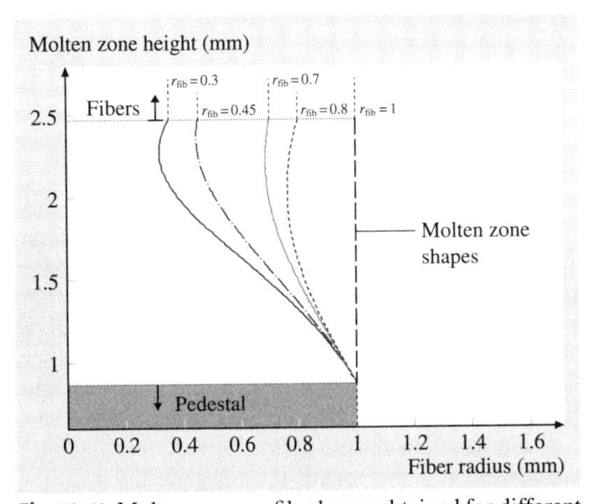

Fig. 13.11 Molten zone profile shapes obtained for different fiber–pedestal diameter reductions, obtained by numerical solution of the minimization of the surface energy of the liquid phase (the pedestal radius was normalized to 1) (after [13.19])

Solving (13.9) and returning to the dimensional form we obtain (13.10) [13.49]

$$x(y) = \frac{y^3}{6C^2} + \frac{\lambda y^2}{2C} + C_0 y + C_1 C \,, \quad (13.10)$$

where λ, C_0, and C_1 are constants (given in terms of the floating zone height, pedestal diameter, and molten zone surface area), C is the capillary constant (defined as $C = (2\sigma/\rho g)^{1/2}$), σ is the surface tension, ρ is the liquid density, and g is the acceleration due to gravity [13.48].

When minimization of the energy takes place, considering an isotropic system, another way to look at the problem is to realize that equilibrium of the molten zone is reached when there is also balance of the surface tensions at the liquid–solid–gas junction. This junction defines what is called the wetting angle and it is possible to express the contact angle by

$$\cos \varphi_0 = \frac{\sigma_{l,g}^2 + \sigma_{s,g}^2 - \sigma_{s,l}^2}{2\sigma_{l,g}\sigma_{s,g}} \,, \quad (13.11)$$

where $\sigma_{l,g}$, $\sigma_{s,g}$, and $\sigma_{s,l}$ are the liquid–gas, solid–gas, and solid–liquid surface tensions, respectively.

When the wetting angle reaches this equilibrium value, the growth process is stable and there should be no diameter fluctuations. However in real crystal growth processes we are faced with nonideal situations which cause fluctuations of the floating zone shape, thus modifying the contact angle. Again for the LHPG technique, there is always the need for good diameter control to obtain high-quality fibers.

13.3.4 Growth Under Controlled Atmosphere

The growth atmosphere and the nature of the solid and liquid phases are also factors that influence the molten zone profile. The effects of external pressure on the melt properties and equilibrium of solid and liquid phases began to be studied in the first decades of the 20th century [13.50]. Fiber growth under different atmospheres using laser-heated systems was first explored in the pioneering work of *Haggerty* [13.11]. He grew fibers under air, Ar, Cl_2, H_2, and CH_3 alone or in combination. It was noticed that the mechanical strength of the fiber was dependent on the growth atmosphere used. *Ardila* et al. reported fiber pulling under controlled high external pressure [13.29]. In this work they modified the LHPG technique to grow crystal fibers in a system that could handle up to 15 atm of oxidizing, reducing or inert gas and studied the profile of the liquid phase. They used various materials, such as $Ba_{0.77}Ca_{0.23}TiO_3$ (BCT), $CaMoO_4$

(CMO), Ca_2FeMoO_6 (CFMO), and $Ca_{1-x}Sr_xMoO_3$ (CSMO) with $0 \leq x \leq 1$ [13.49, 51]. The values for the capillary constant and surface tension coefficient for all of these compounds were quite similar.

Phomsakha and coworkers studied the effect of the atmosphere on Al_2O_3 single-crystal fiber pulling [13.23]. They report that the best atmosphere is He gas, but if the pressure is increased to 15–20 Torr the fiber transmission decreases. They found that the optimum pressure was 5 Torr of He. Lower pressure values allow for vaporization that leaves oxygen in the melt, which is the main cause of the formation of microvoids. Other gases, such as N_2, Ar, and air presented lower transmission measurements. *Wu* et al. [13.52] grew Ti-doped sapphire fibers under various atmosphere (N_2, H_2, Ar) in order to improve the Ti concentration in the fiber. They found that the use of an N_2 atmosphere increased the Ti fiber concentration when compared with experiments performed in air atmosphere, although the fibers presented dark regions on the surface and in the interior. In pure H_2 atmosphere, quite a large amount of white fog-like matter volatilised from the melting zone and was deposited onto the surfaces of the fiber and the furnace wall. The cross section of grown fiber was not circular and the grown fiber was useless [13.52]. However, a mixture of H_2 and Ar gas resulted in fibers without inclusions or bubbles inside. This last mixture allowed the researchers to obtain high-quality laser elements.

13.3.5 Dopant Distribution

As in any crystal growth process, the control of the dopant distribution along the crystalline matrix is desirable for many technological applications such as laser hosts and periodic poled ferroelectric crystals. *Sharp* et al. developed a model based on mass conservation for fiber pulling [13.53], shown in (13.12)

$$\frac{C_f(z)}{C_{S0}} = k_{eff}\left[C'e^{-\beta\frac{z}{v}} + \frac{\alpha'}{\beta}\left(1 - e^{-\beta\frac{z}{v}}\right) + \frac{\alpha'(C'-1)}{(\beta-\gamma)}\left(e^{-\gamma\frac{z}{v}} - e^{-\beta\frac{z}{v}}\right)\right] . \quad (13.12)$$

Assuming that $\alpha' = \pi r_{fib}^2 v/V$; $\beta = (\alpha' k_{eff} + 1/\tau_{ev})$; $\gamma = (v r_{fib}^2)/(\eta r_{ped})$, $C_f(z)$ is the fiber concentration at a point z, C_{S0} is the source concentration, k_{eff} is the effective distribution coefficient, C' is the fractional loss of dopant from the source rod at $z = 0$ due to evaporation, r_{fib}, r_{ped}, v, η, τ_{ev}, and V are the fiber radius, pedestal radius, pulling rate, characteristic length (not usually susceptible to measurement), the evaporation constant, and the molten zone volume, respectively [13.53]. With this model, *Sharp* et al. could describe the observed concentration profile for $Ti{:}Al_2O_3$ fiber growth. Some CW laser applications at high power levels may require periodically doped crystals, as in the case of periodically poled $LiNbO_3{:}Hf$ and $LiNbO_3{:}Nd^{3+}$ (PPLN) [13.54]. The intrinsic characteristics of the growing process (high pulling rates, unidimensional growth-like process, and small molten volumes) usually led to the growth of homogenous dopant distribution in the crystal fiber.

However it is possible to intentionally change this distribution profile by modifying the growth conditions, such as introducing a sinusoidal temperature fluctuation at the molten zone [13.54]. In this case the dopant concentration can be expressed as a function of growth time, by applying also the mass conservation of the dopant. It is assumed that the dopant is supplied from the incoming pedestal. Dopant loss results from evaporation from the surface of the molten zone and from dopant being taken up by the growing fiber. Thus, the evaporation may be characterized by an evaporation time constant τ_{ev} defined as the time it takes for the concentration of dopant in the melt to decrease by a factor of e^{-1} when the fiber and pedestal are in a stationary state, i.e., at zero growth rate. Hence, the dopant concentration in the melt, considering the oscillation effects, can be expressed by (13.13)

$$\frac{dC_m}{dt} = \frac{\pi r_{ped}^2 Cs(t) v_{ped}}{V} - \frac{\pi r_{ped}^2 k_{eff} C_m \frac{dF_p(t)}{dt}}{V} - \left[\frac{\pi r_{fib}^2 k_{eff} C_m v_{fib}}{V} + \left(\frac{\pi r_{fib}^2 k_{eff} C_m \frac{dF_f(t)}{dt}}{V}\right)\right] - \left(\frac{C_m}{V}\frac{dV}{dt}\right) - \frac{1}{\tau_{ev}} C_m , \quad (13.13)$$

where C_m is the melt concentration, $Cs(t)$ is the initial pedestal concentration due to the instabilities at the startup of the melting process [13.54], r_{fib} and r_{ped} are the fiber and pedestal radii, respectively, $k_{eff} = C_f/C_m$ is the effective distribution coefficient, C_f is the fiber concentration, v_{ped} and v_{fib} are the pedestal and fiber pulling rates, respectively, V is the molten zone volume that was approximated to a conical section. $F_p(t)$ and $F_f(t)$ are functions for the periodical input oscillations, which were $A_p \sin(\omega t)$ and $A_f \sin(\omega t)$ for the pedestal melting and fiber growth, respectively [13.54]. The total molten zone oscillation amplitude is $A_p + A_f$.

Equation (13.13) can be numerically solved. The experimental input data required for the program were oscillation amplitude, r_{fib}, r_{ped}, k_{eff}, v_{ped}, v_{fib}, and fre-

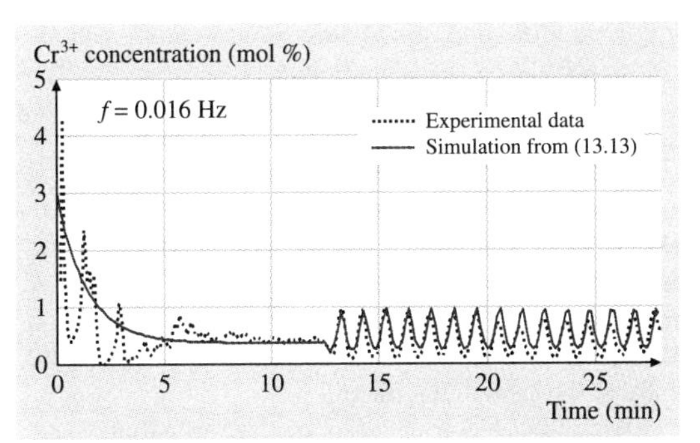

Fig. 13.12 Axial concentration profile measurement (*dashed lines*) for a frequency of 16 mHz molten zone oscillations, and corresponding simulation (*continuous line*) (after [13.33])

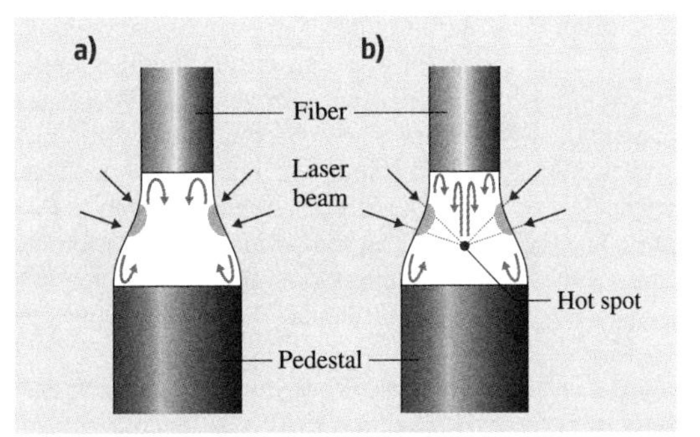

Fig. 13.13a,b Two possible configurations for the convection pattern inside the molten zone in a fiber-pulling process: (**a**) Laser radiation absorption in the opaque melt occurs at the liquid surface, leading to Marangoni convection, and (**b**) semitransparent melt, leading to the creation of a hot spot inside the molten zone at the focus point, generating flow competition with the Marangoni convection

quency. The evaporation time constant τ_{ev} was used to minimize the difference between the calculated and measured Cr^{3+} concentration profile. The typical τ_{ev} values lay within a range of 3–4 min for several experiments, similarly to the Al_2O_3:Ti fiber growth performed by *Sharp* et al. [13.53] without periodic oscillations. A good match between the measured and the calculated concentration profile with an oscillating molten zone can be verified when they are plotted together, as shown in Fig. 13.12. This means that the growth interface position corresponded directly with the periodic input program. With this new approach to single-crystal fiber growth, it is possible to predict and set the distance between two dopant concentration maxima. Also it is possible to pull fibers with alternating uniform and periodic concentration doping profiles regions.

Up to now it has been consider that the pulling of the fibers is unidimensional. Although for the thermal and composition axial profile this is a good approximation, the radial distribution must be considered to explain some of the observed growth aspects. *Liu* et al. recently observed that Al_2O_3:Mg fibers present defects at their centers [13.55]. The radial Mg concentration showed that the center of the fiber is richer in Mg atoms. They proposed that this high concentration is due to Marangoni convection that creates a flow from the surface towards the center. This thermocapillary convection causes the flow to drift from the hot side (near the center, at the external part of the floating zone) to the cold side (at the solid–melt interface) along the melt–gas interface and then back from the cold side to the hot side near the axial center. This is highlighted as the main cause of excess dopant at the fiber center. Constitutional supercooling is then the responsible for the creation of fiber defects at the center. Figure 13.13 illustrates two possible configurations for the fluid flow in the molten zone during fiber pulling. Figure 13.13a represents a molten zone opaque to the laser source, and thus all energy is absorbed in the liquid surface, creating a flow as *Liu* et al. described [13.55]. Figure 13.13b illustrates the condition for a material that is semitransparent to the applied laser source. In this condition a hot spot may be formed in the liquid center, creating competition between the two liquid flows. *Erdei* et al. [13.56] suggested that the configuration shown in Fig. 13.13b may be the cause of the radial segregation of Ba in the SBN fiber growth process. In the same work, the authors suggested that the radial dependence of the effective distribution coefficient can be expressed by

$$k_{eff}(r) = \frac{k_0}{k_0 + (1 - k_0)e^{-v\frac{\delta(r)}{D}}} , \qquad (13.14)$$

where $k_{eff}(r)$ is the radial effective distribution coefficient, k_0 is the equilibrium distribution coefficient, v is the pulling speed, D is the diffusion coefficient of the element in the liquid phase, and $\delta(r)$ is the radial thickness distribution of the diffusion layer that forms ahead of the solidification interface during stationary growth.

13.3.6 Pulling Crystalline Fibers Under Electric Field

During crystallization of ionic melts, *Uda* et al. [13.21] have shown that an electric field will be present in both the liquid and solid, arising for at least two reasons: differential partitioning of opposite-valence ions to place a net charge of one sign on the liquid boundary layer (crystallization electromotive force, EMF) and a Seebeck coefficient produced by the temperature dependence of the equilibrium ion concentration. The thermoelectric potential difference $\Delta\varphi$ between a location at temperature T_1 in the solid and a location at temperature T_2 in the liquid is given by [13.21]

$$\Delta\varphi = \alpha_s(T_1 - T_i) + \alpha_l(T_i - T_2) + \alpha_i v \,, \qquad (13.15)$$

where v is the pulling rate, T_i is the interface temperature, α_i is the crystallization EMF coefficient, and α_s and α_l are the thermoelectric coefficients for the solid and liquid, respectively. The electric field generated at the interface during fiber pulling of some ferroelectric compounds is sufficient to orient the ferroelectric domains in situ, as in the case of $LiNbO_3$ [13.19].

Tiller et al. [13.58] investigated the effect of a strong interface field during the growth of TiO_2-based crystals pulled by LHPG technique. According to the authors, when the field-driven flux is appreciably greater than the interface partitioning flux, a stationary-state solute profile in the solid of an anomalous nature can be maintained. A steady-state solute profile requires the following to hold [13.58]

$$\left(\frac{D_s C_s}{kT}\right) q_s E_s \gg V k_0 C_l^i \,, \qquad (13.16)$$

where D_s, C_s, q_s, and E_s are the diffusion coefficient, solute concentration, effective solute charge, and electric field, respectively, while k_0 is the solute distribution coefficient and C_l^i is the concentration in the liquid phase at the growth interface.

Using the electrical-assisted laser floating zone technique (EALFZ), *Carrasco* et al. studied the effect of a direct electrical current applied through the solidification interface for superconducting oxide materials [13.34]. In the EALFZ process the fiber can be grown with a direct current (DC) by connecting the positive and negative poles, respectively, to the seed and feed rod samples, or by a reverse current (RC) if the negative and positive poles are connected to the seed and feed rod samples, respectively. The application of a DC current during the EALFZ process can

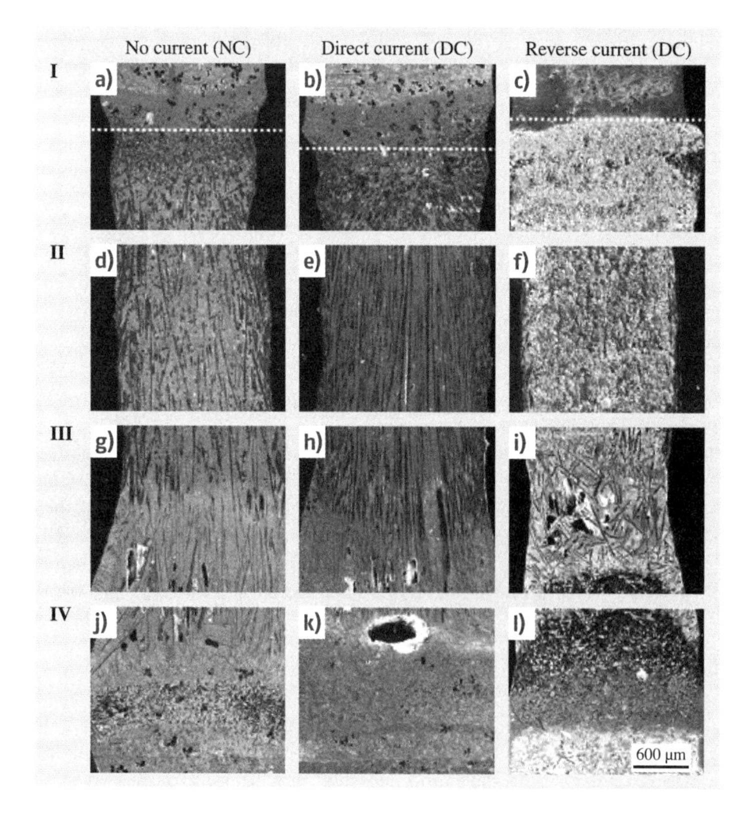

Fig. 13.14a–l Scanning micrograph of longitudinal sections of BSCCO fiber grown at $R = 15$ mm/h without current (NC fiber) **(a,d,g,j)** with direct current (DC fiber), **(b,e,h,k)** and with reverse current (RC fiber) **(c,f,i,l)**. Region I – seeding initial transient. Region II – fiber steady-state region. Region III – frozen melt. Region IV – melting interface. The *dotted line* corresponds to the seed–fiber interface (after [13.57])

drive the system towards equilibrium, since it drastically changes the solidification features, namely the solute ion distribution, phase equilibrium, and crystal growth kinetics [13.57] (Fig. 13.14). Accordingly, important changes in phase structure, crystal morphology, and grain alignment are observed. The presence of the electric field through the solid–liquid interface accelerates the solute ions until a drift velocity $R'(R' = \mu E)$ is reached. In this way, the original Burton–Prim–Slichter (BPS) theory equation for the effective distribution coefficient is modified to [13.34, 59]

$$k_{\text{eff}} = \frac{1+\frac{R'}{R}}{1+\left[\frac{1}{k_0}\left(1+\frac{R'}{R}\right)-1\right]\exp\left[\left(-\frac{R\delta}{D}\right)\left(1+\frac{R'}{R}\right)\right]}, \quad (13.17)$$

where R is the interface velocity, k_0 is the equilibrium distribution coefficient, D is the diffusion coefficient of the element in the liquid phase, and δ is the thickness of the diffusion layer that forms ahead of the solidification interface during stationary growth. Thus, the effective distribution coefficient k_{eff} is a set of two competition processes:

1. The rejection/acceptation of the solute for $k_0 < 1$ and $k_0 > 1$, respectively, by the solid–liquid interface
2. The mobility difference of the solute and solvent ions under the applied electric field in the diffusion layer.

When ions move towards the solid–liquid interface due to the application of a direct current (R' positive) the electrical field will increase k_{eff}. On the contrary, a decrease in k_{eff} is observed for the condition of R' negative, when the ions move away from the interface. It is important to point out that, in the presence of an electrical field and for a given value of interface velocity R, the effective distribution coefficient can exhibit values over a wide range, even outside of the normal range situated between k_0 and unity [13.34]. Useful examples of the effect of the electric current on k_{eff} were given by *Pfann* [13.60]:

1. The refinement of an ingot from an element for which k_0 is close to unity, by increasing the absolute value of $(1-k_{\text{eff}})$
2. Segregation hindrance by making k_{eff} close to unity
3. Simultaneous zone refining even when solutes have k_0 values lying on opposite sides of the unity, forcing them to move to the same side of an ingot.

An external alternating electric field can also be applied to a LHPG fiber during the pulling process without physical contact with the growing fiber. This is performed in order to produce periodic inversion of ferroelectric domains, such as in the method developed by Brenier et al. described in Sect. 13.1. *Lee* et al. showed that assuming a uniform electric field E in the fiber cross-section, the accumulated E field, $E_{\text{acc}}(\bar{z})$ that the grown fiber experiences at position $\bar{z}$ during the poling process can be expressed by the convolution integral [13.61]

$$E_{\text{acc}}(\bar{z}) = E_0 \int_{-\infty}^{+\infty} \text{rect}\left(\frac{(\bar{z}-z)}{W_e}-\frac{1}{2}\right)\sin\left(\frac{2\pi fz}{v}\right)\text{d}z\,, \quad (13.18)$$

where E_0 is the electric field at start point in the center of the poling region, W_e is the effective poling length, f is the applied electric field frequency, rect is the rectangular function (also known as the rectangle function, rect function, unit pulse, or the normalized boxcar function) [13.61], and v is the pulling rate. With this process, magnesium-doped lithium niobate can be periodic poled with controllable distance between the inverted domains.

13.4 Fiber Growth Aspects

As stated previously, the LHPG technique is a very powerful tool for producing single crystals of various compounds, as shown in Fig. 13.15. Table 13.2 lists the different compounds produced using the LHPG technique. The following sections present practical features and aspects of the pulling process using the laser-heated pedestal growth technique, such as pulling of stoichiometric fibers, incongruently melt/evaporating materials, and directional solidification on eutectic compounds.

13.4.1 Congruent Melting Fibers: The Search for Stoichiometry

Congruently melting materials are the easiest type of compound to pull successfully in single-crystal form.

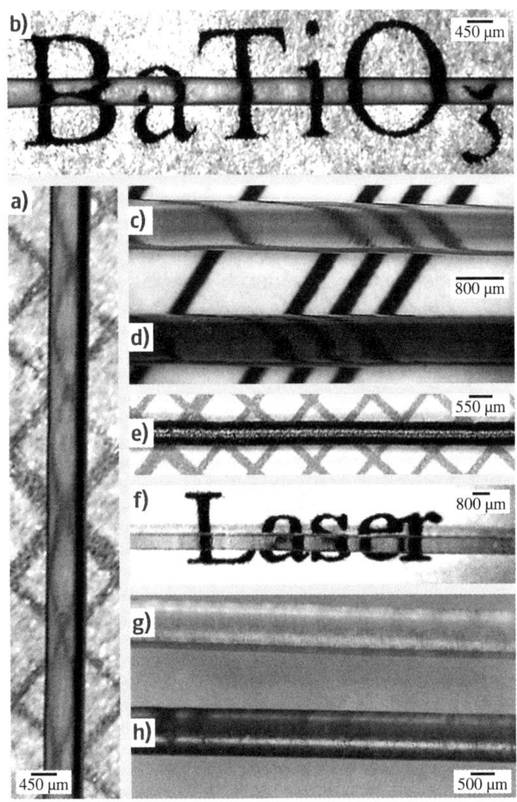

Fig. 13.15a–h Single-crystal fibers of various compounds pulled by the LHPG technique: (**a**) $SrTiO_3$, (**b**) $BaTiO_3$, (**c**) $GdAlO_3:Nd^{3+}$, (**d**) $GdAlO_3:Cr^{3+}$, (**e**) $La_{0.67}Ca_{0.33}MnO_3$, (**f**) $LaVO_4$, (**g**) Al_2O_3 (sapphire), and (**h**) $Al_2O_3:Cr^{3+}$ (ruby)

However, the stoichiometric line in the phase diagram is in reality a region (termed the existence region), i. e., it may extend over a certain homogeneity range [13.62]. Stoichiometry variations can impair the electric, ferroelectric, magnetic, optical, and other characteristics of these single crystals. For many compounds of interest, congruently melting compositions have been detected which are nonstoichiometric, thereby giving rise to unwanted effects such as thermodynamic instabilities, precipitates, intrinsic impurities, and deviations of the distribution coefficient of cations from unity [13.62].

Due to their technological importance, Li(Nb, Ta)O_3, YVO_4, and SBN ($Sr_{0.61}Ba_{0.39}NbO_3$) have been studied in order to achieve stoichiometry composition along the fiber length. Besides the composition variation of metal elements along the crystal growth axis, those compounds also present oxygen deficiency. Since the congruency originates from the defect structure of niobates, tantalates or vanadates, the oxygen vacancy also plays an important role in the growth of highly homogeneous crystals. In other words, metal oxides having different oxygen stoichiometry will possess different congruent compositions [13.56]. Another effect of the oxygen-deficient phases is that they can form solid solutions with their pentavalent variant, as was observed for YVO_4 by *Erdei* et al. [13.80]. These solid-solution formations cannot be eliminated by simple melt growth techniques. The oxygen deficiency can also slightly modify the Li:Nb, Li:Ta or Y:V stoichiometries in $LiNbO_3$, $LiTaO_3$, and YVO_4 during the growth processes. In this way, it is possible to verify that there is not one general precise value for the exact congruent composition in the niobate, tantalate or vanadate families. *Erdei* et al. identified that slightly oxygen-deficient SBN connected with Nb_2O_{5-x} creates a solid solution with the pentavalent niobium–SBN system [13.56, 62].

In order to grow high-quality single-crystal fibers with near-stoichiometric composition and high homogeneity along the fiber length it is possible to distinguish three basic strategies to achieve the goal of stoichiom-

Table 13.2 Pure and doped fibers pulled by laser-heated pedestal growth technique

Compound	Doping	Applications/characteristics	References
Al_2O_3	Cr^{3+}	Temperature sensor	[13.63–68]
	C	Radiation dosimetry	[13.69]
	Pure	Optical transmission	[13.28, 70–72]
	Pure	Mechanical	[13.73, 74]
	Ti^{3+}	Graded-index fiber – laser action	[13.52, 75]
	Yb^{3+}	Spectroscopy	[13.76]
	Er^{3+}, Yb^{3+}	Temperature sensor	[13.77, 78]
	Mg	Mechanical	[13.55]
	Pure	Growth process/synthesis	[13.23]
	Ti	Growth process/synthesis	[13.53, 79]

Table 13.2 (cont.)

Compound	Doping	Applications/characteristics	References
Eutectic Al_2O_3-$Y_3Al_5O_{12}$	Pure	Mechanical	[13.81–84]
	Pure	Growth process/synthesis	[13.85]
Eutectic Al_2O_3-$Y_3Al_5O_{12}$-ZrO_2	Pure	Mechanical	[13.86]
Eutectic Al_2O_3-ZrO_2	Pure	Mechanical	[13.87–90]
	Pure	Mechanical	[13.84, 91–93]
	Y (eutectic)	Growth process/synthesis	[13.94]
$Ba(Mg_{0.33}ME_{0.67})O_3$, ME = Ta, Nb	Pure	Dielectric	[13.95]
		Spectroscopy	[13.96]
		Spectroscopy	[13.97]
		Growth process/synthesis	[13.98]
$Ba(Ti_{1-x}Zr_x)O_3$	Pure	Ferroelectric	[13.99]
	Pure	Dielectric	[13.100, 101]
$Ba_{1-x}M_xTiO_3$	M = Sr	Dielectric	[13.102–104]
	M = Ca	Growth process/synthesis	[13.105, 106]
$Ba_2NaNb_5O_{15}$	Nd^{3+}	Growth process/synthesis	[13.107, 108]
	Nd^{3+}	Spectroscopy	[13.109–114]
	Yb^{3+}	Spectroscopy	[13.115]
Ba_2NdNbO_6	Pure	Spectroscopy	[13.116]
$BaTiO_3$	Pure	Photorefractive effect	[13.117]
	Pure	Growth process/synthesis	[13.118–120]
$Bi_{12}MEO_{20}$; ME =Si, Ti	Pure	Photorefractive effect	[13.121]
		Optical activity	[13.122]
		Growth process/synthesis	[13.46, 123–127]
Bi-Sr-Ca-Cu-O	Pure	Superconductivity	[13.34, 128–130]
	Pure	Superconductivity	[13.57, 131–134]
	Pure	Superconductivity	[13.135–138]
	Pure	Superconductivity	[13.139–143]
	Pure	Magnetic	[13.144–148]
	Ag	Superconductivity	[13.149, 150]
	Pb	Superconductivity	[13.151, 152]
		Growth process/synthesis	
	Pure		[13.153–157]
	Pure		[13.32, 158–161]
	Pure		[13.162–166]
	Pure		[13.40, 167–170]
	Ti		[13.171]
	Ag		[13.172, 173]
	Pb		[13.174–177]
$Ca_{1-x}Yb_xF_{2+x}$	Yb^{3+}	Spectroscopy	[13.178]
$Ca_3(VO_4)_2$	Nd^{3+}	Spectroscopy	[13.179]
	Er^{3+}	Growth process/synthesis	[13.51]
$Ca_{9.5+0.5x}((PO_4)_{6-x})((BO_3)_x)((BO_2)_{1-x}O_x)$	Yb^{3+}	Spectroscopy	[13.180]

Table 13.2 (cont.)

Compound	Doping	Applications/characteristics	References
$CaAl_4O_7$	Ce^{3+}	Spectroscopy	[13.181]
	Eu^{2+}, Nd^{3+}	Spectroscopy	[13.182]
	Tb^{3+}, Ce^{3+}	Spectroscopy	[13.183]
CaF_2	Yb^{3+}	Spectroscopy/laser action	[13.184, 185]
CaM_2O_6; M = Nb, Ta	Nd^{3+}	Spectroscopy	[13.186]
	Nd^{3+}	Spectroscopy	[13.187]
		Growth process/synthesis	
	Pure		[13.188]
	Pure		[13.189]
$CaMoO_4$	Pure	Growth process/synthesis	[13.190]
$CaMoO_4$-$SrMoO_4$	Pure	Growth process/synthesis	[13.39]
$CaTiO_3$	Pure	Growth process/synthesis	[13.191]
Co, Fe, and Co-Fe alloy	Pure	Growth process/synthesis	[13.192]
Dy_2O_3	Pure	Refraction index	[13.193]
$EuAlO_3$	Ti^{3+}-Ti^{4+}	Growth process/spectroscopy	[13.194]
Gd_2O_3	Eu^{3+}	Luminescence	[13.195]
	Nd^{3+}	Spectroscopy	[13.196]
	Er^{3+}	Spectroscopy	[13.197]
	Yb^{3+}	Spectroscopy	[13.31]
$Gd_3Sc_2Al_3O_{12}$	Cr^{3+}	Graded-index fiber	[13.75]
$GdAlO_3$	Er^{3+}	Spectroscopy	[13.198]
	Ti^{3+}-Ti^{4+}	Growth process/spectroscopy	[13.194]
Eutectic $GdAlO_3$-Al_2O_3	Pure	Growth process/synthesis	[13.36]
$GdTaO_4$	Pure	Growth process/synthesis	[13.199]
$GdTaO_4$-$RETaO_4$;	Pure	Growth process/synthesis	
RE = Er, Yb		X-ray optics	[13.200]
		X-ray optics	[13.201]
$K_2NdNb_5O_{15}$	Pure	Spectroscopy	[13.116]
K_2O-WO_3 binary system	Pure	Growth process/synthesis	[13.202]
$K_3Li_{2-x}ME_{5+x}O_{15+2x}$;	Pure	Dielectric	[13.203]
ME = Nb, Ta	Pure	Ferroelectric	[13.204]
	Nd^{3+}	Second-harmonic generation	[13.107, 108, 205]
		Growth process/synthesis	
	Pure		[13.206–210]
	Pure		[13.25]
$La_{1-x}ME_xMnO_3$;	Pure	Magnetic	[13.211–213]
ME = Sr, Ca		Magnetic	[13.212–214]
		Growth process/synthesis	[13.215]
$LaAlO_3$		Growth process/synthesis	
	Pure		[13.38]
	Cr^{3+}		[13.33, 54]
$LaLuO_3$	Ce	Spectroscopy	[13.216]
La-Sr-Cu-O	Pure	Superconductivity	[13.217]
$Li_{1-x}Nb_{1-x}W_xO_3$	Nd^{3+}	Growth process/synthesis	[13.218]
LiB_3O_5	Pure	Growth process/synthesis	[13.219]

Table 13.2 (cont.)

Compound	Doping	Applications/characteristics	References
$LiNbO_3$	Pure	Ferroelectric	[13.220, 221]
	Pure	Spectroscopy	[13.222–225]
	Nd:MgO	Second-harmonic generation	[13.226]
	Fe	Spectroscopy	[13.227, 228]
	Er^{3+}:Sc_2O_3	Visible and IR luminescence	[13.229]
	Nd^{3+}:Sc_2O_3	Spectroscopy	[13.230]
	Yb^{3+}:Sc_2O_3	Spectroscopy/laser action	[13.231]
		Growth process/synthesis	
	Pure		[13.30, 41, 232–235]
	Pure		[13.236–239]
	Fe		[13.240]
	MgO		[13.27, 61, 241–244]
	Mg, Zn		[13.245]
	Mg, Ti		[13.246]
	Cr		[13.21]
$LiTaO_3$	Pure	Growth process/synthesis	[13.247]
$LiYF_4$	Yb^{3+}	Spectroscopy/laser action	[13.248]
		Growth process/synthesis	
	Nd^{3+}		[13.249]
	Tm^{3+}		[13.249]
Lu_2O_3	Pure	Refraction index	[13.193]
	Yb^{3+}	Spectroscopy/laser action	[13.31]-[13.231]
$Lu_4Al_2O_9$	Ce	Spectroscopy	[13.216]
Mg_2SiO_4	Cr	Growth process/synthesis	[13.250]
$MgAl_2O_4$	Pure	Mechanical	[13.251]
		Growth process/synthesis	
	Ni		[13.252]
MgO-Nb_2O_5 binary system: $(Mg, Nb)O_{2.42}$, $Mg_5Nb_4O_{15}$, $Mg_4Nb_2O_9$, $MgNb_2O_6$	Pure	Growth process/synthesis	[13.253]
Eutectic $MgTiO_3$-$CaTiO_3$	Pure	Growth process/synthesis	[13.254]
MTi_2O_7; M = La, Nd	Pure	Piezoelectricity	[13.255]
		Piezoelectricity	[13.255]
NaF	U, Cu	Photoluminescence	[13.256]
Nb_2O_5:MO_2	M = Ti	Dielectric	[13.257–260]
	M = Si	Dielectric	[13.260]
$RE_{1-x}La_xVO_4$; RE = Y, Gd	Nd^{3+}	Spectroscopy	[13.261, 262]
	Tm^{3+}	Spectroscopy	[13.263]
		Growth process/synthesis	
	Pure		[13.80, 264–266]
	Nd^{3+}		[13.267–271]
	Nd^{3+}		[13.267, 268]
$RETiNbO_6$	RE = Nd; Pr; Er	Spectroscopy	[13.272–274]
Sc_2O_3	Pure	Refraction index	[13.193]
	Yb^{3+}	Spectroscopy/laser action	[13.231, 275]
$ScTaO_4$	Pure	Growth process/synthesis	[13.276]

Table 13.2 (cont.)

Compound	Doping	Applications/characteristics	References
$Sr(Al_{0.5}M_{0.5})O_3$	M = Ta; Nb	Dielectric	[13.95]
	M = Ta; Nb	Spectroscopy	[13.277]
		Growth process/synthesis	
	M = Ta; Nb		[13.278]
Sr_2RuO_4	Pure	Growth process/synthesis	[13.279]
$SrAl_2O_4$	Ce^{3+}	Photoconductivity	[13.181, 280]
	Eu^{2+}, Dy^{3+}	Spectroscopy	[13.182]
	Tb^{3+}, Ce^{3+}	Phosphorescence	[13.281]
Sr-Ca-Ti-O	Pure	Dielectric	[13.282]
$SrHfO_3$	Pure	Growth process/synthesis	[13.283]
$SrTiO_3$	Pure	Growth process/synthesis	[13.37, 118]
$SrVO_3$	Pure	Growth process/synthesis	[13.29, 284]
$Sr_xBa_{1-x}Nb_2O_6$	Pure	Dielectric	[13.285–287]
	Ce	Photorefractive	[13.288–290]
	Nd	Spectroscopy	[13.291]
		Growth process/synthesis	
	Pure		[13.22, 56, 292–294]
Ta_2O_5	Ti	Spectroscopy	[13.295]
TiO_2	Pure	Growth process/synthesis	[13.296]
UNi_2Al_3	Pure	Growth process/synthesis	[13.26]
UPt_3	Pure	Growth process/synthesis	[13.26]
Y_2O_3	Tb^{3+}, Ce^{3+}	Spectroscopy	[13.297, 298]
	Eu^{3+}	Spectroscopy	[13.299–303]
	Yb^{3+}	Spectroscopy/laser action	[13.31, 298, 304–306]
	Ho^{3+}	Spectroscopy	[13.298, 304]
	Er^{3+}	Spectroscopy	[13.304]
	Tm^{3+}	Spectroscopy/laser action	[13.306]
	Tb^{3+}	Spectroscopy/laser action	[13.306]
	Tm^{3+}, Pr^{3+}	Spectroscopy	[13.298]
	Er^{3+}, Yb^{3+}	Spectroscopy/laser action	[13.307]
$Y_{2-x}Sc_xO_3$	Eu^{3+}	Spectroscopy	[13.308]
$Y_3Al_5O_{12}$	Tm^{3+}	Fluorescence/laser action	[13.309, 310]
	Nd^{3+}	Optical sensor/thermotherapy	[13.311–314]
	Nd^{3+}	Laser action	[13.315]
	Ca, Cr	Coloration/laser action	[13.42, 316]
	Cr^{4+}, Cr^{3+}	Fluorescence	[13.311, 317]
	Cr^{3+}	Fluorescence/temperature sensor	[13.318]
	Er^{3+}	Fluorescence/temperature sensor	[13.319]
	Pr^{3+}	Spectroscopy	[13.320]
	Yb^{3+}	Spectroscopy	[13.321]
		Growth process/synthesis	
	Cr^{4+}		[13.35, 322, 323]
	Ti^{3+}		[13.324]
$Y_3Fe_5O_{12}$	Pure	Growth process/synthesis	[13.325–328]
$YAlO_3$	Nd^{3+}	Laser action	[13.329]
Yb_2O_3	Pure	Refractive index	[13.193]

Table 13.2 (cont.)

Compound	Doping	Applications/characteristics	References
Y-Ba-Cu-O	Pure	Superconductivity	[13.134, 330]
	Pure	Magnetic	[13.331]
		Crystal growth/synthesis	
	Pure		[13.332, 333]
$ZnLiNbO_4$	Pure	Growth process/synthesis	[13.334]
ZrO_2:Y_2O_3	Pure	IR waveguide	[13.308]
	Er^{3+}, Pr^{3+}	Photoluminescence	[13.335]
	Er^{3+}	Mechanical	[13.336, 337]
	Pure	Growth process/synthesis	[13.338]
Eutectic ZrO_2-CaO-NiO	Pure	Mechanical	[13.93]
		Growth process/synthesis	[13.339]
Eutectic ZrO_2-MgO	Pure	Mechanical	[13.93]
$\alpha(Ba_{1-x}Sr_x)Nb_2O_6$:$\beta(Na_{1-y}K_y)$ NbO_3	Pure	Dielectric	[13.340]
β-BaB_2O_4	Pure	Growth process/synthesis	[13.341, 342]
$(1-x)Pb(Mg_{0.33}Nb_{0.67})O_3$-$xPbTiO_3$	Pure	Growth process/synthesis	[13.343]
$(Gd_{1-x}Nd_x)_2(SiO_4)O$	Pure	Growth process/synthesis	[13.344]
$(La_{1-x}Nd_x)_{9.33}\{(SiO_4)_6O_2\}_{0.67}$	Pure	Growth process/synthesis	[13.344]
$(Lu_{1-x}Nd_x)_2Si_2O_7$	Pure	Growth process/synthesis	[13.344]
2-adamantylamino-5-nitropyridine (AANP)	Organic	Growth process/synthesis	[13.24]
$A_2B_2O_7$ (A = La, Nd, Ca, B = Ti, Nb)	Pure	Dielectric	[13.345]
Miscellaneous:			
Compiled information (reviews)			[13.346–351]
Several oxides and fluorides matrix			[13.352–356]
$La_{0.7}Sr_{0.3}MnO_3$			[13.357]
$LiNbO_3$	–	–	[13.358]
Mixed-oxide perovskites			[13.359]
Al_2O_3			[13.360]
Simulation/modeling	–	–	[13.361–363]

etry control: seed compositions, self-adjusting melt composition, and pulling rate.

The seed composition strategy was already discussed in the seeding aspects of the LHPG technique (Sect. 13.2.1). Another strategy is to allow the system to self-adjust its melt composition. In the case of SBN fiber pulling, Erdei et al. observed that, after a transient composition at the beginning of fiber growth, the melt self-adjusted its composition and a high-quality fiber was obtained. Although a slightly off-congruent source rod composition was used the *composition-control system* of the LHPG technique modifies both the floating zone composition and the highly complex segregation effects, and can produce fibers that are homogeneous in the growth direction [13.56]. However, the intensive convection flows generated in the molten zone are highlighted as responsible for the radial composition variations, as described earlier in Sect. 13.3.5. In another work, Erdei et al. also prepared YVO_4 single-crystal fibers, and the authors identified that the Y/V stoichiometry ratio was mainly caused by deficiency of oxygen due to vanadium oxide dissociation. However, it was *Huang* et al. [13.364] who was able to produce high-quality single-crystal fibers from stoichiometric-composition pedestals. The authors successfully applied pulling rates above 1 mm/min in order to avoid vanadium oxide dissociation and in this way obtained stoichiometric YVO_4 single-crystal fibers. The same strategy was used by *Burlot* et al. [13.247] for growth of $LiTaO_3$ in order to avoid loss of Li_2O. The best results were obtained for pulling speeds of 0.7 mm/min. *Nagashio* et al. [13.232] also used high pulling rates in the LHPG technique to obtain near-stoichiometric $LiNbO_3$ fibers. They found that high pulling rates (4.1 mm/min) associated with a small molten zone length were the best conditions to avoid Li_2O evaporation, and thus to obtain stoichiometric fibers. Also, according to the *Burton–Prim–Slichter* (BPS) theory, the increased pulling rate leads the effective distribution coefficient (k_{eff}) to approach the value of 1 [13.59].

13.4.2 Incongruently Melting and Evaporating Fibers

An important characteristic of the LHPG technique is the possibility of growing incongruently melting and evaporating materials. Growth of incongruently melting bulk oxide single crystal is usually achieved by high-temperature solution growth, which is a slow and complicated technique. The source material in this technique must be rich in one or more elements to compensate for the incongruent melting. However in the LHPG system it has been shown that it is possible to grow such materials, $Bi_{12}TiO_{20}$ (BTO) for example, without the need for enrichment with excess Bi_2O_3 in the source composition. In other words, it is possible to grow incongruently melting materials from a source with the same composition as the grown fiber. According to *Feigelson* [13.8], the liquid composition in such growth experiments changes gradually until it naturally reaches the composition necessary for crystal growth, similar to the effect observed for SBN fiber growth (Sect. 13.4.1). This implies that, at the beginning of the process, the crystallizing solid must have a nonstoichiometric composition to allow the liquid phase to change gradually. With the LHPG technique it was possible to growth single-crystal fibers of BTO compound with 300–1200 μm diameter and up to 70 mm length (Fig. 13.16). The pulling rates typically used in this process were 6–18 mm/h, much higher than the few millimeters per day available with high-temperature solution methods.

Fig. 13.16 The side of a $Bi_{12}TiO_{20}$ single crystal fiber grown from a stoichiometric pedestal, evidencing the natural face on the $[1\bar{1}0]$ crystallographic direction

By conveniently controlling the pulling and feeding ratio, it is also possible to grow materials with incongruent evaporation, such as Sr_2RuO_4. The conventional method for growing Sr_2RuO_4 single crystals was floating zone melting, but due to the high evaporation rates of Ru_2O_3, only single crystals of a few millimeters had been grown. Using the LHPG technique, and with a prereacted powder and then extruded source material with $SrRuO_3$ composition, it was possible to growth Sr_2RuO_4 single-crystal fibers up to 30 mm in length and 0.8–1.0 mm in diameter [13.279]. The best results were obtained with pulling rates of 0.3 mm/min with a feeding (source) rate of 0.4 mm/min. This again was only possible because of the high temperature gradients at the growth interface, allowing the growth velocities to be higher than in the conventional floating zone technique, and that a small amount of material stays at a high temperature for a short period of time, minimizing the ruthenium oxide evaporation.

13.4.3 Eutectic Fibers

The properties of materials are dependent not only on their composition, but also on their microstructure. In this way, phase distribution, size, and shape as well as interface characteristics play a crucial role in determining the behavior of composites [13.93]. The eutectic microstructure can be separated into three basic morphologies: lamellar, fibrous, and what is known as *Chinese script*.

The entropy of melting plays a very important role in determining the final microstructure of a eutectic. In 1966, *Jackson* and *Hunt* [13.365] demonstrated that irregular microstructures such as the *Chinese script* are formed during faceted–nonfaceted growth, in which one phase has high while the other has low entropy of melting. Eutectic compounds, despite the complexity of their microstructure, obey the following relationship [13.365]

$$\lambda_e^2 v = C', \qquad (13.19)$$

where λ_e is the mean spacing between the phases, v is the pulling rate, and C is a constant.

The Al_2O_3-$ZrO_2(Y_2O_3)$ system, called ZA, is one of the eutectic systems that have been systematically studied by the LHPG technique [13.87, 90, 92]. *Farmer* and *Sayir* [13.90] studied the fracture strength of 68 mol % Al_2O_3 in the ZA hypoeutectic composition

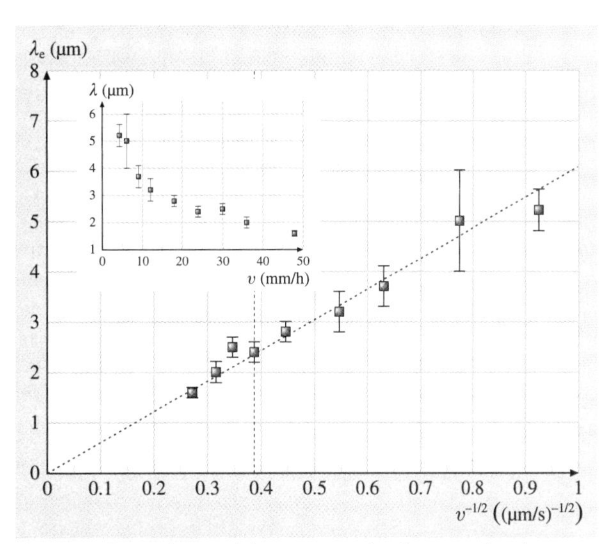

Fig. 13.17 Dependence of the mean spacing between the phases on the pulling rate of $Al_2O_3/GdAlO_3$. Only the *Chinese script* microstructure is present in all fibers on the *right side* of the *vertical line* (after [13.36])

at varying Y_2O_3 content in fibers ranging from 1.0 to 1.8 mm in diameter. Their experiments showed that the Y_2O_3 composition has a major influence on the eutectic microstructure. Pure ZA eutectics have a lamellar microstructure at a pulling rate of 40 mm/h, although with the inclusion of Y_2O_3 the melt became constitutionally supercooled and a planar growth front in not maintained. Under conditions of large undercooling, the leading phase Al_2O_3 facets on the *r*-planes $\{1\bar{1}02\}$ and a transition from lamellar to ZrO_2 rod morphology occur. In this new condition, the fracture strength increases from 0.7 to 1 GPa. This increase is more strongly influenced by partial stabilization of the ZrO_2 than by the change in the phase morphology. With 1.1 mol % or less of Y_2O_3 the observed morphology shows excess Al_2O_3 accommodated in Al_2O_3-rich colony boundary regions and formation of Al_2O_3-rich facets on the fiber exterior in small-diameter (1.0 mm) fibers. With Y_2O_3 composition greater than 1.1 mol %, the system is highly supercooled and primary Al_2O_3 dendrites nucleate. The authors also observed, for compositions in the range 3.5–7.6 mol % Y_2O_3, the formation of large amounts of $Y_3Al_5O_{12}$ within the colony boundary [13.90]. *Peña* et al. [13.94] pulled fibers with 1.5 mm diameter with a composition of 9 mol % Y_2O_3. The authors reported that they were successful in producing fiber with highly homogeneous morphology. This was accomplished by the use of very low pulling rate (10 mm/h), which in other words, increased the value of the ratio G/v, avoiding the constitutional supercooling condition. *Francisco* et al. [13.87] studied the influence of the processing conditions on the eutectic microstructure and reported a transition from coupled to dendrite growth at about 50 mm/h with axial gradient of 6×10^5 K/m, regardless of the rotation speed used.

Another eutectic explored by the LHPG system is Al_2O_3-YAG. *Pastor* et al. [13.82] pulled 1 mm diameter fibers from pedestals containing 81.5 mol % Al_2O_3 and 18.5 mol % Y_2O_3, the known eutectic composition. The fibers showed a *Chinese script*-type homogeneous microstructure. Due to the characteristics of its complex microstructure the hardness was fairly isotropic and the longitudinal strength of eutectics pulled at low and medium rates remained practically constant up to 1700 K.

The substitution of YAG by $GdAlO_3$ produced another interesting aluminate eutectic system ($Al_2O_3/GdAlO_3$) with the *Chinese script* microstructure. *Andreeta* et al. [13.36] reported the pulling of $Al_2O_3/GdAlO_3$ eutectic fibers by LHPG technique using unreacted pedestals with a mixture of 77 mol %

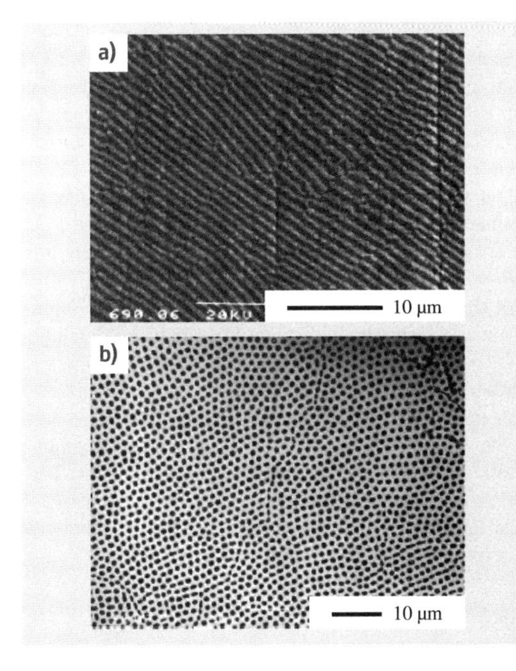

Fig. 13.18a,b SEM photographs of (**a**) $ZrO_2(CaO)$–NiO, and (**b**) transverse section of ZrO_2–MgO (after [13.93]) ◄

Al_2O_3 and 23 mol% Gd_2O_3. The average fiber diameter was 600 μm and the pulling rate ranged from 4.2 to 48.0 mm/h. In transversal analysis of the fibers, it could be observed that for pulling rates higher than 24 mm/h a transition from homogeneous to a complex regular microstructure appeared, with circular regions with larger $GdAlO_3$ phases, and even cellular arrangements with a fibrous pattern inside. Using pulling rates lower than 24 mm/h and with temperature gradients of 6×10^5 K/m, the high value of G/v prevented constitutional supercooling and fibers without the complex regular microstructure could be obtained. The Jackson and Hunt relationship [13.365] that holds for this compound was determined to be $\lambda^2 v = 40\,\mu m^3/s$ (Fig. 13.17).

Orera et al. [13.93] explored the optical properties of the eutectics with regular microstructure to be used as waveguide devices (Fig. 13.18). The eutectics have atomic-scale interfaces, producing sharp refractive index transitions between the phases, which lead to diffraction, interference, polarization effects, etc. Orera et al. also studied the waveguide effect of the fibrous microstructure of the eutectics. The authors showed that the CaF_2–MgO eutectics can produce single-mode waveguides in the third optical window (1500–1600 nm). They also reported the cutoff wavelength, i.e., the most energetic light that can propagate in single-mode form, for the CaO–ZrO_2 (λ_c = 1700 nm), Al_2O_3–ZrO_2 (λ_c = 500 nm), and CaF_2–MgO (λ_c = 1300 nm) eutectic systems. Since in the calcium fluoride eutectic light is guided by MgO fibers, it forms an optical-fiber bunch integrated in a crystal matrix which can be used for optical image transfer, giving a density higher than 40 000 pixels/mm^2.

13.5 Conclusions

The laser-heated pedestal growth technique (LHPG) has been presented as a powerful research tool for crystal growth of new and conventional compounds. In order to understand the state of the art of this technique, we report a historical background that led us to the beginning of the 20th century, when Baker and da Costa Andrade, one of Laue's and Bragg's coworkers, developed the first metallic single-crystal wires. One of the greatest breakthroughs in fiber research was, however, achieved by Fejer and coworkers with the introduction of an optical component (the reflaxicon) to yield what is nowadays called the LHPG technique. This can be verified by the large increase in the number of papers published on the subject after the publication of Fejer and Feigelson's papers in the 1980s. The publication dynamics for laser-heated fiber growth over the last two decades shows that this technique is already well established, with more than 20 papers produced each year and more than 80 new and conventional different crystallographic matrices being produced. It must be noted that in this chapter only fiber pulling by LHPG technique was considered, which means that other fiber-pulling techniques such as the pulling down, melt extraction, and vapor reaction techniques were not included, which would certainly further increase the volume of scientific production.

The trends in the laser-heated pedestal technique can be separated into three basic lines. The first obvious trend is basic research into the fiber-pulling process, since the technique allows visualization of the molten zone and crystallization interfaces. In this way, it is possible to gain access to many solidifying parameters such as temperature gradients, interface shape, and contact angle. These are important, especially for very high melting point (> 2500 °C) compounds, and are virtually impossible to measure in other conventional, crucible techniques. The second line is the miniaturization of some of today's devices by substituting bulk crystals with crystalline fiber. Along these lines it is possible to include the design of completely new devices based on the shape properties of fibers, such as high-temperature thermometers and infrared (IR) waveguides. Finally, with the approach of the data storage limit in semiconductor technology that could lead to holographic storage, and the need for new and efficient scintillators, ferroelectric compounds, and solid-state laser hosts, the most promising use of the laser-heated pedestal growth in the near future is its association with the combinatorial approach for new materials development. The LHPG technique can be used to test the possibility to grow large single crystals of the chosen compound and look more deeply into its physical properties.

References

13.1 B.B. Baker: On the stretching and breaking of sodium and potassium, Proc. Phys. Soc. Lond. **25**, 235 (1912)
13.2 E.N.C. da Andrade: Regular surface markings stretched wires of soft metals, Philos. Mag. **27**, 869 (1914)
13.3 P.P. Ewald (Ed.): *Fifty Years of X-ray Diffraction* (Oosthoek, Utrecht 1962)
13.4 P. Rudolph: What do we want with fiber crystals? An introductory overview, Adv. Mater. Res. **6**, 1–46 (2004)
13.5 H.J. Scheel: Historical aspects of crystal growth technology, J. Cryst. Growth **211**, 1–12 (2000)
13.6 J. Czochralski: Ein neues Verfahren zur Messung der Kristallisationsgeschwindigkeit der Metalle, Z. Phys. Chem. **92**, 219 (1918), in German
13.7 E.V. Gomperz: Untersuchungen an Einkristalldrähten, Z. Phys. A **8**, 184–190 (1922), in German
13.8 R.S. Feigelson: *Growth of Fiber Crystals. Crystal Growth of Electronic Materials* (North-Holland, Amsterdam 1985)
13.9 H.E. Labelle Jr., A.I. Mlavsky: Growth of sapphire filaments from the melt, Nature **216**, 574–575 (1967)
13.10 H.E. Labelle Jr., A.I. Mlavsky: Growth of controlled profile crystals from the melt: Part I – Sapphire filaments, Mater. Res. Bull. **6**, 571 (1971)
13.11 J.S. Haggerty: Production of fibers by a floating zone fiber drawing technique, Final Report NASA-CR-120948 (1972)
13.12 B. Cockayne, D.B. Gasson: The machining of oxides using gas lasers, J. Mater. Sci. **5**, 837 (1970)
13.13 C.A. Burrus, J. Stone: Single-crystal fiber optical devices: A Nd:YAG fiber laser, Appl. Phys. Lett. **26**, 318 (1975)
13.14 J. Stone, C.A. Burrus: Nd:Y_2O_3 single-crystal fiber laser: Room temperature CW operation at 1.07 and 1.35 μm wavelength, J. Appl. Phys. **49**, 2281 (1978)
13.15 C.A. Burrus, L.A. Coldren: Growth of single crystal sapphire-clad ruby fibers, Appl. Phys. Lett. **31**, 383 (1977)
13.16 M.M. Fejer, R.L. Byer, R.S. Feigelson, W. Kway: Growth and characterization of single crystal refractory oxide fibers, advances in Infrared Fibers II, SPIE, Proc. 2nd Meet. Los Angeles (Bellingham 1982), A83-46621 22-74
13.17 M.M. Fejer, J.L. Nightingale, G.A. Magel, R.L. Byer: Laser-heated miniature pedestal growth apparatus for single-crystal optical fibers, Rev. Sci. Instrum. **55**, 1791–1796 (1984)
13.18 W.R. Edmonds: The reflaxicon: A new reflective optical element and some applications, Appl. Opt. **12**, 1940 (1973)
13.19 M.M. Fejer: Single crystal fibers: Growth dynamics and nonlinear optical interactions. Ph.D. Thesis (Stanford Univ., Stanford 1986)
13.20 M.M. Fejer, G.A. Magel, R.L. Byer: High speed high-resolution fiber diameter variation measurement system, Appl. Opt. **24**, 2362 (1985)
13.21 S. Uda, W.A. Tiller: The influence of an interface electric field on the distribution coefficient of chromium in $LiNbO_3$, J. Cryst. Growth **121**, 93–110 (1992)
13.22 Y. Sugiyama, I. Hatakeyama, I. Yokohama: Growth of *a* axis strontium barium niobate single crystal fibers, J. Cryst. Growth **134**, 255–265 (1993)
13.23 V. Phomsakha, R.S.F. Chang, N. Djeu: Novel implementation of laser heated pedestal growth for the rapid drawing of sapphire fibers, Rev. Sci. Instrum. **65**, 3860–3861 (1994)
13.24 A. Yokoo, S. Tomaru, I. Yokohama, H. Itoh, T. Kaino: A new growth method for long rod-like organic nonlinear optical crystals with phase-matched direction, J. Cryst. Growth **156**, 279–284 (1995)
13.25 T. Imai, S. Yagi, Y. Sugiyama, I. Hatakeyama: Growth of potassium tantalate niobate single crystal fibers by the laser-heated pedestal growth method assisted by a crystal cooling technique, J. Cryst. Growth **147**, 350–354 (1995)
13.26 E. Brueck, H.J. Gelders, B.J. Harrison, A.A. Menovsky: Laser-heated fibre pedestal growth under UHV conditions, J. Cryst. Growth **166**, 394–397 (1996)
13.27 A. Brenier, G. Foulon, M. Ferriol, G. Boulon: The laser-heated-pedestal growth of $LiNbO_3$:MgO crystal fibres with ferroelectric domain inversion by in situ electric field poling, J. Phys. D **30**, L37–L39 (1997), rapid communication
13.28 R.K. Nubling, J.A. Harrington: Optical properties of single-crystal sapphire fibers, Appl. Opt. **36**, 5934–5940 (1997)
13.29 D.R. Ardila, J.P. Andreeta, C.T.M. Ribeiro, M.S. Li: Improved laser-heated pedestal growth system for crystal growth in medium and high isostatic pressure environment, Rev. Sci. Instrum. **70**, 4606–4608 (1999)
13.30 D.R. Ardila, L.B. Barbosa, J.P. Andreeta: Bifocal spherical mirror for laser processing, Rev. Sci. Instrum. **72**, 4415–4418 (2001)
13.31 L. Laversenne, Y. Guyot, C. Goutaudier, M.T. Cohen-Adad, G. Boulon: Optimization of spectroscopic properties of Yb^{3+}-doped refractory sesquioxides: cubic Y_2O_3, Lu_2O_3 and monoclinic Gd_2O_3, Opt. Mater. **16**, 475–483 (2001)
13.32 M.R.B. Andreeta, E.R.M. Andreeta, A.C. Hernandes, R.S. Feigelson: Thermal gradient control at the solid–liquid interface in the laser-heated pedestal growth technique, J. Cryst. Growth **234**, 759–761 (2002)
13.33 M.R.B. Andreeta, L.C. Caraschi, A.C. Hernandes: Automatic diameter control system applied to the

laser heated pedestal growth technique, Mater. Res. **6**, 1 (2003)
13.34 M.F. Carrasco, M.R. Soares, V.S. Amaral, J.M. Vieira, R.F. Silva, F.M. Costa: Bi-Sr-Ca-Cu-O superconducting fibers processed by the laser floating zone technique under different electrical current intensities, Supercond. Sci. Technol. **19**, 373–380 (2006)
13.35 C.Y. Lo, K.Y. Huang, J.C. Chen, C.Y. Chuang, C.C. Lai, S.L. Huang, Y.S. Lin, P.S. Yeh: Double-clad Cr^{4+}:YAG crystal fiber amplifier, Opt. Lett. **30**, 129–131 (2005)
13.36 E.R.M. Andreeta, M.R.B. Andreeta, A.C. Hernandes: Laser heated pedestal growth of $Al_2O_3/GdAlO_3$ eutectic fibers, J. Cryst. Growth **234**, 782–785 (2002)
13.37 D.R. Ardila, M.R.B. Andreeta, S.L. Cuffini, A.C. Hernandes, J.P. Andreeta, Y.P. Mascarenhas: Single-crystal $SrTiO_3$ fiber grown by laser heated pedestal growth method: Influence of ceramic feed rod preparation in fiber quality, Mater. Res. **1**, 11 (1998)
13.38 M.R.B. Andreeta, E.R.M. Andreeta, A.C. Hernandes: Laser-heated pedestal growth of colorless $LaAlO_3$ single crystal fiber, J. Cryst. Growth **275**, e757–e761 (2005)
13.39 L.B. Barbosa, D.R. Ardila, E.M. Kakuno, R.H. Camparin, C. Cusatis, J.P. Andreeta: Processing of crystals with controlled lattice parameter gradient by the LHPG technique, J. Cryst. Growth **250**, 67–71 (2003)
13.40 Z. Lu, L.V. Moulton, R.S. Feigelson, R.J. Raymakers, P.N. Peszkin: Factors affecting the growth of single crystal fibers of the superconductor $Bi_2Sr_2CaCu_2O_8$, J. Cryst. Growth **106**, 732–741 (1990)
13.41 C.Y. Chen, J.C. Chen, Y.J. Lai: Investigations of the growth mechanism of stoichiometric $LiNbO_3$ fibers grown by the laser-heated pedestal growth method, J. Cryst. Growth **275**, e763–e768 (2005)
13.42 S. Ishibashi, K. Naganuma, I. Yokohama: Cr,Ca: $Y_3Al_5O_{12}$ laser crystal grown by the laser-heated pedestal growth method, J. Cryst. Growth **183**, 614–621 (1998)
13.43 H.S. Carslaw, J.C. Jaeger: *Conduction of Heat in Solids* (Oxford Univ. Press, London 1959)
13.44 J.C. Brice: The cracking of Czochralski grown crystals, J. Cryst. Growth **42**, 427 (1977)
13.45 M.R.B. Andreeta: Implantação da técnica de crescimento de Cristais por fusão a laser e a preparação de fibras monocristalinas óxidas. Master Thesis (Universidade de São Paulo, São Paulo 1996), in Portuguese
13.46 V.V. Prokofiev, J.P. Andreeta, C.J. de Lima, M.R.B. Andreeta, A.C. Hernandes, J.F. Carvalho, A.A. Kamshilin, T. Jääskeläinen: The relation between temperature gradients and structural perfection of single-crystal $Bi_{12}SiO_{20}$ and $Bi_{12}TiO_{20}$ fibers grown by the LHPG method, Opt. Mater. **4**, 433–436 (1995)
13.47 K.M. Kim, A.B. Dreeben, A. Schujko: Maximum stable zone length in float-zone growth of small diameter sapphire and silicon crystals, J. Appl. Phys. **50**, 4472 (1979)
13.48 M. Saitou: Shape and stability of a floating liquid zone between two solids, J. Appl. Phys. **82**, 6343–6345 (1997)
13.49 D.R. Ardila, L.V. Cofre, L.B. Barbosa, J.P. Andreeta: Study of floating zone profiles in materials grown by the laser-heated pedestal growth technique under isostatic atmosphere, Cryst. Res. Technol. **3**(9), 855–858 (2004)
13.50 A.L. Greer: Too hot to melt, Nature **404**, 134 (2000)
13.51 C.T.M. Ribeiro, D.R. Ardila, J.P. Andreeta, M.S. Li: Effects of isostatic oxygen pressure on the crystal growth and optical properties of undoped and Er^{3+}-doped $Ca_3(VO_4)_2$ single-crystal fibres, Adv. Mater. Opt. Electron. **10**, 9–15 (2000)
13.52 L.S. Wu, A.H. Wang, J.M. Wu, L. Wei, G.X. Zhu, S.T. Ying: Growth and laser properties of Ti:sapphire single-crystal fibers, Electron. Lett. **31**, 1151–1152 (1995)
13.53 J.H. Sharp, T.P.J. Han, B. Henderson, R. Illingworth, I.S. Ruddock: Dopant incorporation in single-crystal fibre growth by the laser-heated miniature pedestal growth technique, J. Cryst. Growth **131**, 457–464 (1993)
13.54 M.R.B. Andreeta, L.C. Caraschi, F. Agulló-Rueda, A.C. Hernandes: Periodic doping in single crystal fibers grown by laser-heated pedestal growth technique, J. Cryst. Growth **242**, 395–399 (2002)
13.55 C.M. Liu, J.C. Chen, C.H. Chiang, L.J. Hu, S.P. Lin: Mg-doped sapphire crystal fibers grown by laser-heated pedestal growth method, Jpn. J. Appl. Phys. **45**, 194–199 (2006)
13.56 S. Erdei, L. Galambos, I. Tanaka, L. Hesselink, L.E. Cross, R.S. Feigelson, F.W. Ainger, H. Kojima: Inhomogeneities and segregation behavior in strontium-barium niobate fibers grown by laser-heated pedestal growth technique – Part II, J. Cryst. Growth **167**, 670–680 (1996)
13.57 M.F. Carrasco, R.F. Silva, J.M. Vieira, F.M. Costa: Electrical field freezing effect on laser floating zone (LFZ)-grown $Bi_2Sr_2Ca_2CU_4O_{11}$ superconducting fibers, Supercond. Sci. Technol. **17**, 612–619 (2004)
13.58 W.A. Tiller, C.T. Yen: Some consequences of a strong interface field-effect operating during the growth of TiO_2-alloy crystals from the melt, J. Cryst. Growth **109**, 120 (1991)
13.59 J.A. Burton, R.C. Prim, W.P. Slichter: The distribution of solutes in crystals grown from the melt: Part I: Theoretical, J. Chem. Phys. **21**, 1987 (1953)
13.60 W.G. Pfann, R.S. Wagner: Simple method of measuring stress relaxation, Trans. Metall. Soc. AIME **224**, 1083 (1962)
13.61 L.M. Lee, C.C. Kuo, J.C. Chen, T.S. Chou, Y.C. Cho, S.L. Huang, H.W. Lee: Periodical poling of MgO doped lithium niobate crystal fiber by modulated pyroelectric field, Opt. Commun. **253**, 375–381 (2005)

13.62 S. Erdei, F.W. Ainger: Trends in the growth of stoichiometric single crystals, J. Cryst. Growth **174**, 293 (1997)

13.63 Y.T. Wang, D.S. Wang, W.Q. Ge, L.C. Cui: A sapphire fibre thermal probe based on fast Fourier transform and phase-lock loop, Chin. Phys. **15**, 975–979 (2006)

13.64 H.C. Seat, J.H. Sharp, Z.Y. Zhang, K.T.V. Grattan: Single-crystal ruby fiber temperature sensor, Sens. Actuators A **101**, 24–29 (2002)

13.65 K.T.V. Grattan, Z.Y. Zhang, T. Sun, Y. Shen, L. Tong, Z. Ding: Sapphire-ruby single-crystal fibre for application in high temperature optical fibre thermometers: Studies at temperatures up to 1500 °C, Meas. Sci. Technol. **12**, 981–986 (2001)

13.66 Y. Shen, Y. Wang, L. Tong, L. Ye: Novel sapphire fiber thermometer using fluorescent decay, Sens. Actuators A **71**, 70–73 (1998)

13.67 H.C. Seat, J.H. Sharp: Dedicated temperature sensing with *c* axis oriented single crystal ruby (Cr^{3+}:Al_2O_3) fibers: Temperature and strain dependences of R-line fluorescence, IEEE Trans. Instrum. Meas. **53**, 140–154 (2004)

13.68 Y.H. Shen, L.M. Tong, Y.Q. Wang, L.H. Ye: Sapphire-fiber thermometer ranging from 20 to 1800 °C, Appl. Opt. **38**, 1139–1143 (1999)

13.69 D. Bloom, D.R. Evans, S.A. Holmstrom, J.C. Polf, S.W.S. McKeever, V. Whitley: Characterization of Al_2O_3 single crystals grown by the laser-heated pedestal growth technique for potential use in radiation dosimetry, Radiat. Meas. **37**, 141–149 (2003)

13.70 R.K. Nubling, J.A. Harrington: Single-crystal laser-heated pedestal-growth sapphire fibers for Er:YAG laser power delivery, Appl. Opt. **37**, 4777–4781 (1998)

13.71 E.K. Renwick, E.E. Robertson, I.S. Ruddock, R. Illingworth: Optical transmission properties of single crystal fibres, Opt. Commun. **123**, 477–482 (1996)

13.72 A.G. Sinclair, G. McCormack, J.H. Sharp, I.S. Ruddock, R. Illingworth: Uniaxial crystalline fibers – optical methods for determining their physical characteristics, Meas. Sci. Technol. **4**, 1501–1507 (1993)

13.73 G.N. Merberg, J.A. Harrington: Optical and mechanical-properties of single-crystal sapphire optical fibers, Appl. Opt. **32**, 3201–3209 (1993)

13.74 H.F. Wu, A.J. Perrotta, R.S. Feigelson: Mechanical characterization of single-crystal α-Al_2O_3 fibers grown by the laser-heated pedestal growth technique, J. Mater. Sci. Lett. **10**, 1428–1429 (1991)

13.75 J.H. Sharp, R. Illingworth, I.S. Ruddock: Graded-index characteristics in single-crystal fibers, Opt. Lett. **23**, 109–110 (1998)

13.76 J.K. Krebs, U. Happek: Yb^{3+} energy levels in α-Al_2O_3, J. Lumin. **94**, 65–68 (2001)

13.77 H.C. Seat, J.H. Sharp: Er^{3+}+Yb^{3+}-codoped Al_2O_3 crystal fibres for high-temperature sensing, Meas. Sci. Technol. **14**, 279–285 (2003)

13.78 J.H. Sharp, C.W.P. Shi, H.C. Seat: Er-doped sapphire fibre temperature sensors using upconversion emission, Meas. Control **34**, 170 (2001)

13.79 J.H. Sharp, T.J.P. Han, B. Henderson, R. Illingworth: Instability in the growth of Ti:Al_2O_3 single-crystal fibres, J. Cryst. Growth **140**, 79–83 (1994)

13.80 S. Erdei, G.G. Johnson, F.W. Ainger: Growth studies of YVO_4 crystals (II). Changes in Y-V-O-stoichiometry, Cryst. Res. Technol. **29**, 815–828 (1994)

13.81 J. Ramírez Rico, A.R. Pinto Gómez, J. Martínez Fernández, A.R. de Arellano López, P.B. Oliete, J.I. Peña, V.M. Orera: High-temperature plastic behavior of Al_2O_3-$Y_3Al_5O_{12}$ directionally solidified eutectics, Acta Mater. **54**, 107–3116 (2006)

13.82 J.Y. Pastor, J. Llorca, A. Salazar, P.B. Oliete, I. Francisco, J.I. Peña: Mechanical properties of melt-grown alumina-yttrium aluminum garnet eutectics up to 1900 K, J. Am. Ceram. Soc. **88**, 1488 (2005)

13.83 A. Laidoune, H. Lahrach, Y. Kagamitani, K. Lebbou, F. Carrillo, C. Goutaudier, O. Tillement: Growth of polycrystalline fibers with eutectic composition $Al_2O_3/Y_3Al_5O_{12}$ for composite reinforcement, J. Phys. IV **113**, 129–134 (2004)

13.84 A. Salazar, J.Y. Pastor, J. Llorca, J.I. Peña, I. Francisco, P.B. Oliete: Mechanical properties of Al_2O_3-$ZrO_2(Y_2O_3)$ and Al_2O_3-YAG eutectic composites processed by laser-heating floating zone, Bol. Soc. Esp. Ceram. Vidr. **44**, 193–198 (2006), in Spanish

13.85 C.S. Frazer, E.C. Dickey, A. Sayir: Crystallographic texture and orientation variants in Al_2O_3-$Y_3Al_5O_{12}$ directionally solidified eutectic crystals, J. Cryst. Growth **233**, 187–195 (2001)

13.86 J.I. Peña, M. Larsson, R.I. Merino, I. de Francisco, V.M. Orera, J. Lorca, J.Y. Pastor, A. Martín, J. Segurado: Processing, microstructure and mechanical properties of directionally-solidified Al_2O_3-$Y_3Al_5O_{12}$-ZrO_2 ternary eutectics, J. Eur. Ceram. Soc. **26**, 3113–3121 (2006)

13.87 I. Francisco, R.I. Merino, V.M. Orera, A. Larrea, J.I. Peña: Growth of $Al_2O_3/ZrO_2(Y_2O_3)$ eutectic rods by the laser floating zone technique: Effect of the rotation, J. Eur. Ceram. Soc. **25**, 1341–1350 (2005)

13.88 J.Y. Pastor, J. Lorca, P. Poza, I. de Francisco, R.I. Merino, J.I. Peña: Mechanical properties of melt-grown Al_2O_3-$ZrO_2(Y_2O_3)$ eutectics with different microstructure, J. Eur. Ceram. Soc. **25**, 1215–1223 (2005)

13.89 N.R. Harlan, R.I. Merino, J.I. Peña, A. Larrea, V.M. Orera, C. González, P. Poza, J. Llorca: Phase distribution and residual stresses in melt-grown Al_2O_3-$ZrO_2(Y_2O_3)$ eutectics, J. Am. Ceram. Soc. **85**, 2025 (2002)

13.90 S.C. Farmer, A. Sayir: Tensile strength and microstructure of Al_2O_3-ZrO_2 hypo-eutectic fibers, Eng. Fract. Mech. **69**, 1015–1024 (2002)

13.91 J.Y. Pastor, P. Poza, J. Llorca, J.I. Peña, R.I. Merino, V.M. Orera: Mechanical properties of directionally solidified Al_2O_3-$ZrO_2(Y_2O_3)$ eutectics, Mater. Sci. Eng. A **308**, 241–249 (2001)

13.92 J.A. Pardo, R.I. Merino, V.M. Orera, J.I. Peña, C. González, J.Y. Pastor, J. Llorca: Piezospectroscopic study of residual stresses in Al_2O_3-ZrO_2 directionally solidified eutectics, J. Am. Ceram. Soc. **83**, 2745–2752 (2001)

13.93 V.M. Orera, R.I. Merino, J.A. Pardo, A. Larrea, J.I. Peña, C. González, P. Poza, J.Y. Pastor, J. Llorca: Microstructure and physical properties of some oxide eutectic composites processed by directional solidification, Acta Mater. **48**, 4683–4689 (2000)

13.94 J.I. Peña, R.I. Merino, N.R. Harlan, A. Larrea, G.F. Fuente, V.M. Orera: Microstructure of Y_2O_3 doped Al_2O_3-ZrO_2 eutectics grown by the laser floating zone method, J. Eur. Ceram. Soc. **22**, 2595–2602 (2002)

13.95 A.S. Bhalla, R.A. Guo: Design of dielectric substrates for high T_c superconductor films, Acta Phys. Pol. A **92**, 7–21 (1997)

13.96 I.G. Siny, R.W. Tao, R.S. Katiyar, R.A. Guo, A.S. Bhalla: Raman spectroscopy of Mg-Ta order-disorder in $BaMg_{1/3}Ta_{2/3}O_3$, J. Phys. Chem. Solids **59**, 181–195 (1998)

13.97 R.L. Moreira, M.R.B. Andreeta, A.C. Hernandes, A. Dias: Polarized micro-raman spectroscopy of $Ba(Mg_{1/3}Nb_{2/3})O_3$ single crystal fibers, Cryst. Growth Des. **5**, 1457–1462 (2005)

13.98 R.A. Guo, A.S. Bhalla, L.E. Cross: $Ba(Mg_{1/3}Ta_{2/3})O_3$ single-crystal fiber grown by the laser-heated pedestal growth technique, J. Appl. Phys. **75**, 4704–4708 (1994)

13.99 Z. Yu, R.Y. Guo, A.S. Bhalla: Growth of $Ba(Ti_{1-x}Zr_x)O_3$ single crystals by the laser heated pedestal growth technique, J. Cryst. Growth **233**, 460–465 (2001)

13.100 Z. Yu, R.Y. Guo, A.S. Bhalla: Growth of $Ba(Ti_{1-x}Zr_x)O_3$ single crystal fibers by laser heated pedestal growth technique, Ferroelectr. Lett. Sect. **27**, 113–123 (2000)

13.101 Z. Yu, R.Y. Guo, A.S. Bhalla: Dielectric behavior of $Ba(Ti_{1-x}Zr_x)O_3$ single crystals, J. Appl. Phys. **88**, 410–415 (2000)

13.102 D. Garcia, R. Guo, A.S. Bhalla: Dielectric properties of $Ba_{1-x}Sr_xTiO_3$ single crystal fibers grown by laser heated pedestal growth technique, Integr. Ferroelectr. **42**, 57–69 (2002)

13.103 D. Garcia, R. Guo, A.S. Bhalla: Growth and properties of $Ba_{0.9}Sr_{0.1}TiO_3$ single crystal fibers, Mater. Lett. **42**, 136–141 (2000)

13.104 D. Garcia, R.Y. Guo, A.S. Bhalla: Field dependence of dielectric properties of BST single crystals, Ferroelectr. Lett. Sect. **27**, 137–146 (2000)

13.105 J.C. Chen, C.Y. Chen: Growth of $Ba_{1-x}Ca_xTiO_3$ single-crystal fibers by a laser heated pedestal method, J. Cryst. Growth **236**, 640–646 (2002)

13.106 L.B. Barbosa, D.R. Ardila, J.P. Andreeta: Crystal growth of congruent barium calcium titanate by LHPG, J. Cryst. Growth **231**, 488–492 (2001)

13.107 G. Foulon, M. Ferriol, A. Brenier, M.T. Cohen-Adad, M. Boudeulle, G. Boulon: Nonlinear single-crystal fibers of undoped or Nd^{3+}-doped niobates: Growth by LHPG, spectroscopy and second harmonic generation, Opt. Mater. **8**, 65–74 (1997)

13.108 G. Foulon, A. Brenier, M. Ferriol, M.T. Cohen-Adad, G. Boulon: Nonlinear single-crystal fibers of Nd^{3+}-doped niobates ($Ba_2NaNb_5O_{15}$ and $K_3Li_{2-x}Nb_{5+x}O_{15+2x}$) Grown by LHPG, spectroscopy and self-frequency doubling, J. Lumin. **72–74**, 794–796 (1997)

13.109 I. Noiret, J. Schamps, J. Lamiot, G. Boulon, A. Brenier: Phase transitions in the 5 at% Nd^{3+}-doped $Ba_2NaNb_5O_{15}$ self-doubling laser crystal, Phys. Rev. B **69**, 104110 (2004)

13.110 G. Foulon, A. Brenier, M. Ferriol, T. Cohen-Adad, G. Boulon: Laser heated pedestal growth and spectroscopic properties of neodymium doped $Ba_2NaNb_5O_{15}$ single crystal fibers, Chem. Phys. Lett. **249**, 381 (1996)

13.111 M. Ferriol, G. Foulon, A. Brenier, G. Boulon: Phenomenological investigation of inhomogeneities in Nd^{3+}-doped $Ba_2NaNb_5O_{15}$ single-crystal fibres grown by the laser-heated pedestal growth technique, J. Mater. Sci. **33**, 1227–1232 (1998)

13.112 G. Foulon, M. Ferriol, A. Brenier, M.T. Cohen-Adad, G. Boulon: Growth by LHPG, structure and spectroscopy of Nd^{3+}-doped $Ba_2NaNb_5O_{15}$ nonlinear single-crystal fibres, Acta Phys. Pol. A **90**, 63–72 (1996)

13.113 G. Foulon, M. Ferriol, A. Brenier, G. Boulon, S. Lecocq: Obtention of good quality $Ba_2NaNb_5O_{15}$ crystals: Growth, characterization and structure of Nd^{3+}-doped single-crystal fibres, Eur. J. Solid State Inorg. Chem. **33**, 673–686 (1996)

13.114 M. Ferriol: Crystal growth and structure of pure and rare-earth doped barium sodium niobate (BNN), Prog. Cryst. Growth Charact. Mater. **43**, 221–244 (2001)

13.115 F.C. Romo, C. Goutaudier, Y. Guyot, M.T. Cohen-Adad, G. Boulon, K. Lebbou, A. Yoshikawa, T. Fukuda: Yb^{3+}-doped $Ba_2NaNb_5O_{15}$ (BNN) growth, characterization and spectroscopy, Opt. Mater. **16**, 199–206 (2001)

13.116 X. Qi, H.G. Gallagher, T.P.H. Han, B. Henderson, R. Illingworth, I.S. Ruddock: Laser heated pedestal growth and spectroscopic properties of $K_2NdNb_5O_{15}$ and Ba_2NdNbO_6 crystals, Chem. Phys. Lett. **264**, 623–630 (1997)

13.117 F. Ito, K.I. Kitayama: Observation of the photorefractive effect in single-domain $BaTiO_3$ crystal fiber, Appl. Phys. Lett. **61**, 2144–2146 (1992)

13.118 M. Saifi, B. Dubois, E.M. Vogel: Growth of tetragonal $BaTiO_3$ single crystal fibers, J. Mater. Res. **1**, 452 (1996)

13.119 Y.C. Lee, J.C. Chen: The effects of temperature distribution on the barium titanate crystal growth in an LHPG system, Opt. Mater. **12**, 83–91 (1999)
13.120 J.C. Chen, Y.C. Lee, S.P. Lin: A new technique to eliminate the 90° in $BaTiO_3$ crystal fibers, Jpn. J. Appl. Phys. **39**, 1812–1814 (2000)
13.121 E. Nippolainen, E. Raita, V.V. Prokofiev, A.A. Kamshilin, T. Jääskeläinen: Photorefractive fibers for adaptive correlation filtering of a speckle-pattern displacement, Opt. Mater. **14**, 1–4 (2000)
13.122 R.M. Ribeiro, A.B.A. Fiasca, P.A.M. dos Santos, M.R.B. Andreeta, A.C. Hernandes: Optical activity measurements in the photorefractive $Bi_{12}TiO_{20}$ single crystal fibers, Opt. Mater. **10**, 201–205 (1998)
13.123 J.C. Chen, L.T. Liu, C.C. Young: A study of the growth mechanism of bismuth silicon oxide during LHPG method, J. Cryst. Growth **199**, 476–481 (1999)
13.124 V.V. Prokofiev, J.P. Andreeta, C.J. Lima, M.R.B. Andreeta, A.C. Hernandes, J.F. Carvalho, A.A. Kamshilin, T. Jääskeläinen: Microstructure of single-crystal sillenite fibers, Radiat. Eff. Defects Solids **134**, 209–211 (1995)
13.125 V.V. Prokofiev, J.P. Andreeta, C.J. de Lima, M.R.B. Andreeta, A.C. Hernandes, J.F. Carvalho, A.A. Kamshilin, T. Jääskeläinen: Growth of single-crystal photorefractive fibers of $Bi_{12}SiO_{20}$ and $Bi_{12}TiO_{20}$ by the laser-heated pedestal growth method, J. Cryst. Growth **137**, 528–534 (1994)
13.126 V.V. Prokofiev, J.P. Andreeta, C.J. de Lima, M.R.B. Andreeta, A.C. Hernandes, J.F. Carvalho, A.A. Kamshilin, T. Jääskeläinen: The influence of themperature-gradients on structural perfection of single-crystal sillenite fibers grown by the LHPG method, Opt. Mater. **4**, 521–527 (1995)
13.127 V.V. Prokofiev, A.A. Kamshilin, T. Jääskeläinen: The formation and stability of the molten zone in single-crystal fiber growth by the LHPG method, Adv. Cryst. Growth Mater. Sci. Forum **203**, 71–75 (1996)
13.128 M.F. Carrasco, V.S. Amaral, R.F. Silva, J.M. Vieira, F.M. Costa: Annealing time effect on Bi-2223 phase development in LFZ and EALFZ grown superconducting fibres, Appl. Surf. Sci. **252**, 4957–4963 (2006)
13.129 M.F. Carrasco, J.H. Monteiro, V.S. Amaral, R.F. Silva, J.M. Vieira, F.M. Costa: The effect of annealing temperature on the transport properties of BSCCO fibres grown by LFZ and EALFZ, Mater. Sci. Forum **514–516**, 338–342 (2006)
13.130 M.F. Carrasco, A.B. Lopes, R.F. Silva, J.M. Vieira, F.M. Costa: Enhancement of Bi-2223 phase formation by electrical assisted laser floating zone technique, J. Phys. Chem. Solids **67**, 416–418 (2006)
13.131 M.F. Carrasco, V.S. Amaral, J.M. Vieira, R.F. Silva, F.M. Costa: The effect of current direction on superconducting properties of BSCCO fibres grown by an electrically assisted laser floating zone process, Supercond. Sci. Technol. **19**, 15–21 (2006)
13.132 A. Leyva, C.M. Cruz, M. Mora, K. Shtejer, J.C. Díez, L.A. Angurel, I. Piñera, Y. Abreu: The effects of ^{137}Cs and ^{60}Co γ radiation on the magnetic susceptibility of BSCCO textured thin rods, Nucl. Instrum. Methods Phys. Res. Sect. B **239**, 281–285 (2005)
13.133 M.F. Carrasco, F.M. Costa, R.F. Silva, F. Gimeno, A. Sotelo, M. Mora, J.C. Díez, L.A. Angurel: Textured Bi-Sr-Ca-Cu-O rods processed by laser floating zone from solid state or melted precursors, Physica C **415**, 163–171 (2004)
13.134 X. Wang, G.W. Qiao: Electron-microscopic studies on BSCCO and YBCO superconducting wires made by the laser heated pedestal growth technique, Physica C **185**, 2443–2444 (1991)
13.135 A. Salazar, J.Y. Pastor, J. Llorca, E. Natividad, F.J. Gimeno, L.A. Angurel: Effect of thermal cycling on the strength and superconducting properties of laser floating zone textured Bi-2212 rods, Physica C **384**, 443–450 (2003)
13.136 E. Natividad, J.A. Gomez, L.A. Angurel, A. Salazar, J.Y. Pastor, J. Llorca: Influence of the post-annealing cooling rate on the superconducting and mechanical properties of LFZ textured Bi-2212 rods, Supercond. Sci. Technol. **15**, 1512–1518 (2002)
13.137 E. Martinez, T.J. Hughes, C. Díez, L.A. Angurel, Y. Yang, C. Beduz: Self-field AC losses of textured $Bi_2Sr_2CaCu_2O_{8+\delta}$ thin rods, Physica C **310**, 71–75 (1998)
13.138 F.M. Costa, R.F. Silva, J.M. Vieira: Influence of epitaxial growth on superconducting properties of LFZ Bi-Sr-Ca-Cu-O fibres. Part II. Magnetic susceptibility and transport properties, Physica C **289**, 171–176 (1997)
13.139 H. Yusheng, Z. Jincang, H. Aisheng, W. Jingsong, H. Yujing: Bi-based superconducting fibers prepared from amorphous materials, Supercond. Sci. Technol. **4**, S154–S156 (1991)
13.140 G.W. Qiao, J.S. Zhang, J.G. Huang, M. Jiang, Y.L. Ge, Y.Z. Wang, Z.Q. Hu: An investigation of melt-textured high T_c superconductor wires made by laser heated pedestal growth technique, Physica C **162–164**, 907–908 (1989)
13.141 G.F. de la Fuente, D. Beltrán, R. Ibáñez, E. Martinez, A. Beltrán, A. Segura: Crystal fibers of Bi-Sr-Ca-Cu-O materials grown by the laser floating zone method, J. Less Common Met. **150**, 253–260 (1989)
13.142 J.Z. Sun, R.S. Feigelson, D. Gazit, D. Fork, T.H. Geballe, A. Kapitulnik: Properties of high-T_c oxide fibers from laser heated pedestal growth, IEEE Trans. Magn. **25**, 2014–2016 (1989)
13.143 R.S. Feigelson, D. Gazit, D.K. Fork, T.H. Geballe: Superconducting Bi-Ca-Sr-Cu-O fibers grown by the laser-heated pedestal growth method, Science **240**, 1642–1645 (1988)
13.144 E. Martínez, L.A. Angurel, J.C. Díez, A. Larrea, M. Aguiló, R. Navarro: Grain texture and bulk magnetic anisotropy correlation in polycrystalline

$Bi_2Sr_2CaCu_2O_{8+\delta}$ thin rods, Physica C Appl. **333**, 93–103 (2000)
13.145 E. Martínez, L.A. Angurel, J.C. Díez, F. Lera, R. Navarro: Magnetic relaxation of highly textured $Bi_2Sr_2CaCu_2O_{8+\delta}$ polycrystalline fibres, Physica C **271**, 133–146 (1996)
13.146 E.R. Yacoby, Y. Yeshurun, D. Gazit, R.S. Feigelson: Magnetic Irreversibility in $Bi_2Sr_2CaCu_2O_8$ fibers irradiated by neutrons, Phys. Rev. B **50**, 13027–13030 (1994)
13.147 Y.L. Ge, Y.N. Jiao, Y.S. Yang, C.S. Liu, Q.M. Liu, Z.Q. Hu: Bi system superconducting fibers by the laser-heated pedestal growth method in stable magnetic-field, Chin. Sci. Bull. **39**, 475–479 (1994)
13.148 A. Badía, Y.B. Huang, G.F. de la Fuente, M.T. Ruiz, L.A. Angurel, F. Lera, C. Rillo, R. Navarro: Magnetic and electric transport properties of Ag/(Bi,Pb)-Sr-Ca-Cu-O superconducting fibres, Cryogenics **32**, 969–974 (1992)
13.149 M. Mora, A. Sotelo, H. Amaveda, M.A. Madre, J.C. Díez, L.A. Angurel, G.F. de la Fuente: Ag addition effect on laser textured Bi-2212 samples, Bol. Soc. Esp. Ceram. Vidr. **44**, 199–203 (2005)
13.150 A. Sotelo, M. Mora, M.A. Madre, H. Amaveda, J.C. Díez, L.A. Angurel, G.F. de la Fuente: Ag distribution in thick Bi-2212 floating zone textured rods, J. Eur. Ceram. Soc. **25**, 2947–2950 (2005)
13.151 A. Sotelo, M. Mora, M.A. Madre, H. Amaveda, J.C. Diez, L.A. Angurel, M.C. Mayoral: Study of the variation of the *E–I* curves in the superconducting to normal transition of Bi-2212 textured ceramics by Pb addition, Bol. Soc. Esp. Ceram. Vidr. **45**, 228–232 (2006)
13.152 Y. Huang, G.F. de la Fuente, A. Sotelo, A. Badia, F. Lera, R. Navarro, C. Rillo, R. Ibañez, D. Beltran, F. Sapiña, A. Beltran: $(Bi,Pb)_2Sr_2Ca_2Cu_3O_{10+\delta}$ superconductor composites: Ceramics vs. fibers, Physica C **185–189**, 2401–2402 (1991)
13.153 D. Gazit, P.N. Peszkin, R.S. Feigelson: Growth of Bi-based superconducting ribbons, J. Cryst. Growth **98**, 541–544 (1989)
13.154 M.C. Mayoral, J.M. Andrés, M.T. Bona, L.A. Angurel, E. Natividad: Approximation to the laser floating zone preparation of high temperature BSCCO superconductors by DSC, Thermochim. Acta **409**, 157–164 (2004)
13.155 E. Natividad, J.C. Díez, L.A. Angurel, J.M. Andrés, A.C. Ferrando, M.C. Mayoral: Radial changes in the microstructure of LFZ-textured Bi-2212 thin rods induced by stoichiometry modifications, Physica C **383**, 379–387 (2003)
13.156 E. Natividad, J.C. Díez, J.I. Peña, L.A. Angurel, R. Navarro, J.M. Andrés, A.C. Ferrando: Correlation of radial inhomogeneties and critical current at 77 K in LFZ Bi-2212 textured thin rods, Physica C **372–376**, 1051–1054 (2002)
13.157 F.M. Costa, R.F. Silva, J.M. Vieira: Phase transformation kinetics during thermal annealing of LFZ Bi-Sr-Ca-Cu-O superconducting fibers in the range 800 – 870 °C, Physica C **323**, 23–41 (1999)
13.158 E. Martínez, L.A. Angurel, J.C. Díez, R. Navarro: Analysis of the length scales in the induced critical currents of $Bi_2Sr_2CaCu_2O_{8+y}$ thick fibres, Physica C **289**, 1–21 (1997)
13.159 D. Gazit, P.N. Peszkin, L.V. Moulton, R.S. Feigelson: Influence of growth rate on the structure and composition of float zone grown $Bi_2Sr_2CaCu_2O_8$ superconducting fibers, J. Cryst. Growth **98**, 545–549 (1989)
13.160 F.M. Costa, R.F. Silva, J.M. Vieira: Influence of epitaxial growth on superconducting properties of LFZ Bi-Sr-Ca-Cu-O fibres. Part I. Crystal nucleation and growth, Physica C **289**, 161–170 (1997)
13.161 L. Yang, F.M. Costa, A.B. Lopes, R.F. Silva, J.M. Vieira: On the half unit cell intergrowth of $Bi_2Sr_2Ca_3Cu_4O_{12}$ with other superconducting phases in two-step annealed LFZ fibres, Physica C **398**, 31–36 (2003)
13.162 H. Miao, J.C. Díez, L.A. Angurel, J.I. Peña, G.F. de la Fuente: Phase formation and microstructure of laser floating zone grown BSCCO fibers: Reactivity aspects, Solid State Ion. **101–103**, 1025–1032 (1997)
13.163 F.M. Costa, A.P. Gonçalves, C. Abilio, M.M. Godinho, M. Almeida, J.M. Vieira: Crystallization process, phase chemistry and transport-properties of superconducting fibers prepared by the LFZ method followed by isothermal annealing, Physica C **235**, 513–514 (1994)
13.164 J. Zhang, A. He, Y. He, Y. Huo, F. Liu, S. Cao: Preparation, structure and properties of Bi-based superconducting fibers grown by laser-floating-zone-growth method, Appl. Supercond. **1**, 1987–1993 (1993)
13.165 C.J. Kim, M.R. De Guire, C.J. Allen, A. Sayir: Growth and characterization of Bi-Sr-Ca-Cu-O superconducting fibers, Mater. Res. Bull. **26**, 29–39 (1991)
13.166 D. Gazit, R.S. Feigelson: Laser-heated pedestal growth of high T_c Bi-Sr-Ca-Cu-O superconducting fibers, J. Cryst. Growth **91**, 318–330 (1998)
13.167 J.S. Zhang, J.G. Huang, M. Jiang, Y.L. Ge, Y.Z. Wang, G.W. Qiao, Z.Q. Hu: Preparation of Bi-Sr-Ca-Cu-O superconductors by laser floating zone melting technique, Mater. Lett. **8**, 46–48 (1989)
13.168 Y.L. Ge, Y.N. Jiao, Y.S. Yang, C.S. Liu, Q.M. Liu, Z.Q. Hu: Bi system superconducting fibers by the laser-heated pedestal growth method in stable magnetic-field, Chin. Bull. **39**, 475–479 (1994)
13.169 G.F. de la Fuente, R. Navarro, F. Lera, C. Rillo, J. Bartolome, A. Badia, D. Beltran, R. Ibanez, A. Beltran, E. Sinn: LFZ growth of (Bi,Pb)-Sr-Ca-Cu-O superconducting fibers, J. Mater. Res. **6**, 699–703 (1991)
13.170 J.M. Brenner, R.S. Feigelson, D. Gazit, P.N. Peszkin: Effects of heat-treatment on the superconducting properties of $Bi_2Sr_2CaCu_2O_x$ fibers produced by the

laser-heated pedestal growth method, Mater. Sci. Eng. B **5**, 351–357 (1990)
13.171 M. Mora, L.A. Angurel, J.C. Díez, R.J. Drost, P.H. Kes: Microstructural changes of LFZ Bi-2212 thin rods due to Ti addition, Physica C **372–376**, 1179–1182 (2002)
13.172 A. Larrea, E. Snoeck, A. Badía, G.F. de la Fuente, R. Navarro: Microstructure, interfaces and magnetic behaviour of thick Ag/BSCCO composite fibres, Physica C **220**, 21–32 (1994)
13.173 D. Gazit, P.N. Peszkin, R.S. Feigelson, J. Sun, T.H. Geballe: Preparation of high temperature superconductor-metal wire composites, Mater. Res. Bull. **24**, 467–474 (1989)
13.174 G.F. de la Fuente, M.T. Ruiz, A. Sotelo, A. Larrea, R. Navarro: Microstructure of laser floating zone (LFZ) textured (Bi,Pb)-Sr-Ca-Cu-O superconductor composites, Mater. Sci. Eng. A **173**, 201–204 (1993)
13.175 E. Snoeck, A. Larrea, C. Roucau: Microstructure of (Bi,Pb)-Sr-Ca-Cu-O fibres study by electron microscopy, Physica C **198**, 129–136 (1992)
13.176 H. Yusheng, H. Yujing, L. Menglin, M. Sining, C. Liying, W. Ying, Z. Jincang, H. Aisheng, W. Jinsong, Y. Xiaohua: Fabrication, characterization and welding of Bi(Pb)-Sr-Ca-Cu-O superconducting fibres, Supercond. Sci. Technol. **4**, 158–164 (1991)
13.177 J.C. Zhang, A.S. He, Y.J. Huo, J.S. Wang, Y.S. He: Bi(Pb)-Sr-Ca-Cu-O Superconducting fibers without postgrowth heat-tretament, Chin. Phys. **12**, 174–179 (1992)
13.178 M. Ito, C. Goutaudier, Y. Guyot, K. Lebbou, T. Fukuda, G. Boulon: Synthesis and spectroscopic characterization of Yb^{3+} in $Ca_{1-x}Yb_xF_{2+x}$ crystals, J. Phys. IV **119**, 201–202 (2004)
13.179 L.H.C. Andrade, D.R. Ardlla, J.P. Andreeta, M.S. Li: Optical properties of Nd^{3+}-doped $Ca_3(VO_4)_2$ single crystal fiber, Opt. Mater. **22**, 369–375 (2003)
13.180 R. Ternane, G. Boulon, Y. Guyot, M.T. Cohen-Adad, M. Trabelsi-Ayedi, N. Kbir-Ariguib: Crystal growth, structural and spectroscopic characterization of undoped and Yb^{3+}-doped oxyboroapatite fibers, Opt. Mater. **22**, 117–128 (2003)
13.181 D.D. Jia: Relocalization of Ce^{3+} 5d electrons from host conduction band, J. Lumin. **117**, 170–178 (2006)
13.182 W. Jia, H. Yuan, L. Lu, H. Liu, W.M. Yen: Crystal growth and characterization of Eu^{2+}, Dy^{3+}:$SrAl_2O_4$ and Eu^{2+}, Nd^{3+}:$CaAl_2O_4$ by the LHPG method, J. Cryst. Growth **200**, 179–184 (1999)
13.183 D. Jia, W. Jia, X.J. Wang, W.M. Yen: Quenching of thermo-stimulated photo-ionization by energy transfer in $CaAl_4O_7$:Tb^{3+},Ce^{3+}, Solid State Commun. **129**, 1–4 (2004)
13.184 G. Boulon, M. Ito, C. Goutaudier, Y. Guyot: Advances in growth of fiber crystal by the LHPG technique. Application to the optimization of Yb^{3+}-doped CaF_2 laser crystals, J. Cryst. Growth **292**, 230–235 (2006)
13.185 M. Ito, C. Goutaudier, Y. Guyot, K. Lebbou, T. Fukuda, G. Boulon: Crystal growth, Yb^{3+} spectroscopy, concentration quenching analysis and potentiality of laser emission in $Ca_{1-x}Yb_xF_{2+x}$, J. Phys. Condens. Matter **16**, 1501–1521 (2004)
13.186 A.S.S. Camargo, R.A. Silva, J.P. Andreeta, L.A.O. Nunes: Stimulated emission and excited state absorption in neodymium-doped $CaNb_2O_6$ single crystal fibers grown by the LHPG technique, Appl. Phys. B **80**, 497–502 (2005)
13.187 A.S.S. Camargo, C.R. Ferrari, A.C. Hernandes, L.A.O. Nunes: Structural and spectroscopic characteristics of neodymium doped $CaTa_2O_6$ single crystal fibres grown by the laser heated pedestal growth technique, J. Phys. Condens. Matter **16**, 5915–5923 (2004)
13.188 C.R. Ferrari, A.S.S. Camargo, L.A.O. Nunes, A.C. Hernandes: Laser heated pedestal growth and optical characterization of $CaTa_2O_6$ single crystal fiber, J. Cryst. Growth **266**, 475–480 (2004)
13.189 R.A. Silva, A.S.S. Camargo, C. Cusatis, L.A.O. Nunes, J.P. Andreeta: Growth and characterization of columbite $CaNb_2O_6$ high quality single crystal fiber, J. Cryst. Growth **262**, 246–250 (2004)
13.190 L.B. Barbosa, D.R. Ardila, C. Cusatis, J.P. Andreeta: Growth and characterization of crack-free scheelite calcium molybdate single crystal fiber, J. Cryst. Growth **235**, 332–337 (2002)
13.191 Y.J. Jiang, R.Y. Guo, A.S. Bhalla: Growth and properties of $CaTiO_3$ single crystal fibers, J. Electroceram. **2**, 199–203 (1998)
13.192 S. Hayashi, W.L. Kway, R.S. Feigelson: Pulling microcrystals of Co, Fe and Co-Fe alloys from their melts, J. Cryst. Growth **75**, 459–465 (1986)
13.193 O. Medenbach, D. Dettmar, R.D. Shannon, R.X. Fischer, W.M. Yen: Refractive index and optical dispersion of rare earth oxides using a small-prism technique, J. Opt. A **3**, 174–177 (2001)
13.194 L. Merkle, H.R. Verdum, U. Braunch, G.F. de la Fluente, E. Behrens, L.M. Thomas, T.H. Allik: Growth and characterization of the spectra of $EuAlO_3$:Ti and $GdAlO_3$:Ti, J. Opt. Soc. Am. **6**, 2342–2347 (1990)
13.195 C. Louis, K. Lebbou, M.A. Flores-Gonzalez, R. Bazzi, B. Hautefeuille, B. Mercier, S. Roux, P. Perriat, C. Olagnon, O. Tillement: Correlation of the structure and the luminescence properties of Eu^{3+}-doped Gd_2O_3 oxide between fiber single crystal and the nano-size powders, J. Cryst. Growth **265**, 459–465 (2004)
13.196 A. Brenier, G. Boulon: Laser heated pedestal growth and spectroscopic investigations of Nd^{3+}-doped Gd_2O_3 single crystal fibres, J. Lumin. **82**, 285–289 (1999)
13.197 D. Jia, L. Lu, W.M. Yen: Erbium energy levels relative to the band gap of gadolinium oxide, Opt. Commun. **212**, 97–100 (2002)
13.198 A. Brenier, A.M. Jurdyc, H. Verweij, M.T. Cohen-Adad, G. Boulon: Up-conversion dynamics in $GdAlO_3$:Er^{3+} single crystal fibre, Opt. Mater. **5**, 233–238 (1996)

13.199 R.A. Silva, G. Tirao, C. Cusatis, J.P. Andreeta: Growth and structural characterization of M-type $GdTaO_4$ single crystal fiber, J. Cryst. Growth **274**, 512–517 (2005)

13.200 R.A. Silva, G. Tirao, C. Cusatis, J.P. Andreeta: Growth and characterization of single crystal fiber with controlled concentration gradient in $GdTaO_4$-$ErTaO_4$ system, J. Cryst. Growth **294**, 447–451 (2006)

13.201 R.A. Silva, G. Tirao, C. Cusatis, J.P. Andreeta: Growth and X-ray characterization of $Gd_xYb_{1-x}TaO_4$ ($0 \leq x \leq 1$) single crystals with large lattice spacing gradient, J. Cryst. Growth **277**, 308–313 (2005)

13.202 E. Gallucci, C. Goutaudier, G. Boulon, M.T. Cohen-Adad: Growth of $KY(WO_4)_2$ single-crystal: Investigation of the rich WO_3 region in the K_2O-Y_2O_3-WO_3 ternary system. The K_2O-WO_3 binary system, Eur. J. Solid State Inorg. Chem. **34**, 1107–1117 (1997)

13.203 S.A. Amin, R. Guo, A.S. Bhalla: Dielectric and thermal expansion properties of LHPG grown potassium lithium niobate single crystals, Ferroelectr. Lett. Sect. **25**, 37–44 (1999)

13.204 M. Matsukura, T. Takeyama, T. Karaki, M. Adachi: Domain structures in $K_3Li_{2-x}Nb_{5+x}O_{15+2x}$ single-crystal fibers produced by the laser-heated pedestal growth technique, Jpn. J. Appl. Phys. **40**, 5783–5785 (2001)

13.205 G. Foulon, A. Brenier, M. Ferriol, G. Boulon: Nonlinear laser crystal as a blue converter: laser heated pedestal growth, spectroscopic properties and second harmonic generation of pure and Nd^{3+}-doped $K_3Li_{2-x}Nb_{5+x}O_{15+2x}$ single crystal fibres, J. Phys. D Appl. Phys. **29**, 3003–3008 (1996)

13.206 M. Matsukura, J. Murakami, T. Karaki, M. Adachi: Diameter control of $K_3Li_{2-x}Nb_{5+x}O_{15+2x}$ single-crystal fibers, Jpn. J. Appl. Phys. **39**, 5658–5661 (2000)

13.207 M. Matsukura, T. Karaki, T. Takeyama, T. Fujii, M. Adachi: Growth of $K_3Li_{2-x}Nb_{5+x}O_{15+2x}$ single-crystal fibers, Jpn. J. Appl. Phys. **38**, 5638–5640 (1999)

13.208 M. Ferriol, G. Boulon: Potassium lithium niobate: single-phase domain boundary in the 30 mol K_2O isopleth of the ternary system Li_2O-K_2O-Nb_2O_5 and characterization of single-crystal fibers, Mater. Res. Bull. **34**, 533–543 (1999)

13.209 M. Matsukura, Z.M. Chen, M. Adachi, A. Kawabata: Growth of potassium lithium niobate single-crystal fibers by the laser-heated pedestal growth method, Jpn. J. Appl. Phys. **36**, 5947–5949 (1997)

13.210 M. Ferriol, G. Foulon, A. Brenier, M.T. Cohen-Adad, G. Boulon: Laser heated pedestal growth of pure and Nd^{3+}-doped potassium lithium niobate single-crystal fibers, J. Cryst. Growth **173**, 226–230 (1997)

13.211 D.L. Rocco, R.A. Silva, A. Magnus, G. Carvalho, A.A. Coelho, J.P. Andreeta, S. Gama: Magnetocaloric effect of $La_{0.8}Sr_{0.2}MnO_3$ compound under pressure, J. Appl. Phys. **97**, 10M317 (2005)

13.212 G.J. Snyder, R. Hiskes, S. Carolis, M.R. Beasley, T.H. Geballe: Intrinsic electrical transport and magnetic properties of $La_{0.67}Ca_{0.33}MnO_3$ and $La_{0.67}Sr_{0.33}MnO_3$ MOCVD thin films and bulk material, Phys. Rev. B **53**, 14434–14444 (1996)

13.213 E. Martínez, L.A. Angurel, J.C. Díez, F. Lera, R. Navarro: Magnetic relaxation of highly textured $Bi_2Sr_2CaCu_2O_{8+\delta}$ polycrystalline fibres, Physica C **271**, 133–146 (1996)

13.214 C.A. Cardoso, F.M. Araujo-Moreira, M.R.B. Andreeta, A.C. Hernandes, E.R. Leite, O.F. de Lima, A.W. Mombru, R. Faccio: Physical properties of single-crystalline fibers of the colossal-magnetoresistance manganite $La_{0.7}Ca_{0.3}MnO_3$, Appl. Phys. Lett. **83**, 3135–3137 (2003)

13.215 F. Büllesfeld, F. Ritter, W. Assmus: Crystal growth and twins in $La_{1-x}Sr_xMnO_3$, J. Magn. Magn. Mater. **226–230**, 815–817 (2001)

13.216 L. Zhang, C. Madej, C. Pedrini, B. Moine, C. Dujardin, A. Petrosyan, A.N. Belsky: Elaboration and spectroscopic properties of new dense cerium-doped lutetium based scintillator materials, Chem. Phys. Lett. **268**, 408–412 (1997)

13.217 P.A. Morris, B.G. Bagley, J.M. Tarascon, L.H. Green, G.W. Hull: Melt growth of high-critical-temperature superconducting fibers, J. Am. Ceram. Soc. **71**, 334 (1988)

13.218 G. Foulon, A. Brenier, M. Ferriol, A. Rochal, M.T. Cohen-Adad, G. Boulon: Laser-heated pedestal growth and optical properties of Nd^{3+}-doped $Li_{1-x}Nb_{1-x}W_xO_3$ single-crystal fibers, J. Lumin. **69**, 257–263 (1996)

13.219 Y.Y. Ji, S.Q. Zhao, Y.J. Huo, H.W. Zhang, M. Li, C.O. Huang: Growth of lithium triborate (LBO) single-crystal fiber by the laser-heated pedestal growth method, J. Cryst. Growth **112**, 283–286 (1991)

13.220 Y.J. Lai, J.C. Chen, K.C. Liao: Investigations of ferroelectric domain structures in the $MgO{:}LiNbO_3$ fibers by LHPG, J. Cryst. Growth **198–199**, 531–535 (1999)

13.221 Y.S. Luh, R.S. Feigelson, M.M. Fejer, R.L. Byer: Ferroelectric domain structures in $LiNbO_3$ single-crystal fibers, J. Cryst. Growth **78**, 135–143 (1986)

13.222 E.K. Renwick, I.S. Ruddock: Birefringent and nonlinear optical assessment of single crystal lithium niobate fibres, J. Phys. D Appl. Phys. **38**, 3387–3390 (2005)

13.223 S.Z. Yin: Lithium niobate fibers and waveguides: Fabrications and applications, Proc. IEEE **87**, 1962–1974 (1999)

13.224 J.J. Carey, E.K. Renwick, I.S. Ruddock: Optical fibre polariser based on GRIN single crystal fibre, Electron. Lett. **35**, 1486–1488 (1999)

13.225 E.K. Renwick, M.P. MacDonald, I.S. Ruddock: GRIN single crystal fibres, Opt. Commun. **151**, 75–80 (1998)

13.226 W.X. Que, Y. Zhou, Y.L. Lam, Y.C. Chan, C.H. Kam, Y.J. Huo, X. Yao: Second-harmonic generation using an *a* axis $Nd{:}MgO{:}LiNbO_3$ single crystal fiber

with Mg-ion in diffused cladding, Opt. Eng. **39**, 2804–2809 (2000)
13.227 P. Bourson, M. Aillerie, M. Cochez, M. Ferriol, Y. Zhang, L. Guilbert: Characterization of iron substitution process in Fe:$LiNbO_3$ single crystal fibers by polaron measurements, Opt. Mater. **24**, 111–116 (2003)
13.228 M. Cochez, M. Ferriol, P. Bourson, M. Aillerie: Influence of the dopant concentration on the OH^- absorption band in Fe-doped $LiNbO_3$ single-crystal fibers, Opt. Mater. **21**, 775–781 (2003)
13.229 R. Burlot, R. Moncorgé, G. Boulon: Visible and infrared luminescence properties of Er^{3+}-doped Sc_2O_3:$LiNbO_3$ crystal fibers, J. Lumin. **72–74**, 135–138 (1997)
13.230 R. Burlot, R. Moncorgé, G. Boulon: Spectroscopic properties of Nd^{3+} doped Sc_2O_3:$LiNbO_3$ crystal fibers, J. Lumin. **72–74**, 812–815 (1997)
13.231 G. Boulon: Yb^{3+}-doped oxide crystals for diode-pumped solid state lasers: Crystal growth, optical spectroscopy, new criteria of evaluation and combinatorial approach, Opt. Mater. **22**, 85–87 (2003)
13.232 K. Nagashio, A. Watcharapasorn, R.C. DeMattei, R.S. Feigelson: Fiber growth of near stoichiometric $LiNbO_3$ single crystals by the laser-heated pedestal growth method, J. Cryst. Growth **265**, 190–197 (2004)
13.233 Y.J. Lai, J.C. Chen: Effects of the laser heating and air bubbles on the morphologies of *c* axis $LiNbO_3$ fibers, J. Cryst. Growth **231**, 222–229 (2001)
13.234 J.C. Chen, Y.C. Lee: The influence of temperature distribution upon the structure of $LiNbO_3$ crystal rods grown using the LHPG method, J. Cryst. Growth **208**, 508–512 (2000)
13.235 J.C. Chen, Y.C. Lee, C. Hu: Observation of the growth mechanisms of lithium niobate single crystal during a LHPG process, J. Cryst. Growth **174**, 313–319 (1997)
13.236 J.C. Chen, C. Hu: Measurement of the float-zone interface shape for lithium niobate, J. Cryst. Growth **149**, 87–95 (1995)
13.237 S. Uda, W.A. Tiller: Microbubble formation during crystallization of $LiNbO_3$ melts, J. Cryst. Growth **152**, 79–86 (1995)
13.238 Y.S. Luh, M.M. Fejer, R.L. Byer, R.S. Feigelson: Stoichiometric $LiNbO_3$ single-crystal fibers for nonlinear optical applications, J. Cryst. Growth **85**, 264–269 (1987)
13.239 C. Hu, J.C. Chen: Experimental observation of interface shapes in the float zone of lithium niobate during a CO_2 laser melting, Int. J. Heat Mass Trans. **39**, 3347–3352 (1996)
13.240 Y.J. Lai, J.C. Chen: The influence of heavy iron-doping on $LiNbO_3$ fibers and their growth, J. Cryst. Growth **212**, 211–216 (2000)
13.241 M. Ferriol, A. Dakki, M.T. Cohen-Adad, G. Foulon, A. Brenier, G. Boulon: Growth and characterization of MgO-doped single-crystal fibers of lithium niobate in relation to high temperature phase equilibria in the ternary system Li_2O-Nb_2O_5-MgO, J. Cryst. Growth **178**, 529–538 (1997)
13.242 W. Que, Y. Zhou, Y. Lam, Y. Chan, C. Kam, L. Zhang, X. Yao: Magnesium-ion diffusion to lithium niobate single-crystal fiber with MgF_2 as diffusion source, Jpn. J. Appl. Phys. **38**, 5137–5142 (1999)
13.243 W. Que, S. Lim, L. Zhang, X. Yao: Characteristics of lithium niobate single-crystal fiber with magnesium-ion-indiffused cladding, J. Am. Ceram. Soc. **80**, 2945 (1997)
13.244 A. Dakki, M. Ferriol, M.T. Cohen-Adad: Growth of MgO-doped $LiNbO_3$ single-crystal fibers: Phase equilibria in the ternary system Li_2O-Nb_2O_5-MgO, Eur. J. Solid State Inorg. Chem. **33**, 19–31 (1996)
13.245 Z.L. Feng, Y. Liao, Z.F. Jiao, X. Wang: Preparation of $LiNbO_3$:Zn+Mg single crystal fibers, Cryst. Res. Technol. **31**, K27–K28 (1996)
13.246 H. Li, Z. Feng, X. Wang, L. Wang, Z. Jiao, X. Xu: Growth and characterization of $LiNbO_3$:Mg+Ti monocrystalline fibers, Cryst. Res. Technol. **30**, 763–765 (1995)
13.247 R. Burlot, M. Ferriol, R. Moncorge, G. Boulon: Li_2O evaporation during the laser heated pedestal growth of $LiTaO_3$ single-crystal fibers, Eur. J. Solid State Inorg. Chem. **35**, 1–8 (1998)
13.248 G. Boulon, Y. Guyot, M. Ito, A. Bensalah, C. Goutaudier, G. Panczer, J.C. Gacon: From optical spectroscopy to a concentration quenching model and a theoretical approach to laser optimization for Yb^{3+}-doped $YLiF_4$ crystals, Mol. Phys. **102**, 1119–1132 (2004)
13.249 L.B. Shaw, R.S.F. Chang: Rare-earth doped YLF grown by the laser-heated pedestal growth technique, J. Cryst. Growth **112**, 731–736 (1991)
13.250 W. Jia, L. Lu, B.M. Tissue, W.M. Yen: Valence and site occupation of chromium ions in single-crystal forsterite fibers, J. Cryst. Growth **109**, 323–328 (1991)
13.251 J. Sigalovsky, J.S. Haggerty, J.E. Sheehan: Growth of spinel single-crystal fibers by the laser-heated floating-zone technique and their characterization as high-temperature reinforcements, J. Cryst. Growth **134**, 313–324 (1993)
13.252 L.H. Wang, M.H. Hon, L.F. Schneemeyer, G.A. Thomas, W.L. Wilson: Growth of single crystal fibers for 3 μm optical amplifiers by the laser-heated pedestal growth method, Mater. Res. Bull. **33**, 1793–1799 (1998)
13.253 E. Bruck, R.K. Route, R.J. Raymakers, R.S. Feigelson: Crystal-growth of compounds in the MgO-Nb_2O_5 binary-system, J. Cryst. Growth **128**, 842–845 (1993)
13.254 Y. Jiang, R. Guo, A.S. Bhalla: LHPG grown crystal fibers of $MgTiO_3$-$CaTiO_3$ eutectic system, J. Phys. Chem. Solids **59**, 611–615 (1998)
13.255 J.K. Yamamoto, A.S. Bhalla: Piezoelectric properties of layered perovskite $A_2Ti_2O_7$ (A = La and Nd) single-crystal fibers, J. Appl. Phys. **70**, 4469–4471 (1991)

13.256 V.Y. Ivanov, A.N. Tcherepanov, B.V. Shul'gin, T.S. Koroleva, C. Pédrini, C. Dujardin: Photoluminescence properties of NaF:U,Cu bulk and fiber crystals, Opt. Mater. **28**, 1123–1127 (2006)

13.257 H. Manuspiya, R. Guo, A.S. Bhalla: Nb_2O_5-based oxide ceramics and single crystals-investigation of dielectric properties, Ferroelectr. Lett. Sect. **31**, 157–166 (2005)

13.258 H. Manuspiya, R. Guo, A.S. Bhalla: Nb_2O_5-based oxide ceramics and single crystals-investigation of dielectric properties, Ceram. Int. **30**, 2037–2041 (2004)

13.259 H. Choosuwan, R. Guo, A.S. Bhalla: Dielectric behaviors of $Nb_2O_5(0.95)$:$0.05TiO_2$ ceramic and single crystal, Mater. Lett. **54**, 269–272 (2002)

13.260 H. Choosuwan, R. Guo, A.S. Bhalla, U. Balachandran: Growth studies of $(Nb_2O_5)(1-x)$:$xTiO_2$ and $(Nb_2O_5)(1-x)$:$xSiO_2$ single crystals and their dielectric behaviors, Ferroelectrics **262**, 1285–1293 (2001)

13.261 F.S. Ermeneux, C. Goutaudier, R. Moncorgé, M.T. Cohen-Adad, M. Bettinelli, E. Cavalli: Comparative optical characterization of various Nd^{3+}:YVO_4 single crystals, Opt. Mater. **13**, 193–204 (1999)

13.262 A.S.S. de Camargo, L.A.O. Nunes, D.R. Ardila, J.P. Andreeta: Excited-state absorption and 1064-nm end-pumped laser emission of Nd:YVO_4 single-crystal fiber grown by laser-heated pedestal growth, Opt. Lett. **29**, 59–61 (2004)

13.263 A.S.S. de Camargo, M.R.B. Andreeta, A.C. Hernandes, L.A.O. Nunes: 4.8 μm emission and excited state absorption in LHPG grown $Gd_{0.8}La_{0.2}VO_4$:Tm^{3+} single crystal fibers for miniature lasers, Opt. Mater. **28**, 551–555 (2006)

13.264 C.H. Huang, J.C. Chen: Nd:YVO_4 single crystal fiber growth by the LHPG method, J. Cryst. Growth **229**, 184–187 (2001)

13.265 S. Erdei, G.G. Johnson, F.W. Ainger: Growth-studies of YVO_4 crystals (II) – changes in Y-V-O stoichiometry, Cryst. Res. Technol. **29**, 815–828 (1994)

13.266 S. Erdei, F.W. Ainger: Crystal growth of YVO_4 using the LHPG technique, J. Cryst. Growth **128**, 1025–1030 (1993)

13.267 M.R.B. Andreeta, A.S.S. de Camargo, L.A.O. Nunes, A.C. Hernandes: Transparent and inclusion-free $RE_{1-x}La_xVO_4$ (RE = Gd,Y) single crystal fibers grown by LHPG technique, J. Cryst. Growth **291**, 117–122 (2006)

13.268 A.S.S. Camargo, L.A.O. Nunes, M.R.B. Andreeta, A.C. Hernandes: Near-infrared and upconversion properties of neodymium-doped $RE_{0.8}La_{0.2}VO_4$ (RE = Y,Gd) single-crystal fibres grown by the laser-heated pedestal growth technique, J. Phys. Condens. Matter **14**, 13887–13889 (2002)

13.269 D.R. Ardila, A.S.S. Camargo, J.P. Andreeta, L.A.O. Nunes: Growth of yttrium orthovanadate by LHPG in isostatic oxygen atmosphere, J. Cryst. Growth **233**, 253–258 (2001)

13.270 L. Sangaletti, B. Allieri, L.E. Depero, M. Bettinelli, K. Lebbou, R. Moncorgé: Search for impurity phases of Nd^{3+}:YVO_4 crystals for laser and luminescence applications, J. Cryst. Growth **198–199**, 454–459 (1999)

13.271 C. Goutaudier, F.S. Ermeneux, M.T. Cohen-Adad, R. Moncorgé, M. Bettinelli, E. Cavalli: LHPG and flux growth of various Nd:YVO_4 single crystals: A comparative characterization, Mater. Res. Bull. **33**, 1457–1465 (1998)

13.272 X. Qi, T.P.J. Han, H.G. Gallagher, B. Henderson, R. Illingworth, I.S. Ruddock: Optical spectroscopy of $PrTiNbO_6$, $NdTiNbO_6$ and $ErTiNbO_6$ single crystals, J. Phys. Condens. Matter **8**, 4837–4845 (1996)

13.273 X. Qi, R. Illingworth, H.G. Gallagher, T.P.J. Han, B. Henderson: Potential laser gain media with the stoichiometric formula $RETiNbO_6$, J. Cryst. Growth **160**, 111–118 (1996)

13.274 X. Qi, T.P.J. Han, H.G. Gallagher, B. Henderson, R. Illingworth, I.S. Ruddock: Optical spectroscopy of $PrTiNbO_6$, $NdTiNbO_6$ and $ErTiNbO_6$ single crystals, J. Phys. Condens. Matter **8**, 4837–4845 (1996)

13.275 G. Boulon, A. Brenier, L. Laversenne, Y. Guyot, C. Goutaudier, M.T. Cohen-Adad, G. Métrat, N. Muhlstein: Search of optimized trivalent ytterbium doped-inorganic crystals for laser applications, J. Alloys Compd. **341**, 2–7 (2002)

13.276 D. Elwell, W.L. Kway, R.S. Feigelson: Crystal growth of a new tetragonal phase of $ScTaO_4$, J. Cryst. Growth **71**, 237–239 (1985)

13.277 R.W. Tao, A.R. Guo, C.S. Tu, I. Siny, R.S. Katiyar: Temperature dependent Raman spectroscopic studies on microwave dielectrics $Sr(Al_{1/2}Ta_{1/2})O_3$ and $Sr(Al_{1/2}Nb_{1/2})O_3$, Ferroelectr. Lett. Sect. **21**, 79–85 (1996)

13.278 R.Y. Guo, A.S. Bhalla, J. Sheen, F.W. Ainger, S. Erdei, E.C. Subbarao, L.E. Cross: Strontium aluminum tantalum oxide and strontium aluminum niobium oxide as potential substrates for HTSC thin-films, J. Mater. Res. **10**, 18–25 (1995)

13.279 D.R. Ardila, M.R.B. Andreeta, S.L. Cuffini, A.C. Hernandes, J.P. Andreeta, Y.P. Mascarenhas: Laser heated pedestal growth of Sr_2RuO_4 single-crystal fibers from $SrRuO_3$, J. Cryst. Growth **177**, 52–56 (1997)

13.280 D.D. Jia, X.J. Wang, W. Jia, W.M. Yen: Temperature-dependent photoconductivity of Ce^{3+}-doped $SrAl_2O_4$, J. Lumin. **119**, 55–58 (2006)

13.281 D. Jia, R.S. Meltzer, W.M. Yen, W. Jia, X. Wang: Green phosphorescence of $CaAl_2O_4$:Tb^{3+},Ce^{3+} through persistence energy transfer, Appl. Phys. Lett. **80**, 1535–1537 (2002)

13.282 M.H. Lente, J. de Los, S. Guerra, J.A. Eiras, T. Mazon, M.R.B. Andreeta, A.C. Hernandes: Microwave dielectric relaxation process in doped-incipient ferroelectrics, J. Eur. Ceram. Soc. **25**, 2563–2566 (2005)

13.283 M.R.B. Andreeta, A.C. Hernandes, S.L. Cuffini, J.A. Guevara, Y.P. Mascarenhas: Laser heated pedestal growth of orthorhombic $SrHfO_3$ single crystal fiber, J. Cryst. Growth **200**, 621–624 (1999)
13.284 D.R. Ardila, J.P. Andreeta, H.C. Basso: Preparation, microstructural and electrical characterization of $SrVO_3$ single crystal fiber, J. Cryst. Growth **211**, 131–317 (2000)
13.285 C. Huang, A.S. Bhalla, R. Guo, L.E. Cross: Dielectric behavior of strontium barium niobate relaxor ferroelectrics in ceramics and single crystal fibers, Jpn. J. Appl. Phys. **45**, 165–167 (2006)
13.286 J.K. Yamamoto, D.A. McHenry, A.S. Bhalla: Strontium barium niobate single-crystal fibers-optical and electrooptic properties, J. Appl. Phys. **70**, 3215–3222 (1991)
13.287 C. Huang, A.S. Bhalla, R. Guo, L.E. Cross: Dielectric behavior of strontium barium niobate relaxor ferroelectrics in ceramics and single crystal fibers, Jpn. J. Appl. Phys. **45**, 165–167 (2006)
13.288 M. Miyagi, Y. Sugiyama, S. Yagi, I. Hatakeyama: Photorefractive properties of Ce-doped strontium barium niobate single-crystal fibers at 830 nm, Jpn. J. Appl. Phys. **33**, L1417–L1419 (1994)
13.289 Y. Sugiyama, S. Yagi, I. Yokohama, I. Hatakeyama: Holographic recording in cerium doped strontium barium niobate *a* axis single crystal fibers, Jpn. J. Appl. Phys. **31**, 708–712 (1992)
13.290 Y. Sugiyama, I. Yokohama, K. Kubodera, S. Yagi: Growth and photorefractive properties of *a* axis and *c* axis cerium-doped strontium barium niobate single-crystal fibers, IEEE Photonics Technol. Lett. **3**, 744–746 (1991)
13.291 J.J. Romero, M.R.B. Andreeta, E.R.M. Andreeta, L.E. Bausá, A.C. Hernandes, J.G. Solé: Growth and characterization of Nd-doped SBN single crystal fibers, Appl. Phys. A **78**, 1037–1042 (2004)
13.292 L. Galambos, S. Erdei, I. Tanaka, L. Hesselink, L.E. Cross, R.S. Feigelson, F.W. Ainger, H. Kojima: Inhomogeneities and segregation behavior in strontium-barium niobate fibers grown by laser-heated pedestal growth technique – Part I, J. Cryst. Growth **166**, 660–669 (1996)
13.293 J.K. Yamamoto, S.A. Markgraf, A.S. Bhalla: $Sr_xBa_{1-x}Nb_2O_6$ single crystal fibers: Dependence of crystal quality on growth parameters, J. Cryst. Growth **123**, 423–435 (1992)
13.294 J.P. Wilde, D.H. Jundt, L. Galambos, L. Hesselink: Growth of $Sr_{0.61}Ba_{0.39}Nb_2O_6$ fibers: New results regarding orientation, J. Cryst. Growth **114**, 500–506 (1991)
13.295 P.S. Dobal, R.S. Katiyar, Y. Jiang, R. Guo, A.S. Bhalla: Structural modifications in titania-doped tantalum pentoxide crystals: A Raman scattering study, Int. J. Inorg. Mater. **3**, 135–142 (2001)
13.296 T. Kotani, H.L. Tuller: Growth of TiO_2 single crystals and bicrystals by the laser-heated floating-zone method, J. Am. Ceram. Soc. **81**, 592 (1998)
13.297 D.D. Jia, X.J. Wang, W.M. Yen: Delocalization, thermal ionization, and energy transfer in singly doped and codoped $CaAl_4O_7$ and Y_2O_3, Phys. Rev. B **69**, 235113 (2004)
13.298 Y. Guyot, R. Moncorgé, L.D. Merkle, A. Pinto, B. McIntosh, H. Verdun: Luminescence properties of Y_2O_3 single crystals doped with Pr^{3+} or Tm^{3+} and codoped with Yb^{3+}, Tb^{3+} or Ho^{3+} ions, Opt. Mater. **5**, 127–136 (1996)
13.299 M.A. Flores-Gonzalez, K. Lebbou, R. Bazzi, C. Louis, P. Perriat, O. Tillement: Eu^{3+} addition effect on the stability and crystallinity of fiber single crystal and nano-structured Y_2O_3 oxide, J. Cryst. Growth **277**, 502–508 (2005)
13.300 G.P. Flinn, K.W. Jang, J. Ganem, M.L. Jones, R.S. Meltzer, M. Macfarlane: Sample-dependent optical dephasing in bulk crystalline samples of $Y_2O_3:Eu^{3+}$, Phys. Rev. B **49**, 5821–5827 (1994)
13.301 K. Jang, I. Kim, S. Park: Optical dephasing of Eu^{3+} in yttrium oxide crystals, J. Phys. Soc. Jpn. **67**, 3969–3971 (1998)
13.302 M.J. Sellars, R.S. Meltzer, P.T.H. Fisk, N.B. Manson: Time-resolved ultranarrow optical hole burning of a crystalline solid: $Y_2O_3:Eu^{3+}$, J. Opt. Soc. Am. B **11**, 1468–1473 (1994)
13.303 G.P. Flinn, K.W. Jang, J. Ganem, M.L. Jones, R.S. Meltzer, R.M. Macfarlane: Anomalous optical dephasing in crystalline Y_2O_3-Eu^{3+}, J. Lumin. **58**, 374–379 (1994)
13.304 F. Auzel, G. Baldacchini, L. Laversenne, G. Boulon: Radiation trapping and self-quenching analysis in Yb^{3+}, Er^{3+}, and Ho^{3+} doped Y_2O_3, Opt. Mater. **24**, 103–109 (2003)
13.305 G. Boulon, L. Laversenne, C. Goutaudier, Y. Guyot, M.T. Cohen-Adad: Radiative and non-radiative energy transfers in Yb^{3+}-doped sesquioxide and garnet laser crystals from a combinatorial approach based on gradient concentration fibers, J. Lumin. **102–103**, 417–425 (2003)
13.306 C. Goutaudier, F.S. Ermeneux, M.T. Cohen-Adad, R. Moncorgé: Growth of pure and RE^{3+}-doped Y_2O_3 single crystals by LHPG technique, J. Cryst. Growth **210**, 693–698 (2000)
13.307 L. Laversenne, S. Kairouani, Y. Guyot, C. Goutaudier, G. Boulon, M.T. Cohen-Adad: Correlation between dopant content and excited-state dynamics properties in Er^{3+}-Yb^{3+}-codoped Y_2O_3 by using a new combinatorial method, Opt. Mater. **19**, 59–66 (2002)
13.308 K.W. Jang, R.S. Meltzer: Homogeneous and inhomogeneous linewidths of Eu^{3+} in disordered crystalline systems, Phys. Rev. B **52**, 6431–6439 (1995)

13.309 Y.H. Shen, W.Z. Zhao, J.L. He, T. Sun, K.T.V. Grattan: Fluorescence decay characteristic of Tm-doped YAG crystal fiber for sensor applications, investigated from room temperature to 1400 °C, IEEE Sens. J. **3**, 507–512 (2003)

13.310 R.S.F. Chang, H. Hara, S. Chaddha, S. Sengupta, N. Djeu: Lasing performance of a Tm-YAG minirod grown by the laser-heated pedestal growth technique, IEEE Photonics Technol. Lett. **2**, 695–696 (1990)

13.311 Y.H. Shen, S.Y. Chen, W.Z. Zhao, J. Chen, L.H. Ye, J.G. Gu, K.T.V. Grattan: Growth characteristics and potential applications in optical sensors of composite Cr^{4+}:yttrium-aluminum-garnet (YAG)-Nd^{3+}:YAG crystal, Rev. Sci. Instrum. **74**, 1187–1191 (2003)

13.312 S.B. Zhang, Z.C. Ding, M.Y. Dong, B.H. Zhou, L.M. Tong: Study of optical-properties of Nd-YAG single-crystal optical fiber, Chin. Phys. **12**, 428–432 (1992)

13.313 L.M. Tong, D. Zhu, Q.M. Luo, D.F. Hong: A laser pumped Nd^{3+}-doped YAG fiber-optic thermal tip for laser thermotherapy, Lasers Surg. Med. **30**, 67–69 (2002)

13.314 L.M. Tong, J.Y. Lou, Y.F. Xu, Q.M. Luo, N. Shen, E. Mazur: Highly Nd^{3+}-doped $Y_3Al_5O_{12}$ crystal fiber tip for laser thermotherapy, Appl. Opt. **41**, 4008–4012 (2002)

13.315 C.Y. Lo, P.L. Huang, T.S. Chou, L.M. Lee, T.Y. Chang, S.L. Huang, L.C. Lin, H.Y. Lin, F.C. Ho: Efficient Nd:$Y_3Al_5O_{12}$ crystal fiber laser, Jpn. J. Appl. Phys. **41**, L1228–L1231 (2002)

13.316 B.M. Tissue, W.Y. Jia, L.Z. Lu, W.M. Yen: Coloration of chromium-doped yttrium-aluminum-garnet single-crystal fibers using a divalent codopant, J. Appl. Phys. **70**, 3775–3777 (1991)

13.317 J.C. Chen, C.Y. Lo, K.Y. Huang, F.J. Kao, S.Y. Tu, S.L. Huang: Fluorescence mapping of oxidation states of Cr ions in YAG crystal fibers, J. Cryst. Growth **274**, 522–529 (2005)

13.318 L. Ye, J. Zhang, Y. Shi: Growth and characteristics of Cr^{3+}:YAG crystal fiber for fluorescence decay temperature sensor, Rev. Sci. Instrum. **77**, 054901 (2006)

13.319 Z. Zhang, J.H. Herringer, N. Djeu: Monolithic crystalline fiber optic temperature sensor, Rev. Sci. Instrum. **68**, 2068–2070 (1997)

13.320 X. Wu, W.M. Dennis, W.M. Yen: Temperature dependence of cross-relaxation processes in Pr^{3+}-doped yttrium aluminum garnet, Phys. Rev. B **50**, 6589–6595 (1994)

13.321 A. Yoshikawa, G. Boulon, L. Laversenne, H. Canibano, K. Lebbou, A. Collombet, Y. Guyot, T. Fukuda: Growth and spectroscopic analysis of Yb^{3+}-doped $Y_3Al_5O_{12}$ fiber single crystals, J. Appl. Phys. **94**, 5479–5488 (2003)

13.322 Y.S. Lin, C.C. Lai, K.Y. Huang, J.C. Chen, C.Y. Lo, S.L. Huang, T.Y. Chang, J.Y. Ji, P. Shen: Nanostructure formation of double-clad Cr^{4+}:YAG crystal fiber grown by co-drawing laser-heated pedestal, J. Cryst. Growth **289**, 515–519 (2006)

13.323 J.Y. Ji, P. Shen, J.C. Chen, F.J. Kao, S.L. Huang, C.Y. Lo: On the deposition of $Cr_{3-\delta}O_4$ spinel particles upon laser-heated pedestal growth of Cr:YAG fiber, J. Cryst. Growth **282**, 343–352 (2005)

13.324 T. Kotani, J.K.W. Chen, H.L. Tuller: Striation formation in Ti-doped $Y_3Al_5O_{12}$ fibers grown by the laser heated floating zone method, J. Electroceram. **2**, 113–118 (1998)

13.325 T.C. Mao, J.C. Chen, C.C. Hu: Characterization of the growth mechanism of YIG crystal fibers using the laser heated pedestal growth method, J. Cryst. Growth **282**, 143–151 (2005)

13.326 J.C. Chen, C.C. Hu: Quantitative analysis of YIG, $YFeO_3$ and Fe_3O_4 in LHPG-grown YIG rods, J. Cryst. Growth **249**, 245–250 (2003)

13.327 C.C. Hu, J.C. Chen, C.H. Huang: Effect of pulling rates on the quality of YIG single crystal fibers, J. Cryst. Growth **225**, 257–263 (2001)

13.328 H.J. Lim, R.C. DeMattei, R.S. Feigelson, K. Rochford: Striations in YIG fibers grown by the laser-heated pedestal method, J. Cryst. Growth **212**, 191–203 (2000)

13.329 J.J. Romero, E. Montoya, L.E. Bausá, F. Agulló-Rueda, M.R.B. Andreeta, A.C. Hernandes: Multiwavelength laser action of Nd^{3+}:$YAlO_3$ single crystals grown by the laser heated pedestal growth method, Opt. Mater. **24**, 643–650 (2004)

13.330 A.M. Figueredo, M.J. Cima, M.C. Flemings, J.S. Haggerty, T. Hara, H. Ishii, T. Ohkuma, S. Hirano: Properties of $Ba_2YCu_3O_{7-\delta}$ filaments directionally solidified by the laser-heated floating zone technique, Physica C **241**, 92–102 (1995)

13.331 H. Ishii, T. Hara, S. Hirano, A.M. Figueredo, M.J. Cima: Magnetization behavior and critical current density along the *c* axis in melt-grown YBCO fiber crystal, Physica C **225**, 91–100 (1994)

13.332 M. Jiang, J.G. Huang, Y.Z. Wang, C. Zhang, D.C. Zeng, X. Wang, G.W. Qiao: Solidification characteristics of textured 123 phase in YBCO by laser floating zone leveling (LFZL) method, J. Cryst. Growth **130**, 389–393 (1993)

13.333 X.P. Jiang, J.G. Huang, Y. Yu, M. Jiang, G.W. Qiao, Y.L. Ge, Z.Q. Hu, C.X. Shi, Y.H. Zhao, Y.J. Wang, G.Z. Xu, Y.E. Zhou: The crystal growth of Y-Ba-Cu-O by laser floating zone melting, Supercond. Sci. Technol. **1**, 102–106 (1988)

13.334 M. Ferriol, Y. Terada, T. Fukuda, G. Boulon: Laser heated pedestal growth and characterization of zinc lithium niobate crystals, J. Cryst. Growth **197**, 221–227 (1999)

13.335 F.S. Vicente, A.C. Hernandes, A.C. Castro, M.F. Souza, M.R.B. Andreeta, M.S. Li: Photoluminescence spectrum of rare earth doped zirconia fibre and power excitation dependence, Radiat. Eff. Defects Solids **149**, 153–157 (1999)

13.336 J. Martínez Fernández, A.R. Pínto Gómez, J.J. Quispe Cancapa, A.R. de Arellano López, J. Llorca, J.Y. Pastor, S. Farmer, A. Sayir: High-temperature plastic deformation of Er_2O_3-doped ZrO_2 single crystals, Acta Mater. **54**, 2195–2204 (2006)

13.337 A. Ridruejo, J.Y. Pastor, J. Llorca, A. Sayir, V.M. Orera: Stress corrosion cracking of single-crystal tetragonal $ZrO_2(Er_2O_3)$, J. Am. Ceram. Soc. **88**, 3125 (2005)

13.338 L.M. Tong: Growth of high-quality Y_2O_3-ZrO_2 single-crystal optical fibers for ultra-high-temperature fiber-optic sensors, J. Cryst. Growth **217**, 281–286 (2000)

13.339 J.I. Peña, H. Miao, R.I. Merino, G.F. de la Fuente, V.M. Orera: Polymer matrix synthesis of zirconia eutectics for directional solidification into single crystal fibres, Solid State Ion. **101–103**, 143–147 (1997)

13.340 Y.J. Jiang, R.Y. Guo, A.S. Bhalla: Single crystal growth and ferroelectric properties of $\alpha(Ba_{1-x}Sr_x)Nb_2O_6$:$\beta(Na_{1-y}K_y)NbO_3$ solid solutions, J. Appl. Phys. **84**, 5140–5146 (1998)

13.341 J.G. Hou, D.Y. Tang, C.T. Chen, L.H. Ye, J.Q. Chen, Z.C. Ding: Growth of beta-barium metaborate single-crystal fibers along the phase matching direction, Chin. Phys. Lett. **7**, 568–571 (1990)

13.342 D.Y. Tang, R.K. Route, R.S. Feigelson: Growth of barium metaborate (BaB_2O_4) single crystal fibers by the laser-heated pedestal growth method, J. Cryst. Growth **91**, 81–89 (1988)

13.343 Y.H. Bing, A.S. Bhalla, R.Y. Guo: One-dimensional crystal growth near morphotropic phase boundary $(1-x)Pb(Mg_{1/3}Nb_{2/3})O_3$-$xPbTiO_3$ crystal fibers, Ferroelectr. Lett. Sect. **33**, 7–14 (2006)

13.344 G.F. de La Fuente, L.R. Black, D.M. Andrauskas, H.R. Verdún: Growth of Nd-doped rare earth silicates by the laser floating zone method, Solid State Ion. **32–33**, 494–505 (1989)

13.345 J.K. Yamamoto, A.S. Bhalla: Microwave dielectric-properties of layered perovskite $A_2B_2O_7$ single-crystal fibers, Mater. Lett. **10**, 497–500 (1991)

13.346 P. Rudolph, T. Fukuda: Fiber crystal growth from the melt, Cryst. Res. Technol. **34**, 3–40 (1999)

13.347 R.S. Feigelson: Pulling optical fibers, J. Cryst. Growth **79**, 669–680 (1986)

13.348 R.S. Feigelson: Opportunities for research on single-crystal fibers, Mater. Sci. Eng. B **1**, 67–75 (1988)

13.349 C. Goutaudier, K. Lebbou, Y. Guyot, M. Ito, H. Canibano, A. El Hassouni, L. Laversenne, M.T. Cohen-Adad, G. Boulon: Advances in fibre crystals: Growth and optimization of spectroscopic properties for Yb^{3+}-doped laser crystals, Ann. Chim. **28**, 73–88 (2003)

13.350 R.S. Feigelson, W.L. Kway, R.K. Route: Single-crystal fibers by the laser-heated pedestal growth method, Opt. Eng. **24**, 1102–1107 (1985)

13.351 R.S. Feigelson, W.L. Kway, R.K. Route: Single-crystal fibers by the laser-heated pedestal growth method, Proc. Soc. Photo-Opt. Instrum. Eng. **484**, 133–141 (1984)

13.352 W.M. Yen: Synthesis, characterization, and applications of shaped single crystals, Phys. Solid State **41**, 693–696 (1999)

13.353 W.M. Yen: Rare earth ions as spectroscopic probes of dynamic properties of insulators, J. Alloys Compd. **193**, 175–179 (1993)

13.354 S.M. Jacobsen, B.M. Tissue, W.M. Yen: New methods for studying the optical-properties of metal-ions in solids, J. Phys. Chem. **96**, 1547–1553 (1992)

13.355 B.M. Tissue, L.Z. Lu, M. Li, W.Y. Jia, M.L. Norton, W.M. Yen: Laser-heated pedestal growth of laser and IR-upconverting materials, J. Cryst. Growth **109**, 323–328 (1991)

13.356 C.T. Yen, D.O. Nason, W.A. Tiller: On controlled solidification studies of some TiO_2 binary-alloys, J. Mater. Res. **7**, 980–991 (1992)

13.357 J. Baszynski, W. Kowalski, C.A. Cardoso, F.M. Araujo-Moreira, M.R.B. Andreeta, A.C. Hernandes: Current-induced conductance jumps in mechanically controllable junctions of $La_{0.7}Sr_{0.3}MnO_3$ manganites, Czechoslov. J. Phys. **54**, D39–D42 (2004)

13.358 R.C. Santana, M.C. Terrile, A.C. Hernandes, M.R.B. Andreeta, G.E. Barberis: Electron spin resonance study of Fe^{3+} in $LiNbO_3$ single crystals: Bulk and fibres, Solid State Commun. **103**, 61–64 (1997)

13.359 R.Y. Guo, P. Ravindranathan, U. Selvaraj, A.S. Bhalla, L.E. Cross, R. Roy: Modified mixed-oxide perovskites $0.7Sr(Al_{1/2}B_{1/2})O_3$-$0.3LaAlO_3$ and $0.7Sr(Al_{1/2}B_{1/2})O_3$-$0.3NdGaO_3$ ($B = Ta^{5+}$ or Nb^{5+}) for high T_c superconductor substrate applications, J. Mater. Sci. **29**, 5054–5058 (1994)

13.360 S.L. Fu, J.S. Jiang, J.Q. Chen, Z.C. Ding: The growth of single-crystal fibers directly from source rods made of ultrafine powders, J. Mater. Sci. **28**, 1659–1662 (1993)

13.361 G.W. Young, J.A. Heminger: Modeling the time-dependent growth of single-crystal fibers, J. Cryst. Growth **178**, 410–421 (1997)

13.362 M.J.P. Nijmeijer, D.P. Landau: Simulation of optical fiber growth in three dimensions, Comput. Mater. Sci. **7**, 325–335 (1997)

13.363 L.H. Wang, B.J. Tsay, M.H. Hon: A thermoelastic analysis in a semi-infinite cylindrical single crystal during laser-heated pedestal growth, J. Chin. Inst. Eng. **21**, 101–108 (1998)

13.364 C.H. Huang, J.C. Chen, C. Hu: YVO_4 single-crystal fiber growth by the LHPG method, J. Cryst. Growth **211**, 237–241 (2000)

13.365 K.A. Jackson, J.D. Hunt: Lamellar and rod eutectic growth, Trans. Metall. Soc. AIME **236**, 1129–1142 (1966)

14. Synthesis of Refractory Materials by Skull Melting Technique

Vyacheslav V. Osiko, Mikhail A. Borik, Elena E. Lomonova

This chapter discusses methods of growing refractory oxide single crystals and synthesis of refractory glasses by skull melting technique in a cold crucible. It shows the advantages of radiofrequency (RF) heating of dielectric materials in a cold crucible and points out some specific problems regarding the process of growing crystals by directional crystallization from the melt and by pulling on a seed from the melt. The distinctive features of the method of directional crystallization from the melt are discussed in detail on the example of technology of materials based on zirconia, i. e., cubic single crystals and partly stabilized single crystals. It is shown that the size and quality of crystals are functions of the process conditions, such as thermal conditions under crystallization, growth rate, and chemical composition. We provide an overview of research on the structure, phase composition, and physicochemical properties of crystals based on zirconia. The optical, mechanical, and electric properties of these crystals make them suitable for a number of technical and industrial applications in optics, electronics, materials processing, and medicine. In this chapter, we also consider some problems regarding the synthesis of refractory glasses by skull melting technique. The physicochemical and optical properties of glasses are given and their practical applications in technology are discussed. We note that one of the better developed and most promising applications of skull melting technique is the immobilization of liquid and solid waste (also radioactive waste) into solid-state materials by vitrification.

14.1 Overview

Refractory single crystals have a wide range of applications in modern technology. Successful development of fields and branches such as microelectronics, fiber optics, laser techniques, metallurgy, and mechanical engineering would be impossible without elements based on refractory crystals and glasses as significant components. In order to promote the development of refractory materials technology it is necessary to find solutions to a number of problems relating to obtaining materials of a required chemical composition with predetermined properties, including

- Development of crystal growth methods and material synthesis techniques at temperatures higher than 2000 °C
- Growing single crystals and casting glasses with perfect internal structure
- Obtaining ultrapure materials where the uncontrolled impurity content does not exceed 10^{-4} wt % and in some cases even 10^{-6} wt %
- Controlled synthesis of materials in oxidation–reduction gas environments
- High production efficiency with minimal harm to the environment

The main challenge is to provide a complex solution to these problems, i. e., in the ideal case the developed technology will meet all these requirements. It is our opinion that the technology surveyed in this chapter can provide a complex approach to the problems related to the production of refractory crystalline materials and glasses, even though it also has certain shortcomings. This technology is based on the skull melting (SM) method in a cold crucible (CC), which follows two main principles:

- Keeping the melt in a solid shell (skull) with a chemical composition identical to that of the melt
- An inductive (i. e., contact-free) method of heating the material

These principles were proposed by various scientists a rather long time ago; for instance, the method of melting metal in an arc furnace with a water-cooled metal plate has been known since 1905 [14.1]. Later on there appeared some work on the method of inductive melting of metals and semiconductors in a CC [14.2, 3]. In the 1960s French scientists published their research on the method of inductive melting in CC of specific refractory oxides (Al_2O_3, ZrO_2, Y_2O_3, TiO_2, and UO_2) and complex chemical compounds based on them [14.4–8]. After that a number of publications appeared on the inductive heating of glass in cooled quartz and ceramic crucibles for vitrification of radioactive waste [14.9], as well as on the synthesis of ultrapure glasses in cooled quartz crucibles for fiber-optic communication lines [14.10, 11]. At the same time, independent research started at the Physical Institute of the USSR Academy of Sciences, aiming at developing technology for refractory materials production by the SM method. In the course of this research, appropriate equipment was designed and developed, conditions of melting and crystallization of various chemical compounds were systematically and thoroughly studied, and large-sized single crystals were grown from solid solutions based on zirconia and hafnia, as well as single crystals of rare-earth oxides, alumina, scandia, and yttria [14.12–15].

To date there have been quite a few publications discussing the modeling of inductive melting of dielectric materials in a CC. The basic properties of a thermal balance for so-called internal melting, which is very similar to SM, were discussed in [14.16, 17]. Some theoretical assumptions made in [14.16, 17] were applied to SM [14.18, 19], and this analytical model was used to account for quantitative description of losses in an RF coil and a crucible. The dependency between the form of a melted zone and the parameters of the system was calculated, a comparison with experimental data was made, and the stability of a melted zone was investigated [14.18, 19]; similar research is reported in [14.20–22]. The experience has shown that the new technology has a number of definite advantages and can successfully be used to synthesize refractory materials, including single-crystal growth. Some of these advantages are:

- There is no upper limit on the temperature (up to 3000 °C and higher).
- There is no contact with alien chemical substances and, therefore, the obtained materials are exceptionally pure. The purity of the resulting material can even exceed the purity of the original material in terms of some impurities (either volatile matters or those which are easily segregated during crystallization).
- The process can be carried out in any atmosphere, including an oxidizing one (air or oxygen).
- Any type of chemical reaction with the melt is possible, in both single-phase and multicomponent systems.
- Melting process can be interrupted or carried out continuously, without any restrictions on the melt volume.
- There are no specific requirements on grain-size composition of the initial materials.
- It is a waste-free technology, since the crystalline scrap from previous melting processes and other waste can be remelted.

- It becomes possible to obtain a large volume of melt, which facilitates convective mass and impurities transport and leads to better-quality single crystals.
- The melt composition can be changed during synthesis by adding different components to the melt via its open surface.

The most striking result of the developed SM technology is the industrial technology of growth and production of crystals based on zirconia. This is the only existing method for zirconia crystal production for industrial purposes. Each month several hundred tons of crystals are manufactured, with production concentrated mostly in the USA, China, South Korea, Russia, and Taiwan. More than 90% of grown crystals are used in jewelry and the rest is used in some technical applications. In this chapter we give an overview of experimental and theoretical research on SM in a CC and the results of this research, with special attention paid to the technology of zirconia-based single-crystal growth and the application of this method to glass synthesis.

14.2 Techniques for Growth of Single Crystals in a Cold Crucible

The development of the SM technique of nonmetal materials in a CC promoted progress in the field of technology of refractory materials, in particular single crystals. As has already been mentioned, the technique allows the melt to be held at very high temperature (up to 3000 °C or higher) to keep it in a stable state for crystallization under controlled conditions. During the last 40 years of the development of this technique a large variety of single crystals of refractory oxides and compounds have been grown, their properties have been investigated, and some industrial technologies and special equipment have been developed. The procedure of SM and crystallization from the melt under direct radiofrequency (RF) heating has been described in many studies [14.15, 23–25] and consists of several steps: (1) start melting, (2) formation of the main volume of the melt, (3) setting the melt–solid shell system in equilibrium (homogenization), and (4) crystallization of the melt (Fig. 14.1).

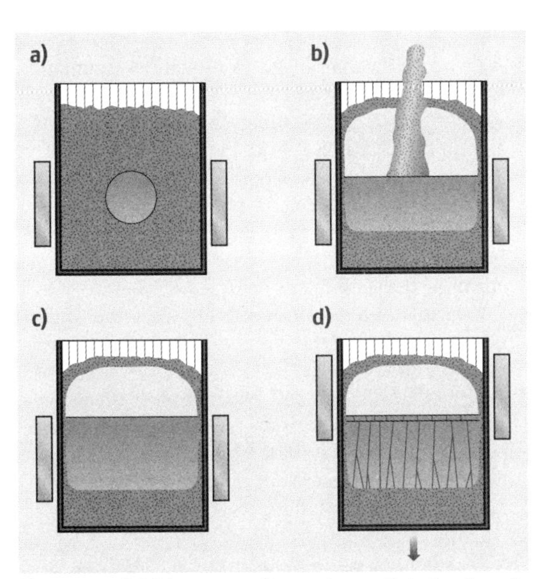

Fig. 14.1a–d Main stages of crystal growth technology in a cold crucible: (**a**) start melting, (**b**) formation of the main volume of the melt, (**c**) homogenization, and (**d**) crystallization

The necessity for start heating of dielectric materials and its practical implementation are discussed in detail in [14.25, 26]. Normally the metal used to initiate the start heating is the same as in the initial oxide charge. The metal and the surrounding charge are heated by the absorption of RF energy and the exothermic oxidation of the metal. In each particular case, the exact contribution of each process depends on the physicochemical properties of the metal used. Moreover, the process is affected by the conditions under which the start melting is carried out, i. e., the composition of the gas atmosphere, the position of the seed metal in the charge, the amount of seed metal, the particle size of the initial charge powder, etc. The time of start melting is largely determined by these factors. Graphite is rather frequently used in start melting because it completely oxidizes in air without any contamination of the melt. Being economical, this technique is very often employed in industrial production of many refractory materials and crystals. As a rule, bar-shaped graphite pieces are linked to obtain a ring with diameter smaller than that of the CC. Graphite is also used for melting complex oxide compounds and glasses, when it is sometimes positioned in the refractory ampoule, made of, for example, quartz, in the form of a heating rod. The rod is inserted into the center of the crucible filled with the initial powder charge. As the initial volume of the melt increases in the crucible, the rod is drawn out by means of various tools.

After start melting is complete, the melt volume should be increased gradually by the melting of ambient solid phase of the charge. The charge is usually composed of either oxide powders or their mixture with previously melted material. The melt volume increases until the charge in a CC is melted completely except for a thin layer of powder adjusting to the walls of the CC. This layer with a thin layer of crystallized melt together forms a polycrystalline solid (shell) or skull. Additional portions of the charge can be gradually poured into the melt until the desired volume of melt is reached. The use of either previously melted or preliminary compacted materials allows the desired volume of the melt to be achieved without additional pouring. In this case all the material is charged into the CC at once and the start melting occurs in the upper part of the charge, while the charged crucible is set at its lowest position. As the melt forms, the crucible is moved upwards to melt the whole charge, until the highest position is reached. This technique is less laborious but more time consuming and, thus, is less economical with respect to energy consumption.

When the volume of the melt reaches the desired value, the melt is exposed for a certain time under a constant energy input in order to achieve thermal and spatial equilibrium between the melt and the skull. The principal feature of the resulting stationary state of the system is a constant volume of the melt, i. e., the immobility of the melt–solid phase (skull) interface. The conditions of the phase equilibrium are discussed in [14.12]. Despite some rough approximations made in the analysis, it provides a simple and clear evaluation of the peculiarities of the phase equilibrium in the melt–solid system in a CC (Fig. 14.2). The following approximations are made in the analysis:

1. The temperature of the melt is equal over the volume, except for a thin layer adjacent to the skull (assuming complete stirring of the melt).
2. The temperature (T_L) at any point of the melt–solid interface is equal to the melting point and does not change.
3. The RF field energy is absorbed by the melt only and the field is uniform inside the inductor.

The amount of heat given off by the melt to the surface of the solid phase (Q_1) can be expressed as $Q_1 = \alpha(T_M - T_L)F\tau$, where α is the heat emission co-

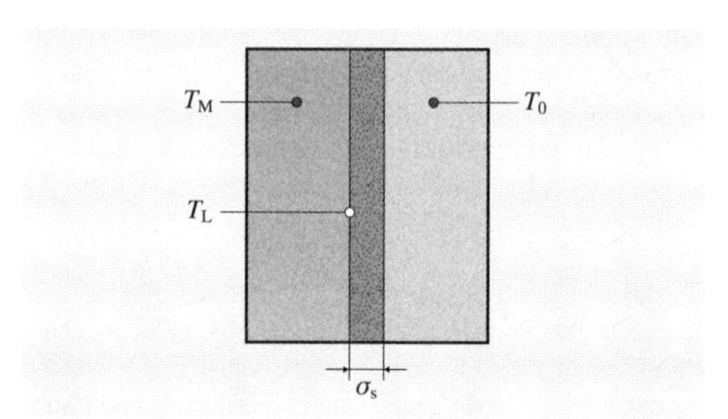

Fig. 14.2 Illustration of melt–solid equilibrium in a cold crucible

efficient, T_M is the temperature of the melt, T_L is the temperature at the solid–liquid interface (which equals the melting point), F is the heat-emitting surface, and τ is time. The amount of heat transported through a solid layer of thickness σ_S and thermal conductivity λ can be expressed by $Q_2 = \lambda/\sigma_S(T_L - T_0)F\tau$, where T_0 is the heat carrier temperature. For stationary conditions $Q_1 = Q_2$, i. e., $(\lambda/\sigma_S)\alpha = (T_M - T_L)/(T_L - L_0)$, where $(T_M - T_L) = \Delta T$ is the parameter characterizing the overheating of the melt. If we assume λ, α, and T_0 to be independent of σ_S, then $\sigma_S \Delta T = \text{const.}$

The above analysis leads to the following conclusions:

- An essential feature of the skull melting technique is the necessary overheating of the melt ΔT, the value of which greatly affects the process of crystal growth in a CC (if $\Delta T \to 0$, then $\sigma_S \to \infty$).
- There is always a solid skull in a CC because complete melting of the solid phase requires infinite overheating (if $\sigma_S \to 0$, then $\Delta T \to \infty$).
- Increasing or decreasing the melt temperature (T_M) changing the skull thickness σ_S.
- Convection of the melt determines the value of the heat emission coefficient α, thus affecting the phase equilibrium.
- Any process in the skull resulting in a change of its thermal conductivity (e.g., sintering or fusion penetration) will disturb the phase equilibrium and therefore cause variation in either the melt temperature (T_M) or the skull thickness (σ_S).

After the thermal equilibrium in a CC has been established, crystallization of the melt takes place. The crystallization can be carried out by various techniques. For producing molten polycrystalline materials a fast mass crystallization is quite suitable. Single-crystal growth is performed either by means of a directional crystallization, which is similar to the Bridgman–Stockbarger technique, or by pulling a seed similarly to in the Czochralski or Kyropoulos growth techniques. Applications of the other methods of single-crystal growth (e.g., crystallization from the flux solution or Stepanov technique) are also possible. The majority of single crystals are produced by either directional crystallization or by pulling a seed from the melt in a CC.

14.2.1 Directional Crystallization of the Melt

Most single crystals grown from a CC are produced by means of directional crystallization of the melt. The choice of this technique is motivated by its relative technical simplicity, which does not require any additional heat sources or complicated design of the thermal unit. Directional crystallization is usually carried out by lowering the crucible containing the melt out of the induction coil at a certain rate. Electromagnetic interaction between the lower part of the charge and the induction coil decreases as this part moves away from the inductor, so the released power also decreases. Gradual cooling then results in the crystallization of the melt at the bottom of the crucible. When using large-volume CCs, it is more convenient to conduct directional crystallization by moving the induction coil upwards, the container being fixed; this prevents technical problems associated with the design of the drive assembly for the heavy (> 1000 kg) charged crucible, and should ensure reliable operation and uniform crystallization rate in order to grow good-quality crystals.

Less frequently, crystallization is conducted by gradual reduction of the power. This procedure is less reliable, as it may lead to spatial instability and uncontrollable solidification of the melt [14.15–18]. Therefore, it is suitable only for crystallization of high-stability melts. The melt stability depends on the electrical and thermophysical properties of the material. Thus, to ensure a reliable and economical process, it is desirable to use melts which have sufficiently high electrical conductivity over a wide range of temperatures and with low thermal conductivity of the solid phase. The feasibility of SM for crystal growth also depends on a number of physicochemical properties of the material used. High evaporation rates of melt components may reduce the volume of the melt, causing the system to become unstable or resulting in uncontrolled crystallization of the melt. However, more frequently, volatility of a certain component causes a deviation from stoichiometry of the grown crystals. If this is the case, it is necessary either to correct the melt composition during growth or to use ambient atmospheres that prevent this deviation.

A CC usually consists of isolated sections and has gaps, so in the case of melt penetration, the melt can leak out, which can disturb the crystallization process. Leakage is prevented by the surface tension of the melt, which is determined by the surface tension coefficient, the size of gaps between the sections of the crucible, and the wettability of the wall. Therefore, it is easier to work with melts, possessing high surface tension. For single-crystal growth, direct crystallization in a CC has a number of advantages compared with other well-known crucible-less techniques (e.g., Verneuil or floating-zone technique); for instance, this method allows a considerable volume of the melt to be created, which promotes convective transfer of the main constituents and impurities and provides better quality of crystals. In addition, this method can be used to grow crystals in any atmosphere. The open surface of the melt makes it possible to vary the melt composition by adding various components and facilitates purification of the melt of volatile impurities and gases. However, single-crystal growth by directional crystallization in a CC faces some general difficulties, mainly associated with peculiarities of the direct RF heating of dielectric materials. With this method of heating the melt itself, being a transient power load in the oscillatory circuit of the RF generator, is a heat-absorbing body. The interaction between the load (melt) and the generator is rather complicated. Melting and crystallization occurring in the skull result in changes of the volume and electrical conductivity of the melt, which affect operation. This is problematic for rigorous control of the input power required to sustain the stability of crystallization. On the one hand, if the input power drops below a certain critical point, this may lead to loss of coupling between the melt and the RF field and, consequently, spontaneous crystallization. On the other hand, if the melt is overheated (i.e., excess input power), this results in melt penetration through the skull, leakage of the melt through the gaps between the CC sections, and disruption of the stability of crystallization (up to complete interruption of the process). Another control problem relates to the absence of visual monitoring of the process and direct measurements of the temperature of the melt and crystals during the growth. The high melting temperatures of the materials usually processed by this technique (such as zirconia, with a melting point of $\approx 3000\,°C$) and the presence of the RF field hinder any temperature measurements, except irradiative ones. Only the melt surface lends itself to direct observations and measurements, when the upper thermal screen (or core) is removed.

Just after the crystallization has been completed, annealing of the crystals is carried out directly in the growth furnace. Annealing is an important stage of the process, necessary to relieve any residual stress in order to prevent fracturing of the crystals. In a conventional growth furnace, there is usually a zone with a certain temperature gradient, where the crystals can slowly cool to the ambient temperature. This, however, becomes impossible when using direct RF heating for crystal growth because the electric conductivity of materials sharply decreases in the process of crystallization. The absorption of the RF field power decreases too, which results in uncontrollable cooling of grown crystals. The cooling rate depends on the volume of the crystallized material (i. e., its thermal inertia). Given that crystals are placed in a water-cooled crucible, the cooling rate of the crystal in zones adjacent to the wall is significantly higher.

The method of directional solidification in a CC essentially differs from the conventional *hot* techniques of crystal growth, e.g., the Bridgman–Stockbarger method, in which nucleation occurs only at the interface between the crystal and a considerable volume of the melt, followed by directional crystallization over a temperature gradient field. In these furnaces a certain gradient of temperature is created and maintained during the growth. The growth usually starts with complete melting of the charge in a crucible, and then, at the stage of initial nucleation, several crystals are formed. Further growth of one of the nuclei at the *crystal–melt* interface is performed by various techniques: a conical bottom, a capillary or constrictions on the bottom, and seeding. The task of growing large and good-quality crystals by this method is solved empirically by appropriate selection of crucible shape, temperature gradients, and the rates of lowering the crucible or cooling of the furnace. In directional crystallization in a CC the melt is in permanent contact with the skull of the same composition formed on the bottom and walls of the crucible. Therefore, polynuclear crystallization takes place. Crystallization begins in the lower part of the crucible from the crystal grains of the skull. As crystallization proceeds, the number of crystals is reduced considerably due to selection according to growth rate. As a result, a bulk of column-shaped single crystals is obtained. It is impossible to use any device to limit the quantity of crystals growing from the bottom and the walls of the crucible because of the very high temperatures and RF field. In general, seed crystals allow to increase the size and decrease the amount of growing crystals, but in practice this is rather problematic. Therefore, the quantity and size of the crystals are determined by the composition of the melt and the growth process conditions.

The peculiar distribution of temperature fields inside a CC is another important feature of the method. On the one hand, the distribution is determined by the RF field energy release in the melt, which depends on the melt properties at a current temperature, and on the other, by heat removal through the walls and bottom of the crucible maintained by the heat carrier (water). Growing good-quality crystals is determined by proper control of the temperature and temperature gradients in the melt and solid phase. Therefore, the distribution of temperature fields in a CC is extremely important. The patterns of temperature distribution both at the surface and inside the oxide melt under direct RF heating, as well as a number of technological conditions influencing these patterns, have been studied [14.27] using glass-forming melts in the temperature range 1000–1500 °C as a model system. The temperature distribution was shown to be determined by both the process parameters (absorbed RF power, position of a CC with respect to the induction coil, and the size and shape of a crucible) and the properties of the melt (temperature dependencies of the viscosity and electric conductivity). Typical temperature distribution patterns in the glass-forming melt under direct RF heating in a CC are presented in Fig. 14.3 for various viscosities. As this figure shows, a decrease in melt viscosity (most of non-glass-forming oxides) results in a signifi-

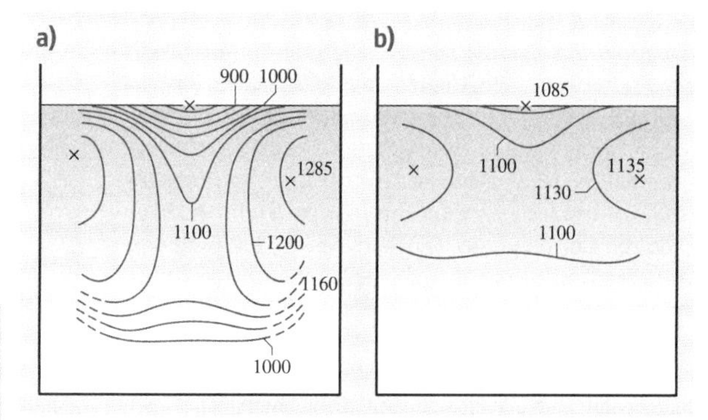

Fig. 14.3a,b Temperature distributions (°C) in oxide melts with different viscosities: **(a)** $45.6Na_2O + 54.4SiO_2$; **(b)** $55.4Na_2O +$ 44., $6SiO_2$

cant decrease of the temperature gradients in the melt, the general pattern of temperature distribution being the same.

A simplified model can help explain the temperature distribution patterns under direct RF heating. Most of the RF energy is released in a certain active layer of the melt, the thickness of which depends on the RF generator frequency and the electric conductivity of the melt, whereas the rest of the melt is heated by conduction of heat from this part. This type of heating can be roughly compared to the one produced by a virtual heat source in the form of a hollow cylinder placed into the melt. The height and diameter of the source are determined by the induction coil height and RF field penetration depth (the thickness of the active layer), respectively. It is clear that, in this model, the radial temperature gradients will be determined by the diameter of the CC and by the input power, and the axial ones by the position of the melt with respect to the induction coil and the heat losses.

For the directional crystallization method of crystal growth, the temperature distribution in the lower part of the melt controlling the crystallization front is of primary importance, but there are no data available for refractory materials. Nevertheless, even on the basis of a model, some conclusions could be made, which were later confirmed in practice. Thus, the maximum temperature is achieved at the level corresponding to the center of the RF coil, while in the upper and lower parts of the melt the temperature is lower due to the heat sink through the lower part of a crucible and heat losses from the surface of the melt. The ratio of heat flows removed from the top and bottom of the crucible determines the position of the maximum temperature zone, related to the melt depth and the values of the axial gradients. At the same time, the axial gradients are higher in the area that corresponds to the RF field energy release (i.e., at the periphery) than in the central zone. Such a temperature distribution results in an increase of temperature in the melt layer above the crystal bulk growing from the bottom with increasing height. Consequently, thermal convection in the melt is hindered. The effect of electrodynamic stirring depends on the applied RF field frequency and is negligible for frequencies of 0.5–10 MHz [14.26]. The most intensive convectional stirring occurs in the upper part of the melt. Radial temperature gradients in the heating zone suppress growth of crystals from the walls of a CC into the melt. However, if the melt is overheated, the crystallization front becomes concave and shifts below the heating zone (i.e., lower than the RF coil). This is associated with intense growth of crystals adjacent the skull and leads to an increase in the total number of the crystals, although the crystals become smaller. In order to suppress this growth, it is desirable to reduce overheating to a minimum and to maintain the heat sink from the bottom so that the crystallization front is either in the active heating zone or as close to it as possible. This is favorable for decreasing the total number of nucleating crystals and improving the quality of single crystals [14.28–31], because it brings the crystallization front near to the zone of intense convective stirring.

The following question thus arises: how to control the temperature distribution in the melt. First, this can be done by changing the pattern of energy absorption in the melt, which is largely determined by the value of the resistivity of the melt, its temperature dependency, and the RF field frequency. The former two parameters are properties of the material itself and so are predefined. The optimal electromagnetic field frequency is also predetermined by the electric properties of the material; it can be varied within a narrow range, but these variations do not significantly affect the temperature fields in the melt. For example, if the resistivity of the melt is 0.1 Ω cm then frequency variation from 5 to 1 MHz results in changing of the penetration depth from ≈ 0.7 to ≈ 1.6 cm. Even for small crucibles (e.g., 10 cm in diameter) this change could not significantly influence the values of temperature gradients. It is also possible to change the electromagnetic field configuration by changing the RF coil design, insertion of additional short-circuited turns, magnetic enhancers, etc. [14.25, 26, 31].

In our opinion, the simplest and most efficient way to control temperature gradients in the melt is to use those means that involve heat removal from the melt through the walls and the bottom of the crucible, as well as from the melt surface. To control heat removal through the walls of a cold crucible and from the melt surface heat shields are used. In order to keep the crucible bottom warmer, a part of the charge at the bottom is not subjected to melting at the stage of formation of the main volume of melt. This part is located below the heating zone and has the same composition as the main charge. This thermal shield decreases heat loss through the crucible bottom, balances the radial temperature gradients in the melt, reduces axial gradients in a growing crystal, decreases electromagnetic losses related to the absorption of energy by a massive metal bottom at start melting, and affects nucleation and degeneration of crystals in polynuclear crystallization. Low axial gradients and uniform thermal properties at the melt–solid interface help to decrease the number of nucleated crystals and to increase the size of individual single crystals.

The upper thermal shields are formed when the melt volume formation is nearly completed. Thermal radiation from the melt surface at high temperatures has high intensity, which leads to fast sintering of the powder charge above the melt. This sintered layer (crust) functions as an upper thermal shield. Such insulating layers of sintered material with composition identical to the melt readily occur in small-diameter cold crucibles. It is much more difficult to form the upper thermal shield in crucibles with larger diameters, although the role of such a shield becomes more important for crystal growth in large-diameter crucibles because the upper thermal shield decreases heat losses associated with radiation, increases the surface temperature, and consequently reduces the probability of spontaneous crystallization in the upper part of the melt, which is frequently the case in refractory melts. Reducing the heat losses results in better performance by decreasing the electricity cost, which is of primary importance for high-capacity installations. Surface heating (skin effect) and high temperature gradients in SM require special technological procedures for balancing thermal fields in the melt in order to achieve uniform melting and stable crystal growth over the whole diameter of the crucible. If the input power is insufficient and efficient upper thermal shielding is absent, the central part of the melt becomes overcooled, which results in the formation of small and imperfect crystals in this area. In the worst case, the material in the central part of a crucible remains unmelted or crystallizes rapidly.

In the absence of shielding, low radial and high axial temperature gradients in the melt and in growing crystals lead to the formation of many small crystals, whereas large-sized single crystals are desirable. It is possible to control the ratio between axial and radial gradients by applying thermal shielding in the top and bottom parts of the melt, thus optimizing the gradients for growing large-sized single crystals. Appropriate selection of the thermal conditions is usually carried out experimentally because of the difficulties in regulating melting and crystallization in this method. Moreover, thermal shields help to increase the mass of the charge in a cold crucible, and, consequently, to slow down the cooling of the crystals after growth to prevent fracturing in large crystal bulks. The shape of a CC influences the temperature distribution inside a crucible, and hence the thermal conditions of crystal growth. If the crucible diameter is increased while the penetration depth and input power are kept constant, then the radial gradients are reduced, leading to the formation of a convex or flat crystallization front. With constant heating parameters (frequency and power) and physical properties of the melt (electric conductivity and viscosity), changing the crucible diameter-to-height ratio affects heat removal from the growing crystals. The ratio of the melt volume to the square of the cooling surface decreases with increasing CC diameter. This reduces energy loss through the walls of a crucible and makes the process more economical. Heat removal from the growing crystals at the bottom mainly occurs through the bottom and side (lateral) walls in the lower part of the charge below the active heating zone. Therefore, the amount of heat removed from the growing crystals is reduced with the increase of crucible diameter for the same reasons: given that the height of this part remains the same, the larger the crucible diameter, the less cooling surface (through which the heat transfers from growing crystals) there is for a unit volume of solid phase.

Thus, there are two important factors for growing crystals from the melt in a CC: first, the power released in the melt should be accurately determined and correlated with the volume of the melt, and second, there is a need for proper control of the heat removal through the bottom of the CC and via the surface of the melt. Optimizing these parameters allows the shape of the crystallization front and its position with respect to the active heating zone to be varied. Unfortunately, as mentioned above, some of the processing parameters in a CC, such as the temperatures in the melt and in the solid phase, the melt volume, and the crystallization rate, are extremely difficult to control. These parameters can only be assessed by the final results of the crystallization after the process has been completed. Some attempts have been made to develop techniques that can provide indirect but reliable data on the melting directly in the course of the process. Continuous monitoring of the RF generator parameters combined with analysis of the generator loading status enables an elucidation of the mechanisms occurring in a CC [14.32–36].

To date a considerable variety of single crystals of simple and complex oxides and oxide solid solutions have been grown by directional crystallization. The conditions of crystallization and the dimensions of the crystals grown by this method are reviewed in [14.23]. It has been shown that the oxides of transition metals have the necessary electrical and thermophysical properties at high temperatures, which makes them suitable for RF heating, melting, maintaining in the molten state, and further crystallization in a CC. CoO, Fe_3O_4, and TiO_2 single crystals up to $1-3\,cm^3$ have been grown [14.37]. The scope of the technique has been demonstrated by magnetite single-crystal growth [14.38], before which good-quality magnetite (Fe_3O_4) single crystals were grown in platinum crucibles by the Bridgman technique. However, this approach was expensive since, following growth, the crucibles were cut to release the crystal, and, moreover, the crucibles were destroyed because of the diffusion of iron into the platinum. Thus, SM has obvious advantages in this case. In order to maintain the oxygen stoichiometry of magnetite it is crucial to create an atmosphere with a certain oxygen activity and to keep it under control. To solve this problem, the process was carried out in a CO/CO_2 buffered atmosphere with a fixed fugacity of O_2. The growth process consisted of two stages. In the first stage the charge is melted in air up to a desirable volume, and in the second stage the chamber is evacuated to allow the gas mixture to flow through it. The melt was kept in this atmosphere for 1.5 h to achieve equilibrium conditions, and then the crucible was lowered. A drop in the growth rate to 7.5 mm/h was mentioned to be necessary to attain the equilibrium between the melt and the gas phase and to grow larger crystals. Similar conditions were required to grow monoxides of some transition metals ($Mn_{1-x}O$, $Co_{1-x}O$, $Fe_{1-x}O$, and $Ni_{1-x}O$) with interesting electrical, optical, and magnetic properties, which can vary with the oxygen stoichiometry. The equipment used for growth in a CC was the same as in the previous case. After studying the effects of atmosphere on the phase composition of single crystals of transition-metal monoxides it was established that the excess of oxygen in the $Fe_{1-x}O$ melt led to the formation of Fe_3O_4 inclusions, whereas metallic inclusions M^0 occurred under a relatively reducing atmosphere. The importance of crystal annealing at subsolidus temperatures, which improves homogeneity of the crystals and completely eliminates the magnetite phase, was also demonstrated in these studies. With an adequate buffer oxygen atmosphere 1 cm-long $(Fe_3O_4)_{1-x}(FeTiO_4)_x$ single crystals were grown. The atmosphere was controlled throughout the process, including the melting stage. Studies on the phase composition of grown crystals proved this method to be applicable to the growth of ferrite crystals of any composition. The same facilities and technology were also used to grow Ln_2NiO_4 (Ln = La, Pr) single crystals. Later on, single crystals of high-temperature superconductors [14.20, 39], several complex oxides (e.g., CeO_2-Y_2O_3) [14.40], oxide eutectics [14.41], and other oxide compounds and compositions were synthesized and studied [14.42].

14.2.2 Crystal Growth by Pulling on a Seed from the Melt in a Cold Crucible

The method of crystal growth by pulling on a seed from the melt (Czochralski technique) in a hot crucible has been widely and successfully used for a variety of oxide crystals. Its application is conditioned by a strictly defined temperature distribution in the melt and at the crystallization front, as well as by a constant temperature at the crystallization front. The process of crystal growth by SM is of a great interest but it is associated with the some principal difficulties discussed above. The temperature distribution in the surface area of the melt is of primary importance for crystal growth by pulling on a seed. The isotherm patterns (Fig. 14.3) show the temperature distribution in the melt. Temperature gradients in the melt can be changed by varying the position of the RF coil with respect to the melt surface or by adjusting the input power. There are various additional factors affecting the formation of temperature gradients in the melt, such as optimizing the shape and the dimensions of a crucible, setting the optimal frequency, using thermal shields or additional heat sources, RF coil design, and some other parameters.

Let us now consider some practical applications of single-crystal growth by pulling on a seed from a CC. The synthesis of corundum and ruby single crystals was the first important scientific and practical result using this technique [14.12, 31, 43]. One of the conditions required for single-crystal growth by pulling on a seed was established to be the following: the specific heat flow from the melt though a crystal has to exceed the specific heat losses from the open surface of the melt. Thermal shields above the melt were used to decrease the heat losses associated with radiation. Besides, a cup-shaped crucible and RF coil were used to improve phase stability and to prevent crystal growth from the skull at the crucible surface caused by changing the input power (Fig. 14.4) [14.31]. The technique was used to grow ruby and corundum single crystals of various orientations, up to 160 mm in length and up to 35 mm in diameter. The pulling rate was in the range of 10–30 mm/h with seed rotation speed of 20–140 rpm.

The method of forming temperature gradients at the melt surface by means of thermal shielding was also applied for growing $SrTiO_3$ [14.44, 45] and $Bi_{12}GeO_{20}$ [14.46] single crystals. The $Bi_{12}GeO_{20}$ crystals were grown in a tubular crucible of 90 mm

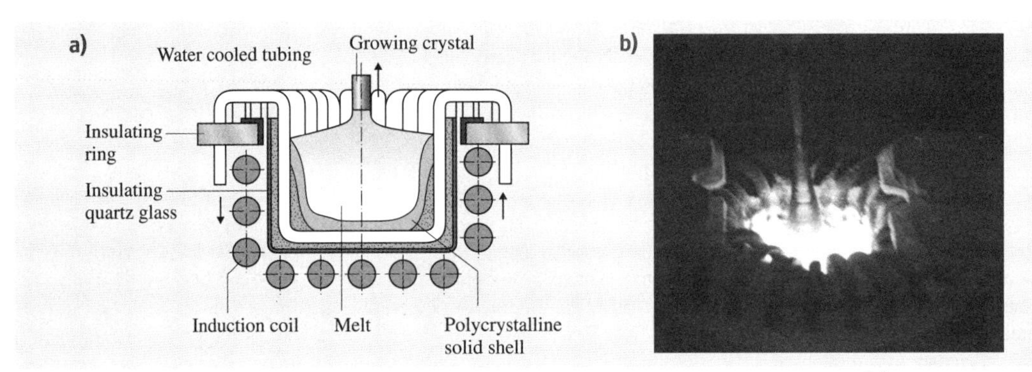

Fig. 14.4a,b Schematic of the unit for growing Al_2O_3 crystals from cup-shaped cold crucible (**a**) and photograph of growing crystal (**b**)

diameter and 60 mm height. The scheme of the heating unit for crystal growth is shown in Fig. 14.5.

The thermal and electrophysical properties of the melt allowed for adjusting the temperature by varying the input power over a wide range of values, while keeping the phase stability undisturbed. The consistency of input power supply was ensured by anode current stabilization. Convectional flows were observed on the melt surface. The intensity of the flows decreased when the input power was decreased, the convection pattern on the surface became regular, and the flows were oriented from the periphery to the center. Thermal shields provided more stable convection patterns, which made the intensity of the flows decrease. Figure 14.6 illustrates the temperature distribution on the melt surface measured by a pyrometer according to the observed convective flow pattern.

The grown crystals were up to 15 mm in cross-section and up to 100 mm in length (Fig. 14.7).

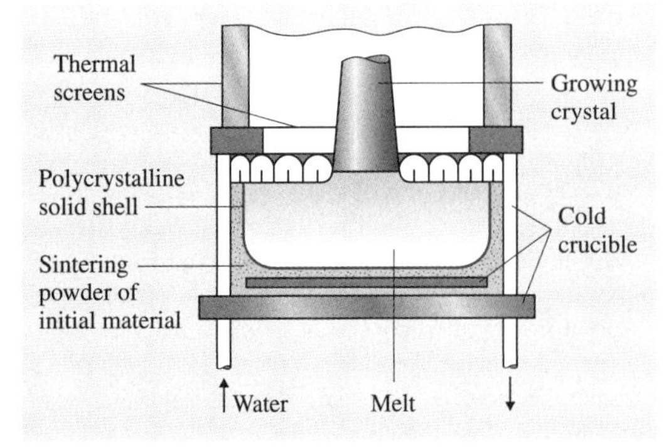

Fig. 14.5 Schematic of the unit for growing $Bi_{12}GeO_{20}$ crystals

In some cases the crystals were faceted while being grown, although crystal edge outlets of the fourth-order symmetry at the crystal surface were more frequently

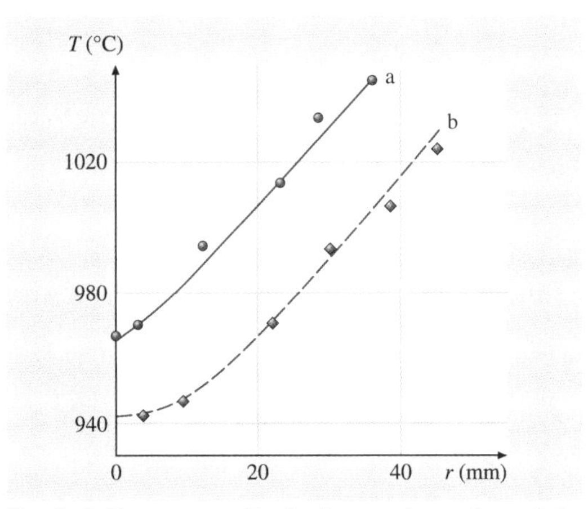

Fig. 14.6 Temperature distribution on the surface of the melt for input power of (a) 2.90 kW and (b) 2.50 kW

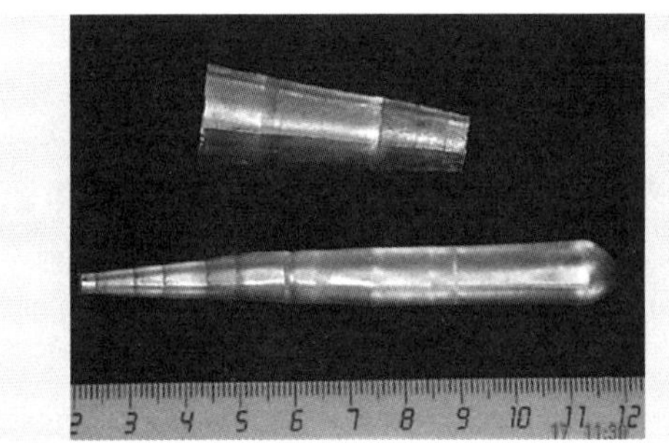

Fig. 14.7 $Bi_{12}GeO_{20}$ crystals

Table 14.1 Concentration of copper in $Bi_{12}GeO_{20}$ crystals grown in a copper water-cooled crucible

Material	Concentration of Cu, wt%
Initial charge	$1\times10^{-2}\pm5\times10^{-4}$
Skull	$9.3\times10^{-2}\pm1.6\times10^{-2}$
Crystal top	$1\times10^{-1}\pm2\times10^{-2}$
Crystal center	$7.5\times10^{-2}\pm1.3\times10^{-2}$
Crystal bottom	$1.4\times10^{-2}\pm2.5\times10^{-3}$

observed. $Bi_{12}GeO_{20}$ crystal grown in platinum crucibles is associated with the occurrence of platinum macroinclusions up to 0.5 mm. These inclusions are due to the chemical activity of the melt, which destroys the crucible surface, and to the metal particles becoming entrapped by a growing crystal. If a CC is used, this problem does not arise. However, X-ray spectral analysis revealed the presence of Cu (the material of a cold crucible) in crystals (Table 14.1). The measured concentration of Cu in the skull and in crystals was somewhat higher than in the initial charge and this was shown to be dependent on a number of technological parameters, such as overheating of the melt during start melting and during crystal detachment, the number of remeltings of the crystallized melt (the recycling factor), and the duration of growth. The presence of Cu did not result in inclusion formation and did not distort the optical homogeneity of crystals. A study of the optical properties of crystals revealed no influence of small copper impurities on the main properties of the material.

Another method for temperature fields formation in the melt was proposed in [14.47] for $Nd_3Ga_5O_{12}$ crystal growth. The growth was associated with significant turbulent convection on the melt surface. To suppress the convection and maintain the necessary radial temperature gradient, a *double-crucible* technique was applied, which consists of the insertion of a split iridium crucible into the melt. On the bottom of the CC there was an iridium ring, which was heated by an additional RF coil. The ring was used to produce a start melt and as an additional heating source to ensure a constant volume of melt during growth. The technique allowed $Nd_3Ga_5O_{12}$ crystals of 35 mm cross-section and 90 mm length to be grown, the quality of which was comparable to that of crystals grown from heated crucibles by Czochralski technique.

An interesting development of the SM technique is presented in [14.48], which proposes to combine direct RF melting with radiative heating of the melt surface. Start melting is carried out by means of three powerful optic concentrators with xenon arc lamps used as light sources. The CC can be moved vertically or rotated. The radiation from these three light sources makes it possible to control the thermal field pattern in the melt by changing the position of the focal spot with respect to the axis of the crucible rotation and by adjusting the radiative power. There are alumina thermal shields and a platinum electric resistance furnace above the melt, which help to decrease the temperature gradients in the pulled crystal and the performance of subsequent annealing. Radiative heating sources in this model allow to form certain thermal patterns in the melt at a certain radiative-to-RF heating power ratio, which facilitates crystal growth by pulling on a seed. The technique was applied to grow $MnZnFe_2O_4$ crystals. The SM method for crystal growth is very promising, primarily for the synthesis of new refractory compounds (in the form of crystals) from the melt. However, the analysis of thermal fluxes in the melt and the development of methods to monitor and control the process to achieve a certain thermal field configuration remain the most complicated problems in this growth technique. This might be the reason why there are so few publications devoted to crystal growth by pulling on a seed from a CC.

14.3 Growth of Single Crystals Based on Zirconium Dioxide

Crystals of pure ZrO_2 are used very rarely because of destructive phase transformations. However, it was shown [14.49] that stabilizing the cubic phase by adding MgO, CaO, Sc_2O_3, Y_2O_3 or CeO_2 could prevent these phase transformations. In this case a metastable cubic solid solution with fluorite structure is formed at room temperature.

The high melting point (2700–2800 °C), the high chemical activity of the melt, and the occurrence of polymorphic transformations made the growth of ZrO_2-based single crystals very difficult. That is why single crystals of pure ZrO_2 were obtained by low-temperature techniques such as flux growth, vapor deposition or hydrothermal method [14.50–54]. These processes are of a very long duration, while the size of the crystals obtained is rather small. Single crystals of solid solutions based on ZrO_2 were grown from the melt by crucible-free techniques (arc and floating-zone melting) [14.55–59]. However, large-scale industrial production of ZrO_2-based single crystals became possible only when the above-described SM method was applied. During the development of the method and its application for single-crystal growth, the first publications on the synthesis of ZrO_2-based single crystals appeared [14.7, 12, 31, 60–62].

Research on some properties of single crystals of cubic solid solutions based on ZrO_2 (CZ) showed that these crystals possess unique optical, mechanical, and electrical characteristics, namely, they are optically isotropic, have high refraction index $n_d = 2.15 - 2.2$, and are transparent in the range 260–7500 nm; these crystals are very hard (8.5 by Moos), have a high melting point (2700–2800 °C), and demonstrate ionic conductivity at temperatures above 300 °C [14.13, 15, 31]. This combination of properties makes CZ crystals promising materials for many technical and industrial applications, including optical, electronic, instrument-making, and high-temperature techniques.

The first practical application of CZ crystals was in jewelry. The similarity of CZ's refractivity index to that of diamond and its high dispersion cause a special play of light under various lighting conditions. These properties make CZ crystals one of the best materials for the imitation of diamond. The possibility of crystal growth of various colors enables the imitation of other natural gemstones, as well as the creation of new gemstones of original colors. Development of the industrial technology, the creation of the first apparatus for crystal production, and the production of the first crystals were accomplished in the Physical Institute of the Academy of Sciences of the USSR (FIAN) as early as the beginning of the 1970s. Exactly this application gave great impetus to the setting up of the industrial production of these crystals and the creation of new equipment for RF heating. Industrial technology for the production of CZ crystals was first elaborated in Russia in FIAN. Mass production of these crystals, which were given the name *fianites*, was already organized at the beginning of the 1970s. In most countries these crystals are known as *cubic zirconia*. World production of CZ crystals for jewelry is inferior only to that of silicon and synthetic quartz crystals, thus occupying the third place.

The following problems remain topical in the modern technology of CZ crystals:

- Enlarging of the color range
- Increasing the size of crystals
- Increasing the optical perfection and uniformity of crystals for technical applications

A little later, in the middle of the 1970s, interest towards high-strength and high-viscosity materials based on ZrO_2 arose. One such material is partly stabilized zirconium dioxide (PSZ) – a solid solution of yttrium oxide or other rare-earth or alkaline-earth oxides in ZrO_2. In 1975 the attention of researchers was concentrated on the creation of two-phase materials characterized by high destruction viscosity due to inherent phase transitions similar to martensitic transformations in steel. Thus, *Harvey* [14.63] obtained and extensively studied a material based on ZrO_2 partially stabilized by CaO. The new material was named *ceramic steel*. Ceramics of even higher characteristics were obtained when Y_2O_3 was used as a stabilizer and the tetragonal phase (TZP) was synthesized [14.64, 65]. The transformational mechanism of hardening was proposed in [14.66]. These results gave great impetus to the development of fundamental research on ZrO_2-based material and studies on its synthesis and applications. An up-to-date review on the transformational mechanism of hardening of constructional ceramics is given in [14.67], where the theoretical aspects of transformational hardening are analyzed and experimental results on the microstructure and mechanical properties of ZrO_2-based materials are presented.

Essential conditions for the preparation of high-strength high-viscosity construction ceramics are small grain size (10–100 nm) and that the residual porosity of the material be close to zero. Methods of nonporous ceramics preparation include high-temperature sintering, sintering under pressure, and hot pressing under isostatic conditions [14.68–70]. In order to obtain high-strength ceramics with high destruction viscosity by these methods one should use ultrafine (grain size $\approx$ 10–200 nm) and highly homogeneous, uniform fused mixture. Nowadays much attention is paid to the preparation of ultrafine oxide particles as starting materials for the production of nonporous ceramics [14.71, 72]. The peculiarities of ceramic materials include randomness of the distribution of initial components, defects in the material structure (which may cause a significant scatter in material properties), and the presence of pronounced grain boundaries, which influence the properties of such materials significantly.

The SM method allows high-strength crack-resistant material with zero porosity without sharp grain boundaries to be obtained by directional crystallization of the melt. The possibility of synthesis under air, the lack of special requirements concerning grain composition and homogeneity of the initial materials, as well as the practically waste-free character of the technology (i. e., the possibility of recycling crystalline waste) make this method very promising for the synthesis of these hard construction materials resistant to aggressive (in particular, oxidizing) gas atmosphere over a broad temperature range.

14.3.1 Crystal Structure of Zirconium Dioxide

Zirconium dioxide has several polymorphic modifications [14.73–75]. Monoclinic (m), tetragonal (t), and cubic (c) modifications of ZrO_2 exist at ambient pressure. The thermal stability range of the m-phase extends to 1160 °C; that of the t-phase extends from 1160 to ≈ 2370 °C; and that of the c-phase extends from ≈ 2370 °C to the melting point of ZrO_2, ≈ 2680 °C. Cubic phase of ZrO_2 is a nondistorted structure of the fluorite type, belonging to the $Fm3m$ space group with lattice parameter $a = 5.07$ Å. The oxygen coordination number is 4, and the zirconium coordination number regarding oxygen positions is 8 [14.76, 77]. Tetragonal phase of ZrO_2 exhibits a slightly distorted fluorite structure, belonging to the $P4_2/nmc$ space group with lattice parameters $a = b = 5.085$ Å, $c = 5.166$ Å, and $a/c = 1.016$ [14.74–76]. The c-phase becomes unstable upon a decrease in temperature and is transformed to the t-modification by means of a small distortion of the fluorite structure. The symmetry of the initial structure is distorted as a result of small translocations of atoms, mainly of oxygen ions. Oxygen ions become shifted from their ideal positions (1/4, 1/4, 1/4) in the fluorite lattice but this does not lead to a change in the coordination number of zirconium. Four oxygen ions located at a distance of 2.065 Å from each other occupy the apices of a tetrahedron; the other four ions occupy distorted tetrahedral positions and are located at a distance of 4.455 Å from each other. On the whole, the lattice of the t-phase displays a minute elongation along the c-axis as compared with the lattice of the c-phase. Monoclinic phase of ZrO_2 belongs to the space group $P2_1/c$, with the following lattice parameters $a = 5.169$ Å, $b = 5.232$ Å, $c = 5.341$ Å, $\beta = 99°15'$, and $Z = 4$ [14.77–79]. The next phase transition, occurring at 1200 °C, causes the transition from the t-form to the m-form. The latter form is stable under normal conditions, widely spread in the Earth's crust, and known as the mineral baddeleyite. Its structure is a result of further distortion of the c-phase of ZrO_2, but the distortion is so pronounced that a completely new structural type emerges. The zirconium coordination number decreases from 8 to 7. Half of the oxygen atoms in the baddeleyite structure are four-coordinated and the others are three-coordinated. It is important to emphasize that the structure of m-ZrO_2 is extremely stable, this being confirmed by the existence of m-ZrO_2 over a broad temperature range and the extreme stability of the mineral baddeleyite under natural conditions. Twinning is characteristic of the monoclinic structure. The twinning plane is constituted by oxygen ions – $\{100\}_m$ or $\{110\}_m$.

Pure α-Y_2O_3 (cubic) has a volume-centered lattice and is isostructural to Mn_2O_3. This structural type can be derived from the CaF_2 structure by deletion of a quarter of all nonmetallic atoms [14.80].

14.3.2 Phase Diagrams of the ZrO_2–Y_2O_3 System

Yttrium oxide is the most widely used stabilizing oxide for the growth of ZrO_2-based crystals. That is why the ZrO_2–Y_2O_3 system will be extensively discussed in this chapter. The first phase diagram for the ZrO_2–$YO_{1.5}$ system was published in 1951 [14.81]. Many publications have appeared since then, but the experimental data obtained are often contradictory and the data for some parts of the phase diagram are not yet sufficiently reliable. The characteristic presence of wide solid-solution regions based on either ZrO_2 or Y_2O_3 was disclosed already during the first studies of the ZrO_2–Y_2O_3 system. The most reliable phase diagrams were plotted at the end of the 1970s [14.82–87] (Fig. 14.8).

These diagrams show that the ordered compound $Zr_3Y_4O_{12}$ formed at Y_2O_3 content of 40 mol % exists along with solid solutions. This compound was first synthesized and described in [14.82]. $Zr_3Y_4O_{12}$ displays a rhombohedral-type symmetry (space group $R3$), is isostructural to UY_6O_{12}, and undergoes incongruent decomposition at 1523 ± 50 K [14.84], i. e., the structure is disordered to yield a fluorite phase.

Special attention was paid to the part of the phase diagram corresponding to zirconia-rich compositions because these materials are of great practical interest. A series of complicated phase transformations depending on the thermal history, grain size or particle size is observed in this part of the system. The presence of metastable states is common. A detailed analysis of these phase transformations is given in [14.88, 89]. A decrease in the temperature of the t–m phase transition accompanied by the increase in yttrium content is a characteristic property of this region of the phase diagram. A narrow two-phase (m–t) region exists in a range of temperatures above the temperatures to which the m-

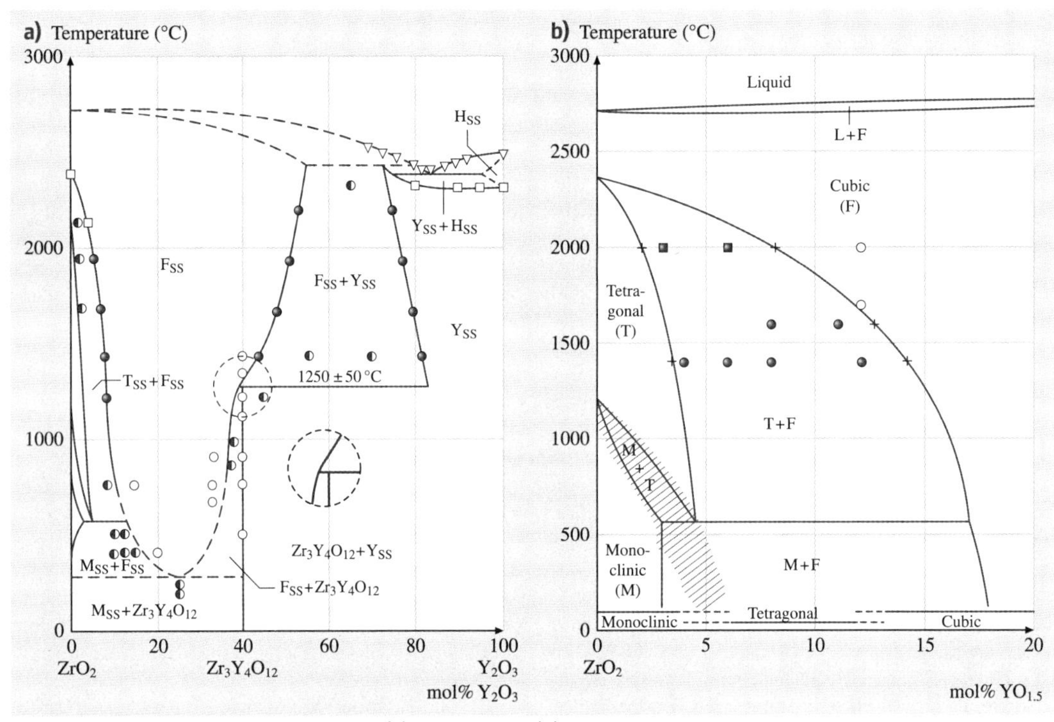

Fig. 14.8a,b ZrO_2–Y_2O_3 phase diagrams: (**a**) after [14.82]; (**b**) after [14.83]

phase corresponds. This two-phase region is followed by the so-called transformable t-phase region. The t-phase, which is transformed into the m-phase upon cooling, exists in the composition range of 0–5 mol % $YO_{1.5}$. However, if the grain size of the ceramic sample is low enough (0.2–1 μm), the transformation of the material with $YO_{1.5}$ content of 3–4 mol % into m-phase may not occur spontaneously at room temperature but takes place upon significant mechanical loading. The complexity and low rate of the m–t diffusion phase transition make the determination of equilibrium phase boundaries in the two-phase m + t region very difficult. Practically, the t ↔ m transition boundary was determined using various methods: differential thermal analysis, dilatometry, acoustic studies, and Raman spectroscopy. The data obtained in these studies were very incoherent, depending on the method of sample preparation, thermal history, presence and size of grains, purity of starting materials, etc. The samples were obtained by oxide mixtures sintering [14.87, 89] or by crystallization of the melt in a CC [14.90–92].

A two-phase region of untransformable tetragonal (t′) and c solid solutions corresponds to higher $YO_{1.5}$ concentrations. Samples containing 4–13 mol % $YO_{1.5}$ undergo a phase transformation into t′-ZrO_2 upon rapid cooling, starting from temperatures corresponding to the temperatures of existence of the c-ZrO_2 solid solution. This phase is named untransformable because it does not turn into the m-ZrO_2 phase. Upon an increase in yttrium oxide content the tetragonal distortion of the structure (c/a) decreases and its lattice parameters become very similar to the cubic parameters of the fluorite lattice. Such phase transitions in metallic systems are well known [14.93]. Their main features are the lack of changes of the chemical composition (the lack of diffusion) and a collective shifting of atoms during the transition (cooperative transformation). This transition is defined mainly by the rate of phase boundary migration. Various models of the phase transition process of this type were proposed, the choice of the model being determined by the temperature [14.94]. In the case of ZrO_2–$YO_{1.5}$ the c ↔ t′ transition occurs in the two-phase region c + t. The nature and properties of the t′-phase as well as its formation from the c-phase upon a phase transition are reviewed in detail in [14.95]. The boundaries of the two-phase region of c + t existence

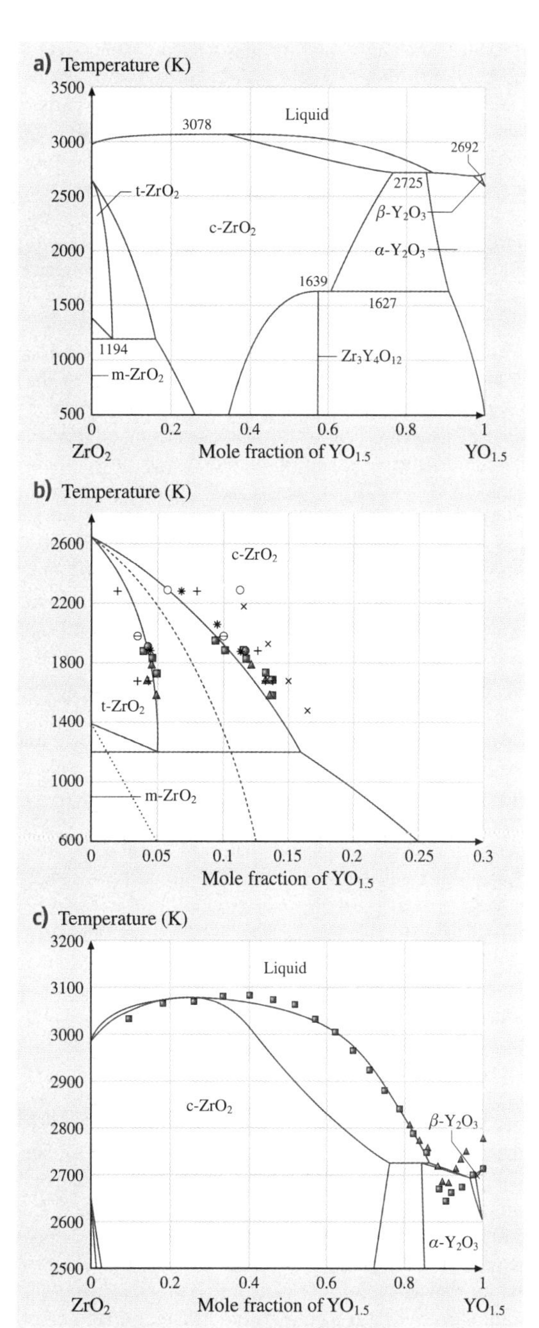

Fig. 14.9a–c Phase diagrams for the ZrO_2-Y_2O_3 system calculated by CALPHAD [14.99]. Experimental data from: (○) [14.81], (+) [14.83], (×) [14.84], (●) [14.100], (▲,■) [14.88], (∗) [14.83], (⊖) [14.101] *dotted line*: t–m transition temperature; *dashed line*: t–c transition temperature ▶

were studied by many researchers [14.81, 84, 87, 95]. The use of various methods of sample synthesis and phase boundary determination causes a scattering of the values. For example, in the temperature range of 1823–2273 K the two-phase region t + c extends from 3–4 mol % to 10–13 mol % $YO_{1.5}$. Further increase of $YO_{1.5}$ concentration to 15 mol % $YO_{1.5}$ and above results in the formation of completely stabilized c solid solutions of fluorite type. The liquidus curves for the whole composition range were first plotted in [14.96]. Eutectic formation was detected at Y_2O_3 concentration of 87.1 mol % and temperature of 2643 K. Melting points of 2983 and 2712 K were determined for pure ZrO_2 and pure Y_2O_3 by the authors of this work. The calculation of phase diagrams (CALPHAD) method was used to give a complete description of the thermodynamic properties and equilibrium phase boundaries of the system ZrO_2–$YO_{1.5}$ [14.97–99]. The calculated phase diagram of the ZrO_2–Y_2O_3 system as well as separate fragments supplemented with points marking the experimental values is presented in Fig. 14.9.

14.3.3 Stabilization of Cubic and Tetragonal Structures in Zirconia-Based Materials

Stability in the whole temperature range including room temperature should be imparted to one of the high-temperature modifications of ZrO_2 in order to grow single crystals from the melt. The ratio of the cation (R_c) and anion (R_a) ionic radii was shown to be the critical parameter for fluorite lattice stability [14.102]. The face-centered anion package in the $Fm3m$ lattice is possible if $R_c/R_a \geq 0.736$. If the parameter value is less than this value, the cubic structure cannot exist because forces of electrostatic repulsion inevitably distort the dense ion package when distances between the anions are small. For ZrO_2 this ratio is 0.66. The cubic lattice can be stabilized if the cation size is increased and the effective anion size is decreased. This may be achieved in two ways: by replacing the zirconium cation by a cation of a larger radius or by creating vacancies in the anionic sublattice (i. e., by introducing cations of lower charge). In order to achieve the required increase in the ratio of the ion radii in the ZrO_2 lattice the lattice parameters should be increased by approximately 4%. In some cases stabilization is also achieved by introduction of oxides with cations having a lower radius than that of the zirconium ion. Therefore, the cation radius is not a decisive factor for the formation of the ZrO_2 c-phase. The stable lattice is the one corresponding to the state of minimum free energy. Apart from the radii of the cation and anion, the character of the interaction between the electron shells of the lattice components is of a great significance for this state. References [14.103, 104] state that the bond between oxygen and the introduced cation must be more heteropolar than the bond between oxygen and the zirconium ion. In this case the fluorite structure is formed. The mechanism of stabilization of high-temperature modifications and the role of the oxygen vacancies formed upon heterovalent replacement of zirconium cations by lower-valency cations of the stabilizing oxide are discussed in several publications [14.105–107]. The degree of stabilization depends on the nature of the stabilizing oxide and its concentration. Oxides structurally similar to ZrO_2 are commonly used. According to crystallochemical conceptions, stable solid solutions can be formed in this case. Such oxides include Y_2O_3, and oxides of rare-earth elements and alkaline-earth elements.

Tetragonal solid solution based on ZrO_2 can be obtained by decreasing the stabilizing oxide concentration. Single-phase samples with t-structure can be prepared only under strictly determined conditions [14.64, 65, 85, 108]. The stability of the t-ZrO_2 structure depends on factors such as density, composition, grain size, length of intergrain boundaries, and annealing conditions. As mentioned above, two forms of t-ZrO_2 exist [14.94, 100, 101, 109]: the t′-ZrO_2, rich in Y_2O_3, and the t-ZrO_2, which is depleted of Y_2O_3 and can undergo a martensitic transition into the m-phase. The t′-phase is predominant in materials obtained by sharp quenching of c-solid solutions. Single crystals grown by directional crystallization of the ZrO_2–3 mol % Y_2O_3 melt composition in a CC have a strongly twinned t-structure [14.110]. Cooling at a low rate facilitates the t → m phase transition, i. e., the t-phase was not the t′-phase in spite of being richer in Y_2O_3 than the equilibrium t-phase. ZrO_2-based ceramics with 2–9 mol % Y_2O_3 stabilizer, obtained by sintering, consists of two or three phases mainly. Monoclinic or t-ZrO_2 can be present in such materials in the form of coherent precipitates in a t- or c-matrix, according to the ratio of the crystallographic parameters of these phases.

The best conditions for preparation of samples consisting only of the t′-phase by directional crystallization of melt in a CC are Y_2O_3 concentration of 3 mol % and high cooling rate (> 400 K/h). These results are consistent with the experimental conditions determined in [14.111]. For the study of the t′-phase, crystals grown by SM were reheated up to 2150 °C (stability region of the c-phase) for 10 min and then quickly cooled to room temperature (cooling time of 60 min) [14.112]. The c → t transformation is accompanied by the formation of a domain structure in the crystals [14.109, 112, 113]. Orientation dispersion and the crystallographic correlations between domains are determined by both the prototype (c-phase) symmetry and the symmetry of the t′-phase formed from the c-phase. Domains form ordered colonies of two alternative variants separated by habitus planes {110}, which are the twinning planes. If the material contains only the t′-phase, then colonies occupy all of its volume, directly adjoining each other. They border on each other along the {110} planes and are elongated along the ⟨111⟩ direction, forming a spiral-like structure. Detailed investigation of the domain structure of the t′-phase was performed in [14.112, 114–116]. It was found that tetragonal domains occupy all of the volume of t′-ZrO_2. Their spatial arrangement must be highly symmetric in order to minimize the energy of coherent deformations [14.112]. Three-dimensional spatial arrange of the colonies in pure t-ZrO_2 obviously differs from that in PSZ, where colonies of t-uniformex domains are inserted into the uniform matrix [14.115, 117]. Colonies and their substructures are a result of the optimal accommodation of spontaneous stress caused by the c → t transition. Upon a c → t phase transition, the c-axis of the elementary cell is slightly elongated. In t′-ZrO_2 containing 3 mol % Y_2O_3 the elongation did not exceed 1%. This is approximately half that of the t-distortion for pure ZrO_2 with $c/a = 1.02$ [14.112]. Correspondingly, the domain c-axes are not orthogonal: the angle between them equals 89.4° in the case of a 1% distortion. It has been found that the domains inside the colony are separated by coherent low-energy twinning planes {110}.

14.3.4 Cubic Zirconia Crystals (Fianits)

As mentioned, the above-described SM method was most successfully applied for the development of an industrial technology for ZrO_2-based single-crystal growth. Crystals up to 60 mm in length and up to 20 mm in cross section had been grown already at the beginning of the 1970s using the technology elaborated in Russia. Such crystals were obtained in the first industrial equipment including a CC of 180–200 mm in diameter.

The power of the RF generator was 60 kW, and the frequency was 5.28 MHz. The weight of the melt was up to 15–20 kg. Further development of the technology was aimed at achieving a larger size and higher quality of the crystals as well as at lowering their price. Using large melt volumes for crystal growth is the simplest and most efficient way of achieving these objectives because it allows:

- Growth of a much larger quantity of crystals during one crystallization cycle at the same crystallization rate. This allows lowering of the production costs.
- Lowering the heat losses through the walls of a CC and the corresponding lowering of the RF generator input power; the ratio of cooled surface area to the melt volume is much less in larger CC.
- Improving the crystallization conditions in order to obtain high-quality crystals. When the melt volume is large, the system has a high thermal inertia which levels out random oscillations of RF output power on the crystallization front.
- Preventing the formation of cracks in large crystals during the cooling of the crystalline ingot. At large melt volumes the cooling rate of the crystals after complete of the crystallization is much lower.

This was confirmed with experience on industrial equipment with a CC of 400 mm in diameter, output power of 160 kW, and RF frequency of 1.76 MHz. The weight of the melt was 80–100 kg. Nowadays large-scale production of single crystals is mainly performed in facilities with CC diameter up to 1000 mm, power of 800 kW, and frequency of 400–800 kHz. The weight of the melt is 600–1500 kg. The weight of individual crystals reaches 15 kg.

Growth of CZ Crystals

The scheme of manufacturing of CZ crystals is shown in Fig. 14.10. The main technological stages are the following: preparation of the initial mixture, loading of the initial mixture into the CC, SM of the material, directional crystallization of the melt, extraction and separation of the crystal ingot, crystal cutting and sorting, and processing of the material and offcuts. The most important stages of the technological process will be discussed below.

The oxides used as starting material are weighed and thoroughly mixed in a ratio corresponding to the predetermined concentration. In the case of mass production, various quantities of crystalline melted material of the same composition from previous melts are added to the initial mixture [14.25, 30, 118]. ZrO_2 and Y_2O_3 are the main raw materials, their purity being of great significance for the crystallization conditions and the degree of structural perfection of the crystals. The presence of some trace contaminants in the oxides used for crystal growth by directional crystallization in the CC is highly undesirable. These contaminants influence both the growth process and the quality of the single crystals obtained. Atomic emission spectroscopy studies have shown that the concentration of contaminants in the crystals obtained is 5–100 times less than that in the starting oxides due to the evaporation of the volatile oxides (namely, oxides of As, Pb, Cu, etc.) and the segregation of impurities (Si, Ti, Al, W, etc.) during directional crystallization [14.25, 30] (Table 14.2).

Requirements concerning the purity of the starting materials depend on the field of application of the crystals. For example, the content of Si, Ti, Al, and W in oxides used to synthesize crystals of high optical uniformity and structural perfection for optical applications must not exceed 10^{-4}–10^{-5} wt %.

Putting the initial materials into the CC is an important technological stage that greatly influences the reproducibility and crystal yield in the cycle, as well as the size and quality of crystals. The method of loading influences the processes of start melting, the formation of the melt volume, and the thermal conditions under which further melting and subsequent crystallization of the melt proceed. As already mentioned, thermal shielding of the melt plays an important role during SM in a CC. The lower thermal shield is formed during crucible loading. As a rule, crystalline melted pieces of the previous processes and the initial powders of the same composition are used to form the shield. The position of the shield is determined experimentally to create appropriate thermal conditions for crystallization. If large quantities of crystalline material are used, circular loading of the CC is often performed [14.118]. Circular loading ensures the formation of a dense, reliable sintered skull. Thus heat losses are decreased and leakage of the melt from the CC through intersection gaps is prevented. As a result, the stability of the melting process becomes higher. In the case of a large-diameter CC this loading method ensures reproducibility of the start melting process and creates the optimal conditions for the formation of large melt volumes required for obtaining large crystals at the lowest power values possible.

During the melting of the initial charge the molten zone approaches the CC walls; emission of radiation through the intersection gaps situated in the upper part of the charge begins and gradually becomes more in-

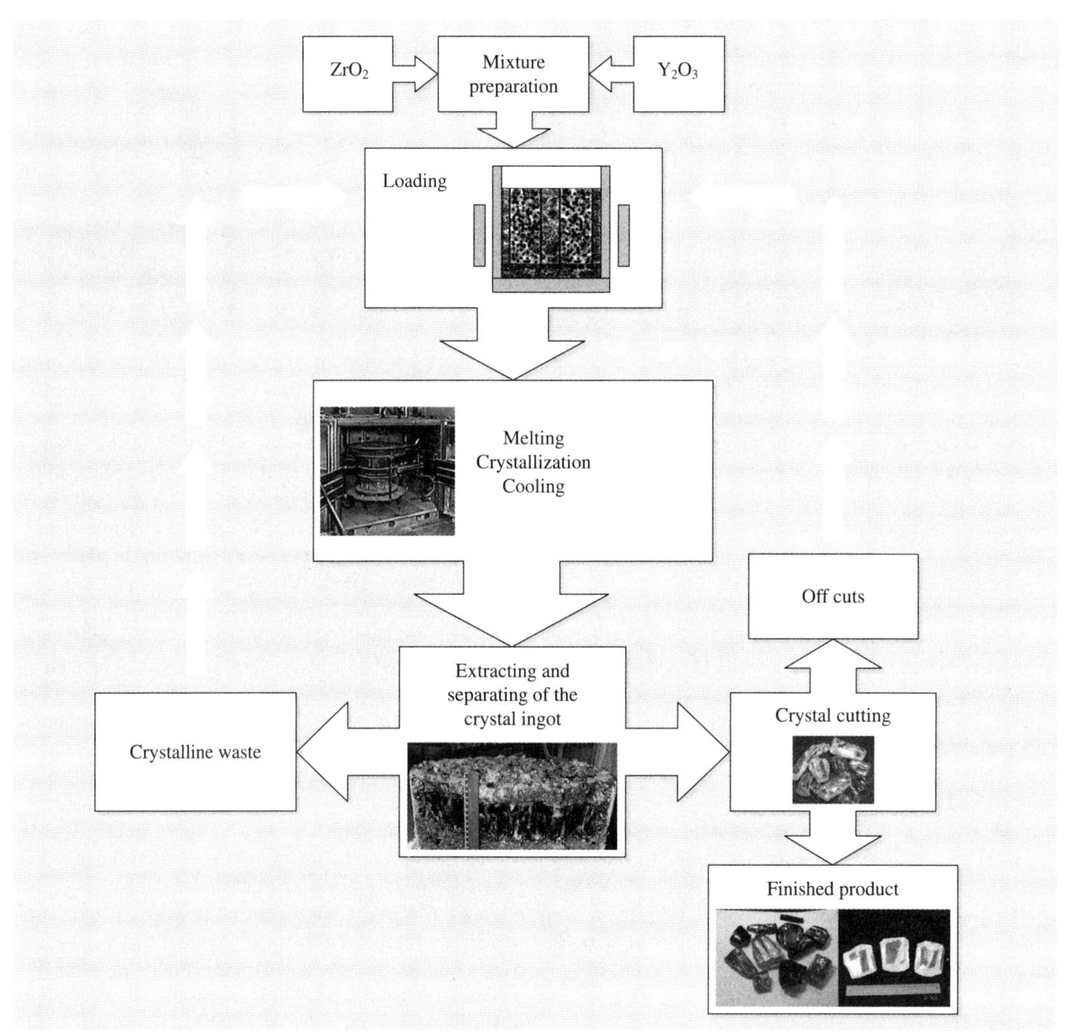

Fig. 14.10 Process flowsheet of the production of crystals based on zirconia

tense. The importance of providing an upper shield for the melt upon crystal growing by directional crystallization from the melt in large CCs was mentioned above. Choice of the size of the upper thermal shield and the method of its formation vary depending on the method of melt volume formation during start melting. If portions of the initial powders are not added during melt volume formation, then the remaining sintered arch of powdered material acts as a heat shield. If initial powders are added during the process of melt formation, then crystalline material (of the same composition) of previous processes can be used for shield formation. This material is loaded after the completion of melt formation. This material does not melt because it remains above the RF coil level during loading; only partial melting of the lower part of the shield may occur. Studies on the initial melt volume formation in the CC have shown [14.33–36] that the same volume can be formed in different ways (Fig. 14.11).

The character of melting depends on the characteristics of the starting material. Studies were carried out for ZrO_2 powders of different dispersity. The experimental data showed that more power is needed for the melting of a less porous (porosity ε_1) powder than for a more

Table 14.2 Results of the analytical determination of impurities concentrations in the initial ZrO_2 and ZrO_2–15 mol. % Y_2O_3 crystals grown at 10 mm/h

Element	Concentration, wt% × 10^{-5} Initial ZrO_2	Crystal		
		Bottom of the crystal	Middle of the crystal	Top of the crystal
Fe	2.0	< 0.1	< 0.5	1.0
Ca	3.0	1.1	1.6	1.8
Si	5.1	0.1	0.5	1.0
Mn	0.2	< 0.1	< 0.1	< 0.1
Cu	1.0	< 0.1	< 0.1	< 0.1
Mg	1.2	0.5	1.1	1.4
Al	0.5	< 0.1	< 0.1	< 0.1
Nb	0.4	0.1	0.1	0.3
W	< 0.1	< 0.01	–	–
Ti	0.5	< 0.1	0.1	0.4
Be	< 0.1	< 0.01	< 0.01	0.01
Sr	4.3	1.5	2.1	3.3

porous one (porosity ε_2). Properties of the initial material (in particular the powder porosity) determine the direction of the melt spreading during the stage of start melting, as well as the structure of the skull. It was confirmed that two modes of melting of a porous dielectric exist: the stationary mode and the quasiperiodic mode. The mode observed is determined by the ratio of the threshold power density (depending on the porosity of the initial powder) and the density of power released in the charge. This experimental data confirmed the theoretical model of induction melting of dielectrics proposed in [14.119]. It was shown that the direction of melt spreading during the initial stage of melting influences the volume and the shape of the molten bath formed, the value of power released in the melt, the value of melt overheating, and the power of heat losses in the charge. The preferable variant of spreading during melting is spreading towards the bottom of the CC. In this case less power is needed for the molten bath formation, the melt overheating is less pronounced, and the melting process is more stable without leakage of melt through the gap between the sections. The mode of initial melt formation is especially important when large CC or charges of large mass are used, or when power is limited. The use of the above-mentioned circular loading method considers the advantages of melt volume formation by spreading of the melt towards the bottom of the CC and thus promotes the formation of large melt

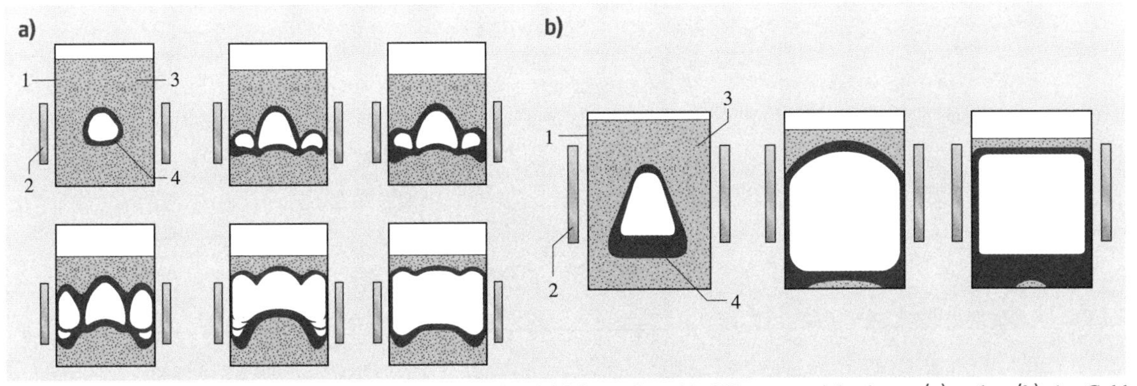

Fig. 14.11a,b Schematic of the melt propagation at melting initial powder with different particle size ε_1 (**a**) and ε_2 (**b**). 1 – Cold crucible; 2 – induction coil; 3 – powder of initial materials; 4 – melt

volumes at minimal power values. After the end of melting, the melt is homogenized to achieve thermal and phase equilibrium.

Crystallization of the melt begins on the bottom of the CC on crystal grains of the solid skull. Only a restricted number of crystals of those initially formed remain during growth. As a result of crystallization, a columnar block of single crystals grows. The size of CZ crystals grown by SM technique is influenced by:

- The conditions of formation of the initial melt volume determining the temperature gradients on the interface, which influence the geometry of the crystallization front and the position of this front relative to the heating zone
- The conditions of nucleation on the skull
- The conditions of degeneration during mass crystallization
- The chemical composition of the initial melt (the nature and concentration of the stabilizing oxide and the presence of impurities)

The melt–solid interface geometry may be concave, flat or convex depending on the conditions of the melt volume formation. The last two types are preferable for crystal growth in the CC as well as for traditional directional crystallization in hot crucibles because they promote a decrease in the number of growing crystals and a consequent increase of crystal size; they also ensure low thermal strains and high purity of the growing crystal. The experience of growing CZ crystals by SM technique shows that larger crystals are formed when the crystallization front is flat or convex, other conditions being equal. In the case of a concave crystallization front the number of crystals formed increases, while their size, especially the size of the crystals in the central part of the ingot, decreases (Fig. 14.12). The actual crystallization front may have a more complex shape and change its geometry during crystallization corresponding to changes in thermal conditions.

Nucleation. The nucleation process determines the size and quantity of growing crystals. Authors [14.120] have described the peculiarities of SM of porous dielectrics during the stage of melting wave propagation, when capillary spreading of the melt takes place inside the pores of the dielectric powder. During this process the melt impregnates the solid phase and crystallizes in it, then slow melting of the crystallized layer takes place, and then the situation cycles. These processes of capillary spreading determine the local structure of the interface on which the single-crystal growth begins. The melt from the heated zone causes partial melting of the upper layer of the lower thermal shield. This layer being nonuniform and porous, the melt penetrates into the surface layer nonuniformly. The photograph of the fragment of the lower part of a crystal (Fig. 14.13) clearly shows that inclusions of unmolten initial powder and bubbles are present in the crystal.

The Laue pattern of this part of the crystal indicated that the atomic planes are heavily distorted near the inclusions but the crystal structure remains single-crystalline. Nonuniformity of the nucleation interface leads to the formation of a great number of crystals and thus decreases the cross-sectional size of isolated crystal. So, the lower thermal shield influences the nucleation process greatly. When this shield is lacking, the thin skull is substantially nonuniform due to the random

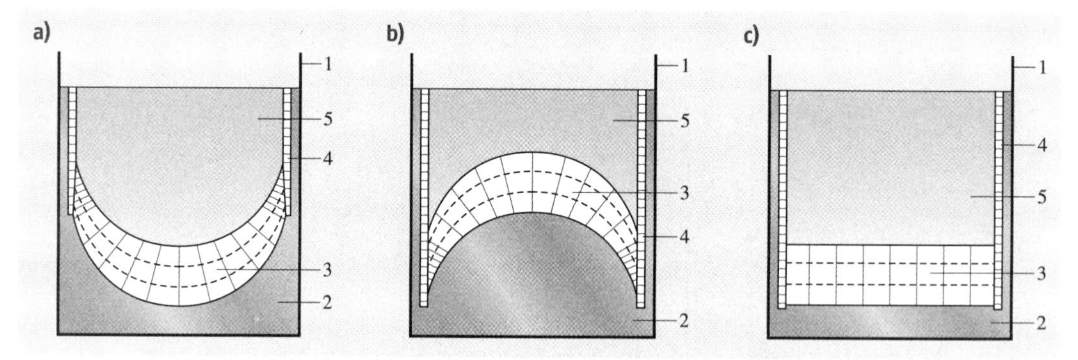

Fig. 14.12a–c Crystal growth in a cold crucible with different shapes of crystallization front: (a) concave crystallization front, (b) convex crystallization front, and (c) flat crystallization front. 1 – cold crucible; 2 – polycrystalline solid shell; 3 – crystals grown from bottom of the crucible; 4 – crystals grown from sides of the crucible; 5 – melt

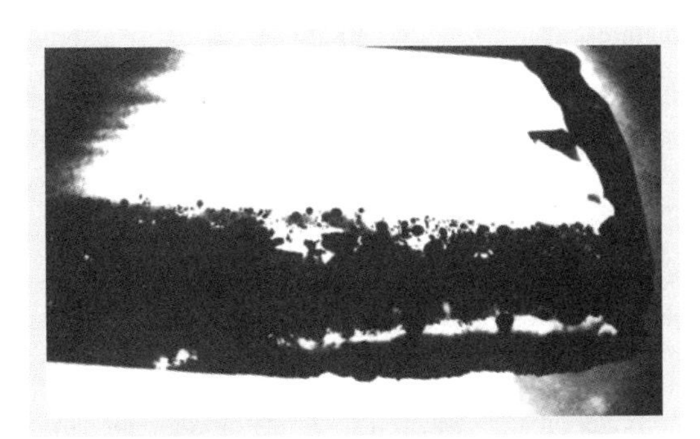

Fig. 14.13 Section of the bottom part of the crystal

character of heat removal in different regions (even direct contact of the melt with the bottom of the CC is possible). This nonuniformity is significantly decreased if a dense heat-insulating layer is formed. This causes a significant decrease in the number of seeded crystals and the increase of the size of crystals already at the nucleation stage. Besides, such a shield allows the use of seeds placed on it. Unfortunately, this seeding process is uncontrolled, has a random character, and requires experimental selection of conditions to increase its repeatability. Nevertheless, crystals were grown on seeds of different orientations [(100), (111), and (101)] and X-ray studies showed that the crystals inherited the crystallographic orientation of the seeds.

Degeneration. The degeneration process consists of a decrease in the number of seeding crystals and is accompanied by an increase in the cross-sectional size of some crystals at the expense of a decrease in the size of other crystals. The problems of competitive growth were analyzed in studies of mass crystallization [14.121–125]; the selection of crystals according to their size was shown to be a result of these processes and so-called *geometric selection*. The mechanism of this process is the following: crystals survive the competition only if their direction of fastest growth is close to the normal of the nucleation surface. As a result, all crystals separated from the nucleation surface by a distance greatly exceeding the mean distance between the nuclei are elongated and their geometric axes are almost parallel to each other: the so-called columnar structure occurs. The formation of such a structure is influenced by only geometric factors, the influence of external conditions (for example, the direction of heat transfer from the growing crystal) being small. It was supposed that the influence of external factors could be only indirect and result in a change of facet growth rate and crystal habitus. Quantitative experiments on geometric selection [14.126, 127] showed that the linear density of the number of surviving crystals $n(h)$ decreases proportionally to the height $n \sim 1/h$. The probabilistic assessment [14.128] yielded the same result for the two-dimensional case and a dependence expressed by the formula $n \sim 1/\sqrt{h}$ for the three-dimensional case. The aspect ratio depends on the crystal habitus in both cases.

A theory of crystal degeneration based on studies of the mass crystallization of metals with a face-centered lattice was proposed by *Tiller* [14.123–125]. It was supposed that grooves formed by boundary surfaces of crystals exist on the interfaces of growing crystals. The depth of the grooves was supposed to depend on the temperature gradients in the solid phase and the melt, as well as on the orientation of the neighboring crystals. The inclination of groove slopes is determined by crystal orientation. Thus a crystal with $\langle 111 \rangle$ orientation relative to the growth axis has a greater slope inclination than a crystal with $\langle hkl \rangle$ orientation, so it will hang over the latter crystal and finally displace it. Such grooves were observed upon growing CZ crystals [14.129]. It was noted that, upon consideration of the degradation process in the framework of Tiller's theory, a decrease of the heat power released in the melt causes a decrease of the mean temperature of the melt and a corresponding decrease of axial gradients at the crystallization front. The decrease in the gradient causes an increase in the depth of the groove between the growing crystals and makes the degeneration process more efficient.

The melt volume influences the size of CZ crystals: larger crystals can be grown in large-volume CCs, exceeding 200 mm in diameter [14.25, 30, 118]. In the case of small crucibles (90–100 mm) the high ratio of the cooling surface area to the melt volume and high heat losses through the CC walls require a higher melt overheating. Significant melt overheating deleteriously influences the degeneration conditions and the shape of the crystallization front [14.129]. The growth rate influences the degeneration process of CZ crystals in the following way: a decrease in the growing rate leads to an increase in the crystal size [14.30, 118]. This corresponds to Tiller's theory of degeneration, which states that the ratio of the rate of displacement of the competing nuclei to the axial growth rate is higher when the growth rates are lower and thus the displacement process becomes more efficient. The influence of melt composition (in particular, the nature and concentration of the stabilizing oxide) on the crystal size was stud-

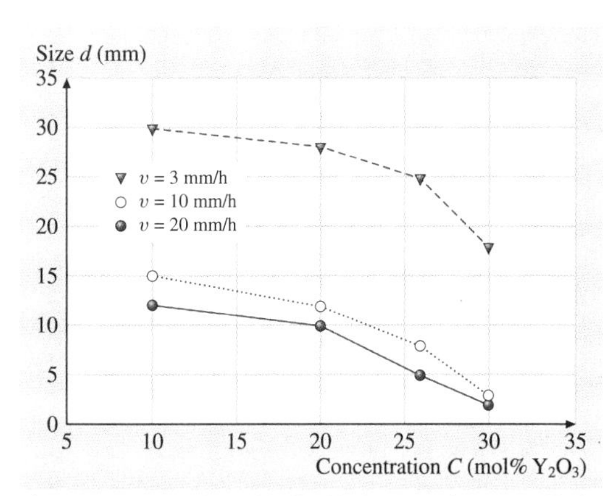

Fig. 14.14 Mean size of ZrO_2-Y_2O_3 crystals versus Y_2O_3 concentration and growth rate

ied in [14.25, 30]. The dependence of the *mean size* of the crystals obtained at different crystallization rates (3, 10, and 20 mm/h) in a CC of 130 mm diameter on the Y_2O_3 concentration is shown in Fig. 14.14. The *mean size* $d = D/n$, where D is the diameter of the ingot and n is the number of crystals in ingot cross-section, was used to compare crystals from different ingots.

The growth conditions were selected for the crystallization front to remain flat throughout the process and for the melt height to be approximately the same for different compositions (the melt height was about ≈ 50 mm, thus being less than the crucible diameter, and this diminished the influence of the degeneration processes). Figure 14.14 demonstrates that an increase in Y_2O_3 concentration (especially above 20 mol %) causes a decrease in the *mean size* of crystals. Consequently, the growth rate must be significantly decreased to maintain an acceptable size of crystals at high Y_2O_3 concentrations.

As mentioned before, impurities that influence the growth of crystals are present in the starting materials. Besides, large quantities of additives may be introduced into the crystals to vary the physicochemical properties of crystals (optical, spectral, electrical, mechanical, and others). These additives significantly influence the crystallization process and, consequently, the size of crystals. It is well known that directional crystallization of multicomponent melts with component distribution coefficients (K) not equal to 1 is accompanied by segregation of the component decreasing the crystallization temperature, into the melt, and its accumulation at the crystallization front (if $K < 1$). In Tiller's theory, grooves on crystal interfaces act as sinks for such admixtures. Thus, the segregation of admixtures inhibits the lateral growth of crystals, and causes an increase in the groove depth and retardation (or complete stop) of the degeneration process. Therefore the introduction of additives, as well as the use of initial mixture with high levels of impurities requires correction of the growing conditions.

Many studies of mass crystallization of metals have been devoted to the problem of preferential orientation of grains in the ingots [14.121–123]. The growth rates anisotropy attempted to give an explanation of the preferential orientation of grains in the ingots: the anisotropy was believed to be due to the anisotropy of the heat conductivity in the crystal. Higher heat conductivity in the temperature gradient direction was suggested to be the cause of the difference in the grain growth rates. However, this is possible only for crystals characterized by anisotropy of heat conductivity, whereas heat conductivity anisotropy is known to be low for most crystals and to be zero for cubic crystals. Experiments on bicrystal growth showed that the equilibrium temperature of the crystal contacting the melt depends on the crystallographic orientation relative to the interface. Dependence of the preferential crystallographic orientation on the growth rate, value of overcooling, and presence of admixtures in the starting material was also noted in these experiments. The problem of preferential crystallographic orientation of CZ crystals grown by SM technique has not been studied in detail. The preferential crystallographic orientation of CZ crystals was shown to be $\langle 110 \rangle$ [14.13, 25, 31, 130, 131]. However, experience shows that the deviation of the crystal growth axis from this direction may often amount to tens of degrees. The correlation structure and morphology of CZ crystals was studied in [14.132]. Single crystals of $M_xZr_{1-x}O_{2-x/2}$ (M = Y, Er, Yb) with $0.3 < x < 0.57$, i.e., crystals with high stabilizer content, were studied. In accordance with the phase diagram, formation of the ordered compound $Zr_3Y_4O_{12}$ was observed at $x = 0.4$. The phenomenon of short-range order was shown to influence the crystal morphology significantly. The interpretation of results was based on the analysis of structural characteristics of ordered rhombohedral phases of the type Y_6UO_{12} by the method of *Hartman* and *Perdok* [14.133]. This method supposes the crystal habitus to be determined by chains of strong bonds. This work clearly demonstrates how the lattice symmetry, the nature of the solid solution, and the crystal growth conditions influence the shape of CZ crystals.

Properties of CZ Crystals

CZ crystals grown by SM technique have been studied rather extensively with regard to chemical and phase compositions, determination of the range of stabilizer concentrations allowing preparation of cubic solid solution crystals from the melt, determination of the phase stability of the crystals, and characterization of growth defects [14.25, 28–31, 61, 134–140]. The structure of crystals was studied in detail by X-ray analysis, Raman spectroscopy, electron paramagnetic resonance, neutron diffraction, etc. [14.91, 113, 141–146]. Lattice parameters and the mechanism of cubic solid solution stabilization were also studied [14.144–149]. A number of investigations have been devoted to the physicochemical properties of crystals: electrical and optical properties [14.150–154], spectral generation characteristics of the activated crystals [14.155–163], and thermophysical and mechanical properties [14.164–177]. Studies of these properties of crystals are of a great significance for practical applications. It should be mentioned that the possibility of using crystals for different applications is to a great extent determined by the perfection of crystals. Therefore types of imperfections appearing in CZ crystals as well as causes of the appearance of imperfections and possible ways of their elimination will be discussed in the following section.

Defects in CZ Crystals

Growth striations, cellular structure, and inclusions of extrinsic phases are imperfections characteristic of CZ crystals [14.28, 29, 138, 139] (Fig. 14.15a–h).

Growth striations. Growth striations are observed in crystals of various compositions as layers with different refraction indices. They appeared as alternating layers perpendicular to the direction of growth of the crystals (Fig. 14.15a). Appearance of striations in ZrO_2–R_2O_3 crystals is indicative of nonuniformity of the components distribution during the crystallization process, in particular of nonuniform distribution of the stabilizing oxide. Local concentration measurements showed that Y_2O_3 concentration fluctuations do not exceed 1 mol %. Such factors as the stabilizing oxide concentration, crystal growth rate, and nature of melt stirring were found to influence the striae density [14.25,28,29]. A considerable increase in the density of the striations was observed at concentrations close to 30 mol % R_2O_3, as well as upon an increase in lowering rate above 16 mm/h. When the lowering rate was decreased from 16 to 1 mm/h (for ZrO_2-8–20 mol % R_2O_3 compositions and CC of 90–200 mm in diameter), the striae density decreased from ≈ 20 to $\approx 8\,cm^{-1}$. When forced stirring by reverse rotation of the crucible was used at low lowering rates, striae density decreased and reached $1\,cm^{-1}$. As is known, nonuniform impurity distributions in crystals is caused by irregular growth rate, resulting in corresponding changes of the effective distribution coefficient and impurity concentration. Growth rate changes may be caused either by peculiarities of the instrumentation or by the processes taking place at the crystallization front. Striations caused by factors of the first type are usually called instrumental while those caused by factors of the second type are called fundamental. Under real conditions, factors of both type usually act simultaneously.

Instrumental striations may be caused by shortcomings of the crystal growth equipment (unstable functioning of the mechanical drives and RF generator output power, oscillations of the coolant flow, etc.) and also by changes of the growth rate caused by temperature oscillations in the melt due to convection. A study of the causes of growth striation formation [14.28, 29] showed that striations are caused by oscillations of the stabilizing oxide concentration (these oscillations are due to constitutional supercooling during growth) and can be eliminated by decreasing the lowering rate to 1–2 mm/h and introducing forced stirring by reverse rotation of the crucible with the melt. It was shown that instrumental instability can cause striations in this case, but this is not the main cause of striations. More uniform crystals are formed upon an increase of the melt volume. Increase of the melt volume causes an increase in the thermal inertia of the crystallization system. High thermal inertia significantly reduces the influence of oscillations of the power supplied by the RF generator and removed by water from a CC, i. e., suppresses the temperature fluctuations, maintaining the stationary state in the system. This results in stabilization of the crystallization front and increases the homogeneity of crystals. This situation was observed upon growing CZ crystals from a CC of more than 400 mm in diameter when the density of the growth striations decreased to $1–2\,cm^{-1}$ upon decreasing the rate to 2–3 mm/h without forced stirring. The determination of the K of components of solid solution in ZrO_2 is very important for growing homogeneous CZ crystals.

Distribution Coefficients of the Solid Solution Components

The K of a component denotes the ratio of concentrations of this component in a liquid and solid phase having a common interface. Actually one usually deals

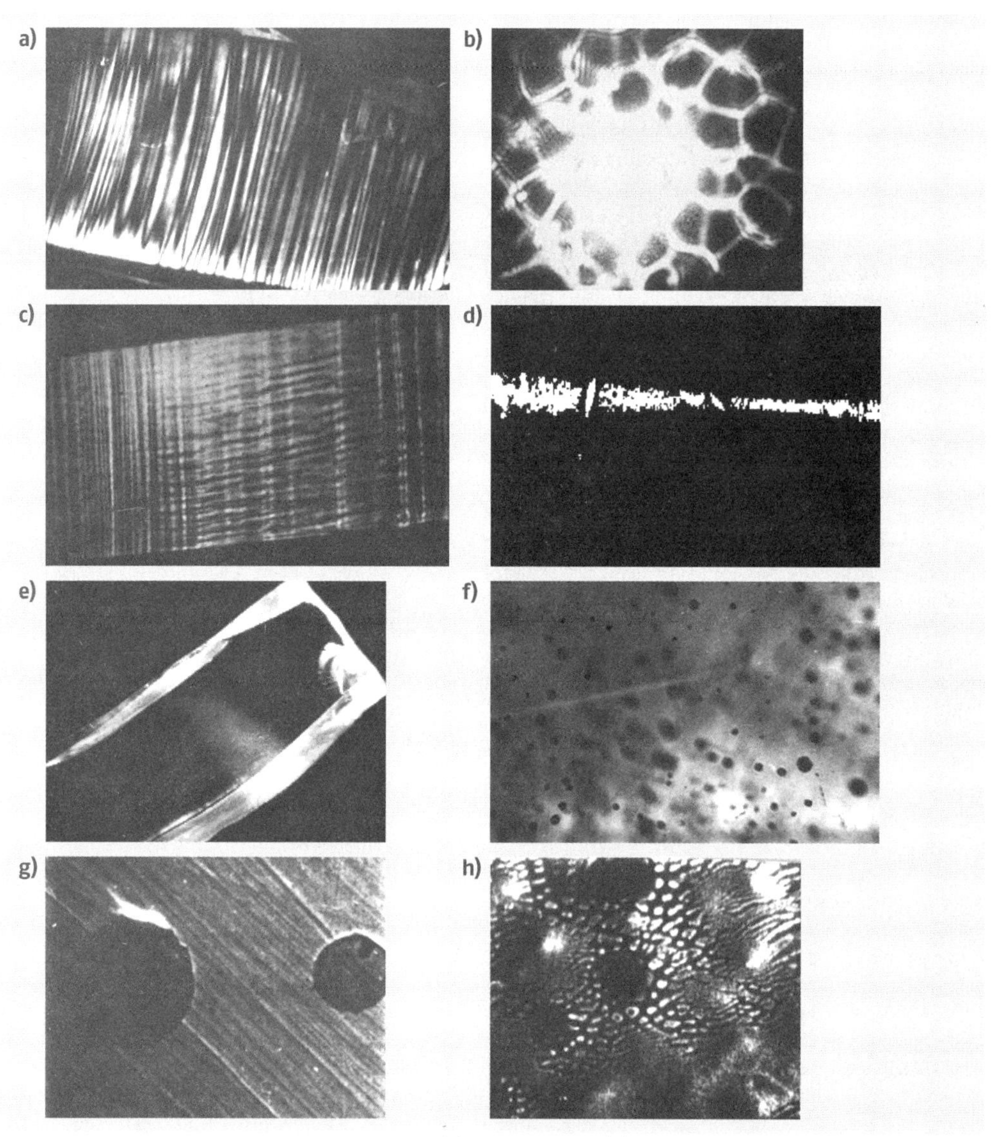

Fig. 14.15a–h Defects in cubic zirconia single crystals: (**a**) growth striae; (**b**) cellular structure; (**c**) growth striae and cellular structure; (**d**) light scattering of laser beam in the crystal; (**e**) light scattering in the defect area; (**f**) inclusions in the boundaries of cells; (**g**) inclusions containing SiO_2; (**h**) SiO_2 inclusions on the surface of ZrO_2–10 mol% Y_2O_3 crystals

with effective distribution coefficients (K_{eff}) which differ from the equilibrium coefficients (K_0). K_0 characterizes the ratio of concentrations corresponding to the solidus and liquidus curves of the phase diagram having a common node. The directional crystallization processes take place with finite rate, so greater or lesser deviations from equilibrium occur. To determine the value of K_{eff}, the distribution of the components along with the axis of the crystal is measured. Quantitative element analysis was performed by electron probe

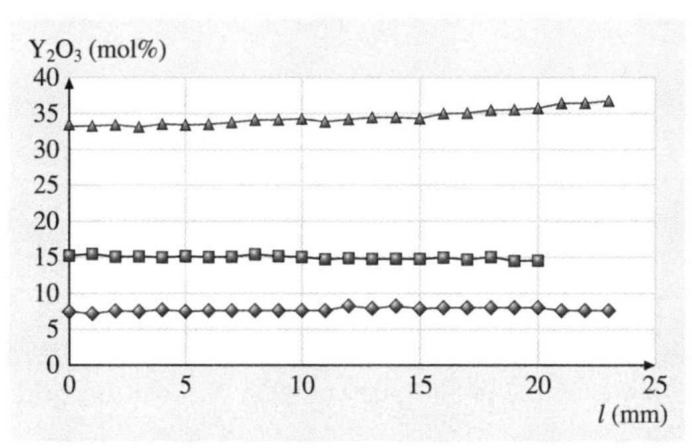

Fig. 14.16 Distribution of Y_2O_3 concentration in ZrO_2-Y_2O_3 crystals of different compositions. Initial composition (mol.% Y_2O_3): (▲) – 35; (□) – 15; (◆) – 7.5

microanalysis [14.25, 28, 30, 31] or by optical spectroscopy [14.134, 135]. The experimental results were approximated by the equation $C(g) = kC_0(1-g)^{k-1}$, which is commonly used to describe the processes of normal directional crystallization [14.25, 28, 30, 31]. An example of measurements of the stabilizing oxide concentrations in ZrO_2–Y_2O_3 crystals of different composition is given in Fig. 14.16.

A more detailed study of the Y_2O_3 distribution curves in crystals taken from different parts of the ingot showed that periodic nonuniformity of crystal composition along the growth direction occurs, resulting from oscillations of the crystal growth rate [14.28]. Waves in concentration curves of crystals taken from different parts of the ingot did not coincide, this being indicative of differences in crystallization conditions along the ingot cross-section (the crystallization front not being flat). Sometimes sudden changes of concentration were observed in the distribution curves. Such changes can occur due to an abrupt change of power supplied by a RF generator, causing a dramatic change of the crystallization rate, or to quick uncontrolled growth caused by constitutional supercooling. As a rule, such sharp concentration changes took place at the end of crystal growth. This is due to accumulation of the displaced admixtures in the melt and enhancement of the constitutional supercooling, which disturb the stable growth of crystals, causing the formation of inclusions and cellular structure. Besides, the coupling between the RF field and the melt is deteriorated when the amount of melt decreases, and this results in quick uncontrolled crystallization of the remaining melt. Changing the lowering rate of a CC in the range of 1–16 mm/h and changing the concentration of the stabilizing oxide weakly influenced the character of the distribution curves. The

Table 14.3 Effective distribution coefficients of Y_2O_3 in the ZrO_2–Y_2O_3 system

C (mol% Y_2O_3)	Growth rate (mm/h)	K_{eff}
5	16	1.08 ± 0.05
8	16	1.06 ± 0.05
10	1	1.12 ± 0.03
10	5	1.05 ± 0.02
10	8	1.02 ± 0.02
10	30	1.07 ± 0.02
15	16	1.02 ± 0.02
20	16	1.00 ± 0.01
30	16	1.01 ± 0.02

values of K_{eff} for different growth rates and compositions are presented in Table 14.3. The values of K_{eff} of Y_2O_3 calculated from these data are close to 1 and only slightly depend on the growth rate and initial concentration of Y_2O_3.

The values of K_{eff} for Gd – Yb oxides are close to those of Y_2O_3 in cubic solid solutions based on ZrO_2 and also slightly depend on the growth rate and initial concentration of stabilizing oxide [14.25, 30]. The values of K_{eff} for rare-earth oxides from the left part of the lanthanide series (Ce–Nd) in cubic solid solutions based on ZrO_2 were found to be less than 1 [14.25] (Table 14.4).

Data on the dependence of K_{eff} in CZ crystals with 12 mol % Y_2O_3 doped with 1.2265 wt % Nd_2O_3 on growth rate are shown in Table 14.5 [14.134]. Using the data obtained, the authors determined that, for Nd_2O_3, $K_0 = 0.426$ and the diffusion layer is approximately 1 mm thick.

Cellular structure [14.25, 28–31] was observed in crystals extracted from the central part of the ingot and, as a rule, in the top part of the crystals (Fig. 14.15b). The top of the crystals exhibiting cellular structure had a characteristic *rugged* appearance; if the cellular structure was highly developed, boundaries between smooth shiny surface and uneven wavy surface could be clearly seen on the sides of the crystals. Regions exhibiting lamellar and cellular structure can alternate in the same crystal (Fig. 14.15c). The appearance of cellular structure is accompanied by a decrease in the striae density. The formation of cellular structure is caused by the destabilization of the smooth crystallization front upon growing [14.178–180]. Most authors consider constitutional supercooling of the melt near the phase interface to be the cause of destabilization of the smooth front and its transformation into the cellu-

Table 14.4 Effective distribution coefficients of Nd_2O_3 and CeO_2 in the ZrO_2-$0Y_2O_3$-Nd_2O_3-CeO_2 system

Composition	V (mm/h)	K_{eff}		
		Nd_2O_3	CeO_2	Y_2O_3
ZrO_2-(12-25 mol %) Y_2O_3-(0.74-4.40 mol %) Nd_2O_3	5	0.62 ± 0.04	–	1.045 ± 0.03
ZrO_2-(9-11 mol %) Y_2O_3-(2-2.5 mol %) Nd_2O_3-1 mol % CeO_2	5	0.58 ± 0.08	0.42 ± 0.04	1.09 ± 0.02
ZrO_2-3 mol % Y_2O_3-0.3 mol % Nd_2O_3-0.4 mol % CeO_2	10	0.87 ± 0.08	0.78 ± 0.02	1.13 ± 0.01

lar type [14.180]. Constitutional supercooling alone was shown to be insufficient for the formation of cellular structure; a threshold value of constitutional supercooling depending on the chemical composition of the melt and the direction of growth had to be achieved. Studies of CZ crystals [14.25, 28–31] showed that the cellular structure appears at the end of the growth stage, most often in crystals situated in the central part of the crystalline ingot. This is connected both with the accumulation of admixtures in the melt before the crystallization front during crystal growth and with the character of the temperature distribution in the CC. Cellular structure, appearing as a *constriction* in the middle of the crystal, was formed at high growth rates (> 10 mm/h) in the absence of stirring.

Oxides of Si, Al, Ti, and W cause the appearance of imperfections deleterious to the optical uniformity of the crystals. Of all the contaminants mentioned, SiO_2 exerts the most negative influence on the growth and perfection of crystals. Even small quantities of SiO_2 promote disruptions of crystal growth and the appearance of cellular structure regions. Decreasing the growth rate and using forced stirring improves the efficiency of impurity segregation and the crystal quality [14.25, 28–30].

Inclusions [14.25, 29–31]. Large isolated inclusions and bubbles are seldom observed in CZ crystals; inclusions formed by light-diffusing particles of 10–0.08 μm in size are observed much more often. Light scattering can be observed either visually or by the scattering of a laser beam passing through the crystal (Fig. 14.15d). The occurrence of light-scattering regions depends on the composition of the melt and the growth conditions. Single crystals may be completely opalescent or exhibit broad opalescent bands in the top central part of the ingot (Fig. 14.15e). Research [14.25, 29–31] showed that band-shaped light-scattering regions coincide with regions of cellular growth. It was established that light-scattering inclusions may be formed during melt crystallization in CZ crystals either on all the phase interface or on the boundaries of the cellular structure (Fig. 14.15f,g). Growth of CZ crystals in a wide range of concentration of the stabilizing oxide at growth rates of 1–2 mm/h showed that light scattering by the whole crystals is characteristic of compositions with 8–12 mol % R_2O_3 (R = Y, Gd) [14.25]. X-ray diffraction patterns of these crystals were not indicative of the appearance of new phases. Scattering in these crystals was caused by particles of the second phase with size of 1–0.1 μm. Under the same growth conditions, increasing the concentration of the stabilizing oxide to above 14 mol % R_2O_3 (R = Y, Yb) or above 16 mol % Gd_2O_3 completely eliminated light scattering in the crystals. The appearance of light-scattering particles may be connected either with the presence of impurities becoming less soluble upon a decrease in temperature (in this case it depends on the melt composition) or with the decomposition of the fluorite-type solid solution. Low concentration of the second phase and the very small size of its particles complicates direct determination of the phase structure and composition. It is worth mentioning that heat treatment of ZrO_2-12 mol % Y_2O_3 crystals at 2100 °C for 3 h under vacuum and subsequent quenching (at cooling rate of 1000 °C/h) led to elimination of light scattering. Large inclusions of the second phase (> 0.5 μm) are an extreme case of inhomogeneity of the crystal and are usually located in regions of highly developed cellular structure. As a rule, such inclusions are caused by the presence of unde-

Table 14.5 Effective distribution coefficients of Nd_2O_3 in the ZrO_2–12 mol % Y_2O_3–1.2265 wt % Nd_2O_3 system

V (mm/h)	K_{eff}
4	0.51 ± 0.016
4	0.52 ± 0.025
6	0.54 ± 0.013
6	0.54 ± 0.013
8	0.61 ± 0.025
8	0.60 ± 0.03
10	0.63 ± 0.006
10	0.68 ± 0.038
12	0.68 ± 0.019
12	0.65 ± 0.013
14	0.73 ± 0.06
14	0.70 ± 0.06

sirable impurities in the initial mixture. The type of inhomogeneity caused by the presence of admixtures is determined by the nature and concentration of the admixture (SiO_2, TiO_2, Al_2O_3, WO_3). Large (4–10 μm in size) silicon-containing inclusions of the second phase were detected in the boundaries of cells (Fig. 14.15f,g). The density of such inclusions in the top part of the crystal varies from 4.5×10^2 to $26\times10^2\,mm^{-3}$.

Electron probe microanalysis of the inclusion composition showed that the concentration of SiO_2 in the inclusions is very high (≈ 60 mol %). The spherical shape of such inclusions and their compositional nonuniformity suggest that trapping of the liquid phase occurs. The melting point of the SiO_2-enriched drops being lower than that of ZrO_2–R_2O_3, these liquid drops become trapped in the solid crystal matrix. The concentration of Y_2O_3 in the inclusions is also higher than in the matrix because R_2O_3, being oxides of basic character, accumulate in the drops due to chemical affinity towards SiO_2, which is more acidic than ZrO_2. The process of chemical differentiation takes place either upon drop formation or upon drop cooling (in the latter case it is due to diffusion from the solid matrix). High content of stabilizing oxide in the inclusions (up to 28 mol % in the case of ZrO_2–12 mol % Y_2O_3) may cause decomposition of the cubic solid solutions and formation of m-ZrO_2 if the concentration of SiO_2 is sufficiently high and the concentration of the stabilizing oxide is low. However, the probability of this process is low if the Y_2O_3 concentration in the cubic crystals is high. Apart from forming inclusions inside the crystal, the second phase can form precipitates on the surface of blocks (Fig. 14.15h). In this case a matted deposit appears on some parts of the crystal facets, which is usually smooth and very shiny [14.25, 29–31].

14.3.5 Growth, Properties, and Application of PSZ Crystals

The SM technology of PSZ crystals is similar to the above-described growth process of CZ crystals. The growing PSZ crystal exhibits a c-structure during the first stage of synthesis from melt; phase transitions in the crystal occur upon cooling of the solid phase. The c→t phase transformation is accompanied by the formation of domain structure in the crystals [14.130–132]. PSZ crystals are used mainly for scientific research on the mechanism of hardening, structural, micro- and nanostructural research, studies on phase transformations, and physicochemical properties [14.109, 112, 116, 149, 181–183]. Although the advantages in mechanical properties (high hardness, strength, and fracture toughness) of PSZ crystals, especially at elevated temperatures, have apparently been demonstrated, practical applications of PSZ as construction material have not yet been reflected in the literature.

Large uniform blocks of material without cracks are required for practical applications. This is why the influence of composition and growth conditions on the size and mechanical properties of crystals was investigated [14.184–187]. The stabilizing oxide concentration range (2.5–4 mol % Y_2O_3) allowing the preparation of large (up to 40 mm in cross-section, up to 120 mm in length) uniform PSZ crystals not containing fractures has been established [14.188]. PSZ crystals of the composition ZrO_2-2.5–4 mol % Y_2O_3 are white and nontransparent, with a smooth shiny surface. The crystal surface becomes rough and matted upon a decrease in Y_2O_3 content. The crystal size is significantly decreased at Y_2O_3 concentration < 1 mol %. Crystals containing 5–8 mol % Y_2O_3 are semitransparent and contain a large number of fractures in the volume. Experiments on the growth of crystals in a CC of 130 mm diameter showed that rate decrease from 40 to 3 mm/h led to a significant increase in the crystal cross-section size (from 3–7 mm to 20–40 mm) [14.185, 186] (Fig. 14.17).

However, growing the crystals at 3 mm/h rate results in a considerable increase in the number of fractures in the crystals. The cooling rate in the crucible of 130 mm diameter being too high for crystals of a large size, residual thermal stresses are probably relieved by means of fracture formation in the crystals. Research into the influence of thermal conditions on crystal growth showed that the quantity of large cross-section crystals is significantly increased upon an increase of the height of the bottom thermal shield. Most crystals have a smooth shiny surface.

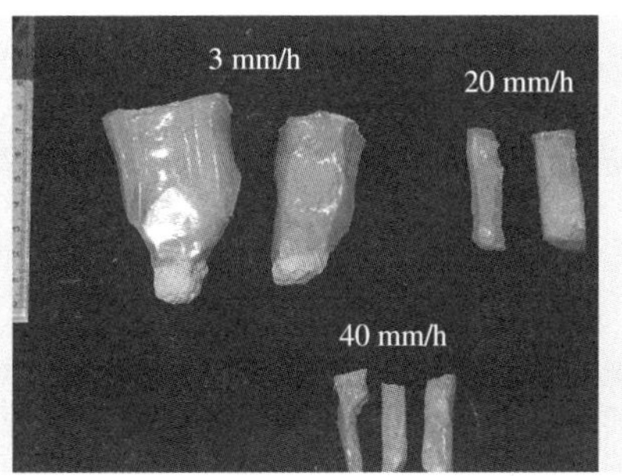

Fig. 14.17 PSZ crystals grown at 3, 20, and 40 mm/h in a cold crucible with 130 mm diameter

Table 14.6 Effective distribution coefficient of Y_2O_3 in the ZrO_2-Y_2O_3 system in PSZ

C (mol% Y_2O_3)	K_{eff} 10 mm/h	20 mm/h	40 mm/h
2.5	1.04 ± 0.01	1.07 ± 0.01	–
3	1.006 ± 0.007	1.025 ± 0.006	1.02 ± 0.02
3.5	1.150 ± 0.009	1.023 ± 0.005	1.027 ± 0.007
4	0.98 ± 0.09	1.001 ± 0.008	1.002 ± 0.009
5	1.03 ± 0.01	–	–
8	1.041 ± 0.007	–	–

Therefore, the most appropriate conditions for the preparation of large, smooth-faceted PSZ crystals are growth rate about 10 mm/h with enlarged bottom thermal shield. Studies of the chemical composition of the crystals showed that microcontaminants from the starting materials are segregated during growth of PSZ crystals in the same way as for CZ crystals. Decreasing the crystallization rate results in a decrease in contaminant concentrations compared with in the initial mixture [14.185]. The distribution of the main components of the solid solution (at crystallization rates of 3–40 mm/h) is fairly uniform, the value of K_{eff} Y_2O_3 in ZrO_2 being close to 1 [14.186] (Table 14.6).

Investigation of the crystal phase composition of as-grown crystals (growth rate 3–10 mm/h) by Raman spectroscopy revealed that molten ZrO_2 was m-phase; crystals containing 2 mol % Y_2O_3 were a mixture of m- and t-phases; crystals containing 2.5 mol % Y_2O_3 contained small m-phase regions, the larger part of the crystal being t-phase; at concentration range 3–5 mol % the crystals were t-phase; at Y_2O_3 concentration of 8–35 mol % the crystals were constituted by a cubic solid solution of fluorite structure [14.185, 186]. X-ray diffraction patterns of powdered samples prepared from PSZ crystals (growth rate 3–10 mm/h) showed that the m-phase is present in crystals containing 2.0–3.5 mol % Y_2O_3 while only the peaks of a highly symmetric phase are present in the pattern of crystals containing 4.0–5.0 mol % Y_2O_3. Apparently, t→m transformation occurred under powdering. Analysis of the same powders by Raman spectroscopy has confirmed that such a transformation in fact took place, indicated by the occurrence of the m-phase in powdered samples containing up to 3.5 mol % Y_2O_3. Differences in the phase composition of bulk and powdered samples show that X-ray diffraction analysis of powders more appropriately reflects the phase composition of materials subjected to intensive mechanical impact, while Raman spectroscopy allows a more detailed determination of the phase composition of bulk crystal samples. It was shown that PSZ crystals grown in a CC of 130 mm in diameter are not sharply quenched, therefore they contain the t-phase as well as the t′-phase, and so the m-phase appears upon mechanical impact [14.186]. The incorporation of admixtures (oxides of Ce, Nd, Tb, Co, etc.) was also shown to influence the number of fractures [14.189–191].

The shape of the crystal growth surface precisely reflects the microstructure of the material, which is influenced by the composition as well as crystallization and annealing conditions. Scanning electron microscopic studies of the as-grown crystals showed that the crystal surface morphology is influenced by the stabilizing oxide concentration in the solid solution [14.185, 186, 189, 191]. Microphotographs of the as-growth crystal surfaces with different content of the stabilizing oxide (Y_2O_3) are shown in Fig. 14.18.

Relief of the growth facets of pure ZrO_2 crystals which had undergone a t→m transformation is typical for martensitic transitions (Fig. 14.18a,b). This transition is accompanied by a significant volume change (up to 5 vol. %) and causes fracturing. A regular microcrack network in the form of rhombs can be clearly seen in Fig. 14.18a. The cracks of each subset are apparently parallel to each other and the angle of intersection is close to 30°. Microcracks can be also seen on the growth surface of ZrO_2–2.0 mol % Y_2O_3 samples (Fig. 14.18c,d), but these microcracks are sufficiently shorter and many of them are warped. Both the elements of the above-described structure and the so-called *tweed* structure characteristic of samples with higher Y_2O_3 content can be seen on the surface. *Tweed* structure is characterized by element intersection angle close to 80–90°. The surface of crystals containing 2.5 mol % Y_2O_3 (Fig. 14.18e,f) is practically devoid of cracks, while the elements of the *tweed* structure occupying the larger part of the crystal surface can be clearly seen. However, only isolated elements of the *tweed* structure are present in some regions of the crystal surface (Fig. 14.18f). The existence of two characteristic types (type 1 and type 2) of growth surface was detected for ZrO_2 crystals containing 3.0 mol % Y_2O_3 (Fig. 14.18g–j). Type 1 is characterized by *smooth* facets (Fig. 14.18g), while type 2 is characterized by *matted* facets (Fig. 14.18h,i). The boundary of the smooth and matted facets is shown in Fig. 14.18j.

Surface morphology of any of these types may be characteristic either for all facets of the crystal block or only for some of them, being present as an individual facet. The surface of type 1 comprises a rather regular structure, which is similar to the *tweed* one. It consists of tiny elongated rectangles (1–10 μm in length and 0.5–3 μm in width) grouped parallel to each other. The angle of orientation of groups relative to each other is close to 80–90°. The surface of type 2 has a more pronounced relief than the one described above and consists of parallel convex rectangles, which are larger than the elements of type 1 structure (up to 50 μm in length and > 5 μm in width). More pronounced relief of the structure forms the matted appearance of the facet. Figure 14.18 clearly shows that smaller structural elements with sizes characteristic of the smooth (type 1) surface coexist with larger surface elements on type 2 surfaces.

Matted surfaces occur in the central and lower part of the crystal bulk, their appearance being dependent on the melt volume and crystal length. At the same time, single-crystalline blocks are often accreted by these facets, so separation of such blocks sometimes leads to destruction of the crystals. It seems that this process is influenced by cooling conditions during crystal growth, which determine the phase composition and microstructure of the crystals. The absence of matted facets on the crystal surface at growth rate of 3 mm/h is worth mentioning. Practically, large crystals are no growing together. There are few such crystals in the ingot, and stress is relieved mainly on the block boundaries and partly in the block volume. The latter process results in crystal fracturing. The matted crystal surface is formed if the initial melt contains significant quantities of contaminants with K value much lower than 1. However, such surfaces differ from that described above by the presence of a characteristic tarnish of the segregated admixtures (Fig. 14.19).

Further increase of the stabilizing oxide concentration results in the appearance of various types of island structures; namely, islands consisting of parallel structural elements without perpendicular crossing appear on a *tweed* structure background in the case of ZrO_2 containing 3.5 mol % Y_2O_3 (Fig. 14.20a,b).

Crystals containing 4.0 mol % Y_2O_3 are characterized by a smooth unstructured surface on which small islands with a pronounced microstructure occur very rarely (Fig. 14.21). The appearance of an island structure can be indicative of incompleteness of the phase transitions occurring during the cooling of the crystal. As a result, Y_2O_3 is unevenly redistributed between two t-phases at temperatures corresponding to the two-phase region of the phase diagram. It is obvious that phases

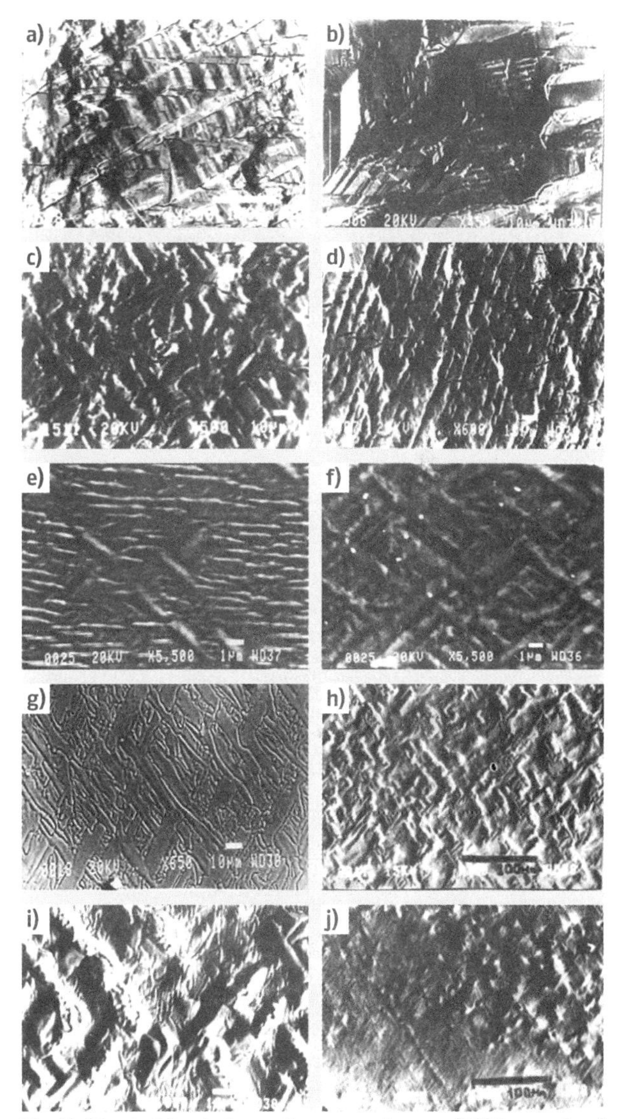

Fig. 14.18a–h Morphology of as-grown facets of PSZ crystals of $(1-x)ZrO_2$-xY_2O_3 mol% compositions **(a,b)** $x = 0$; **(c,d)** $x = 2.0$; **(e,f)** $x = 2.5$; **(g,h)** $x = 3.0$

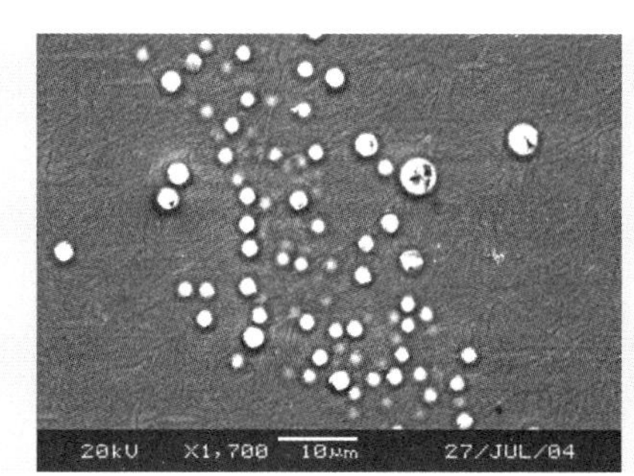

Fig. 14.19 Surface of facets of ZrO_2–3 mol % Y_2O_3 grown at 3 mm/h with SiO_2 inclusions (magnification 1700 ×)

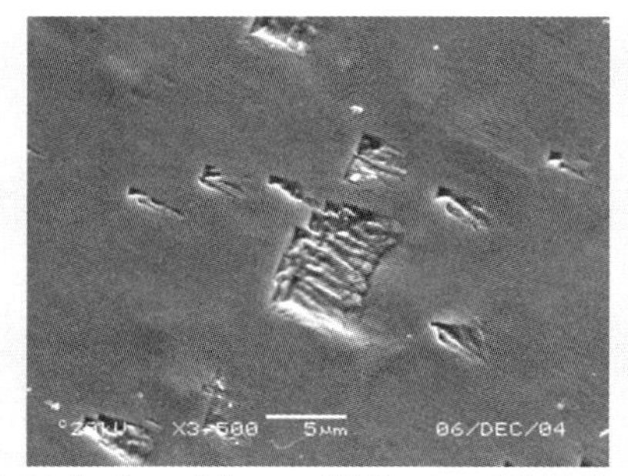

Fig. 14.21 Surface of facets of ZrO_2–4 mol % Y_2O_3 grown at 10 mm/h

with different Y_2O_3 content can differ in microstructure.

Nanostructure of ZrO_2–3 mol % Y_2O_3 crystals was studied by high-resolution transmission electron microscopy (TEM) as well as by x-ray analysis. The latter method was used to estimate the size of coherent scattering regions according to the broadening of the diffraction reflexes [14.112, 114, 115, 186]. A model of ordered domain colonies occupying all the volume of the crystal at the minimum of deformation energy was proposed [14.112]. TEM studies of the ZrO_2–3 mol % Y_2O_3 crystal revealed the presence of colonies formed by *parquet* structure domains (30–100 nm in width, 400–800 nm in length) (Fig. 14.22).

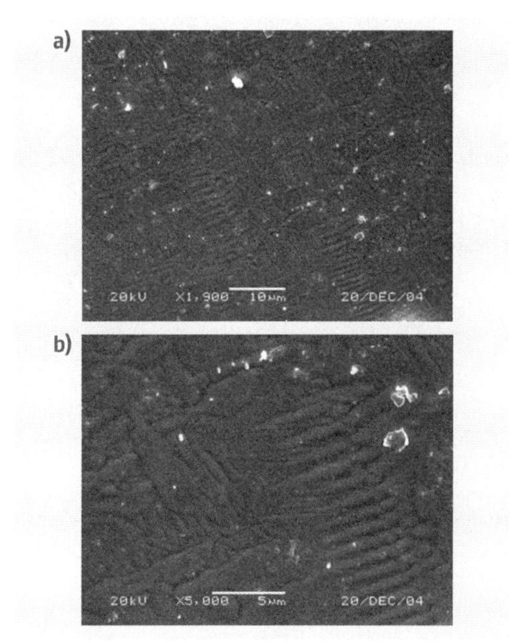

Fig. 14.20a,b Surface of facets of ZrO_2-3.5 mol % Y_2O_3 grown at 10 mm/h.(**a**) 1900 ×; (**b**) 5000 ×

Laminated structure with a period of about 10 nm is observed in the colonies, indicating the presence of a lamellar domain structure in each colony. Their size is estimated to be (0.1–0.5) × (10–20) nm. An estimate of the coherent scattering region size (D) in the ZrO_2–3 mol % Y_2O_3 crystal by X-ray structure analysis yields a value of 87–310 nm, which corresponds rather well to the above estimates of the colony size. A scanning electron microscopic study of the crystal surface microstructure at 25 000 × amplification allowed estimation of the size of the smallest structural elements. The width was estimated to be 70–300 nm and the length was estimated to be 300–900 nm. These estimated values are close to the sizes of colonies observed during the TEM study. Therefore, the microstructure of as-grown facet of the crystals reflects the inner crystal structure (the structure of colonies). Research on

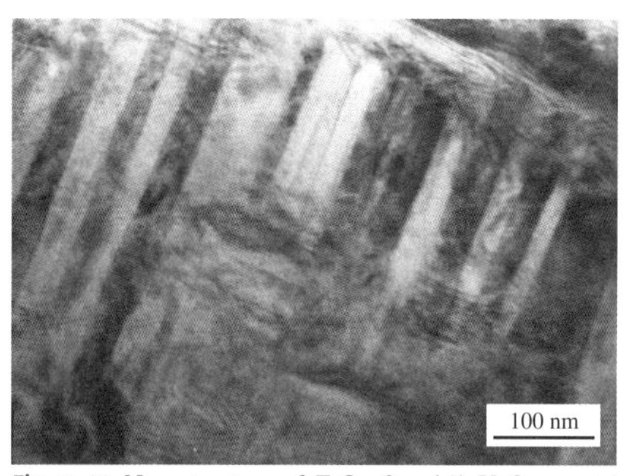

Fig. 14.22 Nanostructure of ZrO_2–3 mol % Y_2O_3 crystal grown at 10 mm/h

Fig. 14.23 Nanostructure of ZrO_2–2.5 mol % Y_2O_3 crystal grown at 10 mm/h

the nanostructure of ZrO_2–2.5 mol % Y_2O_3 crystals showed that the formation of the domain structure just begins at this Y_2O_3 concentration (a system of parallel domains with some elements of the tweed structure is formed) (Fig. 14.23).

Only isolated regions with nonpronounced domain structure were detected in ZrO_2–4 mol % Y_2O_3 crystals (Fig. 14.24).

This corresponds to the results of research on the microstructure of growth surface of PSZ crystals as determined by their composition. The doping of PSZ crystals with additional admixtures in small quantities (≈ 0.1 mol %) affects the microstructure and mechanical properties of the crystals [14.189, 191]. These studies revealed that sample breaking strength is connected with the structure of the crystal surface, which reflects the microstructure of layers adjacent to the surface.

The oxygen-isotope method was used to study the peculiarities of oxygen redistribution during formation

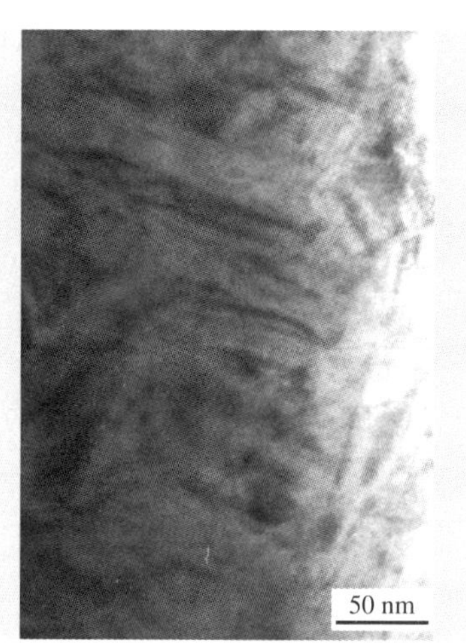

Fig. 14.24 Nanostructure of ZrO_2–4 mol % Y_2O_3 crystal grown at 10 mm/h

and stabilization of the structure of substitutional solid solutions $Zr^{+4}_{1-x}Y^{+3}_{x}O^{-2}_{2-0.5x}V^{0}_{0.5x}$ and further annealing of the crystals under various gas atmosphere. The mobility of oxygen was shown to depend on the chemical composition of the crystal and the conditions of heat treatment. The existence of correlation between oxygen mobility in PSZ crystals and their electrical and mechanical properties was confirmed [14.192]. It was also shown that maximum oxygen mobility is found in PSZ crystals containing 2.5–4.0 mol % Y_2O_3. The optimal composition range of PSZ crystals ensuring the nanocrystalline structure, high mechanical characteristics, and high oxygen mobility was found by physicochemical methods to be ZrO_2–2.5–4.0 mol % Y_2O_3.

Significant anisotropy of strength properties (elastic characteristics, threshold values of strength, deformation properties, specific fracture energy, fracture strength, and hardness) was detected in studies of the dependence of these properties on Y_2O_3 concentration, presence of additives, growth rate, crystallographic orientation, and heat treatment conditions. It should be mentioned that stabilizing oxide concentration exerts a determining influence on strength properties [14.116, 181, 193–197]. The following high values of strength properties were determined for PSZ crystals: dynamic modulus of elasticity of 400–500 GPa, ultimate compression strength of 4000 MPa, ultimate bending strength of 1600 MPa, fracture toughness of 15 MPa $m^{0.5}$, and hardness of 15 GPa.

Results of the studies of the tribological properties of PSZ crystals are presented in [14.195–197].

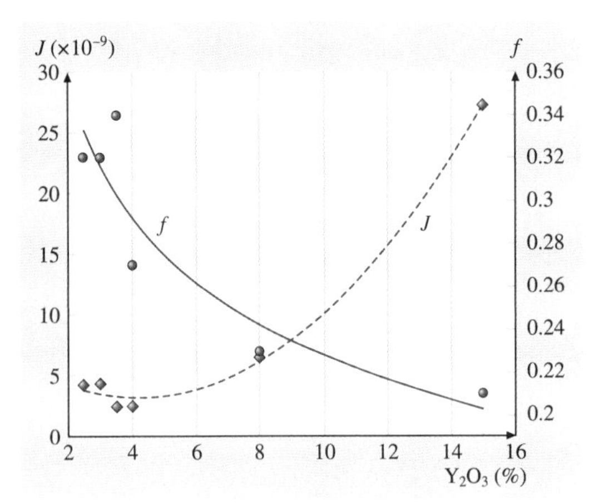

Fig. 14.25 Effect of Y_2O_3 concentration on wearing intensity (J) and friction coefficient (f)

The main tribotechnical properties and the mechanism of wearing of the PSZ crystals were determined in experiments including x-ray and electron microscopic analyses of friction surfaces. PSZ crystals containing 2.5–4.0 mol % Y_2O_3 were the most wear resistance. The intensity of wear was $(2.5-4.3)\times10^{-9}$, the friction coefficient was 0.32–0.34, and microhardness was 11.8–15.08 GPa (assessed with instrumental steel U10A as counterbody) (Fig. 14.25).

The predominant wear mechanism of the PSZ crystals was shown to be a two-stage mechanical (fatigue) mechanism involving the destruction of the surface layer of secondary structures formed upon friction with subsequent destruction of deeper layers of the bulk material. The results obtained show that synthesis of high-strength fracture-proof materials by directional crystallization of melt requires consideration of many factors: the K_{eff} values of the additives used, growth rate, temperature gradients in the melt, the nature and concentrations of the stabilizing oxide and the third additive, and cooling and annealing conditions.

PSZ crystals possess a number of properties giving them advantage over metals and high-strength dielectric materials (including ceramics). Due to their lack of grain boundaries PSZ crystals are characterized by high strength (comparable to the strength of metals), fracture toughness, and hardness. They are also char-

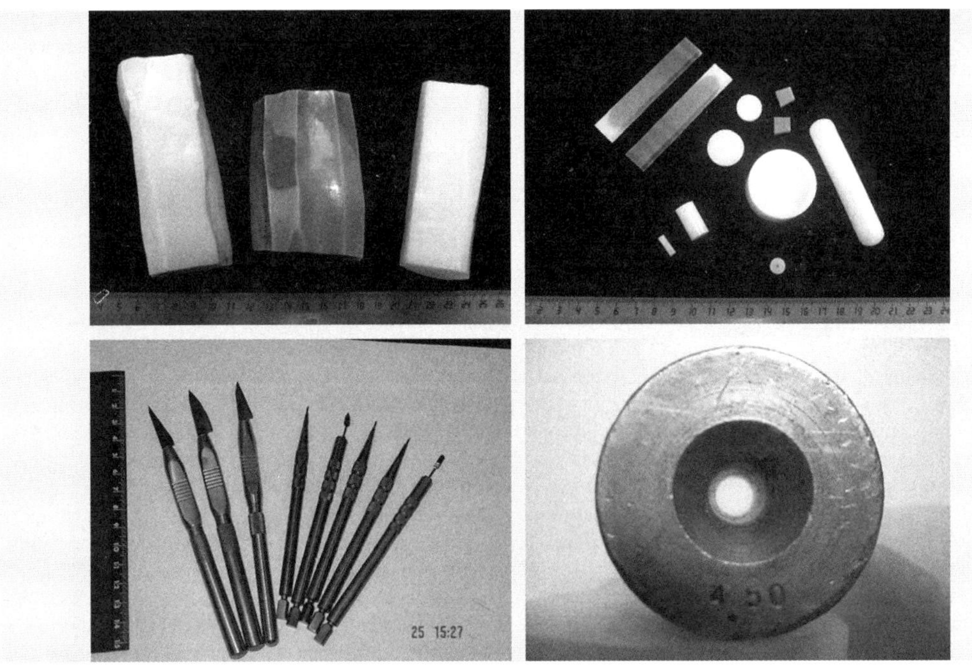

Fig. 14.26 Technical articles and components fabricated from PSZ crystals. PSZ crystals (*top left*); cutting elements, rolls and grinding tool fabricated from PSZ (*top right*); medicine scalpels with cutting edge from PSZ (*bottom left*); wire drawing die with insert from PSZ (*bottom right*)

acterized by low friction coefficient and high abrasive wear resistance, and high resistance to acids, alkali, and water vapor. Deterioration of mechanical characteristics of PSZ crystals placed in oxidizing atmosphere at high temperatures (up to 1400 °C) is much less pronounced than that of metals (which are intensely oxidized) or construction ceramics, which are prone to recrystallization. Besides, the mechanical properties of the crystals are improved upon temperature decrease (at −140 °C the improvement amounts to 70%). Such materials can be used to produce details of appliances functioning under extreme conditions: at high mechanical loadings, in aggressive media, at elevated temperatures, without lubrication, etc. Such appliances include bearings, support prisms, guide plates, and motor valves. Due to the existence of domain nanostructure in PSZ crystals and high mechanical strength of these crystals they may be used in the manufacturing of instruments with very sharp cutting edges, for example, high-quality medical scalpels and instruments for precision machining of various materials (metal, wood, glass, crystals, etc.).

The biological inertness of this material [14.183, 198, 199] enables its use in the manufacturing of implants for medicine. Comparative mechanical and biomedical tests were performed for PSZ crystals, CZ crystals, PSZ ceramics, and Al_2O_3 single crystals [14.182]. PSZ crystals were shown to be somewhat stronger than other materials and displayed the highest fracture toughness. Studies on implants made of ZrO_2-based materials showed that these implants neither provoke an inflammatory reaction in tissues nor cause pathological changes in the internal organs of animals. The results of investigations show that PSZ crystals can be used as a material for various implants. Some articles manufactured from PSZ crystals are shown in Fig. 14.26 to give examples of practical applications of this material.

14.4 Glass Synthesis by Skull Melting in a Cold Crucible

In this section some questions about using SM technique for glass synthesis will be analyzed. Use of this method is justified if the synthesis of a specific type of glass by traditional methods is impossible or very difficult. These glass types include:

- Optical and laser glasses for which high chemical purity is required; for example, in glasses for fiber-optical communication lines the content of transition-metal contaminants must not exceed 10^{-5}–10^{-6} wt %
- Glasses prepared from highly reactive melts and/or materials capable of interacting with the crucible material
- High-temperature glasses of various compositions which must be synthesized at temperatures exceeding 1600 °C.

Let us analyze the peculiarities of the use of this method for glass synthesis.

Glass is an amorphous material, undergoing a smooth transition from solid state to glass-like state upon heating. Glass-forming melts differ from other oxide melts by their high viscosity (depending on the melt composition, the viscosity may be several orders of magnitude higher than that of typical oxide melts) and a smooth temperature dependence of viscosity and electric conductivity. Besides, the absence of a solid–liquid phase transition and the corresponding dramatic change in electric conductivity enhance the phase stability of the melt and make keeping of the melt in a stationary state easier. That is why in the case of induction melting in the cold crucible the skull consists of a thin layer of initial powdered mixture adjacent to crucible walls and a layer of solid glass, the thickness of the latter layer being influenced by the temperature distribution in the melt.

Various types of crucibles are used according to the composition of a specific glass type. Synthesis of silicate glasses was performed in water-cooled quartz crucibles [14.10, 11]. As quartz is a typical dielectric, effective supply of the alternating electromagnetic field energy to the melt without losses in the crucible material is enabled. The use of such crucibles imposes considerable limitations on the temperature of glass synthesis. Other drawbacks of quartz crucibles are the impossibility of repeated use and a requirement for stringent control of the melt temperature to avoid the risks of crucible destruction and water getting into the melt. Metallic crucibles for glass synthesis also have some peculiarities. Aluminum is the preferred crucible material. Beside individual oxides, salts (carbonates, nitrates, oxalates, etc.) are often used as starting materials for glass synthesis. Upon thermal decomposition of salts, interaction of the decomposition products with the crucible material and transfer of small quantities of crucible material into the melt may occur. In the case of aluminum crucibles, a thin layer of Al_2O_3 present on the surface of aluminum prevents the heterogeneous chemical reactions. Besides, Al_2O_3 is often present in glasses and therefore embedding of small amounts of Al_2O_3 into the melt does not cause a significant change in the properties of glass. Aluminum oxide is an optically inactive admixture, so its presence does not influence the optical properties of glass. Metallic crucibles must be dismountable in order to ensure quick extraction of the glass block for further annealing. The use of combined crucibles, i. e., metallic crucibles with thin-walled quartz crucibles inserted into them, has also been reported [14.26].

In the case of glass synthesis, the initial melt volume cannot be created by introducing small pieces of a metal directly into the initial mixture, because this would lead to contamination of the melt by metal and the reduction products dispersed in the melt volume. The high viscosity of glass-forming melts leads to a dramatic decrease in the metal burning-out rate [14.31]. Start melting is usually performed by placing a quartz ampoule filled with pieces of graphite or low-resistance Si into the fusion mixture. Afterwards the ampoule is removed from the melt. To make start melting easier and to prevent the evaporation of low-melting-point components, one should first melt the low-melting-point components and then introduce the refractory ones into the melt; for example, glass synthesis in the system $Li_2O–Al_2O_3–SiO_2–ZrO_2$ was begun with the melting of the $Li_2O–SiO_2$ composition, which melts at 1100–1200 °C [14.200].

Homogenization of the melt is an important technological stage of high-quality optical glass synthesis. Complicated physicochemical processes connected with the melting of individual components of the fusion mixture and the occurrence of solid-phase chemical reactions and reactions in the melt are finished at this stage. Nonuniform temperature distribution in the melt exerts a significant influence on the dissolution of the fusion mixture components not yet involved in chemical reactions and the removal of the previously formed gas bubbles. At first, bubbles are removed from the maximum temperature zone, which is situated at a distance of 10–20 mm from the crucible walls (Fig. 14.3). Intensive convection in the melt, which promotes the removal of gas bubbles and leveling of the chemical composition of the melt, is an important consequence of the existence of high temperature gradients in the melt. Mechanical stirring by a water-cooled stirrer or by gas bubbling through the melt is used to intensify these processes [14.11, 31]. These methods are used in combination with the melt temperature increase, which causes a decrease in viscosity and an increase in the rate of convectional mass transfer.

After glass synthesis in the CC, annealing cannot be performed in the same crucible because, when the temperature of the melt being cooled reaches a value determined by the temperature dependence of the conductivity, the melt ceases to absorb the energy of the RF electromagnetic field. After this, intensive cooling of the glass by the water-cooling crucible walls and glass fracturing occurs. Therefore a method of glass extraction from the crucible with subsequent annealing in the furnace is essential. One of the methods consists of pouring the melt into a mold and further annealing of the glass. However, the risk of contamination of the central part of the glass block by bubbles and nonmolten particles of the fusion mixture from the periphery of the block is very high during this operation. Another method consists of extracting the glass block at a specific temperature, which is determined experimentally for every glass composition. The periphery of the glass block being cooled much quicker than the central part, the extraction temperature must be high enough to prevent fracturing of the peripheral part, but not too high, otherwise softening of the peripheral part and leakage of the viscous glass mass may occur due to heating by the inner hot part. In the case of glass synthesis in cooled quartz crucibles, a crucible with glass is simply transferred into the annealing furnace. Thin-walled crucibles are used to prevent glass fracturing caused by the difference of glass and quartz thermal expansion coefficients. The quartz crucible fractures and exfoliates during annealing, while the glass block remains a monolith.

A significant factor limiting the use of this method for the synthesis of glass with enhanced crystallization ability should be mentioned. The presence of the skull increases the possibility of glass crystallization because the initial mixture crystalline particles are ready nuclei contacting the melt directly. Under certain conditions initial mixture particles may provoke the crystallization of the whole glass volume. The possibility of preparation of glasses with enhanced crystallization ability in the cold crucible depends on a multitude of factors: melt volume, thickness of the viscous glass layer (which is determined by the power released in the melt as well as by the temperature dependence of the viscosity and electrical conductivity of the melt), the cooling rate of melt, and the rates of nucleation and crystal growth. Therefore quantitative assessment of this possibility is very difficult. As for qualitative analysis, it is clear that the crystal nuclei must not be able to grow through the

Table 14.7 Thermal expansion coefficients (K_{te}) and transformation (T_t) and softening (T_g) temperatures in glass of the La_2O_3–Al_2O_3–SiO_2 system [14.201]

Composition (mol%)			K_{te} (20–800 °C)	T_g	T_t
SiO_2	Al_2O_3	La_2O_3	($\times 10^7$, °C^{-1})	(°C)	(°C)
50.1	32.8	17.0	64.0	855	820
55.0	28.4	16.6	62.5	860	825
59.1	34.7	6.3	43.2	895	850
71.6	26.4	2.0	33.7	920	900
72.0	21.5	6.3	38.9	910	865
68.3	25.6	6.1	39.1	905	860

whole melt volume while the temperature of the melt being cooled is in the range of possible crystallization. Otherwise, a block of crystallized glass is formed.

14.4.1 Refractory Glasses of the R_2O_3-Al_2O_3-SiO_2 (R = Y, La, Rare-Earth Element) Systems

Investigation of the properties of glasses formed in this system was performed by several research groups and showed that these glasses possess a combination of valuable properties [14.201–207]. High refractive indices and high mechanical strength combined with high deformation temperature, relatively small values of thermal expansion coefficient (K_{te}), very low electric conductivity, and extremely high resistance to the action of alkalis make these glasses promising materials for use at high temperatures in chemically aggressive media. Besides, glasses formed in this system are devoid of oxides of alkaline or alkaline-earth metals and can be structurally different from traditional alkaline glasses. This undoubtedly makes them objects of scientific interest; for example, a nuclear magnetic resonance study of glasses of the SiO_2–Al_2O_3–Y_2O_3 system revealed the presence of tetra-, penta-, and hexacoordinated Al^{3+} ions in the glass structure, as well as a rather complicated structure of the silicon–oxygen framework [14.208].

Synthesis of glasses of the R_2O_3–Al_2O_3–SiO_2 (R = Y, La, rare-earth element) systems was performed by SM in a CC [14.201, 203]. Synthesis was performed under air at temperature of ≈ 2000 °C in water-cooled metal crucibles of 90–120 mm in diameter. Pieces of graphite were used for initial melting. The glass blocks obtained had a mass of 2–6 kg (depending on the crucible size) and did not contain any inclusions or bubbles. The practically important values of the K_{te} and characteristic temperatures for glasses of various compositions are presented in Table 14.7.

High values of the softening temperatures (855–920 °C) combined with relatively low K_{te} values $(33.7–64)\times 10^{-7}$ make these glasses promising materials for use in electrovacuum technologies for welding and sealing of W and Mo. Apart from being convenient for the preparation of relatively large quantities of glass, the SM technique proved to be very convenient for studies on glass formation in various systems. This method allows obtaining a range of samples of various compositions in one experiment. The glass-formation range in the system La_2O_3–Al_2O_3–SiO_2 was studied in [14.201]. Samples for these investigations were prepared as follows: first, a two-component mixture $Al_2O_3 + SiO_2$ of predetermined composition was melted in the cold crucible. After keeping of the melt at the temperature 1800–2000 °C for 1 h, a mixture

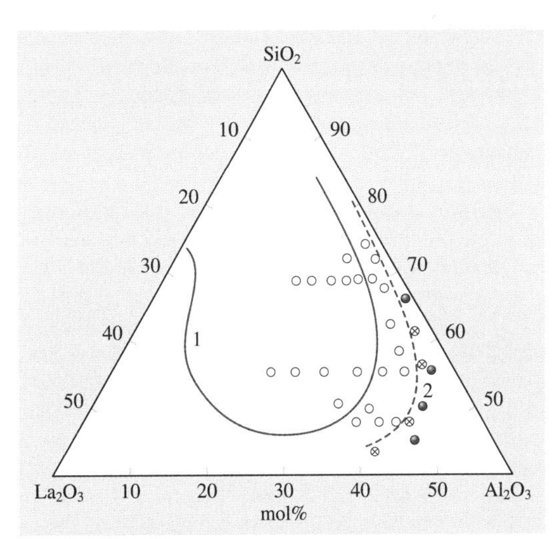

Fig. 14.27 Glass-forming region in the La_2O_3-Al_2O_3-SiO_2 system at different synthesis temperatures: 1 – 1600 °C, 2 – 2000 °C; (○ – glass, ● – sintered, ⊗ – surface crystallized) (after [14.201])

Table 14.8 Compositions and properties of the glasses with high concentrations of rare-earth element ions

Composition (mol%)	Young's modulus ($\times 10^{-8}$ Pa)	Hardness ($\times 10^{-7}$ Pa)	K_{te} ($\times 10^{7}$ °C^{-1})	Thermal conductivity (W/(m K))	Poisson's ratio
$43.2SiO_2$–$28.5Al_2O_3$–$26.4Y_2O_3$–$1.9Nd_2O_3$	1230	830	56.7	1.1	0.259
$29.6SiO_2$–$34.3Al_2O_3$–$36.1Nd_2O_3$	1080	770	62.4	1.0	–
$51.7SiO_2$–$26.0Al_2O_3$–$22.3Nd_2O_3$	1160	800	54.6	0.9	–
$37.6SiO_2$–$32.9Al_2O_3$–$26.6Yb_2O_3$–$2.9Ho_2O_3$	1250	810	–	0.6	–
$47.7SiO_2$–$32.6Al_2O_3$–$19.3Yb_2O_3$–$0.4Er_2O_3$	1270	–	44.0	0.55	–
$48.2SiO_2$–$27.5Al_2O_3$–$24.3Tb_2O_3$	1125	730	55.8	0.7	–
$53.2SiO_2$–$24.4Al_2O_3$–$20.6La_2O_3$–$1.8Nd_2O_3$	990	–	–	0.7	0.15

of $SiO_2 + La_2O_3$ with a predetermined composition was added in small (30–50 g) portions. After the addition of each portion the melt was homogenized for 10 min and then drops of the glass mass weighing 1–3 g were extracted with a quartz rod and quenched on a water-cooled metal surface. After this evaluation of the chemical compositions of the samples by electron probe microanalysis, microscopic analysis and X-ray diffraction analyses were performed. The results of the study compared with the literature data available are presented in Fig. 14.27.

The increase of the glass synthesis temperature to 2000 °C evidently leads to a considerable broadening of the glass-forming range in this system. Thus, at a SiO_2 content of 65–75 mol % the addition of only ≈ 2 mol % of La_2O_3 results in the formation of transparent glass. It is also worth noting that this system allows obtaining glasses with a SiO_2 content less than 50 mol %. These glasses are of interest for structural research because a continuous silicon–oxygen framework cannot exist in them.

Another interesting feature of the glasses of the R_2O_3–Al_2O_3–SiO_2 system is the possibility of obtaining glasses with extremely high concentrations of rare-earth element ions, in particular the Nd^{3+}, Yb^{3+}, and Tb^{3+} ions. It is impossible to obtain such glasses using conventional alkaline glass matrices. The weight of glass blocks obtained amounted to 5 kg, and the rare-earth metal-ion content amounted to 10^{22} cm^{-3}. The properties of some of the glasses synthesized are presented in Table 14.8.

The study of the luminescence spectra of the Sm^{3+} and Eu^{3+} ions present in lanthanalumosilicate glass was performed using the method of selective laser excitation of luminescence and revealed the absence of microphases with drastically different symmetries in the surroundings of the active ions, i. e., the activator ions in the glass were not segregated [14.209, 210]. Glasses synthesized in cold crucibles at high temperatures are characterized by low content of hydroxyl groups OH^-, even without using special dehydration methods (the peak value of absorption coefficient at the wavelength 3 μm is about ≈ 3–4 cm^{-1}) [14.211]. Glasses with Nd^{3+} concentration up to 1.1×10^{22} cm^{-3} are characterized with peak absorption coefficients of 290, 130, 150, and 40 cm^{-1} at wavelengths of 590, 745, 810, and 890 nm, respectively. High thermophysical and mechanical characteristics of such glasses (Table 14.8) combined with their stability at high temperatures make them promising materials for thin, optically dense, selective filters. Glasses with a high content of Tb^{3+} ions are also of considerable practical interest. This ion possesses magnetooptical activity, that is, it can alter the phase of linearly polarized light when magnetic field is applied. Measurements of the magnetooptical properties of the Tb_2O_3–Al_2O_3–SiO_2 glass containing 9.5×10^{21} Tb^{3+} ions/cm^3 showed that the Verdet constant equals −0.33 min/Oe cm at the wavelength of 630 nm. Such a high value of the Verde constant combined with the high value of the laser damage threshold (1.7×10^9 W/cm^2) and high heat conductivity opens wide perspectives for the use of this glass in various Faraday rotator devices in high-power laser systems.

14.4.2 Immobilization of Radioactive Waste in Stable Solid Blocks

During the last few years the SM method was widely applied for the immobilization of liquid and solid wastes (including radioactive waste) in stable solid blocks (phosphate and borosilicate glasses, mineral-like compositions) [14.212]. Improvement of the efficiency of radioactive waste neutralization is one of the main tasks of the atomic power generation industry. Waste recycling is a complex problem. It includes various technological stages (preparation and delivery of waste for recycling, melting, pouring into containers, collection and processing of the gases released, etc.). The melting unit is the most important element of such systems. Until recently all existing programs of waste vitrification relied on the use of traditional melting furnaces with hot crucibles, working in the 1100–1150 °C temperature range. Phosphate and borosilicate glasses are commonly used as matrices for waste immobilization. However, the low operate temperatures of such furnaces impose considerable limitations on the vitrification process. The use of a CC enables working at higher temperatures and removes the restrictions inherent to conventional melting units. Thus, higher temperatures provide broad possibilities for the design and application of new glass-forming and crystalline matrices for waste immobilization, including the possibility of introducing a higher concentration of waste matter into the matrices already being used, and processing of a wider range of waste matters (for example, heat-insulating materials) without the introduction of fluxing additives [14.213]. Other advantages of this technology include the small dimensions of the melting unit and high productivity of waste processing. Research and industrial *cold-crucible* systems are widely used nowadays in Russia, the USA, France, Germany, Italy, and South Korea [14.214].

14.5 Conclusion

This chapter on the application of skull melting of dielectric oxide materials in a cold crucible has illustrated the dynamic development of the method, which provides a basis for industrial-scale technologies as the development and manufacturing of the corresponding equipment has increased. The range of materials synthesized by means of this technique has become increasingly wide. Development of the technique for growth of PSZ crystals and nuclear waste vitrification can be pointed out as two of the most promising directions. Nevertheless, it is apparent that the potential of the method has not been exhausted, particularly in the field of single-crystal growth of refractory compounds.

References

14.1 W.V. Bolton: Das Tantal, seine Darstellung und seine Eigenschaften, Z. Electrochem. **11**, 45–51 (1905), in German

14.2 H.F. Sterling, R.W. Warren: High temperature melting without contamination in cold crucibles, Metallurgia **67**, 301–307 (1963)

14.3 A.A. Neustruev, G.L. Khodorovskii: *Vacuum skull furnaces* (Metallurgija, Moscow 1967), in Russian

14.4 R. Collongues, M. Perez y Jorba: Sur le chauffage et la fusion sans creuset par induction haute frequence de la zircone stabilisee, C. R. Acad. Sci. **257**, 1091–1093 (1963), in French

14.5 B. Gayet, J. Holder, G. Kurka: Fusion du bioxyde d'uraniuum par induction de la haute frequence, Rev. Haut. Temp. Reftact. **1**, 153–157 (1964), in French

14.6 M. Perez y Jorba, R. Collongues: Sur le chauffage et la fusion sans creuset par induction haute frequence de quelques oxydes refractaires, Rev. Haut. Temp. Reftact. **1**, 21–25 (1964), in French

14.7 C. Deportes, B. Lorang, G. Vitter: Sur une amelioration du procede de fusion en auto-creuset des oxydes refractaires, Rev. Haut. Temp. Reftact. **2**, 159–161 (1965), in French

14.8 Y. Roulin, G. Vitter, C. Deportes: Nouveau dispositif de fusion en autocreuset. Fusion d'oxydes refractaires dans une enceinte multitubulaire, Rev. Int. Haut. Temp. Refract. **6**, 153–158 (1969), in French

14.9 J. Moulin, I. Reboux: Nouveaux developpements dans la fusion electrique des verres refracteires, Verres Refract. **26**, 123–127 (1972), in French

14.10 B. Scott, H. Rawson: Techniques for producting low loss glasses for optical fibre communication system, Glass Technol. **14**, 115–124 (1973)

14.11 B. Scott, H. Rawson: Preparation of low loss glasses for optical fibre communication system, Optoelectronics **5**, 285–288 (1973)

14.12 V.I. Aleksandrov, V.V. Osiko, V.M. Tatarintsev: *Report FIAN* (FIAN, Moscow 1968), in Russian

14.13 V.I. Aleksandrov, E.E. Lomonova, A.A. Majer, V.V. Osiko, V.M. Tatarintsev, V.T. Ydovenchik: Physical properties of zirconia and hafnia single crystals, Bull. Lebedev Phys. Inst. (FIAN) **11**, 3–7 (1972), in Russian

14.14 V.I. Aleksandrov, V.V. Osiko, A.M. Prokhorov, V.M. Tatarintsev: Novel method of preparation of refractory single crystals and melted ceramic materials, Herald Rus. Acad. Sci. **12**, 29–36 (1973), in Russian

14.15 V.I. Aleksandrov, V.V. Osiko, A.M. Prokhorov, V.M. Tatarintsev: Synthesis and crystal growth of refractory materials by RF melting in a cold container, Top. Mater. Sci. **1**, 421–480 (1978)

14.16 R.F. Sekerka, R.A. Hartzell, B.J. Farr: Instability phenomena during the RF-heating and melting of ceramics, J. Cryst. Growth **50**, 783–806 (1980)

14.17 R.A. Hartzell, R.F. Sekerka: Mathematical modeling of internal centrifugal zone growth of ceramics and ceramic metal composites, J. Cryst. Growth **57**, 27–42 (1982)

14.18 C. Gross, W. Assmus, A. Muiznieks, G. Raming, A. Muehlbauer, C. Stenzel: Power consumption of skull melting, Part I: analytical aspects and experiments, Cryst. Res. Technol. **34**, 319–328 (1999)

14.19 A. Muiznieks, G. Raming, A. Muehlbauer, C. Gross, W. Assmus, C. Stenzel: Power consumption of skull melting, Part II: numerical calculation of the shape of the molten zone and comparison with experiment, Cryst. Res. Technol. **34**, 329–338 (1999)

14.20 T. Behrens, M. Kudryash, B. Nacke, D. Lopuch, A. Martynow, I. Loginov: Induction skull melting of Y_2O_3–BaO–CuO in a cold crucible, Int. Sci. Coll. Modelling for Electromagnetic Processing (Hannover 2003) pp. 249–254

14.21 B. Nacke, T. Behrens, M. Kudryash, A. Jakovics: Skull melting technology for oxides and glasses, Conf. Fundamental and Applied MHD, Vol. 2 (Riga 2005) pp. 241–244

14.22 A.V. Shkulkov: Attractor formation in induction skull-melting systems and its effect on crystal structure, Bull. Rus. Acad. Sci. Phys. **68**, 1000–1006 (2004)

14.23 V.V. Osiko, M.A. Borik, E.E. Lomonova: Crucible-free methods of growing oxide crystals from the melt, Annu. Rev. Mater. Sci. **17**, 101–122 (1987)

14.24 E.E. Lomonova, V.V. Osiko: Growth of zirconia crystals by skull-melting technique. In: *Crystal Growth Technology*, ed. by H.J. Scheel, T. Fukuda (Wiley, Chichester 2003) pp. 461–484

14.25 Y.S. Kuzminov, E.E. Lomonova, V.V. Osiko: *Refractory Materials from a Cold Crucible* (Nauka, Moscow 2004)

14.26 Y.B. Petrov: *Inductive Melting of Oxides* (Energoatomizdat, Leningrad 1983), in Russian

14.27 V.I. Aleksandrov, A.L. Blinov, M.A. Borik, V.V. Osiko: Investigation of temperature distribution in viscous oxide melts under process of skull melting in a cold crusible, Izv. AS USSR, Inorg. Mater. **19**, 443–447 (1983), in Russian

14.28 V.I. Aleksandrov, S.H. Batygov, V.F. Kalabuhova, S.V. Lavrischev, E.E. Lomonova, V.A. Mizina, V.V. Osiko, V.M. Tatarintsev: Ittria distribution and inhomogenities into cubic ZrO_2–Y_2O_3 crystals, Izv. AS USSR, Inorg. Mater. **16**, 99–104 (1980), in Russian

14.29 V.I. Aleksandrov, S.H. Batygov, V.F. Kalabuchova, S.V. Lavrischev, E.E. Lomonova, V.A. Mizina, V.V. Osiko, V.M. Tatarintsev: Effect SiO_2 on growth and perfection of stabilized ZrO_2, Izv. AS USSR, Inorg. Mater. **16**, 1800–1804 (1980), in Russian

14.30 V.I. Aleksandrov, M.A. Vishnyakova, V.F. Kalabuchova, E.E. Lomonova, V.A. Panov: Growth of zirconia single crystals by direct crystallization in a cold container, Proc. Indian Natl. Sci. Acad. **2**, 133–144 (1991)

14.31 V.I. Aleksandrov, V.V. Osiko, A.M. Prokhorov, V.M. Tatarintsev: Refractory materials preparation by skull melting in a cold container, Rus. Chem. Rev. **47**, 385–427 (1978), in Russian

14.32 V.I. Aleksandrov, V.P. Voitsitskii, E.E. Lomonova, V.V. Osiko, N.P. Khaneev: Study of melting and crystallization processes in a cold container with direct radio-frequency heating, Instrum. Exp. Tech. **3**, 231–234 (1991), in Russian

14.33 V.I. Aleksandrov, V.P. Voitsitskii, E.E. Lomonova, V.V. Osiko, N.P. Khaneev: Study of dielectric materials melting in a cold container with direct radio-frequency heating, Izv. AS USSR, Inorg. Mater. **27**, 983–987 (1991), in Russian

14.34 V.I. Aleksandrov, V.P. Voitsitskii, E.E. Lomonova, V.V. Osiko, N.P. Khaneev: Investigation of melting and crystallization processes in a cold container with direct radio-frequency heating, Izv. AS USSR, Inorg. Mater. **27**, 2167–2171 (1991), in Russian

14.35 V.I. Aleksandrov, V.P. Voitsitskii, E.E. Lomonova, V.V. Osiko, N.P. Khaneev: Control method of melt propogation on a initial stage of direct radio-frequency melting in a cold container, Instrum. Exp. Tech. **4**, 212–217 (1992), in Russian

14.36 V.I. Aleksandrov, V.P. Voitsitskii, E.E. Lomonova, V.V. Osiko, N.P. Khaneev: Monitoring of melt propogation direction on a initial stage of direct radio-frequency melting in a cold container, J. Tech. Phys. **62**, 180–186 (1992), in Russian

14.37 B.T. Melekh, F.F. Andreeva, N.F. Kartenko, I.V. Korkin, V.B. Smirnov: Features of direct induction melting and crystal growth of refractory oxide transition elements, Izv. AS USSR, Inorg. Mater. **18**, 98–101 (1982), in Russian

14.38 R. Aragon, H.R. Harrison, R.H. McCallister, C.J. Sandberg: Skull melting single crystal growth of magnetite (Fe_3O_4)-ulvospinel (Fe_2TiO_4) solid solution members, J. Cryst. Growth **61**, 221–228 (1983)

14.39 V.I. Aleksandrov, A.L. Blinov, M.A. Borik, V.G. Veselago, V.V. Voronov, V.M. Ivanovskaja, V.A. Misina, V.V. Osiko, V.T. Udovenchik, V.A. Fradkov, M.A. Chernikov: Synthesis and crystal growth of $YBa_2Cu_3O_{6.5+x}$ from the melt by direct radio-frequency heating in a cold container, 7 All-Union Conf. on Crystal Growth, Vol. 2 (Moscow 1988), pp. 378–379, in Russian

14.40 M. Hartmanova, E.E. Lomonova, V. Navratil, P. Sutta, F. Kudracik: Characterization of yttria-doped ceria prepared by directional crystallization, J. Mater. Sci. **40**, 5679–5683 (2005)
14.41 L. Mazerolles, D. Michel, M.J. Hÿtch: Microstructures and interfaces in directionally solidified oxide-oxide eutectics, J. Eur. Ceram. Soc. **25**, 1389–1395 (2005)
14.42 W.-J. Jang, K. Imai, M. Hasegawa, H. Takei: Growth and structure of $La_2NiO_{4+\delta}$ ($0.19 \geq \delta \geq 0.12$) single crystals, J. Cryst. Growth **152**, 158–168 (1995)
14.43 V.I. Aleksandrov, V.V. Osiko, V.M. Tatarintsev: Melting of refractory dielectric materials by radio-frequency heating, Instrum. Exp. Tech. **5**, 222–225 (1970), in Russian
14.44 V.I. Aleksandrov, M.A. Vishnyakova, Y.K. Voron'ko, V.F. Kalabuchova, E.E. Lomonova, V.A. Misina, V.V. Osiko: Crystal growth of $SrTiO_3$ by Czochralski method from a cold crusible, Izv. AS USSR, Inorg. Mater. **19**, 104–108 (1983), in Russian
14.45 V.I. Aleksandrov, S.H. Batygov, M.A. Vishnyakova, Y.K. Voron'ko, V.F. Kalabuchova, E.E. Lomonova, V.V. Osiko: Optical properties of $SrTiO_3$ single crystal grown by Czochralski method, Izv. AS USSR, Inorg. Mater. **19**, 265–268 (1983), in Russian
14.46 V.I. Aleksandrov, I.A. Gerasimova, A.V. Kolesnikov, E.E. Lomonova, V.V. Osiko, V.A. Panov, P.A. Makarov, A.V. Archakov, N.G. Gorashchenko, A.A. Mayer: Growth of sillenite (BGO) single crystals from cold container, Russ. J. Inorg. Chem. **35**, 878–883 (1990)
14.47 D. Mateika, R. Lauien, M. Liehr: Czochralski growth by double container technique, J. Cryst. Growth **65**, 237–242 (1983)
14.48 A.M. Balbashov, A.J. Chervonenkis: *Magnetic Materials for Electronics* (Energija, Moscow 1979), in Russian
14.49 O. Ruff, F. Ebert: Contributions on ceramics of high refractory materials. I. The forms of zirconium dioxide, Z. Anorg. Allg. Chem. **180**, 19–41 (1929)
14.50 B.J. Curtis, J.A. Wilkinson: Preparation of mixed oxide crystals by chemical transport, J. Am. Ceram. Soc. **48**, 49–50 (1965)
14.51 V.A. Kysnetsov, O.V. Sidorenko: Crystallization of ZrO_2 and HfO_2 in hydrothermal conditions, Crystallogr. Rep. **13**, 748–749 (1968), in Russian
14.52 A.M. Anthony, V. Loc: Preparation de monocristaux de zircone pure monoclinique, C. R. Acad. Sci. **260**, 1383–1385 (1965), in French
14.53 A.B. Chase, J.A. Osmer: Growth of single crystals of ZrO_2 and HfO_2 from PbF_2, Am. Mineral. **51**, 1811–1888 (1966)
14.54 W. Kleber, L. Ickert, J. Doerschel: Ein Beitrag zum Wachstum von Zirkoniumdioxid-Einkristallen aus Schmelzenlosungen, Krist. Tech. **1**, 237–248 (1966)
14.55 M. Yokoyama, T. Ota, I. Yamai: Flux growth of yttria-stabilized zirconia crystals, J. Cryst. Growth **75**, 630–632 (1986)
14.56 T. Yamakawa, N. Ishizawa, K. Uematsu, N. Mizutani, M. Kato: Growth of yttria partially and fully stabilized zirconia crystals by xenon arc image floating zone method, J. Cryst. Growth **75**, 623–629 (1986)
14.57 M. Yoshimura, T. Hiuga, S. Somiya: Crystal growth of yttria stabilized zirconia (YSZ) under hydrothermal conditions, J. Cryst. Growth **71**, 277–279 (1985)
14.58 T.B. Reed: Growth of refractory crystals using the induction plasma torch, J. Appl. Phys. **32**, 2534–2535 (1961)
14.59 A. Saiki, N. Ishizawa, N. Mizutani, M. Kato: Directional crystal growth of yttria-stabilized zirconia by the arc image floating-zone method, J. Mater. Sci. Lett. **6**, 568–570 (1987)
14.60 D. Michel, M. Perez y Jorba, R. Collongues: Sur l'elaboration de monocristaux de zircone stabiliseret sur certaines proprieles de la solution solide cubique ZrO_2–CaO, C. R. Acad. Sci. C. **266**, 1602–1604 (1968), in French
14.61 D. Michel, M. Perez y Jorba, R. Collongues: Growth from skull-melting of zirconia rare earth oxide crystals, J. Cryst. Growth **43**, 546–548 (1978)
14.62 A.M. Anthony, R. Collongues: Modern methods of growing single crystals of high-melting point oxides. In: *Preparative Methods in Solid State Chemistry*, ed. by P. Hagenmuller (Academic, New York 1972) pp. 147–249
14.63 R.C. Garvi, R.H. Hannink, R.T. Pascoe: Ceramic steel, Nature **258**, 703–704 (1975)
14.64 T.K. Gupta, J.H. Bechtold, R.C. Kuznicki, L.H. Cadoff, B.R. Rossing: Stabilization of tetragonal phase in polycrystalline zirconia, J. Mater. Sci. **12**, 2421–2426 (1977)
14.65 T.K. Gupta: Sintering of tetragonal zirconia and its characteristics, Sci. Sinter. **10**, 205–216 (1978)
14.66 D.L. Porter, A.H. Heuer: Mechanisms of toughening partially stabilized zirconia (PSZ), J. Am. Ceram. Soc. **60**, 183–184 (1977)
14.67 R.H.J. Hannink, P.V. Kelly, B.C. Muddle: Transformation toughening in zirconia-containing ceramics, J. Am. Ceram. Soc. **83**, 461–487 (2000)
14.68 M.E. Mashburg, W.E. Coblenz: Reactio-formed ceramics, J. Am. Ceram. Soc. **67**, 356–363 (1988)
14.69 R. Chaim, M. Hefetz: Effect of grain size on elastic modulus and hardness of nanocrystalline ZrO_2–3 wt%–Y_2O_3 ceramic, J. Mater. Sci. **39**, 3057–3061 (2004)
14.70 K. Tzukuma, K. Heda, M. Shimada: Strength and fracture toughness in isostatically hot pressed composites of Al_2O_3 and Y_2O_3 partially stabilized ZrO_2, J. Am. Ceram. Soc. **68**, C4–C5 (1985)
14.71 M. Hoch, K.M. Nair: Densification characteristics of ultrafine powders, Ceramurg. Int. **2**, 88–97 (1976)
14.72 K.S. Mazdiyasni: Powder synthesis from metal-organic precursors, Ceramurg. Int. **8**, 42–56 (1982)
14.73 C.J. Howard, R.J. Hill, B.E. Reichert: Structures of the ZrO_2 polymorphs at room temperature by

high-resolution neutron powder diffraction, Acta Crystallogr. B **44**, 116–120 (1980)

14.74 G. Teufer: Crystal structure of tetragonal ZrO_2, Acta Crystallogr. **15**, 1187 (1962)

14.75 R.J. Ackermann, S.P. Garg, E.G. Rauh: High-temperature phase diagram for the system Zr-O, J. Am. Ceram. Soc. **60**, 341–345 (1977)

14.76 D.K. Smith, C.F. Cline: Verification of existence of cubic zirconia at high temperature, J. Am. Ceram. Soc. **45**, 249–250 (1962)

14.77 J. McCullough, K. Trueblood: The crystal structure of baddeleyite (monoclinic ZrO_2), Acta Crystallogr. **12**, 507–511 (1959)

14.78 N.V. Belov: Crystallographic structure of baddeleyite, Crystallogr. Rep. **5**, 460–461 (1960), in Russian

14.79 D.K. Smith, H.W. Newkirk: The crystal structure of baddelyite (monoclinic ZrO_2) and its relation to the polymorphism of ZrO_2, Acta Crystallogr. **18**, 983–991 (1965)

14.80 R. Cupres, R. Wollast: Polymorphism conversion of pure zirconia, Ber. Dtsch. Keram. Ges. **40**, 527–532 (1963)

14.81 P. Duwez, F.H. Brown, F. Odell: The zirconia-yttria system, J. Electrochem. Soc. **38**, 356–362 (1951)

14.82 S.P. Ray, V.S. Stubican: Fluorite related ordered compounds in the ZrO_2-CaO and ZrO_2-Y_2O_3 system, Mater. Res. Bull. **12**, 549–556 (1977)

14.83 H.G. Scott: Phase relationships in the yttria-zirconia system, J. Mater. Sci. **10**, 1527–1535 (1975)

14.84 V.S. Stubican, R.C. Hink, S.P. Ray: Phase equilibria and ordering in the system ZrO_2-Y_2O_3, J. Am. Ceram. Soc. **61**, 17–21 (1978)

14.85 V.S. Stubican, G.S. Gorman, J.R. Hellmann, G. Senft: Phase relationships in some ZrO_2 systems. In: *Advances in Ceramics Vol. 12: Science and Technology of Zirconia II*, ed. by N. Claussen, M. Ruhle, A.H. Heuer (Am. Ceram. Soc., Columbus 1984) pp. 96–106

14.86 V.S. Stubican: Phase equilibria and metastabilities in the systems ZrO_2-MgO, ZrO_2-CaO, and ZrO_2-Y_2O_3,. In: *Advances in Ceramics Vol. 12: Science and Technology of Zirconia II*, ed. by N. Claussen, M. Ruhle, A.H. Heuer (Am. Ceram. Soc., Columbus 1984) pp. 71–95

14.87 C. Pascual, P. Duran: Subsolidus phase equilibria and ordering in the system ZrO_2-Y_2O_3, J. Am. Ceram. Soc. **66**, 23–27 (1983)

14.88 M. Rühle, N. Claussen, A.H. Heuer: Microstructural studies of Y_2O_3-containing tetragonal ZrO_2 polycrystals (Y-TZP). In: *Advances in Ceramics Vol. 12: Science and Technology of Zirconia II*, ed. by N. Claussen, M. Rühle, A.H. Heuer (Am. Ceram. Soc., Columbus 1984) pp. 352–370

14.89 A.H. Heuer, M. Rühle: Phase transformations in ZrO_2-containing ceramics. II. The martensitic reaction in t-ZrO_2. In: *Advances in Ceramics Vol. 12: Science and Technology of Zirconia II*, ed. by N. Claussen, M. Rühle, A.H. Heuer (Am. Ceram. Soc., Columbus 1984) pp. 14–32

14.90 C.H. Perry, D.W. Liu, R.P. Ingel: Phase caracterization of partially stabilized zirconia by Raman spectroscopy, J. Am. Ceram. Soc. **68**, 184–187 (1985)

14.91 Y.K. Voron'ko, A.A. Sobol, S.N. Ushakov, L.I. Tsimbal: Tetragonal structure formation in partly stabilized zirconia, Izv. AS USSR, Inorg. Mater. **30**, 803–808 (1994), in Russian

14.92 M.A. Aboimov, M.A. Borik, G.A. Gogotsi, V.F. Kalabuchova, E.E. Lomonova, V.A. Mizina: Phase transitions in crystals of partially stabilized zirconia, Inorg. Mater. **33**, 285–291 (1997)

14.93 M. Hillert: Critical limit for massive transformation, Metall. Mater. Trans. A **33**, 2299–2308 (2002)

14.94 H.G. Scott: Phase relationships in the yttria-rich part of the yttria-zirconia system, J. Mater. Sci. **12**, 311–316 (1977)

14.95 C.A. Ànderson, J. Greggi, T.K. Gupta: Diffusionless transformations in zirconia alloys,. In: *Advances in Ceramics Vol. 12: Science and Technology of Zirconia II*, ed. by N. Claussen, M. Ruhle, A.H. Heuer (Amer. Cer. Soc., Columbus, OH 1984), pp.78–85

14.96 A. Rouanet: Diagrams of solidification and diagrams of high temperature phases in the zirconia-erbium oxide, zirconia-yttrium oxide and zirconia-ytterbium oxide systems, C. R. Acad. Sci. Ser. C **267**, 1581–1584 (1968)

14.97 Y. Du, Z. Jin, P. Huang: Termodynamic assessment in the ZrO_2-$YO_{1.5}$ system, J. Am. Ceram. Soc. **74**, 1569–1577 (1991)

14.98 N.S. Jacobson, Z.-K. Liu, L. Kaufman, F. Zhang: Thermodynamic modeling of $YO_{1.5}$-ZrO_2 system, J. Am. Ceram. Soc. **87**, 1559–1566 (2004)

14.99 M. Chen, B. Hallstedt, L.J. Gauckler: Thermodynamic Modeling of the ZrO_2-$YO_{1.5}$ system, Solid State Ion. **170**, 255–274 (2004)

14.100 V. Lantery, A.H. Heuer, T.E. Mitchell: Tetragonal phase in the system ZrO_2-Y_2O_3. In: *Advances in Ceramics Vol. 12: Science and Technology of Zirconia II*, ed. by N. Claussen, M. Rühle, A.H. Heuer (Am. Ceram. Soc., Columbus 1984), pp.118–130

14.101 T. Yagi, A. Saiki, N. Ishizava, N. Mizutani, M. Kato: Analytical electron microscopy of yttria partitioning in the yttria-partially-stabilized zirconia-crystal, J. Am. Ceram. Soc. **69**, C3–C4 (1986)

14.102 G.B. Bokij: *Crystal Chemistry* (Nauka, Moscow 1971), in Russian

14.103 V.P. Gorelov, S.F. Pal'guev: Examination of oxigen vacancy model in solid solution ZrO_2-Y_2O_3, Izv. AS USSR, Inorg. Mater. **13**, 181–182 (1977), in Russian

14.104 H.H. Möbius: Sauerstoffionenleitende Festelektrolyte und ihre Anwendungsmöglichkeiten, Z. Chem. **2**, 101–106 (1962), in German

14.105 S. Fabris, A.T. Paxton, M.W. Finnis: A stabilization mechanism of zirconia based on oxygen vacancies only, Acta Mater. **50**, 5171–5178 (2002)

14.106 E.V. Stefanovich, A.L. Shluger, C.R.A. Catlow: Theoretical study of the stabilization of the cubic-phase ZrO_2 by impurities, Phys. Rev. B **49**, 11560–11571 (1994)
14.107 S. Fabris, A.T. Paxton, M.W. Finnis: Relative energetics and structural properties of zirconia using a self-consistent tight-binding model, Phys. Rev. B **63**, 094101–13 (2001)
14.108 F.F. Lange: Transformation toughness, Part I: Size effects associated with the thermodynamics of constrained transformation, J. Mater. Sci. **17**, 225–234 (1982)
14.109 D. Michel, L. Mazerolles, M. Perez y Jorba: Fracture of metastable tetragonal zirconia crystals, J. Mater. Sci. **18**, 2618–2628 (1983)
14.110 F.R. Chien, F.J. Ubic, V. Prakash, A.H. Heuer: Stress-induced transformation and ferroelastic deformation adjacent microhardness indents in tetragonal zirconia single crystals, Acta Mater. **46**, 2151–2171 (1998)
14.111 J. Lefevre: Fluorite-type structural modifications in system having a zirconium and hafnium oxide based, Ann. Chem. **8**, 117–149 (1963)
14.112 D. Baither, B. Baufeld, U. Messerschmidt, F.H. Foitzik, M. Rühle: Ferroelasticity of t′-zirconia: I, high electron microscopy studies of the microstructure in polydomain tetragonal zirconia, J. Am. Ceram. Soc. **80**, 1691–1698 (1997)
14.113 Y.K. Voron'ko, M.A. Zufarov, B.V. Ignat'ev, V.V. Osiko, E.E. Lomonova, A.A. Sobol': Raman scattering in tetragonal single crystals ZrO_2-Gd_2O_3 and ZrO_2-Eu_2O_3, Opt. Spectrosc. **51**, 569–571 (1981), in Russian
14.114 K.M. Prettyman, J.-F. Jue, A.V. Virkar, C.R. Hubbard, O.B. Cavin, M.K. Ferber: Hysteresity effects in 3 mol % yttria-doped zirconia (t′-phase), J. Mater. Sci. **27**, 4167–4174 (1992)
14.115 D. Baither, B. Baufeld, U. Messerschmidt: Morphology of tetragonal precipitates in Y_2O_3-stabilized ZrO_2 crystals, Phys. Status Solidi (a) **137**, 569–576 (1993)
14.116 A.H. Heuer, V. Lanteri, A. Dominguez-Rodriguez: High-temperature precipitation hardening of two phase Y_2O_3-partially-stabilized ZrO_2 single crystals: A first report, J. Amer. Cer. Soc, **69**, 285–287 (1986)
14.117 V. Lanteri, T.E. Mitchell, A.H. Heuer: Morphology of tetragonal precipitates in partially stabilized ZrO_2, J. Am. Ceram. Soc. **69**, 564–569 (1986)
14.118 V.I. Aleksandrov, M.A. Borik, M.A. Vishnyakova, V.P. Voitsitskii, V.F. Kalabuchova, E.E. Lomonova, V.V. Osiko: Method of single crystal preparation (modification), RF Patent 2133787 (1999), in Russian
14.119 A.G. Merjanov, V.A. Raduchev, E.N. Rumanov: Heat waves of melting in dielectric crystals, Dokl. AS USSR **253**, 330–334 (1980), in Russian
14.120 E.N. Rumanov: *Heat Wave of Elementary Substance Melting* (Preprint OIHF AS USSR, Chernogolovka, 1982) p. 20, in Russian
14.121 B. Chalmers: *Theory of Solidification* (Metallurgija, Moscow 1968), in Russian
14.122 A. Hellawell, P.M. Herbert: The development of preferred orientations during the freezing of metals and alloys, Proc. R. Soc. London Ser. A **269**, 560–573 (1962)
14.123 W.A. Tiller: Preferred growth direction of metals, J. Met. **9**, 845–855 (1957)
14.124 A. Rosenberg, W.A. Tiller: The relationship between growth forms and the preferred direction of growth, Acta Metall. **5**, 565–573 (1957)
14.125 W.A. Tiller: Solute segregation during ingot solidification, J. Iron Steel Inst. **215**, 447–457 (1959)
14.126 G.G. Lemmlein: Geometric selection in growing crystal agregate, Dokl. AS USSR **48**, 177–180 (1945), in Russian
14.127 A.V. Chubnikov: About geometric selection rule during formation crystal agregate, Dokl. AS USSR **51**, 679–681 (1946), in Russian
14.128 A.N. Kolmogorov: On the Issue of "geometric selection" crystals, Dokl. AS USSR **65**, 681–684 (1949), in Russian
14.129 V.V. Osiko, D.L. Penyaz, N.P. Khaneev: Study of directional crystallization process in a cold container with direct radiofrequency heating, J. Cryst. Growth **128**, 1193–1196 (1993)
14.130 R.P. Ingel, D. Lewis, B.A. Bender, P.W. Rice: Temperature dependence of strength and fracture toughnes of ZrO_2 single crystals, J. Am. Ceram. Soc. **65**, C150–C152 (1982)
14.131 R.P. Ingel, D. Lewis, B.A. Bender, P.W. Rice: Room temperature strength and fracture of ZrO_2-Y_2O_3 single crystals, J. Am. Ceram. Soc. **65**, C108–C109 (1982)
14.132 D. Michel: Relation between morphology and structure for stabilized zirconia crystals. In: *Advances in Ceramics, Vol. 24 Science and Technology of Zirconia III*, ed. by S. Somiya, N. Yamamoto, H. Yanahida (Am. Ceram. Soc., Columbus 1988) pp. 455–461
14.133 P. Hartman, W.G. Perdok: On the relations between structure and morphology of crystals, Acta Crystallogr. **8**, 525–531 (1955)
14.134 H. Römer, K.-D. Luther, W. Assmus: Measurement of the distribution coefficient of neodymium in cubic zirconium dioxide, J. Cryst. Growth **130**, 233–237 (1993)
14.135 H. Römer, K.-D. Luther, W. Assmus: The distribution coefficients of rare earth ions in cubic zirconium dioxide, J. Cryst. Growth **141**, 159–164 (1994)
14.136 H. Römer, K.-D. Luther, W. Assmus: Determination of the distribution coefficients of the rare earth ions Er^{3+} and Nd^{3+} in yttria-stabilized c-ZrO_2 single crystals, Z. Kristallogr. **209**, 311–314 (1994)
14.137 V.B. Glushkova, V.V. Osiko, L.G. Shcherbakova, V.I. Aleksandrov, Y.N. Paputskii, V.M. Tatarintsev: Study of features monocrystalline solid solution in

ZrO_2–Y_2O_3 system, Izv. AS USSR, Inorg. Mater. **13**, 2197–2201 (1977), in Russian
14.138 D.B. Zhang, X.M. He, J.P. Chen, J.C. Wang, Y.F. Tang, B.L. Hu: Research on crystal growth and defects in cubic zirconia, J. Cryst. Growth **79**, 336–340 (1986)
14.139 A. Baermann, W. Guse, H. Saalfeld: Characterization of different (Me,Zr)O_2 single crystals grown by the "skull-melting" technique, J. Cryst. Growth **79**, 331–335 (1986)
14.140 W.-S. Kim, I.-H. Suh, Y.-M. You, J.-H. Lee, C.-H. Lee: Synthesis of yttria-stabilized zirconia crystals by skull-melting method, Neues Jb. Miner. Monatsh. **3**, 136–144 (2001)
14.141 E.V. Alekseev, O.N. Gorshkov, E.V. Chuprunov, V.A. Novikov, A.P. Kasatkin, G.K. Fukin: Investigation into the specific features of the changes in the crystal structure of stabilized zirconia upon thermochemical reduction, Crystallogr. Rep. **51**, 632–635 (2006)
14.142 H. Kahlert, F. Frey, H. Boysen, K. Lassak: Defect structure and diffuse scattering of zirconia single crystals at elevated temperatures and simultaneously applied electric field, Appl. Cryst. **28**, 812–819 (1995)
14.143 J.S. Thorp, A. Aypar, J.S. Ross: Electron spin resonance in single crystal yttria stabilized zirconia, J. Mater. Sci. **7**, 729–734 (1972)
14.144 R.I. Merino, V.M. Orera, O. Povill, W. Assmus, E.E. Lomonova: Optical and electron paramagnetic resonance characterization of Dy^{3+} in YSZ single crystals, J. Phys. Chem. Solids **58**, 1579–1585 (1997)
14.145 R.I. Merino, V.M. Orera, E.E. Lomonova, S.K. Batygov: Paramagnetic electron traps in reduced stabilized zirconia, Phys. Rev. B **52**, 6150–6152 (1995)
14.146 D. Gomez-Garcia, J. Martinez-Fernandez, A. Dominguez-Rodriguez: Recent advances in electron-beam-induced damage models in yttria fully stabilized zirconia single crystals, Philos. Mag. Lett. **81**, 173–178 (2001)
14.147 R.P. Ingel, D. Lewis III: Lattice parameters and density for Y_2O_3-stabilized zirconia, J. Am. Ceram. Soc. **69**, 325–332 (1986)
14.148 V.I. Aleksandrov, G.E. Val'jano, B.V. Lukin, V.V. Osiko, A.E. Rautbort, V.M. Tatarintsev, V.N. Filatova: Crystal structure of stabilized zirconia, Izv. AS USSR, Inorg. Mater. **12**, 273–277 (1978), in Russian
14.149 R.H. Ingel, D. Lewis, B.A. Bender, R.W. Rice: Physical, microstructural and termomechanical properties of ZrO_2 single crystals. In: *Advances in Ceramics Vol. 12: Science and Technology of Zirconia II*, ed. by N. Claussen, M. Rühle, A.H. Heuer (Am. Ceram. Soc., Columbus 1984) pp. 408–414
14.150 A. Cheikh, A. Madani, A. Touati, H. Boussett, C. Monty: Ionic conductivity of zirconia based ceramics from single crystals to nanostructured polycrystals, J. Eur. Ceram. Soc. **21**, 1837–1841 (2001)
14.151 M. Hartmanova, J. Schneider, V. Navratil, F. Kundracik, H. Schulz, E.E. Lomonova: Correlation between microscopic and macroscopic properties of yttria stabilized zirconia. 1. Single crystals, Solid State Ion. **107**, 136–137 (2000)
14.152 D.S. Thorp, H.P. Buckley: The dielectric constants of current-blackened single crystal yttria-stabilized zirconia, J. Mater. Sci. **8**, 1401–1408 (1973)
14.153 S.P.S. Badwal: Electrical conductivity of single crystal and polycrystalline yttria-stabilized zirconia, J. Mater. Sci. **19**, 1767–1776 (1984)
14.154 N. Bonanos, E.P. Butler: Ionic conductivity of monoclinic and tetragonal yttria-zirconia single crystals, J. Mater. Sci. Lett. **4**, 561–564 (1985)
14.155 V.I. Aleksandrov, S.H. Batygov, Y.K. Voron'ko, B.I. Denker, E.E. Lomonova, V.V. Osiko, V.M. Tatarintsev: Coloure centres in cubic ZrO_2 single crystals, Izv. AS USSR, Inorg. Mater. **11**, 664–667 (1975), in Russian
14.156 V.I. Aleksandrov, M.A. Vishnyakova, V.P. Voitsitskii, E.E. Lomonova, M.A. Noginov, V.V. Osiko, V.A. Smirnov, A.F. Umiskov, I.A. Shcherbakov: Fianite (ZrO_2–Y_2O_3:Er^{3+}) laser emitting the 3 μm range, Sov. J. Quant. Electron. **19**, 1555–1556 (1989)
14.157 V.I. Aleksandrov, M.A. Vishnyakova, V.P. Voitsitskii, Y.K. Voron'ko, E.E. Lomonova, V.A. Mizina, A.A. Sobol, S.N. Ushakov, L.I. Tsimbal: Spectroscopic properties solid solution ZrO_2–Y_2O_3 single crystals doped with Cr and Nd, Izv. AS USSR, Inorg. Mater. **26**, 1251–1255 (1990), in Russian
14.158 A.V. Prokof'ev, W. Assmus, A.I. Shelykh, I.A. Smirnov, B.T. Melekh: Absorption edge of zirconium dioxide crystals doped with rare earth ions, Phys. Solid State **38**, 2739–2743 (1996)
14.159 H. Römer, K.-D. Luther, W. Assmus: Coloured zirconia, Cryst. Res. Technol. **29**, 787–794 (1994)
14.160 S.E. Paje, J. Llopis: Photoluminescence decay and time-resolved spectroscopy of cubic yttria-stabilized zirconia, Appl. Phys. A **59**, 569–574 (1994)
14.161 V.I. Aleksandrov, N.A. Abramov, M.A. Vishnyakova, V.F. Kalabukhova, E.E. Lomonova, N.R. Miftiahetdinova, V.V. Osiko: High temperature disproportionation of fianit, Izv. AS USSR, Inorg. Mater. **19**, 100–103 (1983), in Russian
14.162 L. Thome, J. Fradin, J. Jagielski, A. Gentils, S.E. Enescu, F. Garrido: Radiation damage in ion-irradiated yttria-stabilized cubic zirconia single crystals, Eur. Phys. J. Appl. Phys. **24**, 37–48 (2003)
14.163 T. Hojo, H. Yamamoto, J. Aihara, S. Furuno, K. Sawa, T. Sakuma, K. Hojou: Radiation effects on yttria-stabilized zirconia irradiated with He or Xe ions at high temperature, Nucl. Instrum. Methods Phys. Res. B **241**, 536–542 (2005)
14.164 R. Mevrel, J.-C. Laizet, A. Azzopardi, B. Leclercq, M. Poulain, O. Lavigne, D. Demange: Thermal diffusivity and conductivity of $Zr_{1-x}Y_xO_{2-x/2}$ ($x = 0$,

0.084 and 0.179) single crystals, J. Eur. Ceram. Soc. **24**, 3081–3089 (2004)
14.165 J. Martinez-Fernandez, A.R. Pínto Gómez, J.J. Quispe Cancapa, A.R. de Arellano López, J. Llorca, J.Y. Pastor, S. Farmer, A. Sayir: High-temperature plastic deformation of Er_2O_3-doped ZrO_2 single crystals, Acta Mater. **54**, 2195–2204 (2006)
14.166 A.H. Heuer: Indentation induced crackes and the toughness anisotropy of 9.4 mol yttria-stabilised cubic zirconia single crystals, J. Am. Ceram. Soc. **74**, 855–862 (1991)
14.167 A. Dominguez-Rodriguez, K.P.D. Lagerlof, A.H. Heuer: Plastic deformation and solid solution hardening of Y_2O_3-stabilized ZrO_2, J. Am. Ceram. Soc. **69**, 281–284 (1986)
14.168 D. Holmes, A.H. Heuer, P. Pirouz: Dislocation structure around Vickers indents in 9.4 mol Y_2O_3-stabilized cubic ZrO_2 single crystals, Philos. Mag. A **67**, 325–342 (1993)
14.169 A. Pajares, F. Guiberteau, A. Dominguez-Rodriguez, A.H. Heuer: Microhardness and fracture toughness anisotropy in cubic zirconium oxide single crystals, J. Am. Ceram. Soc. **71**, C332–C3331 (1988)
14.170 A. Pajares, F. Guiberteau, A. Dominguez-Rodriguez, A.H. Heuer: Indentation-induced cracks and toughness anisotropy of 9.4 mol%-yttria-stabilized zirconia cubic single crystals, J. Am. Ceram. Soc. **74**, 859–862 (1991)
14.171 K.J. McClellan, A.H. Heuer, L.P. Kubin: Localized yielding during high temperature deformation of Y_2O_3-fully-stabilized cubic ZrO_2 single crystals, Acta Mater. **44**, 2651–2662 (1996)
14.172 F. Guiberteau, F.L. Cumbrera, A. Dominguez-Rodrigues, E. Fries, J. Castaing: X-ray Berg–Barrett topography of the deformation substructure of stabilized zirconium oxide single crystals deformed at 1673 K, J. Appl. Cryst. **27**, 406–410 (1994)
14.173 D. Gomez-Garcia, J. Martinez-Fernandez, A. Dominguez-Rodriguez: Recent advances in electron-beam-induced damage models in yttria fully stabilized zirconia single crystals, Phil. Mag. Lett. **81**, 173–178 (2001)
14.174 M. Bartch, A. Tikhonovsky, U. Messerschmidt: Plastic deformation of yttria stabilized cubic zirconia single crystals II: Plastic instabilities, Phys. Status Solidi (a) **201**, 46–58 (2004)
14.175 G.A. Gogotsi, D.Y. Ostrovoi, E.E. Lomonova: Deformation behavior of cubic ZrO_2 single crystals, Refractories **3**, 15–19 (1992), in Russian
14.176 R.P. Ingel, D. Lewis III: Elastic anisopropy in zirconia single crystals, J. Am. Ceram. Soc. **71**, 265–271 (1988)
14.177 I.L. Chistyi, I.L. Fabelinskij, V.F. Kitaeva, V.V. Osiko, Y.V. Pisalevskij, I.M. Sil'vestrova, N.N. Sobolev: Experimental study of the properties of ZrO_2-Y_2O_3 and HfO_2-Y_2O_3 solid solution, J. Raman Spectrosc. **6**, 183–192 (1977)
14.178 C.F. Bolling, W.A. Tiller: Growth from the melt. II. Cellular interface morphology, J. Appl. Phys. **31**, 2040–2045 (1960)
14.179 V.G. Fomin, M.G. Mil'vidskii, R.S. Beletskaja: Study of heavy doped silicon single crystals with cellular structure, Crystallogr. Rep. **13**, 172–173 (1968), in Russian
14.180 A.N. Kirgintsev, L.I. Isaenko, V.A. Isaenko: *Impurity Distribution Under Direct Crystallization* (Nauka, Novosibirsk 1977), in Russian
14.181 A.H. Heuer, V. Lantery, A. Dominguez-Rodriguez: High-temperature precipitation hardening of Y_2O_3 partially-stabilized ZrO_2 (Y-PSZ) single crystals, Acta Metall. **37**, 559–567 (1989)
14.182 G.A. Gogotsi, E.E. Lomonova, V.G. Pejchev: Strength and fracture toughness of zirconia crystals, J. Eur. Ceram. Soc. **11**, 123–132 (1993)
14.183 G.A. Gogotsi, E.E. Lomonova, Y. Furmanova, I.M. Savitskaya: Zirconia crystals suitable for medicine: 1. Implants, Ceramurg. Int. **20**, 343–348 (1994)
14.184 V.I. Aleksandrov, S.H. Batygov, Y.K. Voron'ko, M.A. Vishnyakova, V.F. Kalabuchova, E.E. Lomonova, V.V. Osiko: Effect Pr_2O_3 on ZrO_2 crystal growth frim melt, Izv. AS USSR, Inorg. Mater. **23**, 349–352 (1987), in Russian
14.185 M.A. Borik, E.E. Lomonova, V.V. Osiko, V.A. Panov, O.E. Porodinkov, M.A. Vishnyakova, Y.K. Voronko, V.V. Voronov: Partially stabilized zirconia single crystals: growth from the melt and investigation of the properties, J. Cryst. Growth **275**, e2173–e2179 (2005)
14.186 V.V. Alisin, M.A. Borik, E.E. Lomonova, A.F. Melshanov, G.V. Moskvitin, V.V. Osiko, V.A. Panov, V.G. Pavlov, M.A. Vishnjakova: Zirconia-bazed nanocrystalline synthesized by directional crystallization from the melt, Mater. Sci. Eng. C **25**, 577–583 (2005)
14.187 M.A. Borik, Y.K. Voron'ko, E.E. Lomonova, V.V. Osiko, V.A. Sarin, G.A. Gogotsi: Mechanical properties of the PSZ crystal grown by skull melting technique: influence of technology conditions. In: *Fracture Mechanics of Ceramics, Vol.13, Crack-Microstructure Interaction, R-curve Behavior, Environmental, Effects in Fracture, and Standardization*, ed. by R.S. Bradt, D. Munz, M. Sakai, V.Y. Shevchenko, K. White (Kluwer/Plenum, New York, Boston, London, Moscow 1999), pp.485–496
14.188 G.A. Gogotsi, E.E. Lomonova, V.V. Osiko: Mechanical characteristics zirconia single crystals for constructional applications, Refractories **8**, 14–17 (1991), in Russian
14.189 G.A. Gogotsi, V.I. Galenko, B.A. Ozerskii, E.E. Lomonova, V.A. Mizina, V.F. Kalabukhova: Prochno strength and fracture toughness of zirconia single crystals with yttrum and terbium oxides, Refractories **6**, 2–8 (1993), in Russian

14.190 S.N. Dub, G.A. Gogotsi, E.E. Lomonova: Hardness and fracture toughness of tetragonal zirconia single crystals, J. Mater. Sci. Lett. **1**, 446–449 (1995)

14.191 G.A. Gogotsi, S.N. Dub, B.A. Ozerskii, D.Y. Ostrovoi, G.E. Khomenko, S.H. Batygov, M.A. Vishnjakova, V.F. Kalabukhova, S.V. Lavrishchev, E.E. Lomonova, V.A. Mizina: Zirconia single crystals with yttrum and terbium oxides, Refractories **7**, 2–10 (1995), in Russian

14.192 V.V. Alisin, K. Amosova, V.P. Voitsitskii, V.V. Voronov, V.A. Grinenko, E.E. Lomonova, N.I. Medvedovskaya, V.I. Ustinov: Influence of temperature on oxygen redistribution in nanocrystal-zirconia-based materials with high mechanical characteristics. In: *Perspektivnie Materiali I Tehnologii: Nanokompozity*, Vol. 2, ed. by A.A. Berlin, I.G. Assovsky (Toruss, Moscow 2006) pp. 183–193

14.193 J. Martinez-Fernandez, M. Jimenez-Melendo, A. Dominguez-Rodriguez: Microstructural evalution and stability of tetragonal precipitates in Y_2O_3-partially stabilized ZrO_2 single crystals, Acta Metall. Mater. **43**, 593–601 (1995)

14.194 J. Martinez-Fernandez, M. Jimenez-Melendo, A. Domiguez-Rodriguez, A.H. Heuer: Microindentation-induced transformation in 3.5 mol yttria-partially-stabilized zirconia single crystals, J. Am. Ceram. Soc. **74**, 1071–1081 (1991)

14.195 V.V. Osiko, V.V. Alisin, M.A. Vishnjakova, Z.V. Ignat'eva, E.E. Lomonova, V.G. Pavlov: Tribological properties of nanocrystalline material based on zirconia, Frict. Wear **26**, 285–289 (2005), in Russian

14.196 V.V. Osiko, V.V. Alisin, M.A. Vishnjakova, Z.V. Ignat'eva, E.E. Lomonova, V.G. Pavlov: Effect Y_2O_3 stabilisator content on tribological properties of nanocristalline material based on zirconia, Zavod. Lab. Diagn. Mashin **4**, 47–52 (2006), in Russian

14.197 K.V. Frolov, V.V. Osiko, V.V. Alisin, M.A. Vishnjakova, Z.V. Ignat'eva, E.E. Lomonova, A.F. Melshanov, G.V. Moskvitin, V.G. Pavlov, M.S. Pugachev: Mechanical and tribological properties of nanocrystalline material based on zirconia, Probl. Mashinostr. Nadezn. Mashin **4**, 3–8 (2006), in Russian

14.198 P. Christel, A. Meunir, M. Heller, Y.P. Torre, C.N. Peille: Mechanical properties and short-term in vivo evaluation of yttrium oxide-partially-stabilized zirconia, J. Biomed. Mater. Res. **23**, 45–61 (1989)

14.199 C. Piconi, G. Maccauro: Zirconia as a ceramic biomaterial, Biomaterials **20**, 1–25 (1999)

14.200 V. Nezhentsev, Y. Petrov, A. Zhilin, O. Dymshits: Use of induction furnaces with a cold crucible for melting hard glasses (review), Steklo Keram. **9**, 9–11 (1986)

14.201 V.I. Aleksandrov, M.A. Borik, G.H. Dechev, N.I. Markov, V.A. Myzina, V.V. Osiko, V.M. Tatarintsev, R.Y. Khodakovskaja: Synthesis and investigations of La_2O_3-Al_2O_3-SiO_2 glasses, Glass Phys. Chem. **6**, 170–173 (1980), in Russian

14.202 B.À. Sakharov, Ò.S. Sedyh, Å.P. Rashevskaja: Study of system Y_2O_3-SiO_2-Al_2O_3, Trudy Giredmet **52**, 83–87 (1974), in Russian

14.203 V.I. Aleksandrov, M.A. Borik, V.B. Glushkova, R.E. Krivosheev, N.I. Markov, V.V. Osiko, V.M. Tatarintsev: Synthesis and some properties of refractory glasses in $R_2O_3 - Al_2O_3 - SiO_2$ system, Glass Phys. Chem. **3**, 177–180 (1977), in Russian

14.204 A. Makishima, Y. Tamura, T. Sakaino: Elastic moduli and refractive indices of aluminosilicate glasses containing Y_2O_3, La_2O_3 and TiO_2, J. Am. Ceram. Soc. **61**, 247–249 (1978)

14.205 A. Makishima, T. Shimohira: Alkaline durability of high elastic modulus aluminosilicate glasses containing Y_2O_3, La_2O_3 and TiO_2, J. Non-Cryst. Solids **39/40**, 661–666 (1980)

14.206 G.E. Malashkevich, V.N. Tadeush, V.V. Kuznetsova, A.K. Cherches, N.I. Bliznyuk, V.G. Mikhalevich, M.B. Rzhevskii: Physicochemical and spectral luminescence properties of a glass based on thf SiO_2-Al_2O_3-La_2O_3-Nd_2O_3 system, J. Appl. Spectrosc. **37**, 926–929 (1982)

14.207 J. Shelby, S. Minton, C. Lord, M. Tuzzolo: Formation and properties of yttrium aluminosilicate glasses, Phys. Chem. Glasses **33**, 93–98 (1992)

14.208 J. Kohli, J. Shelby, J. Frye: A structural investigations of yttrium alumosilicate glasses using ^{29}Si and ^{27}Al magic angle spinning nuclei-magnetic resonance, Phys. Chem. Glasses **33**, 73–78 (1992)

14.209 T.T. Basiev, Y.K. Voron'ko, V.V. Osiko, A.M. Prokhorov: Laser spectroscopy doped crystals and glasses. In: *Spectroscopy of Crystals*, ed. by A.A. Kaplyansky (Leningrad, Nauka 1983) pp. 57–82, in Russian

14.210 M. Weber: Laser exited fluorescence spectroscopy in glass. In: *Laser Spectroscopy of Solids*, ed. by W.M. Yen, P.M. Selzer (Springer, Berlin-Heidelberg, New-York 1981) pp. 189–239

14.211 G.E. Malashkevich, N.N. Ermolenko, V.I. Aleksandrov, M.A. Borik, G.M. Volokhov, A.S. Gigevich, G.A. Denisenko, A.V. Mazovko, V.N. Tadeush: Spectral-luminescent and thermomechanical characteristics of silicate-borate glasses activated with Yb^{3+} and Er^{3+} ions, Izv. AS USSR, Inorg. Mater. **6**, 1053–1055 (1987), in Russian

14.212 Y.I. Matunin, A.V. Demin, T.V. Smelova: Behavior of uranium and rare-earth elements in glasses synthesized in an induction melter with a cold crucible, At. Energy **83**, 795–800 (1997)

14.213 N.D. Musatov, V.G. Pastushkov, P.P. Poluektov, T.V. Smelova, L.P. Sukhanov: Compaction of radioactive thermal-insulation materials and construction debris by melting in a cold crucible, At. Energy **99**, 602–606 (2005)

14.214 K. Guilbeau, A. Giordana, W. Ramsey, N. Shulyak, A. Aloy, R. Soshnikov: Induction-melting technology, Am. Ceram. Soc. Bull. **83**, 38–40 (2004)

15. Crystal Growth of Laser Host Fluorides and Oxides

Hongjun Li, Jun Xu

Following the discovery of the first laser action based on ruby, hundreds of additional doped crystals have been shown to lase. Among those, many crystals, such as $Ti:Al_2O_3$, $Nd:Y_3Al_5O_{12}$, $Nd:YVO_4$, $Yb:Y_3Al_5O_{12}$, $Yb:Ca_5(PO_4)_3F$, and Cr:LiCAF have come to practical application, and are being widely used in scientific research, manufacturing and communication industries, military applications, and other fields of modern engineering. These crystals are mainly oxides and fluorides, which are grown from melt. This chapter reviews the major results obtained during recent years in the growth of various crystalline oxides and fluorides for laser operation, with emphasis on crystals doped with the additional ions Ti^{3+}, Nd^{3+}, and Yb^{3+}. On the other hand, special attention is paid to discuss the elimination of growth defects in these crystals. Limited by the length of this chapter, for each crystal, only outstanding defects are considered herein.

15.1 Crystal Growth of Laser Fluorides and Oxides from Melt

The development of many advanced fields in modern engineering is, to a large extent, governed by the success achieved in the techniques of crystal growth. On the other hand, the demand for new crystals for advanced technologies has stimulated the improvements in growth techniques of these crystals. Certainly, this concerns only a few very important crystals.

In 1960, laser action was first demonstrated on the basis of a ruby laser rod (or Cr^{3+}-doped Al_2O_3) [15.1], indicating the birth of a new subdiscipline of science: the laser technique. In the 1970s, the discovery of Nd^{3+}-doped yttrium aluminum garnet (Nd:YAG) [15.2] crystal prompted the rapid development of the solid-state laser. In the 1980s, the invention of Ti^{3+}-doped sapphire ($Ti:Al_2O_3$) [15.3] made it possible to produce ultra-intense and ultrafast lasers, based on which the intralase technique has been employed widely in both basic and applied science fields. In the 1990s, the successful fabrication of $Nd:YVO_4$ [15.4] led to a new period of development for the laser technique:

the all-solid-state laser. Simultaneously, the advent of cheap diode-laser pump sources brought about the practical application of Yb-doped materials such as Yb:YAG [15.5] and Yb:$Sr_5(PO_4)_3F$ (Yb:S-FAP) [15.6], which have shown great capability in high-power laser systems. Entering the 21st century, research into laser materials is expanding extensively in the forms of single crystal, glass, ceramics, and fiber. Owing to their advantages of simple fabrication processes, lower costs, and greater perfection, laser fiber [15.7] and polycrystalline ceramics [15.8] are challenging the dominant status of traditional materials.

In spite of the hundreds of additional doped crystals that have been shown to lase, few crystals have proven to be useful in practical circumstances. In fact, the gap between demonstration exercises of laser action and the engineering of practical systems is often wide and difficult to bridge.

In order to develop and test new laser crystals, it is first crucial to identify the appropriate means by which to grow and fabricate the crystal. To further optimize the crystal growth conditions, it is necessary to understand the physical properties of the melt and solid. In order to improve the crystal quality, it is important to clarify the formation mechanism of growth defects in the crystal so as to find proper ways to eliminate them. While the methods and issues associated with crystal growth occupy the bulk of this chapter, we begin with a discussion of the important growth techniques and their modifications for laser crystal growth. Our attention is focused on the insights and issues involved in the development, melt-growth, and defect elimination of the main laser oxides and fluorides. Since our approach will entail discussion from the perspective of laser materials, only crystals doped with the most attractive additional ions, such as Ti^{3+}, Nd^{3+}, and Yb^{3+}, will be addressed.

15.1.1 Laser Crystal Growth from the Melt

After the development of the first laser, which was based on a Verneuil-grown ruby, almost all of the classical crystallization techniques have been applied to analyze growth characteristics and provide high-quality laser crystals and the development of new laser crystals. In the 1960s, flux methods were widely used to grow the rare-earth aluminum and gallium garnets, both pure and doped, as well as some of the rare-earth orthoaluminate hosts. The emphasis later was on melt techniques, as they provided both larger crystal size and faster growth rates. Among these, the Czochralski technique (CZ) employing iridium crucibles and radiofrequency (RF) heating was the most adaptable and widely used for the majority of high-temperature laser oxides, while the vertical Bridgman technique (VBT) was commonly used for laser fluorides. On the other hand, static growth techniques, such as the heat-exchange method (HEM) and temperature gradient technique (TGT), have been becoming efficient methods for the growth of large-sized laser crystals, while two other promising techniques, the micro-pulling-down method (μ-PD) and the laser-heated pedestal growth method (LHPG), are especially used for fiber crystal growth.

15.1.2 Czochralski Technique (CZ)

The CZ method is named after *Jan Czochralski* who introduced an early version of the present-day process in 1916, and published it as a method for studying the crystallization rate of metals [15.9]. Further modifications by *Teal* and *Little* [15.10] brought the technique closer to the process known today as the Czochralski, or CZ, method. The application of this technique and its consequent development was stimulated by the invention of the laser in 1960, since when it has been used for oxide crystal production [15.11, 12].

The CZ furnace geometry is relatively simple, as shown in Fig. 15.1, and is usually constructed of either Al_2O_3 or stabilized ZrO_2 ceramics. The crucible can be either Pt or Ir, depending upon the melting point of the material and/or the atmosphere required for the growth process.

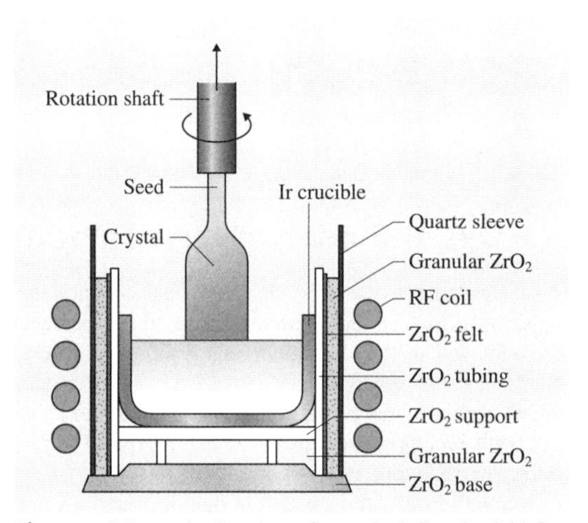

Fig. 15.1 Schematic drawing of a typical Czochralski furnace used for the growth of oxide-based laser crystals

The crucible is heated by means of an RF generator operating in the range of 10–400 kHz. With the development of solid-state RF generators, the trend is toward lower frequencies, preferably in the range 10–30 kHz. These lower frequencies have the distinct advantage of producing more uniform heating of crucible, which is necessary for the growth of large-diameter crystals, and in the case of the higher-melting oxide systems that must use ZrO_2 insulation, avoiding coupling of the RF energy directly into the ceramic parts of the furnace.

The growth and control of the crystal is dependent upon the radial temperature gradient in the liquid and the vertical gradient through the length of the furnace. For smaller growth systems (typically using crucibles less than 75 mm in diameter), the adjustment of the radial and vertical thermal gradients can be easily accomplished by varying the ceramic insulation and its dimension. However, once beyond this size, a suitable liquid radial gradient becomes more difficult to establish, thereby making seeding and the initial cone section of the crystal more difficult to control, hence the need for an automatic diameter control system. Although several methods of diameter control have been developed [15.14–17], the one mostly widely used is that of weighing the crystal while it is growing. In this case, a load cell with a sensitivity of ± 0.1 g is placed in line with the rotation shaft to which the seed crystal is attached. As the crystal is pulled from the melt, the change in weight is used to generate a control signal that modifies the generator output power to the crucible, thereby controlling the diameter through small changes in the liquid temperature. Typical systems are easily capable of maintaining a 75 mm-diameter crystal to within ± 1 mm.

The CZ growth process is, by far, the most commonly used process for the growth of laser materials and is employed over a wider span of materials and melting points than any other melt-growth method, ranging from mixed compounds such as nitrates, through germinates, fluorides, molybdates, tantalates, and garnets, to single oxides such as sapphire and yttria. To circumvent various kinds of problems encountered during crystal growth, numerous modifications have been made to improve the standard CZ growth process.

During the growth of Nd:YAG crystals, the Nd segregation problem always exists. Nd concentration increases as crystal growth proceeds so that, if we want to keep uniformity of concentration within 0.1 at. %, solidification is restricted to 20–30% of melt in crucible. To achieve higher solidification, *Katsurayama* and coworkers [15.13] developed a double-crucible method with automatic powder supply (Fig. 15.2). In this system, material powder is continuously supplied into the crucible to maintain the concentration of Nd in the melt. As a result, the Nd concentration fluctuation in the grown crystal has been reduced to as low as 0.02 at. % when 30% of melt is solidified. On the other hand, *Kanchanavaleerat* et al. [15.18] modified the growth parameters to grow highly doped Nd:YAG crystals with excellent optical quality, as shown in Fig. 15.3, and the laser testing shows that the highly doped Nd:YAG rods have very good efficiency due to low passive losses.

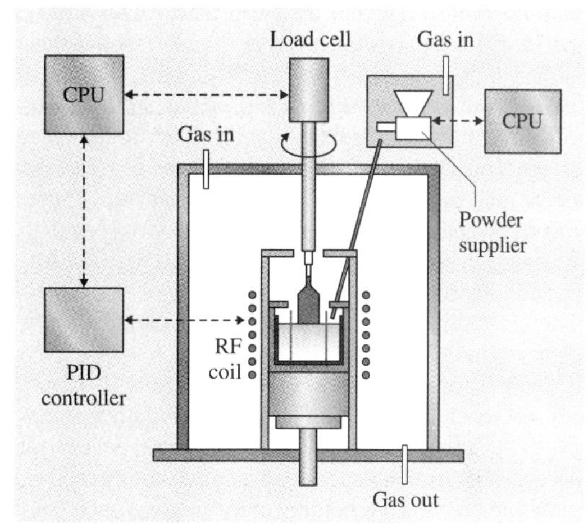

Fig. 15.2 Schematic diagram of crystal growth apparatus with automatic powder supply system [15.13] (CPU – central processing unit, PID – proportional–integral-derivative)

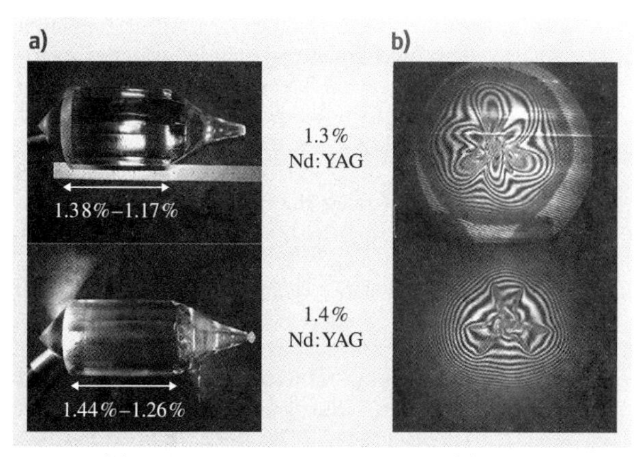

Fig. 15.3 (**a**) Highly doped Nd:YAG crystals, (**b**) Zygo interferogram

Boulon et al. [15.19] indicated that Y_2O_3, Sc_2O_3, and Lu_2O_3 sesquioxides have the highest laser potentialities in the continuous wave (CW) regime. However, due to their high melting points, about 2420 °C, the growth of the sesquioxides is rather difficult. The major problem is that the high temperature restricts the choice of crucible materials, which has to be mechanically stable at the melting point and resistant to chemical reactions with the melt as well as with the surrounding thermal insulation. The results of *Fornasiero* et al. [15.20] reveal that rhenium fulfills these requirements to the greatest extent. Rhenium is sensitive to oxidizing atmospheres but resistant to melts of Al_2O_3 and rare-earth oxides. The melting point of rhenium is 3180 °C. To avoid the react of the hot crucible and the ceramic zirconia insulation at high temperatures, Fornasiero designed a holding construction which consisted of rhenium rods directly welded to the crucible. The crucible was suspended in a thermally insulating tube of zirconia felts so that it was completely surrounded by gas. To reduce the radiation losses from the free melt surface, rings and funnels were inserted into the crucible a few millimeters above the melt. Through the above modifications of the conventional CZ configuration, several crystals with typical length of 5 mm and diameter of 10 mm have been grown [15.20].

Shimamura et al. reported the growth of Ce:LiCAF crystals without either the use of HF gas or the hydrofluorination of raw materials [15.21]. Instead, a growth chamber was evacuated to about 10^{-2} Torr prior to growth, and high-purity Ar (99.9999%) gas was used as a growth atmosphere. Under these conditions, a deposit of a white foreign substance, composed of volatile fluorides and oxyfluoride, was found on the surface of the grown crystals. In order to avoid its formation and to grow high-quality crystals with higher reproducibility, several modifications such as high-vacuum atmosphere prior to growth, use of CF_4 gas instead of Ar, and growth with a low temperature gradient, were required [15.21]. Such modifications are also needed for the growth of other fluorides, e.g., $LiYF_4$ (YLF), $BaLiF_3$, and CaF_2 [15.22, 23].

15.1.3 Temperature Gradient Technique (TGT)

The temperature gradient technique (TGT) is a typical static directional solidification technique, which was invented by *Zhou* et al. [15.24] in 1979; Shanghai Institute of Optics and Fine Mechanics (SIOM) obtained a patent on TGT in 1985 [15.25].

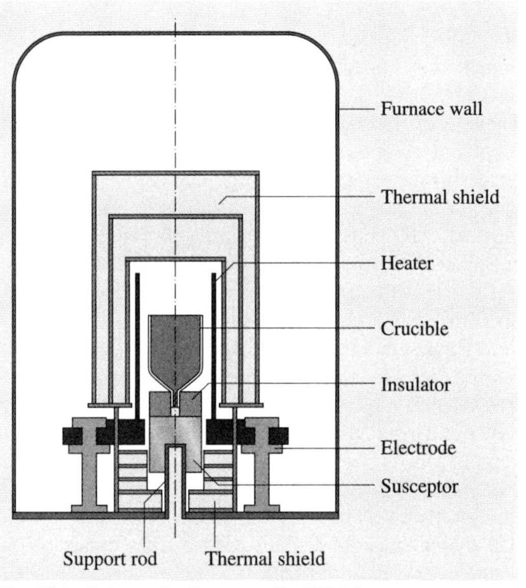

Fig. 15.4 Schematic diagram of a TGT furnace

A schematic diagram of the TGT furnace is shown in Fig. 15.4. It consists of a molybdenum crucible, graphite heating element, and molybdenum heat shields. The cylindrical graphite heating element is designed as an electric circuit with appropriate linear resistance from the top to the middle by making holes with a certain distribution. The cylinder is placed in the graphite electrodes, which are cooled by water. The temperature gradient of the upper part is produced by the linear resistance of the heating element, whereas that in the lower part depends on the extraction of heat by water flowing in the tubes through the electrodes. Besides these, the temperature field near the seed is influenced by the heat conductivity of the water-cooled center rod. In TGT, there are no moving parts. The growth process is accomplished by dropping the temperature at designed rates with a high-precision temperature program controller.

Large-sized laser oxide and fluorides, such as Ti:sapphire [15.26], Nd:YAG [15.27], Yb:$YAlO_3$ (Yb:YAP) [15.28], and Na,Yb:CaF_2 [15.29] shown in Fig. 15.5, have been grown successfully by the TGT method. Taking the growth of Ti:sapphire as an example, the basic growth process of TGT can be described as follows. A cylindrical seed with the selected orientation is placed into the seed hole of the molybdenum crucible, and the high-purity starting materials are placed in the crucible. The furnace is loaded for the growth process, outgassed to $(2-4)\times10^{-3}$ Pa; when the temperature at the bottom of crucible is about 1400 °C, high-purity Ar is fed into furnace; then the material is melted and kept molten for several hours. After the temperature field has stabilized, crystallization is started by slow cooling (1.3–3 K/h). The solid–liquid interface advances upwards as the temperature drops. The whole crystallization process is completed automatically. During the growth process, it is essential that the temperature and thermal field are very stable, and an important factor is the flow stability of the circulating cooling water so that the solid–liquid interface advances with constant velocity.

The typical size of TGT-grown Ti:sapphire boules is $\varnothing 110\times 80\,\text{mm}^2$, the titanium doping levels are between 0.05 and 0.52 wt %, the absorption coefficient at 490 nm is from 1 to 7.5 cm^{-1}, the absorption coefficient can be as high as 10 cm^{-1} in some case (the theoretical limit is 11 cm^{-1} for absorption coefficient at 490 nm), and the figure of merit (FOM) of Ti:sapphire crystals is in the range 150–400. Besides the mentioned advantages, there are other advantages of TGT-grown Ti:sapphire, such as low dislocation density, low scattering, and high perfection, which are very important factors determining laser performance [15.31].

For TGT-grown Nd:YAG crystals, high doping concentration of neodymium is a distinct character. Highly doped (2.5 at. %) Nd:YAG shows high absorption coefficients at the 808 nm laser diode (LD)-pumping wavelength up to 7.55 cm^{-1}, nearly three times higher than 1 at. % Nd:YAG [15.32]. Therefore, a short crystal length (e.g., 1 mm) is preferred, and compact microchip lasers can be constructed by using 2.5 at. %-doped Nd:YAG. Almost the same output has been achieved preliminarily in both (111)-cut 1 mm-long Nd:YAG and *a*-cut 1 mm-long YVO_4 microchip lasers with a very short (9 mm) laser cavity. In particular, the broader and smoothly varying absorption bandwidth allows less stringent requirements on temperature control.

In the growth of $Yb{:}CaF_2$ crystals, it is notable that a lid was employed to seal the crucible to reduce volatilization of the materials. Sometimes, small amounts of PbF_2 or ZnF_2 acting as scavengers were added to the CaF_2 raw materials to eliminate residual moisture prior to growth.

15.1.4 Heat-Exchanger Method (HEM)

The heat-exchanger method (HEM) for growing large sapphire boules was invented by *Schmid* and *Viechnicki* at the Army Materials Research Lab in Watertown, MA, in 1967 [15.33]. The modern implementation of *Schmid* and *Viechnicki*'s heat-exchanger method at Crystal Systems in Salem, MA, is shown in Fig. 15.6 [15.30].

It consists of a water-cooled chamber containing a well-insulated heat zone. A high-temperature heat exchanger is introduced from the bottom of the furnace

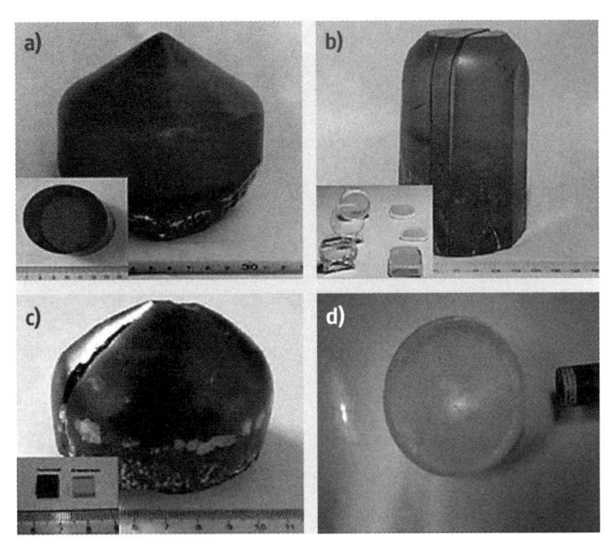

Fig. 15.5a–d Crystal boules and fabricated samples of Ti:sapphire (**a**), Nd:YAG (**b**), Yb:YAP (**c**) and Na, $Yb{:}CaF_2$ (**d**) grown by TGT

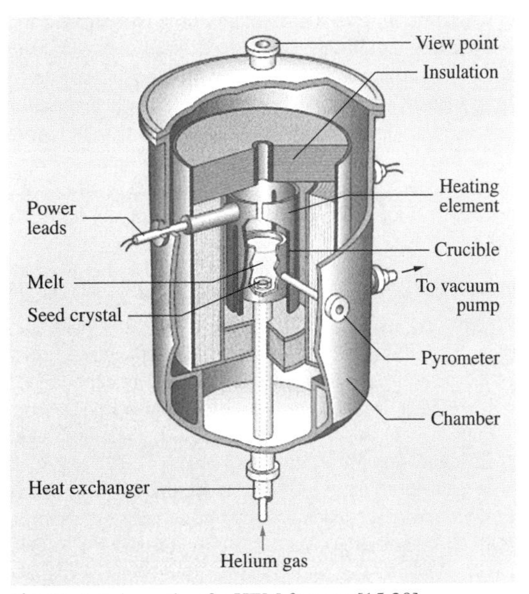

Fig. 15.6 Schematic of a HEM furnace [15.30]

Fig. 15.7 A 10 cm-diameter Ti:Al_2O_3 crystal grown by HEM [15.34]

into the heat zone. The heat exchanger is a closed-end tube with an injection tube through which a controlled flow of coolant gas is introduced. A crucible with a seed crystal positioned at the bottom and loaded with charge is placed on top of the heat exchanger. The furnace chamber can be evacuated with a vacuum pump or backfilled with a gas for controlled atmosphere processing. Heat is supplied by the graphite resistance to melt the charge; the seed is prevented from melting by flowing minimal coolant gas through the heat exchanger. After partial melting of the seed, gas flow is increased to cool the seed and initiate crystallization of melt onto the seed. The furnace is held at constant temperature during growth of the crystal, which proceeds out from the seed in three dimensions. After crystallization is complete, the furnace temperature and the helium flow are decreased and the boule is slowly annealed in situ. The long, slow cool-down produces crystals of the highest quality.

HEM has been successfully utilized for the growth of the world's largest sapphire boules with diameter of 340 mm and mass of 82 kg [15.36]. By this method, more than 95% of the melt can be converted into high-quality crystalline material. HEM was capable of growing laser crystals such as Ti:sapphire (as shown in Fig. 15.7) [15.34], Co:MgF_2 [15.34, 37], Nd:YAG [15.38], and Ti:YAP [15.39].

Recently, a modified HEM technique has been employed in the production of high-melting (about 2400 °C) sesquioxides, such as Yb:Sc_2O_3 and Yb:Y_2O_3 [15.35]. Due to the high melting point (m.p.) of yttria and scandia, rhenium (m.p. 3180 °C) was selected

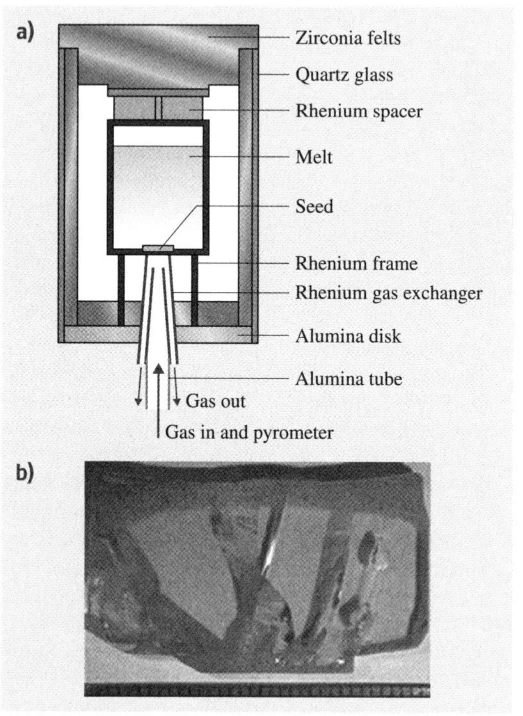

Fig. 15.8a,b Insulation HEM setup for the growth of high-melting sesquioxide crystals (a) and the grown Sc_2O_3 crystal (b) [15.35]

as the crucible material, and the crucible was completely surrounded by gas since no insulation materials could be found that would be stable when in direct contact with the crucible at the required temperature. The growth setup is shown in Fig. 15.8a. During the growth process, the crucible was heated by an RF generator under an atmosphere of 0.01% O_2, 10–15% H_2, and 85–90% N_2 at pressure of 1 bar. Eventually, crystals with vertical volumes of small, single crystals were obtained (Fig. 15.8b), owing to the use of a polycrystalline seed.

15.1.5 Vertical Bridgman Technique (VBT)

The vertical Bridgman technique (VBT) was first used by *Bridgman* in 1925 [15.40] and especially exploited by *Stockbarger* in 1936 and 1949 [15.41, 42]. This technique is commonly known as the Bridgman–Stockbarger method, although sometimes the names of Tammann (1925) and Obreimov (1924) are associated with the technique. *Buckly* [15.43] discussed the histor-

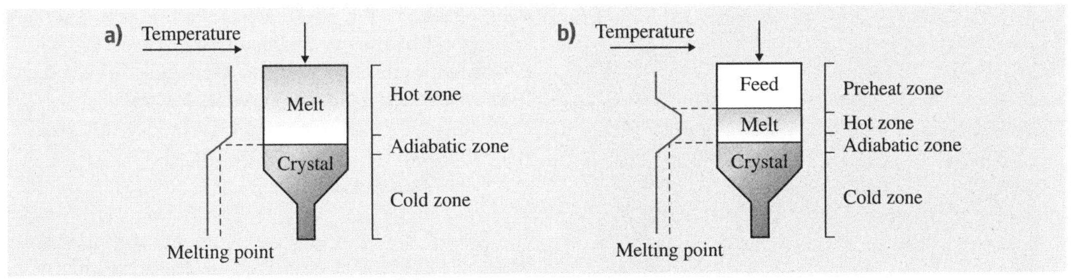

Fig. 15.9a,b A classical (a) and a modified (b) vertical Bridgman process

ical aspects of the technique and assigned the name to the original researcher.

In the Bridgman process, an ampoule containing the materials to be crystallized is translated from a zone hotter than the melting point, through a temperature gradient, to a zone cooler than the melting point in order to solidify the material (Fig. 15.9a). Equivalently, the temperature can be translated through a stationary ampoule by moving the furnace relative to the ampoule.

The vertical Bridgman technique has been applied to the growth of oxide crystals since the late 1960s [15.44]. In apparatus for this purpose a resistively heated tungsten furnace and crucibles (tubes) fabricated from molybdenum are used. Several complex oxide crystals have been grown by this technique, including Nd:YAG, and the first lasing material Nd:$YAlO_3$. This technique has proved to have great potential in both materials research and in the production of oxide laser crystals. Various Bridgman-grown crystals possessing application quality are now available, including Nd:YAG, Er:YAG, Tm^{3+},Cr^{3+}:YAG, Tm^{3+},Cr^{3+},Ho^{3+}:YAG, Er:LuAG, Pr:$YAlO_3$, Pr:$LuAlO_3$, and Nd:$LuAlO_3$ [15.44]. Improvements of crystalline and chemical perfection of crystals, obtained as a result of better understanding of the basic physical possesses occurring during crystallization in this configuration, have resulted in larger-scale use of this technique for oxide crystal growth. In addition, VBT has been widely used for the growth of large-sized commercial CaF_2 crystals for optical applications [15.45]. Certainly, this technique can be modified for the growth of ion-doped CaF_2 crystals or other fluorides for laser applications.

Figure 15.9b illustrates a modification of the Bridgman technique in which a third zone, cold relative to the melting point, is added to the top of the furnace. This is referred to as the moving melt zone or the traveling heater method. In this case the melt volume is small compared with the ampoule. The advantages of this configuration are twofold. First, steady-state conditions may be established in the small melt volume for growth of crystals with nonunity segregation coefficient to level out changes in dopant concentration along the length of the crystal. Second, the upper interface produces a temperature gradient that drives buoyant convection, thereby increasing mixing in the melt. This can be important for growth of multicomponent systems.

15.1.6 Horizontal Bridgman Technique (HBT)

The horizontal Bridgman technique (HBT), also known as the Bagdasarov method, is a method for refractory single-crystal growth proposed by *Bagdasarov* in 1964 [15.46]. In HBT, presented schematically in Fig. 15.10, the crystallizing material (in the form of powder, crystal crackles or ceramics) is placed in a boat-like crucible, melted by moving the crucible through the heating zone, and then crystallized. To obtain a strictly oriented crystal the single crystal seed is mounted on the top of the crucible (boat) and both the moment of seeding and the formation of the growth front are observed visually.

HBT allows repeated crystallizations to be carried out when additional chemical purification of the raw

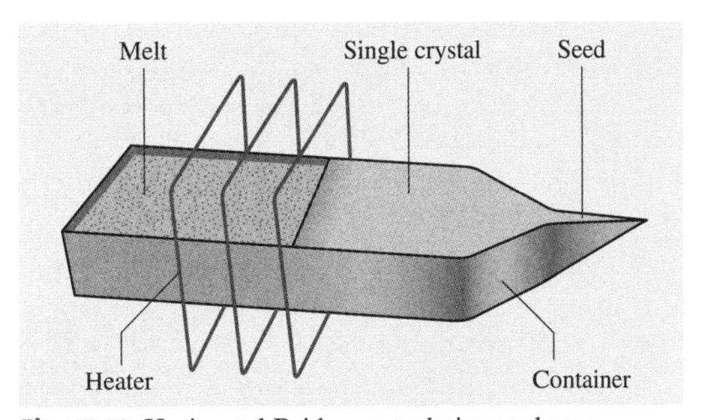

Fig. 15.10 Horizontal Bridgman technique scheme

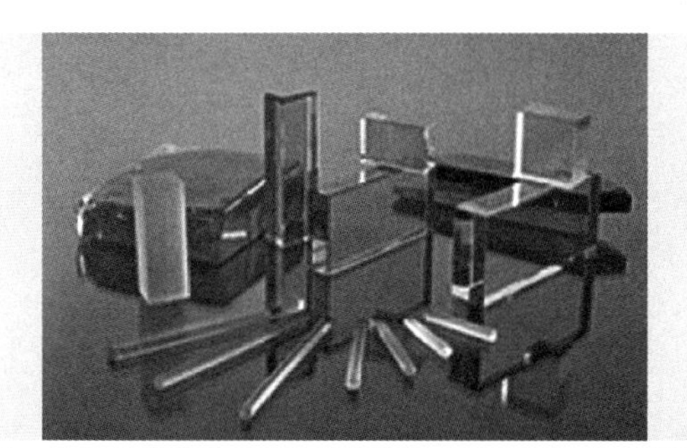

Fig. 15.11 Nd:YAG and Er:YAG crystals grown by HBT [15.46]

material is required. It is also possible to carry out a continuous crystallization process by directed shifting of the crucible echelon through the crystallization zone. With the HBT method it is technically easy to create a controlled temperature field, which is crucial for the growth of high-perfection large-size single crystals. This method makes it possible to obtain large slabs with almost perfect edges and of any given crystallographic orientation.

Initially, this method was designed in order to grow large-size, Nd-doped, high-perfection yttrium–aluminum garnet laser crystals. Later, it turned out to have potential for the growth of yttrium–erbium–aluminum garnet. The crystals have a typical size of $130\times130\times25\,mm^3$ as well as high optical homogeneity, as shown in Fig. 15.11. Recently, it is reported that HBT could be used to produce a new type of Nd:YAG crystal with emission property two times higher than from conventional Nd:YAG crystals [15.46].

15.1.7 Laser-Heated Pedestal Growth Method (LHPG)

Poplawsky [15.48] was the first to initiate crystal growth using a pedestal growth process based on melting materials by the energy created by an image furnace. Then, in 1972, *Haggerty* et al. developed the laser-heated pedestal growth method (LHPG), which was improved later by *Feigelson* [15.49] at Stanford University.

Figure 15.12 illustrates LHPG crystal growth. In this technique a float zone is created at the top of a vertical feed rod by using a focused laser beam. Motion of the float zone is generated by vertical displacement of the feed. During this translation, the feed progressively melts and, behind the float zone, a crystallized rod is formed. The float zone remains in equilibrium with the feed and the crystallized rod due to surface tension. The source rod materials containing the desired host and dopant materials can be used as oriented fiber single crystals or polycrystalline reacted materials prepared by solid-state reaction. A seed crystal, once dipped into the molten zone, is withdrawn at some rate faster than the source material is fed in. By conservation of melt volume, this leads to the crystalline fiber

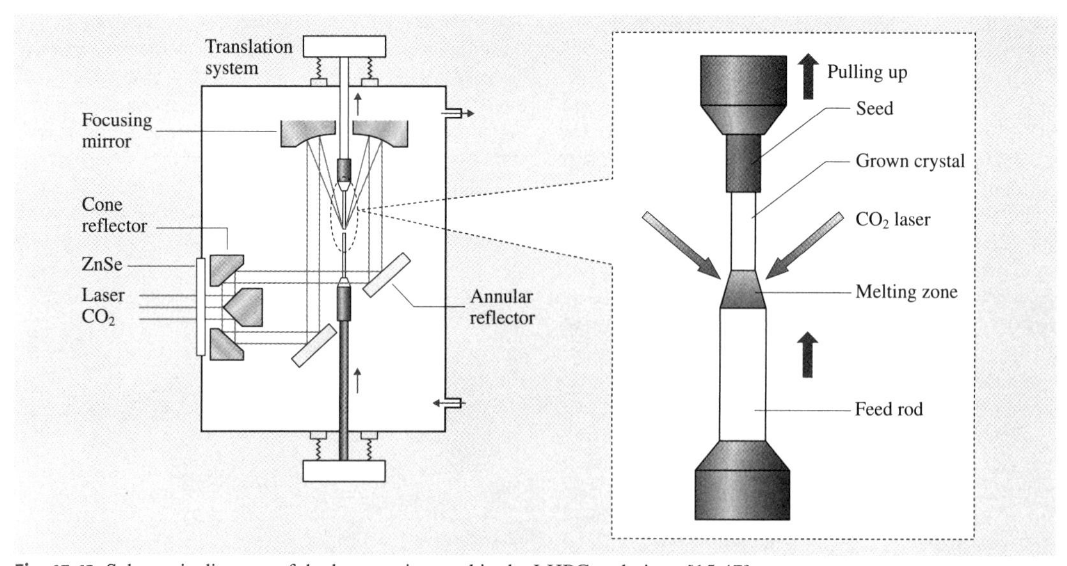

Fig. 15.12 Schematic diagram of the laser optics used in the LHPG technique [15.47]

growing at some constant fraction of the source rod diameter.

A great variety of oxide and fluoride crystal fibers doped with rare-earth and transition metal ion activators in a great range of concentrations have been pulled by the LHPG technique [15.47, 50]. Specially, because of the lack of crucible and the use of a CO_2 laser, the LHPG technique is well adapted to high-refractory crystals such as yttria, scandia, and lutecia sesquioxides, as well as garnets [15.19, 51].

In the combinatorial approach proposed by *Boulon* [15.19], the LHPG technique is applied to a feed rod constituted by two ceramics parts, A and B, with different compositions. When the molten zone moves along the feed rod, there is mutual solubilization of A and B. As the solubilized amounts of A and B vary, the composition of the molten zone changes continuously, inducing a progressive and continuous composition gradient along the crystallized rod. This new approach is an efficient tool for measuring optical, spectroscopic or thermal properties in any type of inorganic optical materials in which either activator ion or nominal composition concentrations can be changed.

15.1.8 Flux Technique (FT)

The flux technique (FT) is based on growth of crystals from a nonaqueous solution. As a rule, a mixture of salts is used as a solvent. In the 1960s, flux methods were widely used to grow the rare-earth aluminum [15.52–54] and gallium [15.55] garnets, both pure and doped, as well as some of the rare-earth orthoaluminate hosts [15.56]. The emphasis later was on melt techniques, as they provided both larger crystal size and higher growth rates. At present, a modified FT method, the top-seeded solution growth method (TSSG), is widely applied to produce those crystals that cannot be grown from melt easily, $Nd:YVO_4$ [15.57], $Yb:KY(WO_4)_2$ (Yb:KYW) [15.58], $Nd:KGd(WO_4)_2$ (Nd:KGW) [15.59], and $Yb:YAl_3(BO_3)_4$ (Yb:YAB) [15.60]. Sometimes, TSSG is also described as a high-temperature solution growth method or a modified Czochralski technique.

Among the Yb^{3+}-doped oxide crystals for diode-pumped solid-state lasers, KYW and KGW were evaluated to be the hosts with the greatest lasing potential in the CW regime [15.19]. FT is obvious the choice for the growth of single crystals such as KGW, since KGW crystal has a phase transition at 1005 °C (KYW at 1014 °C) below its melting point (1075 °C). A typical growth procedure of such crystals is described as follows [15.58].

The starting materials with flux, such as K_2WO_4 or $K_2W_2O_7$, are mixed and placed into the platinum crucible. The fully charged crucible is then placed into the furnace. It is important that the charged crucible be kept at a constant temperature of 1050 °C for 3–4 days to let the solution melt completely and homogeneously. The saturation temperature of the solution is determined exactly by repeated seeding. The seed contacts the melt at a temperature 5 K above the saturation temperature and is kept at constant temperature for half an hour to dissolve the outer surface of the seed. During the growth period, the crystal is slowly cooled at a rate of 1 K/day and rotated at a rate of 4.5 rpm. When the growth process ended, the crystal is pulled up from the melt surface and cooled to room temperature at a rate of 15 K/h.

15.2 Laser Crystal Defects

The main defects which can degrade the performance of melt-grown crystals are now generally recognized to be any which can absorb, reflect, refract or scatter magnetic, optical, acoustic or electrical energy either generated within or incident upon the material. Thus, dislocations, color centers, facet, striations, twins, voids, cellular structures, precipitates, inclusions, and more destructive defects such as cracks are all important defects that have to be eliminated or controlled. Due to the great variety of defects, only those frequently encountered and those specific to the concerned host materials are discussed herein.

15.2.1 Ti:sapphire

Ti:sapphire single crystal is one of the most attractive broadly tunable solid-state laser materials. Both CW and pulsed lasers have been demonstrated with very high efficiency over a tuning range in excess of 300 nm centered at 800 nm. In order to develop this material to meet the needs of current commercial laser systems a variety of growth techniques, such as HEM [15.34], VGF (vertical gradient freeze) [15.61], CZ [15.62], and TGT [15.26], have been used to grow Ti:sapphire crystals. Residual infrared absorption and mosaic structure are the main defects in as-grown Ti:sapphire crystals.

Residual Infrared Absorption

The efficiency at which a laser can operate is fundamentally impacted by its gain-to-loss ratio. As a consequence, the presence of absorptive loss at laser wavelengths can potentially reduce the efficiency, or even render the system inoperable. These parasitic losses generally arise from the presence of unwanted impurities or from unanticipated oxidation states of the laser ion. Parasitic absorption losses can be more harmful to laser performance than an equivalent amount of loss resulting from scattering, since the absorption often leads to additional heating of the medium. Since many solid-state amplifiers operate in the grain regime of 3–30%/cm, losses on the order of 1%/cm can seriously impair laser performance [15.63].

In Ti:sapphire, parasitic losses are mainly induced by the presence of a relatively weak, broad infrared absorption band that peaks between 800 and 850 nm [15.64–67]. *Aggarwal* and coworkers [15.68] proposed that the residual absorption in these crystals is largely due to Ti^{3+}/Ti^{4+} ion pairs, and also indicated that annealing such crystals at high temperatures ($\approx 1600\,°C$) in a reducing atmosphere (an Ar/H_2 mixture) decreased the residual absorption without significantly changing the main absorption, whereas annealing in an oxidizing atmosphere (Ar/O_2 mixture) increased the residual absorption and simultaneously decreased the main absorption. The decrease in the main absorption, which was also observed in earlier experiments on Ti:sapphire crystals grown by the Czochralski method [15.69–71], results from the oxidation of Ti^{3+} ions to Ti^{4+} ions. *Mohapatra* and *Kroger* [15.71] concluded that charge compensation is probably accomplished by the formation of an A1 vacancy for every three Ti^{3+} ions converted to Ti^{4+}. The main absorption could be restored by annealing oxidized samples in a reducing atmosphere [15.69–71].

As noted above, the residual infrared absorption could be eliminated with careful attention to the redox conditions in the melt and also by postannealing techniques [15.72, 73].

Basal Slip

Because of its rhombohedral structure and its resulting anisotropic properties, sapphire exhibits different crystalline habits and structure perfections when growing along different directions [15.74]. Generally it is difficult to grow high-quality sapphire single crystal along the [0001] direction because this orientation is not preserved in the grown crystal [15.74], due to the weakening action of the main slip systems. The slip systems of sapphire reported so far are $(0001)1/3\langle 11\bar{2}0\rangle$ basal slip, $\{11\bar{2}0\}\langle 1\bar{1}00\rangle$ prism slip, and $\{10\bar{1}1\}1/3\langle\bar{1}101\rangle$ pyramidal slip [15.75, 76]. Among those, basal slip is known to be the easiest slip system at elevated temperatures.

In HEM or CZ systems an interface is developed which is convex towards the liquid. With such an interface a boule grown along the [0001] orientation always exhibits solidification stress. So, typically, sapphire is often grown in $[11\bar{2}0]$, $[10\bar{1}0]$, and $[1\bar{1}02]$ orientations to avoid basal slip. For zero-birefringence optics, the [0001] orientation is required, and components can be obtained from $[11\bar{2}0]$- and $[10\bar{1}0]$-oriented boules by fabricating pieces orthogonal to the growth direction [15.77]. However, this limits the size and homogeneity of fabricated components. As will be discussed in Sect. 15.3.2, the TGT system has a rather more stable thermal field than any other system, and the solid–liquid interface in the TGT system is much flatter (slightly convex). This made it easier to grow [0001]-oriented boules [15.78, 79], thereby enhancing the perfection of fabricated components, which results in a good laser performance, as described below [15.31].

The concentration of titanium along the radius in Ti:sapphire crystal grown by TGT is nearly unity. Figure 15.13 shows the distribution of the absorption coefficient at 514 nm ($\alpha_{514\,nm}$) along the growth axis of Ti:sapphire crystal grown by HEM and TGT. It can be seen that the concentration of Ti^{3+} in HEM-grown Ti:sapphire is not high, and the concentration gradi-

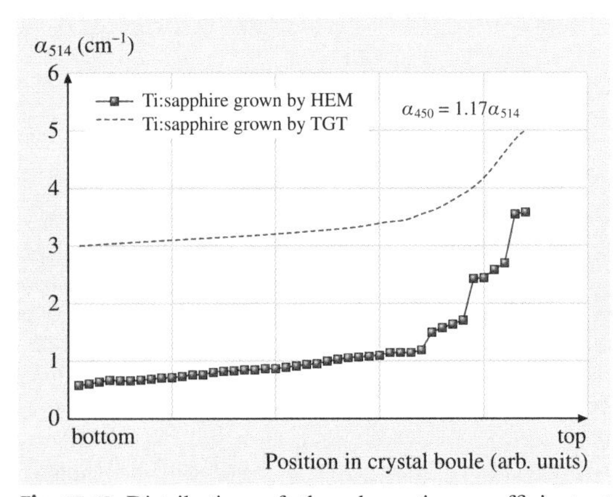

Fig. 15.13 Distribution of the absorption coefficient at 514 nm ($\alpha_{541\,nm}$) along the growth axis of Ti:sapphire crystal grown by HEM and TGT

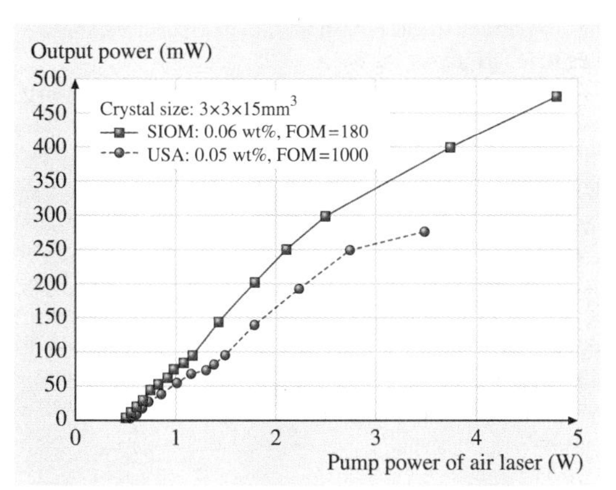

Fig. 15.14 CW laser performance of Ti:sapphire crystals grown by TGT and HEM in the same laser systems without water cooling

ent is large in the highly doped section ($1-3.6\,cm^{-1}$). The concentration of Ti^{3+} in TGT-grown Ti:sapphire is higher, and the absorption coefficient can reach as high as $5\,cm^{-1}$. Figure 15.14 shows the CW laser performance of Ti:sapphire crystals grown by TGT and HEM without water cooling in the same laser system. Although the FOM, mainly determined by the Ti^{3+}/Ti^{4+} ratio in the crystal, of HEM-grown Ti:sapphire is as high as 1000, 5.5 times higher than that of the TGT-grown one, the laser performance (efficiency and output power) of TGT-grown Ti:sapphire crystal is obviously better. These results show that the laser performance of Ti:sapphire depends not only on the FOM, but more on the crystal perfection. Furthermore, ultrashort pulse performance in China (12 fs) and in other counties (8 fs) was also achieved using TGT-grown Ti:sapphire crystals. High gain was obtained in TGT-grown highly doped Ti:sapphire crystals in RIKEN, Japan in 1994, and was 100 times higher than the data reported for Ti:sapphire crystals grown by HEM at that time [15.31].

15.2.2 Nd-Doped Laser Crystals

Nd-ion-based lasers are among the mostly used solid-state laser systems. They find applications in many areas, including the generation of the highest energy per pulse for laser-induced fusion purpose. Nd:YAG is by far the most useful material, owing to its unique combination of excellent thermomechanical properties and high gain cross section at 1.064 μm. Nd:YLF offers lower thermal lensing and longer storage time compared with Nd:YAG, although it is also a much less robust crystal. The Nd:YAP crystal does not exhibit the thermal birefringence problems experienced by Nd:YAG, but crystal growth of this medium has been hampered by several problems, including a tendency to twin. Nd:FAP and $Nd:YVO_4$ crystals possess the important properties of very large emission cross sections.

In spite of possessing many advantages, high-quality Nd-doped laser crystals are not easy to obtain due to the presence of many growth defects such as striations, facets, inclusions, the low distribution coefficient of Nd in host materials, etc.

Striations

Temperature fluctuations can give rise to striations in the growing crystal, which are often detrimental to laser performance [15.80]. These fluctuations can arise both from imperfect temperature/power control in the heating elements as well as from periodic or turbulent flows in the melt.

Striations can be eliminated or reduced in a number of ways. In pure compounds, growth at the congruently melting composition is an obvious solution. However, with deliberately doped or slightly impure materials the lowest temperature gradients and smallest melt depths compatible with crystal diameter control and the avoidance of other defects must be employed in order to limit convection.

Striations represent a useful built-in record of the interface shape at any point in the crystal and are thus widely employed in studying defects and morphology changes related to the solid–liquid interface. Figure 15.15 illustrates the striations observed in CZ and TGT-grown Nd:YAG crystals. It shows that the CZ configuration tends to produce a conical interface shape, while TGT has a flatter shape. Such a discrepancy results mainly from differences in the melt geometries, thermal and boundary conditions, and heating methods.

In CZ growth of laser crystals, the time constant of the system is generally set by the melt and the crys-

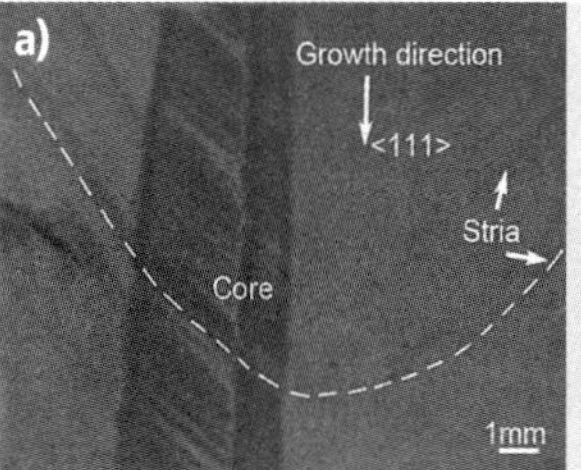

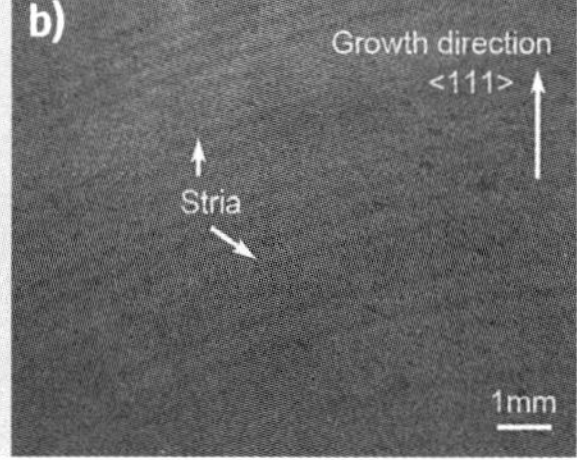

Fig. 15.15a,b Striations in CZ-grown (a) and TGT-grown (b) Nd:YAG crystals, observed by synchrotron radiation topography

tal, since induction heating of the crucible is usually the method of choice. The thin-walled crucible has a negligible impact on the time constant of the furnace. On the other hand, in TGT, VBT or HEM growth, the furnace time constant is often much longer because of the use of massive heating elements (either inductive or resistive). This large mass dampens the inevitable temperature fluctuations from the power supply. Rather than inductively heating the crucible, a relatively massive resistive heater is used that radiates to crucible. The crucible mass can be as large as required. For inductive heating, a massive susceptor can also be used, even for higher-melting materials such as Nd:YAG. A susceptor is a material used for its ability to absorb electromagnetic energy and convert it to heat.

In TGT, VB or HEM growth, with the melt above the crystal, the vertical temperature gradient is stabilizing with respect to natural convection (hot above cold), as opposed to the CZ configurations, which is destabilizing. In this context, stabilizing and destabilizing refer to the tendency to establish buoyancy-driven flows. For systems in which hotter fluid rests on top of colder fluid, such as in TGT, VB, and HEM growth, the lower density associated with the hotter fluid on top means that it will tend to remain in that position; hence this is stable. For systems in which hotter fluid is below colder fluid, such as for CZ growth, the colder upper fluid will tend to fall due to its higher density; this is unstable with respect to convection. In additional to TGT and HEM growth, keeping the crucible and growing crystal stationary will reduce temperature fluctuations resulting from mechanical vibration, and thus further enhance thermal stability in the melt.

In TGT, VB or HEM systems, the temperature gradient can be as small as practically required, depending on thermodynamic considerations. This can be important for controlling thermal stresses or selective evaporation from multicomponent systems. In Czochralski growth, appreciable temperature gradients are required to control the diameter. In fact, *Surek* [15.81] showed by analysis that, when heat transfer effects are neglected isothermally, the CZ process is unstable; small perturbations will produce large fluctuations in crystal diameter.

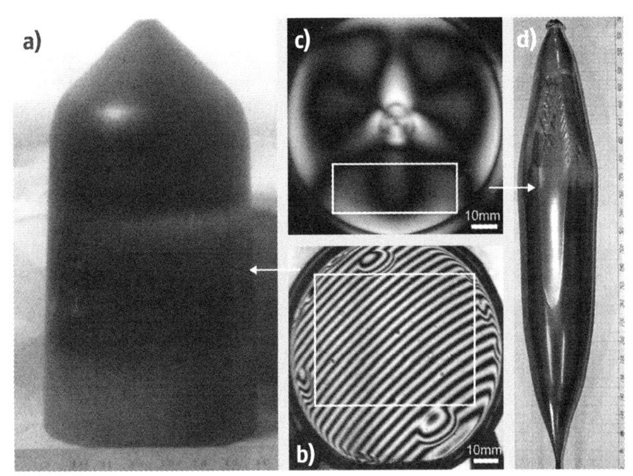

Fig. 15.16a–d Distributions of facet-induced cores in ⌀ 75 mm TGT-grown YAG **(a,b)** and in ⌀ 80 mm CZ-grown Nd:YAG **(c,d)**; the white frames indicate the available aperture for laser elements

Facets

During the growth of Nd:YAG crystal along [111] crystallographic axis under conditions which produce a deep interface, i. e., the growing end of the crystal has an essentially conical shape, convex into the melt, the interface geometry, combined with {211} facets near the tip, generates a nonhomogeneous central core [15.82] surrounded by a six-lobed strain pattern (Fig. 15.16c). The presence of facets limits the size of the slab that can be harvested from the boule and degrades the uniformity of the crystal.

Ming et al. [15.83] indicated that the core will appear at the region where the facet planes are parallel to the solid–liquid interface. Since the interface shapes in TGT and CZ configurations are commonly different, it is certain that the distribution of cores in crystals grown by these methods will be different: one is at the center, while the other is at the periphery (Fig. 15.16).

In the CZ configuration, Nd:YAG naturally grows with a deep interface because of radiative heat loss through the cooler, upper part of the furnace. In effect, the cold part of the furnace pushes the melt/crystal interface down into the melt via this long-range radiative exchange. In the CZ configuration, the interface shape can be controlled to give an approximately planar surface by increasing the crystal rotation rate to a critical value that depends on the crystal diameter [15.84–87], but this approach increases the risk of temperature fluctuations.

For some garnets, e.g. $Dy_3Al_5O_{12}$, the planar interface shape is produced at lower crystal rotation rats as compared with $Y_3Al_5O_{12}$ [15.88], while at higher rates it transforms to a concave one [15.89]. This behavior has been attributed to intensive absorption in the infrared associated with Dy^{3+} ions that overlaps with the maximum wavelength of spectral density of the thermal radiation used for melting [15.88]. In order to adapt this approach to laser crystal growth, impurity ions that are spectroscopically indifferent with respect to the lasing ions must be selected. It has been seen, for example, that additions to YAG melts of small amounts of Pr^{3+}, Sm^{3+} or Zr^{3+} [15.90], producing intensive absorption lines in the infrared region, also affect the shape of the interface.

In the TGT or VBT configurations, the low temperature gradients give an approximately planar solid–liquid interface except near the crucible wall, where the interface becomes slightly convex towards the melt and it is here that the facets occur [15.27, 91]. In the case of the ⟨100⟩ axis, which is normally accepted as the most preferable for Bridgman growth of garnets [15.90], the facets are eliminated further to the crystal peripheral region due the lager angle between the growth axis and the facets plane.

Scattering and Inclusions

The sources of scattering and inclusions can be either extrinsic (impurities) or intrinsic, such as solid-state exsolvation of a compound from a congruently melting, nonstoichiometric crystal.

For small intrinsic defects, often nothing can be done to eliminate them. In general, there are two ways in which small intrinsic defects can be eliminated or reduced in number. Point defects can sometimes be quenched before agglomeration take place. However, quenching implies the presence of large temperature gradients that are often problematic, producing either large thermal stresses or intense melt convection. The second means of elimination is solid-state diffusion of species over macroscopic distances. This mechanism is only applicable for small ions such as lithium or sodium, since the diffusion coefficients must be very large for this to be practical. Large defects, on the order of tens or hundreds of microns, of some intrinsic composition are indicative of other problems in the growth process. These can result from temperature fluctuations leading to large undercooling in portions of the melt, or from insufficient mixing in a multicomponent system, leading to localized composition nonuniformities. In these cases significant changes in growth conditions, or a different growth process, may be in order.

Extrinsic defects are generally controlled by a combination of suitable purification of starting materials and atmosphere control. In the growth of fluorides, this generally involves the use of reactive atmosphere processing using either HF, CF_4, SF_6 or other species to react with oxide impurities to form volatile byproducts [15.92, 93]. In addition, fluorides crystals are often grown in either HF, such as YLF, or in vacuum to further minimize contamination. Sometimes gettering compounds such as PbF_2 are employed to react with oxide impurities prior to growth.

The presence of inclusions in oxide crystals is usually a result of the formation and precipitation of a second phase in the primary liquid phase during the crystal growth process. One example is the formation of an oxygen-deficient phase in many of the gallium-containing garnets. In this case, simply increasing the oxygen partial pressure of the growth atmosphere is sufficient to eliminate this source of inclusions. Similar types of oxygen-deficient phases, usually appearing as submicron particles, i. e. smoke, have been observed in YAG, sapphire, and Y_2SiO_5 (YSO) and can be eliminated by appropriate adjustment of the oxygen partial pressure of the growth atmosphere [15.94].

A second source of inclusions can be associated with the dopant ion, as is the case in the growth of Nd:YAG. In the $Y_3Al_5O_{12}$ system, the distribution coefficient for Nd in YAG is relatively low (less than 0.2). Therefore to achieve approximately a 1 at. % Nd doping level in the crystal, the liquid must contain approximately 5 at. % Nd. Furthermore, since the distribution coefficient is much less than unity, as the growth of the crystal proceeds, the concentration of Nd in the liquid increases. Usually, when only about 20% of the liquid is crystallized, approximately a 20% change in the dopant ion concentration through the length of the crystal is produced. Variations in the dopant concentration beyond this amount often have detrimental effects on the laser performance of the material. Thus a low dopant ion distribution coefficient also imposes another limit on the useful amount of material that can be crystallized from the melt. The desire for large fabricated components therefore enforces the use of large quantities of starting material and large crucibles in CZ growth of Nd:YAG using the batch process. However, the batch process introduces complexity into the growth of a crystal through the progressive decrease in melt height. As the melt height decreases, transfer of heating power to the melt is affected and the strength of convective melt flow and mixing is reduced. The thermal environment of the crystal is also altered as the melt level falls, exposing the crystal to a greater area of the hot crucible wall. This tends to lower the temperature gradient in the melt, which enhances the risk of constitutional supercooling [15.87]. In order to circumvent some of the shortcomings of the standard process, development of an automated CZ growth process was initiated to achieve steady-state conditions through addition to the melt to maintain constant melt height and constant concentration of dopant in the melt [15.19, 87].

On the other hand, the compound $Nd_3Al_5O_{12}$ does not exist among the rare-earth aluminum garnets. Consequently, this system is not a solution between two different rare-earth aluminum garnets, but is a solution between $NdAlO_3$ (a perovskite phase) and $Y_3Al_5O_{12}$ [15.63]. Unfortunately at the required Nd liquid concentration, one is very close to the solubility limit of $NdAlO_3$ in YAG. Small temperature fluctuations from external or internal source, e.g., the heating power or a change in the cooling-water temperature, will easily result in constitutional supercooling, thereby exceeding the solubility limit of $NdAlO_3$ in YAG, which causes localized precipitation of the perovskite phase. Once in the crystal, they cannot be removed by subsequent processing such as annealing.

A similar example occurs in the growth of substituted $Gd_3Ga_5O_{12}$ (GGG) [15.95, 96]. In this case, Mg^{2+}, Ca^{2+}, and Zr^{4+} are added to the liquid in a ratio such that the resulting distribution coefficients are nearly equal to unity. However, this results in a Zr^{4+} liquid concentration (0.65 atoms per formula unit) that is very close to the solubility limit of $Gd_2Zr_2O_7$ in GGG (approximately 0.7 atoms per formula unit). Again, local temperature fluctuations near the growth interface can produce an excess of Zr^{4+} that results in the precipitation of submicron $Gd_2Zr_2O_7$ particles in the liquid, which are trapped by the growing crystal.

In both of these examples, not only are scattering sites produced in the crystal, but generally associated with these inclusions are region of strain [15.97]. In some cases, the induced strain is of sufficient magnitude that it can result in the formation of clusters of dislocations that propagate down the crystal axis and are roughly normal to the growth interface. For a crystal of 25–50 mm in diameter, these dislocations will continue down the crystal axis for approximately another 3–7 cm before reaching the crystal surface, thus further degrading its optical quality. Although the formation and subsequent entrapment of a second phase can be localized, its influence on the optical quality of the crystal can extend far beyond the local environment. As in the case of inclusions, subsequent processing such as annealing will have little, if any, effect on this type of defect. Thus the formation of optical-quality oxide materials must be accomplished during the crystal growth of the material and cannot necessarily be achieved after the growth process is complete.

Both of these cases illustrate the fact that a stable thermal environment is necessary for the growth of optical-quality single crystals suitable for laser applications. Thus much of the effort in the growth of many of these materials is devoted to the design of the growth equipment, its power source, and the furnace geometry.

A third source of scattering sites in oxide materials can result from contamination of the liquid by the crucible material, either Pt or Ir. These scattering centers can vary from submicron to tens of microns in size. Unlike inclusions generated by the precipitation of a second phase, these inclusions always appear as either triangular or pseudohexagonal particles with an aspect ratio of at least 10 : 1. Furthermore, these are flat, thin particles, and tend to align themselves parallel to the growth interface when trapped by the growing crystal. Thus the larger cross section of the particle presents itself to the optical path of the resulting laser rods, which greatly increases the possibility of damage. Both Ir and Pt particles can be formed in the liquid by numerous chemical paths [15.98]; for example, the presence of H_2O in either the starting powder or the growth atmosphere usually results in a high density of particles in many oxide systems. Of the two metals usually used for oxide growth, Ir is more prone to oxidation than is Pt. Consequently, for those oxide systems that require iridium as a crucible, the elimination of Ir particles is always a concern. Special attention must be paid to powder preparation, crucible cleaning and charging, and growth atmosphere to avoid any possible set of conditions that could result in the transport via oxidation of the crucible material into the bulk charge.

15.2.3 Yb-Doped Laser Crystals

The most promising ion that can be used in a non-Nd laser in the same range of emission wavelength is Yb^{3+}. The Yb^{3+} ion has some advantages over the Nd^{3+} ion as a laser-emitting center due to its very simple energy-level scheme, consisting of only two levels: the $^2F_{7/2}$ ground state and the $^2F_{5/2}$ exited state. There is no excited-state absorption to reduce the effective laser cross section, no up-conversion, no concentration quenching. The intense Yb^{3+} absorption lines are well suited for laser-diode pumping near 980 nm and the small Stokes shift between absorption and emission reduces the thermal loading of the material during laser operation. The disadvantage of Yb^{3+} is that the final laser level of the quasi-three-level system is thermally populated, increasing the threshold.

Among new directed searches for novel laser crystals, one important approach is the use of Yb^{3+} active ion in an inertial-fusion energy diode-pumped solid-state laser. $Ca_5(PO_4)_3F$ (FAP) and YAG were soon recognized to be favorable hosts for lasing in the nanosecond pulse regime. This fact was supported by an evaluation of the spectroscopic properties of several Yb^{3+}-doped crystals useful for laser action [15.99,100].

Although, compared with Nd-doped crystals, ytterbium can easily be incorporated into the host materials due to its relatively small ionic radii, many defects induced by doping effects are still present in Yb-doped crystals, and significantly impact on their thermal and spectral properties.

Lattice Distortion

The lattice parameters of YAG and YbAG were measured to be 1.2011597 ± 0.000034 and 1.1937997 ± 0.000054 nm, respectively, and there is only a 1.8% difference in unit-cell size [15.101, 102]. Figure 15.17 shows the Yb^{3+} concentration dependence of lattice parameter. It can be seen that lattice parameter is a linear function of Yb^{3+} concentration, and a linear equation can be obtained as follows:

$$\alpha(x) = 1.20076 - 0.007072x \text{ nm} .$$

From this equation, the densities of crystals with different Yb^{3+} concentration can be estimated from the cell volume and molecular weight of Yb:YAG.

As discussed above, compared with the CZ method, the temperature and thermal field of TGT are very stable and convection that disturbs the solid–liquid interface does not appear. Therefore, Yb^{3+} ions can more easily

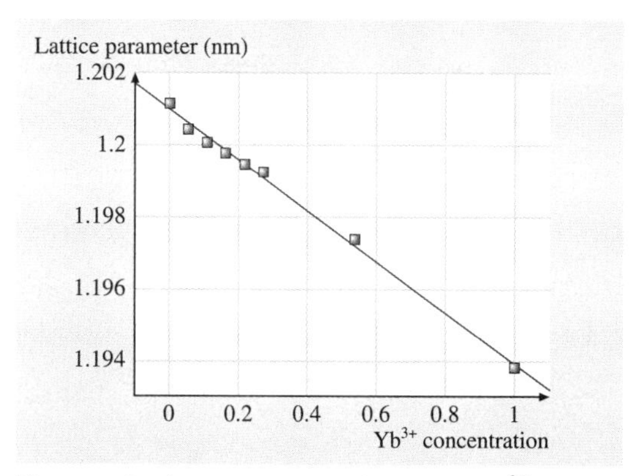

Fig. 15.17 Lattice parameter as a function of Yb^{3+} concentration

substitute at the position of Y^{3+} sites. From the results of cell parameters shown in Table 15.1, it can be seen that the effect of Yb^{3+} on the crystal lattice deformation in TGT-grown Yb:YAG crystals is smaller than that in CZ-grown Yb:YAG crystals. The cell structure of TGT-grown Yb:YAG crystals is steadier.

Different from the typical four-level Nd:YAG laser, the quasi-three-level nature of Yb:YAG requires high pump density in order to overcome the deleterious effect of the lower laser level reabsorption at room temperature; moreover, the laser performance decreases significantly with increasing temperature [15.103]. So it is favorable to keep the temperature of Yb:YAG crystal as low as possible. Obviously, knowledge of the effect of doping on thermal properties of crystal is helpful for both designing laser systems or improving crystal performance.

Figure 15.18a shows that the thermal diffusivity decreases with increasing Yb^{3+} concentration, and values of thermal diffusivity at 50 °C are 1.72×10^{-6}, 1.62×10^{-6}, and 1.54×10^{-6} m^2/s for single crystals with doping level 5 at. %, 10 at. %, and 25 at. %, respectively. Figure 15.18b shows that the specific heat increases as the temperature increases in the measured range, and the variation of temperature has a strong influence on highly doped Yb:YAG crystals [15.104]. The thermal conductivity was calculated from the thermal diffusivity and specific heat capacity, as displayed in Fig. 15.12c, and we can see the apparent influence of Yb^{3+} doping concentration on the thermal conductivity. In Yb:YAG crystals, the main mechanism of heat transfer is by phonons. Yb doping into YAG crystals inevitably induces structural distortion in crystals. The defects in crystals remarkably reduce the phonon mean free path, and the thermal conductivity decreases as Yb doping concentration increases. The deterioration of thermal properties of highly doped Yb:YAG will more easily lead to thermo-optic aberrations, lensing, and birefringence. Therefore, in order to acquire high beam quality and stable laser output from highly doped Yb:YAG media, an efficient cooling system must be adopted [15.105, 106].

The result of *Chenais* et al. [15.107] shows that the thermal conductivity of Yb:GGG, although lower than that of Yb:YAG at weak Yb^{3+} concentrations, becomes

Table 15.1 Cell parameters of Yb:YAG crystals

Growth method	Yb^{3+} concentration (at.%)	α (nm)	β	V (nm^3)
CZ	5.4	1.200424 ± 0.000063	90.0	1.72983
TGT	5.4	1.200704 ± 0.000052	90.0	1.73104

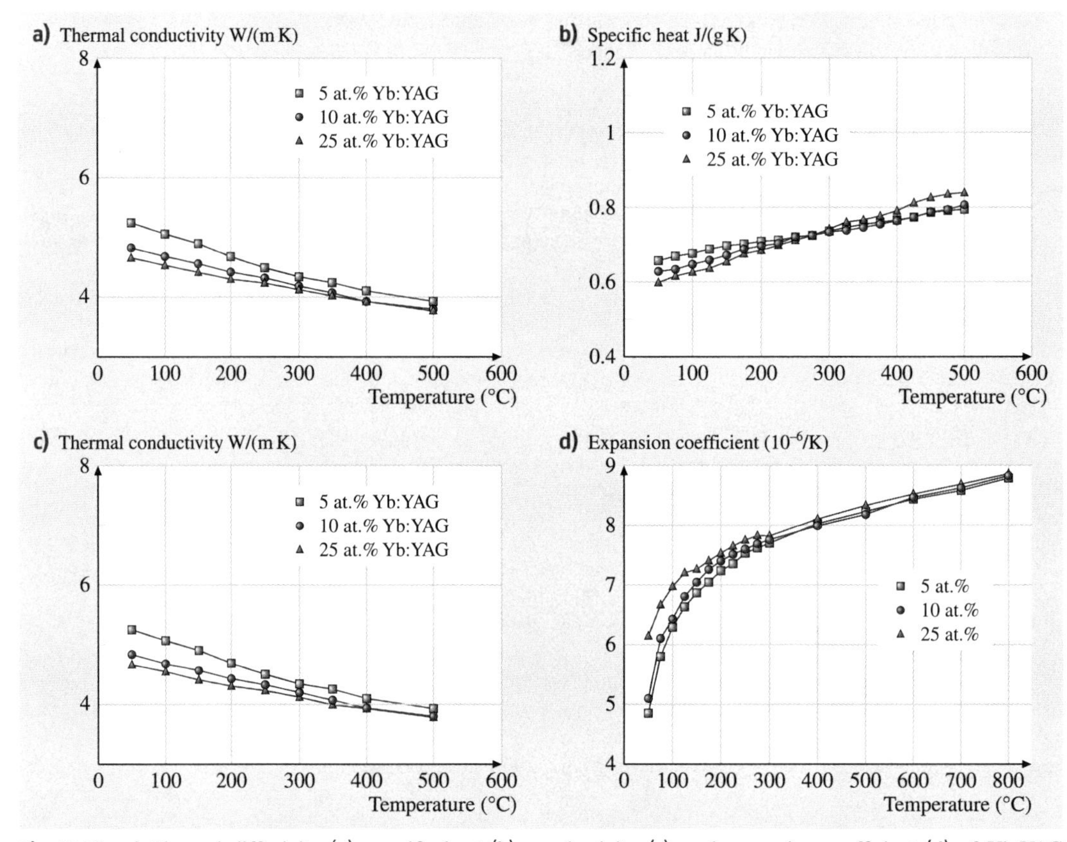

Fig. 15.18a–d Thermal diffusivity (**a**), specific heat (**b**), conductivity (**c**), and expansion coefficient (**d**) of Yb:YAG crystals as a function of temperature for several Yb^{3+} doping concentrations

higher for doping levels above 5×10^{20} ion/cm^3. This means that Yb:GGG exhibits a lower quantum defect and better thermal conductivity, and thermal loading is lower than in Yb:YAG for high doping levels.

Color Centers

Color centers in oxide-based crystals constitute a complex problem, as some centers can be removed by annealing in oxidizing atmosphere [15.108] and others by annealing under reducing or vacuum conditions [15.109, 110]. Color center formation, in general, is attributed to change of lattice defects associated, primarily, with unintentionally introduced impurities.

In CZ-grown Yb:YAG, a particular color center was found [15.111–113], whose absorption spectra are shown in Fig. 15.19. The inert growth atmosphere brought about a lot of oxygen vacancies and formed

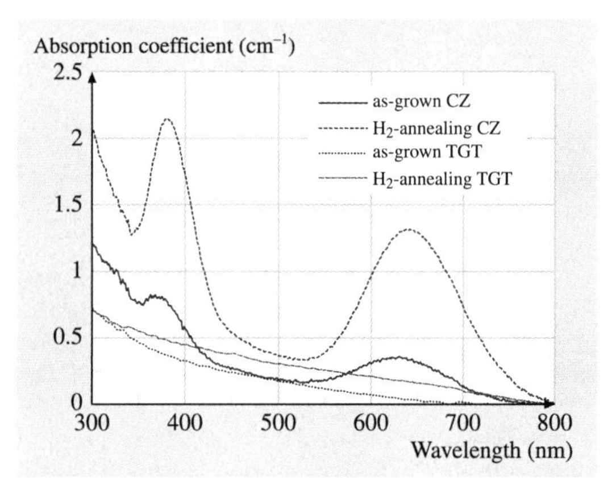

Fig. 15.19 Absorption spectra of color centers in Yb:YAG crystals

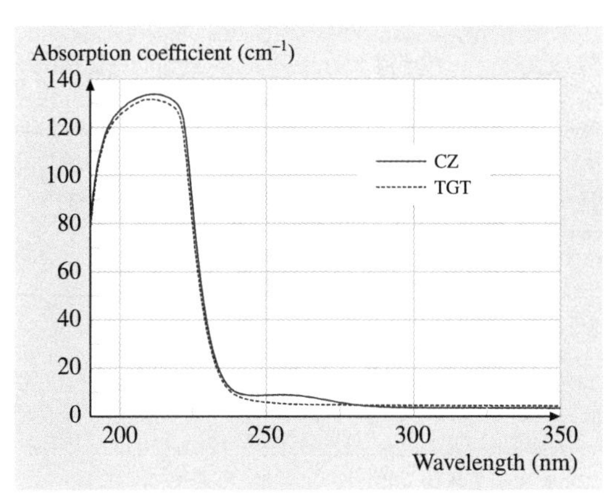

Fig. 15.20 UV absorption spectra of Yb:YAG crystals

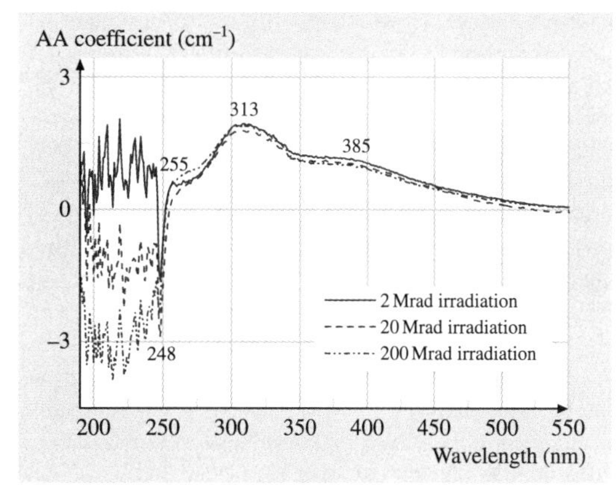

Fig. 15.21 Absorption spectra of color centers in Yb:YAP crystals

Re-F type color centers and Yb^{2+}, which are detrimental to the intrinsic spectroscopic performances of Yb:YAG. They degraded the Yb^{3+} intrinsic absorption at 900–1050 nm and the emission intensity at 1028–1060 nm, and shortened the fluorescence lifetime of Yb^{3+} in YAG host. The color center's two absorption bands are located at wavelengths of 368 and 623 nm, respectively. Each absorption peak increases in intensity after annealing the sample in hydrogen at 1400 °C for 12 h, and the main band positions move from 368 to 381 nm and from 623 to 642 nm, respectively. The absorption spectra of TGT-grown Yb:YAG crystals are also shown in Fig. 15.19. Although also grown in an inert atmosphere, and even when annealing the sample in hydrogen atmosphere, TGT-grown Yb:YAG crystals do not exhibit the same absorption bands. The results indicate that there is no Yb^{2+} and color center absorption in TGT-grown Yb:YAG crystal, which is of great use for the laser performance of Yb:YAG.

The ultraviolet (UV) absorption spectra after annealing in oxygen shows a weak peak around 255 nm in CZ-grown Yb:YAG crystals, as shown in Fig. 15.20. *Chen* et al. [15.114] used optical absorption and electron paramagnetic resonance (EPR) techniques to study iron impurities in YAG crystals and considered that the absorption band at 255 nm was attributable to a Fe^{3+} charge-transfer band that was made up of contributions from substitutional Fe^{3+} ions in octahedral and tetrahedral sites. The role of Fe impurity ions in color center formation has also been demonstrated by other technologists in YAG [15.115], $YAlO_3$ [15.116, 117], and Al_2O_3 [15.118, 119].

For Yb:YAP crystals [15.120], three irradiation-induced color center absorption bands, located at wavelengths 248, 309, and 385 nm, respectively, were observed (Fig. 15.21). The significant additional absorption (AA) peak at 248 nm is believed to be caused by the slight shift of the charge-transfer absorption edge of Yb^{3+} ions induced by the charge density redistribution. Based on the result discussed above, combined with the data reported by *Matkovski* [15.121] for the absorption band at 263 nm corresponding to Fe^{3+} ions in Nd:YAP, Fe^{3+} ions are considered to be responsible for the absorption band at 256 nm in Fig. 15.21. The 313 nm band has been observed by *Matkovski* [15.121] in absorption spectra of Nd:YAP and pure YAP crystals and was attributed to Fe^{2+} in YAP host. Kaczmarek suggested that a 313 nm band in γ-ray-irradiated Cr;Tm;Ho:YAG is most probably correlated to Fe^{2+} ions. Therefore, it is reasonable to

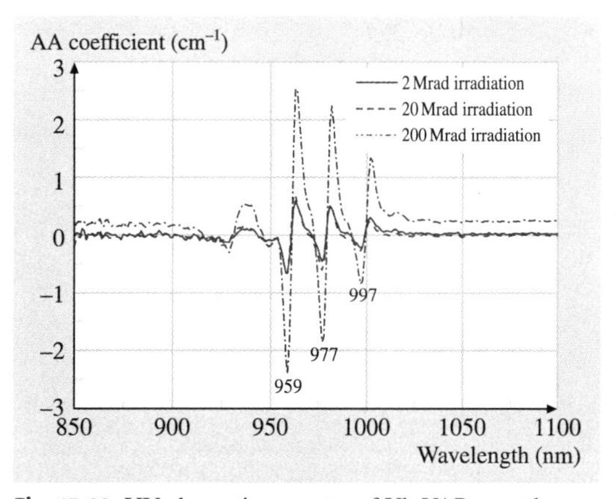

Fig. 15.22 UV absorption spectra of Yb:YAP crystals

believe that the 313 nm bands in Yb:YAP are also associated with the same kind of impurity. The increased AA value of the 313 nm band clearly indicates that the concentration of Fe^{2+} ions increased after γ-ray irradiation. The preexisting Fe^{3+} ions in Yb:YAP captured the electrons produced by γ-ray irradiation and the following interaction takes place: $\gamma + Fe^{3+} \rightarrow Fe^{2+}$. Yb^{3+} ions in Yb:YAP crystal could also capture the free electrons induced by γ-ray irradiation and as a result the transition $Yb^{3+} \rightarrow Yb^{2+}$ takes place, as shown in Fig. 15.22. The broad absorption band centered at about 385 nm shown in Fig. 15.21 has attracted more attention due to its complicated origin. According to the results of *Sugak* et al. [15.122], one broad absorption band located at 385–666 nm occurred in γ-irradiated Nd:YAP and the absorption induced in this region is caused by color centers intrinsic to the YAP lattice. Furthermore, a similar band was also observed in Pr:YAP under γ-ray irradiation [15.123]. *Matkovski* et al. [15.121] found that color centers in this region created in YAP crystals are associated with the crystal host and that their nature does not depend strongly on the type of dopant, even for high dopant concentrations. Because the Yb:YAP crystal was grown in inert atmosphere, a lot of oxygen vacancies are produced in the crystal. In analogy with the charge–recharge process of Fe and Yb ions, these O^{2-} vacancies in Yb:YAP crystal would capture one or two electron(s) to form F^{+} or F centers. It has been shown that cation vacancy would be the most common defect intrinsic to the YAP lattice [15.124]. Consequently, it is most probable that the broad AA band centered at 385 nm is correlated with the cation vacancies and F-type center.

Commonly, color centers may be annealed out or reduced in number at high temperature under oxidizing, reducing or vacuum atmosphere depending on the center's origination. In the case of Yb:YAG, it is notable that annealing at high temperature increases the possibility of scattering formation due to a phase transition occurring at 1369.5 °C, as shown in Fig. 15.23, so a relatively lower temperature should be adopted when annealing.

Fluorescence Quenching

The effective stimulated-emission cross section (σ) and radiative lifetime (τ) are two important parameters for the assessment of a laser crystal [15.125]. Knowledge of both σ and τ is essential when evaluating laser system performance parameters such as saturation intensity and threshold pump power; for example, the threshold pump power is inversely proportional to the product of the effective emission cross section and the radiative lifetime of the lasing crystal [15.126].

Figure 15.24 shows the relationship of measured fluorescence lifetime of Yb:YAG crystals with temperature. At higher temperatures (above 200 K), the fluorescence lifetime increases as the Yb^{3+} concentration increases in the YAG host because of radiative trapping and reabsorption effects. The measured lifetime is longer than the radiative lifetime in these cases. At lower temperatures (below 80 K), the measured fluorescence lifetime of Yb in YAG is nearly the same for each concentration, similar to the results reported by *Patel* et al., [15.127] except for the 30 at. % Yb:YAG crystal. The decreased lifetime ob-

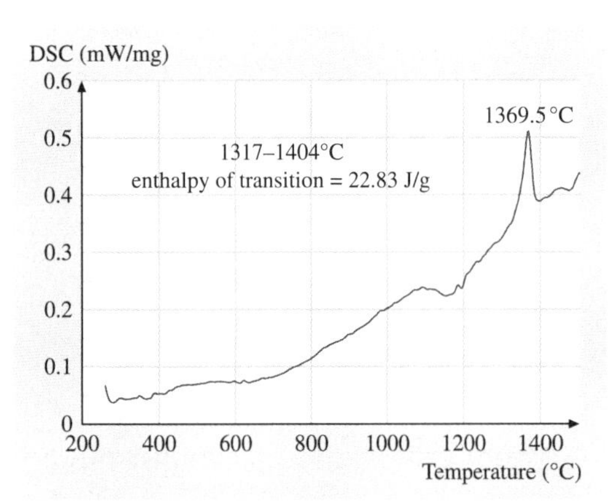

Fig. 15.23 Differential scanning calorimetry (DSC) curve of Yb:YAG

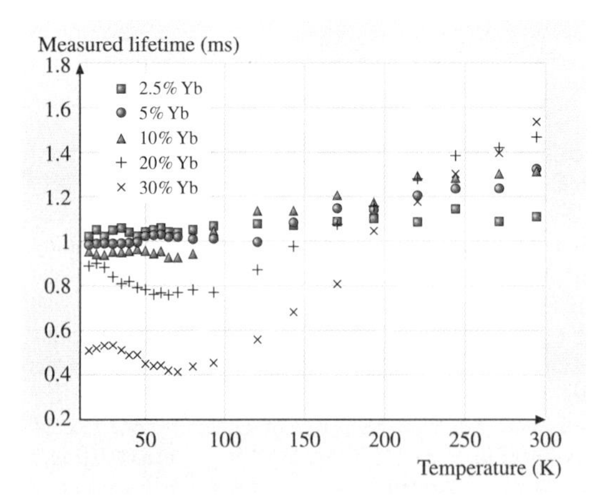

Fig. 15.24 Measured lifetime of Yb:YAG crystals as a function of temperature

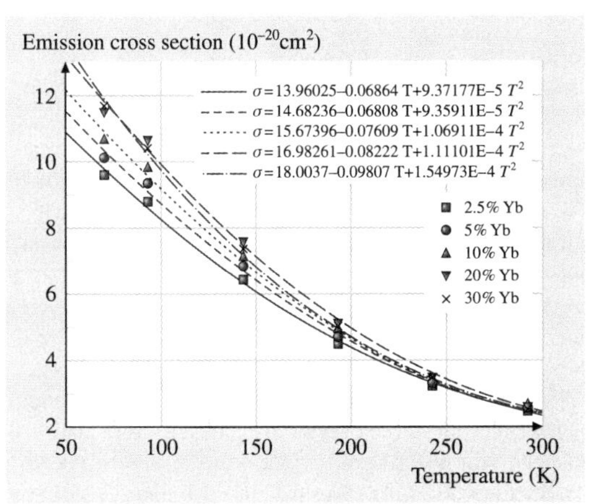

Fig. 15.25 Emission cross sections of Yb:YAG as a function of temperature for five concentrations of Yb; the indicated polynomial fits were obtained by the least-squares method

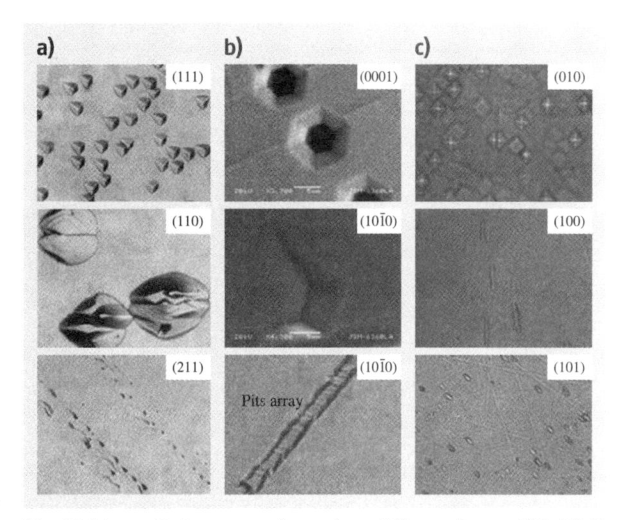

Fig. 15.26a–c Etch patterns formed on different faces of Yb:YAG (**a**), Yb:FAP (**b**), and Yb:YAP (**c**)

served for the 30 at. % crystal at the lowest temperatures may be due to the presence of more impurities in the highly doped crystal, which cause significant fluorescence quenching. These impurities may be other rare-earth ions that are hard to eliminate from the Yb starting material.

Using the measured lifetime at the lowest temperature as the radiative lifetime of different concentrations of Yb:YAG crystals, the effective emission cross section of Yb:YAG crystals can be calculated from the Füchtbauer–Ladenburg (F–L) formula [15.125]. Figure 15.25 shows the dependence of the effective emission cross section of these Yb:YAG crystals on temperature. The emission cross section increases as the temperature decreases.

Dislocations

Dislocation structures in oxides etc. have been widely studied using chemical etching procedures to reveal dislocations as individual pits. Such procedures can be misleading as great care is needed to distinguish dislocation etch pits from etch pits caused by inclusion or surface damage. X-ray topography has provided significant information but a useful innovation has been the application of electron microscopy to such studies because of the inherent resolution improvement at shorter wavelengths.

Generally, the symmetry of an etch pit is in accordance with the symmetry of the crystal face, and the shapes of etch pits on different faces are different for the same crystal [15.128]. The change of etching conditions may also affect etch-pit morphology.

For Yb:YAG crystals [15.129], under the same etching conditions, the pits on the (111) face have two

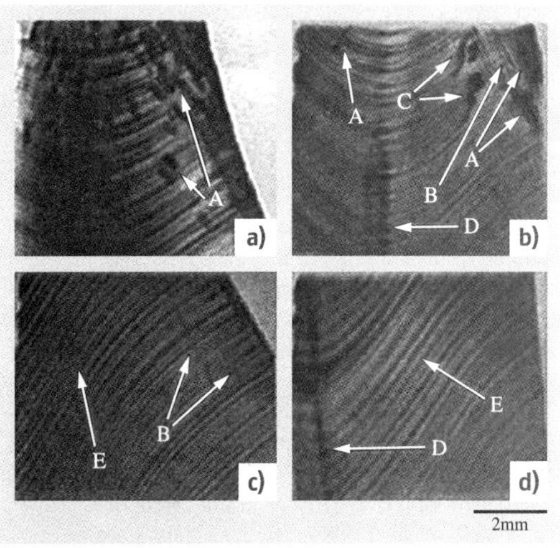

Fig. 15.27a–d Transmission synchrotron topography of (110) slices of Yb:YAG parallel to the [111] growth axis with (121) reflection. (**a–d**) From the upper parts to the middle of the crystal. A: dislocation bundles originated from seed/crystal interfaces; B: dislocation lines; C: dislocation bundles originating from impurity or inclusions; D: core and side core; E: growth striations

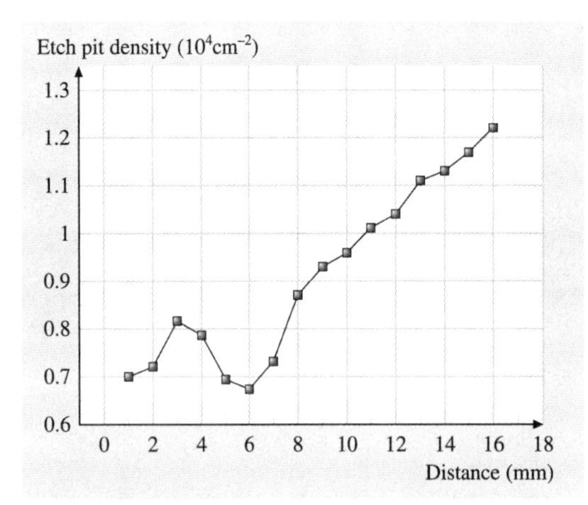

Fig. 15.28 Radial distribution of etch-pit density on (0001) face of Yb:FAP crystal

shapes: one is triangular, and the other is six-sided. The pit pattern on (211) is triangular with a tail and the pit pattern on (110) is a distorted rhombus. In Yb:FAP crystals, etch pits associated with dislocations emerging on the (0001) plane are hexagonal, while those on the $(10\bar{1}0)$ plane have an irregular shape. For Yb:YAP, the etch pits have rhombus shape on the (010) face, haricot-beam shape on the (100) face, and elliptic form on the (101) face. The typical etching patterns are shown in Fig. 15.26.

For Yb:YAG (111) slices, more etch pits can be observed in the initial growth period, which implies that the crystal is imperfect, and with an increase of distance from the seed the number of pits decreases gradually and they are mainly centered at the periphery of the sections. In the middle parts of Yb:YAG crystal, there are few etch pits, which implies that the crystal is perfect. Figure 15.27 shows the transmission synchrotron topography of a (110) slice of Yb:YAG crystal with (121) reflection, displaying typical growth defects in Yb:YAG. This shows that dislocation bundles and dislocation lines distribute towards the crystal periphery in the upper parts of the Yb:YAG crystal. However, in the middle parts of Yb:YAG crystal, they vanish, as shown in Fig. 15.27d. These experimental results are in good agreement with those obtained by chemical etching. Such a rule can also be demonstrated by the radial distribution of etch-pit density along the diameter of Yb:FAP crystal [15.130], as shown in Fig. 15.28.

According to the observations in Fig. 15.27a–c, the growth dislocations may originate from the following sources: (1) dislocations already existing in the seed; (2) dislocations produced by nucleation at the seed–crystal interface, where some defects such as mechanical damage of the seed end, aggregation of impurity particles and inclusions, and thermal shock stress exist, as shown in Fig. 15.27a,b. These dislocations are the main types of source in the as-grown Yb:YAG crystals; (iii) dislocations emerging from impurity particles and inclusions trapped within the crystal during crystal growth. Comparing Fig. 15.27a–c, it can be seen that the propagation direction of the dislocations is perpendicular to the solid–liquid interface and as a result the dislocations are mainly found at the periphery of the upper parts. This rule can be explained by the minimum-energy principle, since the dislocations take the shortest path and locate in the lowest-energy state only when they take this route. *Schmidt* [15.131] found a similar rule in GGG crystal and calculated the propagating path of dislocations with Klapper theory [15.132]. Therefore, in order to obtain high-quality Yb:YAG crystal, it is necessary to choose high-quality seeds free from dislocations and grow the initial part of the crystal with a highly convex solid–liquid interface to eliminate dislocations. Usually, the neck technique is applied to eliminate dislocations originating from the seed or introduced by thermal shock during dipping of a seed crystal into the melt.

15.2.4 Other Activator-Doped Laser Crystals

While flashlamp-pumped ruby lasers are now primarily an artifact of the past, many of the common transition-metal lasers still appear to be a useful system for some specialized applications. Alexandrite, $Cr:LiCaAlF_6$ (Cr:LiCAF), and $Cr:LiSrAlF_6$ (Cr:LiSAF) lasers operate on the vibronic sideband of Cr, and consequently have a much lower threshold than ruby, which renders them much more useful. Cr:LiCAF was found to lase very efficiently. Forsterite ($Mg_2SiO_4:Cr^{4+}$) was found to be an exciting new tunable laser material, and $MgF_2:Co^{2+}$ proved not to require cryogenic cooling if it is pumped on a timescale short compared with the excited-state storage time of 40 μs. Otherwise, several researchers have carefully assessed whether the Mn^{3+} ion might prove to be a useful laser ion [15.63].

While all of the rare-earth ions have been lased, systems based on Er, Tm, and Ho, besides Yb and Nd, have thus far proved to be the most useful. The Cr,Tm:YAG laser material is characterized by a particularly low gain cross section and, as a result, is most efficiently operated in a quasi- or true-CW type of mode. The Cr,Tm,Ho:YAG crystal has been found to be a very efficient flashlamp-pumped system that operates near 2.09 μm, since Cr^{3+} and Tm^{3+} can act as sensitizers of Ho^{3+} to enhance the absorption of flashlamp light. Later, it was noted that the Tm,Ho:YLF crystal is similar to the YAG system. Finally, the Er:YAG crystal has turned out to be useful as a long-wavelength system. The 2.3 μm lasers are particularly useful for novel medical procedures and for applications requiring eye-safe laser output.

During the growth of these crystals, the defects that should be specially mentioned are absorptions at the laser wavelength induced by impurities and the thermal stresses formed in the crystals.

Absorption at Laser Wavelength

As discussed above in Sect. 15.2.1, the presence of absorptive loss at the laser wavelength can strongly degrade the performance of melt-grown crystals. An example similar to the presence of Ti^{4+} in Ti^{3+}:sapphire is the unintended incorporation of Cr^{4+} into tetrahedral Ga-sites of the $Gd_3Sc_2Ga_3O_{12}$ (GSGG) garnet when the crystal is co-doped with Nd^{3+} and Cr^{3+} [15.133]. This problem is complicated by the interplay of numerous growth conditions, such as volatilization of Ga_2O_3 from the melt, oxidation of the crucible, requirement for calcium additives to stabilize growth, and scattering centers under certain conditions. This problem has been particularly vexing because the absorption strength of the tetrahedral Cr^{4+} is orders of magnitude greater than that of the normal Cr^{3+} (on the octahedral Sc^{3+} site), thereby requiring that much less than 1% of the Cr ions be incorporated in the tetravalent state. It is worthwhile to note that, ironically, the Cr^{4+} ion was later recognized to be useful as a saturable absorber in Q-switching applications and then as a laser ion in the forsterite crystal (Mg_2SiO_4). Other examples of the presence of interfering oxidation states of chromium include Cr^{2+} in Cr^{3+}-doped LiCAF and LiSAF [15.134], while Cr^{3+} is found to absorb pump radiation in Cr^{4+}-doped forsterite [15.135].

The growth atmosphere can have a profound effect on the oxidation state of dopants and impurities and hence on the quality of the crystal via the phase diagram. For example, Nd:YLF is generally grown in an HF atmosphere to maintain low concentration of YOF forming from oxide impurities [15.136]. Cr^{4+}:forsterite is grown under oxidizing conditions [15.135], while Ti^{3+}:sapphire is grown under moderately reducing conditions [15.137]. For the gallium-containing garnets, i.e., GGG and GSGG, the formation of a reduced Gd – Ga compound that is insoluble in the molten garnet can be prevented by the addition of 2–3 vol. % oxygen to the growth atmosphere (N_2) [15.98]. In addition to preventing the formation of a second phase, the addition of oxygen to the growth atmosphere reduces the evaporation of Ga_2O_3 from the liquid surface. In general, for most oxide material systems, the addition of small ($\approx$ 500 ppm) to moderate (approximately 2.5% by volume) amounts of oxygen to the growth atmosphere will prevent the partial reduction of the oxide constituents.

Furthermore, during the growth of oxide-based laser materials, the lasing action is dependent upon the dopant ion being in a particular oxidation state, e.g., Cr^{4+} in forsterite [15.138, 139]. Maintaining the correct oxidation state of Cr^{4+} can require a growth atmosphere that is more oxidizing than that required to prevent partial decomposition of the constituent oxides. Postgrowth annealing of the crystal has also been used to convert the dopant ion to the desired valence state. However, the impact of this process is highly dependent upon either cation or anion valencies in the lattice. Therefore, the preferred approach is to produce these crystals by adjusting the initial growth conditions and starting materials to directly yield the desired dopant oxidation state.

The converse can also be true. For Cr,Nd:GSGG and Cr,Nd:GGG, the presence of Cr^{4+} introduces a broad absorption band in the 1 μm region. In this case, it was found that low-level divalent impurities such as Ca^{2+} in the starting materials were charge-compensated in the lattice by the conversion of Cr^{3+} to Cr^{4+}. The elimination of Cr^{4+} in the crystal was achieved by the addition of another tetravalent ion such as Ti^{4+} into the starting composition, thereby providing the growing crystal with a readily available source of another tetravalent ion [15.140, 141].

Thermal Stresses

Thermal stresses result from nonlinear temperature fields in growing crystals, scale linearly with the second derivative of temperature (assuming no external traction forces), and often scale quadratically with the length scale over which they occur [15.142]. The details are actually much more complicated, particularly in noncubic crystals. However, increasing the size of a crystal increases the imposed thermal stress, subject to the degree to which it can be controlled. In CZ growth, radial temperature gradients are required to maintain control over the growth process. Curvature along the growth direction is generally nonlinear, but in principle could be linearized through suitable furnace design to minimize thermal stresses. In TGT, VBT or HEM growth, radial

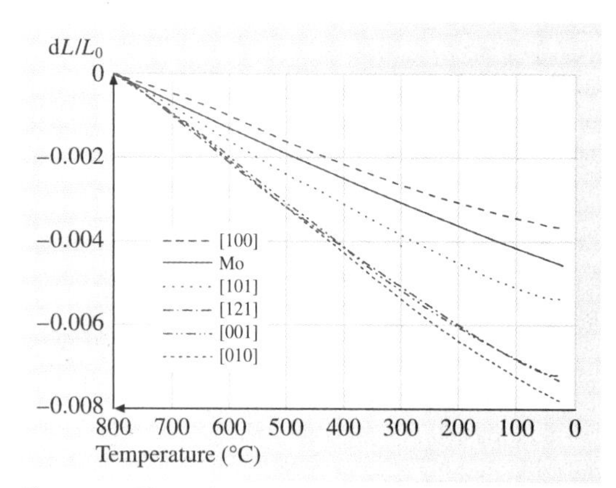

Fig. 15.29 Thermal contraction curves along several typical directions of YAP crystal

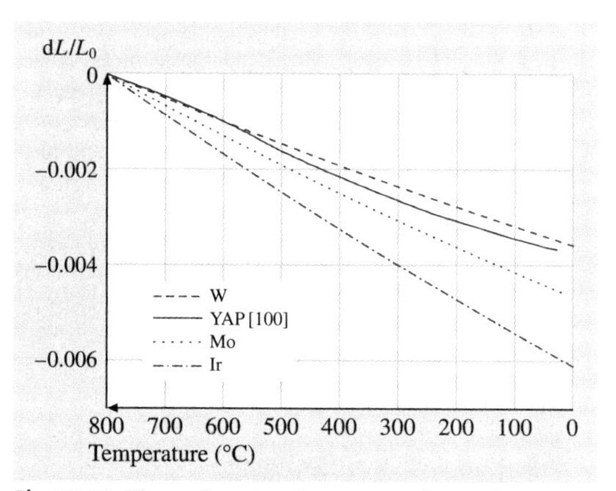

Fig. 15.30 Thermal contraction curves of several common crucible materials in comparison with that along the [100] direction of YAP

(lateral) temperature gradients can be minimized since the crystal takes the shape of the ampoule. However, the ampoule itself can also act as a source of stress in the crystal usually during cool-down.

For robust oxide crystals such as Nd:YAG and Ti:sapphire, scaling up the growth process may ultimately be successful for obtaining larger crystals suitable for large slabs. For fluorides, which have much lower fracture toughness, the thermal stresses present in the CZ process may ultimately limit the sizes that can be grown. The TGT and Bridgman process does not require lateral thermal gradients for stability, and the furnace can be designed to minimize these gradients. As was noted above, however, use of the TGT and Bridgman process can lead to other stress problems due to the different thermal contraction coefficients of the crystal and crucible. A typical example is the cracking and twining defects occurred in TGT-grown YAP crystals using a Mo crucible. YAP possesses a deformed perovskite-like structure, and is highly anisotropic in its coefficient of thermal expansion (CTE), with approximate values of $4.2\times10^{-6}\,K^{-1}$, $11.7\times10^{-6}\,K^{-1}$, and $5.1\times10^{-6}\,K^{-1}$ along the *a*-, *b*-, and *c*-axes, respectively. As shown in Fig. 15.29, since the CTE of the Mo crucible is higher than that along the *a*-axis in YAP, the crystal is compressed along its *a*-axis by the crucible as it cools, hence leading to cracking and twining. Such defects can be reduced or even eliminated by growing YAP crystal using an *a*-oriented seed or tungsten crucible since the tungsten has a lower CTE, as shown in Fig. 15.30.

Cr:LiCAF is also anisotropic in its CTE. Along the *a*-axis the CTE is approximately $2.2\times10^{-6}\,K^{-1}$, while along the *c*-axis the CTE is only $3.6\times10^{-6}\,K^{-1}$. Since the graphite ampoule has a CTE of $8\times10^{-6}\,K^{-1}$, the crystal is compressed along its *c*-axis by the ampoule as it cools. For slab growth, this places the two broad faces of the slab under compression. Materials generally do not fail under compression, and this is probably not a problem for Cr:LiCAF. However, by pinning the crystal (either partially or completely) along the *c*-axis, the slab is put under tension in the orthogonal directions since the CTE is larger than that of graphite. Evidence of failure in tension in the form of long, linear fractures that run the length of slab has been seen by *Atherton* et al. [15.63]. They tried to compensate for this fracture mode by using compressible graphite foam inserts in the ampoule to take up the imposed stress along the *c*-axis. While this clearly helped, it did not eliminate the fracture problem entirely. In addition, there also exist small, nonlinear temperature gradients in the crystal as it cools down, thus creating another type of stress field. Calculations have indicated that these stresses are at least an order of magnitude smaller than the traction-induced stresses, but they may still be important, especially given the experience with the foam inserts.

Should the need for large slabs of Cr:LiCAF (or Cr:LiSAF and other analogues) arise in the future, further modification of the ampoule in the form of lower-CTE graphite, a softer foam insert or an ampoule design that opens as it cools could be utilized to address the traction-induced stress problem. The furnace itself can be modified to further linearize the temperature field imposed on the crystal as it cools to minimize thermal stresses.

15.3 Crystal Growth Techniques Characterization

15.3.1 Czochralski (CZ) Process

The CZ process is by far the most commonly used process for the growth of laser materials. The main CZ characteristics are as follows.

Advantages of CZ

An appropriate thermal field is easy to establish for various crystals with a wider span of materials and melting points than any other melt-growth method. Not only by varying the ceramic insulation and its dimensions but by changing the rotation rate of the crystal, the adjustment of radial and vertical thermal gradients in the melt can be easily accomplished. Moreover, the convenience of the observation of the growing crystal makes it possible to adjust the growth parameter promptly, which is essential for realizing real-time control of the growth process.

Where CZ is used, crystals grow from the free surface of the melt without contacting with crucible. Therefore, defects induced by contact stress and parasitic nucleation can be eliminated, and the selection of crucible material is no longer limited by the mismatch of CTE between crystal and crucible.

In CZ, a neck process can be easily employed to eliminate dislocations originating from the seed and thermal shock. Compared with other methods, CZ can provide a higher growth rate to obtain a relatively large crystal with good perfection.

Disadvantages of CZ

The CZ furnace is usually constructed of either Al_2O_3 or stabilized ZrO_2 ceramics, which contact directly with the crucible. This configuration leads to a weakly oxidizing atmosphere in the furnace. Such conditions may result in the oxidation of crucible and introduce pollutants into the melt.

In CZ growth, appreciable temperature gradients are usually needed to control the diameter of the growing crystal. Otherwise, the space above the crucible is commonly opening due to the presence of the lift mechanism. Such a configuration increases the loss of heat. In addition, the growing crystal can act as a light-conduction tunnel for radiative heat loss. So CZ often has higher temperature gradients, which will introduce thermal stress into the crystal.

15.3.2 Temperature Gradient (TGT)

The TGT method has the following advantages and disadvantages with respect to the CZ technique.

Advantages of TGT

Since the crystal shape is dictated by the crucible, there is no need for sophisticated diameter control based on weight gain or various optical techniques. This aspect is especially significant when growing relatively large sample.

For application where large crystal plates are required, the TGT process can be effectively utilized. This is especially valuable for crystals in which significant scale-up is required.

In TGT growth with the melt above the crystal, the vertical temperature gradient is stabilizing with respect to natural convection (hot above cold), as opposed to the CZ configurations, which is destabilizing, as discussed in Sect. 15.2.2. In addition, the furnace, crucible, and crystal are fixed in space, so this configuration is mechanically simple, which enhances the temperature stability.

In TGT growth, the temperature gradient can be as small as practically required, depending on thermodynamic considerations. This can be important for controlling thermal stresses or selective evaporation from multicomponent systems. In CZ, appreciable temperature gradients are required to control the crystal diameter.

Disadvantages of TGT

Because the crystal is growing in contact with the crucible, two problems can arise. One is spurious nucleation on the crucible wall that gives rise to large-angle grain boundaries. This problem is magnified when insufficient temperature gradients are utilized. The second problem stems from stress imposed on the crystal during cool-down for the cases where the crucible has a large coefficient of thermal expansion for contraction, or where adhesion due to wetting of the crucible wall occurs. The magnitude of this problem is determined by the strength of the crystal and of the adhesion.

Poor mixing in TGT growth stems from the lack of convection in the melt due to the stabilizing temperature gradient discussed above. This aspect is only a disadvantage for growth of multicomponent systems that either grow noncongruently or show selective evaporation of one or more of the components. In this case, convection is absolutely required to homogenize the melt continually.

Blind seeding is a commonly cited disadvantage of the TGT, VBT, and HEM techniques. A combination of diagnostics (usually strategically placed thermocouples) and experience can overcome this disadvantage, even for growth in very low temperature gradients.

15.3.3 Heat-Exchanger Method (HEM)

The HEM process shares most of the advantages and disadvantages of TGT method. The most important differences are noted below.

Advantages of HEM

One of the unique features of HEM is that there are no significant temperature gradients built into the heat zone. The temperature gradients required for crystal growth are controlled by the furnace and heat-exchanger temperatures. For a given material, the furnace temperature determines the temperature gradients in the liquid, and the heat-exchanger temperature determines the temperature gradients in the solid. These temperatures are sensed and controlled externally and, therefore, the temperature gradients can be varied during the growth cycle. This unique feature allows independent control of temperature gradients in the liquid and solid during most of the growth cycle. Once the growth cycle is established, the furnace and heat exchanger temperatures can be readily automated using microprocessors.

Disadvantages of HEM

The HEM process usually uses helium gas as the heat-exchanger medium, which is very expensive. Furthermore, the common growth cycle is very long, so the consumption of helium gas is high.

15.3.4 Vertical Bridgman Technique (VBT)

One of the most distinguishing characters of VBT is the translation of the crucible vertically through a hot zone with a temperature gradient designed to crystallize the materials. This feature leads to both advantages and disadvantages of the growth process relative to TGT.

Advantages of VBT

In VBT growth, only sufficient temperature gradients are needed near the solid–liquid interface, whilst elsewhere a small one is enough. However in TGT growth, a linear temperature gradient is required all along the crucible from the bottom to the top. Apparently, this will increase the difficulty of establishing the thermal field, especially for high-melting-temperature crystals.

For complicated thermodynamic systems, the Bridgman process can be utilized by adding a third zone to the furnace (Fig. 15.9b). If large crystals, especially slabs, are required, this may be the only practical route to growing these crystals.

Disadvantages of VBT

The employment of the lift mechanism will give rise to flow vibration on the solid–liquid interface, and enhance the risk of striation formation, which can degrade crystal performance.

15.3.5 Horizontal Bridgman Technique (HBT)

In the HBT, the crucible is translated horizontally through the hot zone, in contrast with VBT, which leads to some advantages and disadvantages, as noted below.

Advantages of HBT

By using either a boat or a sealed crucible that is not completely filled with starting material, the problem of a crucible imposing stresses on the crystal during cool-down can be overcome. This assumes that either the crystal has a larger coefficient of thermal expansion than the crucible in one or more directions, or that it does not adhere to the crucible during cool-down.

As in this method the melt height is much smaller than the dimension of its surface, this condition is good for effective withdrawal of the impurities at the expense of evaporation. For multicomponent systems, as discussed above, the lack of a stabilizing temperature field can be an advantage. This produces more intense convection that may be crucial for multicomponent systems. Besides, the open surface of the melt provides the opportunity to insert the activating impurity at any stage of crystallization.

HBT possesses the advantage that crystallization is carried out under conditions facilitating the evaporation of extraneous impurities. As a result the chemical purity of the crystal is enhanced on average by one order of magnitude compared with the raw material.

Disadvantages of HBT

The more intense flows generated by the horizontal Bridgman configuration can produce a highly concave interface shape. This can be counteracted by suitable modification of the externally imposed furnace temperature profile, but the process is interactive and definitely not straightforward.

Melt flows in the horizontal configuration can be vigorous, and in addition to complicating control of interface shape, they can also produce striations. This problem can only be overcome by modifying the temperature gradients or the ampoule geometry.

15.3.6 Laser-Heated Pedestal Growth (LHPG)

LHPG and the related float zone growth technique are microvariants of the CZ growth method. Several advantages of LHPG have become apparent, not only in the growth of fibers for applications but, more importantly, as a general way to explore material synthesis and the properties of crystal growth. Other practical advantages of the LHPG method have also become apparent, as follows.

The LHPG relies on surface tension to maintain the integrity of the melt and hence it does not require crucibles, nor does the enclosure containing the fiber growth region possess walls heated to high temperatures as is the case in crystal growth furnaces. Both crucible and furnace surfaces are generally understood to be the primary sources of contamination in normal crystal growth, hence it follows that the absence of these surfaces allows the growth of very pure crystal materials. The impurity levels found in LHPG fibers are solely determined by the purity of the starting materials of the source rod. Furthermore, it is also generally accepted that thermal gradients within the melt container are responsible for introducing stresses and other defects in bulk crystals. Because of this, LHPG pulled fibers can be made practically stress free.

The source rod length as well as the melt volume in LHPG is typically small, of the order of 10 mm and 1 mm^3, respectively. The cost of the chemical compounds required for the growth of single-crystal fibers is, as a consequence, relatively small. Because of this, it is possible to grow fiber crystals of materials which would be prohibitively expensive to grow by traditional methods, especially on a basis. The small volume of the growth area also facilitates the introduction of external perturbations during synthesis of the crystal. The application of an external field to the melt may influence the growing process by encouraging the inclusion of domains or the formation of other stoichiometric combinations.

One of the most attractive features of the LHPG methods is the rapidity with which fibers can be grown by this method. The information feedback made possible by this allows for rapid readjustment of stock compositions and growth conditions for optimized materials.

15.3.7 Flux Technique (FT)

The FT is interesting from the point of view of the peculiarities of crystallization, especially at relatively low temperature. For those crystals that cannot be easily grown from melt, FT may be the most useful method to obtain them. The relatively low growth rate is an obvious disadvantage of this technique.

References

15.1 T.H. Maiman: Stimulated optical radiation in ruby masers, Nature **187**, 493–494 (1960)

15.2 H.G. Danielmeyer, F.W. Ostermayer: Diode-pump-modulated Nd:YAG laser, J. Appl. Phys. **43**, 2911–2913 (1972)

15.3 P.F. Moulton: Spectroscopic and laser characteristics of Ti:Al_2O_3, J. Opt. Soc. Am. B **3**, 125–133 (1986)

15.4 L. De Shazer: Vanadate crystals exploit diode-pump technology, Laser Focus World **30**, 88 (1994)

15.5 P. Lacovara, H.K. Choi, C.A. Wang, R.L. Aggarwal, T.Y. Fan: Room-temperature diode-pumped Yb:YAG laser, Opt. Lett. **16**, 1089–1091 (1991)

15.6 R. Scheps, J.F. Myers, S.A. Payne: End-pumped Yb-doped fluorapatite laser, IEEE Photon. Technol. Lett. **5**, 1285–1288 (1993)

15.7 Y. Jeong, J.K. Sahu, D.N. Payne, J. Nilsson: Ytterbium-doped large-core fiber laser with 1.36 kW continuous-wave output power, Opt. Exp. **12**, 6088–6092 (2004)

15.8 A. Ikesue, T. Kinoshita, K. Kamata, K. Yoshida: Fabrication and optical properties of high-performance polycrystalline Nd:YAG ceramics for solid-state lasers, J. Am. Ceram. Soc. **78**, 1033–1040 (1995)

15.9 J. Czochralski: A new method for the measurement of crystallization rate of metals, Z. Ver. Deutsch. Ing. **61**, 245–351 (1917)

15.10 G.K. Teal, J.B. Little: Growth of germanium crystals, Phys. Rev. **78**, 647 (1950)

15.11 A.E. Paladino, B.D. Roiter: Czochralski growth of sapphire, J. Am. Ceram. Soc. **47**, 465–468 (1964)

15.12 B. Cockayne, M. Chesswas, D.B. Gasson: Single-crystal growth of sapphire, J. Mater. Sci. **2**, 7–11 (1967)

15.13 M. Katsurayama, Y. Anzai, A. Sugiyama, M. Koike, Y. Kato: Growth of neodymium doped $Y_3Al_5O_{12}$ single crystals by double crucible method, J. Cryst. Growth **229**, 193–198 (2001)

15.14 K.J. Gärtner, K.F. Rittinghaus, A. Seeger, W. Uelhoff: An electronic device including a TV-system for controlling the crystal diameter during Czochralski growth, J. Cryst. Growth **13/14**, 619 (1972)

15.15 T.R. Kyle, G. Zydzik: Automated crystal puller, Mater. Res. Bull. **8**, 442–450 (1973)

15.16 R.C. Reinert, M.A. Yatsko: Crystal weighing mechanism for growth monitoring of Czochralski grown crystals, J. Cryst. Growth **21**, 283–286 (1974)

15.17 D.T.J. Hurle: Control of diameter in Czochralski and related crystal growth techniques, J. Cryst. Growth **42**, 473–482 (1977)

15.18 E. Kanchanavaleerat, D. Cochet-Muchy, M. Kokta, J. Stone-Sundberg, P. Sarkies, J. Sarkies, J. Sarkies: Crystal growth of high doped Nd:YAG, Opt. Mater. **26**, 337–341 (2004)

15.19 G. Boulon: Yb^{3+}-doped oxide crystals for diode-pumped solid state lasers: crystal growth, optical spectroscopy, new criteria of evaluation and combinatorial approach, Opt. Mater. **22**, 85–87 (2003)

15.20 L. Fornasiero, E. Mix, V. Peters, K. Petermann, G. Huber: Czochralski growth and laser parameters of RE^{3+}-doped Y_2O_3 and Sc_2O_3, Ceram. Int. **26**, 589–592 (2000)

15.21 K. Shimamura, H. Sato, A. Bensalah, V. Sudesh, H. Machida, N. Sarukura, T. Fukuda: Crystal growth of fluorides for optical applications, Cryst. Res. Technol. **36**, 801–813 (2001)

15.22 S.L. Baldochi, K. Shimamura, K. Nakano, N. Mujilatu, T. Fukuda: Growth and optical characteristics of Ce-doped and Ce:Na-codoped $BaLiF_3$ single crystals, J. Cryst. Growth **200**, 521–526 (1999)

15.23 S. Licia Baldochi, K. Shimamura, K. Nakano, N. Mujilatu, T. Fukuda: Ce-doped $LiYF_4$ growth under CF_4 atmosphere, J. Cryst. Growth **205**, 537–542 (1999)

15.24 F. Cui, Y. Zhou, J. Qiao: Growth of high quality monocrystal sapphire by seed-induced temperature gradient technique (STGT), J. Chin. Ceram. Soc. **8**, 109–113 (1980)

15.25 Y. Zhou: A temperature gradient furnace for growing high-melting crystals, Chin. Patent 85100534.9 (1985)

15.26 Y. Zhou, H. Xia, Z. Huang, M. Lu, P. Deng, J. Qiao, Q. Zhang, B. Hu: Growth of large-size $Ti:Al_2O_3$ tunable laser crystal, SPIE **1627**, 230–233 (1992)

15.27 Y. Zhou: Growth of high quality large Nd:YAG crystals by temperature gradient technique (TGT), J. Cryst. Growth **78**, 31–35 (1986)

15.28 G. Zhao, H. Li, J. Zhu, M. Jie, X. He, J. Xu: The temperature gradient technique (TGT) growth and optical properties of Yb-doped $YAlO_3$ single crystal, J. Cryst. Growth **280**, 483–489 (2005)

15.29 L. Su, J. Xu, H. Li, L. Wen, W. Yang, Z. Zhao, J. Si, Y. Dong, G. Zhou: Crystal growth and spectroscopic characterization of Yb-doped and Yb,Na-codoped CaF_2 laser crystals by TGT, J. Cryst. Growth **277**, 264–268 (2005)

15.30 D.C. Harris: A peek into the history of sapphire crystal growth, Proc. SPIE **5078**, 1–11 (2003)

15.31 J. Dong, P. Deng: Ti:sapphire crystal used in ultrafast lasers and amplifiers, J. Cryst. Growth **261**, 514–519 (2004)

15.32 J. Xu, Y. Zhou, H. Li, P. Deng: Growth of high-doped Nd:YAG laser crystals, Proc. SPIE **3889**, 420–421 (2000)

15.33 F. Schmid, D. Viechnicki: Growth of sapphire disks from the melt by a gradient furnace technique, J. Am. Ceram. Soc. **53**, 528 (1970)

15.34 C.P. Khattak, A.N. Scoville: Growth of laser crystals by heat exchanger method (HEM), SPIE **681**, 58–61 (1986)

15.35 V. Peters, A. Bolz, K. Petermann, G. Huber: Growth of high-melting sesquioxides by the heat exchanger method, J. Cryst. Growth **237-239**, 839–883 (2002)

15.36 C.P. Khattak, P.J. Guggenheim, F. Schmid: Growth of 15-inch diameter sapphire boules, Proc. SPIE **5078**, 47–53 (2003)

15.37 C.P. Khattak, F. Schmid: Growth of large-diameter crystals by HEM for optical and laser application, SPIE **505**, 4–8 (1984)

15.38 J.L. Caslavsky, D. Viechnicki: Melt growth of $Nd:Y_3Al_5O_{12}$ (Nd:YAG) using the heat exchange method (HEM), J. Cryst. Growth **46**, 601–606 (1979)

15.39 C.P. Khattak, F. Schmid, K.F. Wall, R.L. Aggarwal: Growth and characterization of $Ti:YAlO_3$ for turnable solid state laser applications, Proc. SPIE **1104**, 95–99 (1989)

15.40 P.W. Bridgman: Certain physical properties of single crystals of tungsten, antimony, bismuth, tellurium, cadmium, zinc, and tin, Proc. Am. Acad. Arts Sci. **60**, 305–383 (1925)

15.41 D.C. Stockbarger: The Production of large single crystals of lithium fluoride, Rev. Sci. Instrum. **7**, 133–136 (1936)

15.42 D.C. Stockbarger: The production of large artificial fluorite crystals, Disc. Faraday Soc. **5**, 294–299 (1949)

15.43 H.E. Buckley: *Crystal Growth* (Wiley, New York 1951) pp. 71–99

15.44 A.G. Petrosyan: Crystal growth of laser oxides in the vertical Bridgman configuration, J. Cryst. Growth **139**, 372–392 (1994)

15.45 J. Xu, M. Shi, B. Lu, X. Li, A. Wu: Bridgman growth and characterization of calcium fluoride crystals, J. Cryst. Growth **292**, 391–394 (2006)

15.46 K.S. Bagdasarov: http://www.bagdasarovcrystals.com/v1/ (2007)

15.47 G. Boulon, M. Ito, C. Goutaudier, Y. Guyot: Advances in growth of fiber crystal by the LHPG technique. Application to the optimization of Yb^{3+}-doped CaF_2 laser crystals, J. Cryst. Growth **292**, 230–235 (2006)

15.48 R.P. Poplawsky: Ferrite crystals using an arc image furnace, J. Appl. Phys. **33**, 1616 (1962)

15.49 R.S. Feigelson: Pulling optical fibers, J. Cryst. Growth **79**, 669–680 (1986)

15.50 W.M. Yen: Synthesis, characterization, and applications of shaped single crystals, Phys. Solid State **41**, 693–696 (1999)

15.51 L. Laversenne, C. Goutaudier, Y. Guyot, M.T. Cohen-Adad, G. Boulon: Growth of rare earth (RE) doped concentration gradient crystal fibers and analysis of dynamical processes of laser resonant transitions in RE-doped Y_2O_3 ($RE = Yb^{3+}$, Er^{3+}, Ho^{3+}), J. Alloys Compd. **341**, 214–219 (2002)

15.52 R.A. Lefever, J.W. Torpy, A.B. Chase: Growth of single crystals of yttrium aluminum garnet from lead oxide–lead fluoride melts, J. Appl. Phys. **32**, 962–963 (1961)

15.53 R.C. Linares: Substitution of aluminum and gallium in single-crystal yttrium iron garnets, J. Am. Ceram. Soc. **48**, 68–70 (1965)

15.54 L.G. van Uitert, W.H. Grodkiewicz, E.F. Dearborn: Growth of large optical-quality yttrium and rare-earth aluminum garnets, J. Am. Ceram. Soc. **48**, 105–108 (1965)

15.55 J.W. Nielsen: Garnet gemstones, US Patent 3091540 (1963)

15.56 G. Garton, B.M. Wanklyn: The rare earth aluminates, J. Cryst. Growth **1**, 164–166 (1967)

15.57 S. Erdei, B.M. Jin, F.W. Ainger, B. Keszei, J. Vandlik, A. Suveges: Possible trends for the growth of low scattering $Nd{:}YVO_4$ laser crystals; phase relations-growth techniques, J. Cryst. Growth **172**, 466–472 (1997)

15.58 X. Han, G. Wang, T. Tsuboi: Growth and spectral properties of Er^{3+}/Yb^{3+}-codoped $KY(WO_4)_2$ crystal, J. Cryst. Growth **242**, 412–420 (2002)

15.59 C. Pujol, M. Aguil, F. Diaz, C. Zaldo: Growth and characterisation of monoclinic $KGd_{1-x}RE_x(WO_4)_2$ single crystals, Opt. Mater. **13**, 33–40 (1999)

15.60 N.I. Leonyuk, E.V. Koporulina, V.V. Maltsev, O.V. Pilipenko, M.D. Melekhova, A.V. Mokhov: Crystal growth and characterization of $YAl_3(BO_3)_4$ doped with Sc, Ga, Pr, Ho, Tm, Yb, Opt. Mater. **26**, 443–447 (2004)

15.61 E. Fahey, A.J. Strauss, A. Sanchez, R.L. Aggarwal: Growth of $Ti{:}Al_2O_3$ crystals by a gradient-freeze technique. In: *Tunable Solid State Lasers II*, Springer Series in Optical Sciences, Vol. 52, ed. by A.B. Budgor, L. Esterowitz, L.G. DeShazer (Springer, New York 1987) pp. 82–88

15.62 R. Uecker, D. Klimm, S. Ganschow, P. Reiche, R. Bertram, M. Robberg, R. Fornari: Czochralski growth of Ti:sapphire laser crystals, Proc. SPIE **5990**, 599006-1–5990006-9 (2005)

15.63 L.J. Atherton, S.A. Payne, C.D. Brandle: Oxide and fluoride laser crystals, Annu. Rev. Mater. Sci. **23**, 453–502 (1993)

15.64 P.F. Moulton: Spectroscopic and laser characteristics of $Ti{:}Al_2O_3$, J. Opt. Soc. Am. E **3**, 125–133 (1986)

15.65 P. Albers, E. Stark, G. Hube: Continuous-wave laser operation and quantum efficiency of titanium-doped sapphire, J. Opt. Soc. Am. B **3**, 134–139 (1986)

15.66 G.F. Albrecht, J.M. Eggleston, J.J. Ewing: Measurements of $Ti^{3+}{:}Al_2O_3$ as a lasing material, Opt. Commun. **52**, 401–404 (1985)

15.67 R.C. Powell, J.K. Caslavsky, Z. AlShaieb, J.M. Bowen: Growth, characterization, and optical spectroscopy of $Al_2O_3{:}Ti^{3+}$, J. Appl. Phys. **58**, 2331–2336 (1985)

15.68 R.L. Aggarwal, A. Sanchez, M.M. Stuppi, R.E. Fahey, A.J. Strauss, W.R. Rapoport, C.P. Khattak: Residual infrared absorption in As grown and annealed crystals of $Ti{:}Al_2O_3$, IEEE J. Quantum Electron. **24**, 1003–1008 (1988)

15.69 G.A. Keig: Influence of the valence state of added impurity ions on the observed color in doped aluminum oxide single crystals, J. Cryst. Growth **2**, 356–360 (1968)

15.70 T.P. Jones, R.L. Coble, C.J. Mogab: Defect diffusion in single crystal aluminum oxide, J. Am. Ceram. Soc. **52**, 331–334 (1969)

15.71 K.S. Mohapatra, F.A. Kroger: Defect structure of α-Al_2O_3 doped with titanium, J. Am. Ceram. Soc. **60**, 381–387 (1977)

15.72 M.R. Kokta: Processes for enhancing $Ti{:}Al_2O_3$ tunable laser crystal fluorescence by annealing, US Patent 4587035 (1986)

15.73 M.R. Kokta: Processes for enhancing fluorescence of tunable titanium-doped oxide laser crystals, US Patent 4988402 (1991)

15.74 K.S. Bagdasarov, E.R. Dobrovinskaya, V.V. Pishchik, M.M. Chernick, Y.Y. Kovalev, A.S. Gershum, I.F. Zvyagintseva: Low dislocation density single crystals of corundum, Soy. Phys. Crystallogr. **18**, 242–245 (1973)

15.75 C.T. Bodur, J. Chang, A.S. Argon: Molecular dynamics simulations of basal and pyramidal system edge dislocations in sapphire, J. Europ. Ceram. Soc. **25**, 1431–1439 (2005)

15.76 A. Nakamura, T. Yamamoto, Y. Ikuhara: Direct observation of basal dislocation in sapphire by HRTEM, Acta Mater. **50**, 101–108 (2002)

15.77 C.P. Khattak, A.N. Scoville, F. Schmid: Recent developments in sapphire growth by heat exchanger method (HEM), SPIE **683**, 32–35 (1986)

15.78 J. Xu, Y. Zhou, G. Zhou, J. Xu, P. Deng: Producing large ⟨0001⟩-oriented sapphire for optical applications, SPIE **3557**, 11–14 (1998)

15.79 J. Xu, Y. Zhou, G. Zhou, K. Xu, P. Deng, J. Xu: Growth of large-sized sapphire boules by temperature gradient technique (TGT), J. Cryst. Growth **193**, 123–126 (1998)

15.80 J.R. Carruthers: Origins of convective temperature oscillations in crystal growth melts, J. Cryst. Growth **32**, 13–26 (1976)

15.81 T. Surek: Theory of shape stability in crystal growth from the melt, J. Appl. Phys. **47**, 4384–4393 (1976)

15.82 J. Basterfield, M.J. Prescott, B. Cockayne: An X-ray topographic study of single crystals of melt-grown yttrium aluminium garnet, J. Mater. Sci. **3**, 33–40 (1968)

15.83 N. Ming, Y. Yang: Facet and vicinical growth of YAG by Czochralski, Acta Phys. Sin. **28**, 285–295 (1979)

15.84 B. Cockayne, M. Chesswas, D.B. Gasson: The growth of strain-free $Y_3Al_5O_{12}$ single crystals, J. Mater. Sci. **3**, 224–225 (1968)

15.85 G. Zydzik: Interface transitions in Czochralski growth of garnets, Mater. Res. Bull. **10**, 701–707 (1975)

15.86 B. Cockayne, B. Lent: A complexity in the solidification behaviour of molten $Y_3Al_5O_{12}$, J. Cryst. Growth **46**, 371–378 (1979)

15.87 E.W. O'Dell, D.J. Nelson, D. Narasimhan, R.C. Morris, J.E. Marion: Development of a large scale Nd:YAG growth process, SPIE **1223**, 94–102 (1990)

15.88 B. Cockayne, M. Chesswas, D.B. Gasson: Facetting and optical perfection in Czochralski grown garnets and rubies, J. Mater. Sci. **4**, 450–456 (1969)

15.89 Y. Miyazawa, M. Mori, S. Honma: Interface shape transitions in the Czochralski growth of $Dy_3Al_5O_{12}$, J. Cryst. Growth **43**, 541–542 (1978)

15.90 A.G. Petrosyan: Crystal growth of laser oxides in the vertical Bridgman configuration, J. Cryst. Growth **139**, 372–392 (1994)

15.91 P. Deng, J. Qiao, B. Hu, Y. Zhou, M. Zhang: Perfection and laser performances of Nd:YAG crystals grown by temperature gradient technique (TGT), J. Cryst. Growth **92**, 276–286 (1988)

15.92 R.C. Pastor: Effect of RAP purification on materials characterization, J. Cryst. Growth **75**, 54–60 (1986)

15.93 M. Robinson: Processing and purification techniques of heavy metal fluoride glass (HMFG), J Cryst. Growth **75**, 184–194 (1986)

15.94 C.D. Brandle, A.J. Valentino, G.W. Berkstresser: Czochralski growth of rare-earth orthosilicates (Ln_2SiO_5), J. Cryst. Growth **79**, 308–315 (1986)

15.95 D. Mateika, J. Herrnring, R. Rath, C. Rusche: Growth and investigation of $\{Gd_{3-x}Ca_x\}[Ga_{2-y-z}Zr_yGd_z]$ $(Ga_3)O_{12}$ garnets, J. Cryst. Growth **30**, 311–316 (1975)

15.96 D. Mateika, C. Rusche: Coupled substitution of gallium by magnesium and zirconium in single crystals of gadolinium gallium garnet, J. Cryst. Growth **42**, 440–444 (1977)

15.97 D.C. Miller: Defects in garnet substrates and epitaxial magnetic garnet films revealed by phosphoric acid etching, J. Electrochem. Soc. **120**, 678–685 (1973)

15.98 C.D. Brandle, D.C. Miller, J.W. Nielsen: The elimination of defects in Czochralski grown rare-earth gallium garnets, J. Cryst. Growth **12**, 195–200 (1972)

15.99 G. Boulon, A. Brenier, L. Laversenne, Y. Guyot, C. Goutaudier, M.T. Cohen-Adad, G. Metrat, N. Muhlstein: Search of optimized trivalent ytterbium doped-inorganic crystals for laser applications, J. Alloys Compd. **341**, 2–7 (2002)

15.100 D. De Loach, S.A. Payne, L.L. Chase, L.K. Smith, W.L. Kway, W.F. Krupke: Evaluation of absorption and emission properties of Yb^{3+} doped crystals for laser applications, IEEE J. Quantum Electron. **29**, 1179–1191 (1993)

15.101 X. Xu, Z. Zhao, J. Xu, P. Deng: Distribution of ytterbium in Yb:YAG crystals and lattice parameters of the crystals, J. Cryst. Growth **255**, 338–341 (2003)

15.102 P. Lacovara, H.K. Choi, C.A. Wang, R.L. Aggarwal, T.Y. Fan: Room-temperature diode-pumped Yb:YAG laser, Opt. Lett. **16**, 1089–1090 (1991)

15.103 F. Lu, M. Gonga, H. Xueb, Q. Liu, W. Gong: Analysis on the temperature distribution and thermal effects in corner-pumped slab lasers, Opt. Lasers Eng. **45**, 43–48 (2007)

15.104 X. Xu, Z. Zhao, J. Xu, P. Deng: Thermal diffusivity, conductivity and expansion of $Yb_{3x}Y_{3(1-x)}Al_5O_{12}$ (x=0.05, 0.1 and 0.25) single crystals, Solid State Commun. **130**, 529–532 (2004)

15.105 S.A. Payne, R.J. Beach, C. Bibeau, C.A. Ebbers, M.A. Emanuel, E.C. Honea, C.D. Marshall, R.H. Page, K.I. Schaffers, J.A. Skidmore, S.B. Sutton, W.F. Krupke: Diode arrays, crystals, and thermal management for solid-state lasers, IEEE J. Quantum Electron. **3**, 71–81 (1997)

15.106 D.J. Ripin, J.R. Ochoa, R.L. Aggarwal, T.Y. Fan: 300-W cryogenically cooled Yb:YAG laser, IEEE J. Quantum Electron. **41**, 1274–1277 (2005)

15.107 S. Chenais, F. Druon, F. Balembois, P. Georges, A. Brenier, G. Boulon: Diode-pumped Yb:GGG laser: comparison with Yb:YAG, Opt. Mater. **22**, 99–106 (2003)

15.108 J.B. Willis, M. Dixon: Assessment and control of imperfections in crystals for laser devices, J. Cryst. Growth **3-4**, 236–240 (1968)

15.109 V.A. Antonov, P.A. Arsenev, I.G. Linda, V.L. Farshtendiker: Studies of some point defects in $YAlO_3$ and $GdAlO_3$ single crystals, Phys. Status Solidi (a) **5**, K63–K68 (1973)

15.110 B. Cockayne, B. Lent, J.S. Abell, I.R. Harris: Cracking in yttrium orthoaluminate single crystals, J. Mater. Sci. **8**, 871–875 (1973)

15.111 P. Yang, P. Deng, J. Xu, Z. Yin: Growth of high-quality single crystal of 30 at. % Yb:YAG and its laser performance, J. Cryst. Growth **216**, 348–351 (2000)

15.112 H. Qiu, P. Yang, J. Dong, P. Deng, J. Xu, W. Chen: The influence of Yb concentration on laser crystal Yb:YAG, Mater. Lett. **55**, 1–4 (1998)

15.113 X. Xu, Z. Zhao, G. Zhao, P. Song, J. Xu, P. Deng: Comparison of Yb:YAG crystals grown by CZ and TGT method, J. Cryst. Growth **257**, 297–300 (2003)
15.114 C.Y. Chen, G.J. Pogatshnik, Y. Chen, M.R. Kokta: Optical and electron paramagnetic resonance studies of Fe impurities in yttrium aluminum garnet crystals, Phys. Rev. B **38**, 8555–8561 (1988)
15.115 K. Mori: Transient colour centres caused by UV light irradiation in yttrium aluminium garnet crystals, Phys. Status Solidi (a) **42**, 375–384 (1977)
15.116 T.I. Butaeva, K.L. Ovanesyan, A.G. Petrosyan: Growth and spectral investigations of oxide laser crystals with Pr^{3+} ions, Cryst. Res. Technol. **23**, 849–854 (1988)
15.117 R.F. Belt, J.R. Latore, R. Uhrin, J. Paxton: EPR and optical study of Fe in Nd:$YAlO_3$ laser crystals, Appl. Phys. Lett. **25**, 218–220 (1974)
15.118 J. Kvapil, J. Kvapil, J. Kubelka, R. Autrata: The role of iron ions in YAG and YAP, Cryst. Res. Technol. **18**, 127–131 (1983)
15.119 J. Kvapil, B. Perner, M. Košelja, J. Kvapil: Aborption background and laser properties of YAP:Nd, Czech. J. Phys. **40**, 99–108 (1990)
15.120 Y. Dong, G. Zhou, J. Xu, G. Zhao, F. Su, L. Su, H. Li, J. Si, X. Qian, X. Li, J. Shen: Color centers and charge state recharge in γ-irradiated Yb:YAP, Opt. Mater. **28**, 1377–1380 (2006)
15.121 A. Matkovski, A. Durygin, A. Suchocki, D. Sugak, G. Neuroth, F. Wallrafen, V. Grabovski, I. Solski: Photo and gamma induced color centers in the $YAlO_3$ and $YAlO_3$:Nd single crystals, Opt. Mater. **12**, 75–81 (1999)
15.122 D. Sugak, A. Matkovski, D. Savitski, A. Durygin, A. Suchocki, Y. Zhydachevskii, I. Solskii, I. Stefaniuk, F. Wallrafen: Growth and induced color centers in $YAlO_3$-Nd single crystals, Phys. Status Solidi (a) **184**, 239–250 (2001)
15.123 S.M. Kaczmarek: Role of the type of impurity in radiation influence on oxide compounds, Cryst. Res. Technol. **34**, 737–743 (1999)
15.124 V.G. Baryshevsky, M.V. Korzhik, B.I. Minkov, S.A. Smirnova, A.A. Fyodorov, P. Dorenbos, C.W.E. van Eijk: Spectroscopy and scintillation properties of cerium doped $YAlO_3$ single crystals, J. Phys. Condens. Matter **5**, 7893–7902 (1993)
15.125 J. Dong, M. Bass, Y. Mao, P. Deng, F. Gan: Dependence of the Yb^{3+} emission cross section and lifetime on temperature and concentration in yttrium aluminum garnet, J. Opt. Soc. Am. B **20**, 1975–1979 (2003)
15.126 T.Y. Fan, R.L. Byer: Diode laser pumped solid-state laser, IEEE J. Quantum Electron. **24**, 895–912 (1988)
15.127 F.D. Patel, E.C. Honea, J. Speth, S.A. Payne, R. Hutcheson, R. Equall: Laser demonstration of $Yb_3Al_5O_{12}$ (YbAG) and materials properties of highly doped Yb:YAG, IEEE J. Quantum Electron. **37**, 135–144 (2001)
15.128 R. Sang: *Etching of Crystal Theory, Experiment and Application* (North-Holland, Amsterdam 1987) p. 303
15.129 P. Yang, P. Deng, Z. Yin, Y. Tian: The growth defects in Czochralski-grown Yb:YAG crystal, J. Cryst. Growth **218**, 87–92 (2000)
15.130 P. Song, Z. Zhao, X. Xu, P. Deng, J. Xu: Defect analysis in Czochralski-grown Yb:FAP crystal, J. Cryst. Growth **286**, 498–501 (2006)
15.131 W. Schmidt, R. Weiss: Dislocation propagation in Czochralski grown gadolinium gallium garnet (GGG), J. Cryst. Growth **43**, 515–525 (1978)
15.132 H. Klapper, H. Kuppers: Directions of dislocation lines in crystals of ammonium hydrogen oxalate hemihydrate grown from solution, Acta Cryst. A **29**, 495–503 (1973)
15.133 W.F. Krupke, M.D. Shinn, J.E. Marion, J.A. Caird, S.E. Stokowski: Spectroscopic, optical, and thermomechanical properties of neodymium- and chromium-doped gadolinium scandium gallium garnet, J. Opt. Soc. Am. B **3**, 102–114 (1986)
15.134 S.A. Payne, L.L. Chase, L.J. Atherton, J.A. Caird, W.L. Kway, M.D. Shinn, R.S. Hughes, L.K. Smith: Properties and performance of the $LiCaAlF_6$:Cr^{3+} laser material, Proc. SPIE **1223**, 84 (1990)
15.135 P. Pan, H. Zhu, S. Yan, Y. Chai, S. Wang, Y. Hou: Distribution and valence of chromium in forsterite crystals grown by the Czochralski technique, J. Cryst. Growth **121**, 141–147 (1992)
15.136 R. Uhrin, R.F. Belt, V. Rosati: Preparation and crystal growth of lithium yttrium fluoride for laser applications, J. Cryst. Growth **38**, 38–44 (1977)
15.137 M.R. Kokta: Process for enhancing Ti:Al_2O_3 tunable laser crystal fluorescence by controlling crystal growth atmosphere, US Pat 4711696 (1987)
15.138 C.B. Finch, G.W. Clark: Czochralski growth of single-crystal Mg_2SiO_4 (forsterite), J. Cryst. Growth **8**, 307–308 (1971)
15.139 H. Takei, T. Kobayashi: Growth and properties of Mg_2SiO_4 single crystals, J. Cryst. Growth **23**, 121–124 (1974)
15.140 C.D. Brandle, V.J. Fratello, A.J. Valentino, S.E. Stokowski: Effects of impurities and atmosphere on the growth of Cr-doped gadolinium scandium gallium garnet. I, J. Cryst. Growth **85**, 223–228 (1987)
15.141 V.J. Fratello, C.D. Brandle, A.J. Valentino, S.E. Stokowski: Effects of impurities and atmosphere on the growth of Cr-doped gadolinium scandium gallium garnet. II, J. Cryst. Growth **85**, 229–233 (1987)
15.142 B.A. Bolev, J.H. Weiner: *Theory of Thermal Stresses* (Wiley, New York 1960) p. 586

16. Shaped Crystal Growth

Vitali A. Tatartchenko

Crystals of specified shape and size (shaped crystals) with controlled defect and impurity structure have to be grown for the successful development of modern engineering. Since the 1950s many hundreds of papers and patents concerned with shaped growth have been published. In this chapter, we do not try to enumerate the successful applications of shaped growth to different materials but rather to carry out a fundamental physical and mathematical analysis of shaping as well as the peculiarities of shaped crystal structures. Four main techniques, based on which the lateral surface can be shaped without contact with the container walls, are analyzed: the Czochralski technique (CZT), the Verneuil technique (VT), the floating zone technique (FZT), and technique of pulling from shaper (TPS). Modifications of these techniques are analyzed as well. In all these techniques the shape of the melt meniscus is controlled by surface tension forces, i.e., capillary forces, and here they are classified as capillary shaping techniques (CST). We look for conditions under which the crystal growth process in each CST is dynamically stable. Only in this case are all perturbations attenuated and a crystal of constant cross section grown *without any special regulation*. The dynamic stability theory of the crystal growth process for all CST is developed on the basis of Lyapunov's dynamic stability theory. Lyapunov's equations for the crystal growth processes follow from fundamental laws. The results of the theory allow the choice of stable regimes for crystal growth by all CST as well as special designs of shapers in TPS. SCG experiments by CZT, VT, and FZT are discussed but the main consideration is given to TPS. Shapers not only allow crystal of very complicated cross section to be grown but provide a special distribution of impurities. A history of TPS is provided later in the chapter, because it can only be described after explanation of the fundamental principles of shaping. Some shaped crystals, especially sapphire and silicon, have specified structures. The crystal growth of these materials, and some metals, including crystal growth in space, is discussed.

16.1 Definitions and Scope of Discussion: SCG by CST

Modern engineering usually uses device details fabricated from crystals in the shape of plates, rods or tubes. Sometimes the shapes can be more complicated. Traditional ways of the detail fabrication (a growth of a bulk crystal and its machining) bring a loss of expensive material (often up to 90%) as well as an appearance of structure defects. Therefore, crystals of specified shape and size with controlled defect and impurity structure have to be grown. It allows using the crystals as final products with minimal additional machining as well as without one.

The problem of shaped crystal growth seems to be simply solved by profiled container crystallization just as in the case of casting. Indeed, it is possible to find a realization of this idea by the vertical or horizontal Bridgman techniques for growth of silicon, fluoride, sapphire or YAG crystals with different cross section of the crucible used [16.1–3]. But in these cases the crucible material should satisfy a whole set of requirements: it should neither react with the melt nor be wetted by it. Even if all these requirements are satisfied, perfect-crystal growth is not secured: the crucible serves as a source of noncontrolled nucleation as well as internal residual stresses. In addition, if a crucible material is wetted by the melt, the crucible usually has been made from a thin foil and used only once [16.2, 3].

Therefore, the techniques of crystal lateral surface shaping without contact with container walls have to be considered as the candidates for a shaped crystal growth. Since early sixties, both theoretical and practical aspects of the shaped crystal growth by these techniques have been developed. The main information

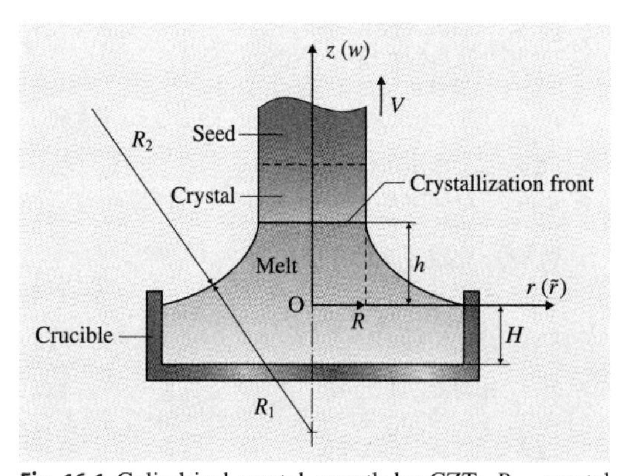

Fig. 16.1 Cylindrical crystal growth by CZT: R – crystal radius; h – crystallization front height; V – crystal pulling rate; H – melt level in crucible; $wO\tilde{r}$ or zOr – coordinate systems; R_1 and R_2 – main radii of liquid surface curvature

Fig. 16.2 Cylindrical crystal growth by VT: H_2 – hydrogen flow; O_2 – mixed oxygen and powder flow; R – crystal radius; l_1 – melt surface position relative to the burner; l – crystallization front position relative to the burner; $h = l - l_1$ is the melt meniscus height; V – crystal displacement rate; ω – crystal rotation rate ▶

Fig. 16.4a–j Melt growth of crystalline rod by TPS **(a–j)**: pulling up **(a–e,h,i)** or lowering down **(f,g,j)** with rate V; shaping on the shaper surfaces **(a,c,g–i)**; shaping on the shaper edges **(b,d–f,j)**; positive melt pressure d **(b,f,g,j)**; negative melt pressure d **(d,e)**; C, G are counters of contact meniscus with shaper; $\boldsymbol{n}$ – normal vector to the shaper walls; d_0 – shaper depth; R – crystal radius; r_0 – edge counter radius; h – crystallization front height; Θ – wetting angle; ψ_0 – growth angle; α_d – meniscus inclination angle with respect to positive r-direction at the point of the contact with the shaper; $\alpha_1 = \pi - \alpha_d$; β – angle of cone shaper wall inclination; r_1 – cone radius on the melt free surface level ▶

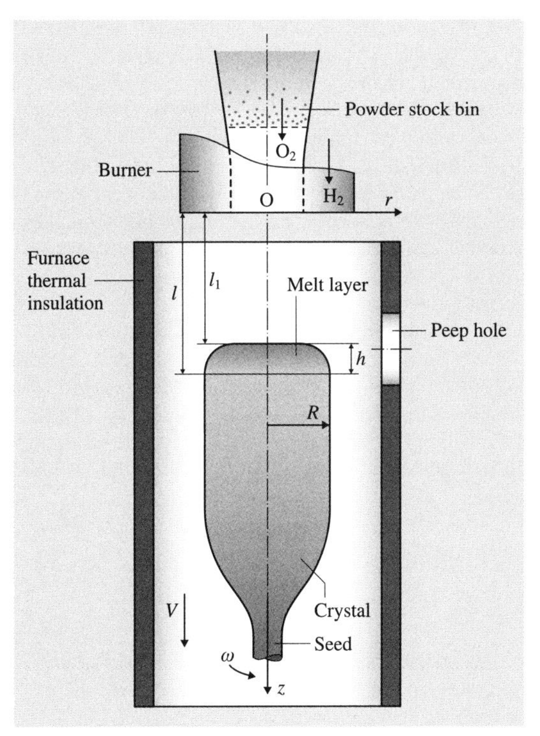

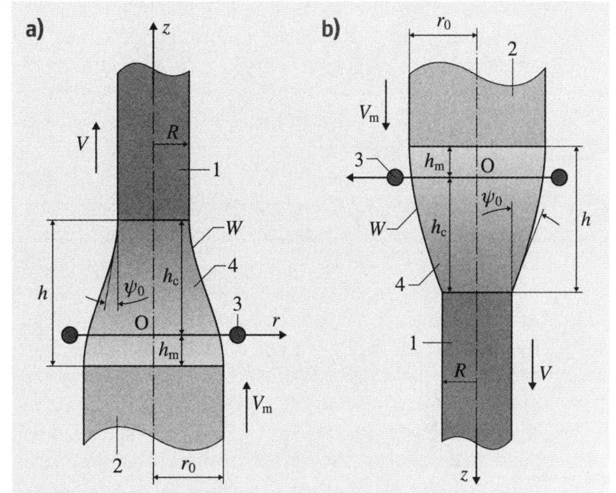

Fig. 16.3a,b Cylindrical crystal growth by FZT. (**a**) Pulling up, (**b**) lowering down: (1) growing crystal with radius R; (2) feeding rod with radius r_0; (3) heater; (4) melted zone; h_c, h_m – positions of crystallization front and melting front relative to the heater, respectively; $h = h_c + h_m$ is the length of the melted zone; W – volume of the melted zone; ψ_0 – growth angle; V – rate of growing crystal displacement relative to the induction heater; V_m – the same for the melting rod

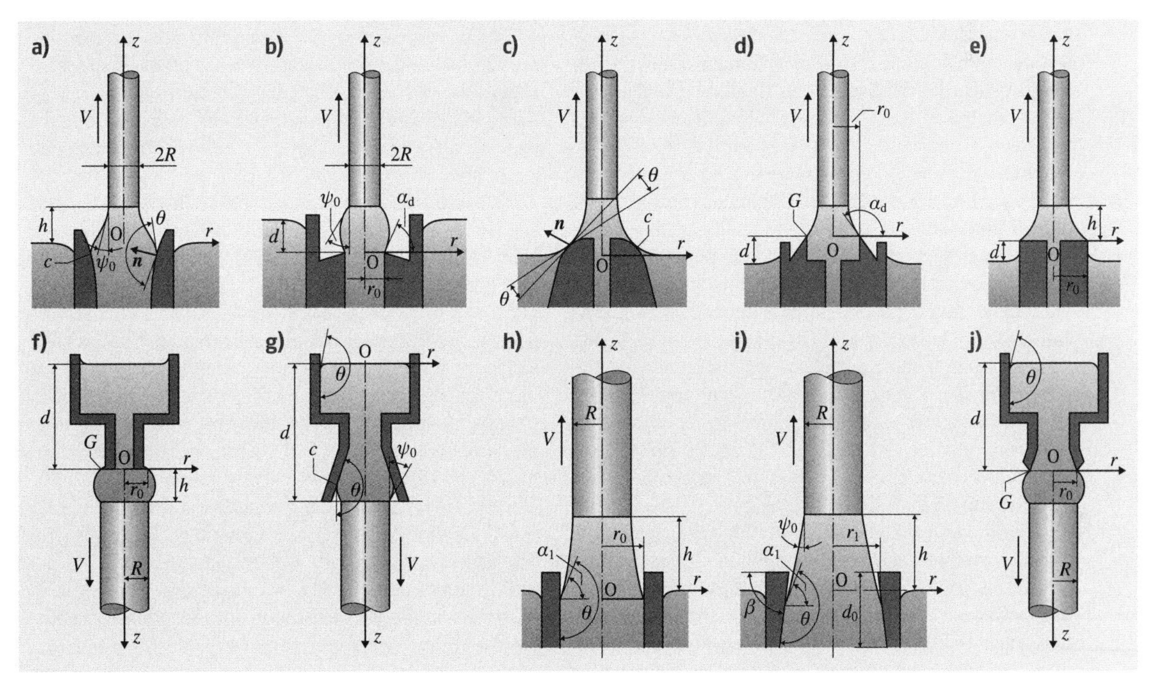

concerned with all of them was published in some books and reviews [16.4–10], in the proceedings of three international conferences [16.11–13] and in hundreds of papers.

The few *classical* techniques of this type are well known: the Czochralski technique (CZT, Fig. 16.1), the Verneuil technique (VT, Fig. 16.2), the floating zone technique (FZT, Fig. 16.3). For all these techniques the shapes and the dimensions of the crystals grown are controlled by the shapes and the dimensions of meniscus of melt existing at the vicinity of interface crystal–melt. A shape of meniscus is controlled by the surface tension forces of the melt – capillary forces. Let the above-mentioned techniques (some *no classical* ones will be added below) be classified as capillary shaping techniques (CST). As a rule, all these techniques are used to grow crystals of nonregular cylindrical shape. On the other hand, all of them have been used for shaped crystal growth as well. For shaped crystal growth it is necessary to guarantee a special shape for the melt meniscus and a dynamical stability of the crystal growth process. But in any case, all above mentioned techniques allow obtaining only simple shaped crystals (cylinders, plates or tubes) and they have to be modified for the growth of more complicated ones. It has to be mentioned that *classical* Kyropoulos technique also belongs to CST but, in fact, it has never been used for the shaped growth because of specific thermal growth conditions [16.8].

Some possible schemes of the modification of *classical* CST are shown on the Fig. 16.4a–j. The schemes Fig. 16.4a–e,h,i differ from CZT by the presence of a shaper at the melt. The schemes Fig. 16.4f,g,j may be classified as FZT modified with lowering down where the melting rod is replaced by a shaper with a melt inside.

16.2 DSC – Basis of SCG by CST

As the crystal is not restricted by crucible walls its cross section depends upon the growth regimes. Any deviations of pulling or lowering (for a down growth) rate, as well as temperature conditions result in changes of the crystal cross section (pinches formation). A lot of defects (an increased amount of inclusions, nonuniform impurity distributions, subgrain formation) are observed at the pinch locations. It is not the pinches themselves, that seem to cause defect formation, but some deviation of the growth conditions (mainly, the crystallization rate) from the optimal ones indicated by a change in the crystal dimensions. Therefore, the stabilization of the crystal cross section as well as the crystallization front position has to be achieved. Why is it very important the stabilization of the crystallization front position? We can see from the schemes of CST (Figs. 16.1–16.4) that, if the crystallization front position is unmoved, a crystal growth speed is exactly equal to the one of pulling or lowering of a seed. A displacement of the crystallization front position changes the real crystal growth, see (16.10). As a result, in spite of the pulling or lowing rate stabilization, there is defect formation.

The modern systems of regulation using weight or crystal diameter detectors allow obtaining the cylindrical crystals by CST. These systems effect with a change of a power or pulling rate. Sometimes, the regulation is not stable and there is a permanent perturbation of the crystallization front position. The solution is to analyze theoretically the dynamic stability of concrete schemes of crystal growth and to select on the basis of this analysis the stable ones. In the dynamically stable system, the perturbations of parameters attenuate because of internal processes and, without any additional active regulation, it is possible to provide crystals of specified shape and of controlled cross section. If the active regulator is included in the system under investigation, this dynamically stable system can improve the shape and the quality of crystals.

For the first time, a comparative theoretical analysis of a dynamic stability of crystallization process for CZT and TPS had been carried out by the author of this chapter in 1971. The very impressive result was published in 1973 [16.14]. It was explained why it is difficult to pull crystals of constant cross section by CZT and easy by TPS: the use of a shaper allows obtaining a dynamic capillary stability of the crystal growth process. In the same year, the first paper concerning the investigation of VT dynamic stability was published [16.15]. In 1974 the investigation of CZT, TPS, VT and FZT stability was presented at the 4th International Conference of Crystal Growth in Japan [16.16]. The analysis of capillary and heat stability in detail using Lyapunovs approach was published in 1976 [16.17]. In 1976 *Surek* published only capillary stability analysis [16.18] repeating our main results for CZT and TPS from [16.14].

16.2.1 Lyapunov Set of Equations

The main results of stability analysis for all CST [16.4–10] were obtained by applying of the Lyapunovs approach [16.19]. With respect to Lyapunov, the crystallization techniques under consideration have to be characterized by a finite number n of the variables (degrees of freedom) X_i which can arbitrarily vary in the process of crystallization. Each CST has to include, as a minimum, crystal dimension R and crystallization front position h as degrees of freedom, i. e. a minimal degree of freedom quantity min n have to be two. Sometimes, it is sufficient for the dynamic stability analysis (CZT, Fig. 16.1; TPS, Fig. 16.4). But for VT, min $n=3$ (Fig. 16.2): R, l, h; for the FZT, min $n=4$ (Fig. 16.3): R, W, h_c, h_m. It has to be mentioned that the min n depends also on the cross section of the crystal to be grown. For instance, for a tube crystal, the internal diameter, as well as the external one, are the degrees of freedom. Therefore, min $n=3$ for CZT and for TPS.

If n exceeds min n, the analysis is more fruitful. Sometimes we can use several iterations. The first iteration can include the stability investigating for min n. After that, one or more variables can be added. For instance, in [16.6, pp. 71–145], the stability of TPS as a system with min n was investigated. As a second step [16.6, pp. 155–159], the melt pressure was added as a third degree of freedom and a complimentary information concerning the influence of pressure perturbation on the stability of growth was obtained. So, we can confirm that min $n \geq 2$ but we can not propose any simple choosing of min n as well as optimal n. For us, in every case, it has been a result of very special investigation.

To realize a mathematical analysis of stability, a set of equations (16.1) for derivation of each degree of freedom X_i with respect to time t as a function of all n degrees of freedom $X_1, \ldots, X_n$, their other (except i) $n-1$ derivatives, time t, and parameters of process C (a temperature of melt, a velocity of pulling, a regime of cooling, etc.) has to be obtained

$$\frac{\mathrm{d}X_i}{\mathrm{d}t} = f_i\left(X_1, X_2, \ldots, X_n, \frac{\mathrm{d}X_1}{\mathrm{d}t}, \frac{\mathrm{d}X_2}{\mathrm{d}t}, \ldots, \frac{\mathrm{d}X_{n-1}}{\mathrm{d}t}, t, C\right), \quad i = 1, 2, \ldots, n . \tag{16.1}$$

To find the explicit function f_i, a set of fundamental laws have to be used. The set should include:

1. The Navier–Stokes equation for a melt with the boundary conditions on the meniscus free surface (the Laplace capillary equation)
2. The continuity equation (the law of crystallizing substance mass conservation)
3. The heat transfer equations for the liquid and the solid phases with the equations of heat balance at the crystallization front and at the melting front as the boundary conditions (the law of energy conservation)
4. The diffusion equation (impurity mass conservation)
5. The growth angle certainty condition
6. Some others.

The set of these equations is general for all the crystallization techniques under consideration (but it does not mean that each time we use all of them) while the specific features of each of crystallization schemes are characterized by the set of boundary conditions and concrete values of the parameters included in the equations.

Equation 16.1 with zero left side corresponds to the system under conditions of equilibrium ($X_i = X_i^0$): the growth of crystals of constant cross section X_1^0 with stationary crystallization front position X_2^0 etc.

$$f_i\left(X_1^0, X_2^0, \ldots, X_n^0, t, C\right) = 0 . \tag{16.2}$$

We are looking for the stable solutions of (16.1). According to Lyapunov [16.19], the solutions of (16.1) are stable if they are stable for the linearized set of equations

$$\frac{\mathrm{d}X_i}{\mathrm{d}t} = \sum_{k=1}^{n} \frac{\partial f_i}{\partial X_k} \delta X_k = \sum_{k=1}^{n} A_{ik} \delta X_k . \tag{16.3}$$

Here, $\delta X_k = X_k - X_k^0$, $\partial f_i / \partial X_k = A_{ik}$, all partial derivatives are taken with $X_k = X_k^0$. The stability of (16.3), in turn, is observed when all the roots S in the characteristic equation (16.4)

$$\det\left(\frac{\partial f_i}{\partial X_k} - S\delta_{ik}\right) = 0 \tag{16.4}$$

have negative real components (δ_{ik} is the Kronecker delta [16.19]). This equilibrium will be unstable if (16.4) has at least one root with a positive real component. If an imaginary number can be found among the roots, additional study including an allowance for the nonlinear terms in (16.3) is required.

Calculation of the time-dependent non stationary functions f_i is usually rather difficult. These difficulties can be avoided using a quasi-stationary approach.

We successfully have been used it in the most of our dynamic stability investigations and can show other examples of the same approach. For instance, *Mullins* and *Sekerka* [16.20] applied it to the temperature and impurity distribution problem while studying the morphological stability of the crystallization-front shape. However, in each particular case, the quasi-stationary approach has to be justified.

A number of constrains imposed on the systems and perturbations occurring in the course of Lyapunov stability study should be noted. Stability is examined over an infinitely long period of time. In this case, the perturbations are considered to be small and are imposed on the initial conditions only, i. e. after the perturbations, the same forces and energy sources affect the system as before the perturbations.

16.2.2 Capillary Problem – Common Approach

Melt Meniscus Shaping Conditions

For the capillary shaping techniques, the crystal cross section is determined by the melt meniscus section formed by the crystallization surface. The melt meniscus shape can be calculated on the basis of the Navier–Stocks equation, the Laplace capillary equation being the free-surface boundary condition.

The full-scale solution of this problem offers considerable mathematical difficulties. Therefore, to simplify the problem formulation, the contributions of various factors of meniscus shaping should be estimated: the inertial forces associated with the melt flow, the capillary forces, the gravity forces, viscous and the thermocapillary forces [16.6, 21]. The relative effect of the first three factors can be estimated by means of dimensionless numbers: the Weber number $\mathrm{We} = \rho V^2 L/\gamma$, characterizing comparative action of the inertial and capillary forces, the Froude number $\mathrm{Fr} = V/(gL)^{1/2}$ characterizing comparative action of the inertial and gravity forces; the Bond number $\mathrm{Bo} = \rho g L^2/\gamma$, characterizing comparative action of the gravity and capillary forces. Here ρ denotes the liquid density, L the liquid meniscus characteristic dimensions, γ the liquid surface tension coefficient, V is the liquid flow rate, and g relates to the gravity acceleration. When the Weber and Froude numbers are small, the melt flow can be neglected. The Bond number defines the region of capillary or gravity force predominance (Fig. 16.5). The effect of the inertial force as compared with the gravity and capillary ones proves to be negligible, if liquid flow rate is considered to

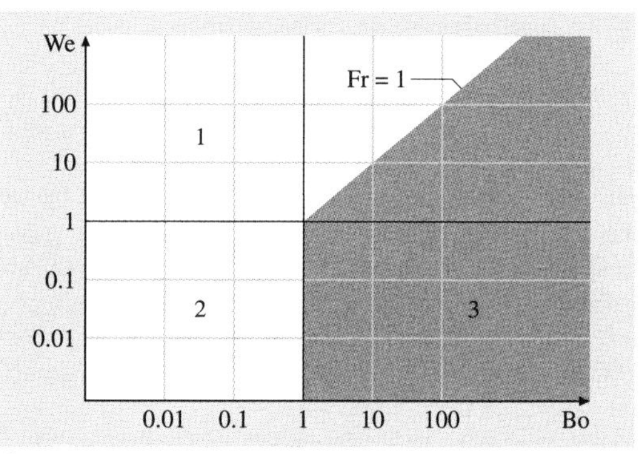

Fig. 16.5 Inertial (1), capillary (2), and gravitational force effects (3) on the melt column shaping (We, Bo, and Fr are characteristic Weber, Bond and Froude numbers, respectively)

be equal to the crystallization rate. Indeed, if we assume that the linear dimensions of the meniscus lie within the range of 10^{-3}–10^{-2} m, $\rho \approx 10^{-3}\,\mathrm{kg\,m^{-3}}$, $\gamma \approx 1\,\mathrm{N\,m^{-1}}$ the liquid meniscus shaping can be examined in the hydrostatic approximation up to the fluid speed of 0.1–$1.0\,\mathrm{m\,s^{-1}}$.

Convective flows, whose rates can substantially exceed the crystallization rate, can occur in a liquid column in addition to the flow associated with crystallization. These flow effects on meniscus shaping and on liquid-phase heat-transfer as well as the influence of the two latter factors can be found in [16.6].

The Meniscus Surface Equation

In the hydrostatic approximation, the equilibrium shape of the liquid surface is described by the Laplace capillary equation [16.22]

$$\frac{\gamma}{R_1} + \frac{\gamma}{R_2} + \rho g w = \mathrm{const}\,. \tag{16.5}$$

Here, R_1 and R_2 denote the main radii of liquid surface curvature. They have to be located in two perpendicular planes. As a rule one of the planes coincides with the diagram plane (R_2, Fig. 16.1) and second one is perpendicular to it (R_1, Fig. 16.1). The w-axis is directed vertically upwards. The value of const. depends upon w-coordinate origin selection and is equal to the pressure p on the liquid in the plane $w = 0$. In particular, if the w-coordinate origin coincides with the plane of the liquid surface, const. $= 0$ (Fig. 16.1).

In this chapter our study will be restricted to considering meniscus possessing axial symmetry (Figs. 16.1–16.4). Such meniscus is obtained during melt pulling of straight circular cylinder, tube-shaped crystals as well as a flat part of ribbon. We will find the equation of such meniscus surface, by introducing the cylindrical coordinates $w, \tilde{r}$. The problem of liquid meniscus shape calculation for an axially symmetric meniscus is reduced to finding the shape of a profile curve $w = f(\tilde{r})$, the liquid surface meniscus being obtained by rotating this curve around the w-axis. Let us introduce the capillary constant a, and pass to the dimensionless coordinates and parameters

$$\left(\frac{2\gamma}{\rho g}\right)^{1/2} = a\,, \quad \frac{w}{a} = z\,,$$
$$\frac{\tilde{r}}{a} = r\,, \quad \frac{pa}{2\gamma} = d\,.$$

This transition means that the capillary constant serves as a linear dimension unit, and the weight of a liquid column of one capillary constant high corresponds to the pressure equal to one unit. The approach allows application of the calculated results to any substance and magnitude of gravity, with the scale changing alone. Then (16.5) takes the form

$$z''r + z'\left(1+z'^2\right) \pm 2(d-z)\left(1+z'^2\right)^{3/2} r = 0\,. \tag{16.6}$$

For large Bond numbers (Bo $\gg$ 1, Fig. 16.5) gravity prevails (this condition corresponds to growing big diameter crystals with $R \geq 5a$) and (16.6) can be simplified

$$z''r \pm 2(d-z)\left(1+z'^2\right)^{3/2} r = 0\,. \tag{16.7}$$

For small Bond numbers (Bo $\ll$ 1, Fig. 16.5) capillarity prevails (this condition corresponds to growing small diameter crystals $R < a$ as well as it was easily satisfied in our TPS experiments on the board of space stations [16.6] when the capillary constant is high) and (16.6) can also be simplified to give

$$z''r + z'\left(1+z'^2\right) \pm 2d\left(1+z'^2\right)^{3/2} r = 0\,. \tag{16.8}$$

Static Stability of the Melt Meniscus as well as the dynamic stability of crystal growth process has to be provided. A presence of static stability means that the melt meniscus exists for all values of crystallization parameters. An analysis of the stability can be realized on the basis of Jacobis equation investigation. For the TPS, this analysis was realized in [16.23, 24].

Growth Angle Certainty – Common Boundary Condition for CST

As the Laplace capillary equation is a second order differential one, formulation of a boundary problem for melt meniscus shape calculation requires assignment of two boundary conditions. The first of them is determined by the structural features of each specific CST and will be analyzed in details below. But second of the boundary conditions (the crystal–melt interface condition) is mutual for all CST. This condition follows from the growth angle certainty.

Let the angle ψ_0 (Figs. 16.3, 16.4a,b), made by the line tangent to the meniscus and the lateral surface of the growing crystal, be called the growth angle. The growth angle should not be confused with the wetting angle. The wetting angle characterizes particular equilibrium relative to the liquid movement along a solid body and is not directly associated with crystallization [16.6]. In the first studies of the CZT and TPS, a crystal of constant cross section was considered to grow in case $\psi_0 = 0$. Judging by a purely geometric diagram of liquid–solid phase conjugation, such assumption is quite natural. However, experimental and theoretical investigations of the crystal growth process showed that the geometrical condition $\psi_0 = 0$ is not satisfied while growing crystals of constant cross sections and ψ_0 is a physical characteristic of the crystals. Particular case $\psi_0 = 0$ is available only for some metals. Experimental determinations of the ψ_0 have included direct measurements as well as indirect calculations. Indirect techniques, as a rule, are more precise [16.6]. In our experiments [16.6, 25], we studied shapes of crystallized drops obtained on the bottom of silicon, germanium and indium antimonide crystals detached from the melts in the process of pulling by the CZT. The values of $\psi_0 = 25 \pm 1°$ for indium antimonide, $\psi_0 = 11 \pm 1°$ for silicon, and $\psi_0 = 12 \pm 1°$ for germanium were obtained. Theoretical investigations have shown that ψ_0 is an anisotropic value as well as it depends on the crystallization speed. In [16.6], a reader can find a discussion of the problem in detail. In this chapter, for the dynamic stability analysis, we will use ψ_0 as a constant value for the crystal to be grown.

16.2.3 The Equation of Crystal Dimension Change Rate

Proceeding from the condition of growth angle certainty, we can obtain an equation for the crystal characteristic dimension change rate $\mathrm{d}R/\mathrm{d}t$ that is common for all CST. On the diagram (Fig. 16.6) a vector

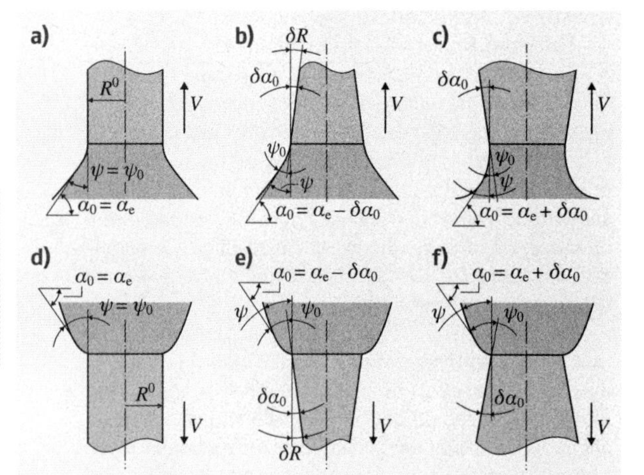

Fig. 16.6a–f Crystal growth by CST: pulling up **(a–c)**; lowering down **(d–f)**; $\alpha_0 = \alpha_e$: growth of constant cross section crystal **(a,d)**; $\alpha_0 < \alpha_e$: growth of widening cross section crystal **(b,e)**; $\alpha_0 > \alpha_e$: growth of narrowing cross section crystals **(c,f)**; V is the crystal displacement rate; other symbols given in the text

$\boldsymbol{R}$ is located within the diagram plane and represents the radius for a straight circular cylinder-shaped crystal. For a plate it is its half-thickness. Now we introduce the angles made by the line tangent to the meniscus on the three-phase line with the horizontal α_0 and with the vertical ψ (the crystal grows in the vertical direction). When the angles α_0 and ψ add up to $\frac{\pi}{2}$, a crystal of constant cross section R^0 grows. In this case, the angle ψ is equal to the growth angle ψ_0 and the value of the angle α_0 is denoted α_e. If $\psi \neq \psi_0$ the crystal lateral surface declines from the vertical on the angle $\psi - \psi_0 = \alpha_0 - \alpha_e = \delta\alpha_0$ the crystal changes its dimension $\delta R = R - R^0$ in accordance with (Fig. 16.6)

$$\frac{\mathrm{d}R}{\mathrm{d}t} = V_c \tan(\delta\alpha_0) = V_c \tan(\alpha_0 - \alpha_e) . \tag{16.9}$$

The angle $\delta\alpha_0$ is the angle of crystal tapering at any moment, the crystallization rate V_c is equal to the difference in rates between pulling and front displacement

$$V_c = V - \frac{\mathrm{d}h}{\mathrm{d}t} . \tag{16.10}$$

Near the stationary state that we need for (16.3), deviations $\mathrm{d}h/\mathrm{d}t$ as well as $\delta\alpha_0$ have to be negligible. Hence, the crystallization rate V_c can be replaced by the rate of the pulling V and $\tan(\delta\alpha_0) \approx \delta\alpha_0$. The angle α_0 together with the meniscus shape as a function of R, h and other parameters can be determined by solving the capillary boundary problem whose equation was discussed above. Assuming that the capillary problem is solved, i.e., the function $\alpha_0(X_1^0, X_2^0, \ldots, X_n^0, t, C) = 0$ is found, (16.9) can be represented by

$$\frac{\mathrm{d}R}{\mathrm{d}t} = \sum_{k=1}^{n} \frac{\partial\alpha_0}{\partial X_k}\delta X_k = \sum_{k=1}^{n} A_{ik}\delta X_k . \tag{16.11}$$

16.2.4 The Equation of the Crystallization Front Displacement Rate

The equation of the crystallization front displacement rate belongs to the set (16.3) and as well as (16.11) is common for all CST. It follows from the heat-balance condition on the crystallization front

$$-\lambda_S G_S(h) + \lambda_L G_L(h) = \zeta V_c . \tag{16.12}$$

Here λ_S and λ_L denote thermal conductivities of the solid and liquid phases respectively, $G_S(h)$ and $G_L(h)$ are the temperature gradients in the solid and liquid phases at the crystallization front, h is a crystallization front position, ζ denotes the latent melting heat of a material unit volume, V_c is the crystallization rate. In accordance with (16.10), we obtain a (16.1)-type equation for $\mathrm{d}h/\mathrm{d}t$

$$\frac{\mathrm{d}h}{\mathrm{d}t} = V - \zeta^{-1}\left[\lambda_L G_L(h) - \lambda_S G_S(h)\right] . \tag{16.13}$$

The equation of (16.3)-type is

$$\frac{\mathrm{d}h}{\mathrm{d}t} = \zeta^{-1}\sum_{k=1}^{n}\left[\lambda_S\left(\frac{\partial G_S}{\partial X_k}\right) - \lambda_L\left(\frac{\partial G_L}{\partial X_k}\right)\right]\delta X_k . \tag{16.14}$$

Now the functions $G_L(h)$ and $G_S(h)$ have to be found, which can be done by solving the Stefans problem – a nonstationary thermal conductivity problem with interface boundary as a heat source.

16.2.5 SA in a System with Two Degrees of Freedom

If only the crystal radius R and the crystallization front position h are regarded as variable parameters, (16.11) and (16.14) look like

$$\frac{\mathrm{d}R}{\mathrm{d}t} = A_{RR}\delta R + A_{Rh}\delta h , \tag{16.15}$$

$$\frac{\mathrm{d}h}{\mathrm{d}t} = A_{hR}\delta R + A_{hh}\delta h , \tag{16.16}$$

where the notations $A_{RR} = -V(\partial\alpha_0/\partial R)$; $A_{Rh} = -V(\partial\alpha_0/\partial h)$; $A_{hR} = \zeta^{-1}[\lambda_S(\partial G_S/\partial R) - \lambda_L(\partial G_L/\partial R)]$;

$A_{hh} = \zeta^{-1}[\lambda_S(\partial G_S/\partial h) - \lambda_L(\partial G_L/\partial h)]$ have been used. The solutions of the set (16.15), (16.16) are

$$\delta R = C_1 \exp(S_1 t) + C_2 \exp(S_2 t)\,, \quad (16.17)$$

$$\delta h = C_3 \exp(S_1 t) + C_4 \exp(S_2 t)\,. \quad (16.18)$$

Here S_1 andS_2 are the roots of the characteristic equation (16.4) that, for our case, reads

$$S^2 - (A_{RR} + A_{hh})\,S + (A_{RR}A_{hh} - A_{Rh}A_{hR}) = 0\,. \quad (16.19)$$

To estimate the stability of the set of equations (16.15) and (16.16) there is no need to solve the equations themselves but the Routh–Gurvitz conditions [16.6] can be used. For the set (16.15) and (16.16) to be stable, it is necessary and sufficient that the coefficients satisfy the inequalities

$$A_{RR} + A_{hh} < 0\,, \quad (16.20)$$

$$A_{RR}A_{hh} - A_{Rh}A_{hR} > 0\,. \quad (16.21)$$

The stability types mentioned are *rough* in the sense that the system stability remains unchanged within a wide range of values of the coefficients $A_{ik}(i, k = R, h)$. If at least one of the inequalities (16.20), (16.21) is replaced by the equality, the roots of the characteristic equation (16.19) are either imaginary or zero. In this case we can not judge the system's stability by its linear approximation, and the nonlinear model should be analyzed.

The coefficients A_{RR} andA_{hh} indicate direct correlation between dR/dt and δR, as well as dh/dt and δh, i. e., self-stability of the parameters. The coefficients A_{Rh} and A_{hR} represent the effect of the change in one value on the rate of change of the other value, i. e., inter-stability of the parameters. It can be concluded from the analysis of (16.20) and (16.21) that the crystal growth system is stable if

$$A_{RR} < 0\,, \quad A_{hh} < 0\,, \quad A_{Rh}A_{hR} < 0\,, \quad (16.22)$$

$$A_{RR} < 0\,, \quad A_{hh} < 0\,, \quad A_{Rh}A_{hR} > 0\,, \quad |A_{RR}A_{hh}| > |A_{Rh}A_{hR}|\,, \quad (16.23)$$

$$A_{RR} < 0\,, \quad A_{hh} > 0\,, \quad |A_{RR}| > |A_{hh}|, \quad A_{Rh}A_{hR} < 0\,, \quad |A_{RR}A_{hh}| < |A_{Rh}A_{hR}|\,, \quad (16.24)$$

$$A_{RR} > 0\,, \quad A_{hh} < 0\,, \quad |A_{RR}| < |A_{hh}|\,, \quad A_{Rh}A_{hR} < 0\,, \quad |A_{RR}A_{hh}| < |A_{Rh}A_{hR}|\,. \quad (16.25)$$

We can see, that negative values of A_{RR} and A_{hh} coefficients to be, is a very important condition for the crystal growth system stability. Below, we will use the following terminology. There is a capillary stability in the system if $A_{RR} < 0$ and there is a heat stability if $A_{hh} < 0$.

16.3 SA and SCG by CZT

16.3.1 Capillary Problem

A solution of the boundary capillary problem allows finding coefficients A_{RR} and A_{Rh}. In [16.6], the reader can find the investigation of the problem in detail. Here are the brief results. The problem includes (16.6) with $d = 0$ and the two boundary conditions

$$\left.\frac{dz}{dr}\right|_{r=R} = -\tan\alpha_0\,, \quad (16.26)$$

$$z|_{r\to\infty} = 0\,. \quad (16.27)$$

The boundary problems numerical solution is presented on the Fig. 16.7 and

$$A_{RR} > 0\,; \quad A_{RR}|_{r\to\infty} \to 0\,, \quad (16.28)$$

$$A_{hR} < 0\,; \quad A_{hR}|_{r\to\infty} \to \approx 4V\,. \quad (16.29)$$

For the numerical calculation the dimensionality of A_{ik} is s^{-1}; all angles have to be measured in radians, the units of length is the capillary constant a and speed of pulling is a/s.

16.3.2 Temperature Distribution in the Crystal–Melt System

A solution of the boundary heat problem allows finding A_{hh} and A_{hR} coefficients. There exists a great number of works dealing with calculation of the temperature fields in the crystal–melt system. They form a group of the Stefan problems in which a crystal–melt boundary is a heat source. However, owing to the variety of growth schemes and presence of a great number of factors that are to be taken into account for the thermal conductivity problems, for instance, complex temperature dependence of the thermophysical characteristics of various matters as well as convective flows in melts, a complete mathematical description of heat patterns during crystal growth is very difficult. Obtaining the solution in its analytical form is usually achieved by significant simplifications. With this end in view, the following equation from [16.26] allowing simple analytical solutions will be used to analyze the heat conditions in the

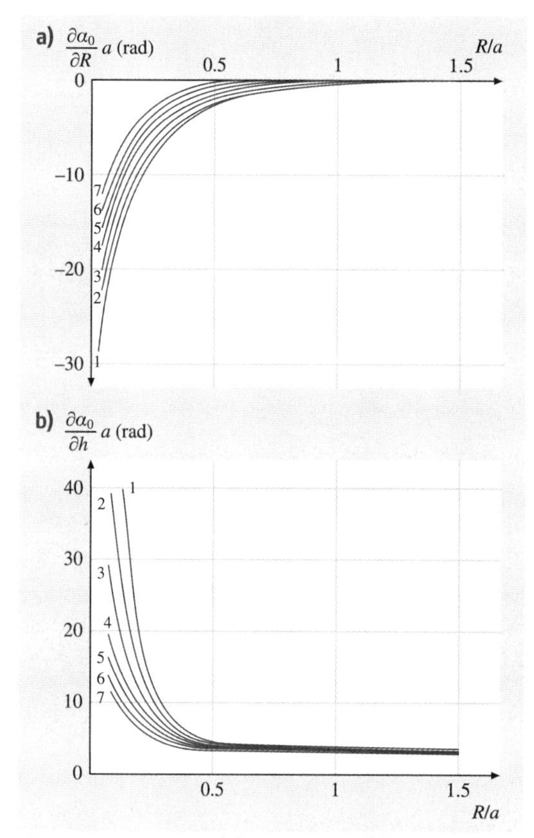

Fig. 16.7a,b For CZT (**a**) $\partial\alpha_0/\partial R$ and (**b**) $\partial\alpha_0/\partial h$ versus crystal radius for various values of the growth angle ψ_0: (1) 15°; (2) 5°; (3) 0°; (4) 10°; (5) 20°; (6) 30°; (7) 40°. Capillary constant a is used as a unit of all dimensions

crystallization process

$$k_i^{-1}\frac{\partial T_i}{\partial t} = \frac{\mathrm{d}^2 T_i}{\mathrm{d}z^2} - Vk_i^{-1}\frac{\mathrm{d}T_i}{\mathrm{d}z} - \mu_i\lambda_i^{-1}F\left(T_i - T_\mathrm{e}\right) . \tag{16.30}$$

Here $i = \mathrm{L}, \mathrm{S}$ ($i = \mathrm{L}$ for a liquid, $i = \mathrm{S}$ for a solid body), T_i denotes the temperature, k_i is the thermal diffusivity coefficient, z is the vertical coordinate, μ_i denotes the coefficient of heat-exchange with the environment, F denotes the crystal (meniscus) cross section perimeter-to-its area ratio, T_e is the environment temperature, λ_i is the thermal conductivity coefficient.

There are few peculiarities for the use of (16.30):

1. It describes the temperature distribution of the crystal–melt system in a one-dimensional approximation that means that the temperature in the crystal (meniscus) cross section is averaged and isotherms are flat.
2. The heat exchange with the environment is allowed for not in the form of the boundary condition but by introducing heat run-offs on the lateral surface in the form of an additional term in the equation.
3. It gives a good description of real temperature distribution for small Biot numbers ($\mathrm{Bi} = \mu_i R\lambda_i^{-1} \ll 1$) – this can be observed during growth of small diameter or thickness wall crystals, for low coefficients of convective heat transfer from the crystal (melt) surface, and high thermal conductivities.
4. Heat exchange is allowed for by the Newton law that means the convective heat exchange is much higher than the heat losses caused by radiation (the heat losses caused by free convection are comparable with the heat losses caused by radiation at the surface temperatures of $\approx 1000\,°\mathrm{C}$ and even higher in case specimen surface blowing is provided).
5. The equation is not available for vacuum pulled refractory materials and this for the radiation heat-exchange, the Stefan–Boltzmann law should be necessarily allowed for which leads to considerable nonlinearity of the problem; in this case a linearization of the crystal–surface radiation law described in [16.6] can be applied that allows (16.30) to be used up to temperatures of 2000 °C.

In Sect. 16.2.1 we discussed a possibility to use a quasi stationary approximation, according to which temperature distribution in the crystal–melt system at any moment of time satisfies the stationary thermal conductivity equation with instantaneous values of all the process parameters. For this approximation to be applied, the time of crystallization front relaxation to the stationary state should be significantly longer than the characteristic time of temperature relaxation. As a rule, this condition is satisfied [16.6]. So, we can use (16.30) with left part set to zero for the calculation of G_S and G_L.

As an example of A_hh and A_hR calculation, we consider a growth of a long crystal (the limiting case is a continuous pulling), with good thermal screening of the melt column provided. So, we use (16.30) with zero left part for crystal as well as for the melt meniscus with $\mu_\mathrm{L} = 0$. Boundary conditions of the problem are the following:

1. The melt temperature at the bottom of the liquid column at the level of the melt free surface is fixed: $T_\mathrm{L}|_{z=0} = T_\mathrm{m}$.

2. The crystallization-front temperature is equal to the melting temperature T_0: $T_L|_{z=h} = T_S|_{z=h} = T_0$.
3. The temperature of the crystal end is equal to the environment temperature: $T_S|_{z\to\infty} = T_e$.

The solution of (16.30) with these boundary conditions [16.6] allows obtaining A_{hh} and A_{hR}

$$A_{hh} = -\zeta^{-1}\lambda_L h^{-2}(T_m - T_0) , \quad (16.31)$$

$$A_{hR} = \zeta^{-1}\mu_s R^{-2}\varsigma_S^{-1}(T_0 - T_e) ; \quad (16.32)$$

where

$$\varsigma_S = \left(\tfrac{1}{4}V^2 k_S^{-2} + 2\mu_s\lambda_S^{-1}R^{-1}\right)^{1/2} .$$

So, on the basis of the capillary and heat boundary problems solution, the signs and the values of CZT growth stability coefficients are found. The signs are $A_{RR} > 0$; $A_{Rh} < 0$; $A_{hR} > 0$; $A_{hh} < 0$, if the melt is superheated ($T_m > T_0$).

16.3.3 SA and Crystal Growth

It can be concluded that:

1. The capillary stability is absent for all diameters of crystal.
2. Heat stability can be realized.
3. It is a chance to satisfy conditions (16.25) and to have the crystal growth stable if the following two inequalities can be fulfilled.

$$\left|\frac{\partial\alpha_0}{\partial R}\right| < \lambda_L V^{-1}\zeta^{-1}h^{-2}(T_m - T_0) , \quad (16.33)$$

$$\left|\frac{\partial\alpha_0}{\partial R}\right| < \left|\frac{\partial\alpha_0}{\partial h}\right| h^2\mu_s R^{-2}\lambda_L^{-1}\varsigma_S^{-1} \times (T_0 - T_e)(T_m - T_0)^{-1} . \quad (16.34)$$

Practically it means that a constant cross section crystal can be grown if the diameter of crystal is bigger than a melt capillary constant and the melt is superheated. If a crystal has a smaller diameter, superheating has to be high as well as a special combination of a process crystal growth parameters given the inequalities (16.33) and (16.34) has to be realized. Our experience of CZT crystal growth without a special diameter regulation as well as the growth of cylindrical crystals of silver [16.27] and of big diameter silicon tubes [16.28] confirms the conclusions of this paragraph. It has to be mentioned a necessity of very good stabilization of the pulling speed: it was shown [16.29] even a stable system can not compensate a sudden speed change and as a result there is a pinch formation.

16.4 SA and SCG by VT

At the beginning of seventies, our attention in some experiments was turned to the fact that growth of corundum crystals of small diameters ≈ 5 mm (they are grown especially as seed crystals) was easy. Practically the growth regime required no correction by an operator. The crystals had smooth surfaces and cylindrical shapes. This stimulated our works on the VT dynamic stability investigation. As was mentioned above, the analysis had been carried out with 3 degrees of freedom for cylindrical crystals (Fig. 16.2) and with 4 ones for tubular ones. The limited volume of the chapter does not allow us to give further details about the theoretical model. The reader can find them in [16.6, 15, 30–32]. Let us notice some peculiarities of the problem formulation for VT and the main conclusions. *First of all*, the mass balance condition in a system including a charge feeder, a melt layer, and a crystal has been used as a third equation for the set (16.2). *Second peculiarity* concerned with a heat problem formulation. As a rule, when formulating the heat problem for crystal growth from melts, the melt temperature has been specified as the boundary condition (Sect. 16.3.2). This boundary condition for the VT does not correspond to the real situation. Crystal displacement in the furnace muffle results in melt-temperature change on its surface. This is a reason why the heat conditions of the technique under consideration will be allowed for by specifying the density Q of the heat flow fed from the burner onto the surface of the melted layer. With the gas flow specified, the density of the heat flow Q depends on the distance between the burner and the level of the melt surface $Q(l)$. The function $Q(l)$ is determined by the burner design and the gas debit.

16.4.1 Practical Results of the Theoretic Analysis

Round Cylindrical Crystals

For the process to be stable the following three conditions have to be fulfilled:

1. The diameter of crystal $2R$ has to be small ($R < a$). We can see that the situation is the opposite one to the CZT.
2. Change in the heat flow density $Q(l)$ along the furnace muffle in the vicinity of the growth zone at the distance of an order of R should not exceed the crystallization heat.
3. The heat flow density value $Q(l)$ in the vicinity of growth zone has to be decreased if the distance between the melt surface and the burner l_1 is increased.

The result of the theoretical investigation explains why growth of corundum crystals of small diameters (≈ 5 mm) is easy. They grow in the dynamically stable regime; it means that there are internal mechanisms for dissipation of perturbations. With respect to our terminology, the capillary stability exists for $R < a$ (for the sapphire melt $a = 6$ mm).

In spite of the capillary stable growth is impossible for the cylindrical round sapphire crystals with the diameters more than 12 mm, the theoretical model allows a minimizing of the crystallization process perturbations while growing big diameter crystals. For this, the previous two conditions have to be fulfilled (evidently, without the crystal dimension limitation) and also:

- The temperature of a muffle wall has to be increased.
- The irregularity of the density distribution of the charge flow falling on the melted layer has to be decreased.

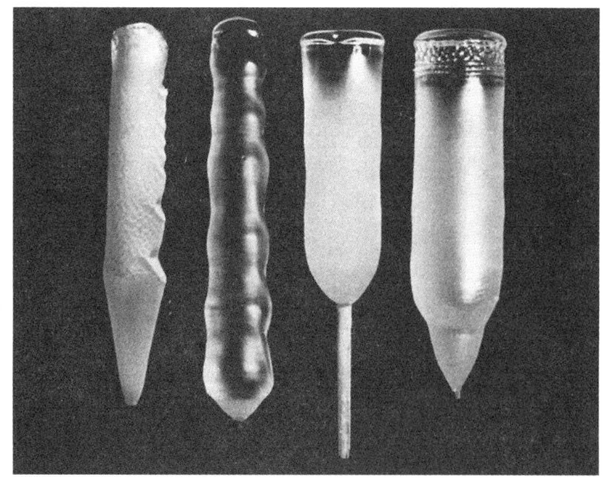

Fig. 16.8 Corundum cylindrical crystals grown by VT in optimized (*two right ones*) and nonoptimized (*two left ones*) regimes

These requirements of the crystallization conditions (we classify them as optimized) are in good agreement with our experimental results [16.15, 30–32]. In the experiments such hydrogen- and oxygen-flow debits in a three-channel burner had set that a crystal grown closer to the burner had a larger diameter. This condition corresponds to the heat-flow density increase when approaching the burner. Preheating the gas before feeding it into the burner and increasing the furnace muffle temperature were also used. As a rule, no parameter control to maintain constant cross section of the crystal was required. Crystals, grown under these conditions, exhibited a smoother surface and improved optical and structural characteristics. In the Fig. 16.8, corundum crystals of 40 mm diameter, grown in 1972 without any automatic control by the author of this chapter with collaborators from the Leningrad State Optical Institute, are presented.

Tube Shaped Crystals

The theoretical analysis states: Crystallization of tubes of arbitrary outer diameters is stable if the tube wall thickness is smaller than some critical thickness. This thickness is smaller than the capillary constant a and depends both on the heat conditions of the process and on the outer diameter of the tube. It increases with the outer diameter increase. In our experiments [16.6, 32] tubes with an outer diameter of 16–25 mm with walls 3–4 mm thick were grown (Fig. 16.9). A crystallization apparatus fitted with a four-channel burner providing charge supply via the central and periphery channels was used. It was experimentally stated the optimal gas distribution in the burner channels: oxygen-hydrogen-oxygen-hydrogen. Crystal growth was initiated from

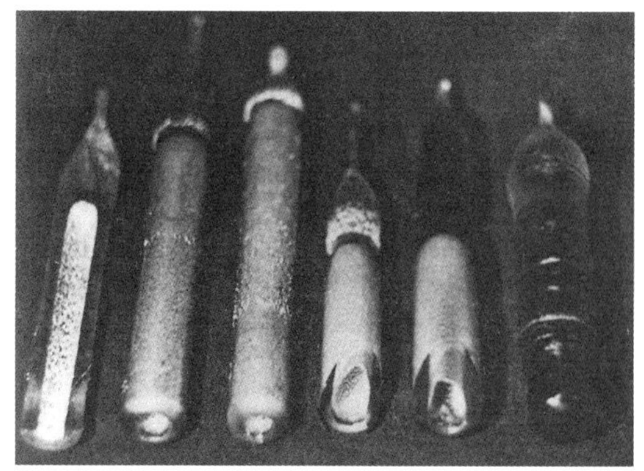

Fig. 16.9 Six corundum single crystals (five tubes and one cylindrical crystal) grown by VT in the same regime

a seed 3–4 mm in diameter. Firstly, a seed cone was grown. The cone was widened by feeding the charge through the central channel, the periphery oxygen flow rate being increased. As soon as the crystal diameter reached the specified value 20–22 mm the charge was fed through the periphery channel. A little later (10–15 min) charge supply from the central tank was cut off and within 30–60 min the rate of the central oxygen flow was reduced. This growth regime has provided smooth transition from a solid crystal to a tube. The sink rate was smoothly increased. After that the process of stationary growth went on. Tubes up to 120 mm in length were grown. Usually no parameter control to maintain constant cross section of the tube was required, i. e., stable growth conditions the existence of which had been theoretically predicted could be attained. Figure 16.9 shows some of the tubes grown in 1981 by the authors of this chapter with collaborators from the Solid State Institute of the Russian Academy of Science. Three of the tubes are cut (one of them along the axe, and two ones obliquely) to show the tube walls. For the comparison, a cylindrical crystal of the same diameter, grown in the same furnace, is shown on the figure. The irregular crystal shape indicates that for it this regime is not optimal from the point of view of stability that is not at variance with our theoretical prediction.

Plate Shaped Crystals

The theoretical results are applicable for the growth of plate shaped crystals: The crystal plates of less than two capillary constants thick (12 mm for sapphire) have to grow stably. Information of corundum crystal plate growth by VT can be found in our review [16.8].

16.4.2 SA-Based Automation of VT

As was shown above, when passing to crystals with diameters exceeding the two capillary constants crystallization stability is lost. Practically it means that the crystallization front position and crystal dimensions are changed during the crystal growth process. In this case, an operator controls the parameters by changing the gas debit, charge feed and crystal sinking rate using his experience and intuition. Automatic system of control provides better result. When developing systems of growing crystal automatic diameters control, a problem of the laws of automatic control of the process parameters under some changes in crystal dimensions arises. Up to publication of our paper [16.33] the required laws of parameters control were defined from the results of empirical search. In [16.33] it was shown that the laws can be performed on the basis of the crystal growth stability analysis. In this instance the controllable parameters side by side with the crystal diameter, the liquid–gas interface position and the melt meniscus height can be required as the degrees-of-freedom of the crystal growth process. For VT the density of the heat flow from the burner Q, (it is regulated by the changing of the gases debit P), the rate of crystal sinking V and the powder charge flow rate Ω can be used as controllable parameters. Usually, P as well as Ω is used as controllable parameters on the stage of crystal widening. But after the crystal has already widened from the seed dimension up to the desired diameter, the control is provided by Ω regulation. In our approach Ω has to be regarded as an additional 4th degree-of-freedom for the cylindrical crystal. But coefficients of the linear equation for W are unknown. They have to be found from the necessary and sufficient conditions of the set of 4 equations stability of (16.3)-type. The problem can have several solutions. Each of them can be used as the regulation law in the control system. In this case our system of crystal growth including the regulator has to be stable. In [16.33] three different Ω change laws, allowed stable growth, were found. Fig. 16.10 illustrates corundum crystals grown with one of the stable laws of Ω regulation and the proportional one. The crystals were grown in 1979 by the author of this chapter with collaborators from the Institutes of Crystallogra-

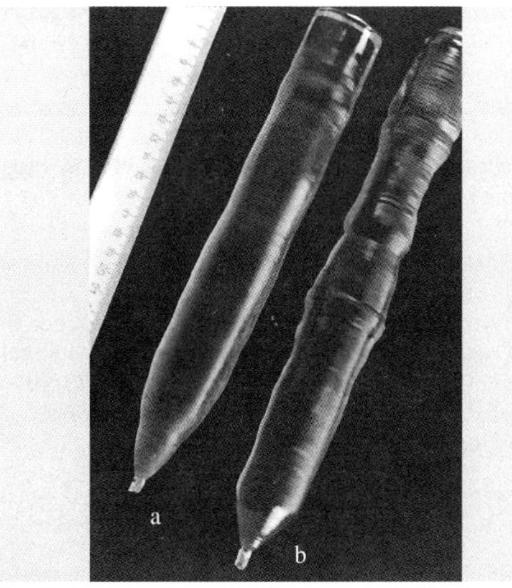

Fig. 16.10 Corundum crystals grown by VT using stable (a) and unstable (b) laws of powder charge control

phy and Solid State Physics of the Russian Academy of Science. A standard industrial Verneuil furnace was used for the experiments. A comparison of crystals from Figs. 16.8 and 16.10 shows that a design of the furnace corresponding optimized, from the point of view of stability, growth conditions can provide the same (or even better) crystal quality as using of automatic system control.

16.5 SA and SCG by FZT

The FZT has been widely used for different materials crystal growth, especially: semiconductors (RF heating), high-melting metals and dielectrics (electron beam, plasma or laser heating). There are a lot of publications concerning a FZT theoretical study but only two aspects have been the investigation topics: a static meniscus stability, melt flows included, and impurity segregation. Only few papers have been devoted the dynamic stability analysis. Reference [16.18] has been the first of them. However, this analysis has not been completed as a heat conditions of crystallization have not been taken into account and a capillary part of the problem has been violently simplified. As a result, in [16.18] a capillary stability proved to exist in all versions of FZT. Our analysis based on the Lyapunovs theory [16.34, 35] is more common and includes the following main points:

1. Min $n = 4$ were chosen (Fig. 16.3).
2. The crystal up pulling as well as down lowering with different diameters ratio of the growing crystal and the rod to be melted were analyzed.
3. Equation (16.11) for $\partial R/\partial t$ was obtained as a result of the capillary boundary problem with the boundary condition of the angle growth certainty on the crystallization front.
4. It has been shown the boundary condition of the angle certainty is not available for the melting front [16.6, 36].
5. The equation for $\mathrm{d}W/\mathrm{d}t$ was obtained from the mass balance of melted and crystallized substances.
6. Equation (16.14) for $\mathrm{d}h_c/\mathrm{d}t$ was obtained as a result of the solution of (16.30) near the crystallization front as well as for $\mathrm{d}h_m/\mathrm{d}t$ near the melting front.

The analysis of the result is rather complicated because four Routh–Gurvitz inequalities have to be simultaneously satisfied. Here are the main conclusions:

1. The capillary stability exists ($A_{RR} < 0$) for big growing crystal and melted rod diameters ($R > a$, $r_0 > a$) with all ratios of them (R/r_0).
2. For small ($R < a$, $r_0 < a$) crystal and rod diameters, $A_{RR} < 0$ if $R > R_{min} \approx \frac{1}{2} r_0$.
3. The biggest negative value of A_{RR} coefficient corresponds to $R = r_0$.
4. The capillary stability exists for both direction of growth (up or down) but down lowering is the preferable one.

As for practical using of FZT for shaped crystal growth, besides widely spreaded round shape rode, the ribbon-to-ribbon (RTR) technique [16.37] has to be mentioned. A ribbon was used for a feeding as well as a ribbon crystal was grown. RTR achieved silicon ribbons for a solar cells application of a width 75 mm, a thickness 0.1 mm, and a 3–9 cm/min growth rate with laser heating being used.

16.6 TPS Capillary Shaping

While analyzing stability of TPS as a system with two degrees of freedom, the equations (16.15)–(16.25) have to be used. Therefore, we can proceed to the analysis of the melt-column shaping conditions in TPS.

16.6.1 Capillary Boundary Problem

For axisymmetrical case, the Laplace capillary equations (16.6)–(16.8) will be used in our analysis. As was mentioned above, each equation is a second order differential one and a boundary problem for a melt meniscus shape calculation strictly requires assignment of two boundary conditions. The first of them is (16.26), common for all CST but the second one is determined by the structural features of each specific TPS. A shaper is used for melt-column shaping in TPS (Fig. 16.4). The functions of the shaper in TPS are wide and we will discuss all of them later. At the moment, we characterize the shaper as a device to control the melt-column shape only. This problem is a fundamental one for a shaped crystal growth by TPS and we will discuss it in detail. In a mathematical description of the problem, a shaper function is to determine the meniscus shape by means of fixation of a capillary problem second boundary condition. For the first time, the characterization of shapers from this point of view was accomplished by us in 1967 [16.38, 39]. In the most cases, the shaper (Fig. 16.4) is characterized by its wall or free edge curvature radius r_0 in the horizontal plane, the angle β made by its wall with the horizontal. The wetting angle Θ formed by the melt and the shaper surface, is a very important shaper characteristic. If this angle exceeds 90°, the shaper material is not wetted by the melt (Fig. 16.4a,b,g–j); if it is smaller, the melt wets the shaper material (Fig. 16.4c–f). Shaping is accomplished either on the surfaces (Fig. 16.4a,c,g–i) or on the sharp edges (Fig. 16.4b,d–f,j) of the shaper. It corresponds to the following boundary conditions of the capillary boundary problem:

Catching Boundary Condition

In case the shaper material is wetted by the melt, the melt is easily caught by its sharp edge. This boundary condition will be termed the *catching condition*. There is a possibility of providing the catching conditions at nonwettable shaper free-edges which will be discussed below. The catching condition means that a counter line on the meniscus surface is fixed by the edge of the shaper, i. e., it coincides with the edge counter of the shaper. It does not matter if the edge counter is internal (Fig. 16.4d,j) or external (Fig. 16.4e,f), if the pulling up (Fig. 16.4d,e) or down lowering (Fig. 16.4f,j) is used for the shape crystal growth. The catching condition has the following mathematical form in the cylindrical coordinate system

$$z|_G = d(r, \phi) . \quad (16.35)$$

Here G is the counter of the shaper edge, d is the distance from the shaper edge to the coordinate plane. In case the shaper is flat and is positioned parallel to the melt plane, $d = \text{const.}$ For axisymmetric flat shaper edges the condition (16.35) has the following form

$$z|_{r=r_0} = \text{const.} = d , \quad (16.36)$$

where r_0 is a shaper edge counter radius. If the coordinate plane coincides with the free level of the melt, d represents the pressure of feeding the melt to the shaper. In this case, the pressure is included in the boundary condition and in the Laplace capillary equation, const. = 0. If the coordinate plane coincides with the shaper edge plane (Fig. 16.4b,d–f,j) the right part of (16.36) is equal zero and the pressure d is included in the Laplace capillary equation (16.5–16.8) as a parameter (const. = d). The pressure being positive, the shaper edges are positioned below the melt free-surface level and vice versa.

Angle Fixation Boundary Condition (the Wetting Condition)

If the melt has a contact with the shaper surface, it makes the wetting angle, Θ with the shaper surfaces (Fig. 16.4a,c,g–i). This boundary condition will be termed the *angle fixation condition or wetting condition*. It can be realized for nonwettable shaper material (Fig. 16.4a,g–i) as well as for wettable one (Fig. 16.4c), for pulling up (Fig. 16.4a,c,h,i) as well as for lowering down (Fig. 16.4g) shaped crystal growth. The condition means that shaper walls fix the meniscus angle on a counter C belonging to the shaper surface by forming the angle of wetting. The condition has the following forms:

In common case, where $\boldsymbol{n}$ denotes the direction of the normal towards the shaper wall, C is a counter of the contact of the meniscus with the shaper walls

$$\left[1+\left(\frac{\partial z}{\partial x}\right)^2+\left(\frac{\partial z}{\partial y}\right)^2\right]^{-1/2} \left.\frac{\partial z}{\partial \boldsymbol{n}}\right|_C = -\cos\Theta . \quad (16.37)$$

The shapers of complicated surfaces (Fig. 16.4a,c) can illustrate this condition although in the figure they represent a particular case of the cylindrical symmetry system (z, r coordinates) whereas, for the common case, the use of x, y, z coordinate system is necessary.

For an axisymmetrical problem (a rod, a tube) on the vertical shaper walls (Fig. 16.4h)

$$\left.\frac{dz}{dr}\right|_{r=r_0} = \tan\left(\Theta - \tfrac{\pi}{2}\right) = -\tan\alpha_1 . \quad (16.38)$$

For circular cone shaper walls formed the angle β with the horizontal (Fig. 16.4i)

$$\left.\frac{dz}{dr}\right|_{r=r_1 - z\cot\beta} = \tan\left(\Theta + \beta - \frac{\pi}{2}\right) = -\tan\alpha_1 . \quad (16.39)$$

Wetting-to-Catching Condition Transition

With the melt pressure increasing, the catching boundary condition at the nonwettable shaper free-edges can be obtained. Figure 16.11a illustrates this transition with the pressure to be increased by gradual shaper immersing into the melt. The diagram is based on [16.40]

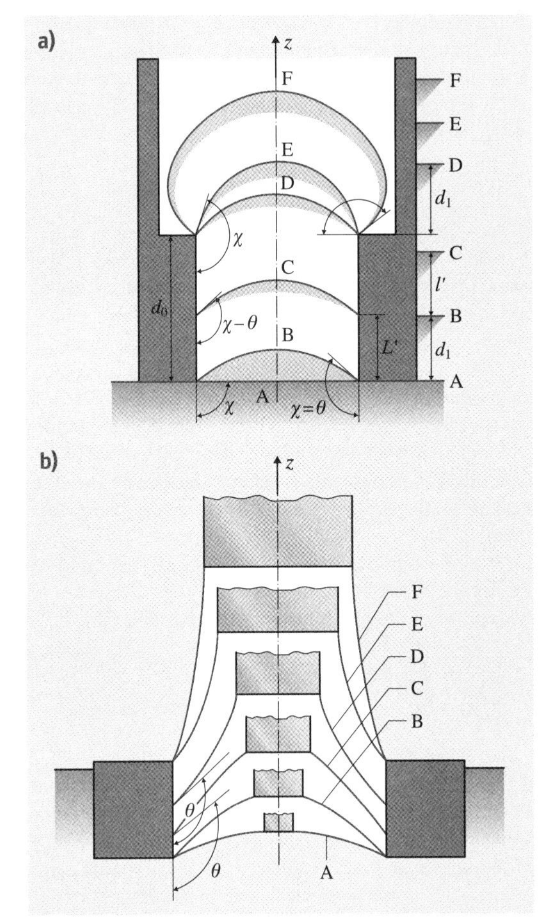

Fig. 16.11a,b Transition of the shaper lower free-edge catching boundary condition (meniscus A) to the wetting condition (menisci B, C, D) and further to the shaper upper free-edge catching condition (menisci E, F); (**a**) pressure changes (the *horizontal lines* denote successive positions of the liquid free surface); (**b**) the seed–shaper dimension ratio changes; χ is the angle between the line tangent to the melt surface and the shaper wall, θ is the wetting angle

describing the particle buoyancy conditions for the flotation processes. A nonwettable shaper, with a hole of d_0 depth, possessing vertical walls is considered. The angle χ between the line tangent to the liquid surface and the shaper wall is introduced. Let us analyze a number of shaper successive positions. *Position* A: The shaper touches the melt with its lower plane. The lower-plane immersion depth is equal to zero. The line tangent to the liquid surface coincides with the liquid surface. Angle χ is $\frac{\pi}{2}$. The catching condition holds at the lower free edge of the shaper. With the shaper being immersed into the liquid, the angle χ increases, and when the shaper lower plane reaches some depth d_1, (*position* B) the angle χ will be equal to the wetting angle Θ, (within the immersion depth range from 0 to d_1 the catching condition holds at the lower free edge). With further shaper immersing into the liquid, χ remains equal to the wetting angle Θ. The liquid–shaper wall contact line goes up by the value of L^1, the shaper immersion depth, the distance between this line and the free surface level remaining equal to d_1 (*position* C). As soon as the immersion depth is equal to d_0+d_1, the liquid–shaper contact line coincides with the shaper sharp edge (*position* D), and with further pressure increase, the catching condition at the shaper upper free edge holds (*position* E). In this case the angle χ will increase until the wetting angle Θ is formed by the liquid and the shaper surface [16.40]. For the horizontal shaper surface $\chi=\frac{\pi}{2}+\Theta$ (*position* F). Further increase in pressure is impossible as it will lead to liquid spreading over the shaper surface and the shaper will not operate properly. But if shaper's free edges are sharp, pressure increase is possible (Fig. 16.4b) up to the loss of the meniscus' static stability.

The presence of a seed or a profile being pulled considerable changes into the conditions of the transition described. It means that the condition at the upper boundary (along the crystal–melt contact line) can affect the character of the condition at the lower boundary (along the melt–shaper contact line). The diagram, that will be proved when solving the boundary problem, is given in Fig. 16.11b. By changing the seed-to-shaper hole dimension ratio alone, the catching boundary condition at the lower free edge (menisci A, B), the wetting condition on the shaper walls (menisci C, D), the catching boundary condition at the upper free edge (menisci E, F) can be achieved. The very important conclusion is the following: the wetting boundary condition means the angle χ is fixed but the counter of the meniscus with the shaper contact is movable. Its position on the shaper walls depends on the melt pressure and the crystal–shaper dimensions ratio.

Certainly, we could change a melt pressure by any other way as well as a growth direction doesnot matter. For instance, by changing a melt pressure or the crystal–shaper edge dimensions ratio, it is possible to realize the scheme of either Fig. 16.4g or j.

A comparison of Fig. 16.11a and b diagrams shows that there are two very different situations: before seeding and after seeding (during pulling). In the second case, even for non wettable shaper walls, the meniscus, changing its curvature with forming a negative pressure inside, raises the melt above the shaper edge, up to the crystallization front (the same situation exists in CZT).

Hereafter it will be shown that the catching boundary condition usually leads to more capillary stability of the process, therefore, the ways of achieving the catching condition at the shaper free edges in TPS should be specified: *Firstly*, a melt-wettable material should be used for the shaper, and the latter should be designed in such a way that the melt could raise up to the shaper free-edges due to capillary forces (Fig. 16.4d–f). *Secondly*, for melt-nonwettable materials, the melt column should be embraced from outside, providing additional pressure on the liquid to make the melt–shaper contact point touch the shaper sharp edge (Fig. 16.4b,j). *Thirdly*, for poorly wettable shaper materials, the crystal–shaper dimensions ratio should be used in order that it could ensure melt column contact with the shaper sharp edges, compare Fig. 16.4g and j.

Influence of the Wetting-Angle Hysteresis on Capillary Boundary Conditions

While analyzing all the capillary effects, the existence of wetting-angle hysteresis should be taken into consideration. The wetting angle hysteresis reveals itself in the fact [16.40] that the wetting angle of liquid run on a solid body is larger than that of liquid run off a solid body. This means that the stationary wetting angle depends upon the process of meniscus formation (on-run or off-run). This results, for example, in the fact that a higher pressure is to be applied to create the catching condition at the shaper free edge than that required to keep it unchanged. In case the catching condition is created by the seed, this condition can remain unchanged in the process of growth, with the clearance between the shaper free edge and the growing crystal changing.

A Comparison of the Catching Boundary Condition and the Angle Fixation One

The comparison can be done on the basis of the previous description of transition one to another. It exhibits a big difference of these two boundary conditions: The catching one fixes a coordinate of the meniscus end counter but an angle of the inclination of the meniscus is not fixed. The wetting one fixes an angle of inclination of the meniscus end counter but not a coordinate. In this case the meniscus is movable (Fig. 16.11b). Its position on the shaper wall depends on the second boundary condition (a seed or growing crystal presence as well as a growing crystal dimension and a growth angle) and can be found as a result of the capillary boundary problem solution. So, for capillary shaping by walls, a shaper has to be designed taking account for this phenomena as well as the angle-wetting hysteresis. We carefully explain this difference because there is a big misunderstanding of this key problem for capillary shaping. Let us analyze some of wrong approaches. For the boundary capillary problem solution, *Stepanov* et al. [16.41] use the idea of *Tsivinskii* from [16.42], where, for the axisymmetrical meniscus described the Laplace equation (16.5), it was suggested to replace $1/R_2$ (Fig. 16.1) by a linear function of the vertical co-

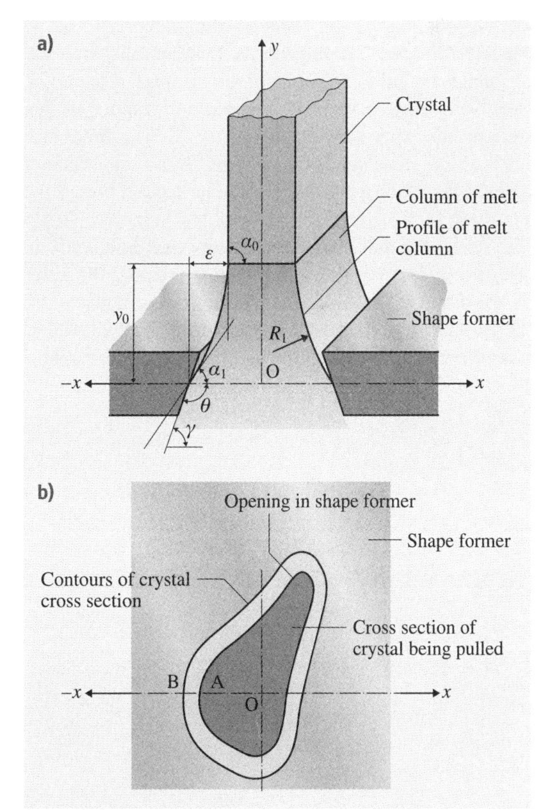

Fig. 16.12a,b Arrangement for producing a crystal with cross section of arbitrary form. Melt column parameters: y_0 represents the melt column height, α is the inclination of the profile curve tangent to the x axis (α_0 at $y = 0$ and α_{01} at $y = y_0$), and R_1 is the radius of curvature of the melt column surface, lying in a plane perpendicular to the tangent (R^0 at $y = 0$ and R^{01} at $y = y_0$) (after [16.41])

ordinate. There are, as minimum, four mistakes in this approach (Fig. 16.12):

1. The axisymmetrical meniscus is applied for the growth of arbitrary cross section crystals
2. A priori, we need to know if the meniscus is concave, convex or concave–convex
3. A priori, for concave–convex meniscus, we need to know the coordinates of the inflection point
4. We need to know y_0, α_0, and α_{01}.

The fourth mistake is the worst. Indeed, the fixation of these parameters means that we need to use three boundary conditions for the second order differential equation: a fixation of the angle growth on the crystal–melt boundary and the catching as well as the wetting boundary conditions on the shaper. But it is nonsense. Authors of this approach had published a lot of papers including cumbersome formulas that never have been used or verified. The main argument of the authors: this approach is applied for CZT – there are a lot of experimental evidences of it. But it is clear, why the approach works for CZT. The second boundary condition for CZT (16.27) confirms the vertical coordinate strives for the zero on the infinity. Automatically it means that the first derivative also strives for the zero on the infinity. So, this is the well known in mathematics peculiarity of the boundary condition on the infinity and, as a result, we have three boundary conditions. For

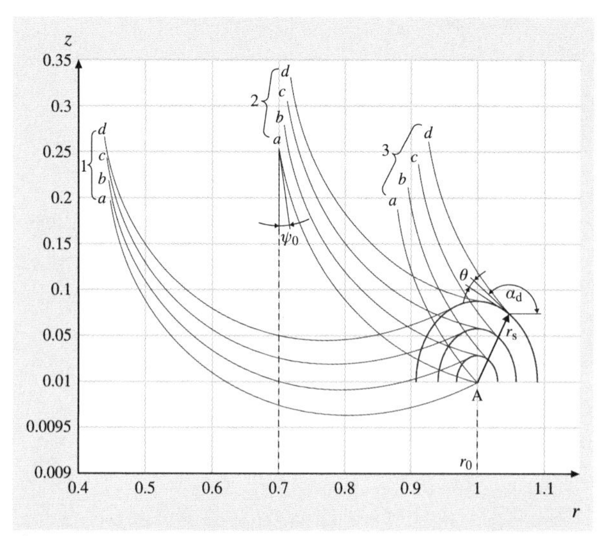

Fig. 16.13 Mid-range Bo. Profile curves $z(r)$, starting from surfaces of toruses (b–d) of different radius r_S and from a sharp edge of the shaper (a) (after [16.23])

the first time we had explained nonapplicability of this approach to the TPS in [16.43] (1969) and later in our reviews [16.5, 6, 9]. But many journals have continued publishing including *Pet'kov* and *Red'kin* paper [16.44] concerned with the investigation of the shaped growth dynamic stability. In our paper [16.45] we once more explained nonapplicability of this approach.

More Precise Definition of the Catching Boundary Condition

Nevertheless, we understood that there is a weak point in our approach: From the formal point of view we are right – mathematical formulation of the problem requires two boundary conditions. But from the physical point of view a melt has to form a wetting angle with a shaper surface. This discrepancy was explained in [16.23]. The paper contents well-grounded mathematical proof but here we only illustrate the main idea. Figure 16.13 presents one specific case: the growth of round cylindrical crystal with respect to scheme Fig. 16.4d. The shaper edges are replaced by circular coaxial tours of four different curvature radiuses r_S (the last $r_S \to 0$). The example corresponds to middle-range bond numbers Bo ≈ 1. The shaper radius $r_0 = 1$. The unit of lengths measurement is the capillary constant a which is equal for silicon 7.6 mm, for sapphire 6 mm, for germanium 4.8 mm, for indium antimonide 3.7 mm [16.6]. It means that for silicon the diameter of the shaper edges circle is 15.2 mm, for sapphire 12 mm. The origin of the z-coordinate is located on the melt free surface. The negative melt pressure $d = -1$, e.g. the melt free surface is located on the distance a lower than the shaper edge (Fig. 16.4). Three set of profile curves (a profile curve is a section of the meniscus by the figure plane) are presented on the Fig. 16.13. Each set consists of four profile curves and is characterized by the same value of the angle α_d-meniscus inclination angle with respect to positive r-direction at the point of the contact with the shaper. The lower end of each profile curve is located on the surface of a torus and forms the wetting angle $\Theta = 10°$ with the surface of the torus. The each torus corresponds to the shaper edge of a different radius of curvature r_S: $r_S = 0.09$ (0.68 mm for Si) for the all three profile curves of set d, $r_S = 0.06$ (0.46 mm for Si) for the curves c, $r_S = 0.03$ (0.23 mm for Si) for curves b, $r_S \to 0$ for curves a. The upper end of each profile curve forms the angle 10° with the vertical. It corresponds to crystal growth of the respective dimension with the growth angle $\psi_0 = 10°$: $R \approx 0.45$ (the crystal of 7 mm diameter for Si) for set 1, $R \approx 0.7$ (the crystal of 10.6 mm diameter for Si) for set 2, $R \approx 0.9$

(the crystal of 16 mm diameter for Si) for set 3. The crystallization front is located on the height (from the plane of the shaper edges) 0.20 (1.5 mm for Si) for sets 1 and 3, 0.30 (2.2 mm for Si) for set 2.

So, on the microlevel, the melt forms the wetting angle with the torus surfaces and we have a normal physical wetting condition. If the crystal changes its dimension the contact point also changes its position on the torus surface but, in any case, its position will be in the vicinity (torus radius r_S) of the point A. With the r_S decreasing, a location of the lower end of the profile curve is more definite near the point A with the coordinates ($r_0 = 1$, $z = 1$). If the torus radius is infinitely small we have, in macroscopic sense, the catching of the meniscus on the sharp shaper edge (in point A). So, the catching condition is just a useful mathematical approach to satisfy a capillary boundary problem. From the physical point of view it defines a wetting boundary condition on the sharp edges of the shaper. It seems, everything is clear. But in spite of the publication of our papers [16.23,45], the history of application of the approach from [16.42] to TPS has not been finished [16.46].

Capillary Boundary Problem Solution

The solution of the problem we are dividing on three parts with respect to the Bond numbers Bo: large, middle-range and small ones. For large and small Bo, the Laplace equation has the forms (16.7) and (16.8), respectively. A solution of the boundary problems for both of them can be obtained in analytical form (sometimes with the using of special functions). For a middle-range Bo a numerical solution is needed. A comparison of the analytical and numerical solutions shows that, with a sufficient accuracy for practice, we can use:

1. The large Bo approximation for the growth of cylindrical round rods or tubes of $10a$ minimal diameter (remind that for the silicon it corresponds 76 mm, for the sapphire 60 mm)
2. The small Bo approximation for the growth of cylindrical round rods or tubes of a maximal diameter (remind that for the silicon it corresponds 7.6 mm, for the sapphire 6 mm) on the Earth surface as well as for the very big diameters crystals growth under microgravity conditions

A solution of a boundary capillary problem allows to obtaining very interesting information concerning a capillary shaping as a function of a shaper design, melt pressure, wetting angle. The information can include:

1. A shape of meniscus-conditions of existence of concave, convex and convexo–concave ones
2. A range of parameters for existence of catching and wetting boundary conditions
3. Design of a shaper and a range of parameters for existence of a meniscus with the definite growth angle ψ_0
4. The same with the fixed crystallization front position
5. Signs and values A_{RR} and A_{Rh} coefficients [16.6, 23,24,47].

Here are few examples of this kind.

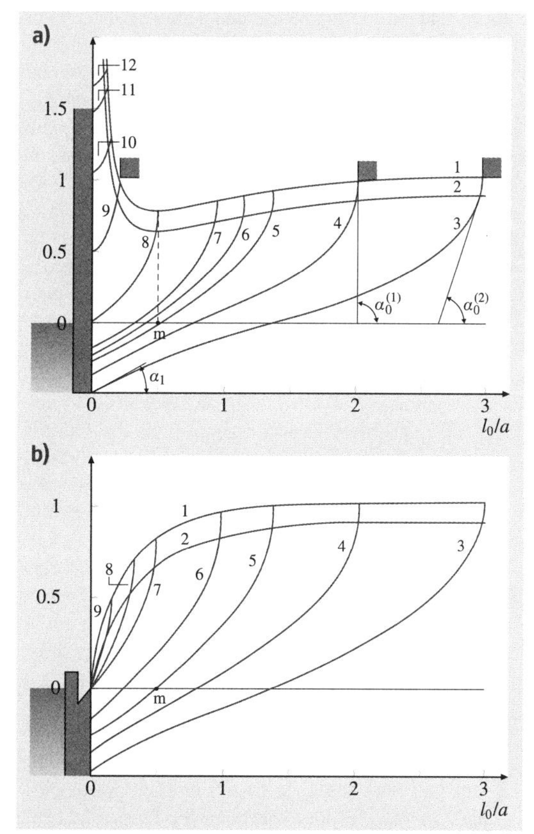

Fig. 16.14a,b Large Bo: (**a**) Wetting boundary condition ($\theta = 135°$); convex–concave (3–7) and concave (8–12) profile curves $z(r)$; boundary 1 ($\psi_0 = 0$) and 2 ($\psi_0 = 15°$) curves $h(R)$; (**b**) transition from the wetting boundary condition (3–6) to the catching condition (7–9) by changing the shaper design

Large Bo: Let us formulate the following crystal growth conditions (Fig. 16.14) which corresponds to the scheme Fig. 16.4h [16.6,47]: To the right, where $r = r_0$, the non wetted shaper wall is located. The line tangent to the melt surface in the melt–shaper contact point makes the wetting angle Θ with the shaper wall. Let us introduce the angle $\alpha_1 = \Theta - \frac{\pi}{2}$. To the left, where $r = R$, the edge of a melt-growing flat crystalline tape or that of a circular cylindrical crystal of a large diameter is situated. Let the angle α_0 (Fig. 16.6a), made by the line tangent to the melt surface at the melt–crystal growing contact point and the negative direction of the r-axis, be specified (while growing crystals of constant cross sections, the angle $\alpha_0 = 1/2\pi - \psi_0$ is the complement of the growth angle ψ_0). Let l_0 denotes the clearance between the crystal edge and the shaper: $l_0 = r_0 - R$. Let us consider what process parameter data can be obtained with such problem formulation. According to our terminology, we have the angle-fixation boundary conditions (or wetting conditions) on both the ends of the r-variation interval: the equation (16.7), boundary conditions (16.26) and (16.38). The analytical solution of the problem was obtained in the Legendre elliptical functions (e.g. [16.6]) and is presented on the Fig. 16.14a for the following parameter values: $\Theta = 135°$ ($\alpha_1 = 45°$), $\psi_0 = 15°$ ($\alpha_0 = 75°$) and $\psi_0 = 0$ ($\alpha_0 = 90°$). The origin of the z-coordinate coincides with the free melt surface. The profile curves 3–12 are the sections of the melt meniscus by the diagram plane. Each profile curve corresponds to the definite distance l_0 between the growing crystal and the shaper wall. The edges of the crystals are located on the one of two boundary curves $h(R)$ corresponding two different growth angles: (1) for $\psi_0 = 0$ and (2) for $\psi_0 = 15°$. Based on this boundary problem solution, the following conformities can be established:

1. With the angle fixation boundary condition satisfied at both the ends of the l_0-variation interval, the vertical coordinates of the liquid–solid phase contact points with respect to the melt free surface are not fixed but depend on the relation between the angles at both the ends of the interval and on the value of the clearance between the shaper and the crystal being pulled. It is a confirmation the scheme Fig. 16.11b.
2. There exists some minimum value m of the clearance between the shaper and the crystal being pulled (for our parameter values it is equal approximately to one capillary constant a) when the meniscus lies both above and below the melt free surface. When this gap is smaller than m, the point of the meniscus contact with the shaper wall is located higher of the melt free surface.
3. The meniscus part, located below the melt surface is convex, the meniscus part, located above the melt surface is concave.
4. A higher crystallization front position corresponds to a smaller growth angle.
5. For the crystal–shaper gap being more than capillary constant a the crystallization front height doesnot exceed capillary constant. For smaller gaps, the crystallization front height can be infinitely tall.
6. Any change in the melt level during pulling will produce the following effect on the crystal dimensions: with the level decreasing, the tape thickness (or the rod diameter) can be kept unchanged only in case the crystallization front is lowered by the same value. With the crystallization front position kept unchanged, the tape thickness or the crystal diameter will decrease with the melt lowering and vice versa. It means that the melt level can be qualified as one of degrees of freedom. We investigated these phenomena (e.g. [16.6]).
7. $|\partial\alpha_0/\partial h| > 0$; $|\partial\alpha_0/\partial R| > 0$ for $l_0 < m$; $|\partial\alpha_0/\partial R| < 0$ for $l_0 > m$. It means that there is a capillary stability ($A_{RR} < 0$) only if the gap between the shaper wall and the pulling crystal is more than capillary constant ($l_0 > m \approx a$).

Now, let us modify the shaper design: let us locate a sharp shaper edge on the melt free surface level. The solution of the new boundary problem is presented on Fig. 16.4b. Let us discuss new results:

1. For $l_0 > m \approx a$, the situation is the same as in previous case.
2. But for $l_0 < m \approx a$, the situation is dramatically changed: the catching boundary condition is realized (this is a second confirmation of the scheme of Fig. 16.11b) and, as a result, either for $l_0 < m \approx a$, $|\partial\alpha_0/\partial R| > 0$, i. e., the capillary stability exists for all range of l_0 variations.
3. On the other hand, a crystallization front $h(R)$ has to be located much lower than in the previous case. In particular, $h(R)|_{R \to r_0} \to 0$.

Middle-ranged Bo: For a middle-ranged Bo (Bo ≈ 1), three profile curves a of the sets 1–3 on the Fig. 16.13 present an example of the numerical solution of the capillary boundary problem with the catching boundary condition that corresponds to Fig. 16.4d Therefore, we use (16.6) and the boundary conditions

(16.26) and (16.36). The results were discussed above but the two peculiarities have to be mentioned here:

1. A middle part of the meniscus, corresponding to the crystal radius $R \approx 0.5$, is located lower than the shaper edge.
2. The highest crystallization front position corresponds to the crystal radius $R \approx 0.7$.

Small Bo: As an example of the capillary boundary problem solution for a small Bo (Bo $\ll 1$), studied in detail in the analytical form with using of the Legendre elliptical functions in [16.6, 47], we show (Fig. 16.15) a melt pressure influence on the shape of profile curves $z(r)$, with the boundary condition of catching in the point r_0 (the shaper radius $r_0 = 0.05$), as well as on the boundary curves $h(R)$ (1–4), corresponding to the growth angle $\psi_0 = 0$. Hence, we use the equation (16.8) and the boundary conditions (16.26) and (16.36). It is very important to mention that in the capillary problem with small Bo it is neglected by the influence of gravity. This is a reason, why these results are applicable for the growth of different size crystals in microgravity conditions. As for the growth in the condition of normal gravity, the results are applicable for filaments growth (for sapphire, for instance, the case under consideration corresponds to growth of a filament of 0.6 mm diameter)

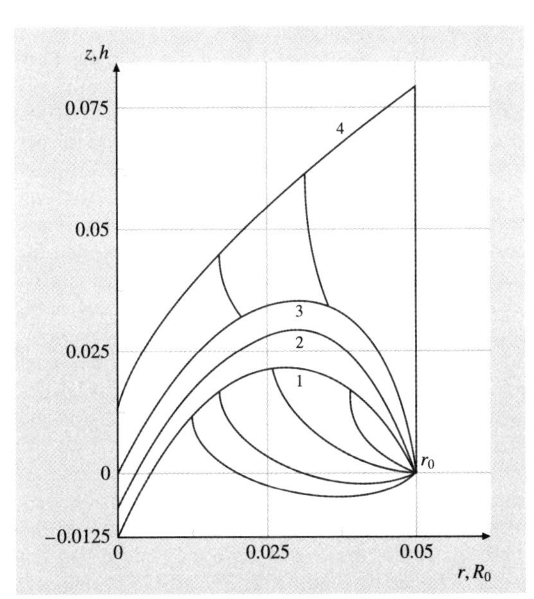

Fig. 16.15 Small Bo: Boundary curves $h(R)$ 1–4 and some profile curves $z(r)$ for shaper with $r_0 = 0.05$, under various pressures d: -10 (1), -5 (2), 0 (3), 10 (4)

with pulling up as well as lowering down. Consequently, the schemes Fig. 16.4b,d–f,j are described in the frame of this model. Here are some peculiarities of the results presented on the Fig. 16.15:

1. The order of magnitude of the crystallization front position is the same as the shaper radius for all values pressures under investigation. In our example, $r_0 = 0.05$. It means that for sapphire, for instance, the crystallization front is located on the distance 0.5–0.6 mm from the plane of sharp shaper edges.
2. For all boundary curves, except 4, $h(R)|_{R \to r_0} \to 0$).
3. The boundary curve 4 is the particular one. From the theoretical study [16.6], it follows that if d corresponds to the value from the formula $2dr_0 = 1$, the boundary curves for all r_0 values (but only from the range of small Bo values) has to have the shape similar to the curve 4. The main particularity of this curve is the following one: $h(R)|_{R=r_0} = 0.5\pi r_0$. Hence, for this particular pressure, if the diameter of pulling crystal is equal to the shaper edge diameter, the very special meniscus in the shape of the right circular cylinder exists.
4. All boundary curves, except 4, have a maximum. The maximum position corresponds to $R = R_m \approx 0.7 r_0$ and it divides all range of the crystal dimensions on two parts for which: $\partial\alpha_0/\partial R > 0$ if $R > R_m$ and $\partial\alpha_0/\partial R < 0$ if $R < R_m$. It means that for d values characterized the formula $2dr_0 < 1$, there is a capillary stability ($A_{RR} < 0$) only if the gap between the shaper edge and the pulling crystal is less than $0.3 r_0$.

16.6.2 Stability Analysis

Signs of capillary coefficients A_{RR} and A_{Rh} are presented on Fig. 16.16. For the stability estimation, heat coefficients (16.31) and (16.32) can be used: $A_{hh} < 0$, $A_{hR} > 0$. With respect to Fig. 16.16, we can choose a shaper design to have a capillary stability: $A_{RR} < 0$, $A_{Rh} < 0$. Therefore, with respect to (16.22) inequalities fulfilled, a dynamic stability of a shaped crystal growth will be provided.

Here we formulated a common problem of stability for growth of arbitrary cross section crystals by TPS. But only the simplest case (a growth of a round cylindrical crystal in the thermal condition described by the one dimensional thermoconductivity equation) was analyzed. Complimentary information on the topic can be found in our reviews [16.6, 9] and papers [16.23, 24]. It includes:

Conditions of crystallization				A_{RR}	A_{Rh}
Method	Bond number	Boundary condition	Parameter values		
CZT					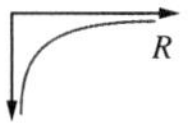
TPS	Small	Catching	$\sin\alpha_0 < 2r_0d$	<0	<0
			$\sin\alpha_0 > 2r_0d$	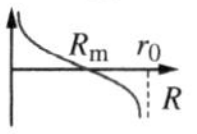	<0
		Catching	$\alpha_0 < \alpha_1$	<0	<0
			$\alpha_0 > \alpha_1$	>0	<0
	Large	Catching	Single- and double-valued meniscus	<0	<0
			Double-valued meniscus $R \approx r_0$	<0	>0
		Angle fixing	$\alpha_0 < \alpha_1$	<0	<0
			$\alpha_0 > \alpha_1,\ r - R > m$	<0	<0
			$\alpha_0 > \alpha_1,\ r - R > m$	>0	<0

Fig. 16.16 CZT and TPS capillary coefficients for different Bo values and different capillary boundary conditions

1. Investigation of the meniscus static stability.
2. Investigation of tubes growth dynamic stability – three degrees of freedom stability problem where internal tube diameter is a third degree of freedom.
3. Investigation of a melt pressure influence on the dynamic stability – three degrees of freedom stability problem where melt pressure is a third degree of freedom.
4. Crystal cross section shape stability. So, some problems are solved but for theory up to now, there have been a lot of problems to be solved. The main of them is the stability of complicated shape crystal growth and an influence of crystallographic anisotropy.

Now, for preliminary analysis of the capillary shaping conditions before the experiments, we use rough estimations. For instance, for ribbon growth, an approximation including three steps, we proposed [16.6]:

1. A solution of the capillary problem with large Bo for the ribbon flat part
2. A solution of the capillary problem with small Bo for the ribbon edge
3. A joining of two solutions on the boundary near ribbon edge with the condition of the same growth angle for two parts.

16.6.3 Experimental Tests of the Capillary Shaping Statements

Growth Angle Certainty

A growth angle certainty is one of the main capillary shaping theory statements. This is a reason, why we carried out special experiments to examine it [16.6, 48]. A growth of thin sapphire crystal from a shaper with 0.8 mm diameter was carried out (Fig. 16.16). By changing the slope between the crystallization front and the surface of the skew shaper by means of a different heating, various boundary conditions were created for left and right sections of the same crystal. In Fig. 16.16a the rod of a constant cross section grows: the angles ψ_1 between the crystallization front and the growth direction are different on the left and on the right crystal sides; constancy of the diameter is provided by maintaining the angle $\alpha_0 = \alpha_e(\alpha_e = \pi/2 - \psi_0)$ all over the perimeter. In Fig. 16.17b on the left $\alpha_0 = \alpha_e$, on the right $\alpha_0 < \alpha_e$, the right side of the crystal widens. In Fig. 16.17c deviation α_0 from the equilibrium value α_e on the right side has led to crystal contraction. In the transient region, the crystal surface is convex on the Fig. 16.17b and concave on the Fig. 16.17c, which corresponds to the capillary stability presence.

Integration of the equation (16.9) in the range of the crystal radius R change from R_0 to the final value R_{01}

gives the following formula

$$\ln D = -\left(\frac{d\alpha_0}{dR}\right) z\,, \tag{16.40}$$

where

$$D = \frac{R - R_{01}}{R_0 - R_{01}}\,.$$

Linear dependence $\ln D$ versus z has to be a proof of the (16.9) correctness. Figure 16.18 presents the date processing from Fig. 16.17b,c.

The Space Experiments. We carried out some model experiments in the microgravity conditions (small Bo) to test some capillary shaping theory statements [16.6, 49–54].

Simulation Experiments. Crystal growth experiments under microgravity conditions in the Space were preceded by simulating the liquid column shape with using immiscible liquids of equal densities [16.6, 49–51]. A meniscus of the alcohol/water solution was formed between two glass tubes, surrounded by the equal-in-density mineral oil (Fig. 16.19). Pressure d in meniscus was equal to the weight of column of the alcohol solution in the upper tube. The lower tube (3) of $2r_0 = 13.12$ mm diameter imitated the shaper, and the upper one (1) of $2R$ in diameter – the crystal. For a right circular cylindrical meniscus (2) existence, a pressure 16.8 dyn/cm^2 was determined experimentally (Fig. 16.19a). A convex meniscus (2) (Fig. 16.19b) was formed under pressure 27.14 dyn/cm^2. The sphere menisci (4) on the ends of the lower tubes (3) were used for the pressure estimation as well as, with a certain pressure, for a surface tension value on the two liquids boundary. The same types of the experiments were carried out by us on the board of a flight laboratory with the 20 s microgravity time.

Crystallization of Copper Under Short-Time Microgravity Conditions. The simulation experiments with liquids were only the first step to estimate the crystal growth real conditions. Moreover, a doubt appeared that a shaped crystal growth under microgravity could be realized because, sometimes, in simulation experiments, the liquid flew on the crystal surface [16.6, 49]. Therefore, we crystallized the copper under the capillary shaping, on a high-altitude rocket, with 20 min time of microgravity [16.6, 50, 51]. The metal has a relatively low melting point (1083 °C), resistance to overloading (it is important for rocket launching as well as for a capsule landing), and its physical-chemical properties are well known.

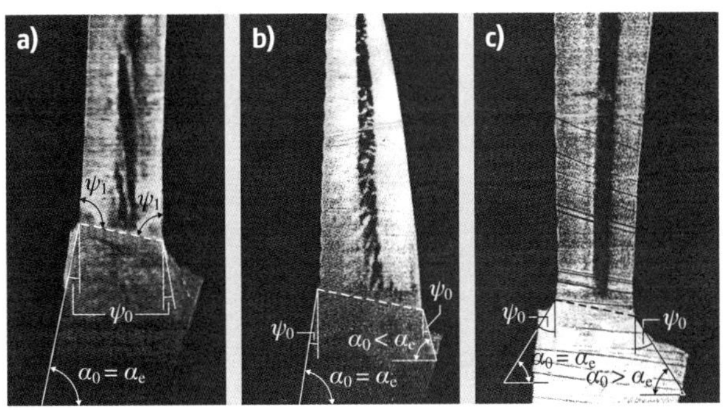

Fig. 16.17a–c TPS growth of a thin sapphire filament, the shaper being skew: (**a**) constant cross section, (**b**) widening from the right, (**c**) constriction from the right

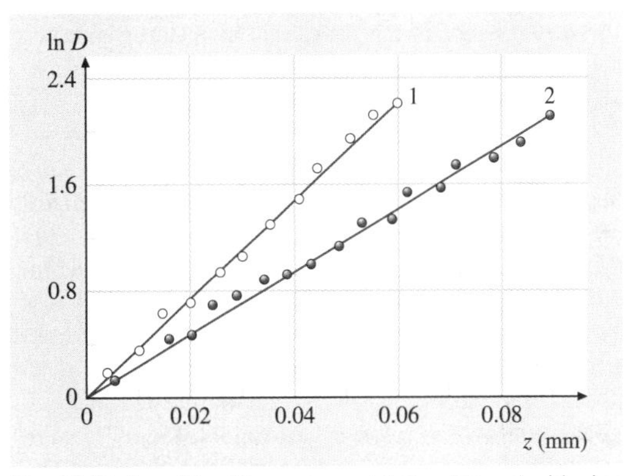

Fig. 16.18 Sapphire filament constriction (1) and widening (2) in the transient range

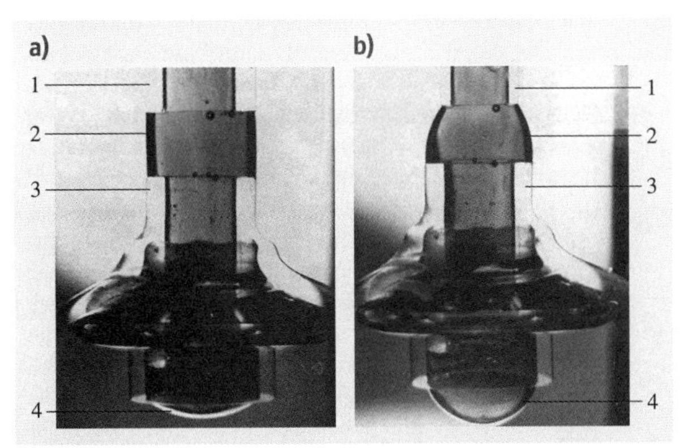

Fig. 16.19a,b Meniscus model of pulling a circular rod under zero-g conditions: (**a**) right circular cylinder, (**b**) convex

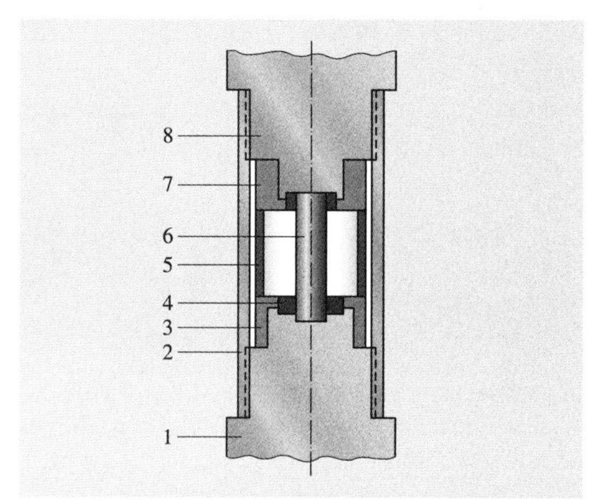

Fig. 16.20 Design of the capsule used to investigate the crystallization process from a melt under microgravity conditions (notations given in the text)

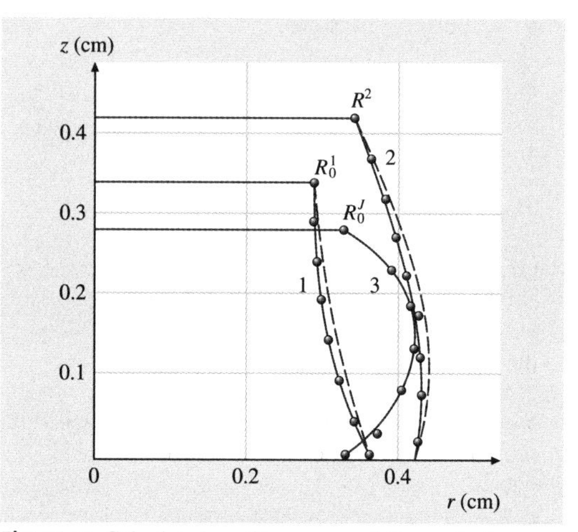

Fig. 16.21 Calculated profile curves of the melted copper columns (*dotted line*) and crystallized specimens (*continuous line*) for different (1, 2, 3) internal meniscus pressures; experimental data for the crystallized specimen shape are indicated by *bullets*

The technique of capillary shaping was preliminary used to produce a rod specimen in an evacuated enclosure. This enables us to reduce the amount of gas in the metal. Specimen (6) in the form of a cylinder, 5–8 mm in diameter, and about 5–6 mm long (Fig. 16.20) was placed between two molybdenum shapers (1) and (8) into which copper (4) was fused preliminary. This guaranteed complete wetting of the shaper during the space experiment. Two graphite guard rings (3) and (7) were introduced to prevent the escape of the melt beyond the sharp lip of the shaper. These rings were supported by the graphite spacer (5), which also acted as a thermal shield. Both shapers were pushed into the coupling tube (2). The above design ensured rigidity, simplicity of assembly, and constant separation between the shapers. Tungsten-rhenium thermocouples were mounted near the ends of the specimens to estimate the temperature distribution along its length. The capsule was inserted into a heating device using the energy of the exothermic chemical reaction. The device did not contain any movable parts. The experiment was carried out by the following way. When a microgravity conditions had achieved, the specimen was completely melted. After that during the microgravity existence, a directional crystallization of the melt column was realized by the heat removing from the upper shaper. On the high-altitude rockets, we had no photographic facilities for recording the crystallization process. Therefore, the shapes of the crystallized specimens had examined experimentally and was compared with the calculated for $\psi_0 = 0$ ones. Figure 16.21 confirms:

1. Our model of capillary shaping is applicable for a crystal growth in the conditions of microgravity.
2. Under recorded by the thermocouples crystallization speed 5–7 mm/min, $\psi_0 = 0$ for Cu crystallization.
3. The melt did not flow on the crystal surface during the crystallization; therefore, the crucible-free zone melting and capillary shaping crystal growth could be realized in the space.

A little later, these crystallization processes were realized in the *shape* and *ribbon* experiments.

Shape Experiment. The main idea of the experiment was to realize a crystal growth by using the right cylindrical melt column which can not exist in the terrestrial condition (except a filament growth). Under microgravity (Figs. 16.15 and 16.19a), such a column of melt is formed if $R = r_0$; $\alpha_0 = \frac{\pi}{2}$ under pressure d, satisfying equality $2dr_0 = 1$, but, reaching the altitude $h \approx \pi r_0$, it loses static stability and transforms first to a meniscus, its profile curve having an ambiguous projection onto the abscissa axis, and then with increasing h, it falls into two independent meniscus. At $h < \pi r_0$ and $d = 0$, a meniscus is a catenoid; with increasing pressure its curvature in the axial cross section decreases, reaching infinity at $2dr_0 = 1$. When pressure increases above this value the meniscus becomes convex (Fig. 16.19b).

A cylindrical meniscus can be used for a crystal growth only at $\psi_0 = 0$, which means that a metal should be chosen for the material grown experimentally (for metals the angle of growth is usually close to zero). In 1984, under microgravity, at *Salyut* orbital space probe, crystallization of indium was carried out by big group of Russian scientists governed by the author of this chapter [16.6, 50, 52]. The advantage of indium is its low melting temperature (156 °C). It is very important for the simplicity of the space furnace and the limited energy power on the space craft board. Here are some other In physical characteristics: a comparatively high density in a solid state (7.28 g/cm^3) which only slightly differs from that of the melt (7.03 g/cm^3), the surface tension of the melt $\gamma = 592\,erg/cm^2$, and capillary constant under terrestrial conditions $a = 0.41$ cm.

Figure 16.22 depicts a scheme of the growth device. A plastic case (1) with a lid (2) has a graphite container (3) filled with indium preliminary. A resistive heater (4), separated by a foam-polyurethane layer (5), is used. The heat is delivered to the container through a copper capsule (6), which also serves for holding a copper cap (7) (shaper itself) and supplied with a hole for leveling the inert gas pressure inside and outside the container (3). The meniscus of melt (8) is formed first between the initial copper rod (9), fastened to a rod (10), and the edge of the shaper (7). The meniscus shape is fixed by a photo camera with an illumination system of windows (11). The most important part of this setup was the system of maintaining a pressure inside of meniscus by means of a melt meniscus formed near the crucible bottom. The pressure depends on the radius R of the graphite container and a wetting angle. The idea had been suggested by the author of this chapter and was used either for the *ribbon* experiments. Figure 16.23 depicts the shape of a drop (a), formed at the edge of the shaper, and of the meniscus of melt (b). Pressure, which had been found from the shape of the drop (Fig. 16.23a), exceeded approximately by 40% that required for formation of a cylindrical column. A reason of non accuracy is evident – there are a lot of factors (a hysteresis included) governed by the wetting angle. Some of them could be found only under real space conditions and could not be taking into account before. Unfortunately, we did not have any possibility to repeat the same experiment with the correction. But the main aim of the experiment was achieved: the typical shape of a meniscus has the ratio $h/r_0 \approx 3.6$ (10 mm) that one order exceeds terrestrial TPS growth conditions. After the seed rod was wetted with the melt, it was pulled at the speed $V \approx 3$ mm/min. The pulling speed of the test indium specimen, grown with the analogous device in terrestrial conditions, could not be elevated above 0.2 mm/min. The flight specimen diameter was ≈ 5.6 mm, whereas it was 2.9–3.2 mm of the test one. No special regulation systems were used, but it is evident that both pulling processes were stable. The both samples were mainly single crystals in structure.

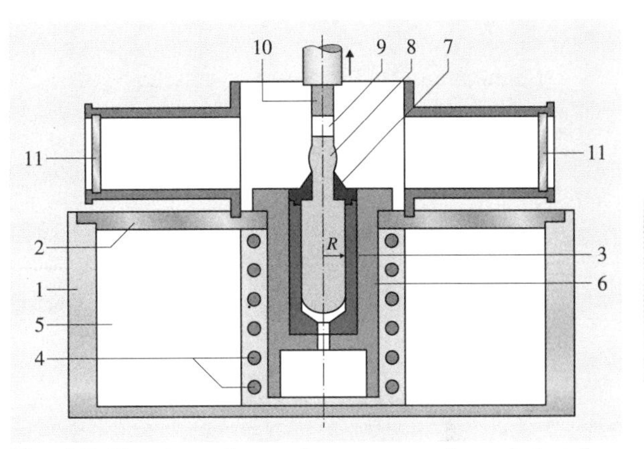

Fig. 16.22 Experimental setup for space crystal growth (notations given in the text) in the *shape* experiment

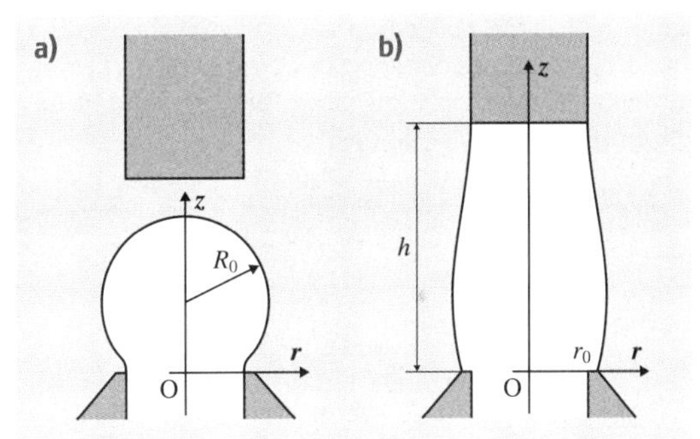

Fig. 16.23a,b Melt drop at the edge of the shaper (**a**) and the meniscus formed thereupon (**b**) in the *shape* space experiment

Ribbon Experiment. The crystallization of ribbons from the melt of Ge and GaAs in the orbital station *Mir* is the next example of TPS realization at microgravity conditions [16.6]. The melt is placed in a fixed gap between two non wettable flat plates (Fig. 16.24). The right part of the melt is a free surface of the $1/r$ curvature which depends on the wetting angle and the distance $2r_0$ between the plates. It was the same idea,

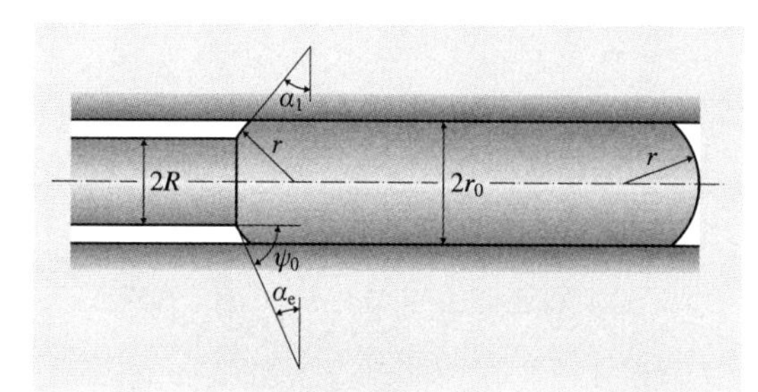

Fig. 16.24 Schematic drawing of tape growth in the *ribbon* space experiment

which before had been realized in the *shape* experiment, to maintain a definite pressure inside the growth meniscus by the curvature of the melt free surface. The growth meniscus projection in the gaps between the walls and the growing crystal are the circular arcs of the same radius r. It is the realization of TPS with the wetting boundary conditions on the walls of the shaper. The employment of unwetted walls has to be successful, if the following inequality is fulfilled

$$\alpha_1 = \Theta - \frac{\pi}{2} > \alpha_0 = \frac{\pi}{2} - \psi_0 \,, \tag{16.41}$$

that is

$$\Theta > \pi - \psi_0 \,.$$

Here Θ is the wetting angle and ψ_0 is the growth angle.

The ribbon thickness $2R$, the capillary coefficients A_{RR} and A_{Rh} are the following ones

$$2R = \frac{2r_0 \sin\alpha_0}{\sin\alpha_1} \,, \tag{16.42}$$

$$A_{RR} = \frac{-V \sin\alpha_1}{r_0 \cos\alpha_0} \,, \tag{16.43}$$

$$A_{Rh} = \frac{-V \sin\alpha_1}{r_0 \sin\alpha_0} \,. \tag{16.44}$$

The both capillary coefficients are negative and there is a capillary stability. For the common dynamic stability estimation, the heat coefficients (16.31) and (16.32) can be used: $A_{hh} < 0$, $A_{hR} > 0$. Therefore, with respect to (16.22) inequalities fulfilled, the dynamic stability of the scheme under investigation is provided.

At the flight experiments, Ge and GaAs ribbons were grown between pyrocarbon plates. Unfortunately, a heating regime was not optimal, the seeds were melted and the ribbon structures were polycrystalline [16.6]. But smooth surfaces and constant thicknesses of the ribbons, grown without any special regulation, is an evidence of the process stability.

16.6.4 Impurity Distribution

Impurity distribution in crystals grown by the Bridgman, CZT and FZT is studied ratter thoroughly, however extension of the mechanisms known to thin-profile growth using TPS can lead to wrong conclusions. Application of the Burton–Prim–Slichter equation [16.55] to calculation of the impurity distribution effective coefficient K requires specification of the boundary diffusion layer thickness. In case this thickness is assumed to be equal to the total height of the meniscus and the shaper capillary channel, $K = 1$ for any impurity [16.56] that does not correspond to reality. The present section gives a model of impurity transfer for the TPS under consideration that allows relating K-values to the parameters of capillary shaping and feeding [16.6, 57]. Figure 16.25a illustrates the case of a thin tape growth.

Stationary process is considerated; it is assumed that the melt in the zone of the meniscus and the capillary channel is not stirred and the conditions of complete stirring are maintained in the crucible. Under the assumptions made, an impurity transfer in the meniscus is described by

$$D\left(\frac{\mathrm{d}^2C}{\mathrm{d}r^2} + r^{-1}\frac{\mathrm{d}C}{\mathrm{d}r}\right) = -V\frac{\mathrm{d}C}{\mathrm{d}r} \,, \tag{16.45}$$

with the following boundary condition at the crystallization front

$$-D\frac{\mathrm{d}C}{\mathrm{d}r}\bigg|_{r=r_0} = V_0\,(1-K_0)\,C(r_0) \,. \tag{16.46}$$

Here C denotes impurity concentration in the melt, D is the diffusion coefficient, K_0 is the impurity distribution equilibrium coefficient, others notations are given in Fig. 16.25a.

The polar coordinate system chosen allows easy specification of the melt-flow rate distribution in the meniscus $V(r) = V_0 r_0 / r$. Then (16.45) is rearranged into the following form

$$\frac{\mathrm{d}^2C}{\mathrm{d}r^2} + r^{-1}\left(1 + \frac{V_0 r_0}{D}\right)\frac{\mathrm{d}C}{\mathrm{d}r} = 0 \,. \tag{16.47}$$

Impurity transfer in the capillary channel is described by the equation

$$D\frac{\mathrm{d}^2C}{\mathrm{d}z^2} = -V_c\frac{\mathrm{d}C}{\mathrm{d}z} \,, \tag{16.48}$$

with the following boundary condition

$$C|_{z\to\infty} \to C_\infty \,. \tag{16.49}$$

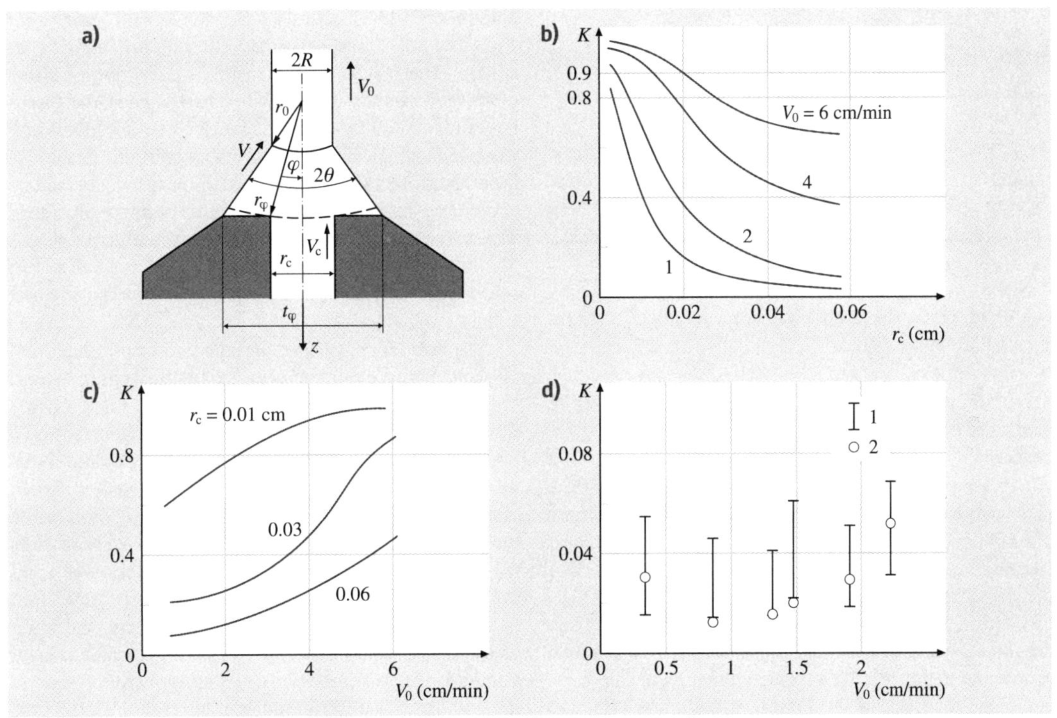

Fig. 16.25a–d Calculating the impurity distribution effective coefficient K (**a**) diagram; (**b**) $K = f(r_c)$ – calculated; (**c**) $K = f(V_0)$ – calculated; (**d**) comparison of experimental (1) and calculated (2) K-values from (16.50) for the In distribution in thin-walled shaped Si crystal

It is assumed that the impurity concentration in the bulk of the melt C_∞ is constant in the process of tape growth. The solution of the problem relates the value of the impurity distribution effective coefficient K to the parameters of the crystallization conditions (V_0 and $2R$), with the conditions of capillary shaping (t_ϕ, r_c, Θ) and the impurity characteristics (D and K_0)

$$K = K_0 \Big/ \left\{ K_0 \left[1 + \left(\frac{\Theta r_c}{\varphi t_\varphi} - 1 \right) \cos\varphi \right] + (1 - K_0)(1 - \cos\varphi) \left(\frac{2R}{t_\varphi} \right) \right\}^{V_0 R / D \sin\Theta} . \quad (16.50)$$

If the width of the capillary channel r_c reaches its maximum value t_φ, equation (16.50) is reduced to

$$K = K_0 \Big/ \left[K_0 (1 + (1 - K_0) \times (1 - \cos\varphi) \left(\tfrac{2R}{t_\varphi} \right) \right]^{V_0 R / D \sin\Theta} . \quad (16.51)$$

Figure 16.25b shows $K = f(r_c)$ calculated for aluminum impurities in silicon. The following values are used: $K_0 = 0.002$, $D = 0.53 \times 10^{-3}\,\mathrm{cm^2\,s^{-1}}$, $2R = 0.03\,\mathrm{cm}$, $t_\varphi = 0.06\,\mathrm{cm}$. Figure 16.25c presents $K = f(V_0)$ calculated from (16.51) for various r_c values. It is obvious that K increases with V_0 (and as a result V_c) increasing. Under actual conditions of growing thin-walled profiled silicon crystals $K_{Al} < 1$. Processing of the data of [16.58] gives the value of $K_{Al} = 0.039$ in silicon, which agrees very well with the value of $K_{Al} = 0.3–0.4$ calculated from (16.50) in accordance with the initial data of [16.58].

Figure 16.26 gives data on sulfur distribution along the axis of a TPS silicon tape obtained by laser emission microanalysis (LEM). The ratio of the sulfur spectral line strength J_S to that of silicon J_{Si} (the ratio is proportional to the concentration of the element analyzed in the silicon matrix) is plotted on the ordinate. The rate of tape pulling was equal to 12 mm/min. Increase of sulfur concentration in the silicon tape in the process of its pulling shows that $K_S < 1$ (it is assumed that

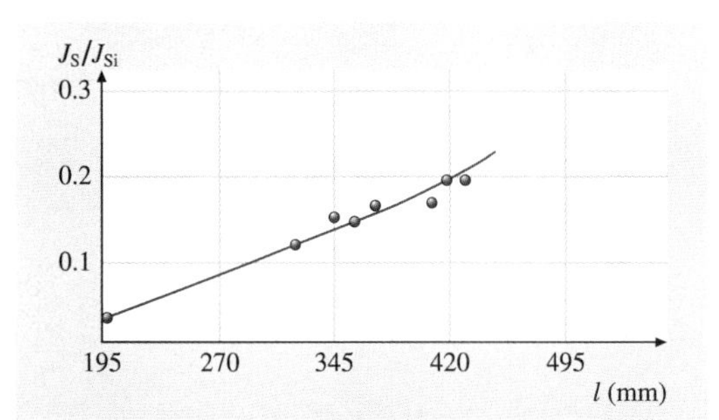

Fig. 16.26 Sulfur distribution along the length of a silicon tape (LEM data)

for sulfur impurity in silicon $K_0 = 0.001 \times 10^{-3}$). While analyzing other impurities, no explicit regularities were observed.

To verify the main equation (16.50) a series of 0.01 mass indium-doped silicon tapes was grown. The mass of each tape-shaped crystal did not exceed 7% of that of silicon charged into the crucible. The tape thickness $2R$, the shaper transverse dimension t_φ, the capillary slot dimension r_c, the growth rate V_0 were measured. Then photometrical spectral lines of indium in the crystals grown and in the crucible residue were drawn. The measurement accuracy of the indium distribution effective coefficient was equal to (40–50%). The results obtained are depicted in Fig. 16.25d. Vertical arrows indicate experimental results, circles show the values calculated from (16.50) in accordance with the above mentioned parameters. The values of $D = 0.52 \times 10^{-3}\,\mathrm{cm^2\,s^{-1}}$, $K_0 = 0.4 \times 10^{-3}$ are assumed for indium impurity in silicon. The results shown in Fig. 16.25d do not allow plotting K versus the growth rate V_0 since the thickness of a silicon tape, $2R$, decreases with V_0 increasing (the value of $2RV_0$ practically was constant in the experiments), and they can only demonstrate satisfactory agreement between calculated and experimental values. The values of K calculated from the Burton–Prim–Slichter equation are equal to 0.8–0.9, i. e., they are some orders of magnitude greater than the experimental values.

Impurity distribution along the widths profiled crystals is to a great extent determined by the technique used to feed the melt to the growth meniscus. In Fig. 16.27a,b resistivity ρ distributions across of silicon tapes, grown under the conditions of the two versions of meniscus melt replenishment, are compared. The shaper shown in Fig. 16.27d possesses one long capillary slot, while the shaper given in Fig. 16.27e has two short slots at its end faces. The same feeding system was used for the sapphire ribbon growth. A diffusion reflection of light (Fig. 16.27c) indicates a bubbles concentration at the center of ribbon. The following explanation of nonuniformity of impurity distribution across the crystal (Fig. 16.27a–c) observed can be offered. In the case of horizontal melt flow in the meniscus from the shaper edges towards its center, impurities with $K < 1$ driven off by the growing crystal accumulate in the central part of the meniscus. Hence, corresponding distribution of

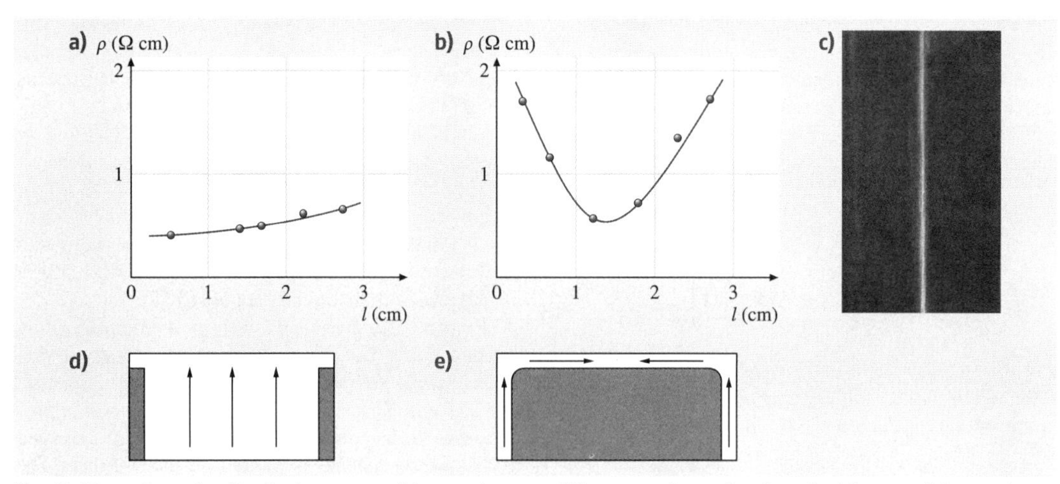

Fig. 16.27a–e Impurity distribution across ribbons using two different versions of melt replenishment of the meniscus: (**a**) silicon ribbon and (**d**) corresponding feeding scheme; (**b**) silicon and (**c**) sapphire ribbon and (**e**) corresponding feeding scheme

capillary channels in the shaper allows controlling impurity distribution at the cross section of crystals being grown.

16.6.5 TPS Definition

On the basis of the theoretical analysis, accomplished above, we can define TPS:

> *TPS is the shape crystal growth technique which uses a solid body (shaper) to define a melt meniscus shape by means of either catching (on edges of the shaper) or wetting (on surfaces of the shaper) capillary boundary condition to obtain the crystal of predominant cross section and impurity distribution as a result of pulling it in a dynamically stable regime.*

This definition follows from our analysis of capillary catching and wetting boundary conditions, completed in 1967 and published for the first time in 1968 [16.38], as well as from our dynamical stability analysis with a capillary shaping, completed in 1970 and published for the first time in 1973 [16.14]. Now, on the basis of this definition, we can analyze a TPS development history.

16.6.6 TPS Brief History

A development of shaped crystal growth for industrial application was begun from the set of papers [16.59–63] published in 1958–1959. The papers [16.59, 60] informed that during 1938–1941, Russian scientist A.V. Stepanov had carried out experiments concerning pulling of shaped polycrystalline and single-crystalline specimens (sheets, tubes and so on) from melts of some metals, especially aluminum and its alloys. The Second World War interrupted these experiments and they have been continued since 1950s in the Physical and Technical Institute of the USSR Academy of Sciences. During 1950–1958 few Stepanovs collaborators, especially postgraduated students *Shakch-Budagov* and *Goltsman*, continued these experiments with low-melting metals and alkaline halides. The papers [16.61–63] described the main results of these experiments. The author of this chapter has been a participant of the below described events because in 1959 he began a scientific activity as the undergraduated student in the Stepanov's laboratory.

In 1963 *Stepanov* formulated his global idea [16.64]:

> *it is necessary to find a way to pull all types of industrial profiles from melts of aluminum, steel and other alloys. It could save energy, and eliminate extruding, rolling, cutting and many other types of mechanical treatments.*

It was a basis of large investigations in the field carried out in the Stepanov's laboratory. Two main directions were being developed:

1. Polycrystalline metals and alloys
2. Semiconductors single crystals.

The author of this chapter was nominated as the head of the first direction. First of all, *Stepanov* and *Tatartchenko* decided to demonstrate the possibilities of the technique as widly as possible. For this, *Tatartchenko* designed an installation for the continuous production of aluminum tubes. During development of the installation, a lot of technical problems were solved [16.65]. The installation was completed in 1963. It contained two connected crucibles: the first for feeding and a second for the pulling of 6 mm diameter tube, automatically wound on a coil. During testing of the installation, specimens of tubes with a length up to 4000 m and a speed of pulling up to 15 m/h were obtained. The installation was made very compact to demonstrate its operation at different exhibitions. In 1964, the installation was shown in Italy. There was a full success: many articles described a new metallurgical technique. After that, a second variant of the installation was developed for the production of the aluminum profiles of different complicated cross sections with lengths up to 3 m [16.10]. The operation of it was demonstrated at exhibitions in Hungary (twice in 1967), in Italy (1968), in Czechoslovakia (1971). Examples of aluminum profiles obtained with this installation are presented in Fig. 16.28. A possibility to use the technique for steel profiles and combined aluminum-steel profiles pulling was either demonstrated by the author of this chapter [16.66] (Fig. 16.29).

In the application of the technique for single crystals pulling, the laboratory intensively was working with the huge industrial germanium project. Some industrial laboratories and plants were included in the project and, at the end of the 60s, in the former Soviet Union, about 85% of the germanium was produced as shaped crystals. For instance, for electronic applications, instead of one crystal by the CZT, 26 round cylindrical crystals with the diameter of 10.0 ± 0.1 mm we pulled, from the same crucible, without any special control [16.10]. The author of this chapter participated in the project by developing the theory described above as well as the application of the theory for the technology (e.g. [16.67]).

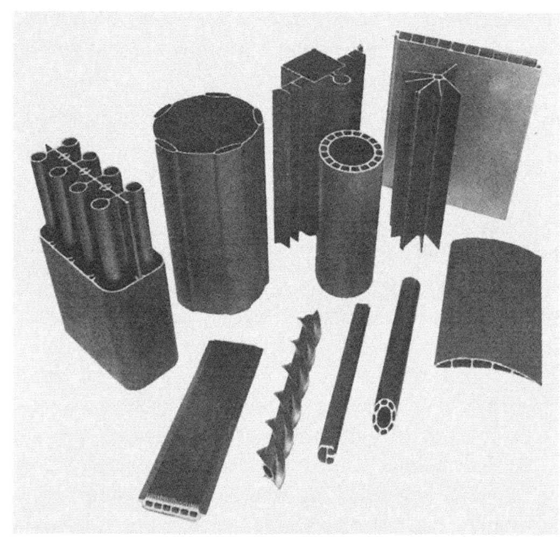

Fig. 16.28 Aluminum heat exchangers and other profiles grown by TPS in Stepanov's laboratory by the author of this chapter with collaborators

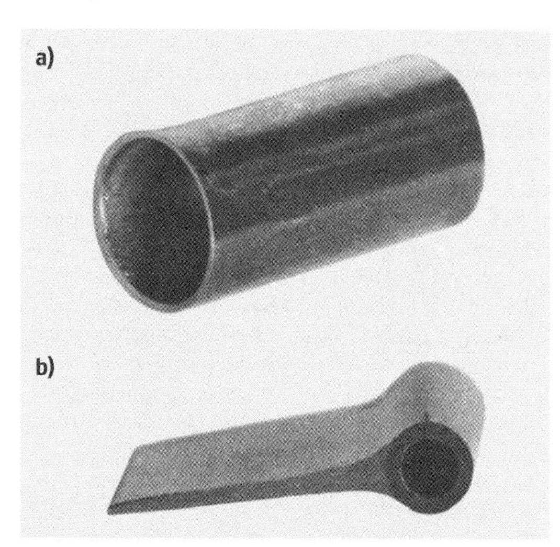

Fig. 16.29a,b Steel tube (**a**) and steel tube with Al rib (**b**) grown by TPS in Stepanov's laboratory by the author of this chapter

Up to 1967 shaped crystal growth was developing only in the former Soviet Union. The publication of the information concerning shaped filaments growth [16.68, 69] and especially edge defined film fed growth (EFG) [16.70] changed the situation drastically: Shaped crystal growth appeared in the USA. It was a beginning of shaped crystal growth spreading in many scientific groups, especially for the sapphire and silicon growth. As for former Soviet Union, the shaped crystal growth direction has continued the development. Since 1968 every year, Russian National Conferences has been organised with publication of the special Proceedings (e.g. [16.38]) as well as special issues of Bulletin of the Academy of Sciences of the USSR, Physical Series (e.g. [16.39]). All variants of pulling techniques from the shaper were named Stepanov's technique. In 1972, A.V. Stepanov died. His laboratory has continued to work. The author of this chapter was invited as the head of the Crystal Growth department in the Solid State Physics Institute of the Academy of Sciences of the USSR. In the frame of the department, the shaped crystal growth laboratory was organized with the stuff of 40 persons. Many theoretical and experimental aspects of shaped growth, Space growth included, were developed there. The main competitor during long time was being the Tyco laboratory where EFG growth was developing for sapphire and silicon. In 1978, the author of this chapter visited the Tyco laboratory and had very interesting discussion with La Belle, the author of EFG patent. That time, it was impossible to imagine that he would work in this laboratory (later Saphikon, and Saint-Gobain Crystals) as the chief scientist twenty four years later.

Now, shaped crystal growth is widely spreaded but there are a lot of titles used, especially EFG and Stepanov's technique. We are sure the situation must be clarified. As for us, since 1980s, we have preferred not to use the title Stepanov's technique neither EFG. We insist to combine all these techniques by the title *technique of pulling from shaper* (TPS). Here are the reasons of that.

A priority in applying of a shaper (holes in plates placed onto the melt surface for shaping melt-pulled crystals) belongs to Gomperz [16.71]. In 1922, he used mica plates floating on melt surfaces to pull Pb, Zn, Sn, Al, Cd, Bi thin filaments through holes in the plates. Sometimes the filaments had a single crystalline structure. In 1923, the same technique was used either for filaments pulling [16.72]. In 1929, for Zn single crystalline filaments growth, for the first time, a single crystalline seed was used [16.73]. In 1929, the technique was titled Czochralski–Gomperz technique [16.74, 75]. In 1928, P. Kapitza, later the Nobel Prize rewarded, used this technique for Bi rods growth [16.76]. So, during 1922–1931, six papers concerned with a shaper using were published. Moreover, the technique was titled as a CZT modification Czochralski–Gomperz technique. Hence, Stepanov does not have a priority. Never in his publications, have we found any of these six references. What about the explanation of the shaper functions? He is one of our coauthors of [16.38, 39] papers, explaining catching and wetting boundary conditions. On the other hand, at the same time his name as a coauthor we can see in the [16.41] publication, the no applicability of which to shaped growth we explained in some our papers, among them with Stepanovs participation [16.43]. This absurd situation does not need any comments [16.7]. In [16.64] Stepanov pretended to have invented:

> *the principle of shape formation from a melt, using capillary forces or some other actions (except of crucible walls) on the melt, of the cross section or an element of a cross section of solid profile its following crystallisation.*

This is a typical *umbrella* formulation which covers CZT, VT, FZT techniques as well as electromagnetic, ultrasonic, inertial and other possibilities of shaping. It can not be discussed seriously. At the same manuscript we read:

> *A shaper should be distinguished from a die. A die is the embodiment of a brute force. A shaper is a more spiritual system. Its aim, first of all, is to provide a delicate effect on the curvature and shape of the mobile column of the liquid melt stretching itself behind the crystal by creating new boundary conditions along its contour.*

We agree that there is a big difference between a die and shaper, but in spite of Stepanovs big role in the TPS spreading [16.7], *new boundary conditions* never have been specified in Stepanov's papers. We think there is sufficient quantity of arguments not to use the title Stepanov's technique. On the contrary, our capillary shaping analysis of TPS, that has been done in 1968 [16.38] is completed: It is impossible to suggest something new, beside catching or wetting if using a shaper. From this point of view EFG technique is simply TPS with catching boundary condition. Probably, this is because of a low level of patent experts that the independent EFG patent exists. Edge defined condition for meniscus was published in our paper [16.39] as catching boundary conditions before the EFG patent appearance. We only agree that capillary feeding was described in the EFG patent for the first time. Certainly, many shaped growth schemes of TPS have peculiarities and we will discuss some of them below.

Now we will describe shaped crystal growth of few materials. The most impressive results of TPS industrial application were obtained for sapphire and silicon. Furthermore structures of shaped sapphire and silicon have a lot of peculiarities.

16.7 TPS Sapphire Growth

Sapphire belongs to the family of corundum crystals. The term corundum designates α-aluminum oxide. Pure corundum crystals are colourless and are named *leucosapphire* ones. But now the name *sapphire* crystals is very often used, although historically the title *sapphire* crystals has been applied to the blue corundum ones. Different colours of corundum can be obtained by the addition of different metallic oxides to aluminum one. Natural and synthetic red ruby contains 1–7% chromium oxides. Synthetic alexandrite is produced by the addition of vanadium oxide. 1% titanium and 2% iron oxides give a blue colour. Nickel oxide gives a yellow colour. An addition of vanadium and cobalt oxides imparts a green colour. Green crystals are also produced if the chromium content exceeds 8%.

The application of TPS for sapphire is an example of the most successful one because of coexistence of few factors [16.8]: First of all, it is a demand for advance techniques of a material with unique physical and chemical properties: high melting point, exceptional hardness, transmission over a wide band of wavelengths, radiation and chemical resistance. Secondly, exceptional hardness of sapphire hinders its machining if it could be grown by other techniques. Thirdly, it has been impossible to growth very big crystals (for instance, optical windows 320 x 500 mm^2 for aircrafts and spacecrafts) by using other crystal growth techniques. Fourthly, chemically resistant and wettable materials (W, Mo, Ir) have been found for shapers.

La Belle and *Mlavsky* were the first who grew shaped sapphire crystals. In [16.68] there was information concerning shaped filament growth without any detail. In [16.69, 70, 77] edge defined film fed growth, EFG, technique was described. EFG corresponds to the scheme Fig. 16.4e, i. e., TPS with a catching boundary condition. A peculiarity of EFG is a capillary feeding. The capillary together with wettable shaper facilitates a seeding. It is important to repeat once more that the capillary from Fig. 16.4e serves only for feeding and does not participate in the capillary shaping: the meniscus, catched on the edge of a shaper, changes its curvature with forming a negative pressure inside, and raises the melt above the shaper edge, up to the crystallization front.

16.7.1 Modifications of TPS

The presence of the shaper in TPS allows many manipulations during growth. The variable shaping technique VST described for the first time in [16.79], gives a possibility to change a cross section of crystal (periodically if it is necessary) during of pulling. Monocrystalline sapphire profiles, grown by TPS and VST by the authors with collaborators in the Solid State Physics Institute of Academy of Sciences of Russia, are presented on the Fig. 16.30. VST is based on controlling the melt mass flow towards the crystallization interface when passing to a new specific cross section shape. One of VST variants described in [16.78] is shown on the Fig. 16.31. The scheme allows not only changing the shape of the crystal but also its composition. It works by the following way: At the beginning both channels 5 and 6 plunged in the melts but, only through capillary channel 5, melt reaches the shaper edge. As a result, during a first stage, a tube (cross section A) grows. Because of a vacuum, formed inside of the growing tube, the other melt through non capillary channel 6 reaches the shaper edges and, during a second stage, the rod of two different compositions grows (section B). During the thirds stage shown on the diagram, the feeding through the channel 5 is eliminated by the crucible lowering (section C). By the similar way, changing of shaper plunging in the melt, different edges or walls of the shaper can operate by turns. As a result, we obtain a variety of shaped crystals, examples of which are presented on the Fig. 16.30. There is very interesting modification of TPS used especially for a dome growth [16.80]. There the seed is rotated around horizontal axes.

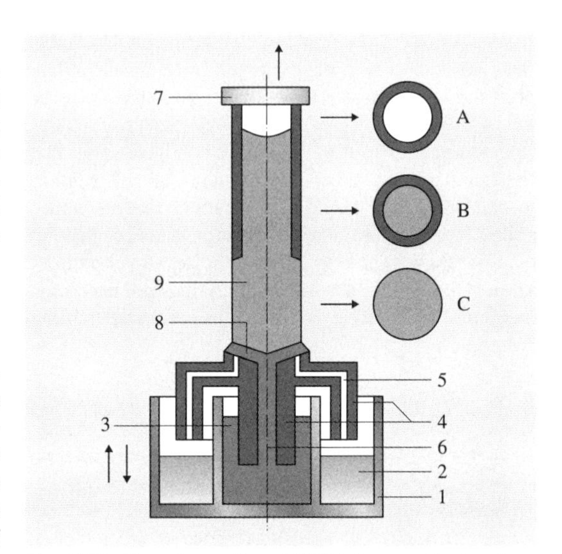

Fig. 16.31 The scheme from [16.78] for growth at various cross sections and various compositions crystals: 1 – crucible; 2 – undoped melt; 3 – doped melt; 4 – shaper; 5 – capillary feeding channel; 6 – noncapillary feeding channel; 7 – seed; 8 – meniscus; 9 – growing crystal; A, B, C – cross sections of the different parts of the growing crystal

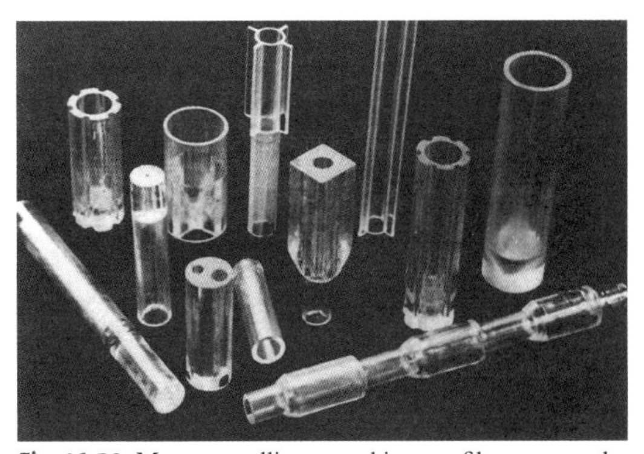

Fig. 16.30 Monocrystalline sapphire profiles grown by TPS and VST

The local shaping technique, LST [16.81], shapes only an element of the crystal that we need to grow. During growth, the horizontal displacement of the seed and shaper with respect to each other in combination with rotation and pulling of the seed allows obtaining a variety of shape crystals. Figure 16.32 presents a tube growth: A shaper of a diameter q shapes an element of the tube of thickness $p = r_2 - r_1$. The seed is rotated with the angle speed ω and is moved up with the speed V. A middle tube radius R is equal to a distance between the shaper center and the axe of the seed rotation. Practically any complicated body of revolution, for instance, a dome with the thin wall, can be obtained by this technique. A dynamic stability investigation of LST was carried out in [16.81]. It has to be mentioned that

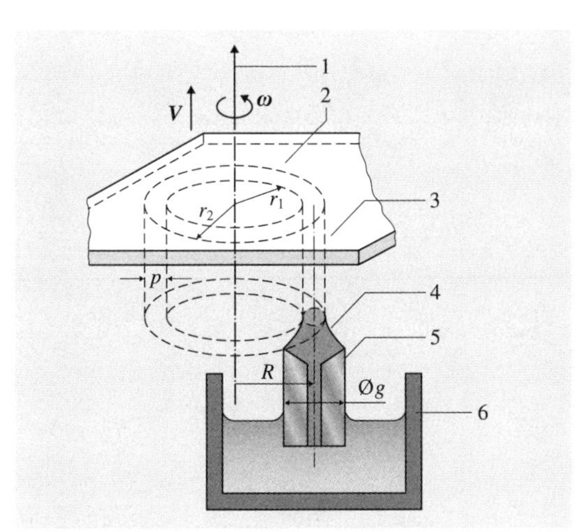

Fig. 16.32 Local shaping technique scheme: 1 – rotation axis, 2 – seed crystal, 3 – growing tube, 4 – melt meniscus, 5 – shaper, 6 – crucible

40 years ago LST was being used in our experiments with aluminum growth as well as a vacuum inside of a tube grown for a seeding. Some specimens are shown on the Fig. 16.28.

16.7.2 Crystal Defects

In compared with other crystal growth techniques, the peculiarities of TPS crystal defect structures are mainly determined by two factors:

1. High growth speed which, as a rule, one or two orders more
2. Small thickness.

Bubbles are the major defects in shaped corundum crystals (e.g. Fig. 16.27c). There are two aspects of the problem concerned with its appearance in the crystal. The first of them is an enrichment of the melt by gas and the bubble formation in the melt. The second one is a capture of the bubbles by grown crystal.

A vacuum treatment of the melt before the crystal growth reduces the bubble contents in the crystals. This is a proof that a dilution of gases in the melt is one of the sources of its gas enrichment. But there is a more important gas source for the corundum melt. In spite of the aluminum oxide is stable under normal conditions; its heating is accompanied with thermal dissociation, evaporation of some reaction products and coagulation of others. Mass spectroscopic analysis shows [16.82] that the products of dissociation of aluminum oxide are O^+, O_2^+, Al^+, AlO^+, Al_2O^+, AlO_2^+, $Al_2O_2^+$. The intensity of thermal dissociation depends greatly on the temperature and environment. It is the most intensive if the hydrogen is presented, less intensive in the vacuum and the least intensive in the inert atmosphere (argon, nitrogen). The VT improves this thesis. One of the first bubble formation versions in VT corundum crystals was concerned with the gas adsorption on the surface of feeding powder particles. But the hypothesis is at variance with the following experimental fact. The γ-Al_2O_3 powder is more friable than α-Al_2O_3 and, therefore, adsorbs more gas on the surface. But corundum crystals grown from γ-Al_2O_3 charge have less bubble inclusions. So, the charge dissociation has to be the main reason of bubble formation because of the hydrogen presence, and α-Al_2O_3 powder is worse because of sticking of α-Al_2O_3 powder particles.

The crucible and shaper presence complicates the situation. Although the problem is not completely understood at this time, the thermal dissociation of the melt because the chemical interaction of the melt with the crucible material seems to be very important cause of the gas appearance in the melt. Here is the list of chemical activity of the main container materials for sapphire crystal growth (from the best to the worst): Ir, W, Mo, Nb, Ta, Zr. As rule, only three first materials are used for the containers. The reaction of Al_2O_3 with refractory metals is generally believed to involve two stages. For example with Mo: $Al_2O_3 \rightarrow Al_2O + O_2$; $Mo + O_2 \rightarrow MoO_2$. In reality, the presence above mentioned aluminum oxide dissociation products does the situation much more complicated. Mass spectroscopic measurements point to the presence of MoO, MoO_2, MoO_3 ions.

The diluted gases and the gas products of the above mentioned reactions are impurities in the melt and assist a gas inclusions formation in the crystals. Here are two possibilities:

1. As rule, dilution of gases is decreased with the temperature. The liberated gases form the bubbles. The crystal catches the bubble existed in the melt.
2. A nuclear bubble forms on the interface, grows and catches by the crystal.

The surface energies estimation [16.83] deduced that the interface is not an effective heterogeneous site for bubble formation. We agree that this deduction is correct for a flat interface but cavities on the interface can be effective bubble nucleation spots. In any case, the melt layer near the interface has to be preferable for the bubble formation because of enrichment of the diluted gas rejected by the growing crystal. Direct experiments of the melted zone quenching [16.84] improved this mechanism of the bubble formation.

As other impurities, bubbles can form striations, parallel to the interface. The striations are concerned with the variations of growth rate and captures of the bubbles situated in the melt layer near the crystallization front. Sometimes it is a result of constitutional super cooling. Sometimes long tube similar bubble can be found in the corundum crystals. A mechanism of this type bubble formation suggested in [16.85, 86].

In each case, the caverns on the interface, formed as a result of morphological stability loss, provokes the bubbles capture, especially for the big growth rate. In our experiments [16.5, 6, 87] with ribbons and tubes grown by TPS in the *C*-direction the following results were obtained. At the crystallization rate of order of 0.2 mm/min the bubble practically were absent. In [16.88] the critical growth rate (0.5 mm/min for sapphire) has been calculated below which a foreign particle is not trapped by the grown crystal. The interface is plane below this rate. At crystallization rates 0.75–1.5 mm/min the bubble distribution is in the shape of a treelike pile-up with a branch diameter of up to 1 mm (Fig. 16.33). The ends of ribbon are bubbles free, probably, because of gas diffusion to the melt surface and leaving of the melt. The same result was obtained in [16.83] where small diameter crystals were bubbles free.

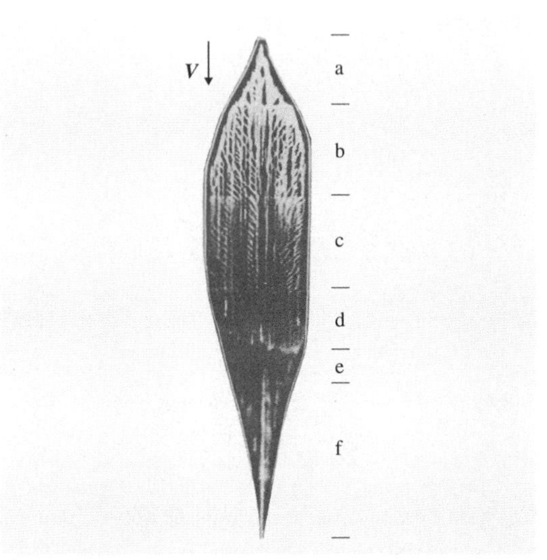

Fig. 16.33 Bubbles in a sapphire ribbon pulled with step changes in the rate V: the *arrow* indicates the growth direction, Vmm/min = 0.75 (a); 1.0 (b); 1.25 (c); 1.5 (d); 1.75 (e); 2.0 (f)

The distribution of bubbles over a specimen section normal to the growth direction is shown in Fig. 16.34a,b. The following explanation is proposed for the bubble distribution observed. The plane interface becomes morphologically unstable. Macroscopic concavities, with a diameter to 1 mm, arise on the crystallization front; these concavities are characterized by an enhanced trapping of bubbles. We suppose that the change of position of the concavities on the crystallization front with time depends on the pattern of the convective flows in the melt that leads to the bubble distribution shown in Figs. 16.33 and 16.34a. This process is in accordance with the *Mullins* and *Sekerka* [16.20] theory for the lowest frequencies. When growing shaped sapphire crystals at rates of 1–3 mm/min the interface has, as a rule, a convex central part and concave peripheral sections that traps the bubbles (Fig. 16.34b).

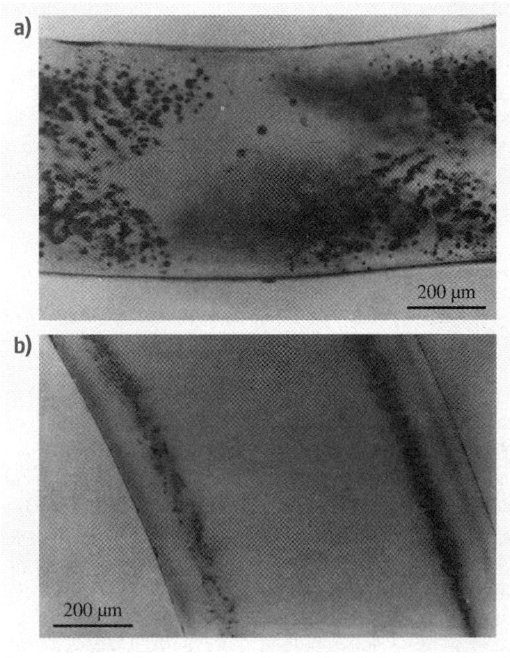

Fig. 16.34a,b Cross section of sapphire ribbon pulled with rate of V (mm/min) = 0.75 (**a**); 2.0 (**b**)

As the growth rate is increased further, the entire solid–liquid interface becomes unstable. The crystallization front becomes faceted, with the new interface shape becoming stabilized by the faceting. Figure 16.35

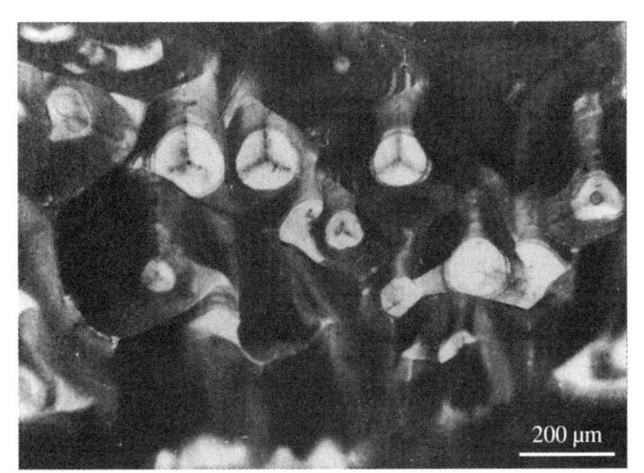

Fig. 16.35 Interface of decanted sapphire ribbon grown with $V = 3\,\mathrm{mm/min}$

shows the decanted crystallization front of a tube faceted by rhombohedral planes. The bubbles distribution in the crystal is a hexagonal pattern and corresponds to the positions of grooves on the interface. Always the bubble agglomerations correspond to cavity positions. The decanted crystallization front of a tube faceted by the $\{11\bar{2}0\}$ and $\{10\bar{1}1\}$ planes is shown in Fig. 16.36a. From geometrical considerations

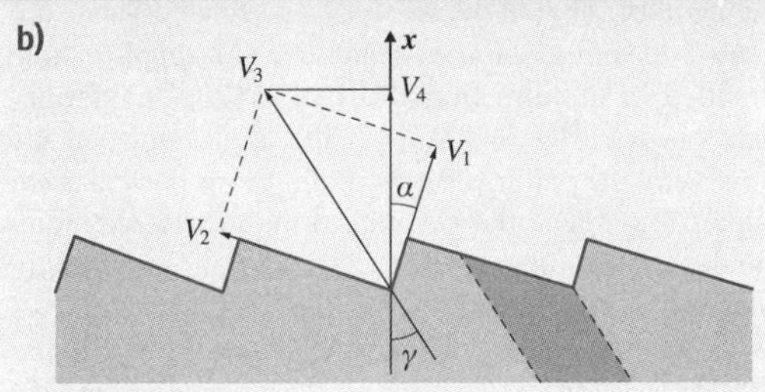

Fig. 16.36 (**a**) View and (**b**) schematic of the decanted crystallization front of a tube pulled at $V > 5\,\mathrm{mm/min}$

(Fig. 16.36b), it is apparent that the angle γ between the crystal pulling direction and the direction of the band of void pile-up depends on the ratio of two mentioned faces growth rates.

Solid inclusions are the second type of inclusions. Container material gives metal particle inclusions, sometimes as a result of complicated chain of chemical reactions mentioned above – volatile oxides formation, gas transport in the melt and decomposition of oxides with metal inclusion formation. Experiments show that the small particle of container metal (Ir, Mo), as rule, presents at the crystal. Innumerable experimental data demonstrate that the density of small dimensions (about $10\,\mu\mathrm{m}$) inclusions (bubbles and metal particles) is a few orders of magnitude greater than of large dimensions (about $1000\,\mu\mathrm{m}$) ones. The low growth speed and low gradients favor to crystal growth with minimum inclusions. Here is an example (American 80/50 standard) of bubbles and solid inclusions requirement for window-grade corundum: maximal allowable size is $500\,\mu\mathrm{m}$, and maximal allowable sum of diameters within any 20 mm – diameter circle is $1000\,\mu\mathrm{m}$. As rule, modern technology of TPS sapphire meets this requirement.

Dislocations, Low Angle Grain Boundaries, Internal Stresses

Four possible mechanisms of dislocation appearance in the corundum crystals can be postulated:

1. Intergrowth from seed
2. Formation on rough defects (inclusions, grain boundaries, twins) during growth
3. Condensation of vacancies during growth and cooling
4. Plastic deformation during growth and annealing.

Theoretically all dislocations can be identified. The dislocation identification of the first and the second types is evident. The third type dislocations, as rule, have a circle shape because they are a result of collapse of flat round discs obtained by a condensation of nonequilibrium vacancies. The geometry of the last type dislocations depends on two plastic deformation glissile systems $\{0001\}\ \langle 11\bar{2}0\rangle$ and $\{11\bar{2}0\}\ \langle 10\bar{1}0\rangle$ in the corundum crystals. The first source can be eliminated by necks formation. The second one – by crystal growth without rough defects. The third mechanism concerned with a fundamental physical phenomena, but their density are not significant for the corundum crystals. As to the plastic deformation, it is a main source of dislocations in the corundum crystals.

The defect structure of TPS crystals is formed as a result of complex interactions of growth, in situ annealing and cooling. A systematic investigation of the defect structure of TPS grown sapphire crystals was carried out both by the optical polarization method and by the technique of widely diverging x-ray beams [16.5, 6, 87]. The tube samples with diameters of 4–40 mm and wall thickness in the range 0.5–3 mm were investigated.

It has been established that, in the absence of low-angle grain boundaries propagating from the seed, the initial part of the crystal does not contain subgrain (or low-angle grain) boundaries. Then, as the crystal grows the dislocation density increases and subgrain boundaries are formed. The disorientation of adjacent subgrains increases with distance from the seed, reaches a certain limit, and then decreases slightly with further growth. The decrease is probably associated with a rearrangement of the subgrain structure; i. e. subgrains with large disorientations branch into a series of subgrain boundaries with smaller disorientations. In addition, there is a decrease in the density of subgrain boundaries which are at large angles to the growth direction. The subgrain boundaries which remain are approximately parallel to the growth direction; the density of these boundaries (with disorientation as high as $5-10^{\circ}$ stays constant.

It should be noted that, at high crystallization rates, crystals which are free of low-angle grain boundaries can sometimes be grown. This is probably explained by the fact that the time spent by the growing crystal in the plastic zone is not long enough for polygonization processes to occur. The presence of bubbles is a source of additional stresses which gives rise to dislocation generation and boundary formation. Sub grain boundaries are frequently observed to form along planes of bubbles pile-up. So, we can conclude that the quality of crystals at the beginning, at lower temperature gradient, was better.

Our experience of sapphire crystals growth by different techniques, TPS included, allows a statement that only temperature gradient less than 2 K/mm gives the possibility to grow the crystals without low angle boundaries and the dislocation density on the level $\leq 10^3\ \mathrm{cm}^{-2}$. Certainly, the dislocation density is a relative characteristic of crystal because the crystal quality can be characterized only taking into account, as minimum, six dependent parameters:

1. The dislocation density
2. The density of low angle boundaries
3. The degree of subgrain disorientation
4. Residual stresses value
5. Twins
6. Impurity inhomogeneities.

Twins

In the temperature lower than 800 °C single crystal sapphire is essentially brittle and is not deformed by usual dislocation mechanism. A twinning is only mode of deformation. The following two main rhombohedral twin systems in sapphire were identified: $\{01\bar{1}2\}\ \langle0\bar{1}11\rangle$, $\{0\bar{1}14\}\ \langle02\bar{2}1\rangle$.

Twinning process consists of two distinct stages: the twin nucleation and the twin growth. The nucleation stress is higher than growth stress. In [16.89] the nucleation stress for 600 °C was determined experimentally. It is high and strictly depends on surface treatment. It was found as $13-18\ \mathrm{kg/mm^2}$ for polished specimens and $36-41\ \mathrm{kg/mm^2}$ for polished and heat treated ones. During growth the stress of twin formation has to be average. It corresponds to the residual stress $\approx 20\ \mathrm{kg/mm^2}$, where the twins can be more often found.

Faceting, Inhomogeneities of Impurity

Faceted growth is a normal mode of some growth techniques. The singular faces appearance on the interface in CZT is not very important for sapphire growth but assists nonhomogeneity in ruby growth because the coefficient of chrome distribution is different for faceted and nonfaceted interface. But for TPS faceting may be serious defects.

The geometrical form of TPS tubes can differ slightly from ideal for reasons of a crystallographic nature. According to the Curie theorem, in the process of growth, there is an interaction between the crystal symmetry and that of the medium (i. e. the thermal environment in which the crystal grows). In the grown crystal, only those elements of symmetry are exhibited which are common both for the crystal and for the medium. That is the reason why grown crystals are faceted depending on the orientation. If a sapphire tube is pulled in the directions $\langle10\bar{1}0\rangle$, $\langle11\bar{2}0\rangle$ or intermediate ones with $\rho = 90^{\circ}$ (ρ is the angle between the $\boldsymbol{C}$-axis and the pulling direction), the close-packed basal plane (0001) becomes parallel to the growth direction and facets the tube [16.5, 6, 87]. If ρ differs slightly from 90°, a steplike faceting by the basal plane appears on the lateral surface of the tube. Tubes grown in the [0001] direction are faceted on the outside by $\{11\bar{2}0\}$ planes and on the inside by $\{10\bar{1}0\}$ prismatic planes. Faceting is reduced if the temperature gradient at the interface increases. The faceting can be explained because of supercooling that singular face needs to grow with the same speed as an isothermal part of interface.

Impurity inhomogeneities are important for ruby crystals. There are macroscopic and microscopic periodic inhomogeneities (striations) along axes of growth. Macroscopic ones concerned with the impurity concentration changing during the growth because of the coefficient of distribution that is not equal of one. TPS allows growing of crystals with macroscopically homogeneous distribution of impurity. Microscopic inhomogeneity concerned with the periodic changing of growth parameters or constitutional supercooling.

Growth Direction

The corundum crystal growth in the direction of ***C***-axis is very difficult for all techniques of the melt growth, except TPS. TPS allows growing of all profile crystals in ***C***-direction without any problems. And what is more, if we grow corundum filaments without seeds, the spontaneous orientation of the filament coincides with ***C***-axis. What is the reason? For many melt growth techniques the growth in the ***C***-direction requires an appearance of the singular ***C***-plane on the interface. ***C***-plane has the lowest growth speed. TPS crystals, the filaments especially, grow with high speed. The interface loses morphologic stability and is faceted by the rhombohedral faces, as it is shown on Fig. 16.35. So, ***C***-plane does not participate in the growth. As for filament orientation, the orientation of ***C***-axes is the most preferable orientation for its growth with respect to Curie principle. On the other hand, the filament interface also is faceted by the rhombohedral faces.

16.7.3 Applications

Special Windows

Application for modern airborne optical reconnaissance systems is one of the most impressive fields of TPS sapphire using [16.8]. Practically TPS does not have any competitors here. Below are the evidences that TPS sapphire meets all requirements for this application:

1. The maximal dimensions achieved are $315 \times 480\,mm^2$ [16.90]. It is necessary up to 750 mm diameter.
2. It has *high optical transmission* in the 3–5 μm wavelength atmospheric transmission window.
3. Sapphire has *a hardness 9 on the Moose scale* (the hardness of diamond is 10). It is the most durable commercially available infrared window material. It has the best resistance to erosion by rain and sand of any available window materials.
4. It also has *excellent thermal shock resistance*. But its thermal shock resistance is limited by loss of mechanical strength at high temperature from $70\,kg/mm^2$ at room temperature to $20\,kg/mm^2$ at 600 °C. Doping or ion implantation with Mg, Ti can double the compressive strength at 600 °C. Heat treatment at 1450 °C in an air atmosphere enriched with oxygen increases compressive strength by 1.5 times. Neutron irradiation with 1×10^{22} neutrons increases the ***C***-axes compressive strength by a factor of 3 at 600 °C.
5. It is available routinely with *minimal optical scatter*.
6. *High refractive index uniformity* is achieved because of simple oxide composition. For stringent optical applications, ***C***-axes optics is preferred as this is zero birefringence orientation.

Domes

The requirements for rocket nose cones correspond mentioned above ones. The traditional technique of dome production is a mechanical treatment of big sapphire crystals. The effort of production concerned with minimization of treatment by a growth of crystals having a shape close to dome. TPS has suggested some successful examples of this kind [16.80, 81, 91].

Substrates

An application of sapphire in electronics as substrate for silicon on sapphire devices was rather large at 1970s. In that case the face $\{1\bar{1}02\}$ was used. The requirements were not very rigid. But since 1990s, sapphire has been becoming the main substrate material for blue and white laser diodes. The face {0001} is used for the epitaxy. The requirements to the wafers concerning a crystal structure and orientation as well as polishing quality are very high. Sometimes special misorientation of wafer is used to get better deposited layer. It is evident, that this TPS application is very promising. But practically there are no publications concerning this topic, except rare ones [16.92], probably because of technological secrets.

Construction Material

Sapphire is increasingly becoming the material of choice for engineers faced with the design challenges of extreme conditions, such as those found in high-temperature, high-pressure or aggressive chemical environment. There are a lot of examples of this TPS sapphire product using in the former Saphikon (now

Saint-Gobain Crystals) catalog. The industrial technique of welding of sapphire pieces is developed there. Sapphire tubes, plates and more complicated assemblies are used a superior alternative to quartz, alumina, and silicon carbide: in semiconductor processing application (plasma containment tubes, process gas injectors, thermocouple protectors); in spectroscopy and chemical and biological analysis; lamps and lamp envelopes (high intensity lamps, flash-lamps, ultraviolet sterilizations); GaAs backer/carriers; mail sorting optical windows.

16.8 TPS Silicon Growth

Silicon is the second example of the TPS successful industrial application. But this case is not similar to the sapphires one. All numerous attempts to grow shaped Si crystals of electronic grade quality have not been successful because shaped silicon is characterized by the presence of a defect structure influencing its electronic properties. At the same time, a quality of big surface thin sheets obtained is acceptable for no expensive solar cells industrial production. We carried out a complex investigation of shaped ribbons and tubes defect structures [16.93, 94]. The aims of these investigations were to understand why it is difficult to obtain a single crystal structure as well as what is an electric activity of different defects. Its influence on the lifetime of secondary charge carriers is of particular importance, since the efficiency of solar elements is determined primarily by the lifetime of the secondary carriers. It is well known that inhomogeneity of the properties results from the inhomogeneous distribution of electrically active defects. The processes occurring at the solid–liquid (S–L) interface essentially affect formation of defects and their electronic properties as well. The simultaneous study of electric and photoelectric properties of sheet and tube crystals, its defect structure and the S–L interface was carried out. The graphite and the quartz were used as materials for crucibles and the graphite – for shapers.

16.8.1 Shaped Silicon Structure

Shaped silicon crystals possess characteristic defects of their crystalline structures including flat boundaries (most often those of the twinning type), dislocations and pileups thereof as well as SiC particles and particle-aggregations. As a result of twinning along intersecting planes, which can be observed at the initial stage of growth, a stable and quasi-equilibrium structure characterized by existence of defect areas (flat boundaries) perpendicular to the tape plane and parallel to the pulling direction. In this case the silicon tape surface orientation is {110} and the crystallographic axis ⟨211⟩ coincides with the direction of pulling. Such orientation is formed and maintained irrespective of the seed orientation and can be attained directly at the seed–crystal contact boundary in case the seed orientation is {110} ⟨211⟩ or a stable-structure section of the previously grown silicon tape is used as a seed. Figure 16.37 gives a photograph of the surface of a silicon tape of a stable defect structure: SiC inclusion surrounded with dislocation pileups can be observed on the tape surface.

As a result of x-ray structure analysis it was established that the orientation of silicon-tape surfaces can deflect within the range of 15° from {110}. Considerable dimensional fragments of defect structures identical to those of tapes are observed in silicon tubes (Fig. 16.38a). The tube-surface orientation in the vicinity of such a fragment deflects from {110} due to surface curvature. This change in orientation does not cause additional structure defects up to the values equal to 15° and as soon as that value reaches 15° surface orientation abruptly changes because there appears *a severely defect spot* that provides an indispensable turn of the crystal orientation in such a way so that the system of twin-boundary planes should be again approximately perpendicular to the tube surface. Either grain boundaries of a general type or dislocation pileups can act as such defect areas (Fig. 16.38b)

Fig. 16.37 Defect structure of a silicon tape: selective chemical etching pattern, 120×; an SiC inclusion is located in the *center*; the label "H.p." indicates the direction of crystal growth

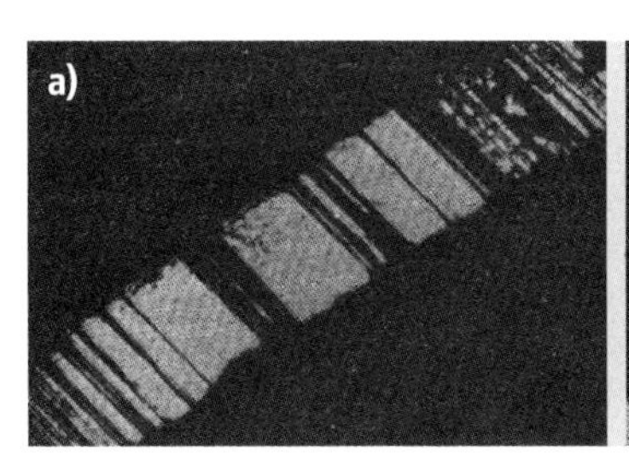

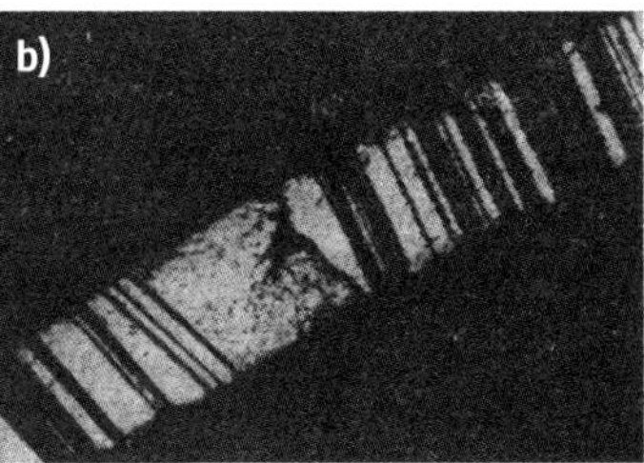

Fig. 16.38a,b Defect structures of silicon tubes (cross-sectional fragments): selective chemical etching pattern, 200×: (**a**) typical structure; (**b**) defect area providing rotation of the system of twin-boundary planes

Apparently, the number of *severely defect spots* and correspondingly the number of fragments of steady structures is determined by the necessity of closing the system of fragments into a cylinder. Since the angular magnitude of an arc of 30° corresponds to the variation of the tube surface orientation of 15° from {110}, the overall number of defect spots is equal to at least twelve, which is experimentally proved. It should also be noted that this linear length of *severely defect spots* along the perimeter of a tube depends little on its diameter, therefore the relation of the volume of *severely defect spots* to crystal volume should quickly decrease with the tube diameter increasing.

Defect structures of silicon tapes were studied by the TEM. With a ≈ 200 magnification, the defect structure pattern proved to be similar to that of selective etching (Fig. 16.37). It was observed that the width of defective areas is equal to ≈ 5 μm and the width of monocrystalline regions between them varies from 10–500 μm. Interpretation of fine structures of defect areas, that look like dark lines parallel to the direction of pulling in the pattern of selective chemical etching (Fig. 16.37) revealed that each area represents a set of microtwins 40–200 Å wide (Fig. 16.39). Small-angle disorientation of monocrystalline sections adjacent to the defect area is caused by such set of microtwins. Disorientation measured by the Kikuchi-line technique varied from 40 to 4°. Besides defect areas of twinning nature, other types of defects were observed in 2% of cases. Those included multilayer lattice defects, small-angle dislocation boundaries as well as separate dislocations and inclusions.

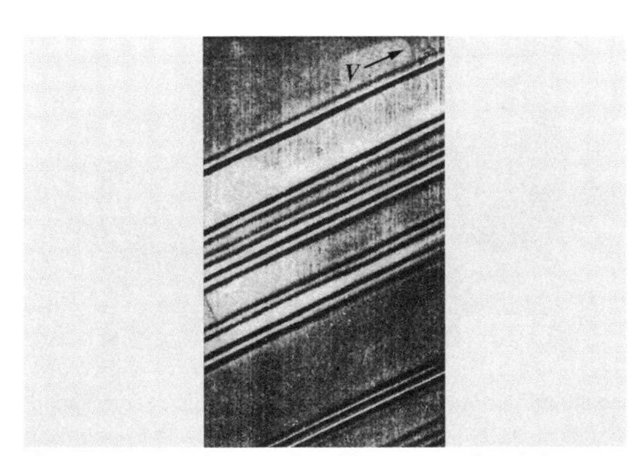

Fig. 16.39 TEM image of a silicon tape: flat defect area structure in the form of a system of microtwins; magnification 160 000; *V* denotes the direction of crystal growth

To interpret the defect structure and macroscopic pattern of profiled silicon crystal growth, a model, according to which the crystallization front tends to be shaped by the most slowly growing crystal faces {111} while high rates are provided by availability of inlet angles with their vertexes coinciding with twin boundaries, was postulated (Fig. 16.40a). To check the model offered, experiments on investigation of the crystallization front shape visualized by impulse changes in the crystal-pulling rate were carried out. In this case local changes in crystal thickness follow the changes in crystallization front shape, so does the horizontal hatching observed on the surface of profiled silicon crystals (Fig. 16.40d).

Crystallization front shaping by faces {111} proved to be observed only in the vicinity of high-energy boundaries of a general type and at twin boundaries of higher orders, e.g., {111}–{115}. Crystallization front deflection from the flat one does not exceed 2 μm in the coherent-twin region. The overall crystallization front area made by faces {111} is not large; the major part of the crystallization front follows the crystallization isotherm and corresponds to the face close to {112} and in separate cases to {110} (Fig. 16.40b,c) each of which can grow according to the normal mechanism, i. e., both layer-by-layer and normal mechanisms of growth take place when shaped crystals grow.

The specific features of the crystallization front reviled might be associated with the impurity influence on the face free energy or on the growth kinetics. Therefore it can be assumed that front shaping and perhaps inlet angle formation at twin boundaries are not necessary for high rates of profiled silicon crystal growth to be put

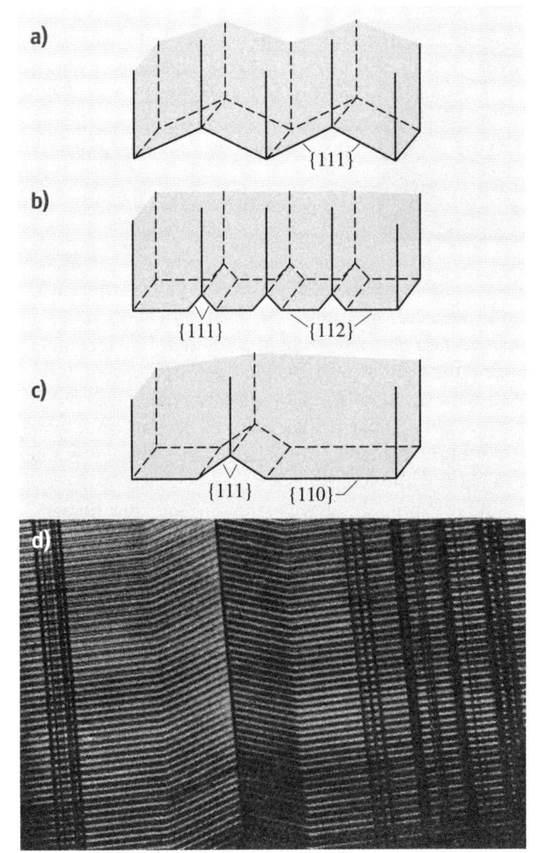

Fig. 16.40 (**a**) Crystallization front cut by {111} planes in accordance with the postulated model; (**b,c**) experimentally observed crystallization front shape; (**d**) correspondence between the crystallization front shape and the horizontal hatching on the surface of profiled crystals

into practice. It can be confirmed by the fact that in separate cases silicon tapes with wide (up to 10–15 mm) monocrystalline areas can be grown.

It was noticed that in the process of silicon tube growth the melt meniscus height strongly influences the character of defect formation in crystals. With the crystallization front in a tube sinking, the number of boundaries of a general type increases, i. e., profile defect structure depends on the interaction efficiency in the crystallization front–shaper system. The shaper influence is mainly determined by the distortions (thermal and capillary) introduced by the silicon carbide layer formed on the operating free edges of a graphite shaper when it interacts with the silicon melt.

Besides such indirect influence on the structure of profiles, silicon carbide entering the crystal subsurface layer in the form of SiC particles greatly affects the quality of the material produced. SiC microcrystals measuring $25-30\,\mu\mathrm{m}$, produced in the melt meniscus, form a carbide layer growing on the shaper free edges. When bulges appear on the layer the meniscus becomes distorted and on the crystal surface there appear distinctive furrows and in separate cases bulges along the direction of pulling that break when SiC particles and aggregations thereof results from random fluctuations of the crystallization front when particles reaching the size of the meniscus escape the carbide layer. The number and the size of simultaneously entrapped particles depend on the amplitude of front fluctuations and on the mean height of the meniscus. The mechanism of carbide inclusion entrapment offered is confirmed by the results of silicon tube growth when the number of SiC particles entrapped sharply decreases for a high meniscus since the number of particles whose dimensions reach the meniscus height decreases.

The number of carbide inclusions in shaped silicon crystals can be decreases in several ways: by using crucibles made of graphite of high density; by replacing the

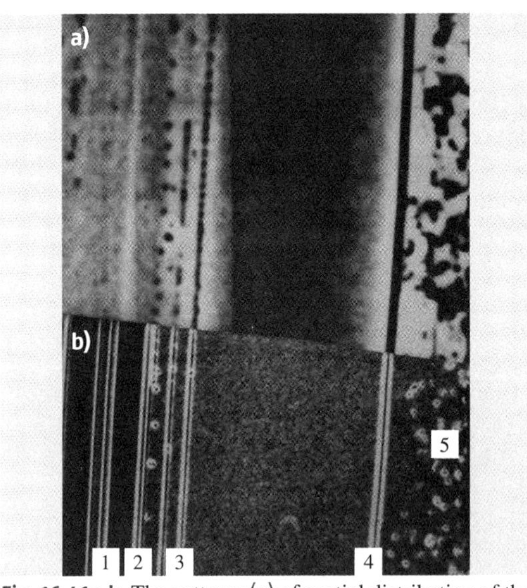

Fig. 16.41a,b The patterns (**a**) of spatial distribution of the secondary carrier current (EBIC, *dark regions* correspond to increased recombination) and (**b**) selective chemical etching; recombination boundaries: (1) inactive, (2) intermittent active, (3) low active, (4) high active, and (5) active dislocations

graphite container for crucibles with a protective layer-coated container; by maintaining a high melt meniscus (easy to secure when growing closed profiles). Besides, experiments on alloying shaped silicon with rare-earth elements showed that no SiC inclusions can be found on profile surfaces for any meniscus height in case at least 0.05 mass % of gadolinium dopant is introduced into silicon melt.

16.8.2 Local Electronic Properties of Shaped Silicon

The electric and photoelectric properties of crystals have been studied by local methods at $T = 300\,\mathrm{K}$: electron-beam induced current (EBIC), spreading resistance (SR), light-beam induced voltage (LBIV) and light-beam induced current (LBIC). EBIC–method is the widely used one. In the case of the SR method, the local conductivity of the sample was detected by an acicular probe displaced on the crystal surface with 8 μm step. In this case the spatial resolution of the method was 30 μm. The LBIV method was used for measuring the potential difference at the edges of a sample locally illuminated by a scanning light beam of 5 μm in diameter. This method made it possible to reveal the built-in electric fields caused by the defects of the crystal structure. In order to apply the LBIC method, a p-n junction was prepared or tin and indium oxide film was grown on the silicon surface, and then the current of minority carriers generated by a method analogous to the LBIV

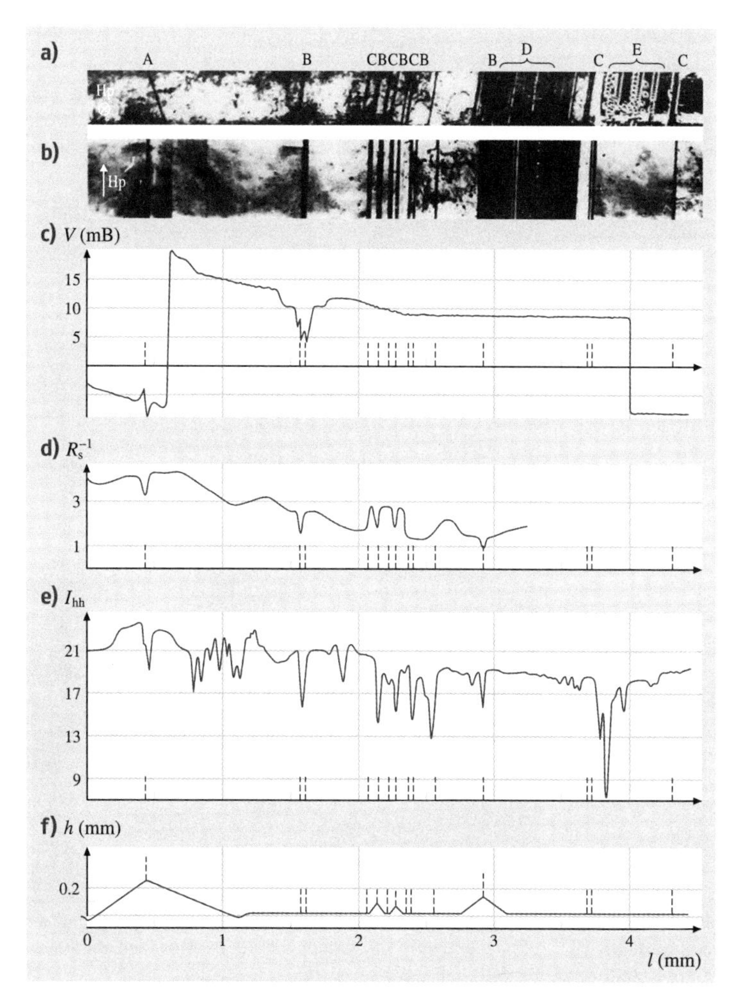

Fig. 16.42 Local study of the defect structure (**a,b**), electric properties (**c–e**), and S–L interface shape (**f**) of a silicon ribbon: (**a**) defect structure of the transverse cross section of an etched silicon ribbon, A–E indicate different types of defects, the label "Hp" indicates the growth direction; (**b**) structural defects on the surface of an etched ribbon, the label "Hp" indicates the growth direction; (**c**) distribution of photo-EMF obtained by the LBIV method across the ribbon; (**d**) spatial distribution of crystal conductivity, obtained by the spreading resistance method (arbitrary unit); (**e**) spatial distribution of minority carriers across the ribbon obtained by the LBIC method (arbitrary unit); (**f**) shape of the S–L interface crystal is shown above the curve, and the melt is below; at $l \geq 2\,\mathrm{mm}$ the height h of the S–L interface distortion is expanded by a factor of 10. The *dashed lines* in (**c–f**) show the locations of some twin boundaries in (**b**)

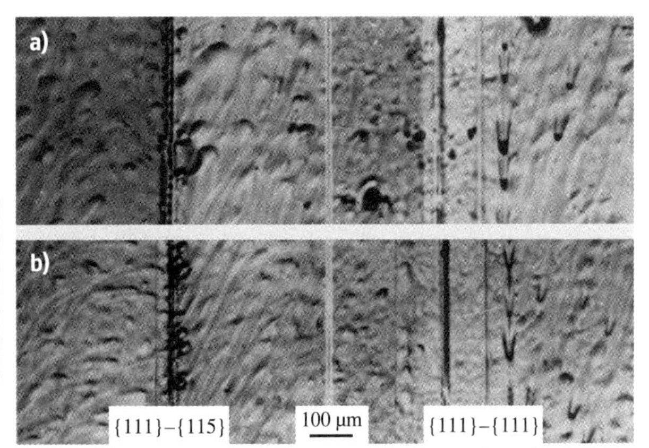

Fig. 16.43a,b Morphology of the melting crystal surface in the region of twin boundaries {111}–{115} at $l = 0.43$ mm, and {111}–{111} at $l = 2.2$ mm. The distances from the melting front are (**a**) 5 mm and (**b**) 3 mm

method was recorded. Effective centers of recombination were revealed by the LBIC method.

The spatial distribution of crystal defects was analyzed by the method of selective chemical etching. In addition, the structure of boundaries was examined with a scanning electron microscope. The morphology of the S–L interface was investigated as it is written above (Fig. 16.40d). Preliminary EBIC investigations of shaped crystals show that the electrical activity of twin boundaries as centers of increased recombination of nonequilibrium charge carriers differs (Fig. 16.41): there are electrically inactive boundaries, which are evidently coherent, boundaries with discontinuous activity, and boundaries of moderate and high activity. In the latter case, the boundaries are evidently incoherent. The combined results are presented on the Fig. 16.42 analyzed in detail in [16.94]. Here are brief results. The most important are defects whose electrical activity, as recombination centers of nonequilibrium carriers, is constant all along the silicon ribbon. It can be seen from comparing the LBIC spectra (Fig. 16.42e) with the pattern of selective chemical etching (Fig. 16.42b), that there are three types of such defects, at least. They are narrow bands of submicrons defects, accumulations or agglomerations of point defects with the coordinate $l = 0.8$–1.2 mm extended along the crystal growth direction. Dislocation etch pits are not visible here. The rows of dislocations denoted by index E in Fig. 16.42a and some twin boundaries are highly electrically active. Twin boundaries are the main defects. Its electric activity, as shown by the analysis in detail, depends on structural properties as well as enriching by impurities.

We have assumed that the decreased melting temperature of boundaries enriched with impurities may affect the S–L interface structure. Therefore, experiments were performed on melting shaped silicon crystals. It has turned out that melting starts on the surface and melt drops are formed in the vicinity of impurity inclusions. Fig. 16.43 shows that their concentration at the boundary with the coordinate $l = 0.43$ mm (Fig. 16.42) is of order of $400\,\mathrm{cm}^{-1}$, which is in accordance with a density of inclusions equal to $10^5\,\mathrm{cm}^{-2}$ found along this boundary after crystal cleavage. At the distance 5 mm from the melting front of reference single crystalline regions, separate drops coalesce and make up a uniform melt band of $10\,\mu\mathrm{m}$ in width. The process of melting thus proceeds mainly along the boundary, and a crystal can be divided into two parts by melting.

So, the main defects of the shaped silicon are the following ones: SiC inclusions, block boundaries, acute-angle boundaries, monocrystalline sections with

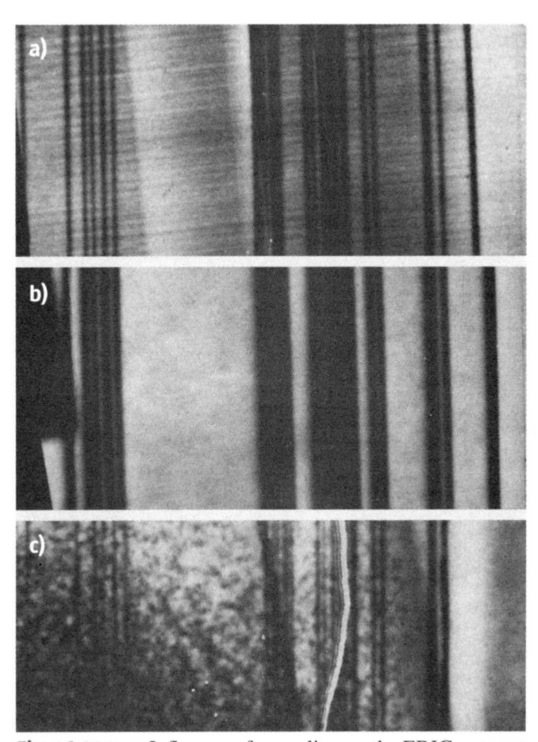

Fig. 16.44a–c Influence of annealing on the EBIC contrast of a silicon strip: (**a**) initial sample; (**b**) annealed at 450 °C for 50 h; (**c**) annealed at 1000 °C for 50 h (130 ×)

reduced lifetime, incoherent twin boundaries, dislocations, and coherent twin boundaries. The electrical activity of the structural defects falls in the order listed: in the region of the SiC inclusion, the lifetime of secondary carriers $\tau \le 10^{-8}$ s, increasing to a magnitude of the order of 10^{-6} s in monocrystalline regions of Si crystals with a resistivity of the order of 1 Ω cm.

The concentration of electrically active defects may possibly be reduced at the stage of crystal growth by improving the procedure. Decreasing of electrical activity of crystal defects in postgrowth treatment is a promising method of increasing the efficiency of solar elements. It has been established that annealing crystals at temperature of the order of 500 °C is a largely ineffective. In the EBIC spectra shown in Fig. 16.44, the regions of increased recombination of nonequilibrium carriers in the vicinity of the boundaries are seen to be only slightly broaden. However, annealing at 1000 °C leads to an increase in lifetime of the secondary carriers as a result of the decrease in electrical activity of the boundaries.

16.8.3 TPS Silicon Growth

The first experiments in the field of shaped silicon growth, especially for solar cells, were carried out more than thirty years ago. More than twenty different techniques have been tested. Information concerning this activity with numerous bibliographies can be found in [16.11–13]. The recent situation was analyzed in [16.95] review where four shaped silicon growth techniques are estimated as promising. There are TPS-EFG production of octagon profiles of 5 m length with 8 faces of 12.5 cm width and 0.3 mm thickness: RWE Schott Solar produces more than 200 000 kg/year with solar cell efficiency 14%. There is no any sawing loss because a laser cutting is used. In the frame of R&D project, a growth of tubes with the diameter of 0.5 m was realized [16.96].

Three other techniques are the following ones:

1. Evergreen Solar uses the string ribbon technique – a vertical growth from free melt surface between two metal strings

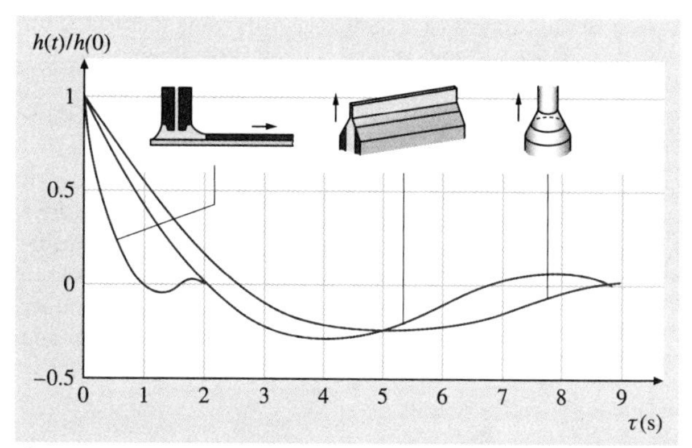

Fig. 16.45 Calculated curves of attenuation of crystallization front perturbation for a silicon tape

2. AstroPower uses the Silicon Film technique – a horizontal growth on a substrate with separation of the ribbon from the substrate
3. ECN uses RGS – the horizontal growth on a substrate when the substrate is used as a part of the solar cell.

We suggested this variant of shaped silicon growth 22 years ago with the name two shaping elements technique, TSET [16.97]. The scheme of the technique can be seen on the Fig. 16.45. When a graphite cloth is used as the substrate we title the technique the silicon on cloth, SOC.

At the conclusion of the silicon shaped growth discussion, let us show how the stability theory can be used for a crystal growth process characterization. On the Fig. 16.45 the curves of attenuations of crystallization front perturbation are presented. The special experiments were carried out for TSET. The relocation time was calculated as a function of the growth speed. It was found that it is decreased with the speed increasing up to the growth speed 6 cm/min and is not changed with the farter speed increasing. Two speeds (1.5 and 6 cm/min) were used in the experiments. The surface quality of the ribbons grown with the speed 6 cm/min was much better. The amplitude of the roughness was five times less.

16.9 TPS Metals Growth

What are the results of TPS application for metals and alloys? On the basis of our big experience in the field, we can conclude: when we find a material for the shaper (both wetted or nonwetted, but chemically inert) we can pull profiles from pure metals and eutectic alloys. The profiles have good quality surfaces (sometimes like a mirror). It is easy to grow low-melting metal single crystals by this technique. But there is no any technical application. As for refractory metal single crystals, for instance Mo that is used for foil production, it is very difficult to find a material for shaper.

Polycrystalline profiles could be used for many applications, for instance heat exchangers from alloys of aluminum (Fig. 16.28). But when we need alloys of high mechanical properties, the composition of alloys is complicated. There is a big gap between the temperatures of liquidus and solidus. Dendrite crystals appear in the meniscus and the problem of profile pulling is rather difficult both from point of view of surface quality and internal structure. So, TPS is difficult to compete with extrusion. On the other hand, the structure and mechanical properties of profiles are better than ones of cast profiles. We think the history is not finished yet. The problem of industrial application is not easy and a lot of efforts will need from numerous researchers. First of all alloys of special compositions have to be developed. Our success in the TPS application for the special steel [16.66] (Fig. 16.29a) and never published before stainless steel with aluminum rib (Fig. 16.29b) were promising but unfortunately have not been continued.

16.10 TPS Peculiarities

1. The crystals have the shape that we need for the most rational practical using.
2. The technique can be applied for any matters if a material for the shaper (both wetted or nonwetted, but chemically inert) is found.
3. The range of crystal dimensions is large: filaments of 0.02 mm diameter; tubes of 0.5 m diameter; plates $480 \times 320 \times 10\,mm^3$.
4. A growth speed, as rule, is much higher than in other crystal growth techniques.
5. Crystal, as rule, has a special structure concerned with the interface faceting.
6. Shaper influences on the interface shape.
7. The separation of growth zone from the melt in the crucible allows realization of continuous feeding by the raw material during growth process. As a result, we can have a short time of the melt presence before the growth, if it is necessary. Periodical change of the doping also is possible.
8. A distribution of impurities along the axis of crystal is more uniform. Indeed, if a coefficient of distribution of impurities is not equal to one during crystal growth a quantity of impurities changes along an axis of crystals in the most of crystal growth techniques. A solution of this problem can be found by localization of the melt at the zone of growth without stirring it with other melt volume. This situation we have, for instance, in using of additional crucible of small volume for CZT. The more effective result gives the capillary feeding at the TPS.
9. The using of special systems of capillaries for feeding allows controlling a distribution of impurities at the cross section of crystal.
10. A combination of doped and undoped parts is achieved in the same crystal.
11. The dynamic stability theory, developed for TPS, was successfully used for other CST as well as for analysis of cylindrical pores (negative crystals) growth and the radial instability of vapor whisker growth [16.5, 6].

References

16.1 A. Horowitz, S. Biderman, D. Gazit, Y. Einav, G. Ben-Amaz, M. Weiss: The growth of dome-shaped sapphire crystals by the gradient solidification method (GSM), J. Cryst. Growth **128**, 824–828 (1993)

16.2 D. Petrova, O. Pavloff, P. Marinov: Optical quality and laser characteristics of YAG:Nd crystals grown by the Bridgman–Stockbarger method, J. Cryst. Growth **99**, 841–844 (1990)

16.3 C.S. Bagdasarov: Problems of synthesis of refractory optic crystals. In: *Rost Kristallov*, Vol. 11, ed. by E.I. Givargizov (Transl. Growth of Crystals, Consultants Bureau, New York 1978) pp. 169–195, in Russian

16.4 V.A. Tatartchenko: Growth of shaped crystals from the melt. In: *Encyclopedia of Materials: Science and Technology*, ed. by K.H.J. Buschow, R.W. Cahn, M.C. Flemings, B. Ilschner, E.J. Kramer, S. Mahajan, P. Veyssière (Elsevier Science, Amsterdam 2001) pp. 3697–3703

16.5 V.A. Tatartchenko: Shaped crystal growth. In: *Handbook of Crystal Growth*, Vol. 2b, ed. by

D.T.J. Hurle (North-Holland, Amsterdam 1994) pp. 1011–1110

16.6 V.A. Tatartchenko: *Shaped Crystal Growth* (Kluwer Academic, London 1993)

16.7 V.A. Tatartchenko: The life of Alexander Stepanov and a brief history of shaped crystal growth, J. Jpn. Assoc. Cryst. Growth, **28**, 55–59 (2001)

16.8 V.A. Tatartchenko: Sapphire crystal growth and applications. In: *Bulk Crystal Growth of Electronic, Optical and Optoelectronic Materials*, ed. by P. Capper (Wiley, London 2005) pp. 299–338

16.9 V.A. Tatartchenko: *Stable Crystal Growth* (Nauka-Science, Moscow 1988), in Russian

16.10 P.I. Antonov, L.M. Zatulovskiy, A.S. Kostigov, D.I. Levinzon, S.P. Nikanorov, V.V. Peller, V.A. Tatartchenko, V.S. Juferev: *Preparation of Shaped Single Crystals and Products by Stepanov's Technique* (Nauka-Science, Moscow 1981), in Russian

16.11 G.W. Cullen, T. Surek, P.I. Antonov (Eds.): Shaped crystal growth, J. Cryst. Growth **50**(1), 1–396 (1980)

16.12 J.P. Kalejs, T. Surek, V.A. Tatartchenko (Eds.): Shaped crystal growth, J. Cryst. Growth **82**(1/2), 1–268 (1987)

16.13 J.P. Kalejs, T. Surek (Eds.): Shaped crystal growth, J. Cryst. Growth **104**(1), 1–199 (1990)

16.14 V.A. Tatartchenko: Influence of capillary phenomena on the stability of the crystallization process during the pulling of shaped specimens from the melt, Phys. Chem. Mater. Treat. **6**, 47–53 (1973), in Russian

16.15 V.A. Tatartchenko, G.I. Romanova: Stability of crystallization by Verneuile technique, Sov. Single Cryst. Tech. **2**, 48–53 (1973), in Russian

16.16 V.A. Tatartchenko: Stability of profiled pattern crystallization from the melt, 4th Int. Conf. Cryst. Growth Abstr. (Tokyo 1974) pp. 521–522

16.17 V.A. Tatartchenko, E.A. Brener: Stability of crystallization from a melt with a capillary shaper, Izv. Akad. Nauk SSSR, Ser. Fiz. **40**, 1456–1467 (1976), in Russian – Translation in Bull. Acad. Sci. USSR Phys. Ser. **40**, **7**, 106–115 (1976)

16.18 T. Surek: Theory of shape stability in crystal growth from the melt, J. Appl. Phys. **47**, 4384–4393 (1976)

16.19 G.A. Korn, T.M. Korn: *Mathematical Handbook for Scientists and Engineers* (McGraw-Hill, New York 1961) pp. 282–284

16.20 V.W. Mullins, F.R. Sekerka: Stability of the flat interface during crystallization of dilute binary alloys, J. Appl. Phys. **35**, 444–455 (1964)

16.21 V.A. Tatartchenko: Survey of quantitative analyses of the effects of capillary shaping on crystal growth, J. Cryst. Growth **82**, 74–80 (1987)

16.22 L.D. Landau, E.M. Lifchits: *Mecanique des Fluids* (Mir, Moscow 1971) p. 289

16.23 V.A. Tatartchenko, V.S. Uspenski, E.V. Tatartchenko, J.P. Nabot, T. Duffar, B. Roux: Theoretical Model of Crystal Growth Shaping Process, J. Cryst. Growth **180**, 615–626 (1997)

16.24 V.A. Tatartchenko, V.S. Uspenski, E.V. Tatartchenko, B. Roux: Theoretical investigation of crystal growth shaping process with the wetting boundary condition, J. Cryst. Growth **220**, 301–307 (2000)

16.25 G.A. Satunkin, V.A. Tatartchenko, V.I. Shaitanov: Determination of the growth angle from the shape of a crystal lateral face and solidified separation drops, J. Cryst. Growth **50**, 133–139 (1980)

16.26 H.S. Carslaw, J.C. Jaeger: *Conduction of Heat in Solids* (Clarendon, Oxford 1959) p. 148

16.27 K.J. Bachmann, H.J. Kirsch, K.J. Vetter: Programmed Czochralski growth of metals, J. Cryst. Growth **7**, 290–295 (1970)

16.28 A.A. Alioshin, N.I. Bletscan, S.I. Bogatyriov, V.N. Fedorenko: Silicon furnace components for microelectronic applications fabricated from shaped silicon tubes, J. Cryst. Growth **104**, 130–135 (1990)

16.29 V.A. Tatartchenko, A.I. Saet: Thermal regime of profile pulling from melt, J. Eng. Phys. **13**(2), 255–258 (1967), in Russian – Translation in J. Eng. Phys. USA

16.30 G.I. Romanova, V.A. Tatartchenko, N.P. Tichonova: Stability of crystallization at the crystal growth by the gas flame technique, Trudi Goi **54**(188), 10–13 (1976), in Russian – Sov. Proc. State Opt. Inst., Leningrad

16.31 V.A. Borodin, E.A. Brener, V.A. Tatartchenko: Investigation of the crystallization process in the Verneuile technique, Cryst. Res. Technol. **16**, 1187–1197 (1982)

16.32 V.A. Borodin, E.A. Brener, T.A. Steriopolo, V.A. Tatartchenko, L.I. Chernishova: Growing single crystal corundum tubes by Verneuile technique in stable conditions, Cryst. Res. Technol. **16**, 1199–1207 (1982)

16.33 V.A. Borodin, E.A. Brener, V.A. Tatartchenko, V.I. Gusev, I.N. Tsigler: Automation of the Verneuile technique on the basis of stability analysis, J. Cryst. Growth **52**, 505–508 (1981)

16.34 E.A. Brener, G.A. Satunkin, V.A. Tatartchenko: Some aspects of the macroscopic theory of oriented crystallization from the melt. IV. The floating zone techniques, Acta Phys. Acad. Sci. Hung. **47**, 159–165 (1979)

16.35 E.A. Brener, G.A. Satunkin, V.A. Tatartchenko: Macroscopic theory of crystallization from a melt with capillary shaping. The floating-zone method. In: *Growth of Crystals*, Vol. 14, ed. by E.I. Givargizov (Nauka Science, Moscow 1983) pp. 153–158, in Russian – Translation in *Growth of Crystals*, Vol. 14 (Consultants Bureau, New York 1987), 181–186

16.36 E.A. Brener, V.A. Tatartchenko, V.E. Fradkov: Differences between the mass transfer processes at a crystal surface near a three-phase line during crystallization and melting, Kristallografiya **27**, 205–206 (1982), in Russian – Translation in Sov. Phys. Crystallogr. **27** 127–128 (1982)
16.37 I.A. Lesk, A. Baghdadi, R.W. Gurtler, R.J. Ellis, J.A. Wiese, M.G. Coleman: Ribbon-to-ribbon crystal growth, Proc. 12th IEEE Photovolta. Spec. Conf. (IEEE, New York 1976) pp. 163–167
16.38 V.A. Tatartchenko, A.I. Saet, A.V. Stepanov: The shape of liquid column on producing of specified shape product by crystallization of melt, Proc. 1st Conf. Stepanov's Growth Semiconduct. Single Cryst. (Ioffe Phys.-Techn. Inst., Leningrad 1968) pp. 83–97, in Russian
16.39 V.A. Tatartchenko, A.I. Saet, A.V. Stepanov: Boundary conditions of capillary shaping at crystallization from melts, Izv. Akad. Nauk SSSR, Ser. Fiz. **33**, 1954–1959 (1969), in Russian – Translation in Bull. Acad. Sci. USSR, Phys. Ser. USA
16.40 N.K. Adam: *The physics and Chemistry of surfaces* (Clarendon, Cambridge 1930)
16.41 S.V. Tsivinskii, P.I. Antonov, A.V. Stepanov: Melt column form in crystal pulling process, Sov. Phys. Tech. Phys. **15**, 274–277 (1970), American translation of Sov. Zh. Tek. Fiz.
16.42 S.V. Tsivinskii: Application of capillary phenomena theory to manufacturing articles of predominant shape directly from the melt by Stepanov's method, J. Eng. Phys. **5**(9), 59–63 (1962), in Russian – Translation in J. Eng. Phys. USA
16.43 V.A. Tatartchenko, A.V. Stepanov: Calculation of the height of liquid column in crystal growth from melts, Izv. Akad. Nauk SSSR Ser. Fiz. **33**, 1960–1962 (1969), in Russian – Translation in Bull. Acad. Sci. USSR, Phys. Ser. USA
16.44 I.S. Pet'kov, B.S. Red'kin: Stability analysis of movable menisci, J. Cryst. Growth **104**, 20–22 (1990)
16.45 V.A. Tatartchenko, J.P. Nabot, T. Duffar, E.V. Tatarchenko, B. Roux: Some problems of stability of movable meniscus, J. Cryst. Growth **148**, 415–420 (1995)
16.46 L.L. Kuandykov, P.I. Antonov: Shaped melt column optimal choice on the basis of an equilibrium growth angle value, J. Cryst. Growth **222**, 852–861 (2001)
16.47 V.A. Tatartchenko: Capillary shaping in crystal growth from melts, I. Theory, J. Cryst. Growth **37**, 272–284 (1977)
16.48 V.A. Tatartchenko, G.A. Satunkin: Capillary shaping in crystal growth from melts, II. Experimental results for sapphire, J. Cryst. Growth **37**, 285–288 (1977)
16.49 V.A. Tatartchenko, S.K. Brantov: A model for a capillary shaper under conditions of weightlessness, Izv. Akad. Nauk SSSR, Ser. Fiz. **40**, 1468–1484 (1976), in Russian – Translation in Bull. Acad. Sci. USSR Phys. Ser. USA **407**, 116-129 (1976)
16.50 Y.A. Osipian, V.A. Tatartchenko: Crystal growth from melt by capillary shaping technique, Adv. Space Res. **8**(12), 16–34 (1988)
16.51 M.S. Agafonov, G.A. Gavrilov, L.I. Gubina, V.A. Kislov, L.V. Leskov, Y.B. Levin, V.V. Savichev, V.A. Tatartchenko, N.I. Timofeeva: Crystallization from the melt by the Stepanov method under zero-gravity conditions, Izv. Akad. Nauk SSSR, Ser. Fiz. **43**, 1935–1939 (1979), in Russian – Translation in Bull. Acad. Sci. USSR, Phys. Ser. **43**(9), 124–128 (1979)
16.52 V.A. Tatartchenko, S.K. Brantov, L.V. Leskov, V.L. Levtov, M.S. Agafonov: Experimental results of the crystallization of indium by the Stepanov method in conditions of micro-gravitation, Izv. Akad. Nauk SSSR, Ser. Fiz. **49**, 708–710 (1985), in Russian – Translation in Bull. Acad. Sci. USSR, Phys. Ser. **49**(4), 77–79 (1985)
16.53 V.A. Tatartchenko: Stability of melt crystal growth under microgravity condition, Adv. Space Res. **11**(7), 307–321 (1991)
16.54 L.A. Slobozhanin, M.A. Svechkareva, V.A. Tatartchenko: Stability of melt meniscus during growth of crystals by the technique of pulling from shaper under zero-gravity conditions, J. Cryst. Growth **133**, 273–280 (1993)
16.55 J.A. Burton, P.C. Prim, W.P. Slichter: The distribution of solute in crystals grown from the melt. Part I. Theoretical, J. Chem. Phys. **21**, 1987–1991 (1953)
16.56 J.C. Swartz, T. Surek, B. Chalmers: EFG (edge-defined, film-fed growth) process applied to the growth of silicon ribbons, J. Electron. Mater. **4**, 255–264 (1975)
16.57 S.K. Brantov, V.A. Tatartchenko: On effective coefficient of impurity distribution at the Stepanov thin-walled crystal growth (EFG), Cryst. Res. Technol. **18**, K59–K64 (1983)
16.58 J.P. Kalejes: Impurity redistribution in EFG, J. Cryst. Growth **44**, 329–334 (1978)
16.59 A.V. Stepanov: A new technique of production of sheets, of tubes, of rods with different cross sections from a melt, J. Tech. Phys. **29**, 382–393 (1959), in Russian – Translation in Sov. Phys. Tech. Phys. USA
16.60 A.V. Stepanov: A new technique of products fabrication from a melt, Mech. Eng. Bull. **11**, 47–50 (1959), in Russian
16.61 A.L. Shakch-Budagov, A.V. Stepanov: A new technique of production of sheets, of tubes, of rods with different cross sections from a melt, J. Tech. Phys. **29**, 394–405 (1959), in Russian – Translation in Sov. Phys. Tech. Phys. USA
16.62 B.M. Goltsman, A.V. Stepanov: A technique of production of sheets and tubes directly from a melt of aluminum and its alloys, Bull. Acad. Sci.

USSR, Metall. Fuel Ser. **5**, 49–53 (1959), in Russian
16.63 B.M. Goltsman: A pulling of crystalline sheets and tubes from a melt, Opt. Mech. Ind. **11**, 45–46 (1958), in Russian
16.64 A.V. Stepanov: *The Future of Metals Machining* (Lenizdat, Leningrad 1963), in Russian
16.65 V.A. Tatartchenko, A.V. Stepanov: The installation for investigation of crystallization parameters influence on the structure properties of crystallized specimens. In: *Machines and Devices for Testing of Materials*, ed. by G.V. Kurdiumov (Metallurgy, Moscow 1971), in Russian
16.66 V.A. Tatartchenko, V.V. Vakhrushev, A.S. Kostygov, A.V. Stepanov: Profile shaping in iron alloys of high carbon content, Izv. Akad. Nauk SSSR, Ser. Fiz. **35**, 511–513 (1971), in Russian – Translation in Bull. Acad. Sci. USSR, Phys. Ser. USA
16.67 G.V. Sachkov, V.A. Tatartchenko, D.I. Levinzon: Control of the capillary shaping process for single crystals grown from a melt, Izv. Akad. Nauk SSSR, Ser. Fiz. **37**, 2288–2291 (1997), in Russan – Translation in Bull. Acad. Sci. USSR, Phys. Ser.
16.68 H.E. La Belle, A.I. Mlavsky: Growth of sapphire filaments from the melt, Nature **216**, 574–575 (1967)
16.69 H.E. La Belle, A.I. Mlavsky: Growth of controlled profile crystals from the melt. Part 1. Sapphire filaments, Mater. Res. Bull. **6**, 571–580 (1971)
16.70 H.E. La Belle, A.I. Mlavsky: Growth of controlled profile crystals from the melt. Part 2. Edge-defined film fed growth (EFG), Mater. Res. Bull. **6**, 581–590 (1971)
16.71 E.V. Gomperz: Untersuchungen an Einkristalldrähten, Z. Phys. **8**, 184–190 (1922), in German
16.72 H. Mark, M. Polanyi, E. Schmid: Vorgänge bei der Dehnung von Zinkkristallen, Z. Phys. **12**, 58–77 (1923), in German
16.73 E. Grüneisen, E. Goens: Untersuchungen an Metallkristallen, Z. Phys. **26**(4/5), 235–273 (1924), in German
16.74 A.G. Hoyem, E.P.T. Tyndall: An experimental study of the growth of zinc crystals by the Czochralski-Gomperz method, Phys. Rev. **33**, 81–89 (1929)
16.75 A.G. Hoyem: Some electrical properties of spectroscopically pure zinc crystals, Phys. Rev. **38**, 1357–1371 (1931)
16.76 P. Kapitza: The study of the specific resistance of bismuth crystals and its change in strong magnetic field and some allied problems, Proc. R. Soc. Lond. Ser. A, **119**(782), 358–443 (1928)
16.77 H.E. LaBelle: EFG, the invention and application to sapphire growth, J. Cryst. Growth **50**, 8–16 (1980)
16.78 V.N. Kurlov, S.V. Belenko: The growth of sapphire shaped crystals with continuously modulated dopants, J. Cryst. Growth **191**, 779–782 (1998)
16.79 V.A. Borodin, V.V. Sidorov, T.A. Steriopolo, V.A. Tatartchenko: Variable shaping growth of refractory oxide shaped crystals, J. Cryst. Growth **82**, 89–94 (1986)
16.80 J.W. Locher, H.E. Bennet, P.C. Archibald, C.T. Newmyer: Large diameter sapphire dome: Fabrication and characterization, Proc. SPIE **1326**, 2–10 (1990)
16.81 V.A. Borodin, V.V. Sidorov, S.N. Rassolenko, T.A. Steriopolo, V.A. Tatartchenko, T.N. Yalovets: Local shaping technique and new growth apparatus for complex sapphire products, J. Cryst. Growth **104**, 69–76 (1990)
16.82 K.S. Bagdasarov: Synthesis of large single crystals of corundum. In: *Ruby and Sapphire*, ed. by L.M. Belyaev (National Bureau of Standards, Washington, DC 1980) pp. 15–38, Translation from Russian
16.83 M. Saito: Gas-bubble formation of ruby single crystals by floating zone melt growth with an infrared radiation convergence type heater, J. Cryst. Growth **71**, 664–672 (1985)
16.84 M. Saito: Growth process of gas-bubble in ruby single crystals by floating zone method, J. Cryst. Growth **74**, 385–390 (1986)
16.85 A.V. Zhdanov, G.A. Satunkin, V.A. Tatartchenko, N.N. Talyanskaya: Cylindrical pores in a growing crystal, J. Cryst. Growth **49**, 659–660 (1980)
16.86 V.A. Tatartchenko: Cylindrical pores in a growing crystal, J. Cryst. Growth **143**, 294–300 (1994)
16.87 V.A. Tatartchenko, T.N. Yalovets, G.A. Satunkin, L.M. Zatulovsky, L.P. Egorov, D.Y. Kravetsky: Defects in shaped sapphire crystals, J. Cryst. Growth **50**, 335–340 (1980)
16.88 A.A. Chernov, D.T. Temkin, A.M. Melnikova: Theory of inclusion capture at the melt crystal growth, Kristallographia (Sov. Crystallography) **21**(4), 652–660 (1976)
16.89 E. Savrun, C. Toy, W.D. Scott, D.C. Harris: Is sapphire inherently weak in compression at high temperature?, Proc. SPIE **3705**, 12–16 (1999)
16.90 J.W. Locher, S. Zanella, H. Bates, V.A. Tatartchenko: Production of large sapphire crystals by EFG, Abstracts 14th Int. Conf. Cryst. Growth (Grenoble 2004) p. 560
16.91 V.N. Kurlov, B.M. Epelbaum: Fabrication of near-net-shaped sapphire domes by noncapillary shaping method, J. Cryst. Growth **167**, 165–180 (1997)
16.92 V.S. Yuferev, V.M. Krymov, L.L. Kuandykov, S.I. Bakholdin, Y.G. Nosov, I.L. Shulpina, P.I. Antonov: The growth of sapphire ribbons with basal facet surface, J. Cryst. Growth **275**, 785–790 (2005)
16.93 J. Shi, G. Geitos, N.V. Abrosimov, S.K. Brantov, S.A. Erofeeva, V.A. Tatartchenko: Some structural and electrophysical properties of silicon ribbons produced by Stepanov's method, Izv. Akad. Nauk SSSR, Ser. Fiz. **43**, 1992–1994 (1979), in Russian – Translation in Bull. Acad. Sci. USSR, Phys. Ser. USA **43**, 164–165 (1979)

16.94 N.V. Abrosimov, A.V. Bazhenov, V.A. Tatartchenko: Growth features and local electronic properties of shaped silicon, J. Cryst. Growth **82**, 203–208 (1987)

16.95 J.P. Kalejs: An overview of new developments in crystalline silicon ribbon material technology for solar cells, Proc. 3rd World Conf. Photovolta. Energy Convers. (Osaka 2003)

16.96 D. Garcia, M. Ouellette, B. Mackintosh, J.P. Kalejs: Shaped crystal growth of 50 cm diameter silicon thin-walled cylinders by EFG, J. Cryst. Growth **225**, 566–571 (2001)

16.97 S.K. Brantov, B.M. Epelbaum, V.A. Tatartchenko: Unidirectional growth of silicon layers on a graphitized fabric substrate, Mater. Lett. **2**, 274–277 (1984)